Integral Transform Methods

in Science & Engineering

including Boundary Value Problems and Fourier Series

CBS Engineering Series

Integral Transform Methods

in Science & Engineering

including Boundary Value Problems and Fourier Series

Useful for

- UG and PG students of Mathematics, Physics and Engineering.
- GATE and many other Entrance and Competitive Examinations.

DR. SUDHIR KUMAR PUNDIR

M.Sc., M.Phil, NET, Ph.D.
Associate Professor
Department of Mathematics
S.D. (P.G.) College,
Muzaffarnagar (U.P.)

CBS

CBS Publishers & Distributors Pvt. Ltd.

New Delhi • Bengaluru • Chennai • Kochi • Kolkata • Mumbai
Hyderabad • Nagpur • Patna • Pune • Vijayawada

Integral
Transform
Methods
in Science & Engineering

ISBN: 978-93-86310-91-0

First Edition: 2017

Copyright © Author

Published by **Satish Kumar Jain** and produced by **Varun Jain** for

CBS Publishers & Distributors Pvt. Ltd.,
4819/XI Prahlad Street, 24 Ansari Road, Daryaganj, New Delhi - 110002
delhi@cbspd.com, cbspubs@airtelmail.in • www.cbspd.com
Ph.: 23289259, 23266861, 23266867 • Fax: 011-23243014

Corporate Office: 204 FIE, Industrial Area, Patparganj, Delhi - 110 092
Ph: 49344934 • Fax: 011-49344935
E-mail: publishing@cbspd.com • publicity@cbspd.com

Branches:
- *Bengaluru:* 2975, 17th Cross, K.R. Road, Bansankari 2nd Stage,
  Bengaluru - 70 • Ph: +91-80-26771678/79 • Fax: +91-80-26771680
  E-mail: cbsbng@gmail.com, bangalore@cbspd.com
- *Chennai:* No. 7, Subbaraya Street, Shenoy Nagar, Chennai - 600030
  Ph: +91-44-26681266, 26680620 • Fax: +91-44-42032115
  E-mail: chennai@cbspd.com
- *Kochi:* Ashana House, 39/1904, A.M. Thomas Road, Valanjambalam,
  Ernakulum, Kochi • Ph: +91-484-4059061-65
  Fax: +91-484-4059065 • E-mail: cochin@cbspd.com
- *Kolkata:* 6-B, Ground Floor, Rameshwar Shaw Road, Kolkata - 700014
  Ph: +91-33-22891126/7/8 • E-mail: kolkata@cbspd.com
- *Mumbai:* 83-C, Dr. E. Moses Road, Worli, Mumbai - 400018
  Ph: +91-9833017933, 022-24902340/41 • E-mail: mumbai@cbspd.com

Representatives:

- Hyderabad: 0-9885175004
- Patna: 0-9334159340
- Vijayawada: 0-9000660880
- Nagpur: 0-9021734563
- Pune: 0-9623451994

Printed at:
Neekunj Print Process, Delhi

Preface

The book entitled "INTEGRAL TRANSFORM METHODS in SCIENCE AND ENGINEERING" has been written to meet the requirements of Mathematics, Physics and Engineering students. Besides, it will also be very useful for students preparing for various competitive examinations.

The contents of this book are designed from the curricula offered by various universities across the country. An attempt has been made to give comprehensive knowledge of the subject. The topic Laplace, Fourier, Mellin, Hankel and Z-transforms have been discuss in detail. In addition to these topics, the book contain Heat, wave, Laplace equations and boundary value problems along with Fourier series. The book consist nine chapters. In each chapter an ample amount of theory is given which is supported by solved examples followed by graded exercises along with their answers. A lot of objective questions and self assessment test are given at the end of each chapter.

I express my gratitude to the authors and publishers of various books I consulted during the preparation of the book.

I wish to sincerely thank **Sh S.K. Jain**, Managing Director, CBS Publishers and Distributors, New Delhi for his encouragement and help in bringing out this publication in a present nice form.

My special thanks to Sh. B.M. Singh, Sh. Sunil Dutt, Sh. Puneet Verma and entire team of CBS Publishers and Distributors, New Delhi whose encouragement and unstinted support enabled me to complete my book. Mr. Peeyush Goel, M/s Dreamshapers also deserve special mention for nice type setting.

I must also record my appreciation due to my wife Dr. Rimple, daughter Rijuta and son Shrish for their understanding and love during the long period that I have taken to complete this book.

Above all I am thankful to The Almighty God, without whose grace nothing is possible for any one.

Readers are welcomed to point out errors, if any and send their valuable suggestions for improving the quality of the book.

Dr. SUDHIR KUMAR PUNDIR

email : skpundir05@yahoo.co.in

Contents

Chapter

0

Pre-Requisite

(A) Some Series Expansions

1. $e^x = 1 + x + \dfrac{x^2}{2!} + \dfrac{x^3}{3!} + \ldots$

2. $(1+x)^n = 1 + nx + \dfrac{n(n-1)}{2!}x^2 + \ldots$

3. $\sin x = x - \dfrac{x^3}{3!} + \dfrac{x^5}{5!} - \dfrac{x^7}{7!} + \ldots$

4. $\cos x = 1 - \dfrac{x^2}{2!} + \dfrac{x^4}{4!} - \dfrac{x^6}{6!} + \ldots$

5. $\tan x = x + \dfrac{x^3}{3} + \dfrac{2x^5}{15} + \ldots$

6. $a^x = 1 + x \log a + \dfrac{(x \log a)^2}{2!} + \ldots$

7. $\log(1+x) = x - \dfrac{x^2}{2} + \dfrac{x^3}{3} - \dfrac{x^4}{4} + \ldots$

8. $\sin^{-1} x = \dfrac{x^3}{6} + \dfrac{3}{40}x^5 + \ldots$

(B) Some Useful Limits

1. $\lim\limits_{\theta \to 0} \dfrac{\sin \theta}{\theta} = 1$

2. If n is any integer or fraction then $\lim\limits_{x \to a} \dfrac{x^n - a^n}{x - a} = na^{n-1}$, $x \ne a$.

3. If m and n are rational numbers then $\lim\limits_{x \to a} \dfrac{x^m - a^m}{x^n - a^n} = \dfrac{m}{n}a^{m-n}(a > 0)$

4. $\lim\limits_{x \to \infty} \left(1 + \dfrac{x}{n}\right)^n = e^x$

5. $\lim\limits_{x \to \infty} \dfrac{\sin x}{x} = 0$

6. $\lim\limits_{x \to 0} \log|x| = -\infty$

7. $\lim\limits_{x \to 0} \sin\dfrac{1}{x} =$ a finite quantity lying between 1 and –1.

8. $\lim\limits_{\theta \to 0} \cos \theta = 1$

9. $\lim\limits_{n \to \infty} \left(1 + \dfrac{x}{n}\right)^p = 1$, if p is finite

10. $\lim\limits_{n \to \infty} \left(1 + \dfrac{x}{n}\right)^{n+p} = e^x$, if p is prime

11. $\lim\limits_{x \to 0} \left(1 + \dfrac{x}{a}\right)^{a/x} = e$

12. $\log_e e = 1, \log_e 1 = 0, \log_e \infty = \infty, \log_e 0 = -\infty$

(C) SOME TRIGONOMETRIC FORMULAE

1. $\sin 2x = 2\sin x \cos x$

2. $2\cos^2\dfrac{x}{2} = 1 + \cos x$

3. $2\sin^2\dfrac{x}{2} = 1 - \cos x$

4. $\tan x = \dfrac{2\tan\dfrac{x}{2}}{1 - \tan^2\dfrac{x}{2}}$

5. $\cos x = \dfrac{1 - \tan^2\dfrac{x}{2}}{1 + \tan^2\dfrac{x}{2}}$

6. $\sin x = \dfrac{2\tan\dfrac{x}{2}}{1 + \tan^2\dfrac{x}{2}}$

7. $\tan^{-1} x - \tan^{-1} y = \tan^{-1}\dfrac{x - y}{1 + xy}$

8. $2\tan^{-1} x = \tan^{-1}\dfrac{2x}{1 - x^2}$

9. $3\tan^{-1} x = \tan^{-1}\dfrac{3x - x^3}{1 - 3x^2}$

10. $\sin 3x = 3\sin x - 4\sin^3 x$

11. $\cos 3x = 4\cos^3 x - 3\cos x$

12. $\cosh x = \dfrac{e^x + e^{-x}}{2}$

13. $\sinh x = \dfrac{e^x - e^{-x}}{2}$

14. $\cosh^2 x - \sinh^2 x = 1$

15. $\cosh 2x = \cosh^2 x + \sinh^2 x$

16. $\cosh 2x = 2\cosh^2 x - 1 = 1 + 2\sinh^2 x$

17. $\operatorname{sech}^2 x = 1 - \tanh^2 x$

18. $\operatorname{cosech}^2 x = \coth^2 x - 1$

19. $\sinh^{-1} x = \log[x + \sqrt{x^2 + 1}]$

20. $\cosh^{-1} x = \log[x + \sqrt{x^2 - 1}]$

21. $\tanh^{-1} x = \dfrac{1}{2}\log\left(\dfrac{1 + x}{1 - x}\right)$

22. $\sin C + \sin D = 2\sin\dfrac{C + D}{2}\cos\dfrac{C - D}{2}$

23. $\sin C - \sin D = 2\cos\dfrac{C + D}{2}\sin\dfrac{C - D}{2}$

24. $\cos C + \cos D = 2\cos\dfrac{C + D}{2}\cos\dfrac{C - D}{2}$

25. $\cos C - \cos D = 2\sin\dfrac{C + D}{2}\sin\dfrac{D - C}{2}$

26. $2\sin A\cos B = \sin(A + B) + \sin(A - B)$

27. $2\cos A\sin B = \sin(A + B) - \sin(A - B)$

28. $2\cos A\cos B = \cos(A + B) + \cos(A - B)$

29. $2\sin A\sin B = \cos(A - B) - \cos(A + B)$

30. $\sin^2 A - \sin^2 B = \sin(A + B)\sin(A - B)$

31. $\cos^2 A - \cos^2 B = -\sin(A + B)\sin(A - B)$

(D) SOME CONCEPT AND RESULTS OF BETA AND GAMMA FUNCTION

Beta Function : The definite integral

$$\int_0^1 x^{m-1}(1 - x)^{n-1}\,dx \quad \forall m > 0, n > 0$$

is known as Beta function and is denoted by $B(m, n)$.

Gamma Function : The definite integral

$$\int_0^\infty e^{-x} x^{n-1} dx \quad \forall n > 0$$

is known as Gamma function and is denoted by $\Gamma(n)$.

PROPERTIES

1. $\Gamma(1) = 1$

2. $\Gamma(n+1) = n\Gamma(n), n > 0$

3. $\Gamma(n) = \int_0^1 \left(\log\frac{1}{x}\right)^{n-1} dx, n > 0$

4. $B(m,n) = \int_0^\infty \dfrac{x^{n-1} dx}{(1+x)^{m+n}} = \int_0^\infty \dfrac{x^{n-1} dx}{(1+x)^{m+n}}, m > 0, n > 0$

5. $B(m,n) = \dfrac{1}{2}\int_0^\infty \dfrac{x^{m-1} + x^{n-1}}{(1+x)^{m+n}} dx$

6. $B(m,n) = \int_0^1 \dfrac{x^{m-1} + x^{n-1}}{(1+x)^{m+n}} dx$

7. $\int_0^\infty \dfrac{x^{m-1} dx}{(ax+b)^{m+n}} = \dfrac{B(m,n)}{a^m b^n}$

8. $\int_0^{\pi/2} \dfrac{\sin^{2m-1}\theta\cos^{2n-1}\theta d\theta}{(a\sin^2\theta + b\cos^2\theta)^{m+n}} = \dfrac{B(m,n)}{2a^m b^n}$

9. $\int_0^1 \dfrac{x^{m-1}(1-x)^{n-1} dx}{(x+a)^{m+n}} = \dfrac{B(m,n)}{a^n(1+a)^m}$

10. $\int_b^a (x-b)^{m-1}(a-x)^{n-1} dx = (a-b)^{m+n-1} B(m,n)$

11. $\int_b^a \dfrac{x^{m-1}(1-x)^{n-1}}{\{a+(b-a)x\}^{m+n}} dx = \dfrac{1}{a^n b^m} B(m,n)$

(E) RELATION BETWEEN BETA AND GAMMA FUNCTIONS

1. $B(m,n) = \dfrac{\Gamma(m)\Gamma(n)}{\Gamma(m+n)}, m > 0, n > 0$

2. $\Gamma(n)\Gamma(1-n) = \pi / \sin n\pi, 0 < n < 1$

3. $\Gamma(1+n)\Gamma(1-n) = (n\pi) / \sin n\pi$

4. $\Gamma(1/2) = \sqrt{\pi}$

5. $\int_0^\infty e^{-x^2} dx = \dfrac{\sqrt{\pi}}{2}$

6. $B(m, n) = B(m+1, n) + B(m, n+1), m > 0, n > 0$

7. $\int_0^{\pi/2} \sin^{2m-1}\theta\cos^{2n-1}\theta d\theta = \dfrac{\Gamma(m)\Gamma(n)}{2\Gamma(m+n)} = \dfrac{B(m,n)}{2}, m > 0, n > 0$

8. $\int_0^{\pi/2} \sin^p\theta\cos^q\theta d\theta = \dfrac{\Gamma\left(\dfrac{p+1}{2}\right)\Gamma\left(\dfrac{q+1}{2}\right)}{2\Gamma\left(\dfrac{p+q+2}{2}\right)}, p > -1, q > -1$

9. $\int_0^{\pi/2} \sin^p \theta\, d\theta = \int_0^{\pi/2} \cos^p \theta\, d\theta = \dfrac{1 \cdot 3 \cdot 5 \ldots (p-1)}{2 \cdot 4 \cdot 6 \ldots p} \dfrac{\pi}{2}$, if p is even positive integer

$$= \frac{2 \cdot 4 \cdot 6 \ldots (p-1)}{1 \cdot 3 \cdot 5 \ldots p}, \text{ if } p \text{ is odd positive integer}$$

10. $\int_0^{\pi/2} \sin^{p-1} \theta \cos^{q-1} \theta\, d\theta = \dfrac{\Gamma(p/2)\Gamma(q/2)}{2\Gamma\left(\dfrac{p+q}{2}\right)}$

11. $\int_0^{\pi/2} \sin^{p-1} \theta\, d\theta = \int_0^{\pi/2} \cos^{p-1} \theta\, d\theta = \dfrac{\Gamma(p/2)\Gamma(1/2)}{2\Gamma\left(\dfrac{p+1}{2}\right)}$

$$= \frac{\sqrt{\pi}}{2} \frac{\Gamma(p/2)}{\Gamma\left(\dfrac{p+1}{2}\right)} \text{ as } \Gamma\left(\frac{1}{2}\right) = \sqrt{\pi}$$

VVVVVV

Chapter 1

The Laplace Transforms

1.1 INTRODUCTION

An integral of the type $\int_{-\infty}^{\infty} k(p,t)F(t)dt$ is defined as the integral transform of $F(t)$, provided it is convergent. It is denoted by $f(p)$ or $T[F(t)]$, *i.e.*,

$$f(p) = TF(t) = \int_{-\infty}^{\infty} k(p,t)F(t)dt \qquad \text{(Rohilkhand–1991)}$$

REMARK

- The function $k(p,t)$ appearing in the integral is called kernel of the transform. Here p is a parameter and is independent of t. Also p may be real or complex number.

1.2 THE LAPLACE TRANSFORM

Definition 1. *If $F(t)$ be a function of t defined for all values of t, then Laplace transform of $f(t)$, denoted by $L\{F(t)\}$ or $f(p)$ is defined by*

$$L\{F(t)\} = f(p) = \int_0^\infty e^{-pt}\, F(t)\, dt \qquad \qquad ...(1)$$

(Meerut–1987, 88, 92, 99; Rajasthan–1983; Rohilkhand–1988, 91, 99, 2000; MDU–2006, 10)

REMARKS

- If the integral (1) converges for some value of p, then only the Laplace transform of $f(t)$ exists otherwise not.
- L is called Laplace transform operator.

Definition 2. *A function $f(x)$ is said to be exponential order a as $x \to \infty$ if $\lim\limits_{x\to\infty} e^{-ax} f(x) = a$ finite quantity, i.e., for a given positive integer n, if a real number M such that $|e^{-ax} f(x)| < M,\ \forall\, x \geq n$ which can be written as $f(x) = O(e^{-ax}),\ x \to \infty$.*

Definition 3. *A function $f(x)$ is called sectionally continuous over the closed interval $x_1 \leq x \leq x_2$ if the closed interval can be divided into a finite number of subintervals $a \leq x \leq b$ such that*

  (i) *$f(x)$ is continuous in the closed interval (a, b).*

  (ii) $\lim\limits_{x\to a+0} f(x)$ *and* $\lim\limits_{x\to b-0} f(x)$ *both exist.*

Definition 4. *A function which is sectionally (or piecewise) continuous over every finite interval in the range $t \geq 0$ and ω of exponential order as $t \to \infty$ is called a function of class A.*

1.3 LINEARITY PROPERTY OF LAPLACE TRANSFORMATION

(Meerut–1986, 87; Rohilkhand–1999; MDU–2006)

THEOREM 1.   ***The Laplace transformation is a linear transformation, i.e.,***
$L\{a_1 F_1(t) + a_2 F_2(t)\} = a_1 L\{F_1(t)\} + a_2 L\{F_2(t)\}$ if $a_1, a_2 \in \mathbf{R}$.

PROOF.   We know that $L = \{F(t)\} = \int_0^\infty e^{-pt} F(t)\, dt$.

Therefore, $L\{a_1 F_1(t) + a_2 F_2(t)\} = \int_0^\infty e^{-pt} \{a_1 F_1(t) + a_2 F_2(t)\}\, dt$

$$= a_1 \int_0^\infty e^{-pt} F_1(t)\, dt + a_2 \int_0^\infty e^{-pt} F_2(t)dt$$

$$= a_1\, L\{F_1(t)\} + a_2 L\{F_2(t)\}.$$

1.4 EXISTENCE OF LAPLACE TRANSFORM

THEOREM-1.   *If F(t) is a function which is piecewise continuous on every finite interval in the range $t \geq 0$ and satisfies $|F(t)| \leq M\, e^{at}$ for all $t \geq 0$ and for constant a and M then the Laplace transform of f(t) exists for all $p > a$.*

(Meerut–1987, 92; Agra–1988)

PROOF.   We know that $L\{F(t)\} = \int_0^\infty e^{-pt} F(t)\, dt = \int_0^{t_0} F(t) e^{-pt} dt + \int_{t_0}^\infty F(t) e^{-pt} dt$ ...(1)

Now, $\int_0^{t_0} F(t)\, e^{-pt} dt$ exists because $F(t)$ is sectionally continuous on every finite interval $0 \leq t \leq t_0$

Also, $\left| \int_{t_0}^\infty F(t) e^{-pt} dt \right| \leq \int_{t_0}^\infty |F(t) e^{-pt}|\, dt \leq \int_{t_0}^\infty e^{-pt} M\, e^{at}\, dt$ $[\because |F(t)| \leq M e^{at}]$

$$= \int_{t_0}^\infty e^{(a-p)t} M\, dt = M \left[\frac{e^{-(p-a)t}}{-(p-a)} \right]_{t_0}^\infty = \frac{M}{p-a} e^{-(p-a)t_0}, \text{ if } p > a$$

$$\Rightarrow \left| \int_{t_0}^\infty F(t)\, e^{-pt}\, dt \right| \leq \frac{M}{p-a} e^{-(p-a)t_0}, \text{if } p > a.$$

Now, $\dfrac{M e^{-(p-a)t_0}}{p-a}$ can be made small as we please by taking t_0 sufficiently large.

Hence, from (1), we conclude that $L\{f(t)\}$ exists for all $p < a$.

REMARK

* The above conditions are sufficient but not necessary for the existence of the Laplace transform. If these conditions are satisfied, the Laplace transform must exist. If these conditions are not satisfied, the Laplace transform may or may not exist.

1.5 LAPLACE TRANSFORM OF SOME ELEMENTARY FUNCTIONS

(i) $F(t) = 1$ (Purvanchal–2003; Meerut–2004; Kanpur–2003)

SOLUTION.   We have $L\{F(t)\} = \int_0^\infty e^{-pt} F(t)\, dt$ ...(1)

Here $F(t) = 1$.

Therefore, from (1) $L\{1\} = \int_0^\infty e^{-pt} \cdot 1\, dt = \left[-\dfrac{e^{-pt}}{p} \right]_0^\infty = \dfrac{1}{p}, \; p > 0$

Hence, $L\{1\} = \dfrac{1}{p}.$

(ii) $F(t) = t^n$ (Meerut–1986, 1991, 2004; Andhra–1980, 90, 2000, 04; Osmania–2004; Purvanchal–1994)

SOLUTION.   We have $L\{F(t)\} = \int_0^\infty e^{-pt} F(t)\, dt$

$\Rightarrow$ $L\{t^n\} = \int_0^\infty e^{-pt}\, t^n\, dt = \int_0^\infty e^{-pt} \cdot t^{(n+1)-1}\, dt$

$$= \frac{\Gamma(n+1)}{p^{n+1}} = \frac{n!}{p^{n+1}}, \; p > 0 \quad \left[\because \int_0^\infty e^{-u}\, u^n du = \Gamma(n+1) \right]$$

Hence, $L\{t^n\} = \dfrac{n!}{p^{n+1}}.$

(iii) $F(t) = t$

SOLUTION.   We have $L\{F(t)\} = \int_0^\infty e^{-pt} \cdot t\, dt = \left[-\dfrac{1}{p} t\, e^{-pt} \right]_0^\infty + \dfrac{1}{p} \int_0^\infty e^{-pt} dt$

$$= \frac{1}{p^2}, \; p > 0$$

(iv) $F(t) = e^{at}$
(Meerut–1991, 2004; Andhra–1990; Purvanchal–1989, 96)

SOLUTION. We have $L\{F(t)\} = \int_0^\infty e^{-pt} e^{at} \, dt = \int_0^\infty e^{-(p-a)t} \, dt$.

If $p \le a$, integral diverges. For $p > a$, the integral converges. Hence, for $p > a$,

$$L\{e^{at}\} = \int_0^\infty e^{-(p-a)t} \, dt = \left[-\frac{e^{-(p-a)t}}{p-a} \right]_0^\infty = 0 + \frac{1}{p-a}$$

$$= \frac{1}{p-a}, \quad p > a.$$

(v) $F(t) = \sin at$
(Raj.–1983; Kanpur–1984, 95, 2002, 03, 04; Meerut–1991, 2005, 08; Purvanchal–1989; MS Univ.(T.N.)–2007)

SOLUTION. We have
$$L\{\sin at\} = \int_0^\infty e^{-pt} \sin at \, dt = \left[\frac{e^{-pt}(-p\sin at - a\cos at)}{p^2 + a^2} \right]_0^\infty$$

$$\left[\because \int e^{ax} \sin bx \, dx = e^{ax} \frac{[a\sin bx - b\cos bx]}{a^2 + b^2} \right]$$

$$= \frac{a}{p^2 + a^2}, \quad p > a$$

Hence, $L\{\sin at\} = \dfrac{a}{p^2 + a^2}$.

(vi) $F(t) = \cos at$
(Meerut–1991; Rohilkhand–2007; Kanpur–1995; MS Univ.(T.N.)–2007)

SOLUTION. We know that
$$\int e^{ax} \cos bx \, dx = \frac{e^{ax}(a\cos bx + b\sin bx)}{a^2 + b^2}$$

Therefore, we have
$$L\{\cos at\} = \int_0^\infty e^{-pt} \cos at \, dt = \left[\frac{e^{-pt}(-p\cos at + a\sin at)}{a^2 + p^2} \right]_0^\infty$$

$$= \frac{p}{p^2 + a^2}, \quad p > 0.$$

(vii) $F(t) = \sinh at$
(Meerut–1980, 83, 1991; 2009; Rohilkhand–1988; Andhra–1990; MS Univ.(T.N.)–2007)

SOLUTION. Consider $L\{\sinh at\} = L\left\{ \dfrac{e^{at} - e^{-at}}{2} \right\} = \dfrac{1}{2} L\{e^{at}\} - \dfrac{1}{2} L\{e^{-at}\}$ [Using (iv)]

$$= \frac{1}{2} \cdot \frac{1}{p-a} - \frac{1}{2} \cdot \frac{1}{p+a} = \frac{a}{p^2 - a^2}$$

Hence, $L\{\sinh at\} = \dfrac{a}{p^2 - a^2}$.

(viii) $F(t) = \cosh at$
(Meerut–1978, 80, 83; Rohilkhand–1988; Andhra–1990; Purvanchal–2001; Osmania–2004; Sagar–2004; MS Univ.(T.N.)–2007)

SOLUTION. Consider

$$L\{\cosh at\} = L\left[\frac{1}{2}(e^{at} + e^{-at}) \right] = \frac{1}{2} L\{e^{at}\} + \frac{1}{2} L\{e^{-at}\}$$

$$= \frac{1}{2}\cdot\frac{1}{p-a} + \frac{1}{2}\cdot\frac{1}{p+a}, \quad p > a \text{ and } p > -a$$

$$= \frac{p}{p^2 - a^2}, \quad p > |a|$$

Hence, $L\{\cosh at\} = \dfrac{p}{p^2 - a^2}$.

✎ RECAPITULATIONS: LAPLACE TRANSFORM OF ELEMENTARY FUNCTIONS

S. No.	F(t)	L[F(t)]	S. No.	F(t)	L[F(t)]
1.	1	$\dfrac{1}{p}, p > 0$	5.	sin at	$\dfrac{a}{p^2 + a^2}, p > a$
2.	t^n (n is a positive integer)	$\dfrac{n!}{p^{n+1}}, p > 0$	6.	cos at	$\dfrac{p}{p^2 + a^2}, p > |a|$
3	$t^a \ (a > -1)$	$\dfrac{\Gamma(a+1)}{p^{a+1}}, p > 0$	7.	sinh at	$\dfrac{a}{p^2 - a^2}, p > |a|$
4.	e^{at}	$\dfrac{1}{p-a}, p > a$	8.	cosh at	$\dfrac{p}{p^2 - a^2}, p > |a|$

Solved Examples

EXAMPLE 1.   *Find the Laplace transform of the function* $F(t) = \dfrac{e^{at} - 1}{a}$.

(Meerut–1988, 89, 92, 94; Purvanchal–2003)

SOLUTION.   We have $L\{F(t)\} = L\left\{\dfrac{e^{at}-1}{a}\right\} = L\left\{\dfrac{1}{a}e^{at} - \dfrac{1}{a}\right\}$

$$= \frac{1}{a}L\{e^{at}\} - \frac{1}{a}L\{1\} = \frac{1}{a}\left(\frac{1}{p-a}\right) - \frac{1}{a}\left(\frac{1}{p}\right) = \frac{1}{p(p-a)}.$$

EXAMPLE 2.   *Find* $L\{(t^2+1)^2\}$.   [Meerut–2009(B.P.)]

SOLUTION.   We have $L\{(t^2+1)^2\} = L\{t^4 + 2t^2 + 1\}$

$$= L\{t^4\} + 2L\{t^2\} + L(1) \qquad \text{(By linearity property)}$$

$$= \frac{4!}{p^5} + 2\cdot\frac{2!}{p^3} + \frac{1}{p} = \frac{24 + 4p^2 + p^4}{p^5}, \quad p > 0.$$

EXAMPLE 3.   *Find* $L\{F(t)\}$ *where* $F(t) = (\sin t - \cos t)^2$.

(Meerut–1986, 92; Bundelkhand–1995; Vikram–2004)

SOLUTION.   Consider $L\{(\sin t - \cos t)^2\} = L\{\sin^2 t + \cos^2 t - 2\sin t \cos t\}$

$$= L\{1 - \sin 2t\} = L\{1\} - L\{\sin 2t\}$$

$$= \frac{1}{p} - \frac{2}{p^2 + 2^2}, \ p > 0 = \frac{p^2 - 2p + 4}{p(p^2 + 4)}, \ p > 0$$

EXAMPLE 4.   *Find* $L\{6\sin 2t - 5\cos 2t\}$.   (Meerut–1999)

SOLUTION.   We have $L\{6\sin 2t - 5\cos 2t\}$

$$= 6L\{\sin 2t\} - 5L\{\cos 2t\}$$

$$= 6\cdot\frac{2}{p^2+2^2} - 5\cdot\frac{p}{p^2+2^2}, \ p>0 = \frac{12-5p}{p^2+4}, \ p>0.$$

EXAMPLE 5.   *Find* $L\{7e^{2t} + 9e^{-2t} + 5\cos t + 7t^3 + 5\sin 3t + 2\}$.

(Kanpur–2003; Agra–1993; Purvanchal–1994)

SOLUTION.   We have $L\{7e^{2t} + 9e^{-2t} + 5\cos t + 7t^3 + 5\sin 3t + 2\}$

$$= 7L\{e^{2t}\} + 9L\{e^{-2t}\} + 5L\{\cos t\} + 7.L\{t^3\} + 5.L\{\sin 3t\} + 2.L\{1\}$$

$$= \frac{7}{p-2} + \frac{9}{p+2} + \frac{5p}{p^2+1} + \frac{4^2}{p^4} + \frac{15}{p^2+9} + \frac{2}{p}$$

$$= \frac{4(4p-1)}{p^2-4} + \frac{5p}{p^2+1} + \frac{42}{p^4} + \frac{15}{p^2+4} + \frac{2}{p}.$$

EXAMPLE 6.   *Find* $L\{2e^{3t} - e^{-3t}\}$.   (Meerut–1983)

SOLUTION.   We have $L\{2e^{3t} - e^{-3t}\} = 2L\{e^{3t}\} - L\{e^{-3t}\} = 2.\dfrac{1}{p-3} - \dfrac{1}{p+3}, \ p>3 \text{ and } p>-3$

$$= \frac{p+9}{p^2-9}, \ p>|3|.$$

EXAMPLE 7.   *Find* $L\{F(t)\}$, *if* $F(t) = \begin{cases} e^t, & 0<t\le 1 \\ 0, & t>1 \end{cases}$ (Meerut–1994; Jhansi–2012)

SOLUTION.   We have $L\{f(t)\} = \int_0^\infty e^{-pt}F(t)\,dt = \int_0^1 e^{-pt}.e^t\,dt + \int_1^\infty e^{-pt}.0\,dt$

$$= \int_0^1 e^{-(p-1)t}.dt = \left[\frac{e^{-(p-1)t}}{-p-1}\right]_0^1$$

$$= \frac{1}{(p-1)}[1 - e^{-(p-1)}], \ p \ne 1.$$

EXAMPLE 8.   *Find* $L\{F(t)\}$, *where* $F(t) = \begin{cases} 0, & 0<t<1 \\ t, & 1<t<2 \\ 0, & t>2 \end{cases}$

(JNTU–2006; WBTU–2005; Lucknow–2010; Purvanchal–1989; Meerut–1978, 80, 91; TDC–1991; Rohilkhand–1997)

SOLUTION.   Since, $F(t)$ is not defined at $t = 0, 1$ and 2.

$$\therefore \quad L\{F(t)\} = \int_0^\infty e^{-pt}F(t)\,dt = \int_0^1 e^{-pt}.0\,dt + \int_1^2 e^{-pt}.t\,dt + \int_2^\infty e^{-pt}.0\,dt$$

$$= \int_1^2 e^{-pt}.tdt = \left[-t\frac{e^{-pt}}{p} - \frac{e^{-pt}}{p^2}\right]_1^2, \ p \ne 0$$

$$= -\left(\frac{2}{p} + \frac{1}{p^2}\right)e^{-2p} + \left(\frac{1}{p} + \frac{1}{p^2}\right)e^{-p}, \ p \ne 0.$$

EXAMPLE 9. *Find the Laplace transform of*

$$F(t) = \begin{cases} t^2, & 0 < t < 2 \\ t-1, & 2 < t < 3 \\ 7, & t > 3 \end{cases}$$

(UPTU–2007, Mumbai–2007)

SOLUTION. We have

$$L[F(t)] = \int_0^\infty e^{-pt} F(t) dt$$

$$= \int_0^2 t^2 e^{-pt} dt + \int_2^3 (t-1) e^{-pt} dt + \int_3^\infty 7 e^{-pt} dt$$

$$= \left[t^2 \frac{e^{-pt}}{(-p)} - 2t \frac{e^{-pt}}{(-p)^2} + 2 \frac{e^{-pt}}{(-p)^3} \right]_0^2 + \left[(t-1) \left(\frac{e^{-pt}}{(-p)} \right) - \frac{e^{-pt}}{(-p)^2} \right]_2^3 + 7 \left[\frac{e^{-pt}}{-p} \right]_3^\infty$$

$$= \left[-4 \left(\frac{e^{-2p}}{p} \right) - 4 \left(\frac{e^{-2p}}{p^2} \right) + \frac{2}{p^3} \right] + \left[2 \left(\frac{e^{-3p}}{-p} \right) - \left(\frac{e^{-3p}}{p^2} \right) + \left(\frac{e^{-2p}}{p} \right) + \frac{e^{-2p}}{p^2} \right] + 7 \left[0 + \frac{e^{-3p}}{p} \right]$$

$$= \frac{2}{p^3} + e^{-2p} \left[-\frac{4}{p} - \frac{4}{p^2} - \frac{2}{p^3} \right] + e^{-3p} \left[-\frac{2}{p} - \frac{1}{p^2} \right] + e^{-2p} \left[\frac{1}{p} + \frac{1}{p^2} \right] + e^{-3p} \left[\frac{7}{p} \right]$$

$$= \frac{2}{p^3} + e^{-2p} \left[-\frac{4}{p} - \frac{4}{p^2} - \frac{2}{p^3} + \frac{1}{p} + \frac{1}{p^2} \right] + e^{-3p} \left[-\frac{2}{p} - \frac{1}{p^2} + \frac{7}{p} \right]$$

$$= \frac{2}{p^3} - \frac{e^{-2p}}{p^3} (2 + 3p + 3p^2) + \frac{e^{-3p}}{p^2} (5p - 1)$$

EXAMPLE 10. *Find* $L\{F(t)\}$, *if* $F(t) = \begin{cases} 1, & 0 < t < 2 \\ t, & t > 2 \end{cases}$.

SOLUTION.

$$L\{F(t)\} = \int_0^\infty F(t) e^{-pt} dt = \int_0^2 1.e^{-pt} dt + \int_2^\infty t e^{-pt} dt$$

$$= \left[\frac{e^{-pt}}{-p} \right]_0^2 + \left[\left(\frac{e^{-pt}}{-p} \right) t - \left(\frac{e^{-pt}}{p^2} \right) .1 \right]_2^\infty$$

$$= -\frac{e^{-2p}}{p} + \frac{1}{p} - \frac{1}{p} \lim_{t \to \infty} \frac{t}{e^{pt}} + \frac{2}{p} e^{-2p} - \frac{1}{p^2} \lim_{t \to \infty} e^{-pt} + \frac{e^{-2p}}{p^2}$$

$$= \frac{1}{p} (1 + e^{-2p}) + \frac{1}{p^2} e^{-2p} - \frac{1}{p} \lim_{t \to \infty} \frac{t}{e^{pt}} - \frac{1}{p^2} \lim_{t \to \infty} e^{-pt}.$$

Now, if $p > 0$, we have

$$\lim_{t \to \infty} e^{-pt} = 0 \text{ and } \lim_{t \to \infty} \frac{t}{e^{pt}}$$ (Form ∞ / ∞)

$$= \lim_{t \to \infty} \frac{1}{p e^{pt}} = 0$$

Hence, $L\{F(t)\} = \frac{1}{p} [1 + e^{-2p}] + \frac{1}{p^2} e^{-2p}, p > 0$.

EXAMPLE 11. *Show that* $L \left\{ \dfrac{1}{\sqrt{\pi t}} \right\} = \dfrac{1}{\sqrt{p}}$.

(Avadh–2014)

SOLUTION. We have $L \left\{ \dfrac{1}{\sqrt{\pi t}} \right\} = \int_0^\infty e^{-pt} . \dfrac{1}{\sqrt{\pi t}} . dt = \dfrac{1}{\sqrt{\pi}} \int_0^\infty e^{-pt} . \dfrac{1}{\sqrt{t}} dt = \dfrac{1}{\sqrt{\pi}} \int_0^\infty e^{-pt} . t^{-1/2} dt$

$$= \frac{1}{\sqrt{\pi}} \int_0^\infty e^{-pt} . t^{-1/2 - 1} \, dt$$

$$= \frac{1}{\sqrt{\pi}} \frac{\boxed{1/2}}{p^{1/2}}, \quad p > 0 \qquad \text{(Using gamma function)}$$

$$= \frac{1}{\sqrt{\pi}} . \frac{\sqrt{\pi}}{p^{1/2}} \qquad\qquad \left[\because \Gamma(\tfrac{1}{2}) = \sqrt{\pi} \right]$$

$$= \frac{1}{\sqrt{p}} .$$

EXAMPLE 12. *Show that* $L \left\{ \dfrac{\cos \sqrt{t}}{\sqrt{t}} \right\} = \sqrt{\left(\dfrac{\pi}{p} \right)} e^{-1/4p}$.

(Mumbai–2009; Agra–1984, 98, 2008; Meerut–1982, 92, 2001, 02, 05; Kanpur–2002, 05; SVTU–2005; Guwahati–2007)

SOLUTION. Here, we have

$$\frac{\cos \sqrt{t}}{\sqrt{t}} = \frac{1}{\sqrt{t}} \left\{ 1 - \frac{1}{2!} (\sqrt{t})^2 + \frac{1}{4!} (\sqrt{t})^4 - \frac{1}{6!} (\sqrt{t})^6 + \ldots \right\}$$

$$= t^{-1/2} - \frac{1}{2!} t^{1/2} + \frac{1}{4!} t^{3/2} - \frac{1}{6!} t^{5/2} + \ldots$$

Therefore,

$$L \left\{ \frac{\cos \sqrt{t}}{\sqrt{t}} \right\} = L\{t^{-1/2}\} - \frac{1}{2!} L\{t^{1/2}\} + \frac{1}{4!} L\{t^{3/2}\} - \frac{1}{6!} L\{t^{5/2}\} + \ldots$$

$$= \frac{\Gamma(\tfrac{1}{2})}{p^{1/2}} - \frac{1}{2!} \frac{\Gamma(\tfrac{3}{2})}{p^{3/2}} + \frac{1}{4!} \frac{\Gamma(\tfrac{5}{2})}{p^{5/2}} - \frac{1}{6!} \frac{\Gamma(\tfrac{7}{2})}{p^{7/2}} + \ldots, p > 0$$

$$= \frac{\sqrt{\pi}}{p^{1/2}} - \frac{1}{1.2} . \frac{\tfrac{1}{2}.\sqrt{\pi}}{p^{3/2}} + \frac{\tfrac{3}{2}.\tfrac{1}{2}.\sqrt{\pi}}{1.2.3.4} . \frac{1}{p^{5/2}} - \frac{\tfrac{5}{2}.\tfrac{3}{2}.\tfrac{1}{2}.\sqrt{\pi}}{1.2.3.4.5.6} . \frac{1}{p^{7/2}} + \ldots$$

$$= \sqrt{\left(\frac{\pi}{p} \right)} \left[1 - \frac{1}{1!} \left(\frac{1}{4p} \right) + \frac{1}{2!} \left(\frac{1}{4p} \right)^2 - \frac{1}{3!} \left(\frac{1}{4p} \right)^3 + \ldots \right]$$

$$= \sqrt{\left(\frac{\pi}{p} \right)} . e^{-1/4p} .$$

EXAMPLE 13. *Evaluate* $L\{\cosh at\}$. (Meerut–1991; MDU–2004)

SOLUTION. Here, we have

$$L\{\cosh at\} = L \left\{ \frac{1}{2} (e^{at} + e^{-at}) \right\}$$

$$= \frac{1}{2} L\{e^{at}\} + \frac{1}{2} L\{e^{-at}\} = \frac{1}{2} \left[\frac{1}{p-a} \right] + \frac{1}{2} \left[\frac{1}{p+a} \right]$$

$$= \frac{1}{2} \left[\frac{p+a+p-a}{p^2 - a^2} \right] = p / (p^2 + a^2)$$

EXAMPLE 14. *Evaluate* $L\{\sin^3 2t\}$. (Delhi–1985)

SOLUTION. Here, we have $L\{\sin^3 2t\}$

We know that $\sin^3 \theta = \dfrac{1}{4} [3 \sin \theta - \sin 3\theta]$

$$\Rightarrow \quad \sin^3 2t = \frac{1}{4}[3\sin 2t - \sin 6t]$$

$$L\{\sin^3 2t\} = \frac{1}{4}L\{3\sin 2t - \sin 6t\} = \frac{3}{4}L\{\sin 2t\} - \frac{1}{4}L\{\sin 6t\}$$

$$= \frac{3}{4}\left[\frac{2}{p^2 + 2^2}\right] - \frac{1}{4}\left[\frac{6}{p^2 + 6^2}\right], p > 0$$

$$= \frac{3}{2}\left(\frac{1}{p^2 + 4}\right) - \frac{3}{2}\left[\frac{1}{p^2 + 36}\right] = 48/[(p^2 + 4)(p^2 + 36)]$$

Additional Solved Examples

EXAMPLE 1.   *Find* $L\{\sin t \cos t\}$ (Meerut–2012)

SOLUTION.   We have

$$L\{\sin t \cos t\} = L\left\{\frac{1}{2}\sin 2t\right\} = \frac{1}{2}L\{\sin 2t\} \qquad (\because \sin 2t = 2\sin t \cos t)$$

$$= \frac{1}{2} \cdot \frac{2}{p^2 + 2^2}, p > 0 \qquad\qquad \left[\because L\{\sin at\} = \frac{a}{p^2 + a^2}\right]$$

$$= \frac{1}{p^2 + 4}, p > 0.$$

EXAMPLE 2.   *Find* $L\{4\cos^2 t\}$.

SOLUTION.   We have $L\{4\cos^2 t\} = 4L[\cos^2 t] = 2L\{1 + \cos 2t\}$ $(\because \cos 2t = 2\cos^2 t - 1)$

$$= 2\left[\frac{1}{p} + \frac{p}{p^2 + 2^2}\right], p > 0$$

$$= \frac{2(2p^2 + 4)}{p(p^2 + 4)}, p > 0 = \frac{4p^2 + 8}{p(p^2 + 4)}, p > 0$$

EXAMPLE 3.   *Find* $L\{\sin^2 at\}$.

(Meerut–1991, 94; Purvanchal–2002, 05; Kanpur–1980; Nagarjuna–2006)

SOLUTION.   We have

$$L(\sin^2 at) = L\left\{\frac{1}{2}(1 - \cos 2at)\right\} = \frac{1}{2}[L\{1\} - L\{\cos 2at\}]$$

$$= \frac{1}{2}\left[\frac{1}{p} - \frac{p}{p^2 + (2a)^2}\right], p > 0 = \frac{2a^2}{p(p^2 + 4a^2)}, p > 0$$

EXAMPLE 4.   *Find* $L\{3\cosh 5t - 4\sinh 5t\}$

SOLUTION.   We have

$$L\{3\cosh 5t - 4\sinh 5t\} = 3L\{\cosh 5t\} - 4L\{\sinh 5t\}$$

$$= 3\left[\frac{p}{p^2 - 5^2}\right] - 4\left[\frac{5}{p^2 - 5^2}\right] = \frac{3p - 20}{p^2 - 25}, p > 5$$

EXAMPLE 5.   *Find* $L\{3t^4 - 2t^3 + 4e^{-3t} - 2\sin 5t + 3\cos 2t\}$

SOLUTION.   We have

$$L\{3t^4 - 2t^3 + 4e^{-3t} - 2\sin 5t + 3\cos 2t\}$$

$$= 3L\{t^4\} - 2L\{t^3\} + 4L\{e^{-3t}\} - 2L\{\sin 5t\} + 3L\{\cos 2t\}$$

$$= 3\frac{4!}{p^5} - 2\frac{3!}{p^4} + 4\frac{1}{p+3} - 2\frac{5}{p^2+5^2} + 3\frac{p}{p^2+2^2}, p > 0 \text{ and } p > -3, \text{ i.e., } p > 0$$

$$= \frac{72}{p^5} - \frac{12}{p^4} + \frac{4}{p+3} - \frac{10}{p^2+25} + \frac{3p}{p^2+4}, p > 0$$

EXAMPLE 6.   Find $L\{e^{-2t} - e^{-3t}\}$ (Meerut–1999)

SOLUTION.   We have $L\{e^{-2t} - e^{-3t}\}$

$$= L\{e^{-2t}\} - L\{e^{-3t}\} \qquad\qquad \text{(By linearly property)}$$

$$= \frac{1}{p+2} - \frac{1}{p+3}, p > -2 \text{ and } p > -3, \text{ i.e., } p > -2$$

$$= \frac{1}{p^2+5p+6}, p > -2$$

EXAMPLE 7.   *Find the Laplace Transform of the function* $F(t)$, *where* $F(t) = \begin{cases} \sin t, & 0 < t < \pi \\ 0, & t > \pi \end{cases}$

(Meerut–1991, 97, 99, 2009; Agra–1982; Ravishankar–2007; Calicut–2004; Madras–2000)

SOLUTION.   We have

$$L\{F(t)\} = \int_0^\infty e^{-pt} F(t)dt = \int_0^\pi e^{-pt} \sin t\, dt + \int_\pi^\infty e^{-pt} 0\, dt$$

$$= \left[\frac{e^{-pt}}{p^2+1}(-p\sin t - \cos t) \right]_0^\pi = \frac{e^{-p\pi}+1}{p^2+1}$$

EXAMPLE 8.   *Evaluate* $L\{F(t)\}$, *if* $F(t) = \begin{cases} (t-1)^2, & t > 1 \\ 0, & 0 < t < 1 \end{cases}$ (Meerut–1991; Agra–1993)

SOLUTION.   We have $L\{F(t)\} = \int_0^\infty e^{-pt} F(t)dt = \int_0^1 0.e^{-kt}dt + \int_1^\infty (t-1)^2 e^{-pt}dt$

$$= \int_1^\infty (t-1)^2 e^{-kt}dt = \int_0^\infty x^2 e^{-p(x+1)}dx,$$

(putting $t - 1 = x$, so that $dt = dx$ and changing the limits)

$$= \int_0^\infty e^{-p}e^{-px}x^2 dx = e^{-p}\int_0^\infty e^{-px}x^{3-1}dx = e^{-p}\frac{\Gamma(3)}{p^3}, p > 0$$

$$= e^{-p}\frac{2!}{p^3} = \frac{2e^{-p}}{p^3}, p > 0$$

[By L' Hospital Rule]

EXAMPLE 9.   *Find* $L\{F(t)\}$, *if* $F(t) = \begin{cases} e^t & 0 < t < 5 \\ 3 & t > 5 \end{cases}$

SOLUTION.   We have $L\{F(t)\} = \int_0^\infty e^{-pt} F(t)dt = \int_0^5 e^{-pt}e^{-t}dt + \int_5^\infty e^{-pt}.3dt$

$$= \int_0^5 e^{-(p-1)t}dt + 3\int_5^\infty e^{-pt}dt = \left[\frac{e^{-(p+1)t}}{(p-1)} \right]_0^5 + 3\left[\frac{e^{-pt}}{p} \right]_5^\infty$$

$$= \frac{1 - e^{-5(p-1)}}{(p-1)} + \frac{3}{p}e^{-5p}, p > 0$$

EXAMPLE 10. Find $L\{F(t)\}$, if $F(t) = \begin{cases} t, & 0 < t < 4 \\ 5, & t > 4 \end{cases}$

SOLUTION. We have $L\{F(t)\} = \int_0^\infty F(t)e^{-kt}dt = \int_0^4 te^{-pt}dt + \int_4^\infty 5e^{-pt}dt$

$$= \left[-t\frac{e^{-pt}}{p} - \frac{1}{p^2}e^{-pt}\right]_0^4 + \left[-\frac{5}{p}e^{-pt}\right]_4^\infty$$

$$= -\frac{4}{p}e^{-4p} - \frac{1}{p^2}(e^{-4p} - 1) + \frac{5}{p}e^{-4p}, p > 0$$

$$= \frac{1 + (p-1)e^{-4p}}{p^2}, p > 0$$

EXAMPLE 11. Find $L\{\sin\sqrt{t}\}$ (Agra–1990, 94, 98, 2000, 01, 02, 07; Kanpur–1999, 2004; SVTU–2004; Avadh–2012; Meerut–1994, 96, 2008, 11; Rohilkhand–2001; Garhwal–1996)

SOLUTION. We have $L\{\sin\sqrt{t}\} = L\left\{\sqrt{t} - \frac{(\sqrt{t})^3}{3!} + \frac{(\sqrt{t})^5}{5!} - \frac{(\sqrt{t})^7}{7!} + \ldots\right\}$

$$= L\left\{t^{1/2} - \frac{t^{3/2}}{3!} + \frac{t^{5/2}}{5!} - \frac{t^{7/2}}{7!} + \ldots\right\}$$

$$= L\{t^{1/2}\} - \frac{1}{3!}L\{t^{3/2}\} + \frac{1}{5!}L\{t^{5/2}\} - \frac{1}{7!}L\{t^{7/2}\} + \ldots$$

$$= \frac{\Gamma\left(\frac{3}{2}\right)}{p^{3/2}} - \frac{1}{3!}\frac{\Gamma\left(\frac{5}{2}\right)}{p^{5/2}} + \frac{1}{5!}\frac{\Gamma\left(\frac{7}{2}\right)}{p^{7/2}} - \frac{1}{7!}\frac{\Gamma\left(\frac{9}{2}\right)}{p^{9/2}} + \ldots$$

$$= \frac{\frac{1}{2}\sqrt{\pi}}{p^{3/2}} - \frac{1}{1.2.3}\frac{\frac{3}{2}\cdot\frac{1}{2}\sqrt{\pi}}{p^{5/2}} + \frac{1}{1.2.3.4.5}\frac{\frac{5}{2}\cdot\frac{3}{2}\cdot\frac{1}{2}\sqrt{\pi}}{p^{7/2}} + \ldots$$

$$= \frac{\sqrt{\pi}}{2p^{3/2}}\left[1 - \frac{1}{1!}\left(\frac{1}{4p}\right) + \frac{1}{2!}\left(\frac{1}{4p}\right)^2 - \frac{1}{3!}\left(\frac{1}{4p}\right)^3 + \ldots\right] = \frac{\sqrt{\pi}}{2p^{3/2}}e^{-1/4p}$$

EXAMPLE 12. Show that t^2 is of exponential order 3.

SOLUTION. We have

$$\lim_{t\to\infty}\{e^{-at}F(t)\} = \lim_{t\to\infty}\left(\frac{t^2}{e^{at}}\right) = \lim_{t\to\infty}\left(\frac{2t}{ae^{at}}\right)$$ [By L′ Hospital Rule]

$$= \lim_{t\to\infty}\frac{2}{a^2e^{at}} = 0 \text{ if } a > 0$$

Therefore, $F(t) = t^2$ is of exponential order.
Now, $|t^2| = t^2 < e^{3t}$ for all $t > 0$.
Hence, the given function is of exponential order 3.

EXAMPLE 13. Show that the function e^{t^2} is not of exponential order as $t \to \infty$.

SOLUTION. We have

$$\lim_{t\to\infty}\{e^{-at}F(t)\} = \lim_{t\to\infty}\{e^{-at}e^{t^2}\} = \lim_{t\to\infty}e^{t(t-a)}$$

$= \infty$ for all values of a.

Therefore, whatever be the value of a, we cannot find a number M such that

$$e^{t^2} < Me^{at}.$$

Hence, the given function is not of exponential order as $t \to \infty$.

EXAMPLE 14. *Show that the Laplace transforms of the function $F(t) = t^n, -1 < n < 0$ exists, although it is not a function of class A.*

SOLUTION. Clearly, $F(t) \to \infty$ as $t \to 0$ from the right, *i.e.*, the function is not piecewise continuous on every finite interval in the range $t \geq 0$.

We have

$$\lim_{t \to \infty} \{e^{-at} F(t)\} = \lim_{t \to \infty} \left(\frac{t^n}{e^{at}}\right) = \lim_{t \to \infty} \frac{1}{t^n e^{at}}$$

$$= \lim_{t \to \infty} \frac{1}{t^m e^{at}}, \text{where } 0 < m < 1$$

$$= 0, \text{ if } a > 0$$

$\therefore \qquad F(t) = t^n$ is of exponential order.

Now, since, $F(t) = t^n$ is not sectionally continuous over every finite interval in the range $t \geq 0$, hence it is not a function of class A. But t^n is integrable from 0 to any positive number t_0

Now, $L\{F(t)\} = \int_0^\infty e^{-pt} F(t) dt = \int_0^\infty e^{-pt} t^n dt = \int_0^\infty e^{-pt} t^{(n+1)-1} dt$

$$= \frac{\Gamma(n+1)}{p^{n+1}}, \text{ if } p > 0 \text{ and } n + 1 > 0, \textit{i.e., } n > -1.$$

Hence, the Laplace transform of t^n, $-1 < n < 0$ exists, although it is not a function of class A.

1.6 TRANSLATION OR SHIFTING THEOREMS

THEOREM 1. **(First Translation or Shifting Theorem). If $f(p)$ is the Laplace transform of $F(t)$, then $f(p-a)$ is the Laplace transforms of $e^{at}F(t)$, i.e., if $L\{F(t)\} = f(P)$, when $p > a$, then $L\{e^{at}F(t)\} = f(p-a)$, $p > a$.**

(Meerut–1981, 82, 86, 89, 91, 2014, 15; Purvanchal–1989, 91)

PROOF. We have, by definition of Laplace transform

$$L\{F(t)\} = f(p) = \int_0^\infty e^{-pt} F(t) dt$$

Therefore, $L\{e^{at} F(t)\} = \int_0^\infty e^{-pt} . e^{at} F(t) dt = \int_0^\infty e^{-(p-a)t} . F(t) dt$

$$= \int_0^\infty e^{-ut} F(t) dt,$$

where $u = p - a > 0$

$$= f(u) \qquad \text{(By definition)}$$

$$= f(p-a).$$

THEOREM 2. **(Second Translation or Heaviside's Shifting Theorem).**

If $L\{F(t)\} = f(p)$ and $G(t) = \begin{cases} F(t-a), & t > a \\ 0, & t < a \end{cases}$ then, $L\{G(t)\} = e^{-ap} f(p)$.

(UPTU–2006, 08; Meerut–1987, 88; Agra–1980)

PROOF. Let $L\{F(t)\} = f(p)$ and $G(t) = \begin{cases} F(t-a), & \text{if } t > a \\ 0, & \text{if } t < a \end{cases}$

Then, $\quad L\{G(t)\} = \int_0^\infty e^{-pt} G(t)\, dt$

$$= \int_0^a e^{-pt} G(t)\, dt + \int_a^\infty e^{-pt} G(t)\, dt$$

$$= \int_0^a e^{-pt} \cdot 0\, dt + \int_a^\infty e^{-pt} F(t-a)\, dt$$

$$= 0 + \int_a^\infty e^{-pt} F(t-a)\, dt$$

Let $t - a = u$, therefore $dt = du$.

If $t = a$, then $u = t - a = a - a = 0$ and if $t = \infty$, then $u = \infty - a = \infty$.

Hence, $L\{G(t)\} = \int_0^\infty e^{-p(u+a)} F(u)\, du = e^{-pa} \int_0^\infty e^{-pu} F(u)\, du = e^{-pa} f(p)$.

THEOREM 3.   **(Change of Scale Property). If** $L\{F(t)\} = f(p)$, **then** $L\{F(at)\} = \dfrac{1}{a} f\left(\dfrac{p}{a}\right)$.

(Meerut–1982, 86, 2006(B.P.); Agra–2007; Avadh–2006, 2010; Purvanchal–2000, 01, 13)

PROOF.   By definition

$$L\{F(at)\} = \int_0^\infty e^{-pt} F(at)\, dt = \int_0^\infty e^{-pu/a} F(u)\, \frac{du}{a} \qquad \text{(where } at = u\text{)}$$

$$= \frac{1}{a} \int_0^\infty e^{-pu/a} F(u)\, du = \frac{1}{a} \int_0^\infty e^{-su} F(u)\, du, \text{ where } s = \frac{p}{a}$$

$$= \frac{1}{a} f(s) = \frac{1}{a} f\left(\frac{p}{a}\right).$$

✍	**RECAPITULATIONS**		
S. No.	**Operation**	$F(t)$	$L\{F(t)\}$
1.	Linearity property	$a_1 F_1(t) + a_2 F_2(t)$	$a_1 L\{F_1(t)\} + a_2 L\{F_2(t)\}$
2.	First translation or Shifting theorem	$e^{ap} F(t)$	$f(p-a)$
3	Second translation or Shifting theorem	$G(t) = \begin{cases} F(t-a), & t > a \\ 0, & t \leqslant a \end{cases}$	$e^{-ap} f(p)$
4.	Change of scale property	$F(at)$	$\dfrac{1}{a} f\left(\dfrac{p}{a}\right)$

Solved Examples

EXAMPLE 1.   Find $L\left\{\dfrac{e^{-at}\, t^{n-1}}{(n-1)!}\right\}$.

SOLUTION.   We have $L\left\{\dfrac{t^{n-1}}{(n-1)!}\right\} = \dfrac{1}{(n-1)!} L[t^{n-1}] = \dfrac{1}{(n-1)!} \dfrac{(n-1)!}{p^n} = \dfrac{1}{p^n}$

Therefore, using first shifting theorem, we have

$$L\left\{e^{-at} \frac{t^{n-1}}{(n-1)!}\right\} = f(p+a) = \frac{1}{(p+a)^n}.$$

EXAMPLE 2.   *Find* $L\{e^t \cos^2 t\}$.   (Agra–2008; Osmania–1992)

SOLUTION.   We have

$$L\{\cos^2 t\} = L\left\{\frac{1}{2}(1+\cos 2t)\right\} = \frac{1}{2}[L\{1\} + L\{\cos 2t\}]$$

$$= \frac{1}{2}\left\{\frac{1}{p} + \frac{p}{p^2+2^2}\right\} = \frac{p^2+2}{p(p^2+4)} = f(p) \text{ (say)}$$

Using first shifting theorem, we get

$$L\{e^t \cos^2 t\} = f(p-1) = \frac{(p-1)^2+2}{(p-1)(p-1)^2+4}$$

$$= \frac{p^2-2p+3}{(p-1)(p^2-2p+5)}$$

EXAMPLE 3.   *Find* $L\{e^{-t}(3\sin 2t - 5\cosh 2t)\}$.   (Meerut–1983, 98; Andhra–1990)

SOLUTION.   We have

$$L\{3\sin 2t - 5\cosh 2t\} = 3.\frac{2}{p^2+2^2} - \frac{5p}{p^2-2^2} = f(p)   \text{(say)}.$$

Using first shifting theorem, we have

$$L\{e^{-t}(3\sin 2t - 5\cosh 2t)\} = f(p+1) = \frac{6}{(p+1)^2+4} - \frac{5(p+1)}{(p+1)^2-4} = \frac{6}{p^2+2p+5} - \frac{5(p+1)}{p^2+2p-3}.$$

EXAMPLE 4.   *Find* $L\{e^{-t}(3\sinh 2t - 5\cosh 2t)\}$.   (Meerut–1981, 84, 87, 88, 2007; Andhra–1990;
Purvanchal–1990; Garhwal–1996, 99, 2003; Amravati–2006; Bilaspur–2004; Gwalior–2003)

SOLUTION.   We have

$$L\{3\sinh 2t - 5\cosh 2t\} = 3\frac{2}{p^2-2^2} - 5\frac{p}{p^2-2^2} = \frac{6-5p}{p^2-4} = f(p)   \text{(say)}$$

Using first shifting theorem, we have

$$L\{e^{-t}(3\sinh 2t - 5\cosh 2t)\} = f(p+1) = \frac{6-5(p+1)}{(p+1)^2-4} = \frac{1-5p}{p^2+2p-3}.$$

EXAMPLE 5.   *Find* $L\{e^t (t+3)^2\}$.   (Meerut–1986; Agra–1998; Rohilkhand–2004)

SOLUTION.   We have

$$L\{(t+3)^2\} = L\{t^2 + 6t + 9\}$$

$$= \frac{2!}{p^3} + 6.\frac{1!}{p^2} + \frac{9}{p} = \frac{2+6p+9p^2}{p^3} = f(p)   \text{(say)}$$

Using first shifting theorem, we have

$$L\{(t+3)^2 e^t\} = f(p-1) = \frac{2+6(p-1)+9(p-1)^2}{(p-1)^3} = \frac{9p^2-12p+5}{(p-1)^3}.$$

EXAMPLE 6. If $L\{\cos^2 t\} = \dfrac{p^2 + 2}{p(p^2 + 4)}$, find $L[\cos^2 at]$.

(UPTU–2006)

SOLUTION. We have $L\{\cos^2 t\} = \dfrac{p^2 + 2}{p(p^2 + 4)}$

By change of scale property, we have

$$L\{\cos^2 at\} = \frac{1}{a} \cdot \frac{\left(\dfrac{p}{a}\right)^2}{\left(\dfrac{p}{a}\right)\left[\left(\dfrac{p}{a}\right)^2 + 4\right]}$$

$$= \frac{1}{p}\left[\frac{p^2 + 2a^2}{\dfrac{p}{a}(p^2 + 4a^2)}\right]$$

$$= \frac{p^2 + 2a^2}{p(p^2 + 4a^2)}.$$

SOME RESULTS OBTAINED BY SHIFTING THEOREM
1. $L(e^{at} \cdot 1) = \dfrac{1}{p - a}, p > a$
2. $L(e^{at} \cdot t^n) = \dfrac{n!}{(p - a)^{n+1}}, n > -1, p > a$
3. $L(e^{at} \sin bt) = \dfrac{b}{(p - a)^2 + b^2}, p > a$
4. $L[e^{at} \cos bt] = \dfrac{p - a}{(p - a)^2 + b^2}, p > a$
5. $L[e^{at} \sinh bt] = \dfrac{b}{(p - a)^2 - b^2}, p > a$
6. $L[e^{at} \cosh bt] = \dfrac{p - a}{(p - a)^2 + b^2}, p > a$

EXAMPLE 7. Given $L\{F(t)\} = \dfrac{p^2 - p + 1}{(2p + 1)^2(p - 1)}$.

Applying the change of scale property, show that $L\{F(2t)\} = \dfrac{p^2 - 2p + 4}{4(p + 1)(p - 2)}$.

SOLUTION. Given that

$$L\{F(t)\} = \frac{p^2 - p + 1}{(2p + 1)^2(p - 1)} = f(p) \text{ (say)}$$

By using change of scale property, we have

$$L\{F(2t)\} = \frac{1}{2} f\left(\frac{p}{2}\right)$$

$$= \frac{1}{2} \cdot \frac{\left(\dfrac{p}{2}\right)^2 - \left(\dfrac{p}{2}\right) + 1}{\left[2 \cdot \left(\dfrac{p}{2}\right) + 1\right]^2 \left(\dfrac{p}{2} - 1\right)} = \frac{p^2 - 2p + 4}{4(p + 1)^2(p - 2)}.$$

EXAMPLE 8. Find $L\{F(t)\}$, where $F(t) = \begin{cases} \cos\left(t - \dfrac{2}{3}\pi\right), & t > \dfrac{2\pi}{3} \\ 0, & t < \dfrac{2\pi}{3} \end{cases}$

(Meerut–1986, 91, 92, 2008; Agra–1983; MDU–2007)

SOLUTION. Let $F(t) = \cos t$

Then, $G(t) = \begin{cases} F\left(t - \dfrac{2\pi}{3}\right), & t > 2\pi/3 \\ 0, & t < 2\pi/3 \end{cases}$

We have $L\{F(t)\} = L\{\cos t\} = \dfrac{p}{p^2+1} = f(p)$ (say)

Using second translation or shifting theorem, we have

$$L\{G(t)\} = e^{\left(-\frac{2\pi}{3}\right).p} \cdot f(p) = e^{-2\pi p/3} \cdot \dfrac{p}{p^2+1}.$$

EXAMPLE 9. Find $L\{G(t)\}$, where $G(t) = \begin{cases} e^{t-a}, & t > a \\ 0, & t < a \end{cases}$.   (UPTU–2008)

SOLUTION. By second shifting theorem, we have

If $L\{F(t)\} = f(p)$ and $G(t) = \begin{cases} F(t-a), & t > a \\ 0, & t < a \end{cases}$. Then, $L\{G(t)\} = e^{-ap} f(p)$

Let $F(t) = e^t$

Then, $L\{F(t)\} = L\{e^t\} = \int_0^\infty e^{-pt} \cdot e^t \, dt = \dfrac{1}{p-1}, p > 1 = f(p)$ (say)

Now, let $G(t) = \begin{cases} F(t-a) = e^{t-a}, & t > a \\ 0, & t < a \end{cases}$

Then, $L\{G(t)\} = e^{-ap} f(p) = \dfrac{e^{-ap}}{p-1}, \ p > 1$.

EXAMPLE 10. Evaluate $L\{e^{-t}\cos^2 t\}$ (Bangalore–1985)

SOLUTION. Here, we have

$$L\left\{e^{-t}\cos^2 t\right\} = L\left\{\frac{1}{2}e^{-t}(1+\cos 2t)\right\} = \frac{1}{2}L\left\{e^{-t}\right\} - \frac{1}{2}L\left\{e^{-t}\cos 2t\right\}$$

$$= \frac{1}{2}\left[\frac{1}{p+1} + \frac{(p+1)}{(p+1)^2+2^2}\right] = \frac{1}{2}\left[\frac{(p^2+2p+5)+(p+1)^2}{(p+1)(p^2+2p+5)}\right]$$

$$= \frac{p^2+2p+3}{(p+1)(p^2+2p+5)}$$

EXAMPLE 11. Evaluate $L\{t^5 e^{3t}\}$ (Kanpur–2004)

SOLUTION. Here, we have

$$L\{t^5\} = \frac{\Gamma(n+1)}{p^{n+1}} \text{ if } n = 5.$$

$\Rightarrow$ $L\{t^5\} = \dfrac{\Gamma(6)}{p^6} = \dfrac{5!}{p^6}$

$\Rightarrow$ $L\{t^5 e^{3t}\} = \dfrac{120}{(p-3)^6}$ as $L\{e^{at} F(t)\} = f(p-a)$.

EXAMPLE 12. Prove that $L\{e^{2t}(\cos 4t + 3\sin 4t)\} = \dfrac{p+10}{p^2-4p+20}$.

SOLUTION. Here, we have

$$L\{e^{2t}(\cos 4t + 3\sin 4t)\}$$

We know that

$$L\{\cos(at)\} = \frac{p}{p^2 + a^2} \text{ and } L\{\sin(at)\} = \frac{p}{p^2 + a^2}$$

So, $L\{\cos 4t + 3\sin 4t\} = \dfrac{p}{p^2 + 4^2} + \dfrac{3(4)}{p^2 + 4^2} = \dfrac{p + 12}{p^2 + 16}$

Since $L\{e^{at} F(t)\} = f(p-a)$, then we get

$$L\{e^{2t}(\cos 4t + 3\sin 4t)\} = \frac{(p-2)+12}{(p-2)^2 + 16} = \frac{p+10}{p^2 - 4p + 20}$$

EXAMPLE 13.   Evaluate $L\{e^{at} \cos bt\}$ (Kanpur–2008)

SOLUTION.   We know that

$$L\{\cos bt\} = \frac{p}{p^2 + b^2} \implies L\{e^{at} \cos bt\} = \frac{p-a}{(p-a)^2 + b^2}$$

$$L\{e^{at} F(t)\} = f(p-a)$$

EXAMPLE 14.   (i) Evaluate $L\{e^{4t} \cosh 5t\}$ (Meerut–1995)

           (ii) $L\{e^{-t}(3 \sin 2t - 5 \cos 2t)\}$ (Bhopal–2004)

SOLUTION.   (i)   It is given that

$$L\{e^{4t} \cosh 5t\} = \frac{p}{(p^2 - 5^2)} = f(p), \text{ say} \qquad\qquad \dots(i)$$

Now, from first shifting theorem, we get

$$L\{e^{4t} \cosh 5t\} = f(p-4) = \frac{(p-4)}{(p-4)^2 - 5^2} = \frac{p-4}{p^2 - 8p - 9}$$

     (ii)   Consider $L\{3 \sin 2t - 5 \cos 2t\}$

$$= 3L\{\sin 2t\} - 5L\{\cos 2t\}$$

$$= 3 \times \frac{2}{p^2 + 2^2} - 5 \times \frac{p}{p^2 + 2^2} = \frac{6 - 5p}{p^2 + 4} = f(p), \text{ say}$$

Now, from first shifting theorem, we get

$$L\{e^{-t}(3\sin 2t - 5\cos 2t)\} = f(p+1) = \frac{6 - 5(p+1)}{(p+1)^2 + 4} = \frac{1 - 5p}{p^2 + 2p + 5}$$

EXAMPLE 15.   Find $L\{e^{t-1} H(t-1)\}$.   (Purvanchal–2004)

SOLUTION.   Let us assume $F(t) = e^t$, then we have

$$L\{F(t)\} = 1/(p-1) = f(p), \text{ say}$$

Now, from second shifting theorem, we get

$$L\{e^{t-1} H(t-1)\} = e^{-p} f(p) = e^{-p}/(p-1)$$

$$[\because L\{F(t)\} = f(p) \implies L\{F(t-1)H(t-a)\} = e^{-ap} f(p)]$$

Additional Solved Examples

EXAMPLE 1.   Find $L\{t^3 e^{-3t}\}$ (Meerut–1984, 2007; Kanpur–2008)

SOLUTION.   We have $L\{t^3\} = \dfrac{3!}{p^4} = \dfrac{6}{p^4} = f(p) \text{ (say)}$

$\therefore$ Using first shifting theorem, we have

$$L\{t^3 e^{-3t}\} = f(p+3) = \frac{6}{(p+3)^4}$$

EXAMPLE 2. *Find $L\{e^{3t}\cos 5t\}$.*

SOLUTION. We have $\qquad L\{\cos 5t\} = \dfrac{p}{p^2+5^2} = \dfrac{p}{p^2+25} = f(p)$

$\therefore$ Using first shifting theorem, we have

$$L\{e^{3t}\cos 5t\} = f(p-3) = \frac{p-3}{(p-3)^2+25} = \frac{p-3}{p^2-6p+34}$$

EXAMPLE 3. *Find $L\{e^{-t}\sin^2 t\}$.*

SOLUTION. We have $\qquad L\{\sin^2 t\} = L\left\{\dfrac{1}{2}(1-\cos 2t)\right\}$

$$= \frac{1}{2}\left[\frac{1}{p} - \frac{p}{p^2+2^2}\right] = \frac{2}{p(p^2+4)} = f(p), \text{ say}$$

$\therefore$ Using first shifting theorem, we have

$$L\{e^{-t}\sin^2 t\} = f(p+1) = \frac{2}{(p+1)\{(p+1)^2+4\}} = \frac{2}{(p+1)(p^2+2p+5)}$$

EXAMPLE 4. *Find $L\{e^t\sin^2 t\}$*

SOLUTION. We have $\qquad L\{\sin^2 t\} = L\left\{\dfrac{1}{2}(1-\cos 2t)\right\}$

$$= \frac{1}{2}\left[\frac{1}{p} - \frac{p}{p^2+2^2}\right] = \frac{2}{p(p^2+4)} = f(p), \text{ say}$$

$\therefore$ Using first shifting theorem, we have

$$L\{e^t\sin^2 t\} = f(p-1) = \frac{2}{(p-1)\{(p-1)^2+4\}} = \frac{2}{(p-1)(p^2-2p+5)}$$

EXAMPLE 5. *Find $L\{e^{-4t}\cosh 2t\}$*

SOLUTION. We have

$$L\{\cosh 2t\} = \frac{p}{p^2-2^2} = \frac{p}{p^2-4} = f(p), \text{ say}$$

$\therefore$ From first shifting theorem, we have

$$L\{e^{-4t}\cosh 2t\} = f(p+4) = \frac{p+4}{(p+4)^2-4} = \frac{p+4}{p^2+8p+12}$$

EXAMPLE 6. *Find $L\{e^{-2t}(3\cos 6t - 5\sin 6t)\}$*

SOLUTION. We have,

$$L\{3\cos 6t - 5\cos 6t\} = 3\frac{p}{p^2+6^2} - 5\frac{6}{p^2+6^2} = \frac{3p-30}{p^2+36} = f(p), \text{ say}$$

$\therefore$ Using first shifting theorem, we have

$$L\{e^{-2t}(3\cos 6t - 5\sin 6t)\} = f(p+2) = \frac{3(p+2)-30}{(p+2)^2+36} = \frac{3p-24}{p^2+4p+40}$$

EXAMPLE 7. *Using first shifting theorem, find the value of* $L\{e^{6t}(t+2)^2\}$ (Meerut–1981, 91)

SOLUTION. We have

$$L\{(t+2)^2\} = L\{t^2 + 4t + 4\} = \frac{2!}{p^3} + 4\frac{1!}{p} + \frac{4}{p} = \frac{2 + 4p + 4p^2}{p^3} = f(p), \text{(say)}$$

∴ Using first shifting theorem, we have

$$L\{e^{6t}(t+2)^2\} = f(p-6) = \frac{2+4(p-6)}{(p-6)^3} \cdot \frac{(p-6)^2}{} = \frac{4p^2 - 44p + 122}{(p-6)^3}$$

EXAMPLE 8. *If* $L\{F(t)\} = f(p)$, *find* $L\{F(t)\cos\omega t\}$

SOLUTION. We have $L\{F(t)\} = f(p)$,

∴ Using first shifting theorem, we h

$$L\{F(t)\cos\omega t\} = L\left\{F(t)\left[\frac{1}{} \quad e^{-i\omega t}\right)\right\}$$

$$^{\omega t}\} + L\{F(t)e^{-i\omega t}\}] = \frac{1}{2}[f(p - i\omega) + f(p + i\omega)]$$

EXAMPLE 9. *Applying change of* ⅄, *find*

(i) $L\{\sinh 3t^1$ (ii) $L\{\cos 5t\}$

SOLUTION. (i) We hav

$$\cdot t\} = \frac{1}{p^2 - 1} = f(p), \text{ say}$$

.ge of scale property

$$L\{\sinh 3t\} = \frac{1}{3}f(p/3) = \frac{1}{3} \cdot \frac{1}{\left(\dfrac{p}{3}\right)^2 - 1} = \frac{3}{p^2 - 9}$$

have $L\{\cos t\} = \dfrac{p}{p^2 + 1}, p > 0 = f(p), \text{ say}$

∴ Using change of scale property, we have

$$L\{\cos 5t\} = \frac{1}{5}f\left(\frac{p}{5}\right) = \frac{1}{5}\frac{p/5}{(p/5)^2 + 1} = \frac{p}{p^2 + 25}, p > 0$$

EXAMPLE 10. *Find* $L\{F(t)\}$, *where* $F(t) = \begin{cases} \sin\left(t - \dfrac{\pi}{3}\right), & t > \pi/3 \\ 0, & t < \pi/3 \end{cases}$

(Meerut–2009(BP); Agra–1978; Rohilkhand–1996, 97)

SOLUTION. Let $\phi(t) = \sin t$

Then $F(t) = \begin{cases} \phi(t - \pi/3), & t > \dfrac{\pi}{3} \\ 0, & t < \dfrac{\pi}{3} \end{cases}$

We have $L\{\phi(t)\} = L\{\sin t\} = \dfrac{1}{p^2 + 1} = f(p), \text{ say}$

∴   Using second shifting theorem, we have

$$L\{F(t)\} = e^{-(\pi/3)p} \cdot f(p) = e^{-\pi p/3} \cdot \frac{1}{p^2 + 1}, p > 0.$$

EXAMPLE 11. *Find L {F(t)}, where* $F(t) = \begin{cases} \sin\left(t - \dfrac{2}{3}\pi\right), & t > 2\pi/3 \\ 0, & t < 2\pi/3 \end{cases}$ (Agra–1997; Osmania–2004)

SOLUTION.   Let $\phi(t) = \sin t$

Then $F(t) = \begin{cases} \phi\left(t - \dfrac{2}{3}\pi\right), & t > \dfrac{2}{3}\pi \\ 0, & t < \dfrac{2}{3}\pi \end{cases}$

We have $L\{\phi(t)\} = L\{\sin t\} = \dfrac{1}{p^2 + 1} = f(p)$, say

∴   Using second shifting theorem, we have

$$L\{F(t)\} = e^{-\left(\frac{2}{3}\pi\right)p} f(p) = e^{-\frac{2\pi p}{3}} \frac{1}{p^2 + 1} = \frac{e^{-2\pi p/3}}{p^2 + 1}, p > 0.$$

EXAMPLE 12. *If* $\{F(t)\} = \dfrac{1}{p}e^{-1/p}$, *show that* $L\{e^{-t}F(3t)\} = \dfrac{e^{-3/(p+1)}}{p+1}$.

SOLUTION.   Given that $L\{F(t)\} = \dfrac{1}{p}e^{-1/p} = f(p)$, say

∴   By the change of scale property, we have

$$L\{F(3t)\} = \frac{1}{3}f(p/3) = \frac{1}{3} \cdot \frac{1}{(p/3)}e^{-1/(p/3)} = \frac{1}{p}e^{-3/p}$$

$$= f_1(p), \text{ say.}$$

∴   Using first shifting theorem,

$$L\{e^{-t}F(3t)\} = f_1(p+1) = \frac{1}{(p+1)}e^{-3/(p+1)}.$$

1.7 LAPLACE TRANSFORM OF DERIVATIVES

THEOREM 1.   *Let* F(t) *be continuous for all* $t \geq 0$ *and be of exponential order as* $t \to \infty$ *and if* F'(t) *is of class A, the Laplace transforms of derivatives* F'(t) *exists when* $p > a$ *and* $L\{F'(t)\} = p L\{F(t)\} - F(0)$.

(Meerut–1987, 91; Agra–1984; Banglore–1997; Nagpur–2005; Madras–2005)

PROOF.   By definition, we have

$$L\{F'(t)\} = \int_0^\infty e^{-pt}F'(t)\, dt = \left[e^{-pt}F(t)\right]_0^\infty + p\int_0^\infty e^{-pt} F(t)\, dt$$

(On integrating by parts)

$$= -F(0) + pL\{F(t)\} \qquad \left[\because \lim_{t\to\infty} e^{-pt} F(t) = 0\right]$$

$$= pL\{F(t)\} - F(0).$$

GENERALIZATION
*   Proceeding same as above, we get

$L\{F''(t)\} = pL\{F'(t)\} - F'(0) = p[p L\{F(t)\} - F(0)] - F'(0)$

$\qquad = p^2 L\{F(t)\} - p F(0) - F'(0) = p^2 f(p) - p F(0) - F'(0).$

THEOREM 2. *If $F(t)$, $F'(t)$, ..., $F^{n-1}(t)$ are continuous for $t \geq 0$ and be of exponential order as $t \to \infty$ and if $F^n(t)$ is of class A and if $L\{F(t)\} = f(p)$, then*

$$L\{F^n(t)\} = p^n f(p) - p^{n-1}F(0) - p^{n-2}F(0)... - pF^{(n-2)}(0) - F^{(n-1)}(0)$$

$$= p^n f(p) - \sum_{r=0}^{n-1} p^{n-1-r} F^r(0)$$

PROOF. Using above theorem, we have

$$L\{F'(t)\} = pL\{F(t)\} - F(0) \quad \text{and} \quad L\{F''(t)\} = p^2 L\{F(t)\} - pF(0) - F'(0)$$

Similarly, we can find

$$L\{F'''(t)\} = pL\{F''(t)\} - F''(0) = p[p^2 L\{F(t)\} - p F(0) - F'(0)] - F''(0)$$

$$= p^3 L\{F(t)\} - p^2 F(0) - pF'(0) - F''(0).$$

Proceeding, similarly, we get

$$L\{F^n(t)\} = p^n L\{F(t)\} - p^{n-1}F(0) - p^{n-2}F'(0) - ... - F^{n-1}(0)$$

$$= p^n L\{F(t)\} - \sum_{r=0}^{n-1} p^{n-1-r} F^r(0).$$

THEOREM 3. *If $F(t)$ is a function of class A and if $L\{F(t)\} = f(p)$, then $L\{t \cdot F(t)\} = -f'(p)$*

(Rohilkhand–1991, 95; Meerut–1987, 90)

PROOF. We know that

$$f(p) = L\{F(t)\} = \int_0^\infty e^{-pt} F(t)\, dt.$$

Therefore,

$$f'(p) = \frac{d}{dp} \int_0^\infty e^{-pt} F(t)dt = \int_0^\infty \frac{\partial}{\partial p} \{e^{-pt}F(t)\}\, dt$$

(By Leibnitz's rule of differentiation under the sign of integral)

$$= -\int_0^\infty t\, e^{-pt} F(t)\, dt = -\int_0^\infty e^{-pt} \{t\, F(t)\}\, dt = -L\{t\, F(t)\}$$

$$\Rightarrow \quad L\{t\, F(t)\} = -f'(p).$$

THEOREM 4. *If $F(t)$ is a function of class A and if $L\{F(t)\} = f(p)$*   **Then,**

$$L\{t^n F(t)\} = (-1)^n \frac{d^n}{dp^n} f(p).$$

(UPTU–2005; Rohilkhand–1992, 94, 2000; Agra–1987; Meerut–1983)

PROOF. We shall prove this theorem by the principle of Mathematical induction.

Step 1. Using previous theorem, we have

$$L\{t\, F(t)\} = (-1)^1 \frac{d}{dp} f(p) \quad \Rightarrow \quad \text{Theorem is true for } n = 1.$$

Step 2. Assume that the theorem is true for a particular value of n say k. Then, we have

$$L\{t^k\, F(t)\} = (-1)^k \frac{d^k}{dp^k} f(p) \quad \Rightarrow \quad \int_0^\infty e^{-pt}\, t^k\, F(t)\, dt = (-1)^k \frac{d^k}{dp^k} f(p).$$

Step 3. Differentiating both sides w.r.t. p, we have

$$\frac{d}{dp} \int_0^\infty e^{-pt}\, t^k\, F(t)\, dt = (-1)^k \frac{d^{k+1}}{dp^{k+1}} f(p).$$

Applying, Leibnitz's rule for differentiation under the sign of integration, we have

$$-\int_0^\infty e^{-pt} \, t^{k+1} \, F(t) \, dt = (-1)^{k+2} \frac{d^{k+1}}{dp^{k+1}} \, f(p)$$

$$\Rightarrow \qquad \int_0^\infty e^{-pt} \{t^{k+1} \, F(t)\} \, dt = (-1)^{k+1} \frac{d^{k+1}}{dp^{k+1}} \, f(p)$$

$$\Rightarrow \qquad L \{t^{k+1} \, F(t)\} = (-1)^{k+1} \frac{d^{k+1}}{dp^{k+1}} \, f(p)$$

$\Rightarrow$ Theorem is true for $n = k+1$.

Hence, by the principle of mathematical induction, it is true for every positive integral value of n.

THEOREM 5. **Let a function $F(t)$ be periodic with period w, so that $F(t + nw) = F(t)$ for $n = 1, 2, 3, \dots$, then $L \{F(t)\} = \dfrac{1}{1 - e^{-pw}} \int_0^w e^{-pt} \, F(t) \, dt$.** (Agra–1982; MDU–2007)

PROOF. We know that

$$L\{F(t)\} = \int_0^\infty e^{-pt} \, F(t) \, dt$$

$$= \int_0^w e^{-pt} \, F(t) \, dt + \int_w^{2w} e^{-pt} \, F(t) \, dt + \dots + \int_{nw}^{(n+1)w} e^{-pt} \, F(t) \, dt + \dots$$

$$= \sum_{n=0}^\infty \int_{nw}^{(n+1)w} e^{-pt} \, F(t) \, dt = \sum_{n=0}^\infty \int_0^w e^{-p(x+nw)} \, F(x + nw) \, dx$$

$$\left[t = x + nw \right]$$

$$= \sum_{n=0}^\infty \int_0^w e^{-px} \, e^{-npw} \, F(x) \, dx$$

$$[\because \text{For periodic function, } f(x + nw) = f(x)]$$

$$= \sum_{n=0}^\infty e^{-npw} \int_0^w e^{-px} \, F(x) \, dx$$

$$= (1 + e^{-pw} + e^{-2pw} + \dots) \int_0^w e^{-px} \, F(x) \, dx$$

$$= \frac{1}{1 - e^{-pw}} \int_0^w e^{-px} \, F(x) \, dx \qquad \left[\because e^{-pw} < 1 \right]$$

THEOREM 6 **(Initial Value Theorem). Let $F(t)$ be continuous for all $t \geq 0$ and be of exponential order as $t \to \infty$ and if $F'(t)$ is of class A, then $\lim\limits_{t \to 0} F(t) = \lim\limits_{p \to \infty} pL\{F(t)\}$.**

PROOF. We know that

$$L\{F'(t)\} = \int_0^\infty e^{-pt} \, F'(t) dt = pL \{F(t)\} - F(0) . \qquad \dots(1)$$

Since $F'(t)$ is sectionally continuous and of exponential order.

Therefore, $\lim\limits_{p \to \infty} \int_0^\infty e^{-pt} \, F'(t) \, dt = 0$

Now, taking limit as $p \to \infty$ in (1), we have

$$0 = \lim_{p \to \infty} pL \{F(t)\} - F(0) \quad \Rightarrow \quad F(0) = \lim_{p \to \infty} pL \{F(t)\}$$

$$\Rightarrow \qquad \lim_{t \to 0} F(t) = \lim_{p \to \infty} pL \{F(t)\} .$$

THEOREM 7   **(Final Value Theorem).** *Let* $F(t)$ *be continuous for all* $t \geq 0$ *and be of exponential order as* $t \to \infty$ *and if* $F'(t)$ *is of class A, then*

$$\lim_{t \to \infty} F(t) = \lim_{p \to 0} pL \{F(t)\}.$$ (Osmania–2004)

PROOF.   We know that

$$L\{F'(t)\} = \int_0^\infty e^{-pt} F'(t) \, dt = pL \{F(t)\} - F(0). \qquad \dots(1)$$

Taking limit as $p \to 0$ in (1), we get

$$\lim_{p \to 0} \int_0^\infty e^{-pt} F'(t) dt = \lim_{p \to 0} [pL \{F(t)\} - F(0)]$$

$$\Rightarrow \qquad \int_0^\infty F'(t) \, dt = \lim_{p \to 0} pL \{F(t)\} - F(0)$$

$$\Rightarrow \qquad \left[F(t)\right]_0^\infty = \lim_{p \to 0} pL \{F(t)\} - F(0)$$

$$\Rightarrow \quad \lim_{t \to \infty} F(t) - F(0) = \lim_{p \to 0} pL \{F(t)\} - F(0)$$

$$\Rightarrow \qquad \lim_{t \to \infty} F(t) = \lim_{p \to 0} pL \{F(t)\}.$$

THEOREM 8   **(Laplace Transform of the Laplace Transform).** *We have*

$$L[L\{F(t)\}] = L\left\{\int_0^\infty e^{-pt}F(t) \, dt\right\} = \int_0^\infty e^{-up}\left\{\int_0^\infty e^{-pt} F(t) \, dt\right\} dp.$$

PROOF.   The area of integration being the whole positive quadrant. Now, changing the order of integration, we get

$$L[L\{F(t)\}] = \int_0^\infty F(t)\left\{\int_0^\infty e^{-p(t+u)}dp\right\} dt = \int_0^\infty F(t)\left\{\left[\frac{e^{-p(t+u)}}{-(t+u)}\right]_{p=0}^\infty\right\} dt$$

$$= \int_0^\infty \frac{F(t)}{t+u} \, dt.$$

THEOREM 9   **(Laplace Transforms of Integrals).** *If* $F(t)$ *is piecewise continuous and satisfies* $|F(t)| \leq Me^{at}, \forall t \geq 0$ *for some constant a and M, then*

$$L\left\{\int_0^t F(x) \, dx\right\} = \frac{1}{p} L \{F(t)\} \qquad (p > 0, \ p > a)$$

PROOF.   Let $F(t)$ be piecewise continuous such that   .

$$|F(t)| \leq M e^{at} \; ; \text{ for some constants } a \text{ and } M. \qquad \dots(1)$$

If (1) holds for some negative value of a, then it also holds for positive value of a. Therefore, suppose that a is positive.

Let $G(t) = \int_0^t F(x) \, dx$.

Then $G(t)$ is continuous.   ($\because$ Integral of an integrable function is continuous)

Now, $|G(t)| \leq \int_0^t |F(x)| \, dx \leq \int_0^t Me^{ax} \, dx$

$$\Rightarrow \qquad |G(t)| \leq \frac{M}{a} (e^{at} - 1), \ a > 0 \qquad \dots(2)$$

Further, $G'(t) = F(t)$, except for points at which $F(t)$ is discontinuous. Therefore, $G'(t)$ is piecewise continuous on each finite interval.

$$\therefore \qquad L\{G'(t)\} = pL \{G(t)\} - G(0) = pL \{G(t)\} \qquad [\because G(0) = 0]$$

$$\Rightarrow \qquad L\{G(t)\} = \frac{1}{p} L \{G'(t)\}$$

$$\Rightarrow \quad L\left\{\int_0^t F(x) \, dx\right\} = \frac{1}{p} L \{F(t)\}.$$

THEOREM 10   **(Division by t). If** $L\{F(t)\} = f(p)$, **then** $L\left\{\dfrac{1}{t}F(t)\right\} = \int_p^\infty f(x)\,dx$ **provided** $\lim\limits_{t\to 0}\left\{\dfrac{1}{t}F(t)\right\}$ **exists.**

(UPTU–2005, 07; Meerut–1991; Agra–1998, 2000; Kanpur–2004; Rohilkhand–1993)

PROOF.   Let $G(t) = \dfrac{1}{t}F(t)$, *i.e.*, $F(t) = t\,G(t)$

Therefore, $L\{F(t)\} = L\{t\,G(t)\} = -\dfrac{d}{dp}L\{G(t)\}$

$\Rightarrow$ $f(p) = -\dfrac{d}{dp}L\{G(t)\}$.

On integrating both sides with respect to p to ∞, we get

$$-\big[L\{G(t)\}\big]_p^\infty = \int_p^\infty f(p)\,dp$$

$\Rightarrow$ $-\lim\limits_{p\to\infty} L\{G(t)\} + L\{G(t)\} = \int_p^\infty f(p)\,dp$

$\Rightarrow$ $0 + L\{G(t)\} = \int_p^\infty f(p)\,dp$,

$$\left(\because \lim_{p\to\infty} L\{G(t)\} = \lim_{p\to\infty}\int_0^\infty e^{-pt}G(t)dt = 0\right)$$

$\Rightarrow$ $L\left\{\dfrac{1}{t}F(t)\right\} = \int_p^\infty f(x)\,dx$.

📖 RECAPITULATIONS

S. No.	Operation	F(t)	L{F(t)}
1.	Differentiation theorem	$F'(t)$	$pf(p) - F(0)$
		$F^n(t)$	$p^n f(p) - \sum\limits_{r=0}^{n-1} p^{n-1-r}F^r(0)$
2.	Multiplication theorem	$t^n F(t)$	$-f'(p)$
		$t\,F(t)$	$(-1)^n \dfrac{d^n}{dp^n}f(p)$
3	Division theorem	$\dfrac{1}{t}F(t)$	$\int_p^\infty f(x)dx$
4.	Integral theorem	$\int_0^t F(x)dx$	$\dfrac{1}{p}f(p)$
5.	Initial value theorem	$\lim\limits_{t\to 0} F(t) = \lim\limits_{p\to\infty} pL\{F(t)\}$	
6.	Final value theorem	$\lim\limits_{t\to\infty} F(t) = \lim\limits_{p\to 0} pL\{F(t)\}$	
7.	Fundamental theorem for Periodic Functions	$L\{F(t)\} = \dfrac{\int_0^T e^{-pt}F(t)dt}{1 - e^{-pT}}$, $F(t)$ is periodic function of period T.	

Solved Examples

EXAMPLE 1.   *Find $L\{t\cos at\}$.*

(Raipur–2005; Meerut–1996, 97; Agra–2012; Kanpur–1982, 2002, 2004, 2015; Purvanchal–2002)

SOLUTION.   We know that

$$L\{\cos at\} = \frac{p}{p^2 + a^2}, \quad p > 0.$$

Therefore, $L\{t\cos at\} = -\dfrac{d}{dp}L\{\cos at\} = -\dfrac{d}{dp}\left(\dfrac{p}{p^2 + a^2}\right) = \dfrac{p^2 - a^2}{(p^2 + a^2)^2}.$

EXAMPLE 2.   *Find $L\{(\sin at - at\cos at)\}$.*

(SVTU–2005; Meerut–2014
Kumaun–2000; Kurukshetra–2004; MDU–2006)

SOLUTION.   Consider $L\{\sin at - at\cos at\} = L\{\sin at\} - aL\{t\cos at\}$

$$= \frac{a}{p^2 + a^2} - a.(-1)\frac{d}{dp}[L\{\cos at\}]$$

$$= \frac{a}{p^2 + a^2} + a\frac{d}{dp}\left(\frac{p}{p^2 + a^2}\right)$$

$$= \frac{a}{p^2 + a^2} + \frac{a(a^2 - p^2)}{(p^2 + a^2)^2} = \frac{2a^3}{(p^2 + a^2)^2}.$$

EXAMPLE 3.   *Find the Laplace transform of the function $F(t) = t\,e^{-t}\sin 2t$*

(UPTU–2002; Kurukshetra–2005, 13; Meerut–1981, 83, 84, 91)

SOLUTION. $L\{\sin 2t\} = \dfrac{2}{p^2 + 4}, L\{e^{-t}\sin 2t\} = \dfrac{2}{(p+1)^2 + 4} = f(p)$ (say)

$$L\{te^{-t}\sin 2t\} = f'(p) = -\frac{d}{dp}\left[\frac{2}{(p+1)^2 + 4}\right]$$

$$= \frac{-2.2\,(p+1)}{[(p+1)^2 + 4]^2} = \frac{4(p+1)}{[(p+1)^2 + 4]^2}.$$

EXAMPLE 4.   *Obtain the Laplace transform of $t^2\,e^t\sin 4t$.*

(UPTU–2002)

SOLUTION. $L\{\sin 4t\} = \dfrac{4}{p^2 + 16}, L\{e^t\sin 4t\} = \dfrac{4}{(p-1)^2 + 16}$

$$L\{te^t\sin 4t\} = -\frac{d}{dp}\left(\frac{4}{p^2 - 2p + 17}\right) = \frac{4(2p-2)}{(p^2 - 2p + 17)^2}$$

$$L\{t^2 e^t\sin 4t\} = -\frac{d}{dp}\left(\frac{4(2p-2)}{(p^2 - 2p + 17)^2}\right)$$

$$= \frac{-4(2p^2 - 4p + 34 - 8p^2 + 16p - 8)}{(p^2 - 2p + 17)^3}$$

$$= \frac{-4(-6p^2 + 12p + 26)}{(p^2 - 2p + 17)^3} = \frac{8(3p^2 - 6p - 13)}{(p^2 - 2p + 17)^3}.$$

EXAMPLE 5. Given $L\{\sin\sqrt{t}\} = \dfrac{\sqrt{\pi}}{2p^{3/2}} e^{-1/4p}$, show that $L\left\{\dfrac{\cos\sqrt{t}}{\sqrt{t}}\right\} = \sqrt{\left(\dfrac{\pi}{p}\right)}.e^{-1/4p}$.

(Mumbai–2009; Agra–2007, 08; Meerut–2001, 05, 08; Purvanchal–2005, 08;
Kashi–2008; Kanpur–2002, 05)

SOLUTION. Let $F(t) = \sin\sqrt{t}$

Then we have $F'(t) = \cos\dfrac{\sqrt{t}}{2\sqrt{t}}$ and $F(0) = 0$

Put all these values in

$$L\{F'(t)\} = pL\{F(t)\} - F(0)$$

We get $L\left\{\dfrac{\cos\sqrt{t}}{2\sqrt{t}}\right\} = pL\{\sin\sqrt{t}\} = p\left[\dfrac{\sqrt{\pi}}{2p^{3/2}} e^{-1/4p}\right] = \dfrac{1}{2}\sqrt{\left(\dfrac{\pi}{p}\right)} e^{-1/4p}$.

Hence, $L\left\{\dfrac{\cos\sqrt{t}}{\sqrt{t}}\right\} = \sqrt{\left(\dfrac{\pi}{p}\right)}.e^{-1/4p}$.

EXAMPLE 6. Show that $L\left\{\dfrac{\sin t}{t}\right\} = \tan^{-1}\dfrac{1}{p}$ and hence find $L\left\{\dfrac{\sin at}{t}\right\}$. Does the Laplace transform

of $\dfrac{\cos at}{t}$ exists? (UPTU–2005; PTU–2010; Meerut–1989, 91, 92, 2011;
Agra–1981, 83, 85; Kanpur–1985; Rohilkhand–2002, 10;
Purvanchal–1989, 2007; SVTU–2004; Guwahati–2004; Kumaun–2001)

SOLUTION. Let $F(t) = \sin t$

Then, $\displaystyle\lim_{t\to 0}\dfrac{F(t)}{t} = \lim_{t\to 0}\dfrac{\sin t}{t} = 1$.

We know that $L\{\sin t\} = \dfrac{1}{p^2 + 1} = f(p)$ (say)

Then we have

$$L\left\{\dfrac{\sin t}{t}\right\} = \int_p^\infty f(x)\,dx = \int_p^\infty \dfrac{dx}{x^2 + 1} = \left(\tan^{-1} x\right)_p^\infty$$

$$= \dfrac{\pi}{2} - \tan^{-1} p = \cot^{-1} p = \tan^{-1}\left(\dfrac{1}{p}\right).$$

Now, $L\left\{\dfrac{\sin at}{t}\right\} = aL\left\{\dfrac{\sin at}{at}\right\} = a.\dfrac{1}{a}\tan^{-1}\left(\dfrac{1}{p/a}\right)$

$$\left[\because L\{f(at)\} = \dfrac{1}{a} f\left(\dfrac{p}{a}\right)\right]$$

$$= \tan^{-1}\left(\dfrac{a}{p}\right).$$

Also, since $L\{\cos at\} = \dfrac{p}{p^2 + a^2} = f(p)$ (say)

Then, $L\left\{\dfrac{\cos at}{t}\right\} = \int_p^\infty \dfrac{x}{x^2 + a^2}\,dx = \left[\dfrac{1}{2}\log(x^2 + a^2)\right]_p^\infty$

$$= \dfrac{1}{2}\lim_{x\to\infty}\log(x^2 + a^2) - \dfrac{1}{2}\log(p^2 + a^2)$$

which does not exist since $\lim\limits_{x \to \infty} \log(x^2 + a^2)$ is infinite.

Therefore, $L\left\{\dfrac{\cos at}{t}\right\}$ does not exist.

EXAMPLE 7. *Find* $L\left\{\dfrac{\sinh t}{t}\right\}$.   (Kanpur–2013)

SOLUTION. Let us assume $F(t) = \sinh t$.

Now, $\lim\limits_{t \to 0}\dfrac{F(t)}{t} = \lim\limits_{t \to 0}\dfrac{\sinh t}{t} = 1$.

Since $L\{\sinh t\} = \dfrac{1}{p^2 - 1} = f(p)$ (say)

$\therefore$ $L\left\{\dfrac{\sinh t}{t}\right\} = \int_p^\infty f(x)\, dx = \int_p^\infty \dfrac{dx}{x^2 - 1}$

$= \left[\dfrac{1}{2}\log\dfrac{x-1}{x+1}\right]_p^\infty = \left[\dfrac{1}{2}\log\dfrac{1-1/x}{1+1/x}\right]_p^\infty$

$= 0 - \dfrac{1}{2}\log\dfrac{p-1}{p+1} = \dfrac{1}{2}\log\dfrac{p+1}{p-1}$

EXAMPLE 8. *If* $F(t) = \dfrac{e^{at} - \cos bt}{t}$, *find the Laplace transform of* $F(t)$.   (UPTU–2003)

SOLUTION. We have $F(t) = \dfrac{e^{at} - \cos bt}{t} = \dfrac{e^{at}}{t} - \dfrac{\cos bt}{t}$

Now, we know that

$L\left(e^{at} - \cos bt\right) = \left(\dfrac{1}{p-a} - \dfrac{p}{p^2 + b^2}\right)$

$\therefore$ $L\left(\dfrac{e^{at} - \cos bt}{t}\right) = \int_p^\infty \left(\dfrac{1}{p-a} - \dfrac{p}{p^2 + b^2}\right) dp$

$= \left[\log(p-a) - \dfrac{1}{2}\log(p^2 + b^2)\right]_p^\infty$

$= \left[\dfrac{2\log(p-a) - \log(p^2 + b^2)}{2}\right]_p^\infty$

$= \dfrac{1}{2}\left[\log(p-a)^2 - \log(p^2 + b^2)\right]_p^\infty$

$= \dfrac{1}{2}\left[\log\dfrac{(p-a)^2}{p^2 + b^2}\right]_p^\infty = \dfrac{1}{2}\left[\log\left[\dfrac{(1 - (a/p))}{\left(1 + (b^2/p^2)\right)}\right]\right]_p^\infty$

$= \dfrac{1}{2}\left[0 - \log\dfrac{(1-(1/p))^2}{\left(1+(b^2/p^2)\right)}\right] = \dfrac{1}{2}\left[\log\dfrac{p^2 + b^2}{(p-a)^2}\right]$

EXAMPLE 9. *Evaluate* $L\{t^3 \cos t\}$ (Meerut–1991, Jhansi–2012)

SOLUTION. Here, we have

$$L\{\cos t\} = \frac{p}{p^2 + 1}, p > 0$$

$$\therefore \quad L\{t^3 \cos t\} = (-1)^3 \frac{d^3}{dp^3}\left(\frac{p}{p^2+1}\right) = -\frac{d^2}{dp^2}\left\{\frac{-p^2+1}{(p^2+1)^2}\right\}$$

$$= \frac{d^2}{dp^2}\left\{\frac{p^2-1}{(p^2+1)^2}\right\} = \frac{d^2}{dp^2}\left\{\frac{p^2+1-2}{(p^2+1)^2}\right\}$$

$$= \frac{d^2}{dp^2}\left\{\frac{1}{p^2+1} - \frac{2}{(p^2+1)^2}\right\}$$

$$= \frac{d}{dp}\left\{\frac{-2p}{(p^2+1)^2} + \frac{8p}{(p^2+1)^3}\right\} = \frac{6p^4 - 36p^2 + 6}{(p^2+1)^4}$$

EXAMPLE 10. *Prove that* $L\{te^{at}\sin(at)\} = \dfrac{2a(p-a)}{(p^2-2ap+2a^2)^2}$ (Kanpur–2010)

SOLUTION. We know that

$$L\{\sin(at)\} = \frac{a}{p^2 + a^2}$$

$$L\{e^{at}\sin(at)\} = \frac{a}{(p-a)^2 + a^2} = \frac{a}{p^2 - 2ap + 2a^2}$$

$$\Rightarrow \quad L\{t \cdot e^{at}\sin at\} = (-1)^1 \frac{d}{dp}\left\{\frac{a}{p^2-2ap+2a^2}\right\} = \frac{2a(p-a)}{(p^2-2ap+2a^2)^2}$$

EXAMPLE 11. *If* $L\left\{2\sqrt{\left(\dfrac{t}{\pi}\right)}\right\} = \dfrac{1}{p^{3/2}}$, *show that* $\dfrac{1}{p^{1/2}} = L\left\{\dfrac{1}{\sqrt{\pi t}}\right\}$ (Meerut–2000)

SOLUTION. Let us assume $F(t) = 2\sqrt{t/\pi}$

Now, $F(0) = 2\sqrt{0/\pi} = 0$

Since, $F'(t) = \dfrac{d}{dt}\left[\dfrac{2}{\sqrt{\pi}}t^{1/2}\right] = \dfrac{2}{\sqrt{\pi}} \times \dfrac{1}{2}t^{-1/2} = \dfrac{1}{\sqrt{(\pi t)}}$

We know that

$$L\{F'(t)\} = pL\{F(t) - F(0)\} \qquad \ldots(1)$$

Putting the value of $F(0)$, $F(t)$ and $F'(t)$ in eq^n (1), we have

$$L\left\{\frac{1}{\sqrt{\pi t}}\right\} = pL\left\{2\sqrt{\left(\frac{t}{\pi}\right)}\right\} - 0 = p \times \frac{1}{p^{3/2}} = \frac{1}{p^{1/2}} \quad \left[\because L\left\{2\sqrt{\left(\frac{t}{\pi}\right)}\right\} = \frac{1}{p^{3/2}}\right]$$

EXAMPLE 12. *Evaluate* $L\{t^2 \sin at\}$ (MDU Rohtak–2003, 05)

SOLUTION. We know that

$$L\{\sin at\} = a/(p^2 + a^2)$$

Now, $L\{t^2 \sin at\} = (-1)^2 \dfrac{d^2}{dp^2}\left(\dfrac{a}{p^2+a^2}\right) = a\dfrac{d}{dp}\left[\dfrac{d}{dp}(p^2+a^2)^{-1}\right]$

$\qquad\qquad = a\dfrac{d}{dp}[-(p^2+a^2)^{-2}\times 2p]$

$\qquad\qquad = -2a\dfrac{d}{dp}\cdot\dfrac{p}{(p^2+a^2)^2} = \dfrac{-2a(p^2+a^2)\times 1 - p\times 2(p^2+a^2)\times 2p}{(p^2+a^2)^4}$

$\qquad\qquad = -2a\dfrac{p^2+a^2-4p^2}{(p^2+a^2)^3} = \dfrac{2a(3p^2-a^2)}{(p^2+a^2)^3}$

EXAMPLE 13. *Prove that* $L\left\{\dfrac{\sin^2 t}{t}\right\} = \dfrac{1}{4}\log\dfrac{p^2+4}{p^2}$ (Kumaun–1999; Meerut–2006; MS Univ.TN–2007)

SOLUTION. We know that

$\qquad\qquad L\{\sin^2 t\} = L\{(1/2)\times(1-\cos 2t)\}$ $\qquad\qquad$ [$\because \sin^2 t = 1/2(1-\cos 2t)$]

$\qquad\qquad\qquad = 1/2\times[L\{1\} - L\{\cos 2t\}]$

$\qquad\qquad L\{\sin^2 t\} = \dfrac{1}{2}\left[\dfrac{1}{p} - \dfrac{p}{p^2+2^2}\right] = f(p),$ say $\qquad\qquad$ ...(1)

Therefore, $L\left\{\dfrac{\sin^2 t}{t}\right\} = \int_p^\infty f(p)dp = \dfrac{1}{2}\int_p^\infty\left[\dfrac{1}{p} - \dfrac{p}{p^2+2^2}\right]dp$ $\qquad$ [From (1)]

$\qquad\qquad = \dfrac{1}{2}\left[\log p - \dfrac{1}{2}\log(p^2+4)\right]_p^\infty = \dfrac{1}{4}\left[\log\dfrac{p^2}{p^2+4}\right]_p^\infty$

$\qquad\qquad = \dfrac{1}{4}\lim_{p\to\infty}\log\dfrac{p^2}{p^2+4} - \dfrac{1}{4}\log\dfrac{p^2}{p^2+4}$

$\qquad\qquad = \dfrac{1}{4}\lim_{p\to\infty}\log\dfrac{1}{1+(4/p^2)} + \dfrac{1}{4}\log\dfrac{p^2+4}{p^2}$

$\qquad\qquad = 0 + (1/4)\log(p^2+4)/p^2 = (1/4)\log\{(p^2+4)/p^2\}$

EXAMPLE 14. *Evaluate* $L\left[\int_0^t\dfrac{\sin x}{x}dx\right]$ (Kumaun–1997; Delhi–1997; Kurukshetra–2006; Purvanchal–2005)

SOLUTION. We know that $L\{\sin t\} = 1/(p^2+1) = f(p),$ say $\qquad\qquad$ ...(1)

Therefore, $L\left\{\dfrac{\sin t}{t}\right\} = \int_p^\infty f(p)dp = \int_p^\infty\dfrac{1}{p^2+1}dp$

$\qquad\qquad = [\tan^{-1} p]_p^\infty$ $\qquad\qquad\qquad\qquad$ [From (1)]

$\qquad\qquad = \tan^{-1}\infty - \tan^{-1}p = (\pi/2) - \tan^{-1}p = \cot^{-1}p$

$\qquad\qquad\qquad\qquad\qquad\qquad$ [$\because \tan^{-1}x + \cot^{-1}x = \pi/2$]

$\qquad L\left[\int_0^t\dfrac{\sin x}{x}dx\right] = \dfrac{\cot^{-1}p}{p}$

EXAMPLE 15. Evaluate $\int_0^\infty te^{-2t} \cos t \, dt$ (Meerut–2000, 06)

SOLUTION. We know that

$$L\{\cos t\} = p/(p^2 + 1) = f(p), \text{ say} \qquad \ldots(1)$$

Therefore, $L\{t \cos t\} = -\dfrac{d}{dp} f(p) = -\dfrac{d}{dp}\left(\dfrac{p}{p^2+1}\right)$

$$= \dfrac{-1 \times (p^2+1) - p \times 2p}{(p^2+1)^2} = \dfrac{p^2-1}{(p^2+1)^2}$$

Now, from the condition of Laplace transform

$$\int_0^\infty e^{-pt}\{t\cos t\}dt = (p^2-1)/(p^2+1)^2 \qquad \ldots(2)$$

Finally, taking limit of both sides as $p \to 2$, we get

$$\int_0^\infty te^{-2t} \cos t \, dt = \dfrac{3}{25}$$

Additional Solved Examples

EXAMPLE 1. Show that $L\{-a \sin at\} = -\dfrac{a^2}{p^2+a^2}$.

SOLUTION. Let $F(t) = -a \sin at$

then $F'(t) = -a^2 \cos at$ and $F''(t) = a^3 \sin at$

so that $F'(0) = -a^2$ and $F(0) = 0$

∴ Using $L\{F''(t)\} = p^2 L\{F(t)\} - pF(0) - F'(0)$,

We have $L\{a^3 \sin at\} = p^2 L\{-a \sin at\} - 0 - (-a^2)$

or $(p^2 + a^2)L\{-a \sin at\} = -a^2$

∴ $L\{-a \sin at\} = -\dfrac{a^2}{p^2+a^2}$

EXAMPLE 2. Evaluate

(i) $L\{t \cosh 3t\}$ (Agra–2000, 02)

(ii) $L\{t \sinh at\}$ (Meerut–2006)

SOLUTION. (i) We have $L\{\cosh 3t\} = \dfrac{p}{p^2-9}, p > 0$

∴ $L\{t \cosh 3t\} = -\dfrac{d}{dp}\left\{\dfrac{p}{p^2-9}\right\} = \dfrac{p^2+9}{(p^2-9)^2}$

(ii) We have $L\{\sinh at\} = \dfrac{1}{p^2-a^2}, p > 0$

∴ $L\{t \sinh at\} = -\dfrac{d}{dp}\left(\dfrac{a}{p^2-a^2}\right) = \dfrac{2ap}{(p^2-a^2)^2}$

EXAMPLE 3. Show that $L\{t^2 \cos at\} = \dfrac{2p(p^2-3a^2)}{(p^2+a^2)^3}, p > 0$.

(Meerut–1987, 90, 92, 93, 96, 2006; Kanpur–1991; Rohilkhand–2011; Agra–2012)

SOLUTION.   Since $L\{\cos at\} = \dfrac{p}{p^2 + a^2}, p > 0$

$$\therefore \quad L\{t^2 \cos at\} = (-1)^2 \frac{d^2}{dp^2}\left\{\frac{p}{(p^2 + a^2)}\right\} = \frac{d}{dp}\left[-\frac{p^2 + a^2}{(p^2 + a^2)^2}\right] = \frac{2p(p^2 - 3a^2)}{(p^2 + a^2)^3}$$

EXAMPLE 4.   *Show tha t $L(t^n e^{at}) = \dfrac{n!}{(p-a)^{n+1}}, p > a$.*   (Purvanchal–2007; Avadh–2011)

SOLUTION.   Since, $L(e^{at}) = \dfrac{1}{p - a}, p > a$

$$\therefore \qquad L(t^n e^{at}) = (-1)^n \frac{d^n}{dp^n}\left(\frac{1}{p - a}\right) = (-1)^n \frac{(-1)^n n!}{(p-a)^{n+1}}$$

$$= \frac{n!}{(p-a)^{n+1}}, p > a$$

EXAMPLE 5.   *Show that $L\{t(3\sin 2t - 2\cos 2t)\} = \dfrac{8 + 12p - 2p^2}{(p^2 + 4)^2}$.*   (Meerut–1986)

SOLUTION.   We have $L\{3\sin 2t - 2\cos 2t\} = 3L\{\sin 2t\} - 2L\{\cos 2t\} = \dfrac{6}{p^2 + 4} - \dfrac{2p}{p^2 + 4} = \dfrac{6 - 2p}{p^2 + 4}$

Hence, $L\{t(3\sin 2t - 2\cos 2t)\} = (-1)\dfrac{d}{dp}\left(\dfrac{6 - 2p}{p^2 + 4}\right) = \dfrac{8 + 12p - 2p^2}{(p^2 + 4)^2}$

EXAMPLE 6.   *Show that $L\{\sin \alpha t + t\cos \alpha t\} = \dfrac{(\alpha + 1)p^2 + (\alpha - 1)\alpha^2}{(p^2 + \alpha^2)^2}$.*   (Meerut–1983)

SOLUTION.   We have $L\{\sin \alpha t + t\cos \alpha t\} = L\{\sin \alpha t\} + L\{t\cos \alpha t\}$ (By linearly property)

$$= L\{\sin \alpha t\} + \left[-\frac{d}{dp}L\{\cos \alpha t\}\right] = \frac{\alpha}{p^2 + \alpha^2} - \frac{d}{dp}\left(\frac{p}{p^2 + \alpha^2}\right)$$

$$= \frac{\alpha}{p^2 + \alpha^2} - \frac{\alpha^2 - p^2}{(p^2 + \alpha^2)^2} = \frac{(\alpha + 1)p^2 + (\alpha - 1)\alpha^2}{(p^2 + \alpha^2)^2}$$

EXAMPLE 7.   *Show that $L\{t^2 - 3t + 2\}\sin 3t = \dfrac{6p^4 - 18p^3 + 126p^2 - 162p + 432}{(p^2 + 9)^3}$.*

(Bundelkhand–2008)

SOLUTION.   We have $L\{(t^2 - 3t + 2)\sin 3t\} = L\{t^2 \sin 3t\} - 3L\{t\sin 3t\} + 2L\{\sin 3t\}$

$$= (-1)^2 \frac{d^2}{dp^2}L\{\sin 3t\} - 3\left[-\frac{d}{dp}L\{\sin 3t\}\right] + 2L\{\sin 3t\}$$

$$= \frac{d^2}{dp^2}\left(\frac{3}{p^2 + 9}\right) + 3\frac{d}{dp}\left(\frac{3}{p^2 + 9}\right) + 2\frac{3}{p^2 + 9}$$

$$= \frac{18p^2 - 54}{(p^2 + 9)^3} + 3\left\{\frac{-6p}{(p^2 + 9)^2}\right\} + \frac{6}{p^2 + 9}$$

$$= \frac{6p^4 - 18p^3 + 126p^2 - 162p + 432}{(p^2 + 9)^3}$$

EXAMPLE 8. *If* $L\{F(t), t \to p\} = f(p)$, *show that* $L\left\{\int_0^t \frac{F(u)}{u} du, t \to p\right\} = \frac{1}{p}\int_p^\infty f(y) dy$. *Hence,*

show that $L\left\{\int_0^t \frac{\sin u}{u} du, t \to p\right\} = \frac{\cot^{-1} p}{p}$.

(Kumaun–1997; Kanpur–1998; Rohilkhand–1990; Meerut–1992; Avadh–2013)

SOLUTION. From Theorem 9, we have

$$L\left\{\int_0^t F(u) du\right\} = \frac{1}{p} f(p) \qquad \qquad \ldots(1)$$

where $\qquad \qquad f(p) = L\{F(t)\}$

Then, $\qquad \qquad L\{G(t)\} = L\left\{\frac{F(t)}{t}\right\} = \int_p^\infty f(y) dy = g(p)$, say

$\therefore$ From (1), we have

$$L\left\{\int_0^t G(u) du\right\} = \frac{1}{p} g(p)$$

or $\qquad \qquad L\left\{\int_0^t \frac{F(u)}{u} du\right\} = \frac{1}{p}\int_p^\infty f(y) dy \qquad \qquad \ldots(2)$

Deduction: Let $\qquad \qquad F(t) = \sin t$

so that $\qquad \qquad f(p) = L\{\sin t\} = \frac{1}{p^2 + 1}$

$\therefore$ From (2), we have

$$L\left\{\int_0^t \frac{\sin u}{u} du\right\} = \frac{1}{p}\int_p^\infty \frac{dy}{y^2 + 1} = \frac{1}{p}\left[\tan^{-1} y\right]_p^\infty = \frac{1}{p}\left(\frac{\pi}{2} - \tan^{-1} p\right) = \frac{1}{p}\cot^{-1} p$$

EXAMPLE 9. *Show that if* $L\{F(t)\} = f(p)$, *then*

$$\int_0^\infty \frac{F(t)}{t} dt = \int_0^\infty F(x) dx , \text{ provided that the integral converges.} \qquad \text{(Rohilkhand–1990)}$$

SOLUTION. From theorem-10, we have

$$L\left\{\frac{F(t)}{t}\right\} = \int_0^\infty e^{-pt} \frac{F(t)}{t} dt = \int_p^\infty f(x) dt$$

$$= \int_p^0 f(x) dx + \int_0^\infty f(x) dx = -\int_0^p f(x) dx + \int_0^\infty f(x) dx$$

Taking limit as $p \to 0$, (assuming the integral converges)

We have $\qquad \int_0^\infty \frac{F(t)}{t} dt = \int_0^\infty f(x) dx$.

1.8 EVALUATION OF INTEGRALS

If $L\{F(t)\} = f(p)$, i.e., $\int_0^\infty e^{-pt} F(t) dt = f(p)$

By taking limit as $p \to 0$, we have $\int_0^\infty F(t) dt = f(0)$, provided the integral is convergent.

1.9 SOME IMPORTANT SPECIAL FUNCTIONS

(i) The sine and cosine integrals. The sine and cosine integrals, which are denoted by $S_i(t)$ and $C_i(t)$ respectively are defined by $S_i(t) = \int_0^t \frac{\sin u}{u} du$ and $C_i(t) = \int_t^\infty \frac{\cos u}{u} du$.

(Kurukshetra–2006)

(ii) Error Function. The error function denoted by $erf\,(t)$, is defined by $erf(t) = \dfrac{2}{\sqrt{\pi}} \int_0^t e^{-u^2}\,du$.

(iii) The Gamma function. If $n > 0$, the gamma function is defined by $\Gamma(n) = \int_0^{\infty} u^{n-1} e^{-u}\,du$.

(iv) Heaviside's unit step function. The unit step function or heaviside's unit function denoted by $H(t - a)$ is defined by

$$H(t-a) = \begin{cases} 0, & t < a \\ 1, & t \geq a \end{cases}.$$

(v) Bessel's functions. $J_n(t) = \dfrac{t^n}{2^n \, \Gamma(n+1)} \left[1 - \dfrac{t^2}{2(2n+2)} + \dfrac{t^4}{2.4(2n+2)(2n+4)} \cdots \right].$

Solved Examples

EXAMPLE 1. *Find* $\int_0^{\infty} \dfrac{(e^{-at} - e^{-bt})}{t}\,dt$.   (Meerut–1983; SVTU–2009; Mumbai–2007; JNTU–2006)

SOLUTION.   Let $F(t) = e^{-at} - e^{-bt}$.
Thus, we have

$$L\{F(t)\} = L\,\{e^{-at}\} - L\{e^{-bt}\} = \frac{1}{p+a} - \frac{1}{p+b} = f(p)\ (\text{say}).$$

Therefore,

$$L\left\{\frac{F(t)}{t}\right\} = \int_p^{\infty} f(x)\,dx = \int_p^{\infty}\left(\frac{1}{x+a} - \frac{1}{x+b}\right) dx$$

$$= \left[\log\left(\frac{x+a}{x+b}\right)\right]_p^{\infty} = \lim_{x\to\infty} \log\frac{x+a}{x+b} - \log\frac{p+a}{p+b}$$

$$= \lim_{x\to\infty} \log\frac{1+a/x}{1+b/x} - \log\frac{p+a}{p+b}$$

$$= 0 - \log\frac{p+a}{p+b} = \log\frac{p+b}{p+a}$$

Therefore, $L\left\{\dfrac{F(t)}{t}\right\} = \int_0^{\infty} e^{-pt}\cdot\dfrac{e^{-at} - e^{-bt}}{t}\,dt = \log\dfrac{p+b}{p+a}$

Hence, taking limit as $p \to 0$, we have

$$\int_0^{\infty} \frac{e^{-at} - e^{-bt}}{t}\,dt = \log\frac{b}{a}.$$

EXAMPLE 2. *Show that* $\int_0^{\infty} t\,e^{-2t}\cos t\,dt = \dfrac{3}{25}$.   (Agra–2003; Meerut–2006)

SOLUTION.   We have $L\,\{t\cos t\} = -\dfrac{d}{dp}\,L\,\{\cos t\}$

$$\Rightarrow \quad \int_0^{\infty} e^{-pt}\cdot t\cos t\,dt = -\frac{d}{dp}\left(\frac{p}{p^2+1}\right) = \frac{p^2-1}{(p^2+1)^2}$$

Taking $p = 2$, we get $\int_0^{\infty} t\,e^{-2t}\cos t\,dt = \dfrac{3}{25}$.

EXAMPLE 3. *Show that*

    (i) $L\{\sinh at \cos at\} = \dfrac{a(p^2 - 2a^2)}{p^4 + 4a^4}$ (SVTU–2006)

    (ii) $L\{\sinh at \sin at\} = \dfrac{2a^2 p}{p^4 + 4a^4}$.   (Kanpur–1992; Meerut–1998)

SOLUTION.   (i) We know that

$$L\{\sinh at\} = \frac{a}{p^2 - a^2} = f(p) \quad \text{(say)}$$

Therefore,

$$L\{e^{iat} \sin at\} = f(p - ia)$$

$$= \frac{a}{(p - ia)^2 - a^2} = \frac{a}{(p^2 - 2a^2) - 2iap}$$

$$= \frac{a\{(p^2 - 2a^2) + 2iap\}}{(p^2 - 2a^2)^2 - (2ipa)^2}$$

$$\Rightarrow \qquad L\{\sinh at(\cos at + i\sin at)\} = \frac{a(p^2 - 2a^2) + 2ia^2 p}{p^4 + 4a^4}$$

$$\Rightarrow L\{\sinh at \cos at\} + iL\{\sinh at \sin at\} = \frac{a(p^2 - 2a^2)}{p^4 + 4a^4} + i\frac{2a^2 p}{p^4 + 4a^4}$$

Equating real and imaginary parts of both the sides, we get

$$L\{\sinh at \cos at\} = \frac{a(p^2 - 2a^2)}{p^4 + 4a^4}$$

and $L\{\sinh at \sin at\} = \dfrac{2a^2 p}{p^4 + 4a^4}$.

EXAMPLE 4. *Find* $L\{erf \sqrt{t}\}$ *and hence prove that* $L\{t \cdot erf(2\sqrt{t})\} = \dfrac{3p + 8}{p^2(p + 4)^{3/2}}$.

                            (Agra–1999; UPTU–2001; Kanpur–1994; Kumaun–1996)

SOLUTION.   We know that

$$erf \sqrt{t} = \frac{2}{\sqrt{\pi}} \int_0^{\sqrt{t}} e^{-x^2} dx$$

$$= \frac{2}{\sqrt{\pi}} \int_0^{\sqrt{t}} \left(1 - x^2 + \frac{x^4}{2!} - \frac{x^6}{6!} + \ldots\right) dx$$

$$= \frac{2}{\sqrt{\pi}} \left[x - \frac{x^3}{3} + \frac{x^5}{10} - \frac{x^7}{42} - \ldots\right]_0^{\sqrt{t}}$$

$$= \frac{2}{\sqrt{\pi}} \left[\sqrt{t} - \frac{t^{3/2}}{3} + \frac{t^{3/2}}{10} - \frac{t^{7/2}}{42} + \ldots\right]$$

so, $L\{erf\sqrt{t}\} = \dfrac{2}{\sqrt{\pi}} \left[\dfrac{\Gamma(3/2)}{p^{3/2}} - \dfrac{\Gamma(5/2)}{3p^{5/2}} + \dfrac{\Gamma(7/2)}{10p^{7/2}} - \dfrac{\Gamma(9/2)}{42p^{9/2}} + \ldots\right]$

$$= \frac{2}{\sqrt{\pi}} \left[\frac{\frac{1}{2}\Gamma(1/2)}{p^{3/2}} - \frac{\frac{3}{2}\cdot\frac{1}{2}\Gamma(1/2)}{3p^{5/2}} + \frac{\frac{5}{2}\cdot\frac{3}{2}\cdot\frac{1}{2}\Gamma(1/2)}{10p^{7/2}} \right.$$

$$\left. - \frac{\frac{7}{2}\cdot\frac{5}{2}\cdot\frac{3}{2}\cdot\frac{1}{2}\Gamma(1/2)}{42p^{9/2}} + \dots \right]$$

$$= \frac{1}{p^{3/2}} - \frac{1}{2}\frac{1}{p^{5/2}} + \frac{1.3}{2.4}\cdot\frac{1}{p^{7/2}} - \frac{1}{2}\frac{3}{4}\frac{5}{6}\cdot\frac{1}{p^{9/2}} + \dots$$

$$= \frac{1}{p^{3/2}}\left[1 - \frac{1}{2}\cdot\frac{1}{p} + \frac{1}{2}\frac{3}{4}\cdot\frac{1}{p^2} - \frac{1}{2}\frac{3}{4}\frac{5}{6}\cdot\frac{1}{p^3} + \dots \right]$$

$$= \frac{1}{p^{3/2}}\left[1 - \frac{1}{2}\cdot\frac{1}{p} + \frac{\left(-\frac{1}{2}\right)\left(-\frac{3}{2}\right)}{2!}\frac{1}{p^2} + \frac{\left(-\frac{1}{2}\right)\left(-\frac{3}{2}\right)\left(-\frac{5}{2}\right)}{3!}\frac{1}{p^3} + \dots \right]$$

$$= \frac{1}{p^{3/2}}\left[1 + \frac{1}{p} \right]^{-1/2} = \frac{1}{p^{3/2}}\left[\frac{p}{p+1} \right]^{1/2} = \frac{1}{p\sqrt{p+1}}$$

Now, $L\left\{ erf\, 2\sqrt{t} \right\} = L\left\{ erf\, \sqrt{4t} \right\}$

$$= \frac{1}{4}\frac{1}{\frac{p}{4}\sqrt{\frac{p}{4}+1}} = \frac{2}{p\sqrt{p+4}} = \frac{2}{p\sqrt{p+4}}$$

Hence, $L\left\{ t.erf\,(2\sqrt{t}) \right\} = -\dfrac{d}{dp}\dfrac{2}{\sqrt{p^3+4p^2}} = -2\left(-\dfrac{1}{2}\right)\left(p^3+4p^2\right)^{-3/2}(3p^2+8p)$

$$= \frac{3p^2+8p}{(p^3+4p^2)^{3/2}} = \frac{3p+8}{p^2(p+4)^{3/2}}.$$

EXAMPLE 5.   *Show that* $L\left\{ \dfrac{\sin^2 t}{t} \right\} = \dfrac{1}{4}\log\left(\dfrac{p^2+4}{p^2} \right).$

(Meerut–1981, 82, 92, 2006; Agra–2004, 09; Rohilkhand–2010)

SOLUTION.   We know that $\sin^2 t = \dfrac{1}{2}(1 - \cos 2t).$

Let $\qquad\qquad F(t) = \sin^2 t = \dfrac{1}{2}(1 - \cos 2t)$

$\Rightarrow \qquad\qquad L\{F(t)\} = \dfrac{1}{2}[L\{1\} - L\{\cos 2t\}]$

$$= \frac{1}{2}\left[\frac{1}{p} - \frac{p}{p^2+4} \right] = f(p) \text{ (say)}$$

Now, $\lim\limits_{t\to 0}\left\{\dfrac{1}{t}F(t)\right\} = \lim\limits_{t\to 0}\left(\dfrac{\sin t}{t}\right).\sin t = 1.0 = 0 \Rightarrow$ limit exists.

Therefore, $L\left\{\dfrac{1}{t}F(t)\right\} = \int_p^\infty f(x)\,dx = \dfrac{1}{2}\int_p^\infty\left(\dfrac{1}{x} - \dfrac{x}{x^2+4}\right)dx$

$$= \dfrac{1}{2}\left[\log x - \dfrac{1}{2}\log(x^2+4)\right]_p^\infty$$

$$= \dfrac{1}{4}\left[\log\dfrac{x^2}{x^2+4}\right]_p^\infty = \dfrac{1}{4}\left[\lim_{x\to\infty}\log\dfrac{x^2}{x^2+4} - \log\dfrac{p^2}{p^2+4}\right]$$

$$= \dfrac{1}{4}\left[\lim_{x\to\infty}\log\dfrac{1}{1+(4/x^2)} - \log\dfrac{p^2}{p^2+4}\right]$$

$$= \dfrac{1}{4}\left[\log 1 - \log\dfrac{p^2}{p^2+4}\right] = \dfrac{1}{4}\left[0 - \log\dfrac{p^2}{p^2+4}\right]$$

$$= -\dfrac{1}{4}\log\dfrac{p^2}{p^2+4} = \dfrac{1}{4}\log\dfrac{p^2+4}{p^2}.$$

EXAMPLE 6. *Show that* $\int_0^\infty \dfrac{\cos at - \cos bt}{t}\,dt = \dfrac{1}{2}\log\left(\dfrac{p^2+b^2}{p^2+a^2}\right).$

(UPTU–2004; Meerut–1996, 2005, 07, 12)

SOLUTION. Let $F(t) = \cos at - \cos bt$

$\Rightarrow \qquad L\{F(t)\} = L\{\cos at\} - L\{\cos bt\}$

$$= \dfrac{p}{p^2+a^2} - \dfrac{p}{p^2+b^2} = f(p) \ \ \text{(say)}$$

Now, $\lim\limits_{t\to 0}\dfrac{F(t)}{t} = \lim\limits_{t\to 0}\dfrac{\cos at - \cos bt}{t}$

$$= \lim_{t\to 0}\dfrac{-a\sin at + b\sin bt}{1} \qquad \text{(By L-Hospital's rule)}$$

$$= 0 \Rightarrow \ \text{limit exist.}$$

Therefore,

$$L\left\{\dfrac{F(t)}{t}\right\} = \int_p^\infty F(x)\,dx = \int_p^\infty\left[\dfrac{x}{x^2+a^2} - \dfrac{x}{x^2+b^2}\right]dx$$

$$= \dfrac{1}{2}\left[\log(x^2+a^2) - \log(x^2+b^2)\right]_p^\infty$$

$$= \dfrac{1}{2}\left[\log\dfrac{x^2+a^2}{x^2+b^2}\right]_p^\infty = \dfrac{1}{2}\lim_{x\to\infty}\log\dfrac{x^2+a^2}{x^2+b^2} - \dfrac{1}{2}\log\dfrac{p^2+a^2}{p^2+b^2}$$

$$= \dfrac{1}{2}\lim_{x\to\infty}\log\dfrac{1+a^2/x^2}{1+b^2/x^2} - \dfrac{1}{2}\log\dfrac{p^2+a^2}{p^2+b^2}$$

$$= 0 - \dfrac{1}{2}\log\dfrac{p^2+a^2}{p^2+b^2} = -\dfrac{1}{2}\log\dfrac{p^2+a^2}{p^2+b^2} = \dfrac{1}{2}\log\dfrac{p^2+b^2}{p^2+a^2}.$$

EXAMPLE 7.   *Find the Laplace transform of* $S_i(t)$.   (Agra–1982; Kumaun–1997; Kurukshetra–2006, 15)

SOLUTION.   By definition, we have

$$S_i(t) = \int_0^t \frac{\sin u}{u}\, du$$

$$= \int_0^t \left(1 - \frac{u^2}{3!} + \frac{u^4}{5!} - \frac{u^6}{7!} + ...\right) du = t - \frac{t^3}{3.3!} + \frac{t^5}{5.5!} - \frac{t^7}{7.7!} + ...$$

Therefore,

$$L\{S_i(t)\} = L(t) - \frac{1}{3.3!} L\{t^3\} + \frac{1}{5.5!} L\{t^5\} - \frac{1}{7.7!} L\{t^7\} + ...$$

$$= \frac{1!}{p^2} - \frac{1}{3.3!}\cdot\frac{3!}{p^4} + \frac{1}{5.5!}\cdot\frac{5!}{p^6} - \frac{1}{7.7!}\cdot\frac{7!}{p^8} + ...$$

$$= \frac{1}{p}\left[\frac{1}{p} - \frac{1}{3}\cdot\frac{1}{p^3} + \frac{1}{5}\cdot\frac{1}{p^5} - \frac{1}{7}\cdot\frac{1}{p^7} + ...\right] = \frac{1}{p}\tan^{-1}\frac{1}{p}.$$

EXAMPLE 8.   *Evaluate* $L\{F(t)\}$ *if,* $F(t)$ *is a function of period* 2π.

SOLUTION.   Since $F(t)$ is a function with period $T = 2\pi$.

Therefore, we have

$$L\{F(t)\} = \frac{\int_0^T e^{-pt} F(t)\, dt}{1 - e^{-pT}} = \frac{\int_0^\pi e^{-pt} \sin t\, dt + \int_\pi^{2\pi} 0.e^{-pt}\, dt}{1 - e^{-2\pi p}}$$

$$= \frac{1}{1 - e^{-2\pi p}} \int_0^\pi e^{-pt} \sin t\, dt$$

$$= \frac{1}{1 - e^{-2\pi p}} \left[\frac{e^{-pt}}{p^2 + 1}(-p\sin t - \cos t)\right]_0^\pi$$

$$= \frac{1}{1 - e^{-2\pi p}} \left[\frac{e^{-p\pi}}{p^2 + 1} + \frac{1}{p^2 + 1}\right] = \frac{1 + e^{-p\pi}}{(p^2 + 1)[1 - (e^{-p\pi})^2]}$$

$$= \frac{1 + e^{-p\pi}}{(p^2 + 1)(1 - e^{-p\pi})(1 + e^{-p\pi})} = \frac{1}{(p^2 + 1)(1 - e^{-p\pi})}$$

EXAMPLE 9.   *Evaluate* $L\left\{\frac{2}{\sqrt{\pi}}\int_0^t e^{-x^2}\, dx\right\}$ (Rohilkhand–2004, Avadh–2012)

SOLUTION.   Here, we have

$$L\left\{\frac{2}{\sqrt{\pi}}\int_0^t e^{-x^2}\, dx\right\} = \frac{2}{\sqrt{\pi}} L\left\{\int_0^t \left(1 - x^2 + \frac{x^4}{2!} - \frac{x^6}{3!} + ...\right) dx\right\}$$

$$= \frac{2}{\sqrt{\pi}} L\left\{t - \frac{t^3}{3} + \frac{t^5}{5.2!} - \frac{t^7}{7.3!} + ...\right\}$$

$$= \frac{2}{\sqrt{\pi}}\left[L\{t\} - L\left\{\frac{t^3}{3}\right\} + L\left\{\frac{t^5}{5.2!}\right\} - L\left\{\frac{t^7}{7.3!}\right\} + ...\right]$$

$$= \frac{2}{\sqrt{\pi}}\left[L\{t\} - \frac{1}{3}L\{t^3\} + \frac{1}{5.2!}L\{t^5\} - \frac{1}{7.3!}L\{t^7\} + ... \right]$$

$$= \frac{2}{\sqrt{\pi}}\left[\frac{1}{p^2} - \frac{1}{3}\cdot\frac{3!}{p^4} + \frac{1}{5.2!}\frac{4!}{p^5} - \frac{1}{7.3!}\cdot\frac{6!}{p^7} + ... \right]$$

$$= \frac{2}{\sqrt{\pi}}\left[\frac{1}{p^2} - \frac{2}{p^4} + \frac{12}{5p^5} - \frac{120}{7p^7} + ... \right].$$

EXAMPLE 10. *Show that* $\int_0^\infty te^{-3t}\sin t\, dt = \dfrac{3}{50}$ (Meerut–2007)

SOLUTION. Here, we have $\int_0^\infty te^{-3t}\sin t$

Since, $\qquad L\{t\sin t\} = -\dfrac{d}{dp}L\{\sin t\}$

Therefore, $\int_0^\infty e^{-pt}t\sin t\, dt = -\dfrac{d}{dp}\left(\dfrac{1}{p^2+1}\right) = \dfrac{2p}{(p^2+1)^2}$

EXAMPLE 11. *Evaluate* $\int_0^\infty J_0(t)dt = 1$ (Agra–1996, 97)

SOLUTION. Here, we have

$$\int_0^\infty J_0(t)dt = 1$$

$$L\{J_0(t)\} = \int_0^\infty e^{-pt}J_0(t)dt = \frac{1}{\sqrt{1+p^2}}$$

Let $p = 0$, then we have

$$\int_0^\infty J_0(t)dt = 1$$

EXAMPLE 12. *If* $C(t) = \int_t^\infty \dfrac{\cos x}{x}dx$, *then show that* $L\{C(t)\} = \dfrac{\log(p^2+1)}{2p}$ (Kanpur–2004)

SOLUTION. Consider $\qquad F(t) = \int_t^\infty \dfrac{\cos x}{x}dx \Rightarrow \lim_{t\to\infty} F(t) = 0$

And also $\qquad F(t) = -\int_\infty^t \dfrac{\cos x}{x}dx$

$$F'(t) = \frac{-\cos t}{t} \qquad \Rightarrow \qquad tF'(t) = -\cos t$$

$\Rightarrow \qquad L\{tf'(t)\} = L\{-\cos t\}$

$-\dfrac{d}{dp}\{pf(p) - F(0)\} = \dfrac{-p}{p^2+1}$

$\Rightarrow \qquad \dfrac{d}{dp}\{pf(p)\} = \dfrac{p}{p^2+1}$. For $f(0) = $ constant

$\Rightarrow \qquad pf(p) = \dfrac{1}{2}\log(p^2+1) + c$

$\Rightarrow \qquad \lim_{p\to 0} pf(p) = \lim_{p\to 0}\dfrac{1}{2}\log(p^2+1) + c = 0 + c$

$\Rightarrow \qquad \lim_{t\to\infty} F(t) = 0 + c$. For $\lim_{t\to\infty} F(t) = \lim_{p\to 0} pf(p)$

$$\Rightarrow \qquad\qquad 0 = 0 + c \ \Rightarrow \ c = 0$$

Then (1) becomes

$$\Rightarrow \qquad\qquad pf(p) = \frac{1}{2}\log(p^2 + 1) \ \Rightarrow \ f(p) = \frac{1}{2p}\log(p^2 + 1)$$

$$\Rightarrow \qquad\qquad L\{F(t)\} = \frac{1}{2p}\log(p^2 + 1)$$

EXAMPLE 13. Find the Laplace transform of $F(t)$,

where $f(t) = \begin{cases} t, & 0 \le t \le 1/2 \\ t-1, & 1/2 \le t \le 1 \\ 0, & t > 1 \end{cases}$　　　　　(Kanpur–2001)

SOLUTION. Let $L\{F(t)\} = f(p)$, then

$$f(p) = \int_0^\infty e^{-pt} F(t)dt$$

$$= \int_0^{1/2} e^{-pt} F(t)dt + \int_{1/2}^1 e^{-pt} F(t)dt + \int_1^\infty e^{-pt} F(t)dt$$

$$= \int_0^{1/2} e^{-pt} \cdot t \cdot dt + \int_{1/2}^1 e^{-pt}(t-1)dt + \int_1^\infty e^{-pt} \cdot 0 \cdot dt$$

$$= \int_0^1 te^{-pt}dt + \int_{1/2}^1 e^{-pt}tdt - \int_{1/2}^1 e^{-pt}dt + 0$$

$$= \int_0^1 e^{-pt}tdt + \frac{1}{p}(e^{-pt})_{1/2}^1$$

$$= \int_0^{-p} e^x \frac{xdx}{p^2} + \frac{1}{p}(a^2 - a), \ a = e^{-p/2}, -pt = x$$

$$= \frac{1}{p^2}\left\{\left((x-1)e^x\right)\right\}_0^{-p} + \frac{1}{p}(a^2 - a)$$

$$= \frac{1}{p^2}\left\{(-p-1)a^2 - (0-1) + a^2p - pa\right\}$$

$$= \frac{1}{p^2}\{1 - ap - a^2\} = \frac{1}{p^2}[1 - pe^{-p/2} - e^{-p}]$$

EXAMPLE 14. Find $L\{f(x)\}$ where $f(x) = \begin{cases} 2x, & 0 \le x \le 5 \\ 1, & x > 5 \end{cases}$　　　(Lucknow–2012)

SOLUTION. Consider $\quad L\{f(x)\} = \int_0^\infty e^{-pt} f(t)dt$

$$= \int_0^5 e^{-px} f(x)dx + \int_5^\infty e^{-px} f(x)dx$$

$$= \int_0^5 2xe^{-px}dx + \int_5^\infty 1 \cdot e^{-px}dx$$

$$= -\frac{9}{p}e^{-5p} - \frac{2}{p^2}e^{-5p} + \frac{2}{p^2}$$

EXAMPLE 15. Find the Laplace transform of $F(t)$, where

$$F(t) = \begin{cases} t/T, & 0 < t < T \\ 1, & t > T \end{cases}$$

SOLUTION. Consider, $L\{F(t)\} = \int_0^\infty e^{-pt} F(t)dt$

$$= \int_0^T e^{-pt} F(t)dt + \int_T^\infty e^{-pt} F(t)dt \ = \int_0^T \frac{t}{T}e^{-pt}dt + \int_T^\infty 1 \cdot e^{-pt}dt$$

$$= \frac{1}{T}\left[\left(\frac{te^{-pt}}{-p}\right) - \frac{1}{p^2}e^{-pt}\right]_{t=0}^{T} - \frac{1}{p}(e^{-pt})_{t=T}^{\infty}$$

$$= \frac{1}{T}\left[\left(\frac{Te^{-pT}}{-p} - 0\right) - \frac{1}{p^2}(e^{-pt} - 1)\right] + \frac{e^{-pT}}{p}$$

$$= \frac{-1}{p^2 T}(1 - e^{-pT})$$

EXAMPLE 16. *Prove that* $L\{J_0(t)\} = 1/(p^2 + 1)^{1/2}$ *and deduce that* $L\{J_0(at)\} = 1/(p^2 + a^2)^{1/2}$.

(Purvanchal–2003; Agra–2001, 02; Kanpur–2004; Rohilkhand–2001)

SOLUTION. Here, we have

$$J_n(t) = \sum_{r=0}^{\infty} \frac{(-1)^r}{r!\,\Gamma(n+r+1)}\left(\frac{t}{2}\right)^{n+2r}$$

Therefore,

$$J_0(t) = \sum_{r=0}^{\infty} \frac{(-1)^r}{r!\,\Gamma(r+1)}\left(\frac{t}{2}\right)^{2r}$$

$$= \sum_{r=0}^{\infty} \frac{(-1)^r}{2^{2r}(r!)^2}t^{2r} = 1 - \frac{t^2}{2^2} + \frac{t^4}{2^2 \cdot 4^2} - \frac{t^6}{2^2 \cdot 4^2 \cdot 6^2} + \ldots$$

∴

$$L\{J_0(t)\} = L\{1\} - \frac{1}{2^2}L\{t^2\} + \frac{1}{2^2 \cdot 4^2}L\{t^4\} - \frac{1}{2^2 \cdot 4^2 \cdot 6^2}L\{t^6\} + \ldots$$

or

$$L\{J_0(t)\} = \frac{1}{p} - \frac{1}{2^2}\frac{2!}{p^3} + \frac{1}{2^2 \cdot 4^2}\frac{4!}{p^5} - \frac{1}{2^2 \cdot 4^2 \cdot 6^2}\frac{6!}{p^7} + \ldots$$

$$= \frac{1}{p}\left\{1 - \frac{1}{2}\left(\frac{1}{p^2}\right) + \frac{1 \cdot 3}{2 \cdot 4}\left(\frac{1}{p^2}\right)^2 - \frac{1 \cdot 3 \cdot 5}{2 \cdot 4 \cdot 6}\left(\frac{1}{p^2}\right)^3 + \ldots\right\}$$

$$= \frac{1}{p}\left(1 + \frac{1}{p^2}\right)^{-1/2} = \frac{1}{(1+p^2)^{1/2}}$$

$$L\{J_0(t)\} = 1/(1+p^2)^{1/2}$$

Finally, using the change of scale property, we get

$$L\{J_0(at)\} = \frac{1}{a} \times \frac{1}{[1 + (p/a)^2]^{1/2}} = \frac{1}{\sqrt{(p^2 + a^2)}}$$

Additional Solved Examples

EXAMPLE 1. *If* $L\left\{2\sqrt{\dfrac{t}{\pi}}\right\} = \dfrac{1}{p^{3/2}}$, *then show that* $\dfrac{1}{p^{1/2}} = L\left\{\dfrac{1}{\sqrt{\pi t}}\right\}$.

(Meerut–2000; UPTU–2004, 05, Madras–2003)

SOLUTION. Let $F(t) = 2\sqrt{\dfrac{t}{\pi}}$, so that $F'(t) = \dfrac{1}{\sqrt{\pi t}}$

We have, $L\{F'(t)\} = pL\{F(t)\} - F(0)$

or $$L\left\{\frac{1}{\sqrt{\pi t}}\right\} = pL\left\{2\sqrt{\frac{t}{\pi}}\right\} - 0 = p\frac{1}{p^{3/2}} = \frac{1}{p^{1/2}}$$

Hence, $$L\{2\sqrt{t/\pi}\} = \frac{1}{\sqrt{p}}$$

EXAMPLE 2. *Find (i) $L\{F(t)\}$ and (ii) $L\{F'(t)\}$ for the function defined by* $F(t) = \begin{cases} 2t, & 0 \le t \le 1 \\ t, & t > 1 \end{cases}$.

(UPTU–2005; Madras–2003)

SOLUTION.   (i) We have, $L\{F(t)\} = \int_0^\infty e^{-pt} F(t)dt$

$$= \int_0^1 e^{-pt} 2t\, dt + \int_1^\infty e^{-pt} t\, dt$$

$$= \left[\left(\frac{e^{-pt}}{-p}\right)2t - \left(\frac{e^{-pt}}{p^2}\right)2\right]_0^1 + \left[\left(\frac{e^{-pt}}{-p}\right)t - \left(\frac{e^{-pt}}{p^2}\right)1\right]_1^\infty$$

$$= -\frac{2e^{-p}}{p} - \frac{2e^{-p}}{p^2} + \frac{2}{p^2} - \frac{1}{p}\lim_{t\to\infty}\frac{1}{e^{pt}} - \frac{1}{p^2}\lim_{t\to\infty}\frac{1}{e^{pt}} + \frac{e^{-p}}{p} + \frac{e^{-p}}{p^2}$$

$$= -\frac{2e^{-p}}{p} - \frac{2e^{-p}}{p^2} + \frac{2}{p^2} + \frac{e^{-p}}{p} + \frac{e^{-p}}{p^2}, p > 0$$

$$\left[\because \text{if } p > 0, \text{then } \lim_{t\to\infty}\frac{1}{e^{pt}} = 0 \text{ and also } \lim_{t\to\infty}\frac{1}{e^{pt}} = 0\right]$$

$$= \frac{2}{p^2} - \left(\frac{1}{p} + \frac{1}{p^2}\right)e^{-p}, p > 0$$

(ii) Now to find $L\{F'(t)\}$

First we observe that F(t) is discontinuous at t = 1

We have $F(1-0) = \lim_{t\to 1-0} F(t)$

$$= \lim_{h\to 0} F(1-h), h \text{ is positive and sufficiently small}$$

$$= \lim_{h\to 0} 2(1-h) = 2$$

and $F(1+0) = \lim_{t\to 1+0} F(t) = \lim_{h\to 0} F(1+h) = \lim_{h\to 0}(1+h) = 1$

Since, $F(1-0) \ne F(1+0)$, therefore $F(t)$ at $t = 1$.

∴ Using the formula for the Laplace transform of the derivative of $F(t)$, we have

$$L\{F'(t)\} = pL\{F(t)\} - F(0) - e^{-p}[F(1+0) - F(1-0)]$$

$$= p\left[\frac{2}{p^2} + \left(\frac{1}{p} + \frac{1}{p^2}\right)e^{-p}\right] - 0 - e^{-p}(1-2)$$

$$= \frac{2}{p} - e^{-p} - \frac{1}{p}e^{-p} + e^{-p} = \frac{2}{p} - \frac{1}{p}e^{-p} = \frac{1}{p}(2 - e^{-p}).$$

EXAMPLE 3.   *Show that* $\int_0^\infty \dfrac{\sin^2 t}{t^2} dt = \dfrac{\pi}{2}.$ (Agra–1999, Rohilkhand–2003, 04)

SOLUTION.   Let $$F(t) = \sin^2 t = \frac{1}{2}(1 - \cos 2t)$$

$\therefore$ $$L\{F(t)\} = \frac{1}{2}\left[\frac{1}{p} - \frac{p}{p^2 + 4}\right] = f(p), \text{ say}$$

Now, $$\lim_{t \to 0}\left[\frac{1}{t}F(t)\right] = \lim_{t \to 0}\left(\frac{\sin t}{t}\right)\sin t = 1.0 = 0.$$

i.e., limit exists.

$\therefore$ $$L\left\{\frac{1}{t}F(t)\right\} = \int_p^\infty f(x)dx = \frac{1}{2}\int_p^\infty \left(\frac{1}{x} - \frac{x}{x^2 + 4}\right)dx$$

$$= \frac{1}{2}\left[\log x - \frac{1}{2}\log(x^2 + 4)\right]_p^\infty = \frac{1}{4}\left[\log\left(\frac{x^2}{x^2 + 4}\right)\right]_p^\infty$$

$$= \frac{1}{4}\left[\lim_{x \to \infty} \log\frac{x^2}{x^2 + 4} - \log\frac{p^2}{p^2 + 4}\right]$$

$$= \frac{1}{4}\left[\lim_{x \to \infty} \log\frac{1}{1 + \left(\dfrac{4}{x^2}\right)} - \log\frac{p^2}{p^2 + 4}\right]$$

$$= \frac{1}{4}\left[\log 1 - \log\frac{p^2}{p^2 + 4}\right] = \frac{1}{4}\left[0 - \log\frac{p^2}{p^2 + 4}\right]$$

$$= -\frac{1}{4}\log\frac{p^2}{p^2 + 4} = \frac{1}{4}\log\frac{p^2 + 4}{p^2}$$

$$L\left\{\frac{\sin^2 t}{t}\right\} = \frac{1}{4}\log\frac{p^2 + 4}{p^2} = f(p), \text{ say}$$

We know that $L\{F(t)\} = f(p)$, then $\int_0^\infty \dfrac{F(t)}{t} dt = \int_0^\infty f(x)dx$ provided the integral converges.

$\therefore$ $$\int_0^\infty \frac{\sin^2 t}{t^2} dt = \int_0^\infty \frac{1}{t}\left(\frac{\sin^2 t}{t}\right)dt = \int_0^\infty \frac{1}{4}\log\frac{x^2 + 4}{x^2} dx \qquad \dots(1)$$

Now let $$I = \frac{1}{4}\int_0^\infty \log\left(\frac{x^2 + 4}{x^2}\right)dx$$

Put $x = 2\tan\theta$, then $dx = 2\sec^2\theta \, d\theta$

$\therefore$ $$I = \frac{1}{4}\int_0^{\pi/2}\left[\log\left(\frac{1}{\sin^2\theta}\right)\right]2\sec^2\theta d\theta = -\int_0^{\pi/2}\sec^2\theta \log\sin\theta d\theta$$

$$= -[\tan\theta \log\sin\theta]_0^{\pi/2} + \int_0^{\pi/2}\tan\theta\cot\theta\, d\theta$$

Integrating by parts by taking $\sec^2\theta$ as the second function, we get

$$I = -\lim_{\theta\to\pi/2}\tan\theta\log\tan\theta + \lim_{\theta\to 0}\tan\theta\log\sin\theta + \int_0^{\pi/2}d\theta$$

$$= 0 + [\theta]_0^{\pi/2} = \pi/2 \cdot$$

$\therefore$ From (1), we have $\int_0^\infty \dfrac{\sin^2 t}{t^2}dt = \dfrac{\pi}{2}\,.$

EXAMPLE 4.   *Show that* $\int_0^\infty \dfrac{\cos 6t - \cos 4t}{t}dt = \log\left(\dfrac{2}{3}\right).$ (Purvanchal–1991; Mumbai–2008; PTU–2006)

SOLUTION.   Let $F(t) = \cos 6t - \cos 4t$

Then $L\{F(t)\} = L\{\cos 6t\} - L\{\cos 4t\} = \dfrac{10}{p^2+6^2} - \dfrac{p}{p^2+4^2} = f(p)$, say

Now $\displaystyle\lim_{t\to 0}\dfrac{F(t)}{t} = \lim_{t\to 0}\dfrac{\cos 6t - \cos 4t}{t} = \lim_{t\to 0}\dfrac{-6\sin at + 4\sin 4t}{1}$

[By L′ Hospital rule]

$\therefore$

$$L\left\{\dfrac{F(t)}{t}\right\} = \int_p^\infty f(x)dx = \int_p^\infty\left[\dfrac{x}{x^2+6^2} - \dfrac{x}{x^2+4^2}\right]dx$$

$$= \dfrac{1}{2}[\log(x^2+6^2) - \log(x^2+4^2)]_p^\infty = \dfrac{1}{2}\left[\log\dfrac{x^2+6^2}{x^2+4^2}\right]_p^\infty$$

$$= \dfrac{1}{2}\lim_{x\to\infty}\log\dfrac{x^2+6^2}{x^2+4^2} - \dfrac{1}{2}\log\dfrac{p^2+6^2}{p^2+4^2}$$

$$= \dfrac{1}{2}\lim_{x\to\infty}\log\dfrac{1+(6^2/x^2)}{1+(4^2/x^2)} - \dfrac{1}{2}\log\dfrac{p^2+6^2}{p^2+4^2}$$

$$= 0 - \dfrac{1}{2}\log\dfrac{p^2+6^2}{p^2+4^2} = -\dfrac{1}{2}\log\dfrac{p^2+6^2}{p^2+4^2} = \dfrac{1}{2}\log\dfrac{p^2+4^2}{p^2+6^2}$$

$$L\left\{\dfrac{1}{t}F(t)\right\} = L\left\{\dfrac{1}{t}(\cos 6t - \cos 4t)\right\} = \dfrac{1}{2}\log\dfrac{p^2+4^2}{p^2+6^2}$$

$\therefore$ $\displaystyle\int_0^\infty e^{-pt}\dfrac{(\cos 6t - \cos 4t)}{t}dt = \dfrac{1}{2}\log\dfrac{p^2+4^2}{p^2+6^2}$

Taking limit as $p \to 0$, we have

$$\int_0^\infty\dfrac{\cos 6t - \cos 4t}{t}dt = \dfrac{1}{2}\log\dfrac{4^2}{6^2} = \dfrac{1}{2}\log\left(\dfrac{2}{3}\right)^2 = \log\left(\dfrac{2}{3}\right)$$

EXAMPLE 5.   *Show that*

(i) $L\{J_0(t)\} = \dfrac{1}{\sqrt{1+p^2}}$ (Agra–1994)

(ii) $L\{tJ_0(at)\} = \dfrac{p}{(p^2+a^2)^{3/2}}\,.$ (Agra–1980, SVTU–2004 ; Rohilkhand–2002)

SOLUTION.　　(i) We know that

$$J_0(t) = 1 - \frac{t^2}{2} + \frac{t^4}{2^2.4^2} - \frac{t^6}{2^2.4^2.6^2} + \dots$$

$$L\{J_0(t)\} = L\{1\} - \frac{1}{2^2}L\{t^2\} + \frac{1}{2^2.4^2}L\{t^4\} - \frac{1}{2^2.4^2.6^2}L\{t^6\} + \dots$$

$$= \frac{1}{p} - \frac{1}{2^2}.\frac{2!}{p^3} + \frac{1}{2^2.4^2}.\frac{4!}{p^5} - \frac{1}{2^2.4^2.6^2}\frac{6!}{p^7} + \dots$$

$$= \frac{1}{p}\left[1 - \frac{1}{2}\left(\frac{1}{p^2}\right) + \frac{1.3}{2.4}\left(\frac{1}{p^2}\right)^2 - \frac{1.3.5}{2.4.6}\left(\frac{1}{p^2}\right)^3 + \dots\right]$$

$$= \frac{1}{p}\left(1 + \frac{1}{p^2}\right)^{-1/2} = \frac{1}{\sqrt{1+p^2}}$$

(ii) Since $L\{F(at)\} = \frac{1}{a}f\left(\frac{p}{a}\right)$, where $f(p) = L\{F(t)\}$

$$\therefore\ L\{J_0(at)\} = \frac{1}{a}.\frac{1}{\sqrt{1+(p/a)^2}} = \frac{1}{\sqrt{p^2+a^2}}$$

Hence,　$L\{tJ_0(at)\} = -\frac{d}{dp}L\{J_0(at)\} = -\frac{d}{dp}\left[\frac{1}{\sqrt{p^2+a^2}}\right] = \frac{p}{(p^2+a^2)^{3/2}}$

EXAMPLE 6.　*Show that* $L\{J_1(t)\} = 1 - \dfrac{p}{\sqrt{p^2+1}}$, *where $J_1(t)$ is the Bessel function of order one*

and hence deduce that $L\{tJ_1(t)\} = \dfrac{1}{(p^2+1)^{3/2}}$.

SOLUTION.　We know that　$J_0'(t) = -J_1(t)$

∴ From $L\{F'(t)\} = pL\{F(t)\} - F(0)$, we have

$$L\{J_1(t)\} = L\{-J_0'(t)\} = -L\{J_0'(t)\}$$

$$= -[pL\{F(t)\} - J_0(0)] = -\left[p\frac{1}{\sqrt{p^2+1}} - 1\right]$$

$$\left[\because \text{From example 5, } L\{J_0(t)\} = \frac{1}{\sqrt{1+p^2}} \text{ and } J_0(0) = 1\right]$$

$$= 1 - \frac{p}{\sqrt{p^2+1}}$$

Deduction.　$L\{tJ_1(t)\} = -\dfrac{d}{dp}L\{J_1(t)\} = -\dfrac{d}{dp}\left\{1 - \dfrac{p}{\sqrt{p^2+1}}\right\} = \dfrac{1}{(p^2+1)^{3/2}}$

EXAMPLE 7.　*Prove that* $L\{J_0(a\sqrt{t})\} = \dfrac{1}{p}e^{-a^2/4p}$.

SOLUTION.　We know that

$$J_0(t) = 1 - \frac{t^2}{2^2} + \frac{t^4}{2^2.4^2} - \frac{t^6}{2^2.4^2.6^2} + \dots$$

$$\therefore \qquad J_0(a\sqrt{t}) = 1 - \frac{a^2 t}{2^2} + \frac{a^4 t^2}{2^2.4^2} - \frac{a^6 t^3}{2^2.4^2.6^2} + \dots$$

$$\Rightarrow \quad L\{J_0(a\sqrt{t})\} = L\{1\} - \frac{a^2}{2^2} L\{t\} + \frac{a^4}{2^2.4^2} L\{t^2\} - \frac{a^6}{2^2.4^2.6^2} L\{t^3\} + \dots$$

$$= \frac{1}{p} - \frac{a^2}{2^2} \frac{1!}{p^2} + \frac{a^4}{2^2.4^2} \frac{2!}{p^3} - \frac{a^6}{2^2.4^2.6^2} \frac{3!}{p^4} + \dots$$

$$= \frac{1}{p}\left[1 - \frac{1}{1!}\left(\frac{a^2}{4p}\right) + \frac{1}{2!}\left(\frac{a^2}{4p}\right)^2 - \frac{1}{3!}\left(\frac{a^2}{4p}\right)^3 + \dots \right] = \frac{1}{p} e^{-a^2/4p}$$

EXAMPLE 8. *Show that* $L\{t.erf(2\sqrt{t})\} = \dfrac{3p+8}{p^2(p+4)^{3/2}}$. (Rohilkhand–2004)

SOLUTION. We know that

$$erf(\sqrt{t}) = \frac{2}{\sqrt{\pi}} \int_0^{\sqrt{t}} e^{-u^2} du = \frac{2}{\sqrt{\pi}} \int_0^{\sqrt{t}}\left(1 - u^2 + \frac{u^4}{2!} - \frac{u^6}{3!} + \dots\right) du$$

$$= \frac{2}{\sqrt{\pi}}\left[u - \frac{u^3}{3} + \frac{u^5}{5(2!)} - \frac{u^7}{7(3!)} + \dots \right]_0^{\sqrt{t}}$$

$$= \frac{2}{\sqrt{\pi}}\left[t^{1/2} - \frac{t^{3/2}}{3} + \frac{t^{5/2}}{5(2!)} - \frac{t^{7/2}}{7(3!)} + \dots \right]$$

$$\Rightarrow L\{erf(\sqrt{t})\} = \frac{2}{\sqrt{\pi}}\left[L\{t^{1/2}\} - \frac{1}{3} L\{t^{3/2}\} + \frac{1}{5(2!)} L\{t^{5/2}\} - \frac{1}{7(3!)} L\{t^{7/2}\} + \dots \right]$$

$$= \frac{2}{\sqrt{\pi}}\left[\frac{\Gamma(3/2)}{p^{3/2}} - \frac{1}{3}\frac{\Gamma(5/2)}{p^{5/2}} + \frac{1}{5(2!)}\frac{\Gamma(7/2)}{p^{7/2}} - \frac{1}{7(3!)}\frac{\Gamma(9/2)}{p^{9/2}} + \dots \right]$$

$$= \frac{2}{\sqrt{\pi}}\left[\frac{\sqrt{\pi}}{2}\frac{1}{p^{3/2}} - \frac{\sqrt{\pi}}{2}.\frac{1}{2}.\frac{1}{p^{5/2}} + \frac{\sqrt{\pi}}{2}.\frac{1.3}{2.4}.\frac{1}{p^{7/2}} - \frac{\sqrt{\pi}}{2}.\frac{1.3.5}{2.4.6}.\frac{1}{p^{9/2}} + \dots \right]$$

$$= \frac{1}{p^{3/2}}\left[1 - \frac{1}{2}.\frac{1}{p} + \frac{1.3}{2.4}.\frac{1}{p^2} - \frac{1.3.5}{2.4.6}.\frac{1}{p^3} + \dots \right]$$

$$= \frac{1}{p^{3/2}}\left(1 + \frac{1}{p}\right)^{-1/2} = \frac{1}{p\sqrt{p+1}}$$

Since $\qquad L\{F(at)\} = \dfrac{1}{a} f\left(\dfrac{p}{a}\right)$, where $f(p) = L\{F(t)\}$

$$\therefore \qquad L\{erf(2\sqrt{t})\} = L\{erf\sqrt{(4t)}\} = \frac{1}{4}\frac{1}{\dfrac{p}{4}\sqrt{\dfrac{p}{4}+1}} = \frac{2}{p\sqrt{p+4}}$$

Hence, $L\{t.erf(2\sqrt{t})\} = -\dfrac{d}{dp} L\{erf 2\sqrt{t}\} = -\dfrac{d}{dp}\left[\dfrac{2}{p\sqrt{p+4}}\right] = \dfrac{3p}{p^2(p+4)^{3/2}}$

EXAMPLE 9. Show that $L\{e^{3t}.erf\sqrt{t}\} = \dfrac{1}{(p-3)\sqrt{p-2}}$.

SOLUTION. From previous example, we have

$$L\{erf\sqrt{t}\} = \frac{2}{p\sqrt{p+1}} = f(p), \text{ say.}$$

By first shifting theorem, we have

$$L\{e^{3t}erf\sqrt{t}\} = f(p-3) = \frac{1}{(p-3)\sqrt{(p-2)}}$$

EXAMPLE 10. Show that $L\{c_i(t)\} = \log\dfrac{p^2+1}{2p}$. (Rohilkhand–2003)

SOLUTION. Since, $L\{c_i(t)\} = L\left\{\int_t^\infty \dfrac{\cos u}{u} du\right\}$

Let $F(t) = \int_t^\infty \dfrac{\cos u}{u} du = -\int_\infty^t \dfrac{\cos u}{u} du$

so that $F'(t) = -\dfrac{\cos t}{t}$ or $tF'(t) = -\cos t$

$\therefore$ $L\{tF'(t)\} = L\{-\cos t\}$

or $-\dfrac{d}{dp}L\{F'(t)\} = -\dfrac{p}{p^2+1}$ or $\dfrac{d}{dp}[pf(p) - F(0)] = \dfrac{p}{p^2+1}$ where $f(p) = L\{F(t)\}$

or $\dfrac{d}{dp}[pf(p)] = \dfrac{p}{p^2+1}$, since $F(0)$ is constant.

Integrating $pf(p) = \dfrac{1}{2}\log(p^2+1) + C$ (constant). ...(1)

Using final value theorem, we have

$$\lim_{p\to 0} pf(p) = \lim_{t\to\infty} F(t) = 0.$$

From (1) as $p \to 0$, we have

$$0 = 0 + C \text{ or } C = 0.$$

$\therefore$ From (1), $pf(p) = \dfrac{1}{2}\log(p^2+1)$

or $f(p) = L\{F(t)\} = L\{C_i(t)\} = \log\left(\dfrac{p^2+1}{2p}\right)$

EXAMPLE 11. If $F(t) = t^2, 0 < t < 2$ and $F(t+2) = F(t)$. Then show that

$$L\{F(t)\} = \frac{-(4p^2+4p+2)e^{-2p+2}}{p^3(1-e^{-2p})}$$ (Vikram–1996; Kanpur–1993)

SOLUTION. Here, $F(t)$ is a periodic function with period $T = 2$

Now, $L\{F(t)\} = \dfrac{\int_0^T e^{-pt}F(t)dt}{1-e^{-2p}} = \dfrac{\int_0^T t^2 e^{-pt}dt}{1-e^{-2p}} = \dfrac{\left(-\dfrac{t^2}{p}e^{-pt}\right)_0^2 + \dfrac{2}{p}\int_0^2 te^{-pt}dt}{1-e^{-2p}}$

$$= \dfrac{\dfrac{4}{p}e^{-2p} + \dfrac{2}{p}\left\{\left(-\dfrac{t^2}{p}e^{-pt}\right)_0^2 + \dfrac{1}{p}\int_0^2 e^{-pt}dt\right\}}{1-e^{-2p}}$$

$$= \frac{-\frac{4}{p}e^{-2p} - \frac{4}{p^2}e^{-2p} + \frac{2}{p^2}\left(-\frac{e^{-pt}}{p}\right)^2_0}{1 - e^{-2p}}$$

$$= \frac{-\frac{4}{p}e^{-2p} - \frac{4}{p^2}e^{-2p} - \frac{2}{p^3}e^{-2p} + \frac{2}{p^3}}{1 - e^{-2p}} = \frac{-(4p^2 + 4p + 2)e^{-2p} + 2}{p^3(1 - e^{-2p})}$$

EXAMPLE 12. *If* $F(t) = \begin{cases} 3t, & 0 < t < 2 \\ t, & 2 < t < 4 \end{cases}$ *and* $F(t)$ *is a periodic function of period 4, then show that*

$$L\{F(t)\} = \frac{3 - 3e^{-2p} - 6pe^{-4p}}{p^2(1 - e^{-4p})}$$

SOLUTION. Here, $F(t)$ is a periodic function with period $T = 4$.
We have

$$L\{F(t)\} = \frac{\int_0^T e^{-pt}F(t)dt}{1 - e^{-pT}} = \frac{\int_0^2 3t e^{-pt}dt + \int_2^4 6e^{-pt}dt}{1 - e^{-4p}}$$

$$= \frac{1}{1 - e^{-4p}}\left[\left(\frac{e^{-pt}}{-p}\right)3t - \left(\frac{e^{-pt}}{p^2}\right)3\right]_0^2 + \frac{1}{1 - e^{-4p}}\left[\frac{6e^{-pt}}{-p}\right]_2^4$$

$$= \frac{1}{1 - e^{-4p}}\left[-\frac{6}{p}e^{-2p} - \frac{3e^{-2p}}{p^2} + \frac{3}{p^2} - \frac{6}{p} - \frac{6}{p}e^{-4p} + \frac{6}{p}e^{-2p}\right] = \frac{3 - 3e^{-2p} - 6pe^{-4p}}{p^2(1 - e^{-4p})}$$

EXAMPLE 13. *Find the Laplace transform of the Heaviside's unit step function* $H(t - a)$.

SOLUTION. We have $H(t - a) = \begin{cases} 0, & t < a \\ 1, & t > a \end{cases}$

$$L\{H(t - a)\} = \int_0^\infty e^{-pt}H(t - a)dt = \int_a^\infty e^{-pt}1dt = \left[\frac{e^{-pt}}{-p}\right]_a^\infty$$

$$= \left[\lim_{t \to \infty}\frac{e^{-pt}}{-p}\right] + \frac{e^{-ap}}{p} = 0 + \frac{e^{-ap}}{p}, p > 0$$

$$= \frac{e^{-ap}}{p}, p > 0$$

EXAMPLE 14. *Show that* $\int_0^\infty \frac{\sin t}{t}dt = \frac{\pi}{2}$. (Agra–1985, 2004; Avadh–2006, 14; Kanpur–2006;

Meerut–2002, 13; Kashi–2007; Purvanchal–2007; Rohilkhand–2009, 10, 14; Nagpur–2005)

SOLUTION. Let $F(t) = \sin t$

So that $f(p) = L\{F(t)\} = L\{\sin t\} = \dfrac{1}{p^2 + 1}$

Now, $L\left\{\dfrac{1}{t}\sin t\right\} = \int_0^\infty e^{-pt}\dfrac{\sin t}{t}dt = \int_p^\infty f(x)dx$

$$= \int_p^\infty \frac{1}{x^2 + 1}dx = [\tan^{-1}x]_p^\infty = \frac{\pi}{2} - \tan^{-1}p$$

Taking limit as $p \to 0$, we have

$$\int_0^\infty \frac{\sin t}{t}dt = \frac{\pi}{2}$$

EXAMPLE 15. Show that $\int_0^\infty \dfrac{e^{-t} - e^{-3t}}{t} dt = \log 3$.

(Meerut–1991, 2012; Agra–2000, 03; Bundelkhand–2005; Rohilkhand–2010)

SOLUTION. Let $F(t) = e^{-t} - e^{-3t}$

Then, $L\{F(t)\} = L\{e^{-t}\} - L\{e^{-3t}\} = \dfrac{1}{p+1} - \dfrac{1}{p+3} = f(p)$, say

$\therefore \quad L\left\{\dfrac{F(t)}{t}\right\} = \int_p^\infty f(x) dx = \int_p^\infty \left(\dfrac{1}{x+1} - \dfrac{1}{x+3}\right) dx$

$= \left[\log\left(\dfrac{x+1}{x+3}\right)\right]_p^\infty = \lim_{x \to \infty} \log\dfrac{x+1}{x+3} - \log\dfrac{p+1}{p+3}$

$= \lim_{x \to \infty} \log\dfrac{(1+1/x)}{(1+3/x)} - \log\dfrac{p+1}{p+3} = 0 - \log\dfrac{p+1}{p+3} = \log\dfrac{p+1}{p+3}$

$\therefore$ Taking limit as $p \to 0$, we have

$\int_0^\infty \dfrac{e^{-t} - e^{-3t}}{t} dt = \log\dfrac{3}{1} = \log 3$.

EXAMPLE 16. Show that $\int_0^\infty t^3 e^{-t} \sin t \, dt = 0$.

(Meerut–1987, 97; Rohilkhand–1996, 97)

SOLUTION. Let $F(t) = t^3 \sin t$

$\therefore \quad L\{F(t)\} = L\{t^3 \sin t\} = (-1)^3 \dfrac{d^3}{dP^3} L\{\sin t\}$

$= -\dfrac{d^3}{dp^3}\left\{\dfrac{1}{p^2+1}\right\} = -\dfrac{d^2}{dp^2}\left[-\dfrac{2p}{(p^2+1)^2}\right] = \dfrac{d}{dp}\left\{\dfrac{2 - 6p^2}{(p^2+1)^3}\right\}$

or $\int_0^\infty e^{-pt} t^3 \sin t \, dt = \dfrac{24(p^2-1)p}{(p^2+1)^4}$

Taking limit as $p \to 1$, we have $\int_0^\infty t^3 e^{-t} \sin t \, dt = 0$.

✎ RECAPITULATIONS: LAPLACE TRANSFORM OF SOME STANDARD FUNCTIONS

S. No.	Function $F(t)$	$L[F(t)]$	S. No.	Function $F(t)$	$L[F(t)]$
1.	$J_0(t)$	$\dfrac{1}{\sqrt{1+p^2}}$	8.	$\dfrac{\cos\sqrt{t}}{\sqrt{t}}$	$\sqrt{\dfrac{\pi}{p}} \cdot e^{-1/4p}$
2.	$J_0(at)$	$\dfrac{p}{(a^2+p^2)^{3/2}}$	9.	$\dfrac{\sin at}{t}$	$\tan^{-1}\dfrac{a}{p}$
3	$\mathrm{erf}(\sqrt{t})$	$\dfrac{1}{p\sqrt{p+1}}$	10.	$\dfrac{\sinh t}{t}$	$\dfrac{1}{2}\log\dfrac{p+1}{p-1}$
4.	$E(t) = \int_t^\infty \dfrac{e^{-x}}{x} dx$	$\dfrac{1}{p}\log(p+1)$	11.	$\dfrac{\sin^2 t}{t}$	$\dfrac{1}{4}\log\dfrac{p^2+4}{p^2}$
5.	$\int_t^\infty \dfrac{\sin u}{u} du$	$\dfrac{1}{p}\tan^{-1}\dfrac{1}{p}$	12.	$\sinh at \cos at$	$\dfrac{a(p^2-2a^2)}{p^4+4a^4}$
6.	$\int_t^\infty \dfrac{\cos x}{x} dx$	$\dfrac{1}{2p}\log(1+p^2)$	13.	$\sinh at \sin at$	$\dfrac{2a^2p}{p^4+4a^4}$
7.	$\sin\sqrt{t}$	$\dfrac{\sqrt{\pi}}{2p^{3/2}} e^{-1/4p}$	14.	$H(t-a)$	$\dfrac{e^{-ap}}{p}; \, p > a$

1.10 THE UNIT STEP FUNCTION

The unit step function, denoted by $H(t-a)$ is defined by

$$H(t-a) = \begin{cases} 0, & t < a \\ 1, & t > a \end{cases}$$

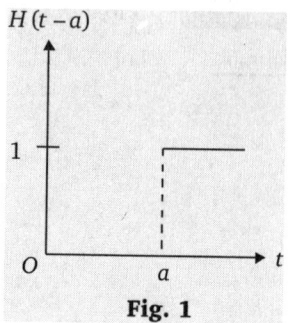

$$\therefore \qquad L\{H(t-a)\} = \int_0^\infty e^{-pt} H(t-a) dt$$

$$= \int_a^\infty e^{-pt} .1 \, dt \; = \left[\frac{e^{-pt}}{-p} \right]_a^\infty$$

$$= \left[\lim_{t \to \infty} \frac{e^{-pt}}{-p} \right] + \frac{e^{-ap}}{p} = 0 + \frac{e^{-ap}}{p}, p > 0$$

$$= \frac{e^{-ap}}{p}, p > 0$$

Fig. 1

THEOREM 1.   *If* $L\{F(t)\} = f(p)$, *then* $L\{F(t-a).H(t-a)\} = e^{-ap} f(p)$.

PROOF.   We have

$$L\{F(t-a).H(t-a)\} = \int_0^\infty e^{-pt} [F(t-a).H(t-a)] \, dt$$

$$= \int_0^a e^{-pt} F(t-a).0 \, dt + \int_a^\infty e^{-pt} F(t-a).(1) dt$$

$$= \int_a^\infty e^{-pt} F(t-a) dt = \int_0^\infty e^{-p(u+a)} F(u) du \quad \text{[Putting } t-a=u]$$

$$= e^{-ap} \int_0^\infty e^{-pu} F(u) \, du = e^{-ap} f(p).$$

THEOREM 2. $L\{F(t).H(t-a)\} = e^{-ap} L\{F(t+a)\} dt$.

PROOF.   We have

$$L\{F(t).H(t-a)\} = \int_0^\infty e^{-pt} [F(t).H(t-a)] \, dt$$

$$= \int_0^a e^{-pt} [F(t).H(t-a)] \, dt + \int_a^\infty e^{-pt} [F(t).H(t-a)] \, dt$$

$$= 0 + \int_0^\infty e^{-pt} . F(t) . (1) dt = \int_0^\infty e^{-p(u+a)} F(u+a) du,$$

[Putting $t - a = u$]

$$= e^{-ap} \int_0^\infty e^{-pu} F(u+a) \, du$$

$$= e^{-ap} \int_0^\infty e^{-pt} F(t+a) dt = e^{-ap} L\{F(t+a)\}$$

Solved Examples

EXAMPLE 1.   *Find the Laplace transform of* $t^2 H(t-3)$.

SOLUTION.   We have $L\{t^2 H(t-3)\} = e^{-3p} L\{(t+3)^2\} = e^{-3p} L\{t^2 + 6t + 9\}$

$$= e^{3p} \left[\frac{2}{p^3} + \frac{6}{p^2} + \frac{9}{p} \right]$$

EXAMPLE 2.   *Express the following function in terms of unit step function and find its Laplace transform :*

$$F(t) = \begin{cases} t-1, & 1 < t < 2 \\ 3-t, & 2 < t < 3 \end{cases}.$$

SOLUTION. We have

$$F(t) = \begin{cases} t-1, & 1 < t < 2 \\ 3-t, & 2 < t < 3 \end{cases}$$

$$= (t-1)[H(t-1) - H(t-2)] + (3-t)[H(t-2) - H(t-3)]$$

$$= (t-1)H(t-1) - (t-1)H(t-2) + (3-t)H(t-2) + (t-3)H(t-3)$$

$$= (t-1)H(t-1) - 2(t-2)H(t-2) + (t-3)H(t-3).$$

$$\therefore \quad L\{F(t)\} = \frac{e^{-p}}{p^2} - 2\frac{e^{-2p}}{p^2} + \frac{e^{-3p}}{p^2}.$$

EXAMPLE 3. *Express the following function in terms of unit step function and find its Laplace transform :* (UPTU–2002)

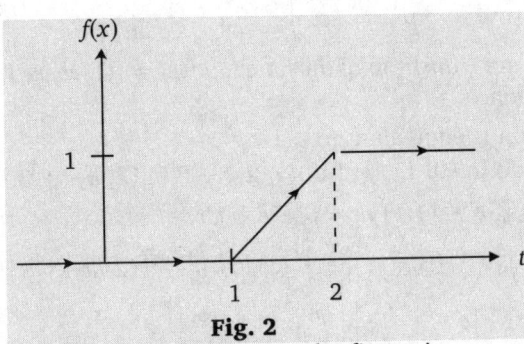

Fig. 2

SOLUTION. The algebraic form of the function in the figure is

$$F(t) = \begin{cases} 0, & 0 < t < 1 \\ t-1, & 1 < t < 2 \\ 1, & 2 < t \end{cases} \qquad ...(1)$$

$$= (t-1)[H(t-1) - H(t-2)] + H(t-2)$$

$$= (t-1)H(t-1) - (t-1-1)H(t-2)$$

$$= (t-1)H(t-1) - (t-2)H(t-2)$$

$$\therefore \quad L\{F(t)\} = L\{(t-1)H(t-1)\} - L\{(t-2)H(t-2)\}$$

$$= \frac{e^{-p}}{p^2} - L\{(t-2)H(t-2)\} = \frac{e^{-p}}{p^2} - \frac{e^{-2p}}{p^2}.$$

Miscellaneous Solved Examples

EXAMPLE 1. *Find* $L\{\sqrt{t}e^{3t}\}$. (PTU–2009)

SOLUTION. Since, $$L\{\sqrt{t}\} = \frac{\overline{(3/2)}}{p^{3/2}} = \frac{(1/2).\sqrt{\pi}}{p^{3/2}}$$

∴ By shifting property, we get

$$L(e^{3t}\sqrt{t}) = \frac{\sqrt{\pi}}{2} = \frac{1}{(p-3)^{3/2}}.$$

EXAMPLE 2. *Find the Laplace transform of the function* $f(t) = |t-1| + |t+1|, t \geq 0$. (SVTU–2009; MDU–2007)

SOLUTION. Given function is equivalent to

$$f(t) = \begin{cases} 2 & 0 \leq t \leq 1 \\ 2t & t \geq 1 \end{cases}$$

$$\therefore \qquad Lf(t) = \int_0^1 e^{-pt}(2)dt + \int_1^\infty e^{-pt}(2t)dt$$

$$= 2\left[\left|\frac{e^{-pt}}{-p}\right|_0^1 + 2\left|\frac{t.e^{-pt}}{-p}\right|_1^\infty - \left|\frac{e^{-pt}}{(-p)^2}\right|_1^\infty\right]$$

$$= 2\left(\frac{e^{-p}}{-p} + \frac{1}{p}\right) + 2\left(\frac{0-e^{-p}}{-p} - \frac{0-e^{-p}}{p^2}\right)$$

$$= \frac{2}{p}\left(1 + \frac{e^{-p}}{p}\right).$$

EXAMPLE 3. *Find the Laplace transform of the function $f(t) = [t]$ where $[\ ]$ stands for the greatest integer function.*
(PTU–2010)

SOLUTION. Given function is equivalent to

$$[t] = 0 \text{ in } (0, 1), + 1 \text{ in } (1, 2), + 2 \text{ in } (2, 3), + 3 \text{ in } (3, 4) + \dots$$

$$\therefore \quad L[F(t)] = \int_0^\infty e^{-pt}[f(t)]dt = \int_0^\infty e^{-pt}[t]dt$$

$$= \int_0^1 e^{-pt}(0)dt + \int_1^2 e^{-pt}(1)dt + \int_2^3 e^{-pt}(2)dt + \int_3^4 e^{-pt}(3)dt + \dots + \infty$$

$$= 0 + \left|\frac{e^{-pt}}{-p}\right|_1^2 + 2\left|\frac{e^{-pt}}{-p}\right|_2^3 + 3\left|\frac{e^{-pt}}{-p}\right|_3^4 + \dots + \infty$$

$$= \frac{-1}{p}[(e^{-2p} - e^{-p}) + 2(e^{-3p} - e^{-2p}) + 3(e^{-4p} - e^{-3p}) + \dots + \infty]$$

$$= \frac{1}{p}(e^{-p} + e^{-2p} + e^{-3p} \dots + \infty) = \frac{1}{p}\left(\frac{e^{-p}}{1-e^{-p}}\right) = \frac{1}{p(e^{-p}-1)}.$$

EXAMPLE 4. *Find $L(erf\, 2\sqrt{t})$.*
(Mumbai–2006)

SOLUTION. Since we know that $L(erf\sqrt{t}) = \dfrac{1}{p(p+1)}$

$$\therefore L(erf\,2\sqrt{t}) = L[erf\sqrt{(4t)}] = \frac{1}{4} \cdot \frac{1}{\dfrac{p}{4}\sqrt{\left(\dfrac{p}{4}+1\right)}} = \frac{2}{p\sqrt{(p+4)}}.$$

EXAMPLE 5. *Find $L\{t^3 e^{-3t}\}$.*
(Meerut–1984, 2007; Kanpur–2008; Kottayam–2005)

SOLUTION. Since $L(e^{-3t}) = \dfrac{1}{p+3}$

$$\therefore \quad L(t^3 e^{-3t}) = (-1)^3 \frac{d^3}{dp^3}\left(\frac{1}{p+3}\right) = -\frac{(-1)^3 \cdot 3!}{(p+3)^{3+1}} = \frac{6}{(p+3)^4}.$$

EXAMPLE 6. *Find $L\left\{(1-e^t)/t\right\}$.*
(Kerala–1985; Madras–2000; Sagar–2004)

SOLUTION. $\because \quad L(1-e^t) = L(1) - L(e^t) = \dfrac{1}{p} - \dfrac{1}{p-1}$

$$\therefore \quad L\left(\frac{1-e^t}{t}\right) = \int_p^\infty \left(\frac{1}{p} - \frac{1}{p-1}\right) dp = \left|\log p - \log(p-1)\right|_p^\infty = \left|\log\left(\frac{p}{p-1}\right)\right|_p^\infty$$

$$= -\log\left[\frac{1}{1-1/p}\right] = \log\left(\frac{p-1}{p}\right).$$

EXAMPLE 7. $Find\ L\left\{\dfrac{\cos at - \cos bt}{t} + t\sin at\right\}.$ (VTU–2010; Meerut–1996; Purvanchal–2004)

SOLUTION. We have $L(\cos at - \cos bt) = \dfrac{p}{p^2+a^2} - \dfrac{p}{p^2+b^2}$ and $L(\sin at) = \dfrac{a}{p^2+a^2}$

$$\therefore \quad L\left(\frac{\cos at - \cos bt}{t}\right) + L(t\sin at)$$

$$= \int_p^\infty \left(\frac{p}{p^2+a^2} - \frac{p}{p^2+b^2}\right) dp - \frac{d}{dp}\left(\frac{a}{p^2+a^2}\right)$$

$$= \left|\frac{1}{2}\log(p^2+a^2) - \frac{1}{2}\log(p^2+b^2)\right|_p^\infty - a\frac{(-2p)}{(p^2+a^2)^2}.$$

$$= \frac{1}{2}\lim_{p\to\infty}\log\frac{p^2+a^2}{p^2+b^2} - \frac{1}{2}\log\frac{p^2+a^2}{p^2+b^2} + \frac{2ap}{(p^2+a^2)^2}$$

$$= \frac{1}{2}\log\left(\frac{1+0}{1+0}\right) - \frac{1}{2}\log\frac{p^2+a^2}{p^2+b^2} + \frac{2ap}{(p^2+a^2)^2}$$

$$= \log\left(\frac{p^2+b^2}{p^2+a^2}\right)^{1/2} + \frac{2ap}{(p^2+a^2)^2}. \qquad [\because \log 1 = 0]$$

EXAMPLE 8. $Find\ L\left\{e^{-t}\int_0^t \dfrac{\sin t}{t}\,dt\right\}.$ (Madras–2006)

SOLUTION. We have $L\{\sin t\} = \dfrac{1}{p^2+1}$

$$L\left(\frac{\sin t}{t}\right) = \int_0^\infty \frac{1}{p^2+1}\,dp = \frac{\pi}{2} - \tan^{-1} p = \cot^{-1} p$$

$$\therefore \quad L\left(\int_0^t \frac{\sin t}{t}\,dt\right) = \frac{1}{p}\cot^{-1} p.$$

Now by shifting property, we get

$$L\left\{e^{-t}\left(\int_0^t \frac{\sin t}{t}\,dt\right)\right\} = \frac{1}{p+1}\cot^{-1}(p+1).$$

EXAMPLE 9. $Find\ L\left\{t\int_0^t \dfrac{e^{-t}\sin t}{t}\,dt\right\}.$ (PTU–2005)

SOLUTION. Since $L\left(\dfrac{\sin t}{t}\right) = \cot^{-1} p$. Therefore $L\left(e^{-t}.\dfrac{\sin t}{t}\right) = \cot^{-1}(p+1)$

and $\quad L\left\{\int_0^t e^{-t} \cdot \dfrac{\sin t}{t} dt\right\} = \dfrac{1}{p} \cot^{-1}(p+1)$

Hence, $\quad L\left\{t\int_0^t e^{-t} \cdot \dfrac{\sin t}{t} dt\right\} = -\dfrac{d}{dp}\left\{\dfrac{\cot^{-1}(p+1)}{p}\right\}$

$$= -\dfrac{p\left[\dfrac{-1}{1+(p+1)^2}\right] - \cot^{-1}(p+1)}{p^2}$$

$$= \dfrac{p + (p^2+2p+2)\cot^{-1}(p+1)}{p^2(p^2+2p+2)}.$$

EXAMPLE 10. Find $L\left\{\int_0^t \int_0^t \int_0^t (t\sin t) dt dt dt\right\}$. $\hfill$ (Mumbai–2006)

SOLUTION. Since, $\qquad L(\sin t) = \dfrac{1}{p^2+1}$

$\therefore \qquad\qquad L(t\sin t) = -\dfrac{d}{dp}\dfrac{1}{(p^2+1)} = \dfrac{2p}{(p^2+1)^2}$

Thus, $L\left\{\int_0^t \int_0^t \int_0^t (t\sin t) dt dt dt\right\} = \dfrac{1}{p^3} L(t\sin t)$

$$= \dfrac{1}{p^3} \cdot \dfrac{2p}{(p^2+1)^2} = \dfrac{2}{p^2(p^2+1)^2}$$

EXAMPLE 11. $\cdot$ Find $L\int_0^\infty t e^{-3t} \sin t \, dt$.

$\hfill$ (Agra–1999; VTU–2007; Rohilkhand–2012; Meerut–2000, 07; Kanpur–2006)

SOLUTION. $\qquad \int_0^\infty t e^{-3t} \sin t \, dt = \int_0^\infty e^{-pt}(t\sin t) dt$, where $p = 3$

$$= L(t\sin t), \text{By def.}$$

$$= (-1)\dfrac{d}{dp}\left(\dfrac{1}{p^2+1}\right) = \dfrac{2p}{(p^2+1)^2} = \dfrac{2\times 3}{(3^2+1)^2} = \dfrac{3}{50}.$$

EXAMPLE 12. Find $\int_0^\infty e^{-t}\left(\dfrac{\cos at - \cos bt}{t}\right) dt$. $\hfill$ (Mumbai–2009)

SOLUTION. Since, $\qquad L(\cos at) = \dfrac{p}{p^2+a^2}$ and $L(\cos bt) = \dfrac{p}{p^2+b^2}$

$\therefore \quad L\left[\dfrac{\cos at - \cos bt}{t}\right] = \int_p^\infty \left(\dfrac{p}{p^2+a^2} - \dfrac{p}{p^2+b^2}\right) dp$

$$= \dfrac{1}{2}\left\{\log\left(\dfrac{p^2+a^2}{p^2+b^2}\right)\right\}_p^\infty = \dfrac{1}{2}\log\left(\dfrac{p^2+b^2}{p^2+a^2}\right)$$

$$\Rightarrow \int_0^\infty e^{-pt}\left(\dfrac{\cos at - \cos bt}{t}\right) dt = \dfrac{1}{2}\log\left(\dfrac{p^2+b^2}{p^2+a^2}\right)$$

Taking $p = 1$, we get

$$\int_0^\infty e^{-t}\left(\frac{\cos at - \cos bt}{t}\right)dt = \frac{1}{2}\log\left(\frac{1+b^2}{1+a^2}\right).$$

EXAMPLE 13. *Evaluate* $L\left(\dfrac{e^t - \cos t}{t}\right)$ (Avadh–2012)

SOLUTION. Since,

$$L(e^t - \cos t) = \frac{1}{p-1} - \frac{p}{p^2+1}$$

$$\Rightarrow \quad L\left(\frac{e^t - \cos t}{t}\right) = \int_p^\infty\left(\frac{1}{x-1} - \frac{x}{x^2+1}\right)dx = \left[\log(x-1) - \frac{1}{2}\log(x^2+1)\right]_{x=p}^\infty$$

$$= \frac{1}{2}\left\{\log\frac{(x-1)^2}{(x^2+1)}\right\}_{x=p}^\infty = \frac{1}{2}\left[0 - \frac{\log(p-1)^2}{(p^2+1)}\right]$$

$$= \frac{1}{2}\log\left\{\frac{p^2+1}{(p-1)^2}\right\}$$

EXAMPLE 14. *Find the Laplace transform of the function (Half wave rectifier):*

$$f(t) = \begin{cases} \sin\omega t & \text{for} \quad 0 < t < \pi/\omega \\ 0 & \text{for} \quad \dfrac{\pi}{\omega} < t < \dfrac{2\pi}{\omega} \end{cases}$$ (Kanpur–2004)

SOLUTION. Here $a = \dfrac{2\pi}{\omega}$ is the period. For periodic function

$$L\{f(t)\} = \frac{1}{k}\int_0^a e^{-pt} f(t)dt \qquad\qquad ...(1)$$

where $\qquad k = 1 - e^{-pa} = 1 - e^{-2\pi p/\omega}$

$$\therefore \quad \int e^{ax}\sin bx\, dx = \frac{e^{ax}(a\sin bx - b\cos bx)}{a^2+b^2}$$

$$\int_0^a e^{-pt} f(t)dt = \int_0^{2\pi/\omega} e^{-pt} f(t)dt$$

$$= \int_0^{\pi/\omega} e^{-pt} f(t)dt + \int_{\pi/\omega}^{2\pi/\omega} e^{-pt}.0\,dt$$

$$= \int_0^{\pi/\omega} e^{-pt}\sin(\omega t)dt$$

$$= \frac{1}{p^2+\omega^2}\left[e^{-pt}(-p\sin\omega t - \omega\cos\omega t)\right]_{t=0}^{\pi/\omega}$$

$$= \frac{1}{p^2+\omega^2}\left[e^{-p\pi/\omega}(-p\sin\pi - \omega\cos\pi) - 1(-p\sin(0\omega).\cos 0)\right]$$

$$= \frac{\omega}{p^2+\omega^2}(1 + e^{-p\pi/\omega})$$

$$Lf(t) = \frac{\omega(1 + e^{-p\pi/\omega})}{(p^2+\omega^2)(1 - e^{-2\pi p/\omega})}$$

EXAMPLE 15. If $L\left\{\dfrac{1}{a^2}(1-\cos at)\right\} = \dfrac{1}{p(p^2+a^2)}$ *prove that* $L\left\{\dfrac{1}{a^2}(1-\cos at)\right\} = \dfrac{3p^2+a^2}{p^2(p^2+a^2)^2}$.

SOLUTION. Here, we have

$$L\left\{\frac{t}{a^2}(1-\cos at)\right\} = (-1)\frac{d}{dp}[L(1-\cos at)]$$

$$= -\frac{d}{dp}\left[\frac{1}{p(p^2+a^2)}\right] = -\frac{d}{dp}\left[\frac{1}{p^3+pa^2}\right]$$

$$= \frac{3p^2+a^2}{(p^3+pa^2)^2} = \frac{3p^2+a^2}{p^2(p^2+a^2)^2}$$

EXAMPLE 16. *Evaluate* $L(5t-2)$. (Kanpur–2002)

SOLUTION. We know that $L\{t^n\} = \dfrac{n!}{p^{n+1}}$, $L(1) = \dfrac{1}{p}$

Thus $L(5t-2) = 5L(t) - 2L(1) = 5 \cdot \dfrac{1!}{p^2} - \dfrac{2}{p} = \dfrac{5}{p^2} - \dfrac{2}{p}$

EXAMPLE 17. *Evaluate* $L\{\sin 2t.\sin 3t\}$ (Kanpur–2012)

SOLUTION. Consider $L\{\sin 2t.\sin 3t\} = \dfrac{1}{2}L\{\cos(3t-2t) - \cos(3t+2t)\}$

$$= \frac{1}{2}L[\cos t - \cos 5t] = \frac{1}{2}\left[\frac{p}{p^2+1} - \frac{p^2}{p^2+25}\right].$$

EXAMPLE 18. *Prove that* $L\{H(t)\} = \dfrac{2(1-e^{-\pi p})}{p^2+4}$ *where* $H(t) = \begin{cases} \sin 2t, & 0 < t < \pi \\ 0, & t > \pi \end{cases}$ (Meerut–2009)

SOLUTION. Consider $L\{H(t)\} = \int_0^\infty e^{-pt}H(t)dt = \int_0^\pi e^{-pt}H(t)dt + \int_\pi^\infty e^{-pt}H(t)dt$

$$= \int_0^\pi e^{-pt}\sin 2t\,dt + \int_\pi^\infty e^{-pt}0 \cdot dt$$

$$= \left[e^{-pt}\frac{(-p\sin 2t - 2\cos 2t)}{p^2+4}\right]_{t=0}^\pi = \frac{2(1-e^{-\pi p})}{p^2+4}$$

EXAMPLE 19. *Find* $L\{tJ_0(at)\}$ (SVTU–2004, 06)

SOLUTION. Consider $L\{tJ_0(at)\} = (-1)^1\dfrac{d}{dp}L\{J_0(at)\}$ ...(1)

$$L\{J_0(at)\} = \frac{1}{a}f\left(\frac{p}{a}\right) \quad \text{where} \quad f(p) = L\{J_0(t)\} \qquad \text{...(2)}$$

Now, $J_n(x) = \dfrac{x^n}{2^n\Gamma(n+1)}\left\{1 - \dfrac{x^2}{2(2n+2)} + \dfrac{x^4}{2\cdot 4(2n+2)(2n+4)}\right\}$

Let $n = 0$, then $x^0 = 2^0 = 1, \Gamma(1) = 1$

$\therefore$ $J_0(x) = \left\{1 - \dfrac{x^2}{2.2} + \dfrac{x^4}{2.4.2.4}\cdots\right\}$

Also, $L\{t^n\} = \dfrac{n!}{p^{n+1}}$, $n = 0, 1, 2, \ldots$

Now. $L\{J_0(t)\} = \left\{ \dfrac{1}{p} - \dfrac{2!}{2^2 \cdot p^3} + \dfrac{4!}{2^2 \cdot 4^2 \cdot p^5} - \cdots \right\}$

$= \dfrac{1}{p} \left\{ 1 - \dfrac{1}{2 \cdot p^2} + \dfrac{(3/2)(1/2)}{2!} \left(\dfrac{1}{p^2} \right)^2 \cdots \right\}$

$= \dfrac{1}{p} \left\{ \left(1 + \dfrac{1}{p^2} \right)^{-1/2} \right\} = \dfrac{1}{p} \left(\dfrac{p^2 + 1}{p^2} \right)^{-1/2} = \dfrac{1}{\sqrt{p^2 + 1}}$

$\therefore \qquad L\{J_0(t)\} = \dfrac{1}{\sqrt{p^2 + 1}}$

Putting this value in eq^n (2), we get

$L\{J_0(at)\} = \dfrac{1}{a} \dfrac{1}{\sqrt{(p/a)^2 + 1}} = \dfrac{1}{\sqrt{p^2 + a^2}}$

From eq^n (1), we get

$L\{tJ_0(at)\} = -\dfrac{d}{dp} \left(\dfrac{1}{\sqrt{p^2 + a^2}} \right) = \dfrac{p}{(p^2 + a^2)^{3/2}}$

EXAMPLE 20. *Find out the Laplace transform of*

(i) $\cos^2 at$ (Kanpur–2005; Purvanchal–2002, 05)

(ii) $\sin^2 t - t$ (MS Univ.(TN)–2007)

SOLUTION.   (i) $L\{\cos^2 at\} = L\{(1/2) \times (1 + \cos 2at)\}$ $\qquad \left[\because \cos^2 at = \dfrac{1 + \cos 2at}{2} \right]$

$= (1/2) \times [L(1) - L\{\cos 2at\}]$

$= \dfrac{1}{2} \left[\dfrac{1}{p} - \dfrac{p}{p^2 + (2a)^2} \right], p > 0$

$= \dfrac{1}{2} \dfrac{4a^2}{p(p^2 + 4a^2)}, p > 0 \ = \dfrac{2a^2}{p(p^2 + 4a^2)}, p > 0$

(ii) $L\{\sin^2 t - t\} = L\{(1/2) \times (1 - \cos 2t) - t\}$

$= (1/2) \times L\{1\} - (1/2) \times L\{\cos 2t\} - L\{t\}$

$= \dfrac{1}{2} \times \dfrac{1}{p} - \dfrac{1}{2} \times \dfrac{p}{p^2 + 4} - \dfrac{1}{p^2}, p > 0$

$= \dfrac{2}{p(p^2 + 4)} - \dfrac{1}{p^2}, p > 0$

EXAMPLE 21. *Find* $L\{\cos^3 x, p\}$ (Osmania–2004; Meerut–1998)

SOLUTION.   We know that
$\cos 3x = 4 \cos^2 x - 3 \cos x$, we get

$\Rightarrow \qquad \cos^3 x = (\cos 3x + 3 \cos x)/4$

$$\therefore \quad L\{\cos^3 x, p\} = L\{\cos 3x\} + (3/4)L\{\cos x\}$$

$$= \frac{1}{4} \times \frac{p}{p^2 + 3^2} + \frac{3}{4} \times \frac{p}{p^2 + 1} = \frac{p}{4}\left[\frac{1}{p^2 + 9} + \frac{3}{p^2 + 1}\right]$$

EXAMPLE 22. *Find the Laplace transform of the following function :*

$$f(t) = \begin{cases} t/a, & 0 < t < a \\ 1, & t > a \end{cases} \qquad \text{(Meerut–1998; Manglore–1997; Rewa–1994)}$$

SOLUTION. We know that $L\{f(t)\} = \int_0^\infty e^{-pt} f(t)dt = \int_0^a e^{-pt} f(t)dt + \int_a^\infty e^{-pt} f(t)dt$ ...(1)

Putting the value of $f(t)$ in eq^n (1), we get

$$= \int_0^a e^{-pt}\left(\frac{t}{a}\right)dt + \int_a^\infty e^{-pt}dt$$

$$= \left[\left(\frac{t}{a}\right)\left(-\frac{1}{p}e^{-pt}\right) - \left(\frac{1}{a}\right)\left(\frac{1}{p^2}e^{-pt}\right)\right]_0^a + \left[-\frac{1}{p}e^{-pt}\right]_a^\infty, p > 0$$

$$= -(1/p)e^{-pa} - (1/ap^2)(e^{-pa} - 1) + (1/p)e^{-pa}$$

$$= (1 - e^{-pa})/ap^2 \text{ where } p > 0$$

MISCELLANEOUS EXERCISE

Find the Laplace transform of :

1. $e^{2t} + 4t^3 - 2\sin 3t + 3\cos 3t$ (JNTU–2003)

2. $\sin 2t \cos 3t$ (Gulberga–2005; Kottayam–2005)

3. $\sin^5 t$ (Mumbai–2007)

4. $e^{2t}(3t^5 - \cos 4t)$ (PTU–2007)

5. $e^{-3t}\sin 5t \sin 3t$ (VTU–2006)

6. $e^{2t}\sin^4 t$ (Mumbai–2007)

7. $\cosh at \sin at$ (Delhi–2002)

8. $\sinh 3t \cos^2 t$ (Madras–2000)

9. $t^2 e^{2t}$ (VTU–2008S)

10. $t\sqrt{(1 + \sin t)}$ (Mumbai–2007)

11. $f(t) = \begin{cases} 4, & 0 \le t \le 1 \\ 3, & t > 1 \end{cases}$ (UPTU–2009)

12. $f(x) = \begin{cases} \sin(x - \pi/3), & x > \pi/3 \\ 0, & x < \pi/3 \end{cases}$

                  (Rajasthan–2006)

13. Find the Laplace transform of the saw-toothed wave of period T, given $f(t) = t/T$ for $0 < t < T$.   (VTU–2007)

14. Find the Laplace transform of the square wave function of period a defined as
$f(t) = k$ when $0 < t < a$
     $= -k$ when $a < t < 2a$.   (VTU–2011)

15. Find the Laplace transform of the triangular wave of period $2a$ given by
$f(t) = t, 0 < t < a$

                 $= 2a - t, a < t < 2a$.
    (Nagarjuna–2008, VTU–2008S, UPTU–2002)

Find the Laplace transform of

16. $t\sin^2 t$ (Mysore–1997; Nagarjuna–2008)

17. $\sin 2t - 2t\cos 2t$ (Anna–2003)

18. $e^{2t}\sin 3t$ (Madras–2003)

19. $te^{-2t}\sin 4t$ (VTU–2008)

20. $t^2 e^{-3t}\sin 2t$ (Madras–2000S)

21. $(e^{-at} - e^{-bt})/t$ (Anna–2005S)

22. $\dfrac{(\sin t \sin 5t)}{t}$ (Mumbai–2008)

23. $(1 - \cos 3t)/t$ (VTU–2006)

24. $(1 - \cos t)/t^2$ (Hazaribag–2008)

25. $2^t + \dfrac{\cos 2t - \cos 3t}{t} + t\sin t$ (VTU–2004)

26. $\int_0^\infty \dfrac{e^{-\sqrt{2}t}\sin nt \sin t}{t}dt$ (Mumbai–2005)

27. $\int_0^\infty te^{-2t}\sin 3t \, dt$ (VTU–2008)

28. Prove that $\int_0^\infty \dfrac{e^{-2t}\sin nt}{t}dt = \dfrac{1}{2}\log 3$.

                      (Mumbai–2008)

29. $L\int_0^t \dfrac{e^t \sin t}{t}dt$

     (PTU–2009S, SVTU–2009, Bhopal–2008)

30. Prove that $L\{2^t\} = 1/(p - \log_e 2)$

                      (Banglore–1997)

31. Find the Laplace transform of the following functions :

(i) e^{4t} (Banglore–2005)

(ii) $t^{2/3}$ (Purvanchal–2004)

(iii) $\cosh 3t$ (Kuvempa–2005)

(iv) $F(t) = \begin{cases} t, & 0 < t < 4 \\ 5, & t > 4 \end{cases}$ (MSU TN–2007)

(v) $F(t) = \begin{cases} \sin 3t, & 0 < t < \pi \\ 0, & t > \pi \end{cases}$ (Banglore–2005)

32. Evaluate the following :

(i) $L\{\sin 2t \sin 3t\}$ (CDLU–2004)

(ii) $L\{t - \sinh 2t\}$ (CDLU–2004)

(iii) $L\{\sin^2 2t\}$ (Karnataka–2005)

(iv) $L\{2 \sin t \sin 3t\}$ (Gulberga–2005)

(v) $L\{\sin^2 4t\}$ (Banglore–2004)

33. Using unit step function, find the Laplace transform of

(i) $(t-1)^2 u(t-1)$

(ii) $\sin t\, u(t-4)$ (UPTU–2008)

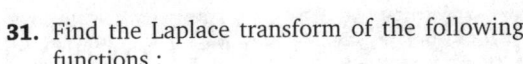

1. $\dfrac{1}{p-2} + \dfrac{24}{p^4} + \dfrac{3(p-2)}{p^4+9}$ **2.** $\dfrac{2(p^2-5)}{(p^2+1)(p^2+25)}$ **3.** $\dfrac{5}{4}\left\{ \dfrac{1}{p^2+1} - \dfrac{3/2}{p^2+9} + \dfrac{1/2}{p^2+25} \right\}$ **4.** $\dfrac{60}{p-2} - \dfrac{p-2}{p^2-4p+20}$

5. $\dfrac{30(p+3)}{(p^2+6p+13)(p^2+6p+73)}$ **6.** $\dfrac{1}{8}\left\{ \dfrac{3}{(p-2)} - \dfrac{4(p-2)}{p^2-4p+8} + \dfrac{p-4}{p^2-8p+32} \right\}$ **7.** $\dfrac{a(p^2+2a^2)}{p^4+4a^4}$

8. $\dfrac{3}{2}\left[\dfrac{1}{p^2-9} + \dfrac{p^2-13}{p^4-10p^2+169} \right]$ **9.** $\dfrac{2}{(p+2)^3}$ **10.** $\dfrac{4(4p^2+4p-1)}{(4p^2+1)^2}$ **11.** $\dfrac{4}{p} - \dfrac{e^{-p}}{p}$ **12.** $\dfrac{e^{-\pi p/3}}{p^2+1}$

13. $(1/p^2 T) - e^{-pT}/p(1-e^{pT})$ **14.** $(a/p)\tanh(ap/2)$ **15.** $(1/p^2)\tanh\dfrac{1}{2}ap$ **16.** $\dfrac{2(3p^2+4)}{p^2(p^2+4)^2}$

16. $\dfrac{2(3p^2+4)}{p^2(p^2+4)^2}$ **17.** $\dfrac{16}{(p^2+4)^2}$ **18.** $\dfrac{6(p-2)}{(p^2-4p+13)^2}$ **19.** $\dfrac{8(p+2)}{p^2+4p+20}$ **20.** $\dfrac{2(p^3+6p^2+9p+2)}{(p^2+4p+5)^3}$

21. $\log\{(p+b)/(p+a)\}$ **22.** $\dfrac{1}{2}\log\{(p^2+36)/(p^2+16)\}$ **23.** $\dfrac{1}{2}\log\left(\dfrac{p^2+9}{p^2}\right)$ **24.** $\cot^{-1}p - \dfrac{1}{2}p\log(1+p^{-2})$

25. $\dfrac{1}{p-\log 2} + \dfrac{2p}{(p^2+1)^2} + \dfrac{1}{2}\log\left(\dfrac{p^2+9}{p^2+4}\right)$ **26.** $\pi/8$ **27.** $12/169$ **29.** $\dfrac{\cot^{-1}(p-1)}{p}$

31. (i) $1/(p-4), p > 4$ (ii) $\dfrac{2\Gamma(2/3)}{3p^{5/3}}, p > 0$ (iii) $p/(p^2-9), p > |3|$

(iv) $\{1 + (p-1)e^{-4p}\}/p^2, p > 0$ (v) $(e^{1-p} - 1 + p)/p - 1, p \neq 1$

32. (i) $(12p)/(p^2+1)(p^2+25), p > 5$ (ii) $(p^2+4)/p^2(4-p^2), p > 2$

(iii) $(p^2+8)/p(p^2+16), p > 0$ (iv) $(6p)/(p^2+4)(p^2+16), p > 4$

(v) $(p^2+32)/p(p^2-64), p > 0$

33. (i) $\dfrac{2e^{-p}}{p^3}$ (ii) $\dfrac{e^{-4p}}{p^3}(4p^2+4p+2)$

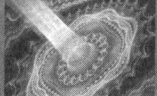

Self Assessment Test

1. Show that $L\{te^{at}\sin at\} = \dfrac{2a(p-a)}{(p^2-2ap+2a^2)^2}$

(Kanpur–1997)

2. Evaluate $L\{te^{-t}\sin^2 t\}$ (Agra–1981)

3. Evaluate $L\{\cos t\}$ (Agra–2001)

4. Evaluate $L\left\{e^{-\lambda t}\dfrac{(\lambda t)^n}{n!}\right\}$ (Agra–2001)

5. Evaluate $L\{\cos^2 at\}$ (Purvanchal–1991, 2005)

6. Show that $L\left\{\dfrac{x}{2a}\sin ax\right\} = \dfrac{p}{(p^2+a^2)^2}$

(Rohilkhand–2004)

7. Show that $L\{e^{ax}x^n\} = \dfrac{n!}{(p-a)^{n+1}}$

(Purvanchal–2007)

8. Show that $L\{e^{ax}\cos bx\} = \dfrac{p-a}{(p-a)^2+b^2}$

(Agra–2004)

9. Evaluate $L\{t^5 e^{3t}\}$ (Kanpur–2004)

10. Show that $L\left\{\int_0^t \dfrac{1-e^{-u}}{u}du\right\} = \dfrac{1}{p}\log\left(1+\dfrac{1}{p}\right)$

(Agra–2003)

11. Find the Laplace transform of

$$F(t) = \begin{cases} e^t, & 0 < t < 1 \\ 0, & t > 1 \end{cases} \quad \text{(UPTU–2004)}$$

12. Find the Laplace transform of $e^t t^{-1/2}$.

(UPTU–2001)

13. Find the Laplace transform of

(i) $t^2 e^t \sin 4t$ (UPTU–2001)

(ii) $te^{-t}\sin 2t$ (UPTU–2002)

14. Find the Laplace transform of

$$F(t) = \begin{cases} 1, & 0 \le t < 1 \\ t, & 1 \le t < 2 \\ t^2, & 2 \le t < \infty \end{cases} \quad \text{(UPTU–2001)}$$

15. If $L\{t\sin\omega t\} = \dfrac{2\omega p}{(p^2+\omega^2)^2}$, evaluate

$L\{\omega t\cos\omega t + \sin\omega t\}$ (UPTU–2001)

16. Find the Laplace transform of

(i) $\int_0^t e^{-t}\cos t\, dt$

(ii) $\int_0^t \dfrac{\sin t}{t}dt$ (UPTU–2001; JNTU–2005;

Meerut–1992; Rohilkhand–1990, 99;

Agra–1982, 85; Kanpur–1998)

17. Show that $\int_0^\infty e^{-t}\dfrac{\sin t}{t}dt = \dfrac{1}{4}\log\dfrac{p^2+4}{p^2}$

(UPTU–2008; VTU–2009; Purvanchal–2008;

Rohilkhand–2009, 11; Kashi–2008; Agra–1996;

Kanpur–1997, 2002; Kurukshetra–2004; Meerut–1998)

18. Find the Laplace transform of $e^{-2t}\sin t\cos 3t$.

(MDU–2006)

19. Show that $\int_0^\infty \cos x^2 dx = \dfrac{1}{2}\sqrt{\dfrac{\pi}{2}}$. (MDU–2005)

ANSWERS

2. $\dfrac{1}{2}\left[\dfrac{1}{(p+1)^2} + \dfrac{d}{dp}\left(\dfrac{p+1}{(p+1)^2+4}\right)\right]$

3. $\dfrac{p}{p^2+1}$, $p > 0$

4. $\dfrac{\lambda^n}{(p+\lambda)^{n+1}}$

5. $\dfrac{p^2+2a^2}{p(p^2+4a^2)}$, $p > 0$

9. $\dfrac{5!}{(p-3)^6}$

11. $\dfrac{e^{1-p}-1}{1-p}$

12. $\dfrac{\sqrt{\pi}}{\sqrt{p-1}}$

13. (i) $\dfrac{8(3p^2-6p-13)}{(p^2-2p+17)^3}$

(ii) $\dfrac{4p+4}{(p^2+2p+5)^2}$

14. $\dfrac{1}{p} + \dfrac{2}{p}e^{-2p} + \dfrac{e^{-p}}{p^2} + \dfrac{3}{p^2}e^{-2p} + \dfrac{2}{p^3}e^{-2p}$

15. $\dfrac{2\omega p^2}{(p^2+\omega^2)^2}$

16. (i) $\dfrac{p+1}{p(p^2+2p+2)}$

(ii) $\dfrac{1}{p}\cot^{-1}(p)$

18. $\dfrac{p^2+4p-4}{(p^2+4p+20)(p^2+4p+8)}$

Objective Evaluations

Fill in the Blanks

1. An integral of the type $\int_{-\infty}^{\infty} k(p,t) F(t)\, dt$ is called _____ of $F(t)$.

2. The function $k(p,t)$ is known as _____ of the transform.

3. If the integral $\int_0^\infty e^{-pt} F(t)\, dt$ _____ for some value of p, then only the Laplace transform of $f(t)$ exists.

4. $L\{1\} = $ _____ .

5. $L\{t^n\} = \dfrac{\overline{}}{p^{n+1}}$.

6. The operation of multiplying $F(t)$ by e^{-pt} and integrating from 0 to ∞ is called _____ .

7. If $f(p)$ is the Laplace transform of $F(t)$, then the Laplace transform of $e^{at} F(t)$ is _____ .

8. The Laplace transform of the function $F(t)$ is _____ .

9. If $F(t)$ is a function of class A and if $\{F(t)\} = f(p)$, then $L\{t \cdot F(t)\} = $ _____ .

True or False

Write 'T' for True and 'F' for False statement.

1. $L\{\cos at\} = \dfrac{p}{p^2 + a^2}$. **(T/F)**

2. $L\{\sin at\} = \dfrac{a}{p^2 + a^2}$. **(T/F)**

3. $L\{e^{at}\} = \dfrac{1}{p - a}$. **(T/F)**

4. $L\{\sinh at\} = \dfrac{p}{p^2 - a^2}$. **(T/F)**

5. $L\{\cosh at\} = \dfrac{p}{p^2 - a^2}$. **(T/F)**

6. The Laplace transform is a linear transform. **(T/F)**

7. If $f(t)$ is a function of class A and if $L\{F(t)\} = f(p)$, then $L\{tF(t)\} = f'(p)$. **(T/F)**

8. The function e^{t^2} is not of exponential order as $t \to \infty$. **(T/F)**

Multiple Choice Questions

Choose the most appropriate one.

1. The Laplace transform of 1 is :
   (a) $1/p$ (b) $1/p^2$
   (c) $1/\sqrt{p}$ (d) none of these

2. The Laplace transform of t is :
   (a) $1/p$ (b) $1/p^2$
   (c) $1/\sqrt{p}$ (d) none of these

3. The Laplace transform of $t^{n-1}/(n-1)!$ is :
   (a) $\dfrac{1}{p_1^{n-1}}$ (b) $1/p^n$
   (c) $\dfrac{1}{p^{n+1}}$ (d) none of these

4. The Laplace transform of $\dfrac{t^{n-1}}{\Gamma(a)}$ is :
   (a) $\dfrac{1}{p^{n-1}}$ (b) $1/p^n$
   (c) $\dfrac{1}{p^{n+2}}$ (d) none of these

5. The Laplace transform of e^{at} is : (Agra–2011)
   (a) $\dfrac{1}{p - a}$ (b) $\dfrac{1}{(p - a)^2}$

   (c) $\dfrac{1}{(p - a)^n}$ (d) none of these

6. The Laplace transform of te^{at} is :
   (a) $\dfrac{1}{p - a}$ (b) $\dfrac{1}{(p - a)^2}$
   (c) $\dfrac{1}{(p - a)^n}$ (d) none of these

7. The Laplace transform of $\dfrac{1}{(n-1)!} t^{n-1} e^{at}$ is :
   (a) $\dfrac{1}{p - a}$
   (b) $\dfrac{1}{(p - a)^2}$
   (c) $\dfrac{1}{(p - a)^n}, n = 1, 2, 3, \ldots$
   (d) none of these

8. The Laplace transform of $\dfrac{1}{a - b}(e^{at} - e^{bt})$ is :
   (a) $\dfrac{1}{(p - a)(p - b)}, (a \neq b)$

(b) $\dfrac{p}{(p-a)(p-b)}, (a \neq b)$

(c) $\dfrac{p}{(p+a)(p+b)}, \quad (a \neq b)$

(d) none of these

9. If $L\{F(t)\} = f(p)$, then $L\{F'(t)\}$ is :

(a) $L\{f'(t)\} = f(p)$

(b) $L\{f'(t)\} = f(p) + f(0)$

(c) $L\{f'(t)\} = pf(p) + f(0)$

(d) $L\{f'(t)\} = pf(p) - f(0)$

10. If $L\{F(t)\} = f(p)$, then $L\{F''(t)\}$ is :

(a) $p^2 f(p) - pF(0) - F'(0)$

(b) $pf(p) - pF(0)$

(c) $f''(p)$

(d) none of these

11. The Laplace transform of $f(t)$ is $f(p)$, then :

(a) $L\{t\, F(t)\} = f(p)$ (b) $L\{tF(t)\} = -f(p)$

(c) $L\{tF(t)\} = f'(p)$ (d) none of these

12. If $u(x,t)$ is a function of two variables x and t

and $L\{u(x,t)\} = U(x,p)$, then $L\left(\dfrac{\partial u}{\partial t}\right) = :$

(a) $pU(x,p) - u(x,0)$ (b) $pU(x,p)$

(c) $u(x,0)$ (d) none of these

13. $L(t^3 e^{-at})$ has the value :   (Bhopal–2008)

(a) $\dfrac{6}{p^4}$ (b) $(p-2)^4$

(c) $\dfrac{6}{(p-2)^4}$ (d) $\dfrac{1}{p^4}$

14. The Laplace transform of $F(t)$ exists for all $p > a$, if   (Rohilkhand–2002)

(a) $F(t)$ is a continuous function

(b) $F(t)$ is a differentiable function

(c) $F(t)$ is a function of class A

(d) none of these

15. The Laplace transform of $\cos at$ is:

(a) $\dfrac{p}{p^2 + a^2}, \; p > 0$

(b) $\dfrac{a}{p^2 + a^2}, \; p > 0$

(c) $\dfrac{a}{p^2 - a^2}, \; p > |a|$

(d) none of these

16. The Laplace transform of $\cosh at$ is:

(Rohilkhand–2003)

(a) $\dfrac{p}{p^2 + a^2}$ (b) $\dfrac{1}{p^2 + a^2}$

(c) $\dfrac{p}{p^2 - a^2}$ (d) $\dfrac{1}{p^2 - a^2}$

17. If $L\{F(t)\} = f(p)$ and $G(t) = \begin{cases} F(t-a), & t > a \\ 0, & t < a \end{cases}$,

then $L\{G(t)\}$ is: (Rohilkhand–2003)

(a) $e^{-ap} f(p)$ (b) $f(p-a)$

(c) $f(ap)$ (d) $\dfrac{1}{a} f\left(\dfrac{p}{a}\right)$

18. If $L\{F(t)\} = f(p)$ then $L\{F(at)\}$ is:

(Agra–2008)

(a) $f\left(\dfrac{p}{a}\right)$ (b) $\dfrac{1}{a} f(p)$

(c) $\dfrac{1}{a} f\left(\dfrac{p}{a}\right)$ (d) none of these

19. The Laplace transform of e^{t^2} is:

(Rohilkhand–2003)

(a) $\dfrac{1}{(p-2)^2}$ (b) $\dfrac{1}{p^2}$

(c) $\dfrac{1}{p-2}$ (d) not existing

20. If $L\left\{\dfrac{\sin t}{t}\right\} = \tan^{-1}\dfrac{1}{p}$ then $L\left\{\dfrac{\sin 6t}{t}\right\}$ is

equal to: (Agra–2010)

(a) $\tan^{-1}\dfrac{1}{p}$ (b) $6\tan^{-1}\dfrac{1}{p}$

(c) $\dfrac{1}{6}\tan^{-1}\dfrac{1}{p}$ (d) $\tan^{-1}\dfrac{6}{p}$

21. The value of $\int_0^\infty \dfrac{\sin^2 t}{t^2}$ is :   (Agra–2009)

(a) $\dfrac{\pi}{4}$ (b) $\dfrac{\pi^2}{2}$

(c) $\dfrac{\pi}{2}$ (d) none of these

22. If $F(t) = \{e^{t-a}, t > a\}$ then $L\{F(t)\}$ is :

(a) $\dfrac{e^{-ap}}{p-1}$ (b) $\dfrac{e^{-ap}}{p+1}$

(c) $\dfrac{e^{ap}}{p-1}$ (d) none of these

23. Laplace transform of $t^{-1/2}$ is:

(a) $\sqrt{\dfrac{p}{\pi}}$

(b) $\sqrt{\dfrac{\pi}{p}}$

(c) $\sqrt{\dfrac{\pi}{p^3}}$

(d) none of these

24. $L(t^p e^{2t})$ has the value :

(a) $\dfrac{1}{(p-2)^4}$

(b) $\dfrac{6}{p^4}$

(c) $\dfrac{1}{p^4}$

(d) $\dfrac{6}{(p-2)^3}$

25. $L\{t^2 e^{-at}\}$ has the value :

(a) $\dfrac{2}{(p+a)^2}$

(b) $\dfrac{2}{(p+a)^{-3}}$

(c) $\dfrac{2}{(p+a)^3}$

(d) none of these

26. If L defines the Laplace transform of a function $L[\sin(at)]$ will be equal to : [GATE(CE)–2008]

(a) $\dfrac{a}{p^2+a^2}$

(b) $\dfrac{a}{p^2-a^2}$

(c) $\dfrac{p}{p^2-a^2}$

(d) $\dfrac{p}{p^2+a^2}$

27. Laplace transform of the function $\sin wt$ is : [GATE(ME)–2003]

(a) $\dfrac{p}{p^2-w^2}$

(b) $\dfrac{w}{p^2-w^2}$

(c) $\dfrac{w}{p^2+w^2}$

(d) $\dfrac{p}{p^2+w^2}$

28. Lapalce transform for the function $f(x) = \cosh(ax)$ is : [GATE(CE)–2009]

(a) $\dfrac{p}{p^2-a^2}$

(b) $\dfrac{p}{p^2+a^2}$

(c) $\dfrac{a}{p^2-a^2}$

(d) $\dfrac{a}{p^2+a^2}$

29. The function $f(t)$ satisfies the differential equation $\dfrac{d^2 f}{dt^2} + f = 0$ and the auxiliary conditions, $f(0) = 0$, $\dfrac{df}{dt}(0) = 4$. The Laplace transform of $f(t)$ is given by : [GATE(ME)–2013]

(a) $\dfrac{4}{p^2+1}$

(b) $\dfrac{2}{p^2+1}$

(c) $\dfrac{4}{p+1}$

(d) $\dfrac{2}{p^2-1}$

30. In what range should $\text{Re}(p)$ remain so that the Laplace transform of the function $e^{(a+2)t+5}$ exists : [GATE(EC)–2005]

(a) $\text{Re}(p) < 2$

(b) $\text{Re}(p) > a + 2$

(c) $\text{Re}(p) > a + 7$

(d) $\text{Re}(p) < a + 5$

31. Evaluate $\int_0^\infty \dfrac{\sin t}{t}\, dt$: [GATE(CE)–2007]

(a) $\pi/2$

(b) π

(c) $\pi/4$

(d) $\pi/3$

32. If $F(p)$ is the Laplace transform of the function $f(t)$, then Laplace transform of $\int_0^1 f(t)\,dt$ is : [GATE(MC)–2007]

(a) $\dfrac{1}{p} F(p) - f(t)$

(b) $\dfrac{1}{p} F(p)$

(c) $\int F(p)\,dp$

(d) none of these

33. The Laplace transform of a function $f(t)$ is $\dfrac{1}{p^2(p+1)}$. The function $f(t)$ is : [GATE(ME)–2010]

(a) $1 + t + e^{-t}$

(b) $-1 + e^{-t}$

(c) $t - 1 + e^{-t}$

(d) $2t + e^t$

ANSWERS

❖ Fill in the Blanks

1. integral transform **2.** kernel **3.** exist **4.** $1/p$ **5.** $n!$ **6.** Laplace transform

7. $f(p-a)$ **8.** $\int_0^\infty e^{-pt} F(t)\,dt$ **9.** $-f'(p)$

❖ True/False

1. T **2.** T **3.** T **4.** F **5.** T **6.** T **7.** F **8.** T

❖ Multiple Choice Questions

1. (a) **2.** (b) **3.** (b) **4.** (b) **5.** (a) **6.** (b) **7.** (c) **8.** (a) **9.** (d)

10. (a) **11.** (b) **12.** (a) **13.** (c) **14.** (c) **15.** (a) **16.** (c) **17.** (a) **18.** (c)

19. (d) **20.** (d) **21.** (c) **22.** (a) **23.** (b) **24.** (d) **25.** (c) **26.** (a) **27.** (c)

28. (a) **29.** (a) **30.** (b) **31.** (a) **32.** (b) **33.** (c)

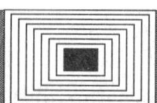

Comprehensive Laplace Transforms Tables

(1) GENERAL FORMULAS

S. No.	$F(x)$	$L(F(x)) = f(p)$
1.	$x^n F(x),\ n = 1, 2, \ldots$	$(-1)^n \dfrac{d^n}{dp^n} f(p)$
2.	$\dfrac{1}{x} F(x)$	$\int_p^\infty f(q)dq$
3.	$e^{ax}F(x)$	$f(p - a)$
4.	$\sinh(ax)F(x)$	$\dfrac{1}{2}[f(p - a) - f(p + a)]$
5.	$\cosh(ax)F(x)$	$\dfrac{1}{2}[f(p - a) + f(p + a)]$
6.	$\sin(\omega x)F(x)$	$\dfrac{1}{2}[f(p - i\omega) - f(p + i\omega)]$
7.	$\cos(\omega x)F(x)$	$\dfrac{1}{2}[f(p - i\omega) + f(p + i\omega)]$
8.	$F(x^2)$	$\dfrac{1}{\sqrt{\pi}} \int_0^\infty e^{-p^2/4t^2} f(t^2)dt$
9.	$F\left(\dfrac{x}{a}\right), a > 0$	$af(ap)$
10.	$x^{\alpha-1}F\left(\dfrac{1}{x}\right), \alpha > -1$	$\int_0^\infty \left(\dfrac{t}{p}\right)^{\alpha/2} J_\alpha(2\sqrt{pt})f(t)dt$
11.	$F(a \sinh x), a > 0$	$\int_0^\infty J_p(at)f(t)dt$
12.	$F(x + a) = F(x)$	$\dfrac{1}{1 - e^{ap}} \int_0^a f(x)e^{-px}dx$
13.	$F(x + a) = -F(x)$	$\dfrac{1}{1 + e^{-ap}} \int_0^a f(x)e^{-px}dx$
14.	$F_x'(x)$	$pf(p) - f(0^+)$
15.	$F_x^n(x)$	$p^n f(p) - \sum\limits_{K=1}^{n} p^{n-K} f_x^{K-1}(0^+)$
16.	$\dfrac{d^n}{dx^n}[x^m F(x)], m \geq n$	$(-1)^m p^n \dfrac{d^m}{dp^m} f(p)$
17.	$\int_0^x F(t)dt$	$\dfrac{1}{p} f(p)$
18.	$\int_0^x (x - t)F(t)dt$	$\dfrac{1}{p^2} f(p)$

19.	$\int_0^x (x-t)^n F(t)dt, n > -1$	$\Gamma(n+1)p^{-n-1}f(p)$
20.	$\int_0^x e^{-a(x-t)}F(t)dt$	$\dfrac{1}{p+a}f(p)$
21.	$\int_0^x \sinh[a(x-t)]F(t)dt$	$\dfrac{af(p)}{p^2-a^2}$
22.	$\int_0^x \sin[a(x-t)]F(t)dt$	$\dfrac{af(p)}{p^2+a^2}$
23.	$\int_0^x F_1(t)F_2(x-t)dt$	$f_1(p)f_2(p)$
24.	$\int_0^x \dfrac{1}{t}F(t)dt$	$\dfrac{1}{p}\int_p^\infty f(q)dq$
25.	$\int_x^\infty \dfrac{1}{t}F(t)dt$	$\dfrac{1}{p}\int_0^p f(q)dq$
26.	$\int_0^\infty \dfrac{1}{\sqrt{t}}\sin(2\sqrt{xt})F(t)dt$	$\dfrac{\sqrt{\pi}}{p\sqrt{p}}f\left(\dfrac{1}{p}\right)$
27.	$\dfrac{1}{\sqrt{x}}\int_0^\infty \cos(2\sqrt{xt})F(t)dt$	$\dfrac{\sqrt{\pi}}{\sqrt{p}}f\left(\dfrac{1}{p}\right)$
28.	$\int_0^\infty \dfrac{1}{\sqrt{\pi x}}e^{-t^2/4x}F(t)dt$	$\dfrac{1}{\sqrt{p}}f(\sqrt{p})$
29.	$F(x) - a\int_0^x F(\sqrt{x^2-t^2})\cdot J_1(atdt)$	$f(\sqrt{p^2+a^2})$
30.	$F(x) + a\int_0^x F(\sqrt{x^2-t^2})\Gamma_1(at)dt$	$f(\sqrt{p^2-a^2})$

(2) FOR POWER LAW FUNCTIONS

S. No.	$F(x)$	$L(F(x)) = f(p)$
1.	1	$1/p$
2.	x	$\dfrac{1}{p^2}$
3.	$x^n,\ n = 1, 2, \ldots$	$\dfrac{n!}{p^{n+1}}$
4.	$\begin{cases} 0; & 0 < x < a \\ 1; & a < x < b \\ 0; & b < x \end{cases}$	$\dfrac{1}{p}(e^{-ap} - e^{-bp})$
5.	$x^{n-1/2},\ n = 1, 2, \ldots$	$\dfrac{1\times 3\times\ldots\times(2n-1)\sqrt{\pi}}{2^n p^{n+1/2}}$
6.	$x^V,\ V > -1$	$\Gamma(V+1)p^{-V-1}$
7.	$(x+a)^V,\ V > -1$	$p^{-V-1}e^{-ap}\Gamma(V+1, ap)$

(3) FOR EXPONENTIAL FUNCTIONS

S. No.	$F(x)$	$L(F(x)) = f(p)$
1.	e^{-ax}	$\dfrac{1}{p+a}$
2.	xe^{-ax}	$\dfrac{1}{(p+a)^2}$
3.	$x^{n-1}e^{-ax}, n > 0$	$\Gamma(n)(b+a)^{-n}$
4.	$\dfrac{1}{x}(e^{-ax} - e^{-bx})$	$\log(p+b) - \log(p+a)$
5.	$\dfrac{1}{x^2}(1 - e^{-ax})^2$	$(p+2a)\log(p+2a) + \log p - 2(p+a)\log(p+a)$
6.	$\sqrt{x} \cdot e^{-a/x}, a \geq 0$	$\dfrac{1}{2}\sqrt{\pi/p^3}(1 + 2\sqrt{ap})e^{-2\sqrt{ap}}$
7.	$\dfrac{1}{\sqrt{x}}e^{-(a/x)}, a \geq 0$	$\sqrt{\dfrac{\pi}{p}}e^{-2/\sqrt{ap}}$
8.	$\dfrac{1}{x\sqrt{x}}e^{-a/x}, a \geq 0$	$\sqrt{\dfrac{\pi}{a}}e^{-2\sqrt{ap}}$

(4) FOR HYPERBOLIC FUNCTIONS

S. No.	$F(x)$	$L(F(x)) = f(p)$
1.	$\sinh(ax)$	$\dfrac{a^2}{p^2 - a^2}$
2.	$\sinh^2(ax)$	$\dfrac{2a^2}{p^3 - 4a^2 p}$
3.	$\dfrac{1}{x}\sinh(ax)$	$\dfrac{1}{2}\log\dfrac{p+a}{p-a}$
4.	$x^{n-1}\sinh(ax), n > -1$	$\dfrac{1}{2}\Gamma(n)[(p-a)^{-n} - (p+a)^{-n}]$
5.	$\sinh(2\sqrt{ax})$	$\dfrac{\sqrt{\pi a}}{p\sqrt{p}}e^{a/p}$
6.	$\dfrac{1}{\sqrt{x}}\sinh^2(\sqrt{ax})$	$\dfrac{1}{2}\pi^{1/2}p^{-1/2}(e^{a/p} - 1)$
7.	$\cosh(ax)$	$\dfrac{p}{p^2 - a^2}$
8.	$\cosh^2(ax)$	$\dfrac{p^2 - 2a^2}{p^3 - 4a^2 p}$

9.	$\sqrt{x}\cosh(2\sqrt{ax})$	$\pi^{1/2}p^{-5/2}\left(\dfrac{1}{2}p+a\right)e^{a/p}$
10.	$\dfrac{1}{\sqrt{x}}\cosh(2\sqrt{ax})$	$\pi^{1/2}p^{-1/2}e^{a/p}$
11.	$\dfrac{1}{\sqrt{x}}\cosh^2(\sqrt{ax})$	$\dfrac{1}{2}\pi^{1/2}p^{-1/2}(e^{a/p}+1)$

(5) For Logarithmic Functions

S. No.	$F(x)$	$L(F(x)) = f(p)$
1.	$\log x$	$-\dfrac{1}{p}(\log p + C), C = 0.5772...$ (The Euler's constant)
2.	$x^n \log x, n = 1, 2, ...$	$\dfrac{n!}{p^{n+1}}\left(1+\dfrac{1}{2}+\dfrac{1}{3}+...+\dfrac{1}{n}-\log p - C\right)$
3.	$\dfrac{1}{\sqrt{x}}\log x$	$-\sqrt{\dfrac{\pi}{p}}(\log 4p + C)$
4.	$(\log x)^2$	$\dfrac{1}{p}\left[(\log x + C)^2 + \dfrac{1}{6}\pi^2\right]$
5.	$e^{-ax}\log x$	$-\dfrac{\log(p+a)+C}{p+a}$

(6) For Trigonometric Functions

S. No.	$F(x)$	$L(F(x)) = f(p)$
1.	$\sin ax$	$\dfrac{a^2}{p^2+a^2}$
2.	$|\sin ax|, a > 0$	$\dfrac{a}{p^2+a^2}\coth\left(\dfrac{\pi p}{2a}\right)$
3.	$\sin^{2n}(ax), n \in \mathbf{N}$	$\dfrac{a^{2n}\cdot(2n)!}{p[p^2+(2a)^2][p^2+(4a)^2]...[p^2+(2na)^2]}$
4.	$x^n\cdot\sin ax, n \in N$	$\dfrac{n!\,p^{n+1}}{(p^2+a^2)^{n+1}}\sum_{0\le 2k<n}(-1)^k C_{n+1}^{2k+1}\left(\dfrac{a}{p}\right)^{2k+1}$
5.	$\dfrac{1}{x}\sin ax$	$\tan^{-1}\left(\dfrac{a}{p}\right)$
6.	$\dfrac{1}{x}\sin^2 ax$	$\dfrac{1}{4}\log(1+4a^2 p^{-2})$
7.	$\dfrac{1}{x^2}\sin^2 ax$	$a\tan^{-1}\left(\dfrac{2a}{p}\right)-\dfrac{1}{4}p\log\left(1+\dfrac{4a^2}{p^2}\right)$

8.	$\cos ax$	$\dfrac{p}{p^2 + a^2}$
9.	$\cos^2 ax$	$\dfrac{p^2 + 2a^2}{p(p^2 + 4a^2)}$
10.	$x^n \cos(ax),\ n \in N$	$\dfrac{n!\, p^{n+1}}{(p^2 + a^2)^{n+1}} \displaystyle\sum_{0 \le 2K \le n+1} (-1)^K C_{n+1}^{2K} \left(\dfrac{a}{p}\right)^{2K}$
11.	$\dfrac{1}{x}[1 - \cos ax]$	$\dfrac{1}{2}\log\left(1 + \dfrac{a^2}{p^2}\right)$
12.	$\dfrac{1}{x}[\cos ax - \cos bx]$	$\dfrac{1}{2}\log \dfrac{p^2 + b^2}{p^2 + a^2}$
13.	$\sqrt{x} \cdot \cos(2\sqrt{ax})$	$\dfrac{1}{2}\pi^{1/2} p^{-5/2}(p - 2a)e^{-a/p}$
14.	$\dfrac{1}{\sqrt{x}}\cos(2\sqrt{ax})$	$\sqrt{\dfrac{\pi}{p}}\, e^{-a/p}$
15.	$\sin ax \sin bx$	$\dfrac{2abp}{[p^2 + (a+b)^2][p^2 + (a-b)^2]}$
16.	$\cos ax \sin bx$	$\dfrac{b[p^2 - a^2 + b^2]}{[p^2 + (a+b)^2][p^2 + (a-b)^2]}$
17.	$\cos ax \cos bx$	$\dfrac{p[p^2 + a^2 + b^2]}{[p^2 + (a+b)^2][p^2 + (a-b)^2]}$
18.	$e^{bx}\sin ax$	$\dfrac{a}{(p-b)^2 + a^2}$
19.	$e^{bx}\cos ax$	$\dfrac{p-b}{(p-b)^2 + a^2}$
20.	$\sin ax \sinh(ax)$	$\dfrac{2a^2 p}{p^4 + 4a^4}$
21.	$\sin ax \cosh(ax)$	$\dfrac{a(p^2 + 2a^2)}{p^4 + 4a^4}$
22.	$\cos ax \sinh(ax)$	$\dfrac{a(p^2 - 2a^2)}{p^4 + 4a^4}$
23.	$\cos ax \cosh(ax)$	$\dfrac{p^3}{p^4 + 4a^4}$

vvvvvv

Chapter

2

The Inverse Laplace Transforms

2.1 INTRODUCTION

If the Laplace transform of a function $F(t)$ is $f(p)$, i.e., if $L\{F(t)\} = f(p)$. Then, $F(t)$ is known as inverse Laplace transform of $f(p)$.

Symbolically, $F(t) = L^{-1}\{f(p)\}$. where, L^{-1} is called the inverse Laplace transformation operator.

For example: If $L\{e^{-2t}\} = \dfrac{1}{p+2}$. Then we can write $L^{-1}\left(\dfrac{1}{p+2}\right) = e^{-2t}$

2.1.1 NULL FUNCTION

A function $N(t)$ of t such that $\int_0^t N(t)\, dt = 0,\quad \forall\, t > 0$ is called the null function.

2.1.2 UNIQUENESS OF INVERSE LAPLACE TRANSFORMS

(Osmania–2004)

Since, we know that the Laplace transform of a null function $N(t)$ is zero. Also, it is clearly that if $L(F(t)) = f(p)$, then also

$$L\{F(t) + N(t)\} = f(p)$$

It follows that we can have two different functions with same Laplace transform.

If we allow null functions, we see that the inverse Laplace transform is not unique. It is unique, however, if we disallow null functions.

2.1.3 LEARCH THEOREM

If we restrict ourselves to functions $F(t)$ which are sectionally continuous in every finite interval $0 \le t \le N$ and of exponential order for $t > N$, then the inverse Laplace transform of $f(p)$. i.e., $L^{-1}\{f(p)\} = F(t)$, is unique.

2.2 SOME INVERSE LAPLACE TRANSFORMS

S. No.	$f(p)$	$L^{-1}[f(p)] = F(t)$	S. No.	$f(p)$	$L^{-1}[f(p)] = F(t)$
(1)	$\dfrac{1}{p}$	1	(6)	$\dfrac{1}{p^2 + a^2}$	$\cos at$
(2)	$\dfrac{1}{p^2}$	t	(7)	$\dfrac{1}{p^2 - a^2}$	$\dfrac{\sinh at}{a}$
(3)	$\dfrac{1}{p^{n+1}}, n = 0, 1, 2\ldots$	$\dfrac{t^n}{n!}$	(8)	$\dfrac{p}{p^2 - a^2}$	$\cosh at$
(4)	$\dfrac{1}{p - a}$	e^{at}	(9)	$\dfrac{\Gamma(a+1)}{p^n}$	$t^a,\ a > -1$
(5)	$\dfrac{1}{p^2 + a^2}$	$\dfrac{\sin at}{a}$	(10)	$\dfrac{1}{(p-a)^2 + b^2}$	$\dfrac{1}{b} e^{at} \sin bt$

(11)	$\dfrac{p-a}{(p-a)^2+b^2}$	$e^{at}\cos bt$	(14)	$\dfrac{p}{(p^2+a^2)^2}$	$\dfrac{1}{2a}t\sin at$
(12)	$\dfrac{1}{(p-a)^2-b^2}$	$\dfrac{1}{b}e^{at}\sinh bt$	(15)	$\dfrac{1}{(p^2+a^2)^2}$	$\dfrac{1}{2a^3}(\sin at - at\cos at)$
(13)	$\dfrac{p-a}{(p-a)^2-b^2}$	$e^{at}\cosh bt$			

2.3 PROPERTIES OF INVERSE LAPLACE TRANSFORMS

(i) (Linearity Property).

(Meerut–1986)

If C_1 and C_2 are any constants while $f_1(p)$ and $f_2(p)$ are the Laplace transform $F_1(t)$ and $F_2(t)$ respectively, then

$$L^{-1}\{C_1 f_1(p) + C_2 f_2(p)\} = C_1 L^{-1}\{f_1(p)\} + C_2 L^{-1}\{f_2(p)\}$$

Proof. We know that

$$L\{C_1 F_1(t) + C_2 F_2(t)\} = C_1 L\{F_1(t)\} + C_2 L\{F_2(t)\} = C_1 f_1(p) + C_2 f_2(p)$$

$$\Rightarrow \quad L^{-1}\{C_1 f_1(p) + C_2 f_2(p)\} = C_1 F_1(t) + C_2 F_2(t) = C_1 L^{-1}\{f_1(p)\} + C_2 L^{-1}\{f_2(p)\}.$$

(ii) First Translation or Shifting Theorem.

If $L^{-1}\{f(p)\} = F(t)$, then $L^{-1}\{f(p-a)\} = e^{at} F(t) = e^{at} L^{-1}\{f(p)\}$

(Meerut–1986, 90; T–1991, 92; Ravishankar–2004; Kanpur–1995; Purvanchal–1993, 94)

Proof. We have

$$f(p) = \int_0^\infty e^{-pt} F(t)\, dt$$

$$\Rightarrow \quad f(p-a) = \int_0^\infty e^{-(p-a)t} F(t)\, dt$$

$$= \int_0^\infty e^{-pt}\{e^{at} F(t)\}\, dt = L\{e^{at} F(t)\}$$

Hence, $\quad L^{-1}\{f(p-a)\} = e^{at} F(t) = e^{at} L^{-1}\{f(p)\}.$

(iii) Second Translation or Shifting Theorem. *If $L^{-1}\{f(p)\} = F(t)$, then $L^{-1}\{e^{-ap} f(p)\} = G(t)$, where*

$$G(t) = \begin{cases} F(t-a), & t > a \\ 0, & t < a \end{cases}$$

(Meerut–1980, 83, 84, 91, 92, 98, 99, 2008; Purvanchal–1995; Kanpur–1994, 96)

Proof. We know that $\quad f(p) = \int_0^\infty e^{-pt} F(t)\, dt$

Therefore, $\quad e^{-ap} f(p) = \int_0^\infty e^{-p(t+a)} F(t)\, dt$

$$= \int_0^\infty e^{-px} F(x-a)dx, \text{ putting } t+a = x \Rightarrow dt = dx$$

$$= \int_0^a e^{-px}.0\, dx + \int_a^\infty e^{-px} F(x-a)\, dx$$

$$= \int_0^a e^{-pt}.0\, dt + \int_a^\infty e^{-pt} F(t-a)\, dt$$

$$= \int_0^\infty e^{-pt} G(t)\, dt = L\{G(t)\}$$

where, $G(t) = \begin{cases} F(t-a), & t > a \\ 0, & t < a \end{cases}$ shows, $L^{-1}\{e^{ap} f(p)\} = G(t)$

(iv) Change of Scale Property.

If $L^{-1}\{f(p)\} = F(t)$, then $L^{-1}\{f(ap)\} = \dfrac{1}{a} F\left(\dfrac{t}{a}\right)$

(Meerut–1989; Agra–1979, 81, 83; Kanpur–1985; Purvanchal–1989; Rohilkhand–1996)

Proof. We know that $f(ap) = \dfrac{1}{a} \int_0^\infty e^{-px} F\left(\dfrac{x}{a}\right) dx \Rightarrow f(ap) = \int_0^\infty e^{-apt} F(t)\, dt$.

Putting $at = x \Rightarrow dt = \dfrac{1}{a} dx$, we get

$$f(ap) = \dfrac{1}{a} \int_0^\infty e^{-px} F\left(\dfrac{x}{a}\right) dx = \dfrac{1}{a} \int_0^\infty e^{-pt} F\left(\dfrac{t}{a}\right) dt$$

[By the property of definite integral]

$$= \dfrac{1}{a} L\left\{F\left(\dfrac{t}{a}\right)\right\} = L\left\{\dfrac{1}{a} F\left(\dfrac{t}{a}\right)\right\}$$

Hence, $L^{-1}\{f(ap)\} = \dfrac{1}{a} F\left(\dfrac{t}{a}\right)$

✎ RECAPITULATIONS

S. No.	Operation	$f(p)$	$L^{-1}\{f(p)\} = F(t)$
1.	Linearity property	$a_1 f_1(p) + a_2 F_2(p)$	$a_1 L^{-1}\{f_1(p)\} + a_2 L^{-1}\{f_2(p)\}$
2.	First translation or Shifting theorem	$f(p - a)$	$e^{at} L^{-1}\{f(p)\}$
3	Second translation or Shifting theorem	$e^{-ap} f(p)$	$G(t) = \begin{cases} F(t-a), & t > a \\ 0, & t < a \end{cases}$
4.	Change of scale property	$f(ap)$	$\dfrac{1}{a} F\left(\dfrac{t}{a}\right)$

Solved Examples

EXAMPLE 1. *Find the inverse Laplace transforms of the following functions :*

(i) $\dfrac{2p+1}{p(p+1)}$ (ii) $\dfrac{3p-8}{4p^2+25}$

SOLUTION. (i) We have

$$L^{-1}\left\{\dfrac{2p+1}{p(p+1)}\right\} = L^{-1}\left\{\dfrac{p+(p+1)}{p(p+1)}\right\}$$

$$= L^{-1}\left\{\dfrac{1}{p+1}\right\} + L^{-1}\left\{\dfrac{1}{p}\right\} = e^{-t} + 1$$

(ii) Here, we have

$$L^{-1}\left\{\dfrac{3p-8}{4p^2+25}\right\} = \dfrac{3}{4} L^{-1}\left\{\dfrac{p}{p^2+\left(\dfrac{5}{2}\right)^2}\right\} - 2L^{-1}\left\{\dfrac{1}{p^2+\left(\dfrac{5}{2}\right)^2}\right\}$$

$$= \frac{3}{4}\cos\left(\frac{5}{2}t\right) - 2.\frac{2}{5}\sin\left(\frac{5}{2}t\right)$$

$$= \frac{3}{4}\cos\left(\frac{5}{2}t\right) - \frac{4}{5}\sin\left(\frac{5}{2}t\right)$$

EXAMPLE 2.   *Show that* $\dfrac{1}{p^{1/2}} = L\left[\dfrac{1}{\sqrt{\pi t}}\right]$ (UPTU 2005; Agra–2007; Meerut–2009, 10)

SOLUTION.   We have to show that $\dfrac{1}{p^{1/2}} = L\left[\dfrac{1}{\sqrt{\pi t}}\right]$

Since, $L^{-1}\left[\dfrac{1}{p^n}\right] = \dfrac{t^{n-1}}{(n-1)!} = \dfrac{t^{n-1}}{\Gamma(n)}$

So, $L^{-1}\left[\dfrac{1}{p^{1/2}}\right] = \dfrac{t^{\frac{1}{2}-1}}{\Gamma(1/2)} = \dfrac{t^{-1/2}}{\Gamma(1/2)} = \dfrac{t^{-1/2}}{\sqrt{\pi}}$

$$L^{-1}\left[\frac{1}{p^{1/2}}\right] = \frac{1}{\sqrt{\pi t}} \Rightarrow \frac{1}{p^{1/2}} = L\left[\frac{1}{\sqrt{\pi t}}\right].$$

EXAMPLE 3.   *Find* $L^{-1}\left\{\dfrac{3p-2}{p^{5/2}} - \dfrac{7}{3p+2}\right\}$

SOLUTION.   We have

$$L^{-1}\left\{\frac{3p-2}{p^{5/2}} - \frac{7}{3p+2}\right\} = 3L^{-1}\left\{\frac{1}{p^{3/2}}\right\} - 2L^{-1}\left\{\frac{1}{p^{5/2}}\right\} - \frac{7}{3}L^{-1}\left\{\frac{1}{p+(2/3)}\right\}$$

$$= 3\frac{t^{1/2}}{\Gamma\left(\dfrac{3}{2}\right)} - 2\frac{t^{3/2}}{\Gamma\left(\dfrac{5}{2}\right)} - \frac{7}{3}e^{\left(-\frac{2}{3}\right)t}$$

$$= 6\sqrt{\left(\frac{t}{\pi}\right)} - \frac{8}{3}t\sqrt{\left(\frac{t}{\pi}\right)} - \frac{7}{3}e^{-2t/3}.$$

EXAMPLE 4.   *Find* $L^{-1}\left\{\dfrac{3}{p^2-3} + \dfrac{3p+2}{p^3} - \dfrac{3p-27}{p^2+9} + \dfrac{6-30\sqrt{p}}{p^4}\right\}.$

SOLUTION.   We have

$$L^{-1}\left\{\frac{3}{p^2-3} + \frac{3p+2}{p^3} - \frac{3p-27}{p^2+9} + \frac{6-30\sqrt{p}}{p^4}\right\}$$

$$= L^{-1}\left\{\frac{3}{p^2-3} + \frac{3}{p^2} + \frac{2}{p^3} - \frac{3p}{p^2+9} + \frac{27}{p^2+9} + \frac{6}{p^4} - \frac{30}{p^{7/2}}\right\}$$

$$= 3L^{-1}\left\{\frac{1}{p^2-(\sqrt{3})^2}\right\} + 3L^{-1}\left\{\frac{1}{p^2}\right\} + 2L^{-1}\left\{\frac{1}{p^3}\right\} - 3L^{-1}\left\{\frac{p}{p^2+3^2}\right\}$$

$$+ 27L^{-1}\left\{\frac{1}{p^2+3^2}\right\} + 6L^{-1}\left\{\frac{1}{p^4}\right\} - 30L^{-1}\left\{\frac{1}{p^{7/2}}\right\}$$

$$= 3.\frac{1}{\sqrt{3}}\sinh\sqrt{3}.t + 3.\frac{t^{2-1}}{(2-1)!} + 2.\frac{t^{3-1}}{(3-1)!} - 3\cos 3t + \frac{27}{3}\sin 3t$$

$$+ 6\frac{t^4 - 1}{(4-1)!} - 30\frac{t^{7/2-1}}{\Gamma\left(\frac{7}{2}\right)}$$

$$= \sqrt{3}\sinh\sqrt{3}t + 3t + t^2 - 3\cos 3t + 9\sin 3t + t^3 - 16t^2\sqrt{\left(\frac{t}{\pi}\right)}.$$

EXAMPLE 5. A function $f(t)$ obey the equation $f(t) + 2\int_0^t f(t)\,dt = \cosh 2t$. Find the Laplace transformation of $f(t)$. [UPTU–2006]

SOLUTION. We have $f(t) + 2\int_0^t f(t)\,dt = \cosh 2t$

Taking Laplace transformation of both the sides, we get

$$L\{f(t)\} + 2L\int_0^t f(t)\,dt = L(\cosh 2t)$$

$$\Rightarrow \qquad F(p) + 2.\frac{1}{p}F(p) = \frac{p}{p^2 - 4}$$

$$\Rightarrow \qquad F(p)\left[1 + \frac{2}{p}\right] = \frac{p}{p^2 - 4} \qquad\qquad \Rightarrow \qquad F(p).\left[\frac{p+2}{p}\right] = \frac{p}{p^2 - 4}$$

$$\Rightarrow \qquad F(p) = \left(\frac{p}{p^2 - 4}\right).\left(\frac{p}{p+2}\right) \qquad \Rightarrow \qquad F(p) = \frac{p^2}{(p^2 - 4)(p+2)}$$

EXAMPLE 6. Show that $L^{-1}\left\{\frac{1}{p}\cos\frac{1}{p}\right\} = 1 - \frac{t^2}{(2!)^2} + \frac{t^4}{(4!)^2} - \frac{t^6}{(6!)^2} + ...$

SOLUTION. $L^{-1}\left\{\frac{1}{p}\cos\frac{1}{p}\right\} = L^{-1}\left\{\frac{1}{p}\left(1 - \frac{(1/p)^2}{2!} + \frac{(1/p)^4}{4!} - \frac{(1/p)^6}{6!} + ...\right)\right\}$

$$= L^{-1}\left\{\frac{1}{p}\right\} - \frac{1}{2!}L^{-1}\left\{\frac{1}{p^3}\right\} + \frac{1}{4!}L^{-1}\left\{\frac{1}{p^5}\right\} - \frac{1}{6!}L^{-1}\left\{\frac{1}{p^7}\right\} + ...$$

$$= 1 - \frac{t^2}{(2!)^2} + \frac{t^4}{(4!)^2} - \frac{t^6}{(6!)^2} + ...$$

EXAMPLE. 7 Evaluate $L^{-1}\left\{\frac{3p-2}{p^2 - 4p + 20}\right\}$. (Meerut–1987, 2011; Purvanchal–1988, 90)

SOLUTION. We have

$$L^{-1}\left\{\frac{3p-2}{p^2 - 4p + 20}\right\} = L^{-1}\left\{\frac{3(p-2) + 4}{(p-2)^2 + 16}\right\} = L^{-1}\left\{\frac{3(p-2)}{(p-2)^2 + 16} + \frac{4}{(p-2)^2 + 16}\right\}$$

$$= 3L^{-1}\left\{\frac{p-2}{(p-2)^2 + 4^2}\right\} + 4L^{-1}\left\{\frac{1}{(p-2)^2 + 4^2}\right\}$$

$$= 3e^{2t}L^{-1}\left\{\frac{p}{p^2 + 4^2}\right\} + 4e^{2t}L^{-1}\left\{\frac{1}{p^2 + 4^2}\right\} = 3e^{2t}\cos 4t + e^{2t}\sin 4t$$

EXAMPLE 8. Evaluate $L^{-1}\left\{\dfrac{p}{(p+1)^{5/2}}\right\}.$ (Meerut–1985, 88)

SOLUTION. $L^{-1}\left\{\dfrac{p}{(p+1)^{5/2}}\right\} = L^{-1}\left\{\dfrac{(p+1)-1}{(p+1)^{5/2}}\right\} = e^{-t}L^{-1}\left\{\dfrac{p-1}{p^{5/2}}\right\}$

$$= e^{-t}L^{-1}\left[\left\{\dfrac{1}{p^{3/2}}\right\} - \dfrac{1}{p^{5/2}}\right] = e^{-t}L^{-1}\left\{\dfrac{1}{p^{3/2}}\right\} - e^{-t}L^{-1}\left\{\dfrac{1}{p^{5/2}}\right\}$$

$$= e^{-t}\dfrac{t^{(3/2)-1}}{\Gamma\left(\dfrac{3}{2}\right)} - e^{-t}\dfrac{t^{(5/2)-1}}{\Gamma(5/2)} = 2e^{-t}\sqrt{\left(\dfrac{t}{\pi}\right)} - \dfrac{4}{3}e^{-t}.t\sqrt{\left(\dfrac{t}{\pi}\right)}$$

$$= \dfrac{2}{3}e^{-t}\sqrt{\left(\dfrac{t}{\pi}\right)}(3-2t)$$

EXAMPLE 9. Evaluate $L^{-1}\left\{\dfrac{1}{(p+2)(p-1)^2}\right\}$ (Meerut–1983, 84)

SOLUTION. $L^{-1}\left\{\dfrac{1}{(p+2)(p-1)^2}\right\} = L^{-1}\left\{\dfrac{1}{(p-1+3)(p-1)^2}\right\}$

$$= e^t L^{-1}\left\{\dfrac{1}{p+3}\cdot\dfrac{1}{p^2}\right\} = e^t L^{-1}\left\{\dfrac{1}{p^2}\left(\dfrac{1}{3} - \dfrac{1}{9}p + \dfrac{1}{9}\dfrac{p^2}{p+3}\right)\right\}$$

(Dividing 1 by $3 + p$ till p^2 is a common factor in the remainder)

$$= e^t L^{-1}\left\{\dfrac{1}{3}\cdot\dfrac{1}{p^2} - \dfrac{1}{9}\cdot\dfrac{1}{p} + \dfrac{1}{9}\cdot\dfrac{1}{(p+3)}\right\}$$

$$= e^t\left(\dfrac{1}{3}t - \dfrac{1}{9} + \dfrac{1}{9}e^{-3t}\right) = \dfrac{1}{9}[(3t-1)e^t + e^{-2t}]$$

EXAMPLE 10. If $L^{-1}\left\{\dfrac{p}{(p^2+1)^2}\right\} = \dfrac{1}{2}t.\sin t$, find $L^{-1}\left\{\dfrac{32p}{(16p^2+1)^2}\right\}$ (Meerut–2001; Kanpur–2010)

SOLUTION. We have

$$L^{-1}\left\{\dfrac{p}{(p^2+1)^2}\right\} = \dfrac{1}{2}t\sin t$$

$$\therefore \quad L^{-1}\left\{\dfrac{ap}{(a^2p^2+1)^2}\right\} = \dfrac{1}{2}\cdot\dfrac{1}{a}\cdot\dfrac{t}{a}\cdot\sin\dfrac{t}{a}$$

$$\Rightarrow \quad L^{-1}\left\{\dfrac{2a^2p}{(a^2p^2+1)^2}\right\} = \dfrac{t}{a}\sin\dfrac{t}{a}$$

Now putting $a = 4$, we get

$$L^{-1}\left\{\dfrac{32p}{(16p^2+1)^2}\right\} = \dfrac{t}{4}\sin\dfrac{t}{4}$$

EXAMPLE 11. Evaluate $L^{-1}\left\{\dfrac{e^{4-3p}}{(p+4)^{5/2}}\right\}$ (Meerut–1984, 86)

SOLUTION. We have

$$L^{-1}\left\{\frac{1}{(p+4)^{5/2}}\right\} = e^{-4t}L^{-1}\left\{\frac{1}{p^{5/2}}\right\} = e^{-4t}\frac{t^{(5/2)-1}}{\Gamma\left(\dfrac{5}{2}\right)} = \frac{4t^{3/2}e^{-4t}}{3\sqrt{\pi}}$$

Therefore,

$$L^{-1}\left\{\frac{e^{4-3p}}{(p+4)^{5/2}}\right\} = e^4 \, L^{-1}\left\{\frac{e^{-3p}}{(p+4)^{5/2}}\right\} = \begin{cases} e^4 \cdot \dfrac{4}{3\sqrt{\pi}}(t-3)^{3/2}e^{-4(t-3)}, & t > 3 \\ 0, & t < 3 \end{cases}$$

$$= \begin{cases} \dfrac{4}{3\sqrt{\pi}}(t-3)^{3/2}e^{-4(t-4)}, & t > 3 \\ 0, & t < 3 \end{cases}$$

$$= \frac{4}{3\sqrt{\pi}}(t-3)^{3/2}e^{-4(t-4)} \cdot H(t-3).$$

EXAMPLE 12. Evaluate $L^{-1}\left(\dfrac{e^{-cp}}{p^2(p+a)}\right),\ c > 0.$

(UPTU–2001 Sp., 2002; Punjab–1996)

SOLUTION. We have

$$L^{-1}\left[\frac{e^{-cp}}{p^2(p+a)}\right] = L^{-1}\left[-\frac{e^{-cp}}{a^2 p} + \frac{e^{-cp}}{ap^2} + \frac{e^{-cp}}{a^2(p+a)}\right] \quad \text{[By Partial Fractions]}$$

$$= L^{-1}\left[-\frac{1}{a^2}\frac{e^{-cp}}{p} + \left(\frac{1}{a}\right)\frac{e^{-cp}}{p^2} + \frac{1}{a^2}\cdot\frac{e^{-c(p+a)}}{e^{-ca}(p+a)}\right]$$

$$= -\frac{1}{a^2}H(t-c) + \frac{1}{a}(t-c)H(t-c) + \frac{1}{a^2 e^{-ca}}\cdot e^{-at}H(t-c)$$

$$= H(t-c)\left[-\frac{1}{a^2} + \frac{1}{a}(t-c) + \frac{1}{a^2}e^{-a(c+t)}\right]$$

Where $H(t-c)$ = unit step function.

EXAMPLE 13. Find a function for which $F(t) = L^{-1}\left\{\dfrac{3}{p} - \dfrac{4e^{-p}}{p^2} + \dfrac{4e^{-3p}}{p^2}\right\}$ (Meerut–1980)

SOLUTION. We have $F(t) = L^{-1}\left\{\dfrac{3}{p} - \dfrac{4e^{-p}}{p^2} + \dfrac{4e^{-3p}}{p^2}\right\}$

$$= 3L^{-1}\left\{\frac{1}{p}\right\} - 4L^{-1}\left[\frac{e^{-p}}{p^2}\right] + 4L^{-1}\left[\frac{e^{-3p}}{p^2}\right] \quad \text{...(1)}$$

Now, $L^{-1}\left[\dfrac{1}{p}\right] = 1,\ L^{-1}\left\{\dfrac{1}{p^2}\right\} = t$

Therefore, $L^{-1}\left\{\dfrac{e^{-p}}{p^2}\right\} = (t-1)H(t-1)$

and $\qquad L^{-1}\left\{\dfrac{e^{-3p}}{p^2}\right\} = (t-3)\,H(t-3)$.

Putting all these values in (1), we get

$$F(t) = 3 - 4(t-1)\,H(t-1) + 4(t-3)H(t-3)$$

EXAMPLE 14.   *Evalute* $L^{-1}\left\{\dfrac{p+1}{p^2+6p+25}\right\}$ (Agra–1985; Meerut–1984, 90, 91, 95; Kanpur–1981)

SOLUTION.   We have

$$L^{-1}\left\{\frac{p+1}{p^2+6p+25}\right\} = L^{-1}\left\{\frac{(p+3)-2}{(p+3)^2+16}\right\} = e^{-3t}L^{-1}\left\{\frac{p-2}{p^2+16}\right\}$$

$$= e^{-3t}\left[L^{-1}\left\{\frac{p}{p^2+4^2}\right\} - 2L^{-1}\left\{\frac{1}{p^2+4^2}\right\}\right]$$

$$= e^{-3t}\left[\cos 4t - \frac{1}{2}\sin 4t\right].$$

EXAMPLE 15.   *Evalute* $L^{-1}\left[\dfrac{e^{-p} - 3e^{-3p}}{p^2}\right]$

SOLUTION.   We have $L^{-1}\left[\dfrac{e^{-p} - 3e^{-3p}}{p^2}\right] = L^{-1}\left[\dfrac{e^{-p}}{p^2} - \dfrac{3e^{-3p}}{p^2}\right]$ ... (1)

We know that $\qquad Lu(t-a) = \dfrac{e^{-ap}}{p}$ and $L\left[(t-a)\,u(t-a)\right] = \dfrac{e^{-ap}}{p^2}$

Using these results in (1), we get

$$L^{-1}\left[\frac{e^{-p} - 3e^{-3p}}{p^2}\right] = (t-1)u(t-1) - 3(t-3)u(t-3).$$

EXAMPLE 16.   *Evalute* $L^{-1}\left\{\dfrac{p+5}{(p+2)(p^2+4)}\right\}$ (Meerut–1988)

SOLUTION.   We have $L^{-1}\left\{\dfrac{p+5}{(p+2)(p^2+4)}\right\}$

$$= L^{-1}\left\{\frac{1}{8}\left(\frac{3}{p+2} - \frac{3p-14}{p^2+4}\right)\right\}$$

$$= \frac{1}{8}\left[3L^{-1}\left\{\frac{1}{p+2}\right\} - 3L^{-1}\left\{\frac{p}{p^2+4}\right\} + 14L^{-1}\left\{\frac{1}{p^2+4}\right\}\right]$$

$$= \frac{1}{8}\,(3e^{-2t} - 3\cos 2t + 7\sin 2t)$$

EXAMPLE 17.   *Show that* $L^{-1}\left\{\dfrac{p}{p^4+p^2+1}\right\} = \dfrac{2}{\sqrt{3}}\sinh\dfrac{t}{2}.\sin\dfrac{1}{2}\sqrt{3}\,t$ (Raipur–2005; Agra–1984, 98, 2014;

SOLUTION.   We have $L^{-1}\left\{\dfrac{p}{p^4+p^2+1}\right\} = L^{-1}\left\{\dfrac{p}{(p^2+1)^2-p^2}\right\}$

$$= L^{-1}\left\{\dfrac{p}{(p^2+p+1)(p^2-p+1)}\right\}$$

$$= L^{-1}\left\{\dfrac{1}{2}\dfrac{(p^2+p+1)-(p^2-p+1)}{(p^2-p+1)(p^2+p+1)}\right\}$$

$$= L^{-1}\left\{\dfrac{1}{2(p^2-p+1)}-\dfrac{1}{2(p^2+p+1)}\right\}$$

$$= \dfrac{1}{2}L^{-1}\left\{\dfrac{1}{\left(p-\dfrac{1}{2}\right)^2+\dfrac{3}{4}}\right\} - \dfrac{1}{2}L^{-1}\left\{\dfrac{1}{\left(p+\dfrac{1}{2}\right)^2+\dfrac{3}{4}}\right\}$$

$$= \dfrac{1}{2}e^{t/2}L^{-1}\left\{\dfrac{1}{p^2+\left(\dfrac{1}{2}\sqrt{3}\right)^2}\right\} - \dfrac{1}{2}e^{-t/2}L^{-1}\left\{\dfrac{1}{p^2+\left(\dfrac{1}{2}\sqrt{3}\right)^2}\right\}$$

$$= \dfrac{1}{2}e^{t/2}\dfrac{2}{\sqrt{3}}\sin\left(\sqrt{3}\cdot\dfrac{t}{2}\right) - \dfrac{1}{2}e^{-t/2}\dfrac{2}{\sqrt{3}}\sin\left(\sqrt{3}\cdot\dfrac{t}{2}\right)$$

$$= \dfrac{1}{\sqrt{3}}(e^{t/2}-e^{-t/2})\sin\left(\sqrt{3}\cdot\dfrac{t}{2}\right) = \dfrac{2}{\sqrt{3}}\sinh\dfrac{t}{2}\sin\left(\sqrt{3}\cdot\dfrac{t}{2}\right).$$

EXAMPLE 18.   *Find* $L^{-1}\left\{\dfrac{2p^3+2p^2+4p+1}{(p^2+1)(p^2+p+1)}\right\}$ (Agra–1988; Kanpur–1987; Ravishankar–2004)

SOLUTION.   We have $L^{-1}\left\{\dfrac{2p^3+2p^2+4p+1}{(p^2+1)(p^2+p+1)}\right\} = L^{-1}\left\{\dfrac{p+2}{p^2+1}+\dfrac{p-1}{p^2+p+1}\right\}$

$$= L^{-1}\left\{\dfrac{p}{p^2+1}\right\} + 2L^{-1}\left\{\dfrac{1}{p^2+1}\right\} + L^{-1}\left\{\dfrac{\left(p+\dfrac{1}{2}\right)-\dfrac{3}{2}}{\left(p+\dfrac{1}{2}\right)^2+\dfrac{3}{4}}\right\}$$

$$= \cos t + 2\sin t + e^{-t/2}L^{-1}\left[\dfrac{p-3/2}{p^2+3/4}\right]$$

$$= \cos t + 2\sin t + e^{-t/2}\left[L^{-1}\dfrac{p}{p^2+\left(\dfrac{\sqrt{3}}{2}\right)^2} - \dfrac{3}{2}L^{-1}\left\{\dfrac{1}{p^2+\left(\dfrac{\sqrt{3}}{2}\right)^2}\right\}\right]$$

$$= \cos t + 2\sin t + e^{-t/2}\left\{\cos\left(\dfrac{1}{2}\sqrt{3}t\right) - \dfrac{3}{2}\cdot\dfrac{2}{\sqrt{3}}\sin\left(\dfrac{1}{2}\sqrt{3}t\right)\right\}$$

$$= \cos t + 2\sin t + e^{-t/2}\left[\cos\left(\dfrac{1}{2}\cdot\sqrt{3}t\right) - \sqrt{3}\sin\left(\dfrac{1}{2}\sqrt{3}t\right)\right]$$

EXAMPLE 19. *Evaluate* $L^{-1}\left\{(p+1)\dfrac{e^{-\pi p}}{p^2+p+1}\right\}$.

(Agra–1981, 99, 2009; Andhra–1990)

SOLUTION. We get

$$L^{-1}\left\{\frac{p+1}{p^2+p+1}\right\} = L^{-1}\left\{\frac{\left(p+\dfrac{1}{2}\right)+\dfrac{1}{2}}{\left[\left(p+\dfrac{1}{2}\right)^2+\dfrac{3}{4}\right]}\right\} = e^{-t/2}L^{-1}\left\{\frac{p+\dfrac{1}{2}}{p^2+\dfrac{3}{4}}\right\}$$

$$= e^{-t/2}L^{-1}\left\{\frac{p}{p^2+\left(\dfrac{\sqrt{3}}{2}\right)^2}\right\} + \frac{1}{2}e^{-t/2}L^{-1}\left\{\frac{1}{p^2+\left(\dfrac{\sqrt{3}}{2}\right)^2}\right\}$$

$$= e^{-t/2}\cos\left(\frac{\sqrt{3}t}{2}\right) + \frac{1}{2}e^{-t/2}\left(\frac{2}{\sqrt{3}}\right)\sin\left(\frac{\sqrt{3}t}{2}\right)$$

$$\therefore \quad L^{-1}\left\{\frac{(p+1)e^{-\pi p}}{p^2+p+1}\right\} = \begin{cases} \dfrac{e^{-(t-\pi)/2}}{\sqrt{3}}\left[\sqrt{3}\cos\dfrac{\sqrt{3}}{2}(t-\pi) \\ \qquad\qquad + \sin\dfrac{\sqrt{3}}{2}(t-\pi)\right], & t>\pi \\ 0\,, & t<\pi \end{cases}$$

$$= \frac{e^{-(t-\pi)/2}}{\sqrt{3}}\left[\sqrt{3}\cos\frac{\sqrt{3}}{2}(t-\pi) + \sin\frac{\sqrt{3}}{2}(t-\pi)\,H(t-\pi)\right].$$

EXAMPLE 20. *Find* $L^{-1}\left\{\dfrac{5p+3}{(p-1)(p^2+2p+5)}\right\}$.

SOLUTION. We have

(UPTU–2004, 05, 14; Rohtak–2009; Rohilkhand–1994; VTU–2014)

$$\frac{5p+3}{(p-1)(p^2+2p+5)} = \frac{1}{p-1} + \frac{2-p}{p^2+2p+5} \qquad \text{[By partial fractions]}$$

$$\therefore L^{-1}\left\{\frac{5p+3}{(p-1)(p^2+2p+5)}\right\}$$

$$= L^{-1}\left\{\frac{1}{p-1} + \frac{2-p}{p^2+2p+5}\right\}$$

$$= L^{-1}\left\{\frac{1}{p-1} - \frac{p+1}{(p+1)^2+4} + \frac{3}{(p+1)^2+4}\right\}$$

$$= L^{-1}\left\{\frac{1}{p-1}\right\} - L^{-1}\left\{\frac{p+1}{(p+1)^2+4}\right\} + 3L^{-1}\left\{\frac{1}{(p+1)^2+4}\right\}$$

$$= e^t - e^{-t}L^{-1}\left\{\frac{p}{p^2+4}\right\} + 3e^{-t}L^{-1}\left\{\frac{1}{p^2+4}\right\}$$

$$= e^t - e^{-t}\cos 2t + \frac{3}{2}e^{-t}\sin 2t = e^t - e^{-t}\left(\cos 2t - \frac{3}{2}\sin 2t\right).$$

EXAMPLE 21. *Show that* $L^{-1}\left(\dfrac{p}{2p^2-8}\right) = \dfrac{1}{2}\cosh 2t$ (Agra–2003; Rohilkhand–2011)

SOLUTION. Here, we have

$$L^{-1}\left(\frac{p}{2p^2-8}\right) = \frac{1}{2}L^{-1}\left(\frac{p}{p^2-4}\right)$$

$$= \frac{1}{2}\cosh 2t$$

EXAMPLE 22. *Evaluate* $L^{-1}(e^{-\sqrt{p}})$ (Agra–2007)

SOLUTION. Here, we have

$$L^{-1}(e^{-\sqrt{p}}) = L^{-1}\left[1 - \sqrt{p} + \frac{p}{2!} - \frac{p^{3/2}}{3!} + \frac{p^2}{4!} - \frac{p^{5/2}}{5!} + \dots\right]$$

$$= L^{-1}(1) - L^{-1}(p^{1/2}) + \frac{1}{2}L^{-1}(p) - \frac{1}{3!}L^{-1}(p^{3/2}) + \frac{1}{4!}L^{-1}(p^2)$$

$$- \frac{1}{5!}L^{-1}(p^{5/2}) + \dots$$

$$= 0 - \frac{(-1)t^{-3/2}}{\sqrt{\pi}} \cdot \frac{1}{2} - \frac{1}{3!}\frac{(-1)^2}{\sqrt{\pi}} \cdot \frac{1}{2}\cdot\frac{3}{2} \cdot t^{-5/2} - \dots$$

$$- \frac{1}{5!}\frac{(-1)^3}{\sqrt{\pi}} \cdot \frac{1}{2}\cdot\frac{3}{2}\cdot\frac{5}{2} \cdot t^{-7/2} - \dots$$

$$= \frac{1}{2\sqrt{\pi}t^{3/2}}\left[1 - \frac{1}{4t} + \frac{(1/4t)^2}{2!} - \frac{(1/4t)^3}{3!} + \dots\right] = \frac{1}{2\sqrt{\pi}t^{3/2}}e^{-1/4t}$$

EXAMPLE 23. *Evaluate* $\dfrac{p+1}{p(p^2+4p+8)}$ (Meerut–1991, 2010)

SOLUTION. Here, we have

$$L^{-1}\left\{\frac{p+1}{p(p^2+4p+8)}\right\} = L^{-1}\left\{\frac{1}{8p} - \frac{p-4}{8(p^2+4p+8)}\right\}$$

$$= \frac{1}{8}L^{-1}\left\{\frac{1}{p}\right\} - \frac{1}{8}L^{-1}\left\{\frac{(p+2)-6}{(P+2)^2+4}\right\}$$

$$= \frac{1}{8} - \frac{1}{8}e^{-2t}L^{-1}\left\{\frac{p-6}{p^2+4}\right\}$$

$$= \frac{1}{8} - \frac{1}{8}e^{-2t}\left[L^{-1}\left\{\frac{p}{p^2+2^2}\right\} - 6L^{-1}\left\{\frac{1}{p^2+2^2}\right\}\right]$$

$$= \frac{1}{8}[1 - e^{-2t}(\cos 2t - 3\sin 2t)]$$

EXAMPLE 24. *Evaluate* (i) $L^{-1}\left\{\dfrac{1}{2p-5}\right\}$ (ii) $L^{-1}\left\{\left(\dfrac{\sqrt{p}-1}{p}\right)^2\right\}$ (Meerut–1991)

SOLUTION. (i) Here, we have

$$L^{-1}\left\{\frac{1}{2p-5}\right\} = L^{-1}\left\{\frac{1}{2(p-5/2)}\right\} = \frac{1}{2}L^{-1}\left\{\frac{1}{p-5/2}\right\} = \frac{1}{2}e^{5t/2}$$

(ii) Here, we have

$$L^{-1}\left\{\left(\frac{\sqrt{p}-1}{p}\right)^2\right\} = L^{-1}\left\{\left(\frac{1}{\sqrt{p}}-\frac{1}{p}\right)^2\right\}$$

$$= L^{-1}\left\{\frac{1}{p}-\frac{2}{p^{3/2}}+\frac{1}{p^2}\right\} = L^{-1}\left\{\frac{1}{p}\right\} - 2L^{-1}\left\{\frac{1}{p^{3/2}}\right\} + L^{-1}\left\{\frac{1}{p^2}\right\}$$

$$= 1 - 2\frac{t^{1/2}}{\Gamma\left(\dfrac{3}{2}\right)} + t = 1 + t - 2\frac{t^{1/2}}{\dfrac{1}{2}\sqrt{\pi}} = 1 + t - 4\sqrt{t/\pi}\,.$$

EXAMPLE 25. *Prove that*

$$L^{-1}\left\{\frac{5}{p^2}+\left(\frac{\sqrt{p}-1}{p}\right)^2-\frac{7}{3p+2}\right\} = 1 + 6t - 4\sqrt{\frac{t}{\pi}} - \frac{7}{3}e^{-2t/3}$$ (Meerut–1996)

SOLUTION. Here, we have

$$L^{-1}\left\{\frac{5}{p^2}+\left(\frac{\sqrt{p}-1}{p}\right)^2-\frac{7}{3p+2}\right\} = L^{-1}\left\{\frac{5}{p^2}+\frac{p-2\sqrt{p}+1}{p^2}-\frac{7}{3}\frac{1}{p+(2/3)}\right\}$$

$$= L^{-1}\left\{\frac{6}{p^2}+\frac{1}{p}-\frac{2}{p^{3/2}}-\frac{7}{3}\cdot\frac{1}{p+(2/3)}\right\}$$

$$= 6L^{-1}\left\{\frac{1}{p^2}\right\}+L^{-1}\left\{\frac{1}{p}\right\}-2L^{-1}\left\{\frac{1}{p^{3/2}}\right\}-\frac{7}{3}L^{-1}\left\{\frac{1}{p+(2/3)}\right\}$$

$$= \frac{6t^{2-1}}{1!}+1-2\frac{t^{3/2-1}}{\Gamma\left(\dfrac{3}{2}\right)}-\frac{7}{3}e^{-2t/3} = 6t+1-4\sqrt{\frac{t}{\pi}}-\frac{7}{3}e^{-2t/3}$$

EXAMPLE 26. *Evaluate* $L^{-1}\left\{\dfrac{3p+2}{4p^2+12p+9}\right\}$ (Meerut–1986, 88, 96, Jhansi–2012)

SOLUTION. Here, we have

$$L^{-1}\left\{\frac{3p+2}{4p^2+12p+9}\right\} = L^{-1}\left\{\frac{3(p+3/2)-5/2}{4(p+3/2)^2}\right\}$$

$$= e^{-3t/2}L^{-1}\left\{\frac{3p-5/2}{4p^2}\right\} = e^{-3t/2}\left[\frac{3}{4}L^{-1}\left\{\frac{1}{p}\right\}-\frac{5}{8}L^{-1}\left\{\frac{1}{p^2}\right\}\right]$$

(From first shifting theorem)

$$= e^{-3t/2}\left[\frac{3}{4}\cdot 1-\frac{5}{8}\cdot\frac{t}{1!}\right] = \frac{1}{8}e^{-3t/2}(6-5t)$$

EXAMPLE 27. *Evaluate* $L^{-1} \dfrac{1}{(8p-27)^{1/3}}$ (Meerut–1998)

SOLUTION. Here, we have

$$L^{-1}\left\{\frac{1}{(8p-27)^{1/3}}\right\} = \frac{1}{2}L^{-1}\left\{\frac{1}{(p-27/8)^{1/3}}\right\}$$

$$= \frac{1}{2}e^{27t/8}L^{-1}\left\{\frac{1}{p^{1/3}}\right\} \qquad \text{(From first shifting theorem)}$$

$$= \frac{1}{2}e^{27t/8}\frac{t^{1/3-1}}{\Gamma(1/3)} = \frac{e^{27t/8}t^{-2/3}}{2\Gamma(1/3)}$$

EXAMPLE 28. *Evaluate* $L^{-1}\left\{\dfrac{p}{(p+3)^{7/2}}\right\}$ (Kanpur–2010; Meerut–1985)

SOLUTION. Here, we have

$$L^{-1}\left\{\frac{p}{(p+3)^{7/2}}\right\} = e^{-3t}L^{-1}\left\{\frac{p-3}{(p-3+3)^{7/2}}\right\} = e^{-3t}L^{-1}\left\{\frac{p-3}{p^{7/2}}\right\}$$

$$= e^{-3t}L^{-1}\left\{\frac{1}{p^{3/2+1}} - \frac{3}{p^{5/2+1}}\right\}$$

$$= e^{-3t}\left\{\frac{t^{3/2}}{\Gamma(5/2)} - \frac{3t^{5/2}}{\Gamma(7/2)}\right\} = \frac{e^{-3t}}{\sqrt{\pi}}\left[\frac{4}{3}t^{3/2} - \frac{8}{5}t^{5/2}\right]$$

EXAMPLE 29. *Find the inverse Laplace transform of* $\dfrac{e^{-\pi p}}{p^2+1}$.   (Kanpur–2008)

SOLUTION. Consider

$$L^{-1}\left\{\frac{1}{p^2+1}\right\} = \sin t = F(t)$$

From second shifting theorem, we get

$$L^{-1}\left\{\frac{e^{-\pi p}}{p^2+1}\right\} = \begin{cases} \sin(t-\pi) & \text{if } t > \pi \\ 0 & \text{if } t < \pi \end{cases} = \sin(t-\pi)H(t-\pi) = -\sin t \cdot H(t-\pi)$$

EXAMPLE 30. *Find inverse Laplace transform of* $\dfrac{1}{\sqrt{p+2}}$ (Lucknow–2012)

SOLUTION. Here, we have

$$L^{-1}\left\{\frac{1}{\sqrt{p+2}}\right\} = e^{-2t}L^{-1}\left\{\frac{1}{\sqrt{(p-2+2)}}\right\} = e^{-2t}L^{-1}\left\{\frac{1}{\sqrt{p}}\right\}$$

$$= e^{-2t}L^{-1}\left\{\frac{1}{p^{-1/2+1}}\right\} = \left\{\frac{e^{-2t}t^{-1/2}}{\Gamma\left(-\dfrac{1}{2}+1\right)} = \frac{e^{-2t}}{\sqrt{\pi t}}\right\}$$

EXAMPLE 31. *Evaluate* $L^{-1}\left\{\dfrac{e^{-3p}}{(p-4)^2}\right\}$ (Agra–2005)

SOLUTION.   Here, $L\{F(t)\} = f(p) = \dfrac{1}{(p-4)^2}$

$$\Rightarrow \qquad F(t) = L^{-1}\{f(p)\} = L^{-1}\left\{\dfrac{1}{(p-4)^2}\right\} = e^{4t}L^{-1}\left\{\dfrac{1}{p^2}\right\} = te^{4t}$$

EXAMPLE 32. *Evaluate* $L^{-1}\left\{\dfrac{6}{2p-3} - \dfrac{3+4p}{9p^2-16} + \dfrac{8-6p}{16p^2+9}\right\}$ (Meerut–1996; Osmania–2004)

SOLUTION.   Here, we have

$$L^{-1}\left\{\dfrac{6}{2p-3} - \dfrac{3+4p}{9p^2-16} + \dfrac{8-6p}{16p^2+9}\right\} \qquad \ldots(1)$$

$$= 6L^{-1}\left\{\dfrac{1}{2p-3}\right\} - 3L^{-1}\left\{\dfrac{1}{9p^2-16}\right\} - 4L^{-1}\left\{\dfrac{p}{9p^2-16}\right\}$$

$$+ 8L^{-1}\left\{\dfrac{1}{16p^2+9}\right\} - 6L^{-1}\left\{\dfrac{p}{16p^2+9}\right\}$$

$$= \dfrac{6}{2}L^{-1}\left\{\dfrac{1}{p-(3/2)}\right\} - \dfrac{3}{9}L^{-1}\left\{\dfrac{1}{p^2-(4/3)^2}\right\} - \dfrac{4}{9}L^{-1}\left\{\dfrac{p}{p^2-(4/3)^2}\right\}$$

$$+ \dfrac{8}{16}L^{-1}\left\{\dfrac{1}{p^2+(3/4)^2}\right\} - \dfrac{6}{16}L^{-1}\left\{\dfrac{p}{p^2+(3/4)^2}\right\}$$

$$= 3e^{3t/2} - \dfrac{1}{3} \times \dfrac{1}{(4/3)}\sinh\dfrac{4t}{3} - \dfrac{4}{9}\cosh\dfrac{4t}{3} + \dfrac{1}{2} \times \dfrac{1}{(3/4)}\sin\dfrac{3t}{4}$$

$$- \dfrac{3}{8}\cos\dfrac{3t}{4}$$

$$= 3e^{3t/2} - \dfrac{1}{4}\sinh\dfrac{4t}{3} - \dfrac{4}{9}\cosh\dfrac{4t}{3} + \dfrac{2}{3}\sin\dfrac{3t}{4} - \dfrac{3}{8}\cos\dfrac{3t}{4}$$

EXAMPLE 33. *Find* $L^{-1}\left\{\dfrac{p^2}{(p+1)(p+2)(p+3)}\right\}$ (SVTU–1997)

SOLUTION.   Here, we have

$$L^{-1}\left\{\dfrac{p^2}{(p+1)(p+2)(p+3)}\right\}$$

Let us assume

$$\dfrac{p^2}{(p+1)(p+2)(p+3)} = \dfrac{A}{(p+1)} + \dfrac{B}{(p+2)} + \dfrac{C}{(p+3)} \qquad \ldots(1)$$

Multiplying both the sides of (1) by $(p + 1)$ and taking limit $p \to -1$, we get

$$A = \lim_{p \to -1} \frac{p^2}{(p+2)(p+3)} = \frac{(-1)^2}{(-1+2)(-1+3)} = \frac{1}{2}$$

Now, multiplying both the sides of (1) by $(p + 2)$ and taking limit $p \to -2$, we get

$$B = \lim_{p \to -2} \frac{p^2}{(p+1)(p+3)} = \frac{(-2)^2}{(-2+1)(-2+3)} = -4$$

Now, multiplying both the sides of (1) by $(p + 3)$ and taking limit $p \to -3$, we get

$$C = \lim_{p \to -3} \frac{p^2}{(p+1)(p+2)} = \frac{(-3)^2}{(-3+1)(-3+2)} = \frac{9}{2}$$

Putting the value of A, B and C in eq^n (1), we get

$$\frac{p^2}{(p+1)(p+2)(p+3)} = \frac{1}{2(p+1)} - \frac{4}{(p+2)} + \frac{9}{2(p+3)}$$

Therefore,

$$\left\{ \frac{p^2}{(p+1)(p+2)(p+3)} \right\} = \frac{1}{2} L^{-1} \left\{ \frac{1}{p+1} \right\} - 4L^{-1} \left\{ \frac{1}{p+2} \right\} + \frac{9}{2} L^{-1} \left\{ \frac{1}{p+3} \right\}$$

$$= \frac{e^{-t}}{2} - 4e^{-2t} + \frac{9e^{-3t}}{2}$$

EXAMPLE 34. *Find the inverse Laplace transform of* $1/p^2(p + 1)^2$.

(Purvanchal–2004; Kumaun–1997; Kanpur–1997)

SOLUTION.   Let us assume

$$\frac{1}{p^2(p+1)^2} = \frac{A}{p} + \frac{B}{p^2} + \frac{C}{(p+1)} + \frac{D}{(p+1)^2} \qquad \ldots(1)$$

Multiplying both the sides of (1) by p^2 and taking $p \to 0$, then

$$A = \lim_{p \to 0} \frac{1}{(p+1)^2} = \frac{1}{(0+1)^2} = 1$$

Now, multiplying both the sides of (1) by $(p + 1)^2$ and taking $p \to -1$, then

$$B = \lim_{p \to -1} \frac{1}{p^2} = \frac{1}{(-1)^2} = 1$$

Then from eq^n (1), we get

$$\frac{1}{p^2(p+1)^2} = \frac{A}{p} + \frac{1}{p^2} + \frac{C}{(p+1)} + \frac{1}{(p+1)^2} \qquad \ldots(2)$$

Now, puting $p = 1$ and $p = -2$ in eq^n (2), we get

$$A + 1 + \frac{C}{2} + \frac{1}{4} = \frac{1}{4} \qquad \text{and} \qquad -\frac{A}{2} + \frac{1}{4} - C + 1 = \frac{1}{4}$$

$$\Rightarrow \qquad 2A + C = -2 \qquad \text{and} \qquad A + 2C = 2$$

So that, $A = -2$, $C = 2$

Then from (2)

$$\frac{1}{p^2(p+1)^2} = -\frac{2}{p} + \frac{1}{p^2} + \frac{2}{(p+1)} + \frac{1}{(p+1)^2}$$

Now, $L\left\{\dfrac{1}{p^2(p+1)^2}\right\} = -2L^{-1}\left\{\dfrac{1}{p}\right\} + L^{-1}\left\{\dfrac{1}{p^2}\right\} + 2L^{-1}\left\{\dfrac{1}{p+1}\right\} + L^{-1}\left\{\dfrac{1}{(p+1)^2}\right\}$

$= -2 + (t/1!) + 2e^{-t} + e^{-t}L^{-1}\{1/p^2\}$

[From first shifting theorem]

$= -2 + t + 2e^{-t} + e^{-t}t$

$= t - 2 + e^{-t}(2 + t)$.

EXAMPLE 35. Evaluate $L^{-1}\{3(1 + e^{-p\pi})/(p^2 + 9)\}$ (Kurukshetra–2004)

SOLUTION. Here, we have

$$L^{-1}\left\{\dfrac{3(1+e^{-p\pi})}{p^2+9}\right\} = L^{-1}\left\{\dfrac{3}{p^2+9}\right\} + L^{-1}\left\{\dfrac{3e^{-p\pi}}{p^2+9}\right\} \qquad \text{...(1)}$$

$$= f(p) = 3(p^2 + 9) \qquad \text{...(2)}$$

Therefore, $F(t) = L^{-1}\{f(p)\} = L^{-1}\left\{\dfrac{3}{p^2+3^2}\right\} = 3 \times \dfrac{1}{3}\sin 3t = \sin 3t$ ...(3)

Now, from second shifting theorem, we get

$$L^{-1}\{e^{-p\pi}f(p)\} = \begin{cases} F(t-\pi), & t > \pi \\ 0, & t < \pi \end{cases}$$

$\Rightarrow \quad L^{-1}\left\{e^{-p\pi}\dfrac{3}{p^2+9}\right\} = \begin{cases} \sin 3(t-\pi), & t > \pi \\ 0, & t < \pi \end{cases}$ [From (1) and (2)]

$= \begin{cases} -\sin 3t, & t > \pi \\ 0, & t < \pi \end{cases} = -\sin 3tH(t-\pi)$ ...(4)

where $H(t - \pi)$ is Heaviside function.
Using eqn (3) and (4), equation (1) reduces

$$L^{-1}\left\{\dfrac{3(1+e^{-p\pi})}{p^2+9}\right\} = \sin 3t - \sin 3tH(t-\pi)$$

Additional Solved Examples

EXAMPLE 1. Find the inverse Laplace transform of the following functions :

(a) $\dfrac{1}{p^4}$ (b) $\dfrac{1}{p^2+4}$ (Meerut–2009BP)

(c) $\dfrac{4}{p-2}$ (d) $\dfrac{1}{\sqrt{p}}$ (Rohilkhand–1995)

(e) $\dfrac{p}{p^2+2} + \dfrac{6p}{p^2-16} + \dfrac{3}{p-3}$ (Osmania–2004)

(f) $\dfrac{2p-5}{p^2-9}$ (Meerut–2013)

(g) $\dfrac{6}{2p-3} - \dfrac{3+4p}{9p^2-16} + \dfrac{8-6p}{16p^2+9}$ (Meerut–1996)

(h) $\dfrac{3(p^2+1)^2}{2p^5} + \dfrac{4p-18}{9-p^2} + \dfrac{(p+1)(2-\sqrt{p})}{p^{5/2}}$ (UPTU–2001, 02)

(i) $\dfrac{1}{p}\sin\dfrac{1}{p}$

SOLUTION. (a) $L^{-1}\left\{\dfrac{1}{p^4}\right\} = \dfrac{t^{4-1}}{3!} = \dfrac{t^3}{6}$

(b) $L^{-1}\left\{\dfrac{1}{p^2+4}\right\} = L^{-1}\left\{\dfrac{1}{p^2+2^2}\right\} = \dfrac{1}{2}\sin 2t$

(c) $L^{-1}\left\{\dfrac{4}{p-2}\right\} = 4L^{-1}\left\{\dfrac{1}{p-2}\right\} = 4e^{2t}$

(d) $L^{-1}\left\{\dfrac{1}{\sqrt{p}}\right\} = \dfrac{t^{1/2-1}}{\Gamma\left(\dfrac{1}{2}\right)} = \dfrac{t^{-1/2}}{\sqrt{\pi}} = \dfrac{1}{\sqrt{\pi t}}$

(e) $L^{-1}\left\{\dfrac{p}{p^2+2} + \dfrac{6p}{p^2-16} + \dfrac{3}{p-3}\right\}$

$\qquad = L^{-1}\left\{\dfrac{p}{p^2+(\sqrt{2})^2}\right\} + 6L^{-1}\left\{\dfrac{p}{p^2-16}\right\} + 3L^{-1}\left\{\dfrac{1}{p-3}\right\}$

$\qquad = \cos\sqrt{2}t + 6\cosh 4t + 3e^{3t}$

(f) $L^{-1}\left\{\dfrac{2p-5}{p^2-9}\right\} = 2L^{-1}\left\{\dfrac{p}{p^2-3^2}\right\} - 5L^{-1}\left\{\dfrac{1}{p^2-3^2}\right\} = 2\cosh 3t - \dfrac{5}{3}\sinh 3t$

(g) $L^{-1}\left\{\dfrac{6}{2p-3} - \dfrac{3+4p}{9p^2-16} + \dfrac{8-6p}{16p^2+9}\right\}$

$\qquad = 3L^{-1}\left\{\dfrac{1}{p-(3/2)}\right\} - 3L^{-1}\left\{\dfrac{1}{9p^2-16}\right\} - 4L^{-1}\left\{\dfrac{p}{9p^2-16}\right\}$

$\qquad\qquad + 8L^{-1}\left\{\dfrac{1}{16p^2+9}\right\} - 6L^{-1}\left\{\dfrac{p}{16p^2+9}\right\}$

$\qquad = 3L^{-1}\left\{\dfrac{1}{p-(3/2)}\right\} - \dfrac{1}{3}L^{-1}\left\{\dfrac{1}{p^2-\left(\dfrac{4}{3}\right)^2}\right\} - \dfrac{4}{9}L^{-1}\left\{\dfrac{p}{p^2-(4/3)^2}\right\}$

$\qquad\qquad + \dfrac{1}{2}L^{-1}\left\{\dfrac{1}{p^2+(3/4)^2}\right\} - \dfrac{3}{8}L^{-1}\left\{\dfrac{p}{p^2+(3/4)^2}\right\}$

$\qquad = 3e^{(3/2)t} - \dfrac{1}{3}\dfrac{1}{(4/3)}\sinh\dfrac{4}{3}t - \dfrac{4}{9}\cosh\dfrac{4t}{3} + \dfrac{1}{2}\dfrac{1}{3/4}\sin\dfrac{3t}{4} - \dfrac{3}{8}\cos\dfrac{3}{4}t$

$\qquad = 3e^{3t/2} - \dfrac{1}{4}\sinh\dfrac{4t}{3} - \dfrac{4}{9}\cosh\dfrac{4t}{3} + \dfrac{2}{3}\sin\dfrac{3t}{4} - \dfrac{3}{8}\cos\dfrac{3}{4}t$

(h) $L^{-1}\left\{\dfrac{3(p^2-1)^2}{2p^5}+\dfrac{4p-18}{9-p^2}+\dfrac{(p+1)(2-\sqrt{p})}{p^{5/2}}\right\}$

$$= L^{-1}\left\{\frac{3}{2}\cdot\frac{1}{p}-3\cdot\frac{1}{p^3}+\frac{3}{2p^5}-4\cdot\frac{p}{p^2-9}+\frac{18}{p^2-9}+\frac{2}{p^{3/2}}-\frac{1}{p}+\frac{2}{p^{5/2}}-\frac{1}{p^2}\right\}$$

$$= \frac{1}{2}L^{-1}\left\{\frac{1}{p}\right\}-3L^{-1}\left\{\frac{1}{p^3}\right\}+\frac{3}{2}L^{-1}\left\{\frac{1}{p^5}\right\}-4L^{-1}\left\{\frac{p}{p^2-3^2}\right\}$$

$$+18L^{-1}\left\{\frac{1}{p^2-3^2}\right\}+2L^{-1}\left\{\frac{1}{p^{3/2}}\right\}+2L^{-1}\left\{\frac{1}{p^{5/2}}\right\}-2L^{-1}\left\{\frac{1}{p^2}\right\}$$

$$= \frac{1}{2}-3\left(\frac{t^2}{2!}\right)+\frac{3}{2}\left(\frac{t^4}{4!}\right)-4\cosh 3t+\left(\frac{18}{3}\right)\sinh 3t$$

$$+2\left\{t^{1/2}\,/\,\Gamma\left(\frac{3}{2}\right)\right\}+2\left\{t^{3/2}\,/\,\Gamma\left(\frac{5}{2}\right)\right\}-(t/11)$$

$$= \frac{1}{2}-\frac{3}{2}t^2+\frac{1}{16}t^4-4\cosh 3t+6\sinh 3t+4\sqrt{t/\pi}+\frac{8}{3}t\sqrt{t/\pi}-t$$

(i) $L^{-1}\left\{\dfrac{1}{p}\sin\dfrac{1}{p}\right\}=L^{-1}\left\{\dfrac{1}{p}\left(\dfrac{1}{p}-\dfrac{(1/p)^3}{3!}+\dfrac{(1/p)^5}{5!}-\dfrac{(1/p)^7}{7!}+...\right)\right\}$

$$= L^{-1}\left\{\frac{1}{p^2}\right\}-\frac{1}{3!}L^{-1}\left\{\frac{1}{p^4}\right\}+\frac{1}{5!}L^{-1}\left\{\frac{1}{p^6}\right\}-\frac{1}{7!}L^{-1}\left\{\frac{1}{p^8}\right\}+...$$

$$= t-\frac{t^3}{(3!)^2}+\frac{t^5}{(5!)^2}-\frac{t^7}{(7!)^2}+...$$

EXAMPLE 2.   *Find the inverse Laplace transform of the following functions :*

(a) $\dfrac{1}{p^2-6p+10}$ (b) $\dfrac{p+b}{(p+b)^2+a^2}$

(c) $\dfrac{3p+7}{p^2+2p-3}$ (Meerut–1976, 91, 94; Bundelkhand–2005; Kanpur–1984, 2005;

Avadh–2003; Purvanchal–1989)

(d) $\dfrac{1}{(p+a)^n}$ (e) $\dfrac{p}{(p+1)^5}$ (Meerut–1985; Osmania–2004)

(f) $\dfrac{p^2-2p+3}{(p-1)^2(p+1)}$

SOLUTION.   (a) $L^{-1}\left\{\dfrac{1}{p^2-6p+10}\right\}=L^{-1}\left\{\dfrac{1}{(p-3)^2+1}\right\}=e^{3t}L^{-1}\left\{\dfrac{1}{10^2+1}\right\}=e^{3t}\sin t$

(b) $L^{-1}\left\{\dfrac{p+b}{(p+b)^2+a^2}\right\}=e^{-bt}L^{-1}\left\{\dfrac{p}{p^2+a^2}\right\}=e^{-bt}\cos at$

(c) $L^{-1}\left\{\dfrac{3p+7}{p^2-2p-3}\right\} = L^{-1}\left\{\dfrac{3(p-1)+10}{(p-1)^2-4}\right\}$

$$= L^{-1}\left\{\dfrac{3(p-1)}{(p-1)^2-4}\right\} + L^{-1}\left\{\dfrac{10}{(p-1)^2-4}\right\}$$

$$= 3L^{-1}\left\{\dfrac{p-1}{(p-1)^2-4}\right\} + 10L^{-1}\left\{\dfrac{1}{(p-1)^2-4}\right\}$$

$$= 3e^t L^{-1}\left\{\dfrac{p}{p^2-2^2}\right\} + 10e^t L^{-1}\left\{\dfrac{1}{p^2-2^2}\right\}$$

$$= 3e^t \cosh 2t + 5e^t \sinh 2t = 4e^{3t} - e^t$$

(d) $L^{-1}\left\{\dfrac{1}{(p+a)^n}\right\} = e^{-at} L^{-1}\left\{\dfrac{1}{p^n}\right\} = e^{-at}\dfrac{t^{n-1}}{\Gamma(n)}$

$$= e^{-at}\dfrac{t^{n-1}}{(n-1)!}, \text{ if } n \text{ is a positive integer.}$$

(e) $L^{-1}\left\{\dfrac{p}{(p+1)^5}\right\} = L^{-1}\left\{\dfrac{(P+1)-1}{(p+1)^5}\right\} = e^{-t}L^{-1}\left\{\dfrac{p-1}{p^5}\right\}$

$$= e^{-t}L^{-1}\left\{\dfrac{1}{p^4}\right\} - e^{-t}L^{-1}\left\{\dfrac{1}{p^5}\right\}$$

$$= e^{-t}\left(\dfrac{t^3}{3!}\right) - e^{-t}\left(\dfrac{t^4}{4!}\right) = \dfrac{e^{-t}(4t^3 - t^4)}{24}$$

(f) $L^{-1}\left\{\dfrac{p^2-2p+3}{(p-1)^2(p+1)}\right\} = L^{-1}\left\{\dfrac{(p-1)^2+2}{(p-1)^2+(p-1+2)}\right\} = e^tL^{-1}\left\{\dfrac{p^2+2}{p^2(p+2)}\right\}$

$$= e^t L^{-1}\left\{\dfrac{1}{p^2}\left(1 - \dfrac{1}{2}p + \dfrac{3p^2}{2(p+2)}\right)\right\},$$

dividing $2 + p^2$ by $2 + p$ till p^2 is a common factor in the remainder

$$= e^t L^{-1}\left\{\dfrac{1}{p^2} - \dfrac{1}{2p} + \dfrac{3}{2(p+2)}\right\}$$

$$= e^t\left[t - \dfrac{1}{2}\cdot 1 + \dfrac{3}{2}e^{-2t}\right] = \left(t - \dfrac{1}{2}\right)e^t + \left(\dfrac{3}{2}\right)e^{-t}$$

EXAMPLE 3. If $L^{-1}\left\{\dfrac{e^{-1/p}}{p^{1/2}}\right\} = \dfrac{\cos 2\sqrt{t}}{\sqrt{\pi t}}$, evaluate $L^{-1}\left\{\dfrac{e^{-a/p}}{p^{1/2}}\right\}, a > 0$. (Kanpur–2012)

SOLUTION. Since $L^{-1}\left\{\dfrac{e^{-1/p}}{p^{1/2}}\right\} = \dfrac{\cos 2\sqrt{t}}{\sqrt{\pi t}}$

$$\therefore \quad L^{-1}\left\{\frac{e^{-1/pk}}{(pk)^{1/2}}\right\} = \frac{1}{k}\frac{\cos 2\sqrt{t/k}}{\sqrt{\pi t/k}} \quad \text{or} \quad L^{-1}\left\{\frac{e^{-1/pk}}{p^{1/2}}\right\} = \frac{\cos 2\sqrt{t/k}}{\sqrt{\pi t}}$$

Taking $k = 1/a$, we have

$$L^{-1}\left\{\frac{e^{-a/p}}{p^{1/2}}\right\} = \frac{\cos 2\sqrt{at}}{\sqrt{\pi t}}$$

EXAMPLE 4.   *Find the inverse Laplace transforms of the following functions :*

(a) $\dfrac{e^{-5p}}{(p-2)^4}$ (Agra–1998; Kanpur–2001; Vikram–2004)

(b) $\dfrac{e^{-4p}}{(p-3)^4}$ (Agra–1982, 92)

(c) $\dfrac{e^{-3p}}{p^3}$ (Meerut–1980) (d) $\dfrac{p+8}{p^2+8p+5}$ (Meerut–1991)

SOLUTION.   (a) Since $L^{-1}\left\{\dfrac{1}{(p-2)^4}\right\} = e^{2t}L^{-1}\left\{\dfrac{1}{p^4}\right\} = e^{2t}\left(\dfrac{t^3}{3!}\right) = \dfrac{1}{6}t^3 e^{2t}$

$$\therefore \quad L^{-1}\left\{\frac{e^{-5p}}{(p-2)^4}\right\} = \begin{cases} 1/6(t-5)^3 e^{2(t-5)}, & t>5 \\ 0, & t<5 \end{cases}$$

$$= \frac{1}{6}(t-5)^3 e^{2(t-5)}H(t-5)$$

(in terms of Heaviside's unit step function.)

(b) Since $L^{-1}\left\{\dfrac{1}{(p-3)^4}\right\} = e^{3t}L^{-1}\left\{\dfrac{1}{p^4}\right\} = \dfrac{1}{6}t^3 e^{3t}$

$$\therefore \quad L^{-1}\left\{\frac{e^{-4p}}{(p-3)^4}\right\} = \begin{cases} \dfrac{1}{6}(t-4)^3 e^3(t-4) , & t>4 \\ 0 & , t<4 \end{cases}$$

$$= \frac{1}{6}(t-4)^3 e^{3(t-4)}H(t-4)$$

(c) $$L^{-1}\left\{\frac{1}{p^3}\right\} = \frac{t^2}{2!} = \frac{1}{2}t^2$$

$$L^{-1}\left\{\frac{e^{-3p}}{p^3}\right\} = \frac{1}{2}(t-3)^2 H(t-3)$$

(d) $$L^{-1}\left\{\frac{p+8}{p^2+8p+5}\right\} = L^{-1}\left\{\frac{(p+4)+4}{(p+4)^2-11}\right\} = e^{-4t}L^{-1}\left\{\frac{p+4}{p^2-11}\right\}$$

$$= e^{-4t}\left[L^{-1}\left\{\frac{p}{p^2-(\sqrt{11})^2}\right\} + 4L^{-1}\left\{\frac{1}{p^2-(\sqrt{11})^2}\right\}\right]$$

$$= e^{-4t}[\cosh\sqrt{11}t + 4\sqrt{11}\sinh(\sqrt{11}t)]$$

EXAMPLE 5. *Show that*

(a) $L^{-1}\left\{\dfrac{pe^{-ap}}{p^2-\omega^2}\right\} = \cosh\omega(t-a)H(t-a), a>0$

(b) $L^{-1}\left\{\dfrac{p+2}{p^2-2p+5}\right\} = e^t\left[\cos 2t + \dfrac{3}{2}\sin 2t\right]$

(Meerut–1991; Rohilkhand–1993, 2000; Kanpur–1994)

(c) $L^{-1}\left\{\dfrac{6p^2+22p+18}{p^3+6p^2+11p+6}\right\} = e^{-t} + 2e^{-2t} + 3e^{-3t}$

(d) $L^{-1}\left\{\dfrac{4p+5}{(p-1)^2(p+2)}\right\} = 3te^t + \dfrac{1}{3}e^t - \dfrac{1}{3}e^{-2t}$ *(Meerut–1983, 87; Garhwal–1996;*

Kurukshetra–2005; AMIE–1997; Banglore–2004; Purvanchal–2005)

(e) $L^{-1}\left\{\dfrac{4p+5}{(p-4)^2(p+3)}\right\} = -\dfrac{1}{7}e^{-3t} + \dfrac{1}{7}e^{4t} + 3te^{4t}$ *(Meerut–1988; Bilaspur–2004)*

(f) $L^{-1}\left\{\dfrac{5p^2-15p-11}{(p+1)(p-2)^3}\right\} = -\dfrac{1}{3}e^{-t} + \dfrac{1}{3}e^{2t} + 4te^{2t} - \dfrac{7}{2}t^2e^{2t}$ *(Meerut–1984)*

(g) $L^{-1}\left\{\dfrac{2p+1}{(p+2)^2(p-1)^2}\right\} = \dfrac{1}{3}t(e^t - e^{-2t})$

(Meerut–1981, 91, 2012; Purvanchal–2002; Kanpur–1995)

(h) $L^{-1}\left\{\dfrac{p}{(p^2-2p+2)(p^2+2p+2)}\right\} = \dfrac{1}{2}\sin t\sinh t$ *(Meerut–2007)*

SOLUTION.

(a) Since, $L^{-1}\left\{\dfrac{p}{(p^2-\omega^2)}\right\} = \cosh\omega t$

$\therefore \quad L^{-1}\left\{\dfrac{pe^{-ap}}{(p^2-\omega^2)}\right\} = \cosh\omega(t-a)H(t-a)$

(b) $L^{-1}\left\{\dfrac{p+2}{p^2-2p+5}\right\} = L^{-1}\left\{\dfrac{(p-1)+3}{(p-1)^2+4}\right\} = e^t L^{-1}\left\{\dfrac{p+3}{p^2+4}\right\}$

$= e^t\left[L^{-1}\left\{\dfrac{p}{p^2+2^2}\right\} + 3L^{-1}\left\{\dfrac{1}{p^2+2^2}\right\}\right] = e^t\left[\cos 2t + \left(\dfrac{3}{2}\right)\sin 2t\right]$

(c) $L^{-1}\left\{\dfrac{6p^2+22p+18}{p^3+6p^2+11p+6}\right\} = L^{-1}\left\{\dfrac{6p^2+22p+18}{(p+1)(p+2)(p+3)}\right\}$

$= L^{-1}\left\{\dfrac{1}{p+1} + \dfrac{2}{p+2} + \dfrac{3}{p+3}\right\} = e^{-t} + 2e^{-2t} + 3e^{-3t}$

(d) We have

$L^{-1}\left\{\dfrac{4p+5}{(p-1)^2(p+2)}\right\} = L^{-1}\left\{\dfrac{1}{3(p-1)} + \dfrac{3}{(p-1)^2} - \dfrac{1}{3(p+2)}\right\}$

$$= \frac{1}{3}L^{-1}\left\{\frac{1}{(p-1)}\right\} + 3L^{-1}\left\{\frac{1}{(p-1)^2}\right\} - \frac{1}{3}L^{-1}\left\{\frac{1}{p+2}\right\}$$

$$= \frac{1}{3}e^t + 3e^t L^{-1}\left\{\frac{1}{p^2}\right\} - \frac{1}{3}e^{-2t} = \frac{1}{3}e^t + 3e^t\left(\frac{t}{1!}\right) - \frac{1}{3}e^{-2t}$$

$$= \frac{1}{3}e^t + 3te^t - \frac{1}{3}e^{-2t}$$

(e) $L^{-1}\left\{\dfrac{4p+5}{(p+3)(p-4)^2}\right\} = L^{-1}\left\{\dfrac{-1}{7(p+3)} + \dfrac{1}{7(p-4)} + \dfrac{3}{(p-4)^2}\right\}$

$$= -\frac{1}{7}e^{-3t} + \frac{1}{7}e^{4t} + 3e^{4t}L^{-1}\left\{\frac{1}{p^2}\right\} = -\frac{1}{7}e^{-3t} + \frac{1}{7}e^{4t} + 3te^{4t}$$

(f) We have

$$L^{-1}\left\{\frac{5p^2 - 15p - 11}{(p+1)(p-2)^3}\right\} = L^{-1}\left\{-\frac{1}{3(p+1)} + \frac{1}{3(p-2)} + \frac{4}{(p-2)^2} - \frac{7}{(p-2)^3}\right\}$$

$$= -\frac{1}{3}L^{-1}\left\{\frac{1}{p+1}\right\} + \frac{1}{3}L^{-1}\left\{\frac{1}{p-2}\right\}$$

$$+ 4L^{-1}\left\{\frac{1}{(p-2)^2}\right\} - 7L^{-1}\left\{\frac{1}{(p-2)^3}\right\}$$

$$= -\frac{1}{3}e^{-t} + \frac{1}{3}e^{2t} + 4e^{2t}L^{-1}\left\{\frac{1}{p^2}\right\} - 7e^{2t}L^{-1}\left\{\frac{1}{p^3}\right\}$$

$$= -\frac{1}{3}e^{-t} + \frac{1}{3}e^{2t} + 4e^{2t}\left(\frac{t}{1!}\right) - 7e^{2t}\left(\frac{t^2}{2!}\right)$$

$$= -\frac{1}{3}e^{-t} + \frac{1}{3}e^{2t} + 4te^{2t} - \frac{7}{2}t^2 e^{2t}$$

(g) We have

$$L^{-1}\left\{\frac{2p+1}{(p+2)^2(p-1)^2}\right\} = \frac{1}{3}t(e^t - e^{-2t}) = L^{-1}\left\{\frac{1}{3(p-1)^2} - \frac{1}{3(p+2)^2}\right\}$$

$$= \frac{1}{3}L^{-1}\left\{\frac{1}{(p-1)^2}\right\} - \frac{1}{3}L^{-1}\left\{\frac{1}{(p+2)^2}\right\}$$

$$= \frac{1}{3}e^t L^{-1}\left\{\frac{1}{p^2}\right\} - \frac{1}{3}e^{-2t}L^{-1}\left\{\frac{1}{p^2}\right\}$$

$$= \frac{1}{3}e^t \cdot t - \frac{1}{3}e^{-2t}t = \frac{1}{3}t(e^t - e^{-2t})$$

(h) We have

$$L^{-1}\left\{\frac{p}{(p^2 - 2p + 2)(p^2 + 2p + 2)}\right\}$$

$$= L^{-1}\left\{\frac{1}{4}\frac{(p^2 + 2p + 2) - (p^2 - 2p + 2)}{(p^2 + 2p + 2)(p^2 - 2p + 2)}\right\}$$

$$= L^{-1}\left\{\frac{1}{4(p^2-2p+2)}-\frac{1}{4(p^2+2p+2)}\right\}$$

$$= \frac{1}{4}L^{-1}\left\{\frac{1}{(p-1)^2+1}\right\}-\frac{1}{4}L^{-1}\left\{\frac{1}{(p+1)^2+1}\right\}$$

$$= \frac{1}{4}e^t L^{-1}\left\{\frac{1}{p^2+1}\right\}-\frac{1}{4}e^{-t}L^{-1}\left\{\frac{1}{p^2+1}\right\}$$

$$= \frac{1}{4}e^t\cdot\sin t-\frac{1}{4}e^{-t}\sin t=\frac{1}{4}(e^t-e^{-t})\sin t=\frac{1}{2}\sinh t\sin t$$

EXAMPLE 6. *Show that*

(a) $L^{-1}\left\{\dfrac{p^2}{p^4+4a^4}\right\}=\dfrac{1}{2a}(\cosh at\sin at+\sinh at\cos at)$ (Agra–2009; Meerut–1985)

(b) $L^{-1}\left\{\dfrac{1}{(p^2+4)(p+1)^2}\right\}=\dfrac{1}{25}\left\{e^{-t}(2+5t)-2\cos 2t-\dfrac{3}{2}\sin 2t\right\}$

(Meerut–1991, 2013)

(c) $L^{-1}\left\{\dfrac{3p^3-3p^2-40p+36}{(p^2-4)^2}\right\}=(5t+3)e^{-2t}-2te^{2t}$ (Meerut–1987)

SOLUTION. (a) We have

$$L^{-1}\left\{\frac{p^2}{p^4+4a^4}\right\}=L^{-1}\left\{\frac{p^2}{(p^2+2a^2)^2-4a^2p^2}\right\}$$

$$= L^{-1}\left\{\frac{p^2}{(p^2-2ap+2a^2)(p^2+2ap+2a^2)}\right\}$$

$$= L^{-1}\left\{\frac{1}{4a}\frac{p(p^2+2ap+2a^2)-p(p^2-2ap+2a^2)}{(p^2+2ap+2a^2)(p^2-2ap+2a^2)}\right\}$$

$$= L^{-1}\left\{\frac{p}{4a(p^2-2ap+2a^2)}-\frac{p}{4a(p^2+2ap+2a^2)}\right\}$$

$$= \frac{1}{4a}L^{-1}\left\{\frac{(p-a)+a}{(p-a)^2+a^2}\right\}-\frac{1}{4a}L^{-1}\left\{\frac{(p+a)-a}{(p+a)^2+a^2}\right\}$$

$$= \frac{1}{4a}e^{at}L^{-1}\left\{\frac{p+a}{p^2+a^2}\right\}-\frac{1}{4a}e^{-at}L^{-1}\left\{\frac{p-a}{p^2+a^2}\right\}$$

$$= \frac{1}{4a}e^{at}\left[L^{-1}\left\{\frac{p}{p^2+a^2}\right\}+aL^{-1}\left\{\frac{1}{p^2+a^2}\right\}\right]$$

$$\qquad -\frac{1}{4a}e^{-at}\left[L^{-1}\left\{\frac{p}{p^2+a^2}\right\}-aL^{-1}\left\{\frac{1}{p^2+a^2}\right\}\right]$$

$$= \frac{1}{4a}e^{at}(\cos at+\sin at)-\frac{1}{4a}e^{-at}(\cos at-\sin at)$$

$$= \frac{1}{2a}\left[\left(\frac{e^{at}+e^{-at}}{2}\right)\sin at + \left(\frac{e^{at}-e^{-at}}{2}\right)\cos at\right]$$

$$= \frac{1}{2a}(\cosh at \sin at + \sinh at \cos at)$$

(b) $L^{-1}\left\{\dfrac{1}{(p^2+4)(p+1)^2}\right\} = L^{-1}\left\{\dfrac{2}{25(p+1)} + \dfrac{1}{5(p+1)^2} - \dfrac{2p+3}{25(p^2+4)}\right\}$

$$= \frac{1}{25}\left[2L^{-1}\left\{\frac{1}{p+1}\right\} + 5L^{-1}\left\{\frac{1}{(p+1)^2}\right\}\right.$$

$$\left. - 2L^{-1}\left\{\frac{p}{p^2+4}\right\} - 3L^{-1}\left\{\frac{1}{p^2+4}\right\}\right]$$

$$= \frac{1}{25}\left[2e^{-t} + 5e^{-t}L^{-1}\left\{\frac{1}{p^2}\right\} - 2\cos 2t - \left(\frac{3}{2}\right)\sin 2t\right]$$

$$= \frac{1}{25}\left[e^{-t}(2+5t) - 2\cos 2t - \left(\frac{3}{2}\right)\sin 2t\right]$$

(c) $L^{-1}\left\{\dfrac{3p^3 - 3p^2 - 40p + 36}{(p^2-4)^2}\right\} = L^{-1}\left\{\dfrac{3p^3 - 3p^2 - 40p + 36}{(p-2)^2(p+2)^2}\right\}$

$$= L^{-1}\left\{-\frac{2}{(p-2)^2} + \frac{3}{(p+2)} + \frac{5}{(p+2)^2}\right\}$$

$$= -2L^{-1}\left\{\frac{1}{(p-2)^2}\right\} + L^{-1}\left\{\frac{1}{p+2}\right\} + 5L^{-1}\left\{\frac{1}{(p+2)^2}\right\}$$

$$= -2e^{2t}L^{-1}\left\{\frac{1}{p^2}\right\} + 3e^{-2t} + 5e^{-2t}L^{-1}\left\{\frac{1}{p^2}\right\}$$

$$= -2e^{2t} + 3e^{-2t} + 5e^{-2t}\cdot t = (5t+3)e^{-2t} - 2te^{2t}$$

2.4 INVERSE LAPLACE TRANSFORM OF DERIVATIVES

THEOREM 1.   If $L^{-1}\{f(p)\} = F(t)$, then $L^{-1}\{f^{(n)}(p)\} = (-1)^n.t^n.F(t)$ (Meerut–1991)

PROOF.   Since we know that $L\{t^n F(t)\} = (-1)^n f^{(n)}(p)$

     Therefore, $t^n F(t) = L^{-1}\{(-1)^n f^{(n)}(p)\} = (-1)^n L^{-1}\{f^{(n)}(p)\}$

     Hence, $L^{-1}\{f^{(n)}(p)\} = (-1)^n t^n F(t)$

2.5 DIVISION BY p

THEOREM 1.   If $L^{-1}\{f(p)\} = F(t)$, then $L^{-1}\left\{\dfrac{f(p)}{p}\right\} = \int_0^t F(u)\,du$

     (Meerut–1990; Rohilkhand–1997, 98; Kanpur–1997)

PROOF.   Since we know that $\dfrac{f(p)}{p} = L\left\{\displaystyle\int_0^t F(u)\,du\right\} \Rightarrow L^{-1}\left\{\dfrac{f(p)}{p}\right\} = \int_0^t F(u)\,du$

2.6 MULTIPLICATION BY POWERS OF p

THEOREM 1. *If $L^{-1}\{f(p)\} = F(t)$ and $F(0) = 0$, then $L^{-1}\{pf(p)\} = F'(t)$*

PROOF. We know that

$$L\{F'(t)\} = pL\{F(t)\} - F(0) = pL[F(t)] = p f(p) \qquad [\because f(0) = 0].$$

Hence, $L^{-1}\{p f(p)\} = F'(t)$

2.7 INVERSE LAPLACE TRANSFORM OF INTEGRALS

THEOREM 1. *If $L^{-1}\{f(p)\} = F(t)$, then $L^{-1}\left[\int_p^\infty f(x)\, dx\right] = \dfrac{F(t)}{t}$.*

PROOF. We know that

$$L\left\{\frac{1}{t} F(t)\right\} = \int_p^\infty f(x)\, dx \text{ provided } \lim_{t \to 0}\left\{\frac{F(t)}{t}\right\} \text{ exists.}$$

Hence, $L^{-1}\left\{\int_p^\infty f(x)\, dx\right\} = \dfrac{F(t)}{t}$.

S. No.	Operation	$f(p)$	$L^{-1}\{f(p)\} = F(t)$
1.	Differentiation theorem	$f^n(p)$	$(-1)^n t^n F(t)$
2.	Integral theorem	$\int_p^\infty f(x)dx$	$\dfrac{1}{t} F(t)$
3	Multiplication theorem	$pf(p)$ $p^n f(p)$	$F'(t)$ $F^n(t)$
4.	Division theorems	$\dfrac{f(p)}{p}$ $\dfrac{f(p)}{p^n}$	$\int_0^t F(x)dx$ $\int_0^t \int_0^t \cdots \int_0^t F(t)(dt)^n$

RECAPITULATIONS

Solved Examples

EXAMPLE 1. *Find $L^{-1}\left\{\dfrac{p}{(p^2 + a^2)^2}\right\}$.* (Agra– 2005, 12; SVTU–2009; VTU–2010; Meerut–1986, 91, 2009;

Rohilkhand–1995; Kashi–2010; Lucknow–2014; Ravishankar–2005; Purvanchal–2010; UPTU–2008)

SOLUTION. We have

$$L^{-1}\left\{\frac{p}{(p^2 + a^2)^2}\right\} = L^{-1}\left\{-\frac{1}{2}\frac{d}{dp}\left(\frac{1}{p^2 + a^2}\right)\right\} = -\frac{1}{2}L^{-1}\left\{\frac{d}{dp}\left(\frac{1}{p^2 + a^2}\right)\right\}$$

$$= -\frac{1}{2}t(-1)L^{-1}\left\{\frac{1}{p^2 + a^2}\right\} = \frac{t}{2a}\sin at$$

EXAMPLE 2. *Evaluate $L^{-1}\left\{\log\left(1 - \dfrac{1}{p^2}\right)\right\}$.* (Meerut–2000, 05)

SOLUTION.   Let us suppose $f(p) = \log\left(1 - \dfrac{1}{p^2}\right) = \log\left(\dfrac{p^2-1}{p^2}\right) = -2\log p + \log(p^2-1)$

$\Rightarrow \qquad\qquad f'(p) = -2\left(\dfrac{1}{p} - \dfrac{p}{p^2-1}\right)$

$\Rightarrow \qquad\qquad L^{-1}\{f'(p)\} = -2(1 - \cosh t) \qquad \Rightarrow \qquad -tL^{-1}\{f(p)\} = -2(1 - \cosh t)$

$\Rightarrow \quad L^{-1}\left\{\log\left(1 - \dfrac{1}{p^2}\right)\right\} = \dfrac{2}{t}(1 - \cosh t).$

EXAMPLE 3.   *Find the function whose Laplace transform is* $\log\left(1 + \dfrac{1}{p}\right)$ (UPTU–2007)

SOLUTION. $L^{-1}\left[\log\left(1 + \dfrac{1}{p}\right)\right] = -\dfrac{1}{t}L^{-1}\left[\dfrac{d}{dp}\log\left(\dfrac{p+1}{p}\right)\right] = -\dfrac{1}{t}L^{-1}\left[\left(\dfrac{p}{p+1}\right)\left(-\dfrac{1}{p^2}\right)\right]$

$\qquad\qquad = -\dfrac{1}{t}L^{-1}\left[-\dfrac{1}{p(p+1)}\right] = -\dfrac{1}{t}L^{-1}\left[\dfrac{1}{p+1} - \dfrac{1}{p}\right]$

$\qquad\qquad = -\dfrac{1}{t}[e^{-t} - 1] = \dfrac{1}{t}[1 - e^{-t}].$

EXAMPLE 4.   *Evaluate* $L^{-1}\left\{\dfrac{p+2}{p^2(p+3)}\right\}$

SOLUTION.   Consider $L^{-1}\left\{\dfrac{p+2}{p^2(p+3)}\right\} = L^{-1}\left\{\dfrac{(p+3)-1}{p^2(p+3)}\right\} = L^{-1}\left\{\dfrac{1}{p^2} - \dfrac{1}{p^2(p+3)}\right\}$

$\qquad\qquad = L^{-1}\left\{\dfrac{1}{p^2}\right\} - L^{-1}\left\{\dfrac{1}{p^2(p+3)}\right\}$ ...(1)

Since, $\qquad L^{-1}\left\{\dfrac{1}{p^2}\right\} = t$ and $L^{-1}\left\{\dfrac{1}{p+3}\right\} = e^{-3t} = F(t)$ (say)

Therefore, we have

$\qquad\qquad L^{-1}\left\{\dfrac{1}{p(p+3)}\right\} = \int_0^t F(x)\,dx$ ...(1)

$\qquad\qquad = \int_0^t e^{-3x}\,dx = \dfrac{1}{3}(1 - e^{-3t}) = F_1(t)$ (say)

Hence, $\qquad L^{-1}\left\{\dfrac{1}{p^2(p+3)}\right\} = \int_0^t F_1(x)\,dx$

$\qquad\qquad = \dfrac{1}{3}\int_0^t (1 - e^{-3x})\,dx = \dfrac{1}{3}t + \dfrac{1}{9}(e^{-3t} - 1)$

Finally, from (1), we have

$\qquad\qquad L^{-1}\left\{\dfrac{p+2}{p^2(p+3)}\right\} = t - \dfrac{t}{3} - \dfrac{1}{9}(e^{-3t} - 1) = \dfrac{2}{3}t - \dfrac{1}{9}e^{-3t} + \dfrac{1}{9}$

EXAMPLE 5. *Find the inverse Laplace transform of* $f(p) = \log \dfrac{p+a}{p+b}$ (UPTU–2001, 03; Anna–2003)

SOLUTION. We have $L^{-1} \log \left[\dfrac{p+a}{p+b} \right] = -\dfrac{1}{t} L^{-1} \left[\dfrac{d}{dp} \log \dfrac{p+a}{p+b} \right]$

$$= -\dfrac{1}{t} L^{-1} \left[\dfrac{d}{dp} \log(p+a) - \dfrac{d}{dp} \log(p+b) \right]$$

$$= -\dfrac{1}{t} L^{-1} \left[\dfrac{1}{p+a} - \dfrac{1}{p+b} \right]$$

$$= -\dfrac{1}{t} \left[e^{-at} - e^{-bt} \right] = \dfrac{1}{t} \left[e^{-bt} - e^{-at} \right].$$

EXAMPLE 6. *Evaluate* $L^{-1} \left\{ \dfrac{1}{p^4(p^2+1)} \right\}$ (Meerut–1990, 99; Agra–1999, 2001; Rohilkhand–1998)

SOLUTION. Since we know that $L^{-1} \left\{ \dfrac{1}{p^2+1} \right\} = \sin t$

Therefore

$$L^{-1} \left\{ \dfrac{1}{p(p^2+1)} \right\} = \int_0^t \sin x \, dx = 1 - \cos t$$

Also, $L^{-1} \left\{ \dfrac{1}{p^2(p^2+1)} \right\} = \int_0^t (1 - \cos x) dx = t - \sin t$

and $L^{-1} \left\{ \dfrac{1}{p^3(p^2+1)} \right\} = \int_0^t (x - \sin x) dx = \dfrac{t^2}{2} + \cos t - 1$

Hence, $L^{-1} \left\{ \dfrac{1}{p^4(p^2+1)} \right\} = \int_0^t \left\{ \dfrac{1}{2} x^2 + \cos x - 1 \right\} dx = \dfrac{1}{6} t^3 + \sin t - t$

EXAMPLE 7. *Evaluate* $L^{-1} \left\{ \dfrac{1}{p(p+1)^3} \right\}$ (Agra–2008; Rohilkhand–1991)

SOLUTION. We know that

$$L^{-1} \left\{ \dfrac{1}{p(p+1)^3} \right\} = L^{-1} \left\{ \dfrac{1}{(p+1-1)(p+1)^3} \right\} = e^{-t} L^{-1} \left\{ \dfrac{1}{(p-1)p^3} \right\} \qquad ...(1)$$

Also, since $L^{-1} \left\{ \dfrac{1}{p-1} \right\} = e^t$

Therefore, $L^{-1} \left\{ \dfrac{1}{p(p-1)} \right\} = \int_0^t e^x dx = (e^t - 1)$

and $L^{-1} \left\{ \dfrac{1}{p^2(p-1)} \right\} = \int_0^t (e^x - 1) dx = e^t - t - 1$

and $L^{-1} \left\{ \dfrac{1}{p^3(p-1)} \right\} = \int_0^t (e^x - x - 1) dx = e^t - \dfrac{1}{2} t^2 - t - 1$

Therefore, from (1), we get

$$L^{-1}\left\{\frac{1}{p(p+1)^3}\right\} = e^{-t}\left\{e^t - \frac{1}{2}t^2 - 1 - t\right\} = 1 - e^{-t}\left(1 + t + \frac{t^2}{2}\right)$$

EXAMPLE 8.   *Obtain the inverse Laplace transformation of* $\cot^{-1}\left(\dfrac{p+3}{2}\right)$ (UPTU–2001 (Sp.), 2002)

SOLUTION.   We know that $L^{-1}[f(p)] = -\dfrac{1}{t}L^{-1}\left[\dfrac{d}{dp}f(p)\right]$

$$\therefore \qquad \left[\cot^{-1}\left(\frac{p+3}{2}\right)\right] = -\frac{1}{t}L^{-1}\left[\frac{d}{dp}\cot^{-1}\left(\frac{p+3}{2}\right)\right]$$

$$= -\frac{1}{t}L^{-1}\left[\frac{-\dfrac{1}{2}}{1+\left(\dfrac{p+3}{2}\right)^2}\right] = \frac{1}{2t}L^{-1}\left[\frac{4}{4+(p+3)^2}\right]$$

$$= \frac{1}{t}L^{-1}\left[\frac{2}{2^2+(p+3)^2}\right] = \frac{1}{t}e^{-3t}L^{-1}\left[\frac{2}{2^2+p^2}\right]$$

$$= \frac{e^{-3t}}{t}\sin 2t\ .$$

EXAMPLE 9.   *Evaluate*

(i) $L^{-1}\left\{\log\left(1 + \dfrac{1}{p^2}\right)\right\}$ (UPTU Q.Bank–2001; Meerut–1987, 96; Agra–1983, 97, 2001, 04)

(ii) $L^{-1}\left\{\dfrac{1}{p}\log\left(1 + \dfrac{1}{p^2}\right)\right\}$ (Garhwal–1996; Agra-1983, Meerut-1987)

SOLUTION.   (i) Let $\qquad f(p) = \log\left(1 + \dfrac{1}{p^2}\right) = -\log\left(\dfrac{p^2}{p^2+1}\right)$

$$= -2\log p + \log(p^2 + 1)$$

Therefore, $\quad f'(p) = -\dfrac{2}{p} + \dfrac{2p}{p^2+1}$

$\Rightarrow \qquad L^{-1}\{f'(p)\} = -2 + 2\cos t \Rightarrow -tL^{-1}\{f(p)\} = -2(1 - \cos t)$

Hence, $L^{-1}\left\{\log\left(1 + \dfrac{1}{p^2}\right)\right\} = \dfrac{2(1 - \cos t)}{t}$

(ii) Since $\quad L^{-1}\left\{\log\left(1 + \dfrac{1}{p^2}\right)\right\} = \dfrac{2(1 - \cos t)}{t}$

Therefore,

$$L^{-1}\left\{\frac{1}{p}\log\left(1 + \frac{1}{p^2}\right)\right\} = L^{-1}\left\{\frac{1}{p}f(p)\right\} = \int_0^t F(x)dx$$

$$= \int_0^t \frac{2}{x}(1 - \cos x)\ dx\ .$$

EXAMPLE 10. *Without expressing $(p-1)/p^2(p^2+1)$ into partial fractions, find its inverse Laplace transform.*
(Lucknow–1998)

SOLUTION. Here, we have

$$L^{-1}\left\{\frac{p-1}{p^2+1}\right\} = L^{-1}\left\{\frac{p}{p^2+1}\right\} - L^{-1}\left\{\frac{1}{p^2+1}\right\} = \cos t - \sin t$$

$$\Rightarrow L^{-1}\left\{\frac{1}{p}\cdot\frac{p-1}{p^2+1}\right\} = \int_0^t(\cos x - \sin x)dx = [\sin x + \cos x]_0^t$$

$$= \sin t + \cos t - 1$$

$$\Rightarrow L^{-1}\left\{\frac{1}{p}\cdot\frac{p-1}{p^2+1}\right\} = \int_0^t(\sin x + \cos x - 1)dx = [-\cos x + \sin x - x]_0^t$$

$$= -\cos t + \sin t - t - (-1) = 1 - t + \sin t - \cos t$$

Additional Solved Examples

EXAMPLE 1. *Evaluate the following inverse Laplace transforms :*

(a) $L^{-1}\left\{\dfrac{p}{(p^2-a^2)^2}\right\}$ (b) $L^{-1}\left\{\dfrac{p}{(p^2-16)^2}\right\}$

(c) $L^{-1}\left\{\dfrac{1}{(p-a)^3}\right\}$ (Meerut–1990)

(d) $L^{-1}\left\{\dfrac{p+1}{(p^2+2p+2)^2}\right\}$ (Agra–1983; Meerut–1998)

(e) $L^{-1}\left\{\dfrac{p^2}{(p^2+4)^2}\right\}$ (Meerut–1981; Agra–2003; UPTU–2001)

SOLUTION. (a) Since

$$\frac{d}{dp}(p^2-a^2)^{-1} = -\frac{2p}{(p^2-a^2)^2} \Rightarrow \frac{p}{(p^2-a^2)^2} = -\frac{1}{2}\frac{d}{dp}\left(\frac{1}{p^2-a^2}\right)$$

Hence

$$L^{-1}\left\{\frac{p}{(p^2-a^2)^2}\right\} = -\frac{1}{2}L^{-1}\left\{\frac{d}{dp}\left(\frac{1}{p^2-a^2}\right)\right\} = -\frac{1}{2}(-1)^1 tL^{-1}\left\{\frac{1}{p^2-a^2}\right\}$$

$$= \frac{1}{2}t\frac{1}{a}\sinh at = \frac{t}{2a}\sinh at$$

(b) Proceed as in part (a). Here $a = 4$.

$$L^{-1}\left\{\frac{p}{(p^2-16)^2}\right\} = \frac{t}{8}\sinh 4t$$

(c) Since $\dfrac{d^2}{dp^2}\left(\dfrac{1}{p-a}\right) = \dfrac{2}{(p-a)^3}$ $\therefore$ $\dfrac{1}{(p-a)^3} = \dfrac{1}{2}\dfrac{d^2}{dp^2}\left(\dfrac{1}{p-a}\right)$

Hence $L^{-1}\left\{\dfrac{1}{(p-a)^3}\right\} = L^{-1}\left\{\dfrac{1}{2}\dfrac{d^2}{dp^2}\left(\dfrac{1}{p-a}\right)\right\} = \dfrac{1}{2}L^{-1}\left\{\dfrac{d^2}{dp^2}\left(\dfrac{1}{p-a}\right)\right\}$

$$= \dfrac{1}{2}(-1)^2 t^2 L^{-1}\left\{\dfrac{1}{p-a}\right\} = \dfrac{1}{2}t^2 e^{at}$$

(d) $L^{-1}\left\{\dfrac{p+1}{(p^2+2p+2)^2}\right\} = L^{-1}\left\{\dfrac{(p+1)}{[(p+1)^2+1]^2}\right\} = e^{-t}L^{-1}\left\{\dfrac{p}{(p^2+1)^2}\right\}$

$$= e^{-t}\dfrac{1}{2}t\sin t = \dfrac{1}{2}te^{-t}\sin t$$

(e) Let $\qquad f(p) = \dfrac{p}{(p^2+4)^2}$

$\therefore \qquad L^{-1}\{f(p)\} = L^{-1}\left\{\dfrac{p}{(p^2+4)^2}\right\} = \dfrac{t}{4}\sin 2t = F(t)$

and $\qquad F(0) = 0$

$\therefore \qquad L^{-1}[pF(p)] = F'(t)$

or $\quad L^{-1}\left\{\dfrac{p^2}{(p^2+4)^2}\right\} = \dfrac{d}{dt}\left(\dfrac{1}{4}t\sin 2t\right) = \dfrac{1}{4}(\sin 2t + 2t\cos 2t)$

EXAMPLE 2. *Show that :*

(a) $L^{-1}\left\{\dfrac{1}{p^3(p+1)}\right\} = 1 - t + \dfrac{t^2}{2} - e^{-t}$

(b) $L^{-1}\left\{\dfrac{1}{p^3(p^2+1)}\right\} = \dfrac{t^2}{2} + \cos t - 1$ $\qquad$ (Agra–1990; Meerut–1987, 92, 2001;

Rohilkhand–1988; UPTU–2004; SVTU–2005; GBTU–2012)

(c) $L^{-1}\left\{\log\dfrac{p+2}{p+1}\right\} = \dfrac{1}{t}(e^{-t} - e^{-2t})$

(d) $L^{-1}\left\{\dfrac{1}{p}\log\dfrac{p+2}{p+1}\right\} = \int_0^t \dfrac{1}{x}(e^{-x} - e^{-2x})dx$ $\qquad$ (SVTU–2004)

(e) $L^{-1}\left\{\dfrac{1}{p}\log\dfrac{p+3}{p+2}\right\} = \int_0^t \dfrac{1}{x}(e^{-2x} - e^{-3x})dx$

SOLUTION. $\quad$ (a) We have, $L^{-1}\left\{\dfrac{1}{p+1}\right\} = e^{-t} = F(t)$ say.

Now, we have

$$L^{-1}\left\{\dfrac{1}{p(p+1)}\right\} = \int_0^t F(x)dx = \int_0^t e^{-x}dx = 1 - e^{-t}$$

$\therefore \qquad L^{-1}\left\{\dfrac{1}{p^2(p+1)}\right\} = \int_0^t (1-e^{-x})dx = t + e^{-t} - 1$

and $\quad L^{-1}\left\{\dfrac{1}{p^3(p+1)}\right\} = \int_0^t (x + e^{-x} - 1)dx = 1 - t + \dfrac{1}{2}t^2 - e^{-t}$

(b) Since $L^{-1}\left\{\dfrac{1}{p^2+1}\right\} = \sin t$

Now, we have

$$L^{-1}\left\{\dfrac{1}{p(p^2+1)}\right\} = \int_0^t \sin x \, dx = 1 - \cos t$$

$$L^{-1}\left\{\dfrac{1}{p^2(p^2+1)}\right\} = \int_0^t (1-\cos t) dt = t - \sin t$$

and $L^{-1}\left\{\dfrac{1}{p^3(p^2+1)}\right\} = \int_0^t (x - \sin x) dx = \dfrac{t^2}{2} + \cos t - 1$

(c) Let $f(p) = \log\dfrac{p+2}{p+1} = \log(p+2) - \log(p+1)$

$\therefore$ $f'(p) = \dfrac{1}{p+2} - \dfrac{1}{p+1}$

$\therefore$ $L^{-1}\{f'(p)\} = e^{-2t} - e^{-t}$

or $L^{-1}\{f'(p)\} = -tL^{-1}\{f(p)\} = e^{-2t} - e^{-t}$

$$L^{-1}\{f(p)\} = L^{-1}\left\{\log\dfrac{p+2}{p+1}\right\} = \dfrac{1}{t}(e^{-t} - e^{-2t})$$

(d) From part (c)

$$L^{-1}\left\{\log\dfrac{p+2}{p+1}\right\} = \dfrac{1}{t}(e^{-t} - e^{-2t}) = F(t)$$

Now, we have

$$L^{-1}\left\{\dfrac{1}{p}\log\dfrac{p+2}{p+1}\right\} = \int_0^t F(x) dx = \int_0^t \dfrac{1}{x}(e^{-x} - e^{-2x}) dx$$

(e) Let $f(p) = \log\dfrac{p+3}{p+2} = \log(p+3) - \log(p+2)$

$\therefore$ $f'(p) = \dfrac{1}{p+3} - \dfrac{1}{p+2}$

$\therefore$ $L^{-1}\{f'(p)\} = e^{-3t} - e^{-2t}$

or $-tL^{-1}\{f(p)\} = e^{-3t} - e^{-2t}$

or $L^{-1}\{f(p)\} = L^{-1}\left\{\log\dfrac{p+3}{p+2}\right\} = \dfrac{1}{t}(e^{-2t} - e^{-3t}) = F(t)$, say

$\therefore$ $L^{-1}\left\{\dfrac{1}{p}\log\dfrac{p+3}{p+2}\right\} = \int_0^t F(x) dx = \int_0^t \dfrac{1}{x}(e^{-2x} - e^{-3x}) dx$

EXAMPLE 3. If $L^{-1}\left\{\dfrac{p}{(p^2+1)^2}\right\} = \dfrac{1}{2}t\sin t$, then show that $L^{-1}\left\{\dfrac{1}{(p^2+1)^2}\right\} = \dfrac{1}{2}(\sin t - t\cos t)$.

SOLUTION.   Given that, $L^{-1}\left\{\dfrac{p}{(p^2+1)^2}\right\} = \dfrac{1}{2}t\sin t = F(t)$, (say)

$$\therefore \qquad L^{-1}\left\{\dfrac{1}{(p^2+1)^2}\right\} = L^{-1}\left\{\dfrac{1}{p}\cdot\dfrac{p}{(p^2+1)^2}\right\} = \int_0^t F(x)\,dx$$

$$= \dfrac{1}{2}\int_0^t x\cdot\sin x\,dx = \dfrac{1}{2}(\sin t - t\cos t)$$

2.8 CONVOLUTION

If $L^{-1}\{f(p)\} = F(t)$ and $L^{-1}\{g(p)\} = G(t)$, where F(t) and G(t) are two functions of class A, then

$$L^{-1}\{f(p).g(p)\} = \int_0^t F(u)\,G(t-u)\,du = F*G$$

*We call F * G the convolution or falting of F and G.*

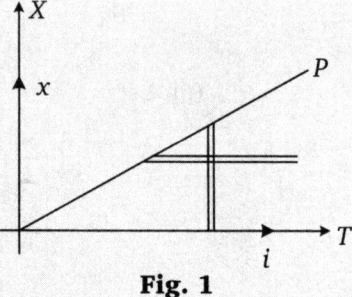

Fig. 1

(UPTU–2002; Agra–1980;
Meerut–2005, 06, 07, 08,11;Rohilkhand–2003;
Guwahati–2000; Nagpur–2005; Purvanchal–2002; MDU–2007)

Proof. $\int_0^t F(x)\,G(t-x)\,dx = H(t)$

Then, $L\{H(t)\} = \int_0^\infty e^{-pt}H(t)\,dt$

$$= \int_0^\infty e^{-pt}\left[\int_0^t F(x)G(t-x)dx\right]dt$$

$$= \int_0^\infty\left[\int_0^t e^{-pt}F(x)G(t-x)dx\right]dt \qquad \ldots(1)$$

The integration being first with respect to x and then t.

The integration (1) is within the region lying below the line *OP* whose equation is $x = t$ and above *OT*, t being taken along *OT* and x along *OX*, with *O* is the origin the axes being perpendicular to each other. If the order of integration is changed, the strip will be taken parallel to *OT*, so that the limits of t are from x to ∞ and of x from 0 to ∞.

Therefore, $L\{H(t)\} = \int_0^\infty dx \int_x^\infty e^{-pt}\,F(x)\,G(t-x)\,dt$

$$= \int_0^\infty e^{-px}F(x)\,dx \int_x^\infty e^{-p(t-x)}\,G(t-x)\,dt$$

Putting $t - x = \theta \Rightarrow dt = d\theta$

$\Rightarrow$ $L\{H(t)\} = \int_0^\infty e^{-px}F(x)\left\{\int_0^\infty e^{-p\theta}G(\theta)\,d\theta\right\}dx$

$$= \int_0^\infty e^{-px}F(x)\,g(p)\,dx = f(p)\,g(p)$$

$\Rightarrow$ $L\left\{\int_0^t F(x)\,G(t-x)dx\right\} = f(p)\,g(p)$

$\Rightarrow$ $\int_0^t F(x)\,G(t-x)\,dx = L^{-1}\{f(p)\,g(p)\} = F*G.$

REMARKS

- $F*G$ is commutative, *i.e.*, $F*G = G*F.$
- $F*G$ is associative.
- $F.*G$ is distributive over addition.

2.9 THE HEAVISIDE EXPANSION FORMULA

THEOREM. *If $F(P)$ and $G(P)$ are polynomials in P, the degree of $F(P)$ being less than that of $G(P)$ and if $G(p) = (p - \alpha_1)(p - \alpha_2)...(p - \alpha_n)$*

where, $\alpha_1, \alpha_2, ..., \alpha_n$ are distinct constants, real or complex, then

$$L^{-1}\left\{\frac{F(p)}{G(p)}\right\} = \sum_{r=1}^{n} \frac{F(\alpha_r)}{G'(\alpha_r)} e^{\alpha_r . t}$$

(Meerut–2006 B.P., 12)

PROOF. By the method of partial fractions, let

$$\frac{f(p)}{G(p)} = \frac{A_1}{p - \alpha_1} + \frac{A_2}{p - \alpha_2} + ... + \frac{A_r}{p - \alpha_r} + ... + \frac{A_n}{p - \alpha_n}$$

Multiplying both sides by $(p - \alpha_r)$ and taking the limit $(p - \alpha_r)$, we get

$$A_r = \lim_{p \to \alpha_r} \frac{F(p)(p - \alpha_r)}{G(p)} = \lim_{p \to \alpha_r} F(p) . \lim_{p \to \alpha_r} \frac{p - \alpha_r}{G(p)}$$

$$= \lim_{p \to \alpha_r} F(p) . \lim_{p \to \alpha_r} \frac{1}{G'(p)} = \frac{F(\alpha_r)}{G'(\alpha_r)} \qquad \text{[By L'Hospital's rule]}$$

Therefore,

$$\frac{F(p)}{G(p)} = \frac{F(\alpha_1)}{G'(\alpha_1)} . \frac{1}{p - \alpha_1} + \frac{F(\alpha_2)}{G'(\alpha_2)} . \frac{1}{p - \alpha_2}$$

$$... + \frac{F(\alpha_r)}{G'(\alpha_r)} . \frac{1}{p - \alpha_r} + ... + \frac{F(\alpha_n)}{G'(\alpha_n)} . \frac{1}{p - \alpha_n}$$

Hence,

$$L^{-1}\left[\frac{F(p)}{G(p)}\right] = \frac{F(\alpha_1)}{G'(\alpha_1)} . e^{\alpha_1 t} + \frac{F(\alpha_2)}{G'(\alpha_2)} e^{\alpha_2 t} +$$

$$... + \frac{F(\alpha_r)}{G'(\alpha_r)} e^{\alpha_r t} + ... + \frac{F(\alpha_n)}{G'(\alpha_n)} e^{\alpha_n t}$$

$$= \sum_{r=1}^{n} \frac{F(\alpha_r)}{G'(\alpha_r)} . e^{\alpha_r . t}$$

✍	RECAPITULATIONS		
S. No.	**Operation**	**$f(p)$**	**$L^{-1}\{f(p)\} = F(t)$**
1.	Convolution theorem	$f(p) \cdot g(p)$	$F * G = \int_0^t F(x) \cdot G(t - x)dx$
2.	Heaviside's expansion theorem	$\dfrac{F(p)}{G(p)}$ degree $F(p)$ < degree $G(p)$	$\sum_{r=1}^{n} \dfrac{F(\alpha_r)}{G'(\alpha_r)} e^{\alpha_r t}$ where $\alpha_r = 1, 2,..., n$ are roots of $G(p) = 0$ and are all distinct

Solved Examples

EXAMPLE 1. *Using convolution theorem, evalute*

$$L^{-1}\left\{\frac{1}{(p-1)(p+2)}\right\}$$ (Agra–1982; Meerut–1987, 92, 2004; Bilaspur–2004; Kanpur–2004)

SOLUTION. We have $L^{-1}\left\{\dfrac{1}{p-1}\right\} = e^t = F_1(t)$ (say)

and $\qquad L^{-1}\left\{\dfrac{1}{p+2}\right\} = e^{-2t} = F_2(t)$ (say)

Using convolution theorem, we have

$$L^{-1}\left\{\frac{1}{p-1}\cdot\frac{1}{p+2}\right\} = F_1 * F_2 = \int_0^t F_1(x)\,F_2(t-x)\,dx$$

$$= \int_0^t e^x e^{-2(t-x)}dx = e^{-2t}\int_0^t e^{3x}dx = \frac{1}{3}(e^t - e^{-2t})$$

EXAMPLE 2. *Use the convolution theorem to find*

$$L^{-1}\left\{\frac{p^2}{(p^2+a^2)^2}\right\}$$

(Agra–2005; Meerut–1996, 98, 2006; Hazaribag–2009; Purvanchal–1990; MDU–2004, 07)

SOLUTION. We know that $L^{-1}\left\{\dfrac{p}{(p^2+a^2)}\right\} = \cos at$

Therefore, by convolution theorem, we have

$$L^{-1}\left\{\frac{p^2}{(p^2+a^2)^2}\right\} = L^{-1}\left\{\frac{p}{p^2+a^2}\cdot\frac{p}{p^2+a^2}\right\}$$

$$= \int_0^t \cos ax\,\cos a(t-x)\,dx$$

$$= \int_0^t \cos ax\,(\cos at\,\cos ax + \sin at\,\sin ax)\,dx$$

$$= \cos at\int_0^t \cos^2 ax\,dx + \sin at\int_0^t \cos ax\,\sin ax\,dx$$

$$= \frac{1}{2}\cos at\int_0^t (1+\cos 2ax)\,dx + \frac{1}{2}\sin at\int_0^t \sin 2ax\,dx$$

$$= \frac{1}{2}\cos at\left[x + \frac{1}{2a}\sin 2ax\right]_0^t + \frac{1}{2}\sin at\left[-\frac{1}{2a}\cos 2ax\right]_0^t$$

$$= \frac{1}{2}\cos at\left[t + \frac{1}{2a}\sin 2a\right] + \frac{1}{4a}\sin at(1 - \cos 2at)$$

$$= \frac{1}{2}t\cos at + \frac{1}{4a}\sin at + \frac{1}{4a}(\sin 2at\,\cos at - \sin at\,\cos 2at)$$

$$= \frac{1}{2}t\cos at + \frac{1}{4a}[(\sin at + \sin(2at - at)]$$

$$= \frac{1}{2a}[at\cos at + \sin at]$$

EXAMPLE 3. Evaluate $L^{-1}\left[\dfrac{p}{(p^2+1)(p^2+4)}\right]$ (UPTU–2001, 02)

SOLUTION. We know that

$$L^{-1}\left[\frac{p}{p^2+1}\right] = \cos x \text{ and } L^{-1}\left[\frac{2}{p^2+2^2}\right] = \sin 2x$$

Therefore,

$$L^{-1}\left[\frac{p}{(p^2+1)(p^2+4)}\right] = \frac{1}{2}L^{-1}\left[\left(\frac{p}{p^2+1}\right)\cdot\left(\frac{2}{p^2+4}\right)\right]$$

$$= \frac{1}{2}\int_0^t \sin 2x \cos(t-x)dx$$

$$= \int_0^t \sin x \cos x \{\cos t \cos x + \sin t \sin x\} \, dx$$

$$= \int_0^t (\sin x \cos^2 x \cos t + \sin^2 x \cos x \sin t) \, dx$$

$$= \left[-\frac{\cos^3 x}{3}\cos t + \frac{\sin^3 x}{3}\sin x\right]_0^t$$

$$= -\frac{\cos^4 t}{3} + \frac{\sin^4 t}{3} + \frac{\cos t}{3} = \frac{1}{3}\left[\sin^4 t - \cos^4 t\right] + \frac{\cos t}{3}$$

$$= \frac{1}{3}(\sin^2 t + \cos^2 t)(\sin^2 t - \cos^2 t) + \frac{\cos t}{3}$$

$$= \frac{1}{3}(\sin^2 t - \cos^2 t) + \frac{\cos t}{3} = -\frac{1}{3}\cos 2t + \frac{\cos t}{3}$$

$$= \frac{1}{3}(\cos t - \cos 2t)$$

EXAMPLE 4. Using convolution theorem, prove that

$$L^{-1}\left[\frac{1}{p^3(p^2+1)}\right] = \frac{t^2}{2} + \cos t - 1$$ (UPTU–2005; VTU–2007; GBTU–2012)

SOLUTION. We have to evaluate $L^{-1}\left[\dfrac{1}{p^3(p^2+1)}\right]$ by convolution theorem.

Now, $L^{-1}\left\{\dfrac{1}{p^3}\right\} = \dfrac{t^2}{2!}$

and $L^{-1}\left\{\dfrac{1}{p^2+1}\right\} = \sin t$

Using convolution theorem, we get

$$L^{-1}\left[\frac{1}{p^3(p^2+1)}\right] = \int_0^t \frac{(t-x)^2}{2!}\sin x \, dx = \frac{1}{2}\int_0^t (t^2 + x^2 - 2tx)\sin x \, dx$$

$$= \frac{1}{2}\left[(t^2 + x^2 - 2tx)(-\cos x) - \int (2x - 2t)(-\cos x)dx\right]$$

$$= \frac{1}{2}\left[-\cos x(t^2 + x^2 - 2tx) + 2\int (x - t)\cos x \, dx\right]_0^t$$

$$= \frac{1}{2}\left[-\cos x(t^2 + x^2 - 2tx) \quad +2(x-t)\sin x + 2\cos x\right]_0^t$$

$$= \frac{1}{2}\left[-\cos x(t^2 + x^2 - 2tx) \quad +2(x-t)\sin x + 2\cos x\right]_0^t$$

$$= \frac{1}{2}\left[-\cos t(t^2 + t^2 - 2t^2) \quad +0 + 2\cos t + t^2\cos 0 - 2\cos 0\right]_0^t$$

$$= \cos t + \frac{t^2}{2} - 1 = \frac{t^2}{2} + \cos t - 1.$$

EXAMPLE 5. *Using the convolution theorem, find*

$$L^{-1}\left[\frac{p^2}{(p^2+a^2)(p^2+b^2)}\right], \quad a \neq b \qquad \text{(UPTU–2004; Mumbai–2007; Bhopal–2008;}$$

UKTU–2011; VTU–2011S)

SOLUTION. We have $\quad L(\cos at) = \dfrac{p}{p^2+a^2}$ and $\quad L(\cos bt) = \dfrac{p}{p^2+b^2}$

Hence, by convolution theorem,

$$L\left[\int_0^t \cos ax \,\cos(bt-x)dx\right] = \frac{p^2}{(p^2+a^2)(p^2+b^2)}$$

Therefore $L^{-1}\left[\dfrac{p^2}{(p^2+a^2)(p^2+b^2)}\right]$

$$= \int_0^t \cos ax \,\cos b(t-x)\,dx$$

$$= \frac{1}{2}\int_0^t \{\cos(ax+bt-bx) + \cos(ax-bt+bx)\}dx$$

$$= \frac{1}{2}\int_0^t \cos\{(a-b)x+bt\}dx + \frac{1}{2}\int_0^t \cos[(a+b)x-bt]dx$$

$$= \frac{\sin at - \sin bt}{2(a-b)} + \frac{\sin at + \sin bt}{2(a+b)} = \frac{a\sin at - b\sin bt}{a^2-b^2}.$$

EXAMPLE 6. *Evaluate* $L^{-1}\left\{\dfrac{1}{\sqrt{p}}\dfrac{1}{(p-a)}\right\}$ *by the convolution theorem.*

(Meerut–1979, 2001; Agra–1979; Andhra–1990; Kumaun–1998)

SOLUTION. We know that

$$L^{-1}\left\{\frac{1}{\sqrt{p}}\right\} = L^{-1}\left\{\frac{1}{p^{1/2}}\right\} = \frac{t^{(1/2)-1}}{\Gamma(1/2)} = \frac{1}{\sqrt{\pi}.\sqrt{t}} = F_1(t) \text{ (say)}$$

Also, $\quad L^{-1}\left\{\dfrac{1}{p-a}\right\} = e^{at} = F_2(t)$

Then, by convolution theorem, we have

$$L^{-1}\left\{\frac{1}{\sqrt{p}\,(p-a)}\right\} = F_1(t) * F_2(t) \ = \int_0^t F_1(x).F_2(t-x)\,dx$$

$$= \int_0^t \frac{1}{\sqrt{\pi}} \cdot \frac{1}{\sqrt{x}} e^{a(t-x)} dx = \frac{e^{at}}{\sqrt{\pi}} \int_0^{\sqrt{at}} \frac{\sqrt{a}}{u} e^{-u^2} \cdot \frac{2u}{a} . du$$

$$\left[\text{By Putting } ax = u^2 \Rightarrow dx = \frac{2u\, du}{a} \right]$$

$$= \frac{e^{at}}{\sqrt{a}} \cdot \frac{2}{\sqrt{\pi}} \int_0^{\sqrt{(at)}} e^{-u^2} du = \frac{e^{at}}{\sqrt{a}} \operatorname{erf} [\sqrt{at})]$$

EXAMPLE 7. *Using convolution theorem, show that*

$$B(m, n) = \int_0^1 x^{m-1}(1-x)^{n-1} dx = \frac{\Gamma(m)\,\Gamma(n)}{\Gamma(m+n)}, \quad m > 0,\ n > 0$$

(Meerut–1982, 83, 85, 94, 2009, 10; Agra–2002; Ravishankar–2004)

SOLUTION. Let $\quad F(t) = \int_0^t x^{m-1}(t-x)^{n-1} dx = \int_0^t F_1(x).F_2(t-x)\, dx$

where, $F_1(t) = t^{m-1} = F_1 * F_2$ and $F_2(t) = t^{n-1}$

Therefore $\quad L\{F(t)\} = L\{F_1 * F_2\}$

$$= L\{F_1(t)\} . L\{F_2(t)\} = L\{t^{m-1}\}.L\{t^{n-1}\}$$

$$= \frac{\Gamma(m)}{p^m} \cdot \frac{\Gamma(n)}{p^n} = \frac{\Gamma(m)\,\Gamma(n)}{p^{m+n}}$$

$$\Rightarrow \qquad F(t) = L^{-1}\left\{ \frac{\Gamma(m) . \Gamma(n)}{p^{m+n}} \right\}$$

$$\Rightarrow \qquad F(t) = \int_0^t x^{m-1}(t-x)^{n-1} dx = L^{-1}\left\{ \frac{\Gamma(m)\,\Gamma(n)}{p^{m+n}} \right\}$$

$$= \Gamma(m) . \Gamma(n) . L^{-1}\left\{ \frac{1}{p^{m+n}} \right\} = \frac{\Gamma(m)\,\Gamma(n)}{\Gamma(m+n)} t^{m+n-1}$$

Let $t = 1$, then we have

$$B(m, n) = \int_0^1 x^{m-1}(1-x)^{n-1} dx = \frac{\Gamma(m)\,\Gamma(n)}{\Gamma(m+n)} .$$

EXAMPLE 8. *Using Heaviside's expansion formula, evaluate* $L^{-1}\left\{ \dfrac{3p+1}{(p-1)(p^2+1)} \right\}$

(Meerut–1983, 91; Agra–1982; Rohilkhand–1989; Kanpur–1986)

SOLUTION. We have $\quad F(p) = 3p+1$

and $\quad G(p) = (p-1)(p^2+1) = (p-1)(p+i)(p-i)$

Clearly, $G(p)$ has 3 distinct zeroes $\alpha_1 = 1$, $\alpha_2 = i$ and $\alpha = -i$

Also, $\quad G'(p) = 3p^2 - 2p + 1$

Using Heaviside's expansion formula, we have

$$L^{-1}\left\{ \frac{3p+1}{(p-1)(p^2+1)} \right\} = \frac{F(1)}{G'(1)} e^t + \frac{F(i)}{G'(i)} e^{it} + \frac{F(-i)}{G'(-i)} e^{-it}$$

$$= \frac{4e^t}{2} + \frac{3i+1}{-(2+2i)} e^{it} + \frac{(-3i+1)}{(-2+2i)} e^{-it}$$

$$= 2e^t - \frac{(3i+1)(1-i)}{2(1+i)(1-i)}e^{i.t} + \frac{(3i-1)(1+i)}{2(1-i)(1+i)}e^{-i.t}$$

$$= 2e^t - \frac{1}{2}(i+2)e^{it} + \frac{1}{2}(i-2)e^{-it}$$

$$= 2e^t - \frac{1}{2}i(e^{it} - e^{-it}) - (e^{it} + e^{-it})$$

$$= 2e^t - \frac{1}{2}.i.2i\,\sin t - 2\cos t = 2e^t + \sin t - 2\cos t.$$

EXAMPLE 9.   *Find* $L^{-1}\left[\dfrac{2p^2 - 6p + 5}{p^3 - 6p^2 + 11p - 6}\right]$.   (UPTU–2004; Agra–2010; Meerut–2005; SVTU–1997)

SOLUTION.   Let $f(p) = 2p^2 - 6p + 5$

and $G(p) = p^3 - 6p^2 + 11p - 6 = (p-1)(p-2)(p-3)$

$\Rightarrow$ $G'(p) = 3p^2 - 12p + 11$

$G(p) = 0$ has three roots 1, 2, 3. So, let $\alpha_1 = 1, \alpha_2 = 2, \alpha_3 = 3$

By Heaviside's inverse formula, we have

$$L^{-1}\left[\frac{F(p)}{G(p)}\right] = \sum_{i=1}^{n} \frac{F(\alpha_i)}{G'(\alpha_i)}e^{t\alpha_i}$$

$$L^{-1}\left[\frac{2p^2 - 6p + 5}{p^3 - 6p^2 + 11p - 6}\right] = \frac{F(\alpha_1)}{G'(\alpha_1)}e^{t\alpha_1} + \frac{F(\alpha_2)}{G'(\alpha_2)}e^{t\alpha_2} + \frac{F(\alpha_3)}{G'(\alpha_3)}e^{t\alpha_3}$$

$$= \frac{F(1)}{G'(1)}e^t + \frac{F(2)}{G'(2)}e^{2t} + \frac{F(3)}{G'(3)}e^{3t}$$

$$= \frac{1}{2}e^t + \frac{1}{-1}e^{2t} + \frac{5}{2}e^{3t} = \frac{1}{2}e^t - e^{2t} + \frac{5}{2}e^{3t}.$$

EXAMPLE 10.   *Show that* $L^{-1}\left\{\dfrac{1}{\sqrt{p^2 + a^2}}\right\} = \int_0^t J_0(ak)\,dx$.

SOLUTION.   Here, we have

$$L^{-1}\left\{\frac{1}{\sqrt{p^2 + a^2}}\right\} = L^{-1}\left\{\frac{1}{p}\left(1 + \frac{a^2}{p^2}\right)^{-1/2}\right\}$$

$$= L^{-1}\left\{\frac{1}{p}\left(1 - \frac{1}{2}\frac{a^2}{p^2} + \frac{1.3}{2.4}\frac{a^4}{p^4} - \frac{1.3.5}{2.4.6}\frac{a^6}{p^6} + ...\right)\right\}$$

$$= L^{-1}\left\{\frac{1}{p} - \frac{a^2}{2}.\frac{1}{p^3} + \frac{1.3}{2.4}\frac{a^4}{p^5} - \frac{1.3.5}{2.4.6}\frac{a^6}{p^7} + ...\right\}$$

$$= 1 - \frac{a^2}{2}.\frac{t^2}{2!} + \frac{1.3}{2.4}\frac{a^4 t^4}{4!} - \frac{1.3.5}{2.4.6}\frac{a^6 t^6}{6!} + ...$$

$$= 1 - \frac{(at)^2}{2^2} + \frac{(at)^4}{2^2.4^2} - \frac{(at)^6}{2^2.4^2.6^2} + ... = J_0(at).$$

EXAMPLE 11. *Evaluate* $L^{-1}\left\{\dfrac{e^{-\sqrt{p}}}{p}\right\}$ *and hence deduce that* $L^{-1}\left\{\dfrac{e^{x\sqrt{p}}}{p}\right\} = erfc\left(\dfrac{x}{2\sqrt{t}}\right)$

(Andhra–1990; Agra–1980, 81, 2000; Kanpur–1996)

SOLUTION. Let $f(p) = e^{-\sqrt{p}}$

Therefore, $F(t) = L^{-1}\{e^{-\sqrt{p}}\}$

$$= L^{-1}\left\{1 - \sqrt{p} + \frac{p}{2!} - \frac{p^{3/2}}{3!} + \frac{p^2}{4!} - \frac{p^{5/2}}{5!} + ...\right\}$$

$$= L^{-1}\{1\} - L^{-1}\{p^{1/2}\} + \frac{1}{2!}L^{-1}\{p\}$$

$$\quad - \frac{1}{3!}L^{-1}\{p^{3/2}\} + \frac{1}{4!}L^{-1}\{p^2\} - \frac{1}{5!}L^{-1}\{p^{5/2}\} + ... \quad ...(1)$$

Now, $L^{-1}\{p^{n+(1/2)}\} = L^{-1}\left\{\dfrac{1}{p^{-n-(1/2)}}\right\} = \dfrac{t^{-n-(3/2)}}{\Gamma\left(-n-\dfrac{1}{2}\right)}, \quad n \in \mathbf{Z}$

$$= \frac{(-1)^{n+1}}{\sqrt{n}}\left(\frac{1}{2}\right)\left(\frac{3}{2}\right)\left(\frac{5}{2}\right)...\left(\frac{2n+1}{2}\right)t^{-n-(3/2)}$$

Now, from (1), we have

$$F(t) = -\frac{(-1)t^{-3/2}}{\sqrt{\pi}}\cdot\frac{1}{2} - \frac{1}{3!}\cdot\frac{(-1)^2}{\sqrt{\pi}}\left(\frac{1}{2}\right)\left(\frac{3}{2}\right)t^{-5/2}$$

$$\quad - \frac{1}{5!}\frac{(-1)^3}{\sqrt{\pi}}\left(\frac{1}{2}\right)\left(\frac{3}{2}\right)\left(\frac{5}{2}\right)t^{-7/2} + ...$$

$$= \frac{1}{2\sqrt{\pi}.t^{3/2}}\left[1 - \frac{1}{4t} + \frac{(1/4t)^2}{2!} - \frac{(1/4t)^3}{3!} + ...\right]$$

$$= \frac{1}{2\sqrt{\pi}.t^{3/2}}e^{-1/4t}.$$

Since $L^{-1}\left\{\dfrac{f(p)}{p}\right\} = \int_0^t F(x)\,dx$ where $F(t) = L^{-1}\{f(p)\}$

Therefore, $L^{-1}\left\{\dfrac{e^{-\sqrt{p}}}{p}\right\} = \int_0^t \dfrac{1}{2\sqrt{\pi}\,x^{3/2}}e^{-1/(4x)}.dx$

$$= -\frac{2}{\sqrt{\pi}}\int_\infty^{1/2\sqrt{t}} e^{-y^2}\,dy \quad \left(\text{where } x = \frac{1}{4y^2} \Rightarrow dx = -\frac{dy}{2y^3}\right)$$

$$= \frac{2}{\sqrt{\pi}}\int_{1/(2\sqrt{t})}^\infty e^{-y^2}\,dy = erfc\left(\frac{1}{2\sqrt{t}}\right)$$

DEDUCTION. We have

$\therefore \qquad L^{-1}\left\{\dfrac{e^{-\sqrt{p}}}{p}\right\} = erfc\left(\dfrac{1}{2\sqrt{t}}\right)$

$$\therefore \qquad L^{-1}\left\{\frac{e^{-\sqrt{(x^2 p)}}}{x^2 p}\right\} = \frac{1}{x^2}\, erfc\left(\frac{1}{2\sqrt{t/x^2}}\right) \qquad \text{(By change of scale property)}$$

or $\qquad L^{-1}\left\{\dfrac{e^{-x\sqrt{p}}}{p}\right\} = erfc\left(\dfrac{x}{2\sqrt{t}}\right)$

EXAMPLE 12. *Show that* $L^{-1}\left\{\dfrac{1}{p^3 + 1}\right\} = \dfrac{t^2}{2!} - \dfrac{t^5}{5!} + \dfrac{t^8}{8!} - \dfrac{t^{11}}{11!} + ...$

SOLUTION. We have $\qquad \dfrac{1}{p^3 + 1} = \dfrac{1}{p^3}\left(1 + \dfrac{1}{p^3}\right)^{-1} = \dfrac{1}{p^3}\left[1 - \dfrac{1}{p^3} + \dfrac{1}{p^6} - \dfrac{1}{p^9} + \dfrac{1}{p^{12}} - ...\right]$

$$= \dfrac{1}{p^3} - \dfrac{1}{p^6} + \dfrac{1}{p^9} - \dfrac{1}{p^{12}} + ...$$

Hence, $\quad L^{-1}\left\{\dfrac{1}{p^3 + 1}\right\} = \dfrac{t^2}{2!} - \dfrac{t^5}{5!} + \dfrac{t^8}{8!} - \dfrac{t^{11}}{11!} + ...$

EXAMPLE 13. *Show that* $\int_0^\infty \cos x^2 dx = \dfrac{1}{2}\sqrt{\dfrac{\pi}{2}}$ (Meerut–1983; Kanpur–1984)

SOLUTION. Let $\qquad F(t) = \int_0^\infty \cos t \,.\, x^2\, dx$

Therefore, $\quad L\{F(t)\} = \int_0^\infty e^{-pt} F(t)\, dt$

$$= \int_0^\infty e^{-pt}\left[\int_0^\infty \cos t.x^2 dx\right] dt = \int_0^\infty \left[e^{-pt}\int_0^\infty \cos t.x^2 dt\right] dx$$

$$= \int_0^\infty L\{\cos tx^2\}\, dx = \int_0^\infty \frac{p}{p^2 + x^4}\, dx$$

$$= \frac{1}{2\sqrt{p}}\int_0^{\pi/2} \frac{d\theta}{\sqrt{(\tan\theta)}}$$

$$= \frac{1}{2\sqrt{p}}\int_0^{\pi/2} \sin^{-1/2}\theta\, \cos^{1/2}\theta\, d\theta$$

$$\left(\text{where } x = \sqrt{p\tan\theta} \;\Rightarrow\; dx = \frac{p\sec^2\theta\, d\theta}{2\sqrt{p\tan\theta}}\right)$$

$$= \frac{1}{2\sqrt{p}}\,\frac{\Gamma\left(\frac{1}{4}\right)\Gamma\left(\frac{3}{4}\right)}{2\,\Gamma(1)} = \frac{1}{2\sqrt{p}}\,\frac{\Gamma\left(\frac{1}{4}\right)\Gamma\left(1 - \frac{1}{4}\right)}{2}$$

$$= \frac{1}{4\sqrt{p}}\cdot\frac{\pi}{\sin\frac{\pi}{4}} = \frac{\pi}{2\sqrt{2p}} \qquad \left[\because \Gamma(p)\,\Gamma(1 - p) = \frac{\pi}{\sin p\pi}\right]$$

Hence, $\qquad F(t) = \dfrac{\pi}{2\sqrt{2}} L^{-1}\left\{\dfrac{1}{p^{1/2}}\right\} = \dfrac{\pi}{2\sqrt{2}} \dfrac{t^{(1/2)-1}}{\Gamma(1/2)} = \dfrac{1}{2}\sqrt{\dfrac{\pi}{2t}}$

$$= \int_0^\infty \cos t \cdot x^2\, dx = \dfrac{1}{2}\sqrt{\dfrac{\pi}{2t}}$$

If $t=1$, then we have $\int_0^\infty \cos x^2\, dx = \dfrac{1}{2}\sqrt{\dfrac{\pi}{2}}$.

EXAMPLE 14. *Show that* $\int_0^\infty e^{-x^2}\, dx = \dfrac{\sqrt{\pi}}{2}$ $\qquad$ (Meerut–1990; Kanpur–1997)

SOLUTION. Let, $\qquad F(t) = \int_0^\infty e^{-tx^2}\, dx$

Then proceed as in example (13), we get

$$L\{F(t)\} = \int_0^\infty L\{e^{-tx^2}\}\, dx = \int_0^\infty \dfrac{dx}{p+x^2}$$

$$= \left(\dfrac{1}{\sqrt{p}}.\tan^{-1}\dfrac{x}{\sqrt{p}}\right)_0^\infty = \dfrac{\pi}{2\sqrt{p}}$$

$\Rightarrow \qquad F(t) = \dfrac{\pi}{2} L^{-1}\left(\dfrac{1}{\sqrt{p}}\right) = \dfrac{\pi}{2}.\dfrac{1}{\sqrt{(\pi t)}} = \dfrac{1}{2}\sqrt{\dfrac{\pi}{t}}$

$\Rightarrow \qquad \int_0^\infty e^{-tx^2}\, dx = \dfrac{1}{2}\sqrt{\left(\dfrac{\pi}{t}\right)}$

Taking $t=1$, we have $\int_0^\infty e^{-x^2}\, dx = \dfrac{1}{2}\sqrt{\pi}$

EXAMPLE 15. *Show that* $L^{-1}\left\{\dfrac{8}{(p^2+1)^3}\right\} = (3-t^2)\sin t - 3t\cos t$

SOLUTION. We know that $L^{-1}\left\{\dfrac{1}{p^2+1}\right\} = \sin t$

$\therefore \quad L^{-1}\left\{\dfrac{1}{(p^2+1)}.\dfrac{1}{(p^2+1)}\right\} = \int_0^t \sin x \sin(t-x)\, dx$

$$= \int_0^t \sin x\,(\sin t\cos x - \cos t\sin x)\, dx$$

$$= \sin t \int_0^t \sin x\cos x\, dx - \cos t \int_0^t \sin^2 x\, dx$$

$$= \dfrac{1}{2}\sin t \int_0^t \sin 2x\, dx - \dfrac{1}{2}\cos t\int_0^t(1-\cos 2x)dx$$

$$= \dfrac{1}{2}\sin t.\dfrac{1}{2}(1-\cos 2t) - \dfrac{1}{2}\cos t\left(t-\dfrac{1}{2}\sin 2t\right)$$

$$= \dfrac{1}{2}\sin t\,\sin^2 t - \dfrac{1}{2}t\cos t + \dfrac{1}{2}\cos t\,\sin t\cos t$$

$\Rightarrow \qquad L^{-1}\left\{\dfrac{1}{(p^2+1)^2}\right\} = \dfrac{1}{2}\sin t - \dfrac{t}{2}\cos t$

$$\Rightarrow \quad L^{-1}\left\{\frac{8}{(p^2+1)^3}\right\} = 8L^{-1}\left\{\frac{1}{(p^2+1)^2}\cdot\frac{1}{(p^2+1)}\right\}$$

$$= 8\int_0^t \left(\frac{1}{2}\sin x - \frac{x}{2}\cos x\right)\sin(t-x)\,dx$$

[By convolution theorem]

$$= 4\int_0^t (\sin x - x\cos x)(\sin t \cos x - \cos t \sin x)\,dx$$

$$= 4\sin t \int_0^t (\sin x \cos x - x\cos^2 x)\,dx$$

$$-4\cos t \int_0^t (\sin^2 x - x\sin x.\cos x)\,dx$$

$$= 2\sin t \int_0^t \{\sin 2x - x(1+\cos 2x)\}\,dx - 2\cos t$$

$$\int_0^t \{(1-\cos 2x) - (x\sin 2x)\}\,dx$$

$$= 2\sin t \left[-\frac{t^2}{2} + \frac{1-\cos 2t}{2} - \frac{t}{2}\sin 2t + \frac{1-\cos 2t}{4}\right]$$

$$-2\cos t \left[t - \frac{1}{2}\sin 2t + \frac{t\cos 2t}{2} - \frac{\sin 2t}{4}\right]$$

$$= -t^2\sin t + \frac{3}{2}\sin t - \frac{3}{2}\sin t \cos 2t - t\sin t \sin 2t$$

$$-2t\cos t + \frac{3}{2}\sin 2t \cos t - t\cos t \cos 2t$$

$$= -t^2\sin t + \frac{3}{2}\sin t + \frac{3}{2}(-\sin t \cos 2t + \sin 2t \cos t)$$

$$-t(\cos 2t \cos t + \sin t \sin 2t) - 2t\cos t$$

$$= (3-t^2)\sin t - 3t\cos t$$

EXAMPLE 16. *Show that* $1*1*1*...*1\,(n\ times) = \dfrac{t^{n-1}}{(n-1)!}$ *where* $n=1,2,3,...$

(Agra–2001; Rohilkhand–2002, 03; Ravishankar–2004)

SOLUTION. Since we know that

$$F*G = \int_0^t F(x)\,G(t-x)\,dx$$

$$\therefore \qquad 1*1 = \int_0^1 1.1\,dx = t$$

Again, $\qquad 1*1*1 = t*1 = \int_0^t x.1\,dx = \dfrac{t^2}{2}$

$$1*1*1 = \left(\frac{t^2}{2}\right)*1 = \int_0^t \frac{x^2}{2}.1\,dx = \left(\frac{x^3}{2.3}\right)_0^t = \frac{t^3}{3!}$$

Proceeding in the similar way, we get

$$1*1*1*...*1\,(n\ times) = \frac{t^{n-1}}{(n-1)!}\ \text{where}\ n=1,2,3,...$$

EXAMPLE 17. *Evaluate* $L^{-1}\left\{\dfrac{1}{(p^2+4)(p+2)}\right\}$ (Meerut–1987)

SOLUTION. Here, we have

$$L^{-1}\left\{\frac{1}{p^2+4}\right\} = \frac{1}{2}\sin 2t$$

$$L^{-1}\left\{\frac{1}{p+2}\right\} = e^{-2t}$$

Using convolution theorem, we have

$$L^{-1}\left\{\frac{1}{(p^2+4)(p+2)}\right\} = \int_0^t \frac{1}{2}\sin 2x . e^{-2(t-x)}dx = \frac{1}{8}[e^{-2t}+\sin 2t - \cos 2t]$$

EXAMPLE 18. *Evaluate* $L^{-1}\left\{\dfrac{p+5}{(p+1)(p^2+1)}\right\}$ (Agra–1986; SVTU–2006)

SOLUTION. Here, we have
$$F(p) = p+5$$

and $$G(p) = (p+1)(p^2+1) = (p+1)(p+i)(p-i)$$

Hence $G(p)$ has 3 distinct zero, *i.e.*, $-1, i, -i$.
Now, $$G'(p) = 3p^2 + 2p + 1$$
Using Heaviside's expansion formula, we have

$$L^{-1}\left\{\frac{p+5}{(p+1)(p^2+1)}\right\} = \frac{F(-1)}{G'(-1)}e^{-t} + \frac{F(i)}{G'(i)}e^{it} + \frac{F(-i)}{G'(-i)}e^{-it}$$

$$= 2e^{-t} - \frac{1}{2}(2+3i)e^{it} + \frac{1}{2}(3i-2)e^{-it}$$

$$= 2e^{-t} - 2.\frac{1}{2}(e^{it}+e^{-it}) - 3i.\frac{1}{2}(e^{it}-e^{-it})$$

$$= 2e^{-t} - 2\cos t + 3\sin t$$

EXAMPLE 19. *Evaluate* $L^{-1}\left\{\dfrac{1}{p^2(p+1)^2}\right\}$ (Purvanchal–2004; Agra–1999, 2001, 04; Kanpur–1997; Kumaun–1997; IAS–1982; Vikram–1996)

SOLUTION. Here, we have

$$L^{-1}\left\{\frac{1}{p^2(p+1)^2}\right\}$$

Since $L^{-1}\left\{\dfrac{1}{p^2}\right\} = t$ so $L^{-1}\left\{\dfrac{1}{(p+1)^2}\right\} = e^{-t}L^{-1}\left\{\dfrac{1}{p^2}\right\} = te^{-t}$

Using convolution theorem, we get

$$L^{-1}\left\{\frac{1}{p^2(p+1)^2}\right\} = \int_0^t xe^{-x}(t-x)dx = \int_0^t (xt-x^2)e^{-x}dx$$

$$= \left[(xt-x^2)\frac{e^{-x}}{-1} - (t-2x)e^{-x} + (-2)(-e^{-x})\right]_0^t$$

$$= te^{-t} + 2e^{-t} + t - 2$$

EXAMPLE 20. *Evaluate* $L^{-1}\left\{\dfrac{p}{(p^2+4)^3}\right\}$

(Agra–2003, 06; Kanpur–2010)

SOLUTION. Here,

$$L^{-1}\left\{\frac{p}{(p^2+4)^2}\right\} = \frac{t\sin 2t}{4} \qquad \Rightarrow \qquad L^{-1}\left\{\frac{1}{p^2+4}\right\} = \frac{1}{2}\sin 2t$$

Using convolution theorem, we get

$$L^{-1}\left\{\frac{p}{(p^2+4)^3}\right\} = L^{-1}\left\{\frac{p}{(p^2+4)^2}\cdot\frac{1}{p^2+4}\right\} = \int_0^t \frac{x\sin 2x}{4}\cdot\frac{1}{2}\sin 2(t-x)dx$$

$$= \frac{1}{8}\int_0^t x\sin 2x(\sin 2t\cos 2x - \cos 2t\sin 2x)dx$$

$$= \frac{\sin 2t}{8}\int_0^t x\sin 2x\cos 2xdx - \frac{\cos 2t}{8}\int_0^t x\sin^2 2xdx$$

$$= \frac{\sin 2t}{16}\int_0^t x\sin 4xdx - \frac{\cos 2t}{16}\int_0^t x(1-\cos 4x)dx$$

$$= \frac{\sin 2t}{16}\left[\frac{-x\cos 4x}{4} + \frac{\sin 4x}{16}\right]_0^t - \frac{\cos 2t}{16}\left[\frac{x^2}{2} - \frac{x\sin 4x}{4} - \frac{\cos 4x}{16}\right]_0^t$$

$$= \frac{-t\sin 2t\cos 4t}{64} + \frac{\sin 2t\sin 4t}{256} - \frac{t^2}{32}\cos 2t$$

$$\qquad\qquad + \frac{t\cos 2t\sin 4t}{64} + \frac{\cos 2t}{256}(\cos 4t - 1)$$

$$= \frac{t}{64}(\sin 4t\cos 2t - \sin 2t\cos 4t) + \frac{1}{256}(\cos 2t\cos 4t$$

$$\qquad\qquad + \sin 2t\sin 4t) - \frac{t^2}{32}\cos 2t - \frac{\cos 2t}{256}$$

$$= \frac{t}{64}\sin(4t-2t) + \frac{1}{256}\cos(4t-2t) - \frac{t^2}{32}\cos 2t - \frac{\cos 2t}{256}$$

$$= \frac{t}{64}(\sin 2t - 2t\cos 2t)$$

EXAMPLE 21. *Using Heaviside expansion formula evaluate*

$$L^{-1}\left\{\frac{2p^2+5p-4}{p^3+p^2-2p}\right\}$$

(Meerut–1986, 87, 88; Kanpur–1999; Avadh–2012)

SOLUTION. Here, we have to evaluate

$$L^{-1}\left\{\frac{2p^2+5p-4}{p^3+p^2-2p}\right\}$$

Since, $F(p) = 2p^2+5p-4$, $G(p) = p^3+p^2-2p = p(p-1)(p+2)$

and $G'(p) = 3p^2+2p-2$

Clearly, $G(p)$ has 3 distinct zeroes, namely $\alpha_1 = 0$, $\alpha_2 = 1$ and $\alpha_3 = -2$.

Now, using Heaviside's expansion formula, we get

$$L^{-1}\left\{\frac{2p^2+5p-4}{p^3+p^2-2p}\right\} = \frac{F(0)}{G'(0)}e^{0\cdot t} + \frac{F(1)}{G'(1)}e^{(t)} + \frac{F(-2)}{G'(-2)}e^{-2t} = 2te^t - e^{-2t}$$

EXAMPLE 22. *Evaluate* $L^{-1}\left\{\dfrac{1}{(p+a)(p+b)}\right\}$ (Lucknow–2012)

SOLUTION.   Let $F(p) = \dfrac{1}{p+a}, G(p) = \dfrac{1}{p+b}$

$\Rightarrow$ $F(t) = e^{-at}, G(t) = e^{-bt}$

So, $L^{-1}\{f(p).g(p)\} = \int_0^t F(u)G(t-u)du = \int_0^t e^{-au}e^{-b(t-u)}du = e^{-bt}\int_0^t e^{u(b-a)}du$

$$= e^{-bt}\left[\frac{e^{u(b-a)}}{b-a}\right]_{u=0}^t = \frac{e^{-bt}}{b-a}[e^{t(b-a)}-1] = \frac{e^{-at}-e^{-bt}}{b-a}$$

EXAMPLE 23. *Using Heaviside's expansion formula, find*

$$L^{-1}\left\{\frac{3p+1}{(p-1)(p^2+1)}\right\}$$ (Meerut–2003, Kanpur–2002, Agra–2009)

SOLUTION.   Here, $F(p) = 3p+1, G(p) = (p-1)(p^2+1),$

$G'(p) = (p^2+1)+(p-1)(2p) = 3p^2-2p+1$

Now, $G(p)$ has three zeroes, *i.e.*, $\alpha_1 = 1, \alpha_2 = i, \alpha_3 = -i$

$$L^{-1}\left[\frac{F(p)}{G(p)}\right] = \sum_{i=1}^3 \frac{F(\alpha_i)}{G'(\alpha_i)}e^{t\alpha_i}$$

$$= \frac{F(1)}{G'(1)}e^t + \frac{F(i)}{G'(i)}e^{it} + \frac{F(-i)}{G'(-i)}e^{-it}$$

$$= \frac{4}{2}e^t + \frac{(3i+1)}{-2(i+1)}e^{it} + \frac{(1-3i)}{-2(1-i)}e^{-it}$$

$$= 2e^t - \frac{1}{2}[(2+i)e^{it}+(2-i)e^{-it}]$$

$$= 2e^t - \frac{1}{2}[2(e^{it}+e^{-it})+i(e^{it}-e^{-it})]$$

$$= 2e^t - (2\cos t - \sin t) = 2e^t + \sin t - 2\cos t$$

Additional Solved Examples

EXAMPLE 1. *Use Convolution Theorem, show that*

(a) $L^{-1}\left\{\dfrac{1}{(p+1)(p-2)}\right\} = \dfrac{1}{3}[e^{2t}-e^{-t}]$

(Lucknow–1993, 97; Ravishankar–2005; Agra–1995; Meerut–2000, 06BP ;

Kanpur–2012; Kumaun–2002)

(b) $L^{-1}\left\{\dfrac{p}{(p^2+a^2)^2}\right\} = \dfrac{1}{2a}t\sin at$ (Purvanchal–2010, 11; UPTU–2008)

(c) $L^{-1}\left\{\dfrac{1}{p(p^2+4)^2}\right\} = \dfrac{1}{16}(1-t\sin 2t-\cos 2t)$ (Lucknow–2011)

(d) $L^{-1}\left\{\dfrac{1}{(p-2)(p^2+1)}\right\} = \dfrac{1}{5}[e^{2t}-2\sin t-\cos t]$ (Meerut–1988, 95)

(e) $L^{-1}\left\{\dfrac{p^2}{(p^2+a^2)^2}\right\} = \dfrac{1}{a}[\cos at+\sin at]$

(f) $L^{-1}\left\{\dfrac{1}{\sqrt{p}(p-1)}\right\} = e^t\,erf(\sqrt{t})$

(g) $\int_0^t \sin u\cos(t-u)du = \dfrac{t}{2}\sin t$ (Meerut–1991)

(h) $\int_0^t J_0(u)J_0(t-u)du = \sin t$ (Meerut–1991)

SOLUTION.   (a) We have $L^{-1}\left\{\dfrac{1}{p+1}\right\} = e^{-t} = F_1(t)$ (say)

$$L^{-1}\left\{\dfrac{1}{p-2}\right\} = e^{2t} = F_2(t) \qquad \text{(say)}$$

∴ By the convolution theorem, we have

$$L^{-1}\left\{\dfrac{1}{(p+1)}\cdot\dfrac{1}{(p-2)}\right\} = \int_0^t F_1(x)\cdot F_2(t-x)dx = \int_0^t e^{-x}e^{2(t-x)}dx$$

$$= \int_0^t e^{2t}\cdot e^{-3x}dx = e^{2t}\left[-\dfrac{1}{3}e^{-3x}\right]_0^t = \dfrac{1}{3}[e^{2t}-e^{-t}]$$

(b) Since $L^{-1}\left\{\dfrac{p}{(p^2+a^2)}\right\} = \cos at = F_1(t)$

and $L^{-1}\left\{\dfrac{1}{(p^2+a^2)}\right\} = \dfrac{1}{a}\sin at = F_2(t)$

∴ By the convolution theorem, we have

$$L^{-1}\left\{\dfrac{p}{(p^2+a^2)^2}\right\} = \int_0^t F_1(x).F_2(t-x)dx = \int_0^t \cos ax\cdot\dfrac{1}{a}\sin a(t-x)dx$$

$$= \dfrac{1}{a}\int_0^t \cos ax.(\sin at\cos ax-\cos at\sin ax)dx$$

$$= \dfrac{1}{a}\sin at\int_0^t \cos^2 ax\,dx - \dfrac{1}{a}\cos at\int_0^t \sin ax\cos ax\,dx$$

$$= \dfrac{1}{2a}\sin at\int_0^t(1+\cos 2ax)dx - \dfrac{1}{2a}\cos at\int_0^t\sin 2ax\,dx$$

$$= \dfrac{1}{2a}\sin at\left[t+\dfrac{\sin 2at}{2a}\right] - \dfrac{1}{2a}\cos at\left(1-\dfrac{\cos 2at}{2a}\right)$$

$$= \dfrac{t\sin at}{2a} + \dfrac{\sin at.2\sin at\cos at}{4a^2} - \dfrac{\cos at.2\sin^2 at}{4a^2}$$

$$= \dfrac{1}{2a}t.\sin at$$

(c) $L^{-1}\left\{\dfrac{1}{p(p^2+4)^2}\right\} = L^{-1}\left\{\dfrac{1}{p^2}\cdot\dfrac{p}{(p^2+4)^2}\right\}$

$\Rightarrow \quad L^{-1}\{f_1(p)\} = L^{-1}\{1/p^2\} = t = F_1(t),\text{ (say)}$

Let $\qquad f_1(p) = \dfrac{1}{p^2}$ and $f_2(p) = \dfrac{p}{(p^2+4)^2}$

and $\quad L^{-1}\{f_2(p)\} = L^{-1}\left\{\dfrac{p}{(p^2+4)^2}\right\} = L^{-1}\left\{-\dfrac{1}{2}\dfrac{d}{dp}\left(\dfrac{1}{p^2+4}\right)\right\}$

$\qquad = -\dfrac{1}{2}L^{-1}\left\{\dfrac{d}{dp}\left(\dfrac{1}{p^2+4}\right)\right\} = -\dfrac{1}{2}(-1)tL^{-1}\left\{\dfrac{1}{p^2+4}\right\}$

$\qquad = \dfrac{1}{2}t\cdot\dfrac{\sin 2t}{2} = \dfrac{1}{4}t\sin 2t = F_2(t),\text{ say}$

By the convolution theorem, we have

$L^{-1}\left\{\dfrac{1}{p^2}\dfrac{p}{(p^2+4)^2}\right\} = \int_0^t F_2(x)F_1(t-x)dx$

$\qquad = \int_0^t\left[\dfrac{x}{4}\sin 2x\right](t-x)dx = \dfrac{t}{4}\int_0^t x\sin 2x\,dx - \dfrac{1}{4}\int_0^t x^2\sin 2x\,dx$

$\qquad = \dfrac{t}{4}\left(-\dfrac{x}{2}\cos 2x + \dfrac{1}{4}\sin 2x\right)_0^t$

$\qquad\qquad -\dfrac{1}{4}\left(-\dfrac{x^2}{2}\cos 2x + \dfrac{x}{2}\sin 2x + \dfrac{1}{4}\cos 2x\right)_0^t$

$\qquad = \dfrac{1}{16}(1 - t\sin 2t - \cos 2t)$

(d) $\qquad L^{-1}\left\{\dfrac{1}{p-2}\right\} = e^{2t} = F_1(t),\ L^{-1}\left\{\dfrac{1}{p^2+1}\right\} = \sin t = F_2(t)$

$\therefore$ By the convolution theorem, we have

$L^{-1}\left\{\dfrac{1}{(p-2)(p^2+1)}\right\} = F_2(t) * F_1(t) = \int_0^t \sin x\,e^{2(t-x)}dx$

$\qquad = e^{2t}\int_0^t e^{-2x}\sin x\,dx = \dfrac{1}{5}e^{2t}\left[e^{-2x}(-2\sin x - \cos x)\right]_0^t$

$\qquad = \dfrac{1}{5}e^{2t}[e^{-2t}(-2\sin t - \cos t) + 1] = \dfrac{1}{5}[e^{2t} - 2\sin t - \cos t]$

(e) We have $\quad L^{-1}\left\{\dfrac{p}{p^2+2^2}\right\} = \cos 2t$

$\therefore$ By the convolution theorem, we have

$L^{-1}\dfrac{p^2}{(p^2+a^2)^2} = L^{-1}\left\{\dfrac{p}{p^2+a^2}\cdot\dfrac{p}{p^2+a^2}\right\} = \int_0^t \cos ax.\cos a(t-x)dx$

$$= \int_0^t \cos ax (\cos at \cos ax + \sin at \sin ax) dx$$

$$= \cos at \int_0^t \cos^2 ax \, dx + \sin at \int_0^t \cos ax \sin ax \, dx$$

$$= \frac{1}{2} \cos at \int_0^t (1 + \cos 2ax) dx + \frac{1}{2} \sin at \int_0^t \sin 2ax \, dx$$

$$= \frac{1}{2} \cos at \left[x + \frac{1}{2a} \sin 2ax \right]_0^t + \frac{1}{2} \sin at \left[-\frac{1}{2a} \cos 2ax \right]_0^t$$

$$= \frac{1}{2} \cos at \left[t + \frac{1}{2a} \sin 2at \right] + \frac{1}{4a} \sin at (1 - \cos 2at)$$

$$= \frac{1}{2} t \cos at + \frac{1}{4a} \sin at + \frac{1}{4a} (\sin 2at \cos at - \sin at \cos 2at)$$

$$= \frac{1}{2} t \cos at + \frac{1}{4a} [\sin at + \sin(2at - at)] = \frac{1}{2a} [at \cos at + \sin at]$$

(f) We have $L^{-1} \left\{ \dfrac{1}{\sqrt{p}} \right\} = L^{-1} \left\{ \dfrac{1}{p^{1/2}} \right\} = \dfrac{t^{1-1}}{\Gamma \left(\dfrac{1}{2} \right)} = \dfrac{1}{\sqrt{\pi} \sqrt{t}} = F_1(t)$, say

and $L^{-1} \left\{ \dfrac{1}{p-1} \right\} = e^t = F_2(t)$, (say).

∴ By Convolution theorem, we have

$$L^{-1} \left\{ \frac{1}{\sqrt{p}(p-1)} \right\} = F_1(t) * F_2(t) = \int_0^t \frac{1}{\sqrt{x}\sqrt{\pi}} e^{(t-x)} dx$$

$$= \frac{e^t}{\sqrt{\pi}} \int_0^{\sqrt{t}} \frac{1}{u} e^{-u^2} 2u \, du$$

(Putting $x = u^2$, so that $dx = 2u du$)

$$= \frac{e^t}{\sqrt{\pi}} 2. \int_0^{\sqrt{t}} e^{-u} \, du = e^t \, erf(\sqrt{t})$$

(g) Let $F(t) = \int_0^t \sin u \cos(t - u) du$

∴ By the convolution theorem, we have

$$L\{F(t)\} = L\{\sin t\}.L\{\cos t\} = \frac{1}{p^2 + 1} \frac{p}{p^2 + 1} = \frac{p}{(p^2 + 1)^2}$$

∴ $F(t) = L^{-1} \left\{ \dfrac{p}{(p^2 + 1)^2} \right\} = \dfrac{t}{2} \sin t$

(h) Let $F(t) = \int_0^t J_0(u) J_0(t - u) du$

∴ Applying the convolution theorem, we have

$$L\{F(t)\} = L\{J_0(t)\} L\{J_0(t)\} = \frac{1}{\sqrt{p^2 + 1}} \cdot \frac{1}{\sqrt{p^2 + 1}} = \frac{1}{(p^2 + 1)}$$

$$F(t) = L^{-1} \left\{ \frac{1}{p^2 + 1} \right\} = \sin t$$

EXAMPLE 2. *Using the Heaviside expansion formula, show that*

(a) $L^{-1}\left\{\dfrac{2p^2-6p+5}{p^3-6p^2+11p-6}\right\} = \dfrac{1}{2}e^t - e^{2t} + \dfrac{5}{2}e^{3t}$

(b) $L^{-1}\left\{\dfrac{p^2-6}{p^3+4p^2+3p}\right\} = -2 + \dfrac{5}{2}e^{-t} + \dfrac{1}{2}e^{-3t}$ (Meerut–1988, 2006)

(c) $L^{-1}\left\{\dfrac{19p+37}{(p+1)(p-2)(p+3)}\right\} = -3e^t + 5e^{2t} - 2e^{-3t}$ (Meerut–1988)

(d) $L^{-1}\left\{\dfrac{1}{p^3-1}\right\} = \dfrac{1}{3}\left[e^t - e^{-t/2}\right]\left\{\cos\left(\dfrac{1}{2}\sqrt{3}t\right) + \sqrt{3}\sin\left(\dfrac{1}{2}\sqrt{3}t\right)\right\}$ (Meerut–1983)

(e) $L^{-1}\left\{\dfrac{1}{p^3+1}\right\} = \dfrac{1}{3}\left[e^{-t} - e^{t/2}\right]\left\{\cos\left(\dfrac{1}{2}\sqrt{3}t\right) - \sqrt{3}\sin\left(\dfrac{1}{2}\sqrt{3}t\right)\right\}$ (Rohilkhand–1989)

SOLUTION.

(a) Here $f(p) = 2p^2 - 6p + 5$

and $G(p) = p^3 - 6p^2 + 11p - 6 = (p-1)(p-2)(p-3)$

$\Rightarrow$ $G'(p) = 3p^2 - 12p + 11$

i.e., $G(p)$ has 3 distinct zeroes $\alpha_1 = 1$, $\alpha_2 = 2$, and $\alpha_3 = 3$.

$\therefore$ By the Heaviside's expansion formula, we have

$L^{-1}\left\{\dfrac{2p^2-6p+5}{p^3-6p^2+11p-6}\right\} = \dfrac{F(1)}{G'(1)}e^t + \dfrac{F(2)}{G'(2)}e^{2t} + \dfrac{F(3)}{G'(3)}e^{3t}$

$= \dfrac{1}{2}e^t - e^{2t} + \dfrac{5}{2} \cdot e^{3t}$

(b) Here $f(p) = p^2 - 6$

and $G(p) = p^3 + 4p^2 + 3p = p(p+1)(p+3)$

$G'(p) = 3p^2 + 8p + 3$

$G(p)$ has 3 distinct zeroes $\alpha_1 = 0$, $\alpha_2 = -1$ and $\alpha_3 = -3$.

$\therefore$ By the Heaviside's expansion formula, we have

$L^{-1}\left\{\dfrac{p^2-6}{p^3+4p^2+3p}\right\} = \dfrac{F(0)}{G'(0)}e^{0 \cdot t} + \dfrac{F(-1)}{G'(-1)}e^{-t} + \dfrac{F(-3)}{G'(-3)}e^{-3t}$

$= -\dfrac{6}{3} + \dfrac{-5}{-2}e^{-t} + \dfrac{3}{6}e^{-3t} = -2 + \dfrac{5}{2}e^{-t} + \dfrac{1}{2}e^{-3t}$

(c) Here $F(p) = 19p + 37$, $G(p) = (p+1)(p-2)(p+3)$

$\therefore$ $G(p)$ has 3 distinct zeroes $\alpha_1 = -1$, $\alpha_2 = 2$ and $\alpha_3 = -3$.

Also $G'(p) = 3p^2 + 4p - 5$

$\therefore$ By the Heaviside's expansion formula, we have

$L^{-1}\left\{\dfrac{19p+37}{(p+1)(p-2)(p+3)}\right\} = \dfrac{F(-1)}{G'(-1)}e^{-t} + \dfrac{F(2)}{G'(2)}e^{2t} + \dfrac{F(-3)}{G'(-3)}e^{-3t}$

$= -3e^{-t} + 5e^{2t} - 2e^{-3t}$

(d) Here $\quad F(p) = 1, G(p) = p^3 - 1 = (p-1)(p^2 + p + 1)$

$$G'(p) = 3p^2$$

$G(p)$ has 3 distinct roots, $1, \dfrac{1}{2}(-1 + \sqrt{3}i), \dfrac{1}{2}(-1 - \sqrt{3}i)$

∴ By the Heaviside's expansion formula, we have

$$L^{-1}\left\{\frac{1}{p^3 - 1}\right\} = \frac{F(1)}{G(1)}e^t + \frac{F\left\{\dfrac{1}{2}(-1 + \sqrt{3}i)\right\}}{G'\left\{\dfrac{1}{2}(-1 + \sqrt{3}i)\right\}}e^{\left\{\frac{1}{2}(-1+\sqrt{3}i)\right\}t}$$

$$+ \frac{F\left\{\dfrac{1}{2}(-1 - \sqrt{3}i)\right\}}{G'\left\{\dfrac{1}{2}(-1 - \sqrt{3}i)\right\}}e^{\left\{\frac{1}{2}(-1-\sqrt{3}i)\right\}t}$$

$$= \frac{1}{3}e^{-t} + \frac{1}{\dfrac{3}{4}(-1+\sqrt{3}i)^2}e^{\frac{1}{2}(-1+\sqrt{3}i)t} + \frac{1}{\dfrac{3}{4}(-1-\sqrt{3}i)^2}e^{\frac{1}{2}(-1-\sqrt{3}i)t}$$

$$= \frac{1}{3}e^{-t} - \frac{2(1-\sqrt{3}i)}{3(1+\sqrt{3}i)(1-\sqrt{3}i)}e^{\frac{1}{2}(-1+\sqrt{3}i)t} - \frac{2(1+\sqrt{3}i)}{3(1-\sqrt{3}i)(1+\sqrt{3}i)}e^{\frac{1}{2}(-1-\sqrt{3}i)t}$$

$$= \frac{1}{3}e^{-t} - \frac{1}{6}(1-\sqrt{3}i)e^{\frac{1}{2}(-1+\sqrt{3}i)t} - \frac{1}{6}(1+\sqrt{3}i)e^{\frac{1}{2}(-1-\sqrt{3}i)t}$$

$$= \frac{1}{3}e^{-t} - \frac{1}{6}e^{-t/2}[(e^{i\sqrt{3}\,t/2} + e^{-i\sqrt{3}\,t/2}) - i\sqrt{3}(e^{i\sqrt{3}\,t/2} - e^{-i\sqrt{3}\,t/2})]$$

$$= \frac{1}{3}\left[e^{-t} - e^{-t/2}\left\{\cos\left(\frac{1}{2}\sqrt{3}t\right) + \sqrt{3}\sin\left(\frac{1}{2}\sqrt{3}t\right)\right\}\right]$$

(e) Here $\quad F(p) = 1$ and $\quad G(p) = p^3 + 1 = (p+1)(p^2 - p + 1)$

$\Rightarrow \qquad G'(p) = 3p^2$

$G(p)$ has 3 distinct zeroes $-1, \dfrac{1}{2}(1 + \sqrt{3}i)$ and $\dfrac{1}{2}(1 - \sqrt{3}i)$

∴ By the Heaviside's expansion formula, we have

$$L^{-1}\left\{\frac{1}{p^3 + 1}\right\} = \frac{F(-1)}{G(-1)}e^{-t} + \frac{F\left\{\dfrac{1}{2}(1 + \sqrt{3}i)\right\}}{G'\left\{\dfrac{1}{2}(1 + \sqrt{3}i)\right\}}e^{\left\{\frac{1}{2}(1+\sqrt{3}i)\right\}t}$$

$$+ \frac{F\left\{\dfrac{1}{2}(1 - \sqrt{3}i)\right\}}{G'\left\{\dfrac{1}{2}(1 - \sqrt{3}i)\right\}}e^{\left\{\frac{1}{2}(1-\sqrt{3}i)\right\}t}$$

$$= \frac{1}{3}e^{-t} + \frac{1}{\frac{3}{4}(1+\sqrt{3}i)^2}e^{\left\{\frac{1}{2}(1+\sqrt{3}i)\right\}t} + \frac{1}{\frac{3}{4}(1-\sqrt{3}i)^2}e^{\left\{\frac{1}{2}(1-\sqrt{3}i)\right\}t}$$

$$= \frac{1}{3}e^{-t} + \frac{2(\sqrt{3}i+1)}{3(\sqrt{3}i-1)(\sqrt{3}i+1)}e^{\left\{\frac{1}{2}(1+\sqrt{3}i)\right\}t} + \frac{2(\sqrt{3}i-1)}{3(\sqrt{3}i+1)(\sqrt{3}i-1)}e^{\left\{\frac{1}{2}(1-\sqrt{3}i)\right\}t}$$

$$= \frac{1}{3}e^{-t} - \frac{1}{6}e^{t/2}[(\sqrt{3}i+1)e^{\sqrt{3}it/2} - (\sqrt{3}i-1)e^{-\sqrt{3}it/2}]$$

$$= \frac{1}{3}e^{-t} - \frac{1}{6}e^{t/2}[\sqrt{3}i(e^{\sqrt{3}it/2} - e^{-\sqrt{3}it/2}) + (e^{\sqrt{3}it/2} + e^{-\sqrt{3}it/2})]$$

$$= \frac{1}{3}\left[e^{-t} - e^{t/2}\left\{-\sqrt{3}\sin\left(\frac{1}{2}\sqrt{3}t\right) + \cos\left(\frac{1}{2}\sqrt{3}t\right)\right\}\right]$$

Miscellaneous Solved Examples

EXAMPLE 1. Find the inverse Laplace transforms of $\dfrac{p+2}{p^2-4p+13}$.

(Jiwaji–2006; Ravishankar–2004; VTU–2008)

SOLUTION.
$$L^{-1}\left(\frac{p+2}{p^2-4p+13}\right) = L^{-1}\left(\frac{p+2}{(p-2)^2+9}\right) = L^{-1}\left[\frac{p-2+4}{(p-2)^2+3^2}\right]$$

$$= L^{-1}\left[\frac{p-2}{(p-2)^2+3^2}\right] + 4L^{-1}\left(\frac{1}{(p-2)^2+3^2}\right)$$

$$= c^{2t}\cos 3t + \frac{4}{3}e^{2t}\sin 3t.$$

EXAMPLE 2. Find $L^{-1}\left\{\dfrac{(p+2)^2}{(p^2+4p+8)^2}\right\}$

(Mumbai–2005)

SOLUTION.
$$L^{-1}\left\{\frac{(p+2)^2}{(p^2+4p+8)^2}\right\} = L^{-1}\left\{\frac{(p+2)^2}{(p^2+4p+4+4)^2}\right\}$$

$$= L^{-1}\left\{\frac{(p+2)^2}{[(p+2)^2+4]^2}\right\} = e^{-2t}L^{-1}\left\{\frac{p^2}{(p^2+4)^2}\right\}$$

$$= e^{-2t}L^{-1}\left\{\frac{p^2+4-4}{(p^2+4)^2}\right\} = e^{-2t}L^{-1}\left\{\frac{1}{p^2+4} - \frac{4}{(p^2+4)^2}\right\}$$

$$= \frac{e^{-2t}\sin 2t}{2} - 4e^{-2t}L^{-1}\left\{\frac{1}{(p^2+4)^2}\right\}$$

$$= \frac{e^{-2t}\sin 2t}{2} - 4e^{-2t}\left\{\frac{1}{4}\left(\frac{\sin 2t}{4} - \frac{t\cos 2t}{2}\right)\right\}$$

$$= e^{-2t}\left\{\frac{\sin 2t}{2} - \frac{\sin 2t}{4} + \frac{t\cos 2t}{2}\right\}$$

$$= e^{-2t}\left\{\frac{\sin 2t}{4} + \frac{t\cos 2t}{2}\right\} = e^{-2t}\left\{\frac{\sin 2t}{4} + \frac{t\cos 2t}{2}\right\}$$

EXAMPLE 3. *Find* $L^{-1}\left\{\dfrac{1}{p(p^2+a^2)}\right\}$

(PTU–2003; Kerala–2001)

SOLUTION. We know that, $L^{-1}\left(\dfrac{1}{p^2+a^2}\right) = \dfrac{1}{a}\sin at$

$\therefore$ $L^{-1}\left\{\dfrac{1}{p(p^2+a^2)}\right\} = \displaystyle\int_0^t \dfrac{1}{a}\sin at\, dt = \dfrac{1}{a^2}[-\cos at]_0^t = \dfrac{(1-\cos at)}{a^2}.$

EXAMPLE 4. *Find* $L^{-1}\left\{\dfrac{p+2}{p^2(p+1)(p-2)}\right\}$

(VTU–2003)

SOLUTION. Here, $L^{-1}\left\{\dfrac{p+2}{(p+1)(p-2)}\right\}$

$$= \dfrac{4}{3}L^{-1}\left(\dfrac{1}{p-2}\right) - \dfrac{1}{3}L^{-1}\left(\dfrac{1}{p+1}\right) = \dfrac{4}{3}e^{2t} - \dfrac{1}{3}e^{-t}$$

Now, $L^{-1}\left\{\dfrac{p+2}{p(p+1)(p-2)}\right\}$

$$= \int_0^t L^{-1}\left(\dfrac{p+2}{(p+1)(p-2)}\right)dt = \int_0^t\left(\dfrac{4}{3}e^{2t} - \dfrac{1}{3}e^{-t}\right)dt$$

$$= \dfrac{2}{3}e^{2t} + \dfrac{1}{3}e^{-t} - 1$$

Therefore, $L^{-1}\left\{\dfrac{p+2}{p^2(p+1)(p-2)}\right\}$

$$= \int_0^t L^{-1}\left\{\dfrac{p+2}{p(p+1)(p-2)}\right\}dt$$

$$= \int_0^t\left(\dfrac{2}{3}e^{2t} + \dfrac{1}{3}e^{-t} - 1\right)dt = \dfrac{1}{3}(e^{2t} - e^{-t} - t)$$

EXAMPLE 5. *Find* $L^{-1}\left\{\dfrac{p+2}{(p^2+4p+5)^2}\right\}.$

(SVTU–2009; PTU–2005)

SOLUTION. We have $L^{-1}\left\{\dfrac{1}{p^2+4p+5}\right\} = L^{-1}\left\{\dfrac{1}{(p+2)^2+1}\right\} = e^{-2t}\sin t$

Now, $L^{-1}\left\{\dfrac{d}{dp}\left(\dfrac{1}{p^2+4p+5}\right)\right\} = (-1)\,t\cdot e^{-2t}\sin t$

$\Rightarrow$ $L^{-1}\left\{\dfrac{-(2p+4)}{(p^2+4p+5)^2}\right\} = -t\cdot e^{-2t}\sin t$

$\Rightarrow$ $L^{-1}\left\{\dfrac{p+2}{(p^2+4p+5)^2}\right\} = \dfrac{1}{2}t\cdot e^{-2t}\sin t$

EXAMPLE 6.   Find $L^{-1}\left\{\log\dfrac{p+1}{p-1}\right\}$.   (Bhopal–2008; SVTU–209; UPTU–2009, 2014; UKTU–2012)

SOLUTION.   If $f(t) = L^{-1}\log\dfrac{p+1}{p-1}$

then, $tf(t) = L^{-1}\left\{\dfrac{-d}{dp}\log\left(\dfrac{p+1}{p-1}\right)\right\} = -L^{-1}\left\{\dfrac{d}{dp}\log(p+1)\right\} + L^{-1}\left\{\dfrac{d}{dp}\log(p-1)\right\}$

$= -L^{-1}\left(\dfrac{1}{p+1}\right) + L^{-1}\left(\dfrac{1}{p-1}\right) = -e^{-t} + e^{t} = 2\sinh t$

Thus $f(t) = (2\sinh t)/t$

EXAMPLE 7.   Find $L^{-1}\left\{\log\dfrac{p^2+1}{p(p+1)}\right\}$ (SVTU–2009; VTU–2008)

SOLUTION.   If $f(t) = L^{-1}\log\dfrac{p^2+1}{p(p+1)}$

Then $tf(t) = L^{-1}\left\{-\dfrac{d}{dp}\log\left(\dfrac{p^2+1}{p(p+1)}\right)\right\}$

$= L^{-1}\left\{\dfrac{d}{dp}\log(p^2+1)\right\} + L^{-1}\left\{\dfrac{d}{dp}\log p\right\} + L^{-1}\left\{\dfrac{d}{dp}\log(p+1)\right\}$

$= -L^{-1}\left(\dfrac{2p}{p^2+1}\right) + L^{-1}\left(\dfrac{1}{p}\right) + L^{-1}\left(\dfrac{1}{p+1}\right) = -2\cos t + 1 + e^{-t}$

Hence, $f(t) = \dfrac{1}{t}(1 + e^{-t} - 2\cos t)$

EXAMPLE 8.   Evaluate $L^{-1}\{1/(p+a)^3\}$ (Purvanchal–1991)

SOLUTION.   We have to evaluate $L^{-1}\{1/(p+a)^3\}$

Let $f(p) = \dfrac{1}{(p+a)} = (p+a)^{-1}$

$\therefore \qquad \dfrac{df}{dp} = -(p+a)^{-2}, \ \dfrac{d^2f}{dp^2} = 2(p+a)^{-2}$

$\Rightarrow \qquad \dfrac{1}{(p+a)^3} = \dfrac{1}{2}\dfrac{d^2f}{dp^2} = \dfrac{1}{2}\dfrac{d^2}{dp^2}\left(\dfrac{1}{p+a}\right)$

$\therefore \qquad L^{-1}\left\{\dfrac{1}{(p+a)^3}\right\} = \dfrac{1}{2}L^{-1}\left\{\dfrac{d^2}{dp^2}\left(\dfrac{1}{p+a}\right)\right\} = \dfrac{1}{2}\left[(-1)^2t^2L^{-1}\left\{\dfrac{1}{p+a}\right\}\right] = \dfrac{1}{2}t^2e^{-at}$

EXAMPLE 9.   Evaluate $L^{-1}\left\{\dfrac{5p^2-15p-11}{(p+1)(p-2)^3}\right\}$ (Meerut–1984, 88)

SOLUTION.   Here, we have

$L^{-1}\left\{\dfrac{5p^2-15p-11}{(p+1)^2(p-2)^3}\right\}$

$$= L^{-1}\left\{\frac{-1}{3(p+1)} + \frac{1}{3(p-2)} + \frac{4}{(p-2)^2} - \frac{7}{(p-2)^3}\right\}$$

$$= \frac{-1}{3}L^{-1}\left\{\frac{1}{p+1}\right\} + \frac{1}{3}L^{-1}\left\{\frac{1}{p-2}\right\} + 4L^{-1}\left\{\frac{1}{(p-2)^2}\right\} - 7L^{-1}\left\{\frac{1}{(p-2)^3}\right\}$$

$$= \frac{-1}{3}e^{-t} + \frac{1}{3}e^{2t} + 4e^{2t}L^{-1}\left\{\frac{1}{p^2}\right\} - 7e^{2t}L^{-1}\left\{\frac{1}{p^3}\right\}$$

$$= \frac{-1}{3}e^{-t} + \frac{1}{3}e^{2t} + 4e^{2t}\frac{t}{1!} - 7e^{2t}\frac{t^2}{2!} = \frac{-1}{3}e^{-t} + \frac{1}{3}e^{2t} + 4te^{2t} - \frac{7}{2}t^2e^{2t}$$

EXAMPLE 10.   *Evaluate* $L^{-1}\left\{\dfrac{3(p^2 + 2p + 3)}{(p^2 + 2p + 2)(p^2 + 2p + 5)}\right\}$.   (Meerut–1996)

SOLUTION.   Here, we have

$$L^{-1}\left\{\frac{3(p^2 + 2p + 3)}{(p^2 + 2p + 2)(p^2 + 2p + 5)}\right\}$$

$$= L^{-1}\left\{\frac{1}{p^2 + 2p + 2} + \frac{2}{p^2 + 2p + 5}\right\}$$

$$= L^{-1}\left\{\frac{1}{p^2 + 2p + 2}\right\} + 2L^{-1}\left\{\frac{1}{p^2 + 2p + 5}\right\}$$

$$= L^{-1}\left\{\frac{1}{(p+1)^2 + 1}\right\} + 2L^{-1}\left\{\frac{1}{(p+1)^2 + 4}\right\}$$

$$= e^{-t}L^{-1}\left\{\frac{1}{p^2 + 1}\right\} + 2e^{-t}L^{-1}\left\{\frac{1}{p^2 + 2^2}\right\} = e^{-t}\sin t + e^{-t}\sin 2t$$

EXAMPLE 11.   *Evaluate* $L^{-1}\left\{\log\dfrac{p+3}{p+2}\right\}$ (Agra–1982; Kanpur–2007)

SOLUTION.   We have $f(p) = \log\dfrac{p+3}{p+2} = \log(p+3) - \log(p+2)$

$$f'(p) = \frac{1}{p+3} - \frac{1}{p+2}$$

Now, taking inverse Laplace transform, we get

$$L^{-1}\{f'(p)\} = L^{-1}\left\{\frac{1}{p+3}\right\} - L^{-1}\left\{\frac{1}{p+2}\right\} = e^{-3t} - e^{-2t}$$

$$\Rightarrow \quad (-1)^1 t L^{-1}\{f(p)\} = e^{-3t} - e^{-2t}$$

Hence, $L^{-1}\left\{\log\dfrac{p+3}{p+2}\right\} = \dfrac{1}{t}(e^{-2t} - e^{-3t})$

EXAMPLE 12.   *Evaluate* $L^{-1}\left\{\dfrac{1}{(p-1)(p+2)(p+4)}\right\}$ (Kanpur–2013)

SOLUTION. Consider

$$L^{-1}\left\{\frac{1}{(p-1)(p+2)(p+4)}\right\} = L^{-1}\left\{\frac{A}{p-1}+\frac{B}{p+2}+\frac{C}{p+4}\right\}$$

$$= L^{-1}\left\{\frac{1}{15(p-1)}-\frac{1}{6(p+2)}+\frac{1}{10(p+4)}\right\}$$

$$= \frac{1}{15}e^t - \frac{1}{6}e^{-2t} + \frac{1}{10}e^{-4t}$$

EXAMPLE 13. *Find the inverse Laplace transform of* $\dfrac{p+2}{p^2-2p+5}$. (Avadh–2013; Kanpur–2008)

SOLUTION. Consider $p^2 - 2p + 5 = (p-1)^2 + 4 = (p-1)^2 + 2^2$

$$L^{-1}\left\{\frac{p+2}{p^2-2p+5}\right\} = L^{-1}\left\{\frac{(p-1)+3}{(p-1)^2+2^2}\right\}$$

$$= L^{-1}\left\{\frac{p-1}{(p-1)^2+2^2}+\frac{3}{(p-1)^2+2^2}\right\}$$

$$= L^{-1}\left\{\frac{p-1}{(p-1)^2+2^2}\right\}+\frac{3}{2}L^{-1}\left\{\frac{2}{(p-1)^2+2^2}\right\}$$

$$= e^t L^{-1}\left\{\frac{p}{p^2+2^2}\right\}+\frac{3}{2}e^t L^{-1}\left\{\frac{2}{p^2+2^2}\right\} = e^t\cos(2t)+\frac{3}{2}e^t\sin(2t)$$

EXAMPLE 14. *Establish the reciprocity between the following pair of Laplace transform.*

$$f(t) = \frac{e^{at}-1}{a}, \quad F(p) = \frac{1}{p(p-a)}$$

SOLUTION. Here, $L\{f(t)\} = L\left\{\dfrac{e^{at}-1}{a}\right\} = \dfrac{1}{a}[L\{e^{at}\}-L\{1\}] = \dfrac{1}{a}\left[\dfrac{1}{p-a}-\dfrac{1}{p}\right]$

$$= \frac{1}{a}\cdot\frac{a}{p(p-a)} = \frac{1}{p(p-a)} = F(p) \qquad \left[\because F(p) = \frac{1}{p(p-a)}\right]$$

$$L\{f(t)\} = F(p)$$

EXAMPLE 15. *Find the inverse Laplace transform of* $\dfrac{1}{p^2(p^2-a^2)}$, *by using the convolution theorem.* (Meerut–1993; Purvanchal–1996)

SOLUTION. We know that $L^{-1}\left\{\dfrac{1}{p^2}\right\} = t$ and $L^{-1}\left\{\dfrac{1}{p^2-a^2}\right\} = \dfrac{\sin at}{a}$

From convolution theorem

$$L^{-1}\left\{\frac{1}{p^2(p^2-a^2)}\right\} = \int_0^t u\frac{1}{a}\sinh a(tu)du = \frac{1}{a}\int_0^t u\sinh(at-au)du$$

$$= \frac{1}{a}\left[\frac{-u}{a}\cosh(at-au)-\frac{\sinh(at-au)}{a^2}\right]_{u=0}^t$$

$$= \frac{1}{a}\left[\frac{-t}{a}\cosh(at-at)-0-\frac{1}{a^2}\{0-\sinh at\}\right]$$

$$= \frac{1}{a}\left[\frac{-t}{a}+\frac{1}{a^2}\sinh at\right] = \frac{1}{a^3}[-at+\sinh at].$$

EXAMPLE 16.   *Evaluate* $L^{-1}\left\{\dfrac{1}{p(p+a)^2}\right\}$

SOLUTION.   Consider

$$L^{-1}\left\{\frac{1}{(p+a)^2}\right\} = -L^{-1}\left\{\frac{d}{dp}\left(\frac{1}{p+a}\right)\right\} = -(-1)^1 t^1 e^{-at} = te^{-at}$$

$$\Rightarrow \; L^{-1}\left\{\frac{1}{p(p+a)^2}\right\} = \int_0^t ue^{-au}du = [-ue^{-au}a - e^{-au}]_{u=0}^t$$

$$= [ate^{-at}+(e^{-at}-1)] = 1-e^{-at}(at+1)$$

EXAMPLE 17.   *Find the inverse Laplace transform of* $\dfrac{5p-2}{p^2(p+2)(p-1)}$ (Purvanchal–1996)

SOLUTION.   We can write, $\dfrac{5p-2}{(p+2)(p-1)} = \dfrac{A}{p+2}+\dfrac{B}{p-1}$

$$\Rightarrow \quad A = \frac{5(-2)-2}{-2-1} = 4, B = \frac{5\cdot 1-2}{1+2} = \frac{3}{3} = 1$$

$$\Rightarrow \quad \frac{5p-2}{(p+2)(p-1)} = \frac{4}{p+2}+\frac{1}{p-1}$$

$$L^{-1}\left\{\frac{5p-2}{(p+2)(p-1)}\right\} = 4L^{-1}\left(\frac{1}{p+2}\right)+L^{-1}\left(\frac{1}{p-1}\right) = 4e^{-2t}+e^t$$

Now, $L^{-1}\left\{\dfrac{5p-2}{p(p+2)(p-1)}\right\} = \int_0^t (4e^{-2u}+e^u)du$

$$= [-2e^{-2u}+e^u]_{u=0}^t = -2e^{-2t}+e^t-(-2+1)$$

$$= 1+e^t-2e^{-2t}$$

$$= L^{-1}\left\{\frac{5p-2}{p^2(p+2)(p-1)}\right\} = \int_0^t (1+e^u-2e^{-2u})du$$

$$= [u+e^u+e^{-2u}]_{u=0}^t = t+e^t+e^{-2t}-(1+1)$$

$$= t+e^t+e^{-2t}-2$$

EXAMPLE 18.   *Find* $L^{-1}\left\{\dfrac{p+1}{p^2(p+2)^3}\right\}$ (Kanpur–2014)

SOLUTION.   We know that $L(t^n) = \dfrac{n!}{p^{n+1}}$, $L\{e^{at}F(t)\} = f(p-a)$

$$\Rightarrow \quad L(t^2) = \frac{2!}{p^3}, L(e^{-2t}t^2) = \frac{2!}{(p+2)^3}$$

and $L^{-1}\left\{\dfrac{1}{(p+2)^3}\right\} = \dfrac{t^2}{2}e^{-2t} = L^{-1}\left\{\dfrac{1}{p(p+2)^3}\right\} = \int_0^t \dfrac{x^2}{2}e^{-2x}dx$

$$= \frac{1}{2}\left[\frac{x^2}{-2}e^{-2x} + \int \frac{e^{-2x}}{2}\cdot 2x\,dx\right]_{x=0}^{t} = \frac{-t^2}{4}e^{-2t} + \frac{1}{2}\int_0^t xe^{-2x}\,dx$$

$$\Rightarrow \quad L^{-1}\left\{\frac{1}{p(p+2)^3}\right\} = \frac{-1}{4}e^{-2t}(t^2+t+1) + \frac{1}{4} \qquad\qquad \ldots(1)$$

Assume that $\qquad A = L^{-1}\left\{\dfrac{p+1}{p^2(p+2)^3}\right\}$

$$\Rightarrow \qquad\qquad A = L^{-1}\left\{\frac{1}{p(p+2)^3}\right\} + L^{-1}\left\{\frac{1}{p^2(p+2)^3}\right\} \qquad\qquad \ldots(2)$$

From eq^n (1)

$$L^{-1}\left\{\frac{1}{p}\cdot\frac{1}{p(p+2)^3}\right\} = \frac{-1}{4}\int_0^t[(x^2+x+1)e^{-2x}-1]dx$$

$$= \frac{t}{4} - \frac{1}{4}\left[\frac{e^{-2x}}{-2}(x^2+x+1) + \frac{1}{2}\int e^{-2x}(2x+1)dx\right]_{x=0}^{t}$$

On integrating by parts, we get

$$L^{-1}\left\{\frac{1}{p^2(p+2)^3}\right\} = \frac{t}{4} + \frac{1}{8}[e^{-2t}(t^2+2t+2)-2] \qquad\qquad \ldots(3)$$

Substituting the value of eq^n (1) & (3) in eq^n (2), we get

$$A = \left\{-\frac{1}{4}e^{-2t}(t^2+t+1) + \frac{1}{4}\right\} + \left\{\frac{t}{4} + \frac{1}{8}e^{-2t}(t^2+2t+2) - \frac{1}{4}\right\} = \frac{t}{4} - \frac{t^2}{8}e^{-2t}$$

EXAMPLE 19. *Find the inverse Laplace transform of*

$$\frac{1}{p^2+4p+13} - \frac{p+4}{p^2+8p+97} + \frac{p+2}{p^2-4p+29} \qquad\qquad \text{(Kurukshetra–2004)}$$

SOLUTION. We have

$$L^{-1}\left\{\frac{1}{p^2+4p+13}\right\} = L^{-1}\left\{\frac{1}{(p+2)^2+9}\right\} = \frac{1}{3}e^{-2t}\sin 3t$$

$$L^{-1}\left\{\frac{p+4}{p^2+8p+17}\right\} = L^{-1}\left\{\frac{p+4}{(p+4)^2+81}\right\} = e^{-4t}\cos 9t$$

$$L^{-1}\left\{\frac{p+2}{p^2-4p+29}\right\} = L^{-1}\left\{\frac{p-2+4}{(p-2)^2+25}\right\}$$

$$= L^{-1}\left\{\frac{p-2}{(p-2)^2+25}\right\} + 4L^{-1}\left\{\frac{1}{(p-2)^2+25}\right\}$$

$$= e^{2t}\cos 5t + 4\cdot\frac{1}{5}e^{2t}\sin 5t$$

Hence, $L^{-1}\left\{\dfrac{1}{p^2+4p+13} - \dfrac{p+4}{p^2+8p+97} + \dfrac{p+2}{p^2-4p+29}\right\}$

$$= \frac{1}{3}e^{-2t}\sin 3t - e^{-4t}\cos 9t + e^{2t}\cos 5t + \frac{4}{5}e^{2t}\sin 5t$$

EXAMPLE 20. *Find the inverse Laplace transform of* $\dfrac{1}{p(p+2)^3}$.   (Kurukshetra–2006)

SOLUTION.   Here, we have

$$L^{-1}\left\{\frac{1}{p(p+2)^3}\right\} = e^{-2t}L^{-1}\left[\frac{1}{(p-2)p^3}\right]$$

$$= e^{-2t}L^{-1}\left[\frac{\left(\dfrac{1}{p-2}\right)}{p^3}\right] = e^{-2t}\int_0^t\int_0^t\int_0^t e^{2t}\,dt$$

$$= e^{-2t}\int_0^t\int_0^t\left(\frac{e^{2t}}{2}\right)_0^t dt = \frac{e^{-2t}}{2}\int_0^t\int_0^t(e^{2t}-1)dt$$

$$= \frac{1}{2}e^{-2t}\int_0^t\left(\frac{e^{2t}}{2}-t\right)_0^t dt = \frac{1}{2}e^{-2t}\int_0^t\left(\frac{e^{2t}}{2}-t-\frac{1}{2}\right)dt$$

$$= \frac{1}{4}e^{-2t}\int_0^t(e^{2t}-2t-1)dt = \frac{1}{4}e^{-2t}\left[\frac{e^{2t}}{2}-t^2-t\right]_0^t$$

$$= \frac{1}{4}e^{-2t}\left[\left(\frac{e^{2t}}{2}-t^2-t\right)-\frac{1}{2}\right] = \frac{1}{4}e^{-2t}\left[\frac{e^{2t}}{2}-t^2-t-\frac{1}{2}\right]$$

EXAMPLE 21. *Find the inverse Laplace transform of* $\log\dfrac{p^2+1}{(p-1)^2}$.   (MDU–2005)

SOLUTION.   Let $$F(p) = \log\frac{p^2+1}{(p-1)^2}.$$

$$L^{-1}[F(p)] = f(t) \Rightarrow L^{-1}\left[-\frac{d}{dp}\{F(p)\}\right] = tf(t)$$

$$\Rightarrow -L^{-1}\frac{d}{dp}\left(\log\frac{p^2+1}{(p-1)^2}\right) = tf(t)$$

$$\Rightarrow -L^{-1}\left[\frac{(p-1)^2}{p^2+1}\times\frac{(p-1)^2 2p-(p^2+1)2(p-1)}{(p-1)^4}\right] = tf(t)$$

$$\Rightarrow -L^{-1}\left[\frac{(p-1)^2}{p^2+1}\times\frac{-2p-2}{(p-1)^3}\right] = tf(t) \Rightarrow L^{-1}\left[\frac{2(p+1)}{(p^2+1)(p-1)}\right] = tf(t) \quad\quad \text{...(A)}$$

Using partial fractions, we can write

$$\frac{2(p+1)}{(p^2+1)(p-1)} = \frac{A}{p-1}+\frac{Bp+C}{p^2+1} \quad\quad \text{...(1)}$$

$$\Rightarrow 2(p+1) = A(p^2+1)+(Bp+C)(p-1) \quad\quad \text{...(2)}$$

Putting $p = 1$, we get

$$4 = 2A \quad \Rightarrow \quad A = 2$$

Now, comparing the coefficient of p^2 of eq^n (2), we get

$$0 = A + B \quad \Rightarrow \quad B = -A = -2$$

Comparing the constant term, we get

$$2 = A - C \qquad \Rightarrow C = 0$$

Putting these values in eq^n (1), we get

$$\frac{2(p+1)}{(p^2+1)(p-1)} = \frac{2}{p-1} + \frac{2p}{p^2+1}$$

Hence, $\quad L^{-1}\left[\dfrac{2(p+1)}{(p^2+1)(p-1)}\right] = 2e^t - 2\cos t$

From eq^n (A), we get

$$f(t) = \frac{1}{t}[2e^t - 2\cos t] = \frac{2}{t}[e^t - \cos t]$$

MISCELLANEOUS EXERCISE

(A) Find the inverse Laplace transform of the following functions.

1. $\dfrac{p}{(2p-1)(3p-1)}$ (VTU–2010)

2. $\dfrac{1}{p^2-5p+6}$ (SVTU–2008)

3. $\dfrac{3p+2}{p^2-p-2}$ (VTU–2010 S)

4. $\dfrac{1}{p(p^2-1)}$ (Nagarjuna–2008)

5. $\dfrac{1-7p}{(p-3)(p-1)(p+2)}$ (BPTU–2005 S)

6. $\dfrac{p}{(p^2-1)^2}$ (Kurukshetra–2005)

7. $\dfrac{p^3}{p^4-a^4}$ (Kurukshetra–2005)

8. $\dfrac{p^2+2p+3}{(p^2+2p+2)(p^2+2p+5)}$ (Mumbai–2008)

9. $\dfrac{a(p^2-2a^2)}{p^4+4a^4}$ (Mumbai–2009)

10. $\dfrac{1}{p^2(p+5)}$ (Madras–2003 S)

11. $\dfrac{p}{a^2p^2+b^2}$ (Madras–2000 S)

12. $\dfrac{p+2}{(p^2+4p+8)^2}$ (Mumbai–2006)

13. $\dfrac{1}{2}\log\left(\dfrac{p^2+b^2}{p^2+a^2}\right)$

(Mumbai–2008; VTU–2008)

14. $\log\dfrac{p^2+1}{(p-1)^2}$

(Madras–2000 S; MDU Rohtak–2005)

15. $\tan^{-1}\left(\dfrac{2}{p}\right)$

(Mumbai–2007; PTU–2005; Chennai–1997)

16. $\cot^{-1}(p)$ (VTU–2005)

17. $p\log\dfrac{p-1}{p+1}$ (Osmania–1992; Madras–1999)

18. $\dfrac{2p+1}{p^2-4}$ (UPTU (SUM)–2009)

19. $\dfrac{e^{-2\pi p}}{p(p^2+1)}$ (GBTU–2011)

20. $\dfrac{p-1}{p^2(p-7)}$ (GBTU (CO)–2011)

21. $\dfrac{14p+10}{49p^2+28p+13}$ (UPTU (SUM)–2007)

22. $\dfrac{p}{p^2+6p+25}$ (GBTU–2011)

23. $\dfrac{p+1}{p^2-6p+25}$ (GBTU–2010)

24. $\log\left(\dfrac{p^2+4p+5}{p^2+2p+5}\right)$ (MTU (SUM)–2011)

25. $\dfrac{1}{p^4+4}$ (UPTU (SUM)–2007)

26. $\dfrac{1}{p(p+1)(p+2)}$ (UPTU (SUM)–2008;

Madras–2005; Mumbai Kamraj–2008)

27. $\log\left(1-\dfrac{a^2}{p^2}\right)$

(UPTU (CO)–2009; Lucknow–1997)

28. $\dfrac{1}{p(p+a)^3}$ (UKTU–2012)

29. $\dfrac{1}{p^2(p^2+1)}$ (UPTU (SUM)–2009)

30. $\dfrac{p^2}{(p^2+\omega^2)^2}$ (GBTU(CO) 2010)

31. $\dfrac{8p}{(p^2+16)(p^2+1)^2}$ (UPTU (CO)–2009)

(B) Use convolution theorm to find:

1. $L^{-1}\left[\dfrac{1}{(p^2+a^2)^2}\right]$

(UPTU (SUM)–2007, UPTU–2009)

2. $L^{-1}\left[\dfrac{16}{(p-2)(p+2)^2}\right]$

(Kanpur–2011; MTU–2011)

3. $L^{-1}\left[\dfrac{p}{(p^2+a^2)^3}\right]$ (UPTU–2008, MTU–2012)

──────── 𝒜NSWERS ────────

(A)

1. $3e^{t/2}+2e^{t/3}$ **2.** $e^{3t}-e^{2t}$ **3.** $\dfrac{1}{3}(8e^{2t}-e^{-t})$ **4.** $\cosh t$ **5.** $e^t+e^{-2t}-2e^{3t}$

6. $\dfrac{1}{2}t\sinh t$ **7.** $\dfrac{1}{2}[\cos at+\cosh at]$ **8.** $\dfrac{1}{3}e^{-t}(\sin t+\sin 2t)$ **9.** $\cos at\sinh at$

10. $\dfrac{1}{25}(e^{-5t}+5t-1)$ **11.** $\dfrac{1}{a^2}\cos\left(\dfrac{bt}{a}\right)$ **12.** $\dfrac{1}{2}te^{-2t}\sin 2t$ **13.** $\dfrac{1}{t}(\cos at-\cos bt)$

14. $\dfrac{2}{t}(e^t-\cos t)$ **15.** $\dfrac{\sin 2t}{t}$ **16.** $\dfrac{\sin t}{t}$ **17.** $\dfrac{2(\sinh t-t\cosh t)}{t^2}$

18. $2\cosh 2t+\dfrac{1}{2}\sinh 2t$ **19.** $1-\cos tu(t-2\pi)$ **20.** $-\dfrac{6}{49}+\dfrac{1}{7}t+\dfrac{6}{49}e^{7t}$

21. $\dfrac{2}{7}e^{-2t/7}\left(\cos\dfrac{3}{7}t+\sin\dfrac{3}{7}t\right)$ **22.** $e^{-3t}\left(\cos 4t-\dfrac{3}{4}\sin 4t\right)$ **23.** $e^{3t}(\cos 4t+\sin 4t)$

24. $\dfrac{2}{t}(e^{-t}\cos 2t-e^{-2t}\cos t)$ **25.** $\dfrac{1}{4}(\sin t\cosh t-\cos t\sinh t)$ **26.** $\dfrac{1}{2}-e^{-t}+\dfrac{1}{2}e^{-2t}$

27. $\dfrac{2}{t}(1-\cosh at)$ **28.** $\dfrac{1}{a^3}-\dfrac{1}{a^3}e^{-at}\left(1+at+\dfrac{a^2t^2}{2}\right)$ **29.** $t-\sin t$

30. $\dfrac{1}{2\omega}\sin\omega t+\dfrac{t}{2}\cos\omega t$ **31.** $\dfrac{60t\sin t-8\cos t+8\cos 4t}{225}$

(B)

1. $\dfrac{1}{2a^3}(\sin at-at\cos at)$ **2.** $e^{2t}-e^{-2t}(1+4t)$ **3.** $\dfrac{t}{8a^3}(\sin at-at\cos at)$

Self Assessment Test

1. Evaluate $L^{-1}\left\{\dfrac{p+2}{p^2-2p+5}\right\}$ (Agra–1996)

2. Evaluate $L^{-1}\left\{\dfrac{1}{p(p^2+2p+2)}\right\}$ (Agra–2010)

3. Evaluate $L^{-1}\left\{\dfrac{3p}{(p^2+16)}-\dfrac{2}{(p^2+16)}\right\}$

(Agra–2013)

4. Find $L^{-1}\left\{\dfrac{(3p+1)}{(p-2)(p^2+1)}\right\}$ (Rohilkhand–2013)

5. Find $L^{-1}\left\{\dfrac{p}{(p+2)^5}\right\}$ (Lucknow–2011)

6. Evaluate $L^{-1}\left\{\dfrac{1}{p^2-6p+14}\right\}$

(Ravishankar–2004)

7. Evaluate $L^{-1}\left\{\dfrac{1}{p(p^2+1)^3}\right\}$ (Ravishankar–2005)

8. Evaluate $L^{-1}\left\{\dfrac{1}{(p+a)(p+b)}\right\}$

(Purvanchal–2001)

9. Evaluate $L^{-1}\left\{\dfrac{1}{(p-1)(p+3)}\right\}$ (Meerut–2006)

Prove the following (10-12):

10. $L^{-1}\left\{\dfrac{p}{2p^2-8}\right\}=\dfrac{1}{2}\cosh 2t$ (Agra–2003)

11. $L^{-1}\left\{\dfrac{p^2}{(p+2)^3}\right\}=e^{-2t}(1-4t+2t^2)$

(Meerut–1998)

12. $L^{-1}\left\{\dfrac{3p-2}{p^2-4p+20}\right\}=3e^{2t}\cos 4t+e^{2t}\sin 4t$

(Meerut–2011; Agra–2014)

13. Find the inverse Laplace transform of the following function :

$\dfrac{2p^2-1}{(p^2+1)(p^2+4)}$ (UPTU–2004)

14. Show that

$$L^{-1}\left\{\dfrac{p^2}{p^4+4a^2}\right\}$$

$$=\dfrac{1}{2a}(\cosh at\sin at+\sinh at\cos at)$$

(Agra–2009, Meerut–1985)

15. Show that

$$L^{-1}\left\{\tan^{-1}\left(\dfrac{2}{p^2}\right)\right\}=\dfrac{2}{t}\sin t\sinh t$$

(Agra–2002, 03; UPTU–2001; VTU–2011;
Mumbai–2005; Jabalpur–2004)

16. Using the Convolution theorem, show that

(i) $L^{-1}\left(\dfrac{p}{(p^2+4)^2}\right)=\dfrac{t}{4}\sin 2t$

(UPTU–2004; GBTU–2010)

(ii) $L^{-1}\left\{\dfrac{1}{(p^2+1)(p^2+9)}\right\}$

$$=\dfrac{1}{8}\left(\sin t-\dfrac{1}{3}\sin 3t\right)$$

(Mumbai–2005)

(iii) $L^{-1}\left\{\dfrac{p}{(p^2+1)(p^2+4)(p^2+9)}\right\}$

$$=\dfrac{1}{12}\cos t-\dfrac{1}{10}\cos 2t+\dfrac{1}{60}\cos 3t$$

(Mumbai–2006)

(iv) $L^{-1}\left\{\dfrac{1}{(p-2)(p+2)^2}\right\}$

$$=\dfrac{1}{16}(e^{2t}-e^{-2t}-4te^{-2t})$$

(Meerut–1996, 97; Mumbai–2009;
Osmania–2004)

(v) $L^{-1}\left\{\dfrac{p}{(p+2)(p^2+9)}\right\}$

$$=\dfrac{1}{13}(3\sin 3t+2\cos 2t-2e^{-2t})$$

(VTU–2008)

(vi) $L^{-1}\left\{\dfrac{1}{(p^2+4p+13)^2}\right\}$

$\qquad = \dfrac{e^{-2t}}{54}(\sin 3t - 3t\cos 3t)$

(Mumbai–2008)

17. Find the inverse Laplace transform of

$\qquad \dfrac{p^2+p-2}{p(p+3)(p-2)}$.   (Kurukshetra–2007)

18. Using Convolution theorem, evaluate

$\qquad L^{-1}\left(\dfrac{1}{p^2(p^2+a^2)}\right)$

ANSWERS

1. $e^t\left(\cos 2t + \dfrac{3}{2}\sin 2t\right)$ **3.** $3\cos(4t) - \dfrac{1}{2}\sin(4t)$ **4.** $\dfrac{1}{5}[7e^{2t} - 7\cos t + \sin t]$

5. $e^{-2t}\left[\dfrac{t^3}{6} - \dfrac{t^4}{24}\right]$ **6.** $e^{3t}\sin t$ **7.** $1 - e^{-t}\left(1 + t + \dfrac{t^2}{2}\right)$ **8.** $\dfrac{e^{-at} - e^{-bt}}{b-a}$

9. $\dfrac{e^{-t} - e^{-3t}}{4}$ **13.** $-\sin t + \dfrac{3}{2}\sin 2t$ **17.** $\dfrac{1}{3} + \dfrac{4}{15}e^{-3t} + \dfrac{2}{5}e^{2t}$

18. $\dfrac{1}{a^3}[at - \sin at]$

Objective Evaluations

❖ Fill in the Blanks

1. A function $N(t)$ of t such that $\int_0^t N(t) = 0 \ \forall t$ is called _____ function.

2. If $L^{-1}[f(p)] = F(t)$, then $L^{-1}\{f(p-a)\} =$ ____

3. If $L^{-1}\{f(p)\} = F(t)$, then $L^{-1}\{f(ap)\} =$ _____ $F\left(\dfrac{t}{a}\right)$.

4. $L^{-1}\left[\dfrac{1}{p}\right] =$ _____.

5. $L^{-1}\left[\dfrac{p}{p^2 + a^2}\right] =$ _____.

6. The inverse Laplace transform of $\dfrac{1}{p-a}, p > a$ is _____.

7. $\int_0^\infty \sin^2 x\, dx =$ _____.

8. If $L^{-1}\{f(p)\} = F(t)$, then $L^{-1}\{e^{-ap} f(p)\} = G(t)$, where $G(t) =$ _____.

9. If $f(p)$ is the Laplace transform of a function $f(t)$, then $F(t)$ is called the _____ Laplace transform of $f(p)$.

10. The inverse Laplace transform of $\dfrac{1}{p^2 - a^2}$ is _____.

11. The inverse Laplace transform of $\dfrac{1}{p^{n+1}}$, $n > -1$ is _____.

12. Let $F(t)$ and $G(t)$ be two functions of $F_1(t)$ and $F_2(t)$ respectively and C_1, C_2 are any two constants, then $L^{-1}\{C_1 f_1(p) + C_2 f_2(p)\} =$ _____.

❖ True or False

Write 'T' for True and 'F' for False statement.

1. $L^{-1}\left(\dfrac{1}{p^2}\right) = t$ **(T/F)**

2. $L^{-1}\left(\dfrac{1}{p^{n+1}}\right) = \dfrac{t^n}{n!}$ **(T/F)**

3. $L^{-1}\left(\dfrac{1}{p^2 - a^2}\right) = \dfrac{\sinh at}{a}$ **(T/F)**

4. $L^{-1}\left(\dfrac{p}{p^2 - a^2}\right) = \cos at$ **(T/F)**

5. $L^{-1}\left(\dfrac{1}{p-a}\right) = e^{at}$ **(T/F)**

6. The value of $L^{-1}\left\{\dfrac{1}{\sqrt{p}}\right\}$ is $\dfrac{1}{\sqrt{\pi t}}$.   **(T/F)**

7. The inverse Laplace transform of $\dfrac{2}{p-5}$ is $2e^{5t}$.   **(T/F)**

8. The inverse Laplace transform of $\dfrac{1}{(p-1)(p+2)}$ is $\dfrac{1}{3}(e^{-t} - e^{2t})$.   **(T/F)**

9. The value of $\int_0^\infty \cos x^2 dx$ is $\sqrt{\dfrac{\pi}{2}}$.   **(T/F)**

10. $L^{-1}\left\{\dfrac{1}{p^2 + a^2}\right\} = \dfrac{1}{a}\sin at$ **(T/F)**

11. The convolution of two functions F and G obeys the commutative law. (Agra–2008)   **(T/F)**

❖ Multiple Choice Questions

Choose the most appropriate one.

1. The function whose Laplace transform is $\dfrac{1}{p^2 + \omega^2}$ is :   (Rohilkhand–2003)
   (a) $\dfrac{1}{\omega}\sin \omega t$ (b) $\cos \omega t$
   (c) $\sin \omega t$ (d) none of these

2. The function whose Laplace transform is $\dfrac{p}{p^2 + a^2}$ is:
   (a) $\cos \omega t$ (b) $\sin \omega t$

(c) $\dfrac{1}{\omega}\cos\omega t$ (d) none of these

3. The function whose Laplace transform is $\dfrac{1}{p^2 - a^2}$ is:

(a) $\dfrac{1}{a}\sinh at$ (b) $\dfrac{1}{a}\cosh at$

(c) $\sinh at$ (d) none of these

4. The function whose Laplace transform is $\dfrac{p-a}{(p-a)^2 + \omega^2}$ is:

(a) $e^{at}\cos\omega t$ (b) $\cos\omega t$

(c) $e^{at}\sin\omega t$ (d) none of these

5. The function whose Laplace transform is $\dfrac{1}{p(p^2 + \omega^2)}$ is :

(a) $\dfrac{1}{\omega^3}(\omega t - \sin\omega t)$

(b) $\dfrac{1}{2\omega^2}(\sin\omega t - \omega t\cos\omega t)$

(c) $\dfrac{1}{2\omega}\sin\omega t$

(d) none of these

6. The function whose Laplace transform is $\dfrac{p}{(p^2 + \omega^2)^2}$ is:

(a) $\dfrac{1}{\omega}(\omega - \sin\omega t)$ (b) $\dfrac{1}{2\omega}\sin\omega t$

(c) $\dfrac{1}{\omega}\sin\omega t$ (d) none of these

7. The function whose Laplace transform is $\dfrac{p^2}{(p^2 + \omega^2)^2}$ is:

(a) $\dfrac{1}{2\omega}(\sin\omega t + \omega t + \cos\omega t)$

(b) $\sin\omega t + \omega t\cos\omega t$

(c) $\sin\omega t$

(d) none of these

8. The function whose Laplace transform is $\dfrac{1}{p^4 - a^4}$ is:

(a) $\sinh at$ (b) $\cosh at$

(c) $\sinh at - \cosh at$ (d) none of these

9. The value of $L^{-1}\left\{\dfrac{1}{(p-1)^3}\right\}$ is:

(a) $\dfrac{1}{2}t^2 e^t$ (b) $\dfrac{1}{2}te^t$

(c) $t^2 e^t$ (d) none of these

10. The value of is $L^{-1}\left\{\dfrac{1}{p^4}\right\}$ is :

(Rohilkhand–2004)

(a) $\dfrac{t^4}{4!}$ (b) $\dfrac{t^3}{3!}$

(c) $\dfrac{t^5}{5!}$ (d) $\dfrac{t^2}{2!}$

11. If $f(p)$ denoted the Laplace transform of the function $F(t)$, then $L^{-1}\{f(ap)\}$ becomes :

(Agra–2003; Rohilkhand–2003, 10)

(a) $\dfrac{1}{a}F\left(\dfrac{t}{a}\right)$ (b) $\dfrac{1}{t}F\left(\dfrac{a}{t}\right)$

(c) $\dfrac{1}{t}F\left(\dfrac{t}{a}\right)$ (d) $\dfrac{1}{a}F\left(\dfrac{a}{t}\right)$

12. The inverse Laplace transform of $f(p) = \dfrac{1}{p} + \dfrac{1}{p+1}$ is :   (Rohilkhand–2008)

(a) $1 + e^{-t}$ (b) e^t

(c) e^{-t} (d) $1 + e^t$

13. $L^{-1}\left\{\dfrac{1}{p-2}\right\}$ is :   (Agra–2005, 08)

(a) e^{-2t} (b) e^t

(c) e^{2t} (d) e^{-t}

14. The value of $t * t * t$ is :   (Agra–2009)

(a) $\dfrac{t^2}{2!}$ (b) $\dfrac{t^6}{5!}$

(c) $\dfrac{t^3}{3!}$ (d) $\dfrac{t^5}{5!}$

15. The value of $1 * 1 * 1$ is :   (Agra–2004)

(a) $\dfrac{t}{2}$ (b) $\dfrac{t^2}{2!}$

(c) $\dfrac{t^2}{3}$ (d) $\dfrac{t^3}{3!}$

16. The value of $L^{-1}\left\{\dfrac{p}{2p^2 - 8}\right\}$ is :   (Agra–2009)

(a) $\dfrac{1}{2}\cosh 2t$ (b) $\dfrac{1}{2}\sinh 2t$

(c) $\cosh 2t$ (d) $\sinh 2t$

17. $L^{-1}\left\{\dfrac{p}{p^2 - a^2}\right\}$ is equal to :   (Agra–2010)

(a) $\sinh at$ (b) $\cosh at$

(c) $\tanh at$ (d) $\coth at$

18. $L^{-1}\left\{\dfrac{1}{\sqrt{p}}\right\}$ is :   (Agra–2007)

(c) $t^2 e^t$ (d) none of these

(a) $\dfrac{1}{\sqrt{\pi p}}$ (b) $\dfrac{1}{\sqrt{\pi^3 t}}$

(c) $\dfrac{1}{\sqrt{\pi t}}$ (d) none of these

19. $L^{-1}\left(\dfrac{1}{p}\right)$ is :   (Rohilkhand–2006)

(a) 1   (b) 2

(c) p (d) none of these

20. If $F(t)$ and $G(t)$ be two functions of class A and let $L^{-1}\{f(p)\} = F(t)$ and $L^{-1}\{g(p)\} = G(t)$, then $L^{-1}\{f(p) \cdot g(p)\} =$ (Agra–2011)

(a) $\int_0^t F(x)G(x)dx$ (b) $\int_0^t F(x)G(t-x)dx$

(c) $\int_0^t F(t-x)G(x)dx$ (d) none of these

21. $L^{-1}\left\{\dfrac{1}{(p+a)^3}\right\}$ is :   (Rohilkhand–2010)

(a) $\dfrac{1}{2}t^2 e^{-at}$ (b) te^{-at}

(c) $\dfrac{1}{2}e^{-at}$ (d) 0

22. If $L^{-1}\{f(p)\} = F(t)$, then $L^{-1}\{f(p-a)\} =$
     (Agra–2011; Rohilkhand–2003)

(a) $e^{-at}F(t)$ (b) $e^{at}Lf(p)$

(c) $e^{at}L^{-1}f(p)$ (d) none of these

23. The inverse Laplace transform of
$f(p) = \dfrac{3}{(p+2)^2 + 9}$ is :   (Rohilkhand–2009)

(a) $e^{2t}\sin 3t$ (b) $e^{3t}\sin 2t$

(c) $e^{2t}\cos 3t$ (d) $e^{3t}\cos 2t$

24. If $L^{-1}\{f(p)\} = F(t)$ and $F(0) = 0$, then $L^{-1}\{pf(p)\} =$ (Agra–2011)

(a) $F(t)$ (b) $F'(t)$

(c) $F''(t)$ (d) none of these

25. The inverse Laplace transform of $\dfrac{1}{(p^2 + p)}$ is :
     [GATE(ME)–2009]

(a) $1 - e^{-t}$ (b) $1 + e^{-t}$

(c) $1 + e^t$ (d) $1 - e^t$

26. Given $L^{-1}\left[\dfrac{3p+1}{p^3 + 4p^2 + (K-3)p}\right]$. If $\lim\limits_{t\to\infty} f(t) = 1$, then value of K is :   [GATE(EC)–2010]

(a) 1   (b) 4

(c) 2   (d) 3

27. The inverse Laplace transform of the function $F(p) = \dfrac{1}{p(p+1)}$ is given by :   [GATE(ME)–2012]

(a) $f(t) = 1 - e^{-t}$ (b) $f(t) = e^{-t}$

(c) $f(t) = \sin t$ (d) none of these

28. A delayed unit step function is defined as
$u(t-a) = \begin{cases} 0 & , \text{ for } t < a \\ 1 & , \text{ for } t \ge a \end{cases}$. Its Laplace transform is :   [GATE(ME)–2004]

(a) $\dfrac{e^{ap}}{p}$ (b) $a.e^{-ap}$

(c) $\dfrac{e^{-ap}}{p}$ (d) $\dfrac{e^{ap}}{a}$

ANSWERS

❖ Fill in the Blanks

1. null   **2.** $e^{at}F(t)$ **3.** $1/a$ **4.** 1   **5.** $\cos at$ **6.** e^{at} **7.** $\dfrac{1}{2}\sqrt{\dfrac{\pi}{2}}$

8. $\begin{cases} F(t-a) & , \quad t > a \\ 0 & , \quad t < a \end{cases}$ **9.** inverse   **10.** $\cosh at$ **11.** $\dfrac{t^n}{\Gamma(n+1)}$

12. $\int_0^t F(x)G(t-x)dx$

❖ True or False

1. T   **2.** T   **3.** T   **4.** F   **5.** T   **6.** T   **7.** T   **8.** F   **9.** F

10. T   **11.** T

❖ Multiple Choice Questions

1. (a)   **2.** (a)   **3.** (a)   **4.** (a)   **5.** (b)   **6.** (b)   **7.** (a)   **8.** (d)   **9.** (a)

10. (b)   **11.** (a)   **12.** (a)   **13.** (c)   **14.** (d)   **15.** (b)   **16.** (a)   **17.** (b)   **18.** (c)

19. (a)   **20.** (b)   **21.** (a)   **22.** (c)   **23.** (a)   **24.** (b)   **25.** (c)   **26.** (b)   **27.** (a)

28. (b)

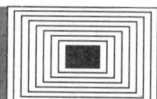

Comprehensive Inverse Laplace Transforms Tables

(1) GENERAL FORMULAS

S. No.	$f(p)$	$L^{-1}\{f(p)\} = F(x)$
1.	$f(p+a)$	$e^{-ax}F(x)$
2.	$f(ap), a > 0$	$\dfrac{1}{a}F\left(\dfrac{x}{a}\right)$
3.	$f(ap+b), a > 0$	$\dfrac{1}{a}e^{(-\frac{b}{a}x)}F\left(\dfrac{x}{a}\right)$
4.	$f(p-a)+f(p+a)$	$2F(x)\cosh(ax)$
5.	$f(p-a)-f(p+a)$	$2F(x)\sinh(ax)$
6.	$e^{-ap}f(p), a \geq 0$	$\begin{cases} 0 & \text{if } 0 \leq x < a \\ F(x-a) & \text{if } a < x \end{cases}$
7.	$pf(p)$	$\dfrac{dF(x)}{dx}$ if $F(+0) = 0$
8.	$\dfrac{1}{p}f(p)$	$\int_0^x F(t)dt$
9.	$\dfrac{1}{p+a}f(p)$	$e^{-ax}\int_0^x e^{at}F(t)dt$
10.	$\dfrac{1}{p^2}f(p)$	$\int_0^x (x-t)F(t)dt$
11.	$f(p)/p(p+a)$	$\dfrac{1}{a}\int_0^x [1-e^{a(x-t)}]F(t)dt$
12.	$f(p)/(p+a)^2$	$\int_0^x (x-t)e^{-a(x-t)}F(t)dt$
13.	$f(p)/(p+a)(p+b)$	$\dfrac{1}{b-a}\int_0^x [e^{-a(x-t)}-e^{-b(x-t)}]F(t)dt$
14.	$f(p)/(p+a^2)b^2$	$\dfrac{1}{b}\int_0^x e^{-a(x-t)}\sin[b(x-t)]F(t)dt$
15.	$\left(\dfrac{1}{p^n}\right)f(p), n = 1,2,...$	$\dfrac{1}{(n-1)!}\int_0^x (x-t)^{n-1}F(t)dt$
16.	$f_1(p)f_2(p)$	$\int_0^x F_1(t)F_2(x-t)dt$
17.	$\dfrac{1}{\sqrt{p}}f\left(\dfrac{1}{p}\right)$	$\int_0^\infty \dfrac{\cos(2\sqrt{xt})}{\sqrt{\pi x}}F(t)dt$
18.	$\dfrac{1}{p\sqrt{p}}f\left(\dfrac{1}{p}\right)$	$\int_0^\infty \dfrac{\sin(2\sqrt{xt})}{\sqrt{\pi x}}F(t)dt$
19.	$\dfrac{1}{p}f\left(\dfrac{1}{p}\right)$	$\int_0^\infty J_0 2\sqrt{xt}\,F(t)dt$
20.	$\dfrac{1}{p}f\left(p+\dfrac{1}{p}\right)$	$\int_0^x J_0(2\sqrt{xt-t^2})F(t)dt$

21.	$f(\sqrt{p})$	$\int_0^\infty \dfrac{t}{2\sqrt{\pi x^3}} e^{(-t^2/4x)} F(t)\,dt$
22.	$\dfrac{1}{\sqrt{p}} f(\sqrt{p})$	$\dfrac{1}{\sqrt{\pi x}} \int_0^\infty e^{(-t^2/4x)} F(t)\,dt$
23.	$f(p+\sqrt{p})$	$\dfrac{1}{2\pi} \int_0^x \dfrac{t}{(x-t)^{3/2}} e^{[-t^2/4(x-t)]} F(t)\,dt$
24.	$f(\sqrt{p^2+a^2})$	$F(x) - a\int_0^x F(\sqrt{x^2-t^2}) J_1(at)\,dt$
25.	$\dfrac{f(\sqrt{p^2+a^2})}{\sqrt{p^2+a^2}}$	$\int_0^x J_0(a\sqrt{x^2-t^2}) F(t)\,dt$
26.	$f(\log p)$	$\int_0^\infty \dfrac{x^t}{\Gamma(t)} F(t)\,dt$
27.	$\dfrac{1}{p} f(\log p)$	$\int_0^\infty \dfrac{x^t}{\Gamma(t+1)} F(t)\,dt$
28.	$f(p-ia) + f(p+ia),\, i^2 = -1$	$2F(x)\cos(ax)$
29.	$i[f(p-ia) - f(p+ia)],\, i^2 = -1$	$2F(x)\sin(ax)$
30.	$\dfrac{df(p)}{dp}$	$-xF(x)$
31.	$\dfrac{d^n f(p)}{dp^n}$	$(-x)^n F(x)$
32.	$\dfrac{p^n d^m f(p)}{dp^m}\quad m \geq n$	$(-1)^m \dfrac{d^n}{dx^n}[x^m F(x)]$
33.	$\int_p^\infty \tilde{f}(q)\,dq$	$\dfrac{1}{x} F(x)$
34.	$\dfrac{1}{p}\int_0^p \tilde{f}(q)\,dq$	$\int_x^\infty \dfrac{F(t)}{t}\,dt$

(2) For Rational Functions

S. No.	$f(p)$	$L^{-1}\{f(p)\} = F(x)$
1.	$\dfrac{1}{p}$	1
2.	$\dfrac{1}{p+a}$	e^{-ax}
3.	$\dfrac{1}{p^2}$	x
4.	$\dfrac{1}{p(p+a)}$	$\dfrac{1}{a}(1-e^{-ax})$
5.	$\dfrac{1}{(p+a)^2}$	xe^{-ax}

6.	$\dfrac{p}{(p+a)^2}$	$(1-ax)e^{-ax}$
7.	$\dfrac{1}{p^2-a^2}$	$\dfrac{1}{a}\sinh(ax)$
8.	$\dfrac{p}{p^2-a^2}$	$\cosh(ax)$
9.	$\dfrac{1}{(p+a)(p+b)}$	$\dfrac{1}{a-b}(e^{-bx}-e^{-ax})$
10.	$\dfrac{p}{(p+a)(p+b)}$	$\dfrac{1}{a-b}(ae^{-ax}-be^{-bx})$
11.	$\dfrac{1}{p^2+a^2}$	$\dfrac{1}{a}\sin(ax)$
12.	$\dfrac{p}{p^2+a^2}$	$\cos(ax)$
13.	$\dfrac{1}{(p+b)^2+a^2}$	$\dfrac{1}{a}e^{-bx}\sin(ax)$
14.	$\dfrac{p}{(p+b)^2+a^2}$	$e^{-bx}[\cos(ax)-\dfrac{b}{a}\sin(ax)]$
15.	$\dfrac{1}{p^3}$	$\dfrac{1}{2}x^2$
16.	$\dfrac{1}{p^2(p+a)}$	$\dfrac{1}{a^2}(e^{-ax}+ax-1)$
17.	$\dfrac{1}{p(p+a)(p+b)}$	$\dfrac{1}{ab(a-b)}(a-b+be^{-ax}-ae^{-bx})$
18.	$\dfrac{1}{p(p+a)^2}$	$\dfrac{1}{a^2}(1-e^{-ax}-axe^{-ax})$
19.	$\dfrac{1}{(p+a)(p+b)(p+c)}$	$\dfrac{(c-b)e^{-ax}+(a-c)e^{-bx}+(b-a)e^{-cx}}{(a-b)(b-c)(c-a)}$
20.	$\dfrac{p}{(p+a)(p+b)(p+c)}$	$\dfrac{a(b-c)e^{-ax}+b(c-a)e^{-bx}+c(a-b)e^{-cx}}{(a-b)(b-c)(c-a)}$
21.	$\dfrac{p^2}{(p+a)(p+b)(p+c)}$	$\dfrac{a^2(c-b)e^{-ax}+b^2(a-c)e^{-bx}+c^2(b-a)e^{-cx}}{(a-b)(b-c)(c-a)}$
22.	$\dfrac{1}{(p+a)(p+b)^2}$	$\dfrac{1}{(a-b)^2}[e^{-ax}-e^{-bx}+(a-b)xe^{-bx}]$
23.	$\dfrac{p}{(p+a)(p+b)^2}$	$\dfrac{1}{(a-b)^2}\{-ae^{-ax}+[a+b(b-a)x]e^{-bx}\}$
24.	$\dfrac{p^2}{(p+a)(p+b)^2}$	$\dfrac{1}{(a-b)^2}\{a^2e^{-ax}+b(b-2a-b^2x+abx)x)e^{-bx}\}$

25.	$\dfrac{1}{(p+a)^3}$	$\dfrac{1}{2}x^2 e^{-ax}$
26.	$\dfrac{p}{(p+a)^3}$	$x\left(1-\dfrac{1}{2}ax\right)e^{-ax}$
27.	$\dfrac{p^2}{(p+a)^3}$	$\left(1-2ax+\dfrac{1}{2}a^2x^2\right)e^{-ax}$
28.	$\dfrac{1}{p^2(p^2+a^2)}$	$\dfrac{1}{a^2}[1-\cos(ax)]$
29.	$\dfrac{1}{p[(p+b)^2+a^2]}$	$\dfrac{1}{a^2+b^2}\{1-e^{-bx}[\cos(ax)+\dfrac{b}{a}\sin(ax)]\}$
30.	$\dfrac{1}{(p+a)(p^2+b^2)}$	$\dfrac{1}{a^2+b^2}[e^{-ax}+\dfrac{a}{b}\sin(bx)-\cos(bx)]$
31.	$\dfrac{p}{(p+a)(p^2+b^2)}$	$\dfrac{1}{a^2+b^2}[-ae^{-ax}+a\cos(bx)+b\cos(bx)]$
32.	$\dfrac{p^2}{(p+a)(p^2+b^2)}$	$\dfrac{1}{a^2+b^2}[a^2e^{-ax}-ab\sin(bx)+b^2\cos(bx)]$
33.	$\dfrac{1}{p^3+a^3}$	$\dfrac{1}{3a^2}e^{-ax}-\dfrac{1}{3a^2}e^{ax/2}[\cos(kx)-\sqrt{3}\sin(kx)]$
34.	$\dfrac{p}{p^3+a^3}$	$-\dfrac{1}{3a}e^{-ax}+\dfrac{1}{3a}e^{ax/2}[\cos(kx)+\sqrt{3}\sin(kx)];k=\dfrac{1}{2}a\sqrt{3}$
35.	$\dfrac{p^2}{p^3+a^3}$	$\dfrac{1}{3}e^{-ax}+\dfrac{2}{3}e^{ax/2}\cos(kx),k=\dfrac{1}{2}a\sqrt{3}$
36.	$\dfrac{1}{(p+a)[(p+b)^2+c^2]}$	$\dfrac{-ae^{-ax}+ae^{-bx}\cos(cx)+ke^{-bx}\sin(cx)}{(a-b)^2+c^2},k=\dfrac{b^2+c^2-ab}{c}$
37.	$\dfrac{p}{(p+a)[(p+b)^2+c^2]}$	$\dfrac{-ae^{-ax}+ae^{-bx}\cos(cx)+ke^{-bx}\sin(cx)}{(a-b)^2+c^2},k=\dfrac{b^2+c^2-ab}{c}$
38.	$\dfrac{p^2}{(p+a)[(p+b)^2+c^2]}$	$\dfrac{-a^2e^{-ax}+(b^2+c^2-2ab)e^{-bx}\cos(cx)+ke^{-bx}\sin(cx)}{(a-b)^2+c^2},$ $k=\dfrac{-ac-bc+ab^2-b^3}{c}$
39.	$\dfrac{1}{p^4}$	$\dfrac{1}{6}x^3$
40.	$\dfrac{1}{p^3(p+a)}$	$\dfrac{1}{a^3}-\dfrac{1}{a^2}x+\dfrac{1}{2a}x^2-\dfrac{1}{a^3}e^{-ax}$
41.	$\dfrac{1}{p^2(p+a)^2}$	$\dfrac{1}{a^2}x(1+e^{-ax})+\dfrac{2}{a^3}(e^{-ax}-1)$
42.	$\dfrac{1}{p^2(p+a)(p+b)}$	$-\dfrac{a+b}{a^2b^2}+\dfrac{1}{ab}x+\dfrac{1}{a^2(b-a)}e^{-ax}+\dfrac{1}{b^2(a-b)}e^{-bx}$

43.	$\dfrac{1}{(p+a)^2(p+b)^2}$	$\dfrac{1}{(a-b)^2}\left[e^{-ax}\left(x+\dfrac{2}{a-b}\right)+e^{-bx}\left(x-\dfrac{2}{a-b}\right)\right]$
44.	$\dfrac{1}{(p+a)^4}$	$\dfrac{1}{6}x^3e^{-ax}$
45.	$\dfrac{p}{(p+a)^4}$	$\dfrac{1}{2}x^2e^{-ax}-\dfrac{1}{6}ax^3e^{-ax}$
46.	$\dfrac{1}{p^2(p^2+a^2)}$	$\dfrac{1}{a^3}[ax-\sin ax]$
47.	$\dfrac{1}{p^4-a^4}$	$\dfrac{1}{2a^3}[\sinh(ax)-\sin ax]$
48.	$\dfrac{p}{p^4-a^4}$	$\dfrac{1}{2a^2}[\cosh(ax)-\cos ax]$
49.	$\dfrac{p^2}{p^4-a^4}$	$\dfrac{1}{2a}[\sinh(ax)+\sin ax]$
50.	$\dfrac{p^3}{p^4-a^4}$	$\dfrac{1}{2a}[\cosh(ax)+\cos ax]$
51.	$\dfrac{1}{(p^2+a^2)^2}$	$\dfrac{1}{2a^3}[\sin(ax)-ax\cos(ax)]$
52.	$\dfrac{p}{(p^2+a^2)^2}$	$\dfrac{1}{2a}x\sin(ax)$
53.	$\dfrac{p^2}{(p^2+a^2)^2}$	$\dfrac{1}{2a}[\sin(ax)+ax\cos(ax)]$
54.	$\dfrac{p^3}{(p^2+a^2)^2}$	$\cos(ax)-\dfrac{1}{2}ax\sin ax$
55.	$\dfrac{1}{[(p+b)^2+a^2]^2}$	$\dfrac{1}{2a^3}e^{-bx}[\sin(ax)-ax\cos(ax)]$
56.	$\dfrac{1}{(p^2-a^2)(p^2-b^2)}$	$\dfrac{1}{a^2-b^2}\left[\dfrac{1}{a}\sinh(ax)-\dfrac{1}{b}\sinh(bx)\right]$
57.	$\dfrac{p}{(p^2-a^2)(p^2-b^2)}$	$\dfrac{\cosh(ax)-\cosh(bx)}{a^2-b^2}$
58.	$\dfrac{p^2}{(p^2-a^2)(p^2-b^2)}$	$\dfrac{a\sinh(ax)-b\sinh(bx)}{a^2-b^2}$
59.	$\dfrac{p^3}{(p^2-a^2)(p^2-b^2)}$	$\dfrac{a^2\cosh(ax)-b^2\cosh(bx)}{a^2-b^2}$
60.	$\dfrac{1}{(p^2+a^2)(p^2+b^2)}$	$\dfrac{1}{b^2-a^2}\left[\dfrac{1}{a}\sin(ax)-\dfrac{1}{b}\sin(bx)\right]$
61.	$\dfrac{p}{(p^2+a^2)(p^2+b^2)}$	$\dfrac{\cos(ax)-\cos(bx)}{b^2-a^2}$

62.	$\dfrac{p^2}{(p^2+a^2)(p^2+b^2)}$	$\dfrac{-a\sin(ax)+b\sin(bx)}{b^2-a^2}$
63.	$\dfrac{p^3}{(p^2+a^2)(p^2+b^2)}$	$\dfrac{-a^2\cos(ax)+b^2\cos(bx)}{b^2-a^2}$
64.	$\dfrac{1}{(p+a)^n}, n=1,2,\dots$	$\dfrac{1}{(n-1)!}x^{n-1}e^{-ax}$
65.	$\dfrac{1}{p(p+a)^n}, n=1,2,\dots$	$a^{-n}[1-e^{-ax}e_n(ax)],\ e_n(z)=1+\dfrac{z}{1!}+\dots+\dfrac{z^n}{n!}$

(3) FOR SQUARE ROOTS

S. No.	$f(p)$	$L^{-1}\{f(p)\} = F(x)$
1.	$\dfrac{1}{\sqrt{p}}$	$\dfrac{1}{\sqrt{\pi x}}$
2.	$\sqrt{p-a}-\sqrt{p-b}$	$\dfrac{e^{bx}-e^{ax}}{2\sqrt{\pi x^3}}$
3.	$\dfrac{1}{\sqrt{p+a}}$	$\dfrac{1}{\sqrt{\pi x}}e^{-ax}$
4.	$\dfrac{1}{p\sqrt{p}}$	$2\sqrt{\dfrac{x}{\pi}}$
5.	$p^{-n-1/2}$	$\dfrac{2^n}{1\times 3\times\dots\times(2n-1)\sqrt{\pi}}x^{n-1/2}$
6.	$(p+a)^{-n-1/2}$	$\dfrac{2^n}{1\times 3\times\dots\times(2n-1)\sqrt{\pi}}x^{n-1/2}e^{-ax}$
7.	$\dfrac{1}{\sqrt{p^2+a^2}}$	$J_0(ax)$
8.	$\dfrac{1}{\sqrt{p^2+ap+b}}$	$e^{(-1/2ax)}J_0[(b-1/4a^2)^{1/2}x]$
9.	$(\sqrt{p^2+a^2}-p)^{1/2}$	$\dfrac{1}{\sqrt{2\pi x^3}}\sin(ax)$
10.	$\dfrac{1}{\sqrt{p^2+a^2}}(\sqrt{p^2+a^2}+p)^{1/2}$	$\dfrac{\sqrt{2}}{\sqrt{\pi x}}\cos(ax)$
11.	$\dfrac{1}{\sqrt{p^2-a^2}}(\sqrt{p^2-a^2}+p)^{1/2}$	$\dfrac{\sqrt{2}}{\sqrt{\pi x}}\cosh(ax)$
12.	$(\sqrt{p^2+a^2}+p)^{-n}$	$na^{-n}x^{-1}J_n(ax)$
13.	$(p^2+a^2)^{-n-1/2}$	$\dfrac{(x/a)^nJ_n(ax)}{1\times 3\times 5\times\dots\times(2n-1)}$

(4) FOR ARBITRARY POWERS

S. No.	$f(p)$	$L^{-1}\{f(p)\} = F(x)$
1.	$(p^2 + a^2)^{-n-1/2}, n > -1/2$	$\dfrac{\sqrt{\pi}}{(2a)^n \Gamma\left(n + \dfrac{1}{2}\right)} x^n J_n(ax)$
2.	$p(p^2 + a^2)^{-n-1/2}, n > 0$	$\dfrac{a\sqrt{\pi}}{(2a)^n \Gamma\left(n + \dfrac{1}{2}\right)} x^n J_{n-1}(ax)$
3.	$\begin{aligned}&[(p^2 + a^2)^{1/2} + p]^n \\ &= a^{-2n}[p - (p^2 - a^2)^{1/2}]^n, n > 0\end{aligned}$	$n a^{-n} x^{-1} J_n(ax)$
4.	$\dfrac{(\sqrt{p^2 + a^2} + p)^{-n}}{(\sqrt{p^2 + a^2})}, n > -1$	$a^{-n} J_n(ax)$

(5) FOR EXPONENTIAL FUNCTIONS

S. No.	$f(p)$	$L^{-1}\{f(p)\} = F(x)$
1.	$p^{-1} e^{-ap}, a > 0$	$\begin{cases} 0, & 0 < x < a \\ 1, & a < x \end{cases}$
2.	$p^{-1}(1 - e^{-ap}), a > 0$	$\begin{cases} 1, & 0 < x < a \\ 0, & a < x \end{cases}$
3.	$p^{-1}(e^{-ap} - e^{-bp}), 0 \le a < b$	$\begin{cases} 0, & 0 < x < a \\ 1, & a < x < b \\ 0, & b < x \end{cases}$
4.	$p^{-2}(e^{-ap} - e^{-bp}), 0 \le a < b$	$\begin{cases} 0, & 0 < x < a \\ x - a, & a < x < b \\ x - b, & b < x \end{cases}$
5.	$(p + b)^{-1} e^{-ap}, a > 0$	$\begin{cases} 0, & 0 < x < a \\ e^{-b(x-a)}, & a < x \end{cases}$
6.	$p^{-1/2} e^{a/p}$	$\dfrac{1}{\sqrt{\pi x}} \cosh(2\sqrt{ax})$
7.	$p^{-3/2} e^{a/p}$	$\dfrac{1}{\sqrt{\pi a}} \sinh(2\sqrt{ax})$
8.	$p^{-5/2} e^{a/p}$	$\sqrt{\dfrac{x}{\pi a}} \cosh(2\sqrt{ax}) - \dfrac{1}{2\sqrt{\pi a^3}} \sinh(2\sqrt{ax})$
9.	$1 - e^{-a/p}$	$\sqrt{\dfrac{a}{x}} J_1(2\sqrt{ax})$
10.	$p^{-1/2} e^{-a/p}$	$\dfrac{1}{\sqrt{\pi x}} \cos(2\sqrt{ax})$
11.	$p^{-3/2} e^{-a/p}$	$\dfrac{1}{\sqrt{\pi a}} \sin(2\sqrt{ax})$

12.	$p^{-5/2}e^{-a/p}$	$\dfrac{1}{2\sqrt{\pi a^3}}\sin(2\sqrt{ax})-\sqrt{\dfrac{x}{\pi a}}\cos(2\sqrt{ax})$
13	$e^{-\sqrt{ap}},a>0$	$\dfrac{\sqrt{a}}{2\sqrt{\pi}}x^{-3/2}e^{(-a/4x)}$
14.	$pe^{-\sqrt{ap}},a>0$	$\dfrac{\sqrt{a}}{8\sqrt{\pi}}(a-6x)x^{-7/2}e^{(-a/4x)}$
15.	$\sqrt{p}e^{(-\sqrt{ap})},a>0$	$\dfrac{1}{4\sqrt{\pi}}(a-2x)x^{-5/2}e^{(-a/4x)}$
16.	$\dfrac{1}{\sqrt{p}}e^{(-\sqrt{ap})},a\ge 0$	$\dfrac{1}{\sqrt{\pi x}}e^{(-a/4x)}$

(6) FOR HYPERBOLIC FUNCTIONS

S. No.	$f(p)$	$L^{-1}\{f(p)\}=F(x)$
1.	$\dfrac{\sinh(a/p)}{\sqrt{p}}$	$\dfrac{1}{2\sqrt{\pi x}}[\cosh(2\sqrt{ax})-\cos(2\sqrt{ax})]$
2.	$\dfrac{\sinh(a/p)}{p\sqrt{p}}$	$\dfrac{1}{2\sqrt{\pi a}}[\sinh(2\sqrt{ax})-\sin(2\sqrt{ax})]$
3.	$\dfrac{1}{p\cosh(ap)},a>0$	$f(x)=\begin{cases}0 & \text{if} \quad a(4n-1)<x<a(4n+1)\\ 2 & \text{if} \quad a(4n+1)<x<a(4n+3)\end{cases}$ $n=0,1,2,\dots\ (x>0)$
4.	$\dfrac{\cosh(a/p)}{\sqrt{p}}$	$\dfrac{1}{2\sqrt{\pi x}}[\cosh(2\sqrt{ax})+\cos(2\sqrt{ax})]$
5.	$\dfrac{\cosh(a/p)}{p\sqrt{p}}$	$\dfrac{1}{2\sqrt{\pi a}}[\sinh(2\sqrt{ax})+\sin(2\sqrt{ax})]$

(7) FOR LOGARITHMIC FUNCTIONS

S. No.	$f(p)$	$L^{-1}\{f(p)\}=F(x)$
1.	$\dfrac{1}{p}\log p$	$-\log x-C$ $C=0.5772\dots$ is the Euler constant
2.	$p^{-n-1}\log p$	$\left(1+\dfrac{1}{2}+\dfrac{1}{3}+\dots+\dfrac{1}{n}-\log x-C\right)\dfrac{x^n}{n!}$ $C=0.5772\dots$ is the Euler constant
3.	$\dfrac{1}{p}(\log p)^2$	$(\log x+C)^2-\dfrac{1}{6}\pi^2,C=0.5772\dots$
4.	$\dfrac{1}{p^2}(\log p)^2$	$x\left[(\log x+C-1)^2+1-\dfrac{1}{6}\pi^2\right]$
5.	$\log\dfrac{p+b}{p+a}$	$\dfrac{1}{x}(e^{-ax}-e^{-bx})$
6.	$\log\dfrac{p^2+b^2}{p^2+a^2}$	$\dfrac{2}{x}[\cos(ax)-\cos(bx)]$

7.	$p \log \dfrac{p^2 + b^2}{p^2 + a^2}$	$\dfrac{2}{x}[\cos(bx) + bx\sin(bx) - \cos(ax) - ax\sin(ax)]$
8.	$\log \dfrac{(p+a)^2 + k^2}{(p+b)^2 + k^2}$	$\dfrac{2}{x}\cos(kx)(e^{-bx} - e^{-ax})$
9.	$p \log\left(\dfrac{1}{p}\sqrt{p^2 + a^2}\right)$	$\dfrac{1}{x^2}[\cos(ax) - 1] + \dfrac{a}{x}\sin(ax)$
10.	$p \log\left(\dfrac{1}{p}\sqrt{p^2 - a^2}\right)$	$\dfrac{1}{x^2}[\cosh(ax) - 1] - \dfrac{a}{x}\sinh(ax)$

(8) FOR TRIGONOMETRIC FUNCTIONS

S. No.	$f(p)$	$L^{-1}\{f(p)\} = F(x)$
1.	$\dfrac{\sin(a/p)}{\sqrt{p}}$	$\dfrac{1}{\sqrt{\pi x}}\sinh(\sqrt{2ax})\sin(\sqrt{2ax})$
2.	$\dfrac{\sin(a/p)}{p\sqrt{p}}$	$\dfrac{1}{\sqrt{\pi a}}\cosh(\sqrt{2ax})\sin(\sqrt{2ax})$
3.	$\dfrac{\cos(a/p)}{\sqrt{p}}$	$\dfrac{1}{\sqrt{\pi x}}\cosh(\sqrt{2ax})\cos(\sqrt{2ax})$
4.	$\dfrac{\cos(a/p)}{p\sqrt{p}}$	$\dfrac{1}{\sqrt{\pi a}}\sinh(\sqrt{2ax})\cos(\sqrt{2ax})$
5.	$\dfrac{1}{\sqrt{p}}e^{(-\sqrt{ap})}\sin(\sqrt{ap})$	$\dfrac{1}{\sqrt{\pi x}}\sin\left(\dfrac{a}{2x}\right)$
6.	$\dfrac{1}{\sqrt{p}}e^{(-\sqrt{ap})}\cos(\sqrt{ap})$	$\dfrac{1}{\sqrt{\pi x}}\cos\left(\dfrac{a}{2x}\right)$
7.	$\tan^{-1}\left(\dfrac{a}{p}\right)$	$\dfrac{1}{x}\sin ax$
8.	$p\tan^{-1}\left(\dfrac{a}{p} - a\right)$	$\dfrac{1}{x^2}[ax\cos(ax) - \sin(ax)]$
9.	$\tan^{-1}\left(\dfrac{2ap}{p^2 + b^2}\right)$	$\dfrac{2}{x}\sin(ax)\cos(x\sqrt{a^2 + b^2})$

VVVVVV

Chapter 3

Applications of Laplace Transform to Solution of Differential Equation and Boundary Value Problems

3.1 INTRODUCTION

Consider a linear differential equation with constant coefficients

$$\frac{d^n y}{dt^n} + A_1 \frac{d^{n-1} y}{dt^{n-1}} + \dots + A_{n-1} \frac{dy}{dt} + A_n y = F(t) \qquad \dots (1)$$

where t is the independent variable and $F(t)$ is a function of t.

Let $\qquad\qquad y(0) = C_1, \; y'(0) = C_2, \dots, y^{n-1}(0) = C_{n-1} \qquad \dots (2)$

be the given initial or boundary conditions, where $C_1, C_2, \dots, C_{n-1}$ are constants. Now, taking the Laplace transform of both sides of (1) and using the conditions given by (2), we get an algebraic equation from which $\bar{y}(p) = L\{y(t)\}$ is determined. The required solution is then obtained by finding the inverse Laplace transform of $\bar{y}(p)$.

REMARKS

- The algebraic equation, obtained above is known as subsidiary equation.
- The above method is easily extended to higher order differential equation.

WORKING PROCEDURE

Step 1. *Taking Laplace transform of both the sides of the given differential equation and use given initial conditions.*

Step 2. *Solve the equation obtained in step (1) for L{y}.*

Step 3. *Taking inverse Laplace transform to find y.*

Solved Examples

EXAMPLE 1. *Solve $\dfrac{d^2 y}{dt^2} + y = 0$ under the condition that $y = 1, \dfrac{dy}{dt} = 0$ when $t = 0$.*

(Meerut–1985, 92; Rohilkhand–2002)

SOLUTION. Here, the given equation is

$$\frac{d^2 y}{dt^2} + y = 0. \qquad \dots (1)$$

Taking the Laplace transform of both sides of the given differential equation, we get

$$L\{y''\} + L\{y\} = 0 \;\Rightarrow\; p^2 L\{y\} - py(0) - y'(0) + L\{y\} = 0$$

$$\Rightarrow \quad (p^2 + 1)\, L\{y\} - p.1 - 0 = 0 \qquad \text{[Using the given conditions]}$$

$$\Rightarrow \qquad\qquad L\{y\} = \frac{p}{p^2 + 1}$$

Therefore, $\qquad y = L^{-1}\left\{\dfrac{p}{p^2 + 1}\right\} = \cos t$.

EXAMPLE 2. *Solve* $(D^2 + 1)y = 6\cos 2t$ *if* $y = 3$, $Dy = 1$ *when* $t = 0$. (Agra–2001;

Meerut–1991, 93, 2012; Kanpur–2012;
Purvanchal–1989, 2000, 02, 07; Kashi–2007; Kurukshetra-2005)

SOLUTION. The given equation can be written as

$$y'' + y = 6\cos 2t$$

Taking the Laplace transform of both the sides of the given differential equation, we get

$$L\{y''\} + L\{y\} = 6L\{\cos(2t)\}$$

$$\Rightarrow \quad p^2 L\{y\} - py(0) - y'(0) + L\{y\} = 6\frac{p}{p^2 + 2^2}$$

$$\Rightarrow \quad (p^2 + 1)\, L\{y\} - 3p - 1 = \frac{6p}{p^2 + 4} \qquad \text{[Using the given conditions]}$$

$$\Rightarrow L\{y\} = \frac{3p}{p^2 + 1} + \frac{1}{p^2 + 1} + \frac{6p}{(p^2 + 1)(p^2 + 4)}$$

$$= \frac{3p}{p^2 + 1} + \frac{1}{p^2 + 1} + \frac{2p[(p^2 + 4) - (p^2 + 1)]}{(p^2 + 1)(p^2 + 4)}$$

$$= \frac{3p}{p^2 + 1} + \frac{1}{p^2 + 1} + 2p\left\{\frac{1}{p^2 + 1} - \frac{1}{p^2 + 4}\right\}$$

$$= \frac{5p}{p^2 + 1} + \frac{1}{p^2 + 1} - \frac{2p}{p^2 + 4}$$

Therefore,

$$y = 5L^{-1}\left\{\frac{p}{p^2 + 1}\right\} + L^{-1}\left\{\frac{1}{p^2 + 1}\right\} - 2L^{-1}\left\{\frac{p}{p^2 + 4}\right\}$$

$$\Rightarrow \quad y = 5\cos t + \sin t - 2\cos 2t$$

EXAMPLE 3. *Using Laplace transforms, find the solution of the initial value problem :*

$$y'' + 9y = 6\cos 3t, \quad y(0) = 2, \quad y'(0) = 0 \qquad \text{(UPTU–2005, 06)}$$

SOLUTION. The given equation can be written as

$$y'' + 9y = 6\cos 3t, \quad y(0) = 2, \quad y'(0) = 0 \qquad \text{... (1)}$$

Taking Laplace transform of (1), we get

$$[p^2 L\{y\} - py(0) - y'(0)] + 9L\{y\} = 6\frac{p}{p^2 + 9}$$

Putting the value of $y(0)$ and $y'(0)$ in (2), we have

$$p^2 L\{y\} - 2p + 9L\{y\} = \frac{6p}{p^2 + 9}$$

$$(p^2 + 9)L\{y\} = 2p + \frac{6p}{p^2 + 9}$$

$$L\{y\} = \frac{2p}{p^2 + 9} + \frac{6p}{(p^2 + 9)^2}$$

$$\Rightarrow \quad y = L^{-1}\left\{\frac{2p}{p^2+9}\right\} + L^{-1}\left\{\frac{6p}{(p^2+9)^2}\right\} = 2\cos 3t + 3L^{-1}\frac{d}{dp}\left[-\frac{3}{p^2+9}\right]$$

$$= 2\cos 3t + t\sin 3t$$

EXAMPLE 4. *Solve* $(D^2+9)y = \cos 2t$ *if* $y(0)=1,\ y\left(\frac{\pi}{2}\right) = -1.$ (UPTU–2002, 06; Bhopal–2008;

Agra–2001, 04, 09, 11; Meerut–1982, 91, 92, 93, 94, 98, 2007, Rohilkhand–1990, 2000, 02, 04,

Garhwal–1996; Bundelkhand–2007; Purvanchal–1990, 2001, 05; Kanpur–2004)

SOLUTION.   The given equation can be written as

$$y''+9y = \cos 2t \qquad \qquad \dots (1)$$

Taking the Laplace transform of both the sides of (1), we get

$$L\{y''\} + 9L\{y\} = L\{\cos 2t\}$$

$$\Rightarrow \quad p^2 L\{y\} - py(0) - y'(0) + 9L\{y\} = \frac{p}{p^2+4}$$

$$\Rightarrow \quad (p^2+9)\,L\{y\} - p - C = \frac{p}{p^2+4},\ \text{where}\ C = y'(0)$$

$$\therefore \quad L\{y\} = \frac{p+C}{p^2+9} + \frac{p}{(p^2+9)(p^2+4)}$$

$$= \frac{p}{p^2+9} + \frac{C}{p^2+9} + \frac{p}{5(p^2+4)} - \frac{p}{5(p^2+9)}$$

Therefore,

$$y = L^{-1}\left\{\frac{p}{p^2+9}\right\} + CL^{-1}\left\{\frac{1}{p^2+9}\right\} + \frac{1}{5}L^{-1}\left\{\frac{p}{p^2+4}\right\} - \frac{1}{5}L^{-1}\left\{\frac{p}{p^2+9}\right\}$$

$$= \cos 3t + \frac{1}{3}C\sin 3t + \frac{1}{5}\cos 2t - \frac{1}{5}\cos 3t$$

$$= \frac{4}{5}\cos 3t + \frac{1}{3}C\sin 3t + \frac{1}{5}\cos 2t \qquad \qquad \dots (2)$$

Now, since $y\left(\frac{\pi}{2}\right) = -1$, therefore, from (1), we have

$$-1 = \frac{4}{5}\cos\frac{3\pi}{2} + \frac{1}{3}C\sin\frac{3\pi}{2} + \frac{1}{5}\cos\pi$$

On solving, we get $C = \frac{12}{5}$

Put this value in (2), we get

$$y = \frac{4}{5}\cos 3t + \frac{4}{5}\sin 3t + \frac{1}{5}\cos 2t\ .$$

EXAMPLE 5.   *Solve using Laplace transform method*

$$y''(t) + 4y'(t) + 4y(t) = 6e^{-t}, \text{with } y(0) = -2,\ y'(0) = 8.$$ (UPTU–2007)

SOLUTION.   The given equation can be written as

$$y''(t) + 4y'(t) + 4y(t) = 6e^{-t}$$

Taking Laplace transform on both sides of the given equation, we get

$$[p^2 L\{y\} - py(0) - y'(0)] + 4[pL\{y\} - y(0)] + 4L\{y\} = \frac{6}{p+1} \qquad ...(1)$$

Putting $y(0) = -2$ and $y'(0) = 8$ in (1), we get

$$[p^2 L\{y\} - p(-2) - 8] + 4[pL\{y\} + 2] + 4L\{y\} = \frac{6}{p+1}$$

$$\Rightarrow \qquad (p^2 + 4p + 4)L\{y\} + 2p = \frac{6}{p+1}$$

$$\Rightarrow \qquad (p^2 + 4p + 4)L\{y\} = -2p + \frac{6}{p+1}$$

$$\Rightarrow \qquad (p+2)^2 L\{y\} = \frac{-2p^2 - 2p + 6}{(p+1)}$$

$$\Rightarrow \qquad L\{y\} = \frac{-2p^2 - 2p + 6}{(p+1)(p+2)^2}$$

Let

$$\frac{-2p^2 - 2p + 6}{(p+1)(p+2)^2} = \frac{A}{p+1} + \frac{B}{p+2} + \frac{C}{(p+2)^2}$$

$$-2p^2 - 2p + 6 = A(p+2)^2 + B(p+1)(p+2) + C(p+1)$$

$$-2 + 2 + 6 = A(-1+2)^2 \Rightarrow A = 6 \qquad \text{[Putting } p = -1]$$

$$-8 + 4 + 6 = C(-2+1) \Rightarrow C = -2 \qquad \text{[Putting } p = -2]$$

Comparing the coefficients of p^2 on both sides, we get

$$-2 = A + B \Rightarrow -2 = 6 + B \Rightarrow B = -8$$

$$L\{y\} = \frac{6}{p+1} + \frac{-8}{p+2} + \frac{2}{(p+2)^2}$$

$$y = L^{-1}\left[\frac{6}{p+1} - \frac{8}{p+2} - \frac{2}{(p+2)^2}\right]$$

Hence,

$$y = 6e^{-t} - 8e^{-2t} - 2e^{-2t}t$$

EXAMPLE 6. *Solve* $(D^3 - 2D^2 + 5D)y = 0$ *given that* $y(0) = 0$, $y'(0) = 1$, $y\left(\dfrac{\pi}{8}\right) = 1$.

(UPTU Q. Bank–2001; Agra–2009, 10; Meerut–1986; Kanpur–1984, 86; Roorkee–1981)

SOLUTION. The given equation can be written as

$$y''' - 2y'' + 5y' = 0 \qquad ... (1)$$

Taking the Laplace transforms of both sides of (1), we get

$$L\{y'''\} - 2L\{y''\} + 5L\{y'\} = 0$$

$$\Rightarrow \quad p^3 L\{y\} - p^2 y(0) - py'(0) - y''(0) - 2[p^2 L\{y\} - py(0) - y'(0)]$$
$$+ 5[pL\{y\} - y(0)] = 0$$

$$\Rightarrow \quad [p^3 - 2p^2 + 5p]\, L\{y\} - p - C - 2(-1) + 5.0 = 0 \text{, where } y''(0) = C$$

$$L\{y\} = \frac{C - 2 + p}{p^3 - 2p^2 + 5p} = \frac{C-2}{p(p^2 - 2p + 5)} + \frac{1}{p^2 - 2p + 5}$$

$$= \frac{C-2}{5p} - \frac{C-2}{5} \cdot \frac{p-2}{p^2 - 2p + 5} + \frac{1}{p^2 - 2p + 5}$$

$$= \frac{C-2}{5p} - \frac{C-2}{5} \cdot \frac{(p-1)-1}{(p-1)^2 + 4} + \frac{1}{(p-1)^2 + 4}$$

$$= \frac{C-2}{5p} - \frac{C-2}{5} \cdot \frac{(p-1)}{(p-1)^2 + 4} + \frac{C+3}{10} \cdot \frac{2}{(p-1)^2 + 4}$$

Therefore,

$$y = \frac{C-2}{5} \cdot L^{-1}\left\{\frac{1}{p}\right\} - \frac{C-2}{5} L^{-1}\left\{\frac{p-1}{(p-1)^2 + 4}\right\} + \frac{C+3}{10} L^{-1}\left\{\frac{2}{(p-1)^2 + 4}\right\}$$

$$= \frac{C-2}{5} - \frac{C-2}{5} e^t \cos 2t + \frac{C+3}{10} e^t \sin 2t \qquad \ldots(2)$$

Now, since $y\left(\frac{\pi}{8}\right) = 1$, therefore $1 = \frac{C-2}{5} - \frac{C-2}{5} e^{\pi/8} \frac{1}{\sqrt{2}} + \frac{C+3}{10} e^{\pi/8} \cdot \frac{1}{\sqrt{2}}$

$$\Rightarrow \quad \frac{7-C}{5} = \frac{e^{\pi/8}}{10\sqrt{2}}(-2C + 4 + C + 3)$$

$$\Rightarrow \quad \left(\frac{7-C}{5}\right) \cdot \left(1 - \frac{e^{\pi/8}}{2\sqrt{2}}\right) = 0$$

$$\Rightarrow \quad C = 7$$

Put this value of C in (2), we get

$$y = 1 + e^t(\sin 2t - \cos 2t).$$

EXAMPLE 7. *Solve* $(D^2 - 3D + 2)y = 1 - e^{2t}$, $y = 1$, $Dy = 0$ *when* $t = 0$.

(Agra–1986; Meerut–1981; Garhwal–1996)

SOLUTION. The given equation can be written as

$$y'' - 3y' + 2y = 1 - e^{2t} \qquad \ldots (1)$$

Taking Laplace transform of both the sides of (1), we get

$$L\{y''\} - 3L\{y'\} + 2L\{y\} = L\{1\} - L\{e^{2t}\}$$

$$p^2 L\{y\} - py(0) - y'(0) - 3[pL\{y\} - y(0)] + 2L\{y\} = \frac{1}{p} - \frac{1}{p-2}$$

$$\Rightarrow \quad (p^2 - 3p + 2)L\{y\} - p + 3 = -\frac{2}{p(p-2)}$$

$$\Rightarrow \quad (p-1)(p-2)L\{y\} = -\frac{2}{p(p-2)} + (p-3)$$

$$\Rightarrow \quad L\{y\} = \frac{p^3 - 5p^2 + 6p - 2}{p(p-1)(p-2)^2} = \frac{p^2 - 4p + 2}{p(p-2)^2}$$

$$= \frac{1}{2p} + \frac{1}{2(p-2)} - \frac{1}{(p-2)^2}$$

Therefore, $y = \dfrac{1}{2}L^{-1}\left\{\dfrac{1}{p}\right\} + \dfrac{1}{2}L^{-1}\left\{\dfrac{1}{p-2}\right\} - L^{-1}\left\{\dfrac{1}{(p-2)^2}\right\}$

Hence, $y = \dfrac{1}{2} + \dfrac{1}{2}e^{2t} - te^{2t}$

EXAMPLE 8.   *Using Laplace transform, solve the following differential equation.*

     $y''(t) + y(t) = 8 \cos t : y(0) = 1, \, y'(0) = -1$ (Agra–2003; Kanpur–1983)

SOLUTION.   Here, we have

$$y''(t) + y(t) = 8 \cos t$$

Assume that $L\{y(t)\} = y(p)$.

Now, taking Laplace transform of both the sides, we get

     $L\{y''(t)\} + L\{y(t)\} = 8\,L\{\cos t\}$

$\Rightarrow$ $p^2 y(p) - py(0) - y'(0) + y(p) = \dfrac{8p}{p^2 + 1}$

Using given boundary conditions

$\Rightarrow$ $p^2 y(p) - p \cdot 1 + 1 + y(p) = \dfrac{8p}{p^2 + 1}$

$\Rightarrow$ $(p^2 + 1)y(p) = p - 1 + \dfrac{8p}{p^2 + 1}$

$\Rightarrow$ $y(p) = \dfrac{p}{p^2 + 1} - \dfrac{1}{p^2 + 1} + \dfrac{8p}{(p+1)^2}$

$\Rightarrow$ $y(t) = L^{-1}\left(\dfrac{p}{p^2 + 1}\right) - L^{-1}\left(\dfrac{1}{p^2 + 1}\right) + 8L^{-1}\left[\dfrac{p}{(p^2 + 1)^2}\right]$

       $= \cos t - \sin t + 8L^{-1}\left\{\dfrac{1}{p^2 + 1} \cdot \dfrac{p}{p^2 + 1}\right\}$

       $= \cos t - \sin t + 8 \sin t * \cos t$

       $= \cos t - \sin t + 8\int_0^t \sin u \cos(t - u)du$

       $= \cos t - \sin t + 4\int_0^t \{\sin t + \sin(2u - t)\}du$

       $= \cos t - \sin t + 4t \sin t - 4\left[\dfrac{\cos(2u - t)}{2}\right]_0^t$

       $= \cos t - \sin t + 4t \sin t - 2[\cos t - \cos(-t)]$

       $= \cos t - \sin t + 4t \sin t.$

EXAMPLE 9.   *Solve* $(D^2 + 6D + 5)y = e^{-t}$, *if* $y(0) = 0, \, y'(0) = 1$.   (Purvanchal–1991)

SOLUTION.   Here, we have

     $(D^2 + 6D + 5)y = e^{-t}$ $\Rightarrow$ $y'' + 6y' + 5y = e^{-t}$

Taking Laplace transform of both the sides, we get

$L\{y''\} + 6L\{y'\} + 5L\{y\} = L\{e^{-t}\}$

$\Rightarrow$ $[p^2 L\{y\} - py\{0\} - y'(0)] + 6[pL\{y\} - y(0)] + 5L\{y\} = 1/(p - 1)$

Using boundary conditions, we get

$\Rightarrow$ $(p^2 + 6p + 5)L\{y\} - 1 = 1/(p + 1)$

$$\Rightarrow \qquad (p+1)(p+5)L\{y\} = 1 + \frac{1}{(p+1)} = \frac{p+2}{p+1}$$

Now, using partial fraction, we get

$$L\{y\} = \frac{p+2}{(p+1)(p+5)} = \frac{3}{16(p+1)} + \frac{1}{4(p+1)^2} - \frac{3}{16}L^{-1}\left\{\frac{1}{p+5}\right\}$$

Taking inverse Laplace transform, we get

$$y = \frac{3}{16}L^{-1}\left\{\frac{1}{p+1}\right\} + \frac{1}{4}L^{-1}\left\{\frac{1}{(p+1)^2}\right\} - \frac{3}{16}L^{-1}\left\{\frac{1}{p+5}\right\}$$

$$= \frac{3}{16}e^{-t} - \frac{1}{4}e^{-t}L^{-1}\left\{\frac{1}{p^2}\right\} - \frac{3}{16}e^{-5t}$$

$$= \frac{3}{16}e^{-t} - \frac{1}{4}e^{-t}\left(\frac{t}{1!}\right) - \frac{3}{16}e^{-5t} \qquad \left(\because L^{-1}\left\{\frac{1}{p^{n+1}}\right\} = \frac{t^n}{n!}\right)$$

$$y = \frac{1}{16}e^{-t}[3 - 4t - 3e^{-4t}].$$

EXAMPLE 10. Solve $(D^2 + D)x = 2$, when $x(0) = 3$, $x'(0) = 1$. (Meerut–2001; Kanpur–2007)

SOLUTION. Here, we have $(D^2 + D)x = 2$

$$x'' + x' = 2$$

Now, taking Laplace transform of both the sides, we get

$$L\{x''\} + L\{x'\} = L\{2\}$$

$$\Rightarrow \quad p^2L\{x\} - px(0) - x'(0) + pL\{x\} - x(0) = 2/p$$

$$\Rightarrow \qquad (p^2 + p)L\{x\} = 3p + 4 + \frac{2}{p}$$

$$\Rightarrow \qquad L\{x\} = \frac{3p^2 + 4p + 2}{p^2(p+1)} = \frac{2}{p} + \frac{2}{p^2} + \frac{1}{p+1}$$

$$\Rightarrow \qquad x = L^{-1}\left\{\frac{2}{p} + \frac{2}{p^2} + \frac{1}{p+1}\right\}$$

$$= L^{-1}\left\{\frac{2}{p}\right\} + L^{-1}\left\{\frac{2}{p^2}\right\} + L^{-1}\left\{\frac{1}{p+1}\right\} = 2 + 2t + e^{-t}.$$

EXAMPLE 11. Solve $(D^2 - D - 6)y = 2$, $t > 0$ if $y = 1$, $Dy = 0$ when $t = 0$. (Agra–2010; Meerut–1986)

SOLUTION. Here, we have

$$(D^2 - D - 6)y = 2$$

$$\Rightarrow \qquad y'' - y' - 6y = 2$$

Taking Laplace transform of both the sides, we get

$$L\{y''\} - L\{y'\} - 6L\{y\} = L\{2\}$$

$$\Rightarrow \quad p^2L\{y\} - py(0) - y'(0) - [pL\{y\} - y(0)] - 6L\{y\} = \frac{2}{p}$$

$$\Rightarrow \qquad p^2L\{y\} - p.1 - 0 - pL\{y\} + 1 - 6L\{y\} = 2/p$$

$$\Rightarrow \quad (p^2 - p - 6)L\{y\} = (2/p) + p - 1$$

$$L\{y\} = \frac{p^2 - p + 2}{p(p^2 - p - 6)} = \frac{p^2 - p + 2}{p(p+2)(p-3)}$$

$$= \left[-\frac{1}{3p} + \frac{4}{5(p+2)} + \frac{8}{15(p-3)} \right]$$

$$y = -1/3L^{-1}\left\{ \frac{1}{p} \right\} + \frac{4}{5}L^{-1}\left\{ \frac{1}{p+2} \right\} + \frac{8}{15}L^{-1}\left\{ \frac{1}{p-3} \right\}$$

$$y = -\frac{1}{3} + \frac{4}{5}e^{-2t} + \frac{8}{15}e^{3t}.$$

EXAMPLE 12. *Solve* $(D^2 - 4D + 5)\,y = 125t^2$, *if* $y = 0 = Dy$, *when* $t = 0$. (Andhra–1990)

SOLUTION. Here, we have

$$(D^2 - 4D + 5)y = 125t^2$$

$$\Rightarrow \qquad y'' - 4y' + 5y = 125t^2$$

Taking Laplace transform of both the sides, we get

$$L\{y''\} - 4L\{y'\} + 5L\{y\} = 125L\{t^2\}$$

$$\Rightarrow \quad p^2 L\{y\} - py(0) - y'(0) - 4[pL\{y\} - y(0)] + 5L\{y\}$$

$$= 125 \cdot (2!/p^3)$$

$$= (p^2 - 4p + 5)L\{y\} = 250p^3$$

$$L\{y\} = \frac{250}{p^3(p^2 - 4p + 5)} = \frac{50}{p^3} + \frac{40}{p^2} + \frac{22}{p} + \frac{2(24 - 11p)}{p^2 - 4p + 5}$$

$$= \frac{50}{p^3} + \frac{40}{p^2} + \frac{22}{p} + 2\left[\frac{2 - 11(p - 2)}{(p - 2)^2 + 1} \right]$$

$$\therefore \qquad y = 50 \cdot \frac{t^2}{2!} + 40 \cdot \frac{t}{1!} + 22 \cdot 1 + 2e^{2t} L^{-1}\left\{ \frac{2 - 11p}{p^2 + 1} \right\}$$

$$y = 25t^2 + 40t + 22 + 2e^{2t}(2\sin t - 11\cos t).$$

EXAMPLE 13. *Find the general solution of the differential equation* $x''(t) + k^2 x(t) = F(t)$; $x(0) = A$, $x'(0) = B$.

(Meerut–1999)

SOLUTION. Here, we have

$$x''(t) + k^2 x(t) = F(t)$$

Taking Laplace transform of both the sides, we get

$$L\{x''\} + k^2 L\{x\} = L\{F(t)\}$$

$$\Rightarrow \quad p^2 L\{x\} - px(0) - x'(0) + k^2 L\{x\} = L\{F(t)\}$$

$$\Rightarrow \qquad (p^2 + k^2)L\{x\} = p(A) + B + f(p) \qquad [\because L\{F(t)\} = f(p)]$$

$$L\{x\} = A\frac{p}{p^2 + k^2} + B\frac{1}{p^2 + k^2} + \frac{1}{p^2 + k^2} f(p)$$

$$x = AL^{-1}\left\{ \frac{p}{p^2 + k^2} \right\} + BL^{-1}\left\{ \frac{1}{p^2 + k^2} \right\} + L^{-1}\left\{ \frac{1}{p^2 + k^2} \cdot f(p) \right\}$$

$$= A\cos kt + (B/k)\sin kt + \{(1/k)\sin kt\} * F(t)$$

Using convolution theorem, we get

$$x(t) = A\cos kt + (B/k)\sin kt + (1/k)\int_0^t \sin k(t - x) \cdot F(x)dx.$$

EXAMPLE 14. Solve $(D^2 - 2D + 2)\, y = 0,\ y = Dy = 1$ when $t = 0,\ D \equiv \dfrac{d}{dt}$

(Agra–1996; Meerut–1983, 84, 86, 2014; Kumaun–2002; Osmania–1992; Kanpur–1967)

SOLUTION.   Here, we have

$$(D^2 - 2D + 2)y = 0$$
$$y'' - 2y' + 2y = 0$$

Taking Laplace transform of both the sides, we get

$$L\{y''\} - 2L\{y'\} + 2L\{y\} = L\{0\}$$
$$\Rightarrow \quad p^2 L\{y\} - py(0) - y'(0) - 2[2L\{y\} - y(0)] + 2L\{y\} = 0$$
$$\Rightarrow \quad p^2 L\{y\} - p - 1 - 2pL\{y\} + 2 + 2L\{y\} = 0$$

$$L\{y\} = \frac{p-1}{p^2 - 2p + 2} = \frac{p-1}{(p-1)^2 + 1}$$

Now, taking inverse Laplace transform of both the sides, we get

$$y = L^{-1}\left\{ \frac{p-1}{(p-1)^2 + 1} \right\} = e^t L^{-1}\left\{ \frac{p}{p^2 + 1} \right\}$$

$$\Rightarrow \qquad y = e^t \cos t.$$

EXAMPLE 15. Solve $(D + 1)^2 y = t$ given that $y = -3$ when $t = 0$ and $y = -1$, when $t = 1$.

(Agra–1985; Meerut–1990, 98, 2009; Rohilkhand–1991; Bundelkhand–2008)

SOLUTION.   Here, we have $(D + 1)^2 y = t$

$$\Rightarrow \qquad (D^2 + 2D + 1)y = t$$
$$\Rightarrow \qquad y'' + 2y' + y = t$$

Now, taking Laplace transform of both the sides, we get

$$L\{y''\} + 2L\{y'\} + L\{y\} = L\{t\}$$

$$\Rightarrow \quad p^2 L\{y\} - py(0) - y'(0) + 2[pL\{y\} - y(0)] + L\{y\} = \frac{1}{p^2}$$

$$\Rightarrow \qquad (p^2 + 2p + 1)L\{y\} - p(-3) - A - 2(-3) = \frac{1}{p^2} \qquad (\because\ y'(0) = A)$$

$$\Rightarrow \qquad (p+1)^2 L\{y\} = \frac{1}{p^2} - 3p - 6 + A$$

$$L\{y\} = \frac{1}{p^2(p+1)^2} - \frac{3p+6}{(p+1)^2} + \frac{A}{(p+1)^2}$$

$$= -\frac{2}{p} + \frac{1}{p^2} + \frac{2}{p+1} + \frac{1}{(p+1)^2} - \frac{3}{p+1} - \frac{3}{(p+1)^2} + \frac{A}{(p+1)^2}$$

$$= -\frac{2}{p} + \frac{1}{p^2} - \frac{1}{p+1} + \frac{A-2}{(p+1)^2}$$

$$y = -2L^{-1}\left\{ \frac{1}{p} \right\} + L^{-1}\left\{ \frac{1}{p^2} \right\} - L^{-1}\left\{ \frac{1}{p+1} \right\} + (A-2)L^{-1}\left\{ \frac{1}{(p+1)^2} \right\}$$

$$= -2.1 + \frac{t}{1!} - e^{-t} + (A-2)e^{-t}L^{-1}\left\{ \frac{1}{p^2} \right\}$$

$$= -2 + t - e^{-t} + (A-2)te^{-t} \qquad\qquad\qquad ...(1)$$

Using given conditions, we get

$$-1 = -2 + 1 - e^{-1} + (A - 2)e^{-4} \quad \Rightarrow \quad A = 3$$

Putting this value in eq^n (1), we get

$$y = -2 + t - e^{-t} + te^{-t}.$$

EXAMPLE 16.   *Using the method of Laplace transform solve the equation*

$$\frac{d^2y}{dt^2} + 4\frac{dy}{dt} + 5y = (\cos t - \sin t)e^{-2t}$$

subject to the conditions $y(0) = 1$, $y'(0) = -3$.   (Agra–2003, 05)

SOLUTION.   Here, we have

$$\frac{d^2y}{dt^2} + 4\frac{dy}{dt} + 5y = (\cos t - \sin t)e^{-2t}$$

$$\Rightarrow \qquad y'' + 4y' + 5y = e^{-2t}\cos t - e^{-2t}\sin t$$

Now, taking Laplace transform of both the sides, we get

$$L\{y''\} - 4L\{y'\} + 5L\{y\} = L\{e^{-2t}\cos t\} - L\{e^{-2t}\sin t\}$$

$$\Rightarrow \quad p^2 L\{y\} - py(0) - y'(0) + 4[pL\{y\} - y(0)] + 5L\{y\}$$

$$= \frac{p+2}{(p+2)^2 + 1^2} - \frac{1}{(p+2)^2 + 1^2}$$

$$\Rightarrow \quad p^2 L\{y\} - p \cdot 1 - (-3) + 4[pL\{y\} - 1] + 5L\{y\}$$

$$= \frac{p+1}{(p+2)^2 + 1}$$

$$\Rightarrow \quad (p^2 + 4p + 5)L\{y\} = \frac{p+1}{(p+2)^2 + 1} + p + 1$$

$$L\{y\} = \frac{p+1}{[(p+2)^2 + 1]^2} + \frac{p+1}{(p+2)^2 + 1}$$

Now, taking inverse Laplace transform of both the sides, we get

$$y = L^{-1}\left\{\frac{p+1}{[(p+2)^2 + 1]^2}\right\} + L^{-1}\left\{\frac{p+1}{(p+2)^2 + 1}\right\}$$

$$= e^{-2t} L^{-1}\left\{\frac{p-2+1}{(p^2+1)^2}\right\} + e^{-2t} L^{-1}\left\{\frac{p-2+1}{p^2+1}\right\}$$

$$ye^{2t} = L^{-1}\left\{\frac{1}{(p^2+1)^2} - \frac{1}{(p^2+1)^2}\right\} + L^{-1}\left\{\frac{p}{p^2+1^2} - \frac{1}{p^2+1^2}\right\}$$

$$= L^{-1}\left\{\frac{p}{(p^2+1)^2}\right\} - L^{-1}\left\{\frac{1}{(p^2+1)^2}\right\} + L^{-1}\left\{\frac{p}{p^2+1^2}\right\} - L^{-1}\left\{\frac{1}{p^2+1^2}\right\}$$

$$= -\frac{1}{2}L^{-1}\left\{\frac{d}{dp}\left(\frac{1}{p^2+1}\right)\right\} - L^{-1}\left\{\frac{1}{(p^2+1)^2}\right\} + \cos t - \sin t$$

$$= \cos t - \sin t - \frac{1}{2}(-1)t^1 L^{-1}\left\{\frac{1}{p^2+1}\right\} - L^{-1}\left\{\frac{1}{(p^2+1)^2}\right\}$$

$$= \cos t - \sin t + \frac{t}{2}\sin t - L^{-1}\left\{\frac{1}{p^2+1}\cdot\frac{1}{p^2+1}\right\}$$

$$= \cos t - \sin t + \frac{t}{2}\sin t - \int_0^t \sin x \sin(t-x)dt$$

Using convolution theorem, we get

$$L^{-1}\left\{\frac{1}{p^2+1}\right\} = \sin t = \cos t - \sin t + \frac{t}{2}\sin t - \frac{1}{2}\int_0^t[\cos(t-2x) - \cos t]dx$$

$$= \cos t - \sin t + \frac{t}{2}\sin t - \frac{1}{2}[\sin t - t\cos t]$$

$$\Rightarrow \quad ye^{2t} = \cos t - \frac{3}{2}\sin t + \frac{t}{2}\sin t + \frac{t}{2}\cos t.$$

EXAMPLE 17. *Solve* $ty'' + y' + 4ty = 0$ *if* $y(0) = 3$, $y'(0) = 0$.   (Agra–1987, 2000, 01, 02)

SOLUTION. Here, we have $ty'' + y' + 4ty = 0$

$$\Rightarrow \quad -\frac{d}{dp}L\{y''\} + L\{y'\} + 4(-1)\frac{d}{dp}L\{y\} = 0$$

$$\Rightarrow \quad -\frac{d}{dp}[p^2 L\{y\} - py(0) - y'(0)] + [pL\{y\} - y(0)] - 4\frac{d}{dp}L\{y\} = 0$$

$$\Rightarrow \quad -\frac{d}{dp}(p^2 z - 3p) + (pz - 3) - 4\frac{dz}{dp} = 0 \text{, where } z = L\{y\}$$

$$\Rightarrow \quad -(p^2+4)\frac{dz}{dp} - pz = 0 \qquad \text{or} \qquad \frac{dz}{z} + \frac{p}{p^2+4}dp = 0$$

On integrating, we have

$$\log z + \frac{1}{2}\log(p^2+4) = \log c_1 \quad \text{or} \qquad z = \frac{c_1}{\sqrt{p^2+4}}$$

$$\Rightarrow \quad L\{y\} = \frac{c_1}{\sqrt{p^2+4}}$$

$$y = c_1 L^{-1}\left\{\frac{1}{\sqrt{p^2+4}}\right\} \qquad \text{or} \qquad y = c_1 J_0(2t)$$

Therefore, $y(0) = 3$ $\therefore$ $3 = c_1 J_0(0) = c_1$ $[\because J_0(0) = 1]$

Hence, $y = 3J_0(2t)$.

EXAMPLE 18. *Solve* $2\dfrac{d^2y}{dt^2} + 5\dfrac{dy}{dt} + 2y = e^{-2t}$, $y(0) = 1$, $y'(0) = 1$.   (Kanpur–2005)

SOLUTION. The given equation can be written as,

$$(2D^2 + 5D + 2)y = e^{-2t},$$

Taking Laplace transform of both the sides, we get

$$2[p^2\bar{y} - py(0) - y'(0)] + 5[p\bar{y} - y(0)] + 2\bar{y} = \frac{1}{p+2}$$

Substituting $y(0) = 1 = y'(0)$, we get

$$2[p^2\bar{y} - p - 1] + 5[p\bar{y} - 1] + 2\bar{y} = \frac{1}{p+2}$$

$$\Rightarrow \qquad (2p^2 + 5p + 2)\bar{y} - 2p - 7 = \frac{1}{p+2}$$

$$\Rightarrow \quad \overline{y} = \frac{1}{(p+2)(2p^2+5p+2)} + \frac{(2p+7)}{2p^2+5p+2}$$

$$\Rightarrow \quad \overline{y} = \frac{1}{(p+2)^2(2p+1)} + \frac{(2p+7)}{(p+2)(2p+1)} \quad \text{...(1)}$$

$$L^{-1}\left\{\frac{1}{(p+2)^2(2p+1)}\right\} = e^{-2t} L^{-1}\left\{\frac{1}{p^2[2(p-2)+1]}\right\}$$

$$= e^{-2t} L^{-1}\left\{\frac{1}{p^2(2p-3)}\right\} \quad \text{...(2)}$$

But
$$\frac{1}{p(2p-3)} = \frac{1}{3}\left[\frac{2}{2p-3} - \frac{1}{p}\right] = \frac{1}{3}\left[\frac{1}{p-\dfrac{3}{2}} - \frac{1}{p}\right]$$

$$\therefore \quad L^{-1}\left\{\frac{1}{p(2p-3)}\right\} = \frac{1}{3}[e^{3t/2} - 1]$$

$$\Rightarrow \quad L^{-1}\left\{\frac{1}{p^2(2p-3)}\right\} = \frac{1}{3}\int_0^t (e^{3x/2} - 1)dx$$

$$\Rightarrow \quad L^{-1}\left\{\frac{1}{p^2(2p-3)}\right\} = \frac{2}{9}(e^{3t/2} - 1) - \frac{1}{3}t \quad \text{...(3)}$$

and
$$\frac{2p+7}{(p+2)(2p+1)} = \frac{4}{2p+1} - \frac{1}{p+2} = \frac{2}{p+\dfrac{1}{2}} - \frac{1}{p+2}$$

$$\Rightarrow \quad L^{-1}\left\{\frac{2p+7}{(p+2)(2p+1)}\right\} = 2e^{-t/2} - e^{-2t} \quad \text{...(4)}$$

Taking inverse Laplace transform of eq^n (1) and substituting the values from (2), (3) and (4), we get

$$y = (2e^{-t/2} - e^{-2t}) + e^{-2t}\left[\frac{2}{9}(e^{3t/2} - 1) - \frac{t}{3}\right]$$

$$= \frac{20}{9}e^{-t/2} - e^{-2t}\left(\frac{11}{9} + \frac{t}{3}\right).$$

EXAMPLE 19.   *Solve the differential equation*
$$tY''(t) + Y'(t) + tY(t) = 0$$
under the conditions that Y(0) = 1 and Y(t) and its derivative have transforms.

SOLUTION.   Let $L\{Y(t)\} = y(p)$,   (Agra–2003, 05; Kanpur–2001)

Now, taking Laplace transform of the given eq^n, we get

$$(-1)^1 \frac{d}{dp}\{p^2 y - pY(0) - Y'(0)\} + \{py - Y(0)\} + (-1)\frac{dy}{dp} = 0$$

Let Y(0) = 1, Y'(0) = c, then

$$-\frac{d}{dp}\{p^2 y - p - c\} + py - 1 - \frac{dy}{dp} = 0$$

$$\Rightarrow \qquad 2py + p^2\frac{dy}{dp} - 1 - py + 1 + \frac{dy}{dp} = 0$$

$$\Rightarrow \qquad py + \frac{dy}{dp}(p^2 + 1) = 0$$

$$\Rightarrow \qquad \frac{dy}{dp} = \frac{-py}{p^2 + 1}$$

$$\Rightarrow \qquad \frac{dy}{y} + \frac{1}{2}\left(\frac{2pdp}{p^2 + 1}\right) = 0$$

On integrating, we get

$$\log y + \frac{1}{2}\log(p^2 + 1) + \text{cons.} = 0$$

$$\Rightarrow \qquad \log y^2(p^2 + 1)a^2 = 0 = \log 1$$
$$= a^2y^2(p^2 + 1) = 1$$

$$y = \frac{1}{a\sqrt{p^2 + 1}}$$

Now, taking inverse Laplace transform, we get
$$aY(t) = J_0(t)$$
Putting $t = 0$, $Y(0) = 1$, $J_0(0) = 1$, then we get $a = 1$
$$Y(t) = J_0(t).$$

EXAMPLE 20. *Solve* $Y''(x) + 9Y(x) = 40e^x$, $Y(0) = 5$, $Y'(0) = -2$. (Kanpur–2010)

SOLUTION. Let $L(Y(t)) = y(p)$

Now, taking inverse Laplace transform of the given eq^n.

$$p^2y - pY(0) - Y'(0) + 9y = \frac{49}{p^2 - 1}$$

Given $\qquad Y(0) = 5$, $Y'(0) = -2$

$$p^2y - 5p + 2 + 9y = \frac{40}{p - 1}$$

$$\Rightarrow \qquad y(p^2 + 9) = \frac{40}{p - 1} - 2 + 5p$$

$$y = \frac{40}{(p - 1)(p^2 + 9)} - \frac{2}{p^2 + 9} + \frac{5p}{p^2 + 9}$$

$$\Rightarrow \qquad L^{-1}\{y\} = L^{-1}\left\{\frac{40}{(p - 1)(p^2 + 9)} - \frac{2}{p^2 + 9} + \frac{5p}{p^2 + 9}\right\}$$

$$\Rightarrow \qquad Y = L^{-1}\left\{\frac{40}{(p - 1)(p^2 + 9)}\right\} - \frac{2}{3}\sin 3x + 5\cos 3x \qquad \ldots(1)$$

Let $\dfrac{1}{(p - 1)(p^2 + 9)} = \dfrac{A}{p - 1} + \dfrac{Bp + c}{p^2 + 9}$ $\qquad \ldots(2)$

Then $\qquad A = \dfrac{1}{1^2 + 9} = \dfrac{1}{10}$

Substituting $p = 0$ in eq^n (2), we get

$$\frac{1}{-9} = -A + \frac{c}{9} = -\frac{1}{10} + \frac{c}{9}$$

$$c = \frac{-1}{10}$$

Now, multiplying eq^n (2) by p and then making $p \to \infty$,
$$a + b = 0$$

$$\Rightarrow \qquad b = -a = -\frac{1}{10} = c$$

$$\therefore \quad L^{-1}\left\{\frac{1}{(p-1)(p^2+9)}\right\} = L^{-1}\left\{\frac{1}{10}\left(\frac{1}{p-1} - \frac{p+1}{p^2+9}\right)\right\}$$

$$= \frac{1}{10}\left[e^x - \cos 3x - \frac{1}{3}\sin 3x\right]$$

Now, eq^n (1) becomes

$$Y = 4\left[e^x - \cos 3x - \frac{1}{3}\sin 3x\right] - \frac{2}{3}\sin 3x + 5\cos 3x$$

Hence, $\qquad Y = 4e^x + \cos 3x - 2\sin 3x.$

EXAMPLE 21.   *Use inversion theorem of Laplace transform to solve $(D^2 + a^2)^2 x = \cos at$, if x, Dx, D^2x, D^3x are zero, when $t = 0$.*

(Meerut–1981; Kanpur–1990)

SOLUTION.   Here, the given equation is
$$(D^4 + 2a^2D^2 + a^4)x = \cos at$$

Now, taking Laplace transform, we get

$$L\{x^4\} + 2a^2 L\{x''\} + a^4 L\{x\} = \frac{p}{p^2 + a^2}$$

$$\Rightarrow \quad p^4\bar{x} - p^3 x(0) - p^2 x'(0) - px''(0) - x'''(0) \qquad \text{[where } \bar{x} = L(x)\text{]}$$

$$+ 2a^2[p^2\bar{x} - px(0) - x'(0)] + a^4\bar{x} = \frac{p}{p^2 + a^2}$$

Given $x = 0$, $Dx = x' = x'' = x$, $D^3x = 0$ when $t = 0$

$$p^4\bar{x} + 2a^2 p^2 \bar{x} + a^4 \bar{x} = \frac{p}{(p^2 + a^2)}$$

$$\Rightarrow \quad \bar{x} = \frac{p}{(p^2 + a^2)^3}$$

$$x = L^{-1}\left\{\frac{p}{(p^2 + a^2)^3}\right\} = \frac{1}{2\pi i}\int_{\gamma - i\infty}^{\gamma + i\infty} \frac{pe^{pt}}{(p^2 + a^2)^3}\, dp$$

$$= \frac{1}{2\pi i} \cdot 2\pi i \text{ sum of residue of } \frac{pe^{pt}}{(p^2 + a^2)^3} \text{ at the poles } p = ai \text{ (pole of order 3)}$$

and at $p = -ai$ (pole of order 3)

$$\therefore \qquad x = R_1 + R_2 \qquad\qquad \text{...(A)}$$

$$\frac{pe^{pt}}{(p^2 + a^2)^3} = \frac{pe^{pt}}{(p - ia)^3(p + ia)^3}$$

$$R_1 = \text{Residue at } p = ia$$

$$= \lim_{p \to ia} \frac{1}{2!}\left[\frac{d^2}{dp^2}\frac{pe^{pt}}{(p + ia)^3}\right] \qquad\qquad \text{...(1)}$$

$$R_2 = \lim_{p \to -ia} \frac{1}{2!} \left[\frac{d^2}{dp^2} \frac{pe^{pt}}{(p-ia)^3} \right]$$ ...(2)

Adding eq^n (1) & (2), we get

$$R_1 + R_2 = \frac{t \sin at - at^2 \cos at}{8a^3}$$

From eq^n (A)

$$x = \frac{t \sin at - at^2 \cos at}{8a^3}.$$

Additional Solved Examples

EXAMPLE 1. *Solve* $\dfrac{dy}{dt} + y = 1,$ *if* $y = 2$ *when* $t = 0.$

SOLUTION. We can write the given equation in the form

$$y' + y = 1.$$ ...(1)

Now taking Laplace transformation of both sides of equation (1), we have

$$L(y') + L(y) = L(1)$$

$$\Rightarrow \quad pL(y) - y(0) + L(y) = \frac{1}{p} \qquad \Rightarrow \quad (p+1)L(y) - y(0) = \frac{1}{p}$$

$$\Rightarrow \quad (p+1)L(y) - 2 = \frac{1}{p} \qquad\qquad [\because y = 2 \text{ at } t = 0]$$

$$\Rightarrow \quad (p+1)L(y) = \frac{2p+1}{p} \quad \Rightarrow \quad L(y) = \frac{2p+1}{p(p+1)}$$

$$\Rightarrow \quad L(y) = \frac{(p+1)+p}{p(p+1)} \quad \Rightarrow \quad L(y) = \frac{1}{p+1} + \frac{1}{p}$$

Taking the inverse Laplace transform of both sides of above equation, we have

$$\Rightarrow \quad y = L^{-1}\left\{ \frac{1}{p+1} + \frac{1}{p} \right\}$$

$$\Rightarrow \quad y = L^{-1}\left(\frac{1}{p+1} \right) + L^{-1}\left(\frac{1}{p} \right)$$

$$\Rightarrow \quad y = e^{-t} + 1$$

which is the required solution.

EXAMPLE 2. *Show that the general solution of the equation* $(D^2 + k^2)y = 0$ *is* $y = C_1 \cos kt + C_2 \sin kt.$

SOLUTION. We can write the given equation in the form as

$$y'' + k^2 y = 0$$

Taking Laplace transformation of both sides, we have

$$L(y'') + k^2 L(y) = 0$$

$$\Rightarrow \quad p^2 L(y) - py(0) - y'(0) + k^2 L(y) = 0$$

$$\Rightarrow \quad (p^2 + k^2)L(y) - py(0) - y'(0) = 0$$

Let $y(0) = A, y'(0) = B$, after putting in equation (1), we have

$$\Rightarrow \quad (p^2 + k^2)L(y) - pA - B = 0$$

$$\Rightarrow \qquad L(y) = \frac{pA}{(p^2 + k^2)} + \frac{B}{p^2 + k^2}$$

Taking the inverse Laplace transformation of both sides, we have

$$\Rightarrow \qquad y = L^{-1}\left(\frac{pA}{(p^2 + k^2)}\right) + L^{-1}\left(\frac{B}{p^2 + k^2}\right)$$

$$\Rightarrow \qquad y = A\cos kt + \frac{B}{k}\sin kt$$

$\Rightarrow \quad y = C_1 \cos kt + C_2 \sin kt$, where $C_1 = A$ and $C_2 = \dfrac{B}{k}$ be arbitrary constant.

EXAMPLE 3. *Solve* $y''(t) + y(t) = t$ *if* $y'(0) = 1, y(\pi) = 0$.

SOLUTION.

(Meerut–1991; Kanpur–2002, 08; Lucknow–2010)

Here, the given equation is

$$y''(t) + y(t) = t \qquad \qquad \dots(1)$$

Taking Laplace transformation of both sides, we have

$$L\{y''(t)\} + L\{y(t)\} = L(t)$$

$$\Rightarrow \qquad p^2 L(y) - py(0) - y'(0) + L(y) = \frac{1}{p^2}$$

$$\Rightarrow \qquad (p^2 + 1)L(y) - py(0) - y'(0) = \frac{1}{p^2}$$

Let $y(0) = A$ and given $y'(0) = 1$.

Thus above equation becomes

$$(p^2 + 1)L(y) - pA - 1 = \frac{1}{p^2}$$

$$\Rightarrow \qquad (p^2 + 1)L(y) = pA + 1 + 1/p^2$$

$$\Rightarrow \qquad L(y) = \frac{pA}{p^2 + 1} + \frac{(p^2 + 1)}{(p^2 + 1)p^2} \quad \Rightarrow \quad L(y) = \frac{pA}{p^2 + 1} + \frac{1}{p^2}$$

Taking inverse Laplace transformation of both sides, we have

$$(y) = y(t) = AL^{-1}\left(\frac{p}{p^2 + 1}\right) + L^{-1}\left(\frac{1}{p^2}\right)$$

$\Rightarrow \qquad y(t) = A\cos t + t$

Given $\qquad y(\pi) = 0 \quad \Rightarrow \quad A\cos \pi + \pi = 0$

$\Rightarrow \qquad A = \pi$

Hence, $\qquad y = \pi \cos t + t$.

EXAMPLE 4. *Solve* $(D^2 - 1)y = a \cosh nt$ *if* $y = Dy = 0$ *when* $t = 0$.

SOLUTION. We can write the given equation in the form of

$$y'' - y = a \cosh nt.$$

Taking Laplace transformation of both sides of equation (1), we have

$$L(y'') - L(y) = aL(\cosh nt)$$

$$\Rightarrow \quad p^2 L(y) - py(0) - y'(0) - L(y) = ap/(p^2 - n^2)$$

$$\Rightarrow \quad (p^2 - 1)L(y) - p(0) - 0 = \frac{ap}{p^2 - n^2} \qquad \text{[Since given } y(0) = 0, y' = 0]$$

$$\Rightarrow \quad (p^2-1)L(y) = \frac{ap}{p^2-n^2}$$

$$\Rightarrow \quad L(y) = \frac{ap}{(p^2-1)(p^2-n^2)} = \frac{ap}{(n^2-1)}\left[\frac{(p^2-1)-(p^2-n^2)}{(p^2-1)(p^2-n^2)}\right]$$

$$= \frac{ap}{(n^2-1)}\left[\frac{1}{p^2-n^2} - \frac{1}{p^2-1}\right] = \frac{a}{(n^2-1)}\left[\frac{p}{p^2-n^2} - \frac{p}{p^2-1}\right]$$

$$\Rightarrow \quad L(y) = \frac{a}{n^2-1}\left[\frac{p}{p^2-n^2} - \frac{p}{p^2-1}\right].$$

Taking inverse Laplace transformation of both sides, we have

$$y = \frac{a}{n^2-1}\left[L^{-1}\left(\frac{p}{p^2-n^2}\right) - L^{-1}\left(\frac{p}{p^2-1}\right)\right]$$

Hence, $\qquad y = \dfrac{a}{n^2-1}(\cosh nt - \cosh t).$

EXAMPLE 5. *Solve* $(D^2+m^2)x = a\cos nt$, $t > 0$ *where* x, Dx *equal to* x_0 *and* x_1 *when* $t = 0$, $n \neq m$.

(Meerut–1980, 94; Rohilkhand–1992)

SOLUTION. We can write the given equation in the form of $x'' + m^2 x = a\cos nt$.

Taking Laplace transformation of both sides, we have

$$L(x'') + m^2 L(x) = aL(\cos nt)$$

$$\Rightarrow p^2 L(x) - px(0) - x'(0) + m^2 L(x) = \frac{ap}{p^2+n^2}$$

$$\Rightarrow (p^2+m^2)L(x) - px_0 - x_1 = \frac{ap}{p^2+n^2} \qquad \text{[Since given } x(0) = x_0, Dx(0) = x_1]$$

$$\Rightarrow L(x) = \frac{p}{p^2+m^2}x_0 + x_1 \cdot \frac{1}{p^2+m^2} + \frac{ap}{(p^2+m^2)(p^2+n^2)}$$

$$\Rightarrow L(x) = x_0 \frac{p}{p^2+m^2} + x_1\frac{1}{p^2+m^2} + \frac{a}{(m^2-n^2)}\left[\frac{p}{p^2+n^2} - \frac{p}{p^2+m^2}\right]$$

Taking inverse Laplace transformation of both sides, we have

$$x = x_0 L^{-1}\left[\frac{p}{p^2+m^2}\right] + x_1 L^{-1}\left[\frac{1}{p^2+m^2}\right] + \frac{a}{m^2-n^2}L^{-1}\left(\frac{p}{p^2+n^2} - \frac{p}{p^2+m^2}\right)$$

Hence, $x = x_0 \cos mt + \dfrac{x_1}{m}\sin mt + \dfrac{a}{m^2-n^2}(\cos nt - \cos mt).$

EXAMPLE 6. *Solve* $(D^2+m^2)y = a\cos nt$, $t > 0$, $y = Dy = 0$, *when* $t = 0$.

(Meerut– 2010; Kumaun–2002; Rohilkhand–1983)

SOLUTION. We can write this equation in the form of

$$y'' + m^2 y = a\cos nt.$$

Taking Laplace transformation of both sides, we have

$$L(y'') + m^2 L(y) = a\cos nt$$

$$\Rightarrow \quad p^2 L(y) - py(0) - y'(0) + m^2 L(y) = \frac{ap}{p^2+n^2}$$

$$\Rightarrow \qquad (p^2 + m^2)L(y) = \frac{ap}{p^2 + n^2} \qquad \text{[Since given } y(0) = y'(0) = 0]$$

$$\Rightarrow \qquad L(y) = \frac{ap}{(p^2 + m^2)(p^2 + n^2)}$$

$$\Rightarrow \qquad L(y) = \frac{a}{m^2 - n^2}\left[\frac{p}{p^2 + n^2} - \frac{p}{p^2 + m^2}\right]$$

Taking inverse Laplace transformation, we have

$$y = \frac{a}{m^2 - n^2}(\cos nt - \cos mt).$$

EXAMPLE 7. Solve $(D^2 + m^2)\, y = a \sin nt$, $t > 0$, where x, Dx equal to x_0 and x_1 when $t = 0$, $n \ne m$.

(Meerut–1999, 83)

SOLUTION. We can write the given equation in the form of

$$y'' + m^2 y = a \sin nt$$

Taking Laplace transformation of both sides, we have

$$L(x'') + m^2 L(x) = aL(\sin nt)$$

$$\Rightarrow \qquad p^2 L(x) - px(0) - x'(0) + m^2 L(x) = \frac{an}{p^2 + n^2}$$

$$\Rightarrow \qquad (p^2 + m^2)L(x) - px_0 - x_1 = \frac{an}{p^2 + n^2} \qquad \text{[Since given } x(0) = x_0, x'(0) = x_1]$$

$$\Rightarrow \qquad L(x) = \frac{px_0}{(p^2 + m^2)} + \frac{x_1}{(p^2 + m^2)} + \frac{an}{(p^2 + m^2)(p^2 + n^2)}$$

$$\Rightarrow \qquad L(x) = \frac{px_0}{(p^2 + m^2)} + \frac{x_1}{(p^2 + m^2)} + \frac{a}{(m^2 - n^2)}\left[\frac{n}{p^2 + n^2} - \frac{n}{p^2 + m^2}\right]$$

Taking inverse Laplace transformation of both sides, we have

$$x = x_0 L^{-1}\left(\frac{p}{p^2 + m^2}\right) + x_1 L^{-1}\left(\frac{1}{p^2 + m^2}\right) + \frac{a}{m^2 - n^2}L^{-1}\left[\frac{n}{p^2 + n^2} - \frac{n}{p^2 + m^2}\right]$$

Hence, $x = x_0 \cos mt + \dfrac{x_1}{m}\sin mt + \dfrac{a}{m^2 - n^2}\left(\sin nt - \dfrac{n}{m}\sin mt\right).$

EXAMPLE 8. Solve $(D + 2)^2 y = 4e^{-2t}$ using $y(0) = -1$ and $y'(0) = 4$.

(Agra–2008; Meerut–2001; Rohilkhand–1995)

SOLUTION. We can write the given equation as follows

$$(D^2 + 4D + 4)y = 4e^{-2t}$$

$$\Rightarrow \qquad y'' + 4y' + 4y = 4e^{-2t}$$

Taking Laplace transformation of both sides, we have

$$L(y'') + 4L(y') + 4L(y) = 4L(e^{-2t})$$

$$\Rightarrow \qquad p^2 L(y) - py(0) - y'(0) + 4pL(y) - 4y(0) + 4L(y) = \frac{4}{p + 2}$$

$$\Rightarrow \qquad (p^2 + 4p + 4)L(y) - (p + 4)y(0) - y'(0) = \frac{4}{p + 2}$$

$\Rightarrow \quad (p^2 + 4p + 4)L(y) + (p+4) - 4 = \dfrac{4}{p+2}$

[Since given $y(0) = -1, y'(0) = 4$]

$\Rightarrow \quad (p+2)^2 L(y) + p = \dfrac{4}{p+2}$

$\Rightarrow \quad L(y) = \dfrac{4}{(p+2)^3} - \dfrac{p}{(p+2)^2}$

Taking inverse Laplace transformation of both sides, we have

$$y = L^{-1}\left[\dfrac{4 - 2p^2}{(p+2)^3}\right]$$

$\Rightarrow \quad y = L^{-1}\left[\dfrac{4 - (p+2)^2 + 2(p+2)}{(p+2)^3}\right]$

$\Rightarrow \quad y = e^{-2t} L^{-1}\left[\dfrac{4}{p^3} + \dfrac{2}{p^2} - \dfrac{1}{p}\right]$

$\Rightarrow \quad y = e^{-2t}[2t^2 + 2t + 1].$

EXAMPLE 9. *Solve* $(D^2 + 6D + 9)y = \sin x$, *where* $y(0) = 1, y'(0) = 0$.

(Meerut–1996; Rohilkhand–1983, 88, 89)

SOLUTION. We can write this equation in the form of

$$y'' + 6y' + 9y = \sin x \qquad \ldots(1)$$

Taking Laplace transformation of both sides of equation (1), we have

$L(y'') + 6L(y') + 9L(y) = L(\sin x)$

$\Rightarrow \quad p^2 L(y) - py(0) - y'(0) + 6pL(y) - 6y(0) + 9L(y) = \dfrac{1}{(p^2+1)}$

$\Rightarrow \quad (p^2 + 6p + 9)L(y) - (p+6)y(0) - y'(0) = \dfrac{1}{p^2+1}$

$\Rightarrow \quad (p+3)^2 L(y) - (p+6) - 0 = \dfrac{1}{(p^2+1)}$

[Since given $y(0) = 1, y'(0) = 0$]

$\Rightarrow \quad L(y) = \dfrac{(p+6)}{(p+3)^2} + \dfrac{1}{(p^2+1)(p+3)^2}$

$\Rightarrow \quad L(y) = \dfrac{1}{(p+3)} + \dfrac{3}{(p+3)^2} + \left[\dfrac{3}{50(p+3)} + \dfrac{1}{10(p+3)^2} - \dfrac{3p-4}{50(p^2+1)}\right]$

$\qquad = \dfrac{1}{50}\left[\dfrac{53}{p+3} + \dfrac{155}{(p+3)^2} - \dfrac{3p}{(p^2+1)} + \dfrac{4}{(p^2+1)}\right]$

Taking inverse Laplace transformation of both sides, we have

$$y = \dfrac{1}{50}\left[53L^{-1}\left(\dfrac{1}{p+3}\right) + 155L^{-1}\left(\dfrac{1}{(p+3)^2}\right) - 3L^{-1}\left(\dfrac{p}{p^2+1}\right) + 4L^{-1}\dfrac{1}{(p^2+1)}\right]$$

$$= \frac{1}{50}\left[53e^{-3x} + 155e^{-3x}L^{-1}\left(\frac{1}{p^2}\right) - 3\cos x + 4\sin x\right]$$

$$= \frac{1}{50}[(53 + 155x)e^{-3x} - 3\cos x + 4\sin x]$$

EXAMPLE 10. *Solve* $(D^2 + 4D + 4)x = \sin \omega t$, $t > 0$ *where* x_0 *and* x_1 *are the values of* x *and* Dx, *when* $t = 0$.

SOLUTION. We can write the given equation in the form of

$$y'' + 4y' + 4x = \sin \omega t \qquad \ldots(1)$$

Taking Laplace transformation of both sides of equation (1), we have

$$L(x'') + 4L(x') + 4L(x) = L(\sin \omega t)$$

$$\Rightarrow \quad p^2 L(x) - px(0) - x'(0) + 4pL(x) - 4x(0) + 4L(x) = \frac{\omega}{p^2 + \omega^2}$$

$$\Rightarrow \quad (p^2 + 4p + 4)L(x) - px(0) - 4x(0) - x'(0) = \frac{\omega}{p^2 + \omega^2}$$

$$\Rightarrow \quad (p+2)^2 L(x) - (p+4)x_0 - x_1 = \frac{\omega}{p^2 + \omega^2}$$

$$\Rightarrow \quad L(x) = \frac{(p+4)x_0 + x_1}{(p+2)^2} + \frac{\omega}{(p^2 + \omega^2)(p+2)^2}$$

$$\Rightarrow \quad + \omega\left[\frac{1}{(4+\omega^2)(p+2)^2} + \frac{4}{(4+\omega^2)^2(p+2)} \right.$$

$$\left. + \frac{(-4p+4-\omega^2)}{(4+\omega^2)^2(p^2+\omega^2)}\right]$$

Taking inverse Laplace transformation of both sides, we have

$$x = x_0 e^{-2t} L^{-1}\left(\frac{p-2}{p^2}\right) + (x_1 + 4x_0)e^{-2t}L^{-1}\left(\frac{1}{p^2}\right)$$

$$+ \omega t^{-1}\left[\frac{1}{(4+\omega^2)(p+2)^2} + \frac{4}{(4+\omega^2)^2(p+2)} + \frac{(-4p+4-\omega^2)}{(4+\omega^2)^2(p^2+\omega^2)}\right]$$

$$= x_0 e^{-2t} L^{-1}\left(\frac{p-2}{p^2}\right) + (x_1 + 4x_0)e^{-2t}L^{-1}\left(\frac{1}{p^2}\right) + \frac{\omega e^{-2t}}{(4+\omega^2)}L^{-1}\left(\frac{1}{p^2}\right)$$

$$+ \frac{4\omega e^{-2t}}{(4+\omega^2)^2}L^{-1}\left(\frac{1}{p}\right) - \frac{4\omega}{(4+\omega^2)^2}L^{-1}\left(\frac{p}{p^2+\omega^2}\right)$$

$$+ \frac{(4-\omega^2)\omega}{(4+\omega^2)^2}L^{-1}\left(\frac{1}{p^2+\omega^2}\right)$$

Hence, $x = e^{-2t}\left[x_0(1-2t) + (x_1 + 4x_0)t + \frac{\omega}{(4+\omega^2)}t + \frac{4\omega}{(4+\omega^2)^2}\right.$

$$\left. - \frac{4\omega}{(4+\omega^2)^2}\cos \omega x + \frac{4-\omega^2}{(4+\omega^2)^2}\sin \omega t\right]$$

EXAMPLE 11. *Solve $(D^2 + 3D + 2)y = 0$, where $y = y_0$ and $Dy = y_1$ at $t = 0$.*

(Meerut–1991; Rohilkhand–2004)

SOLUTION. We can write the given equation in the form of

$$y'' + 3y' + 2y = 0 \qquad \ldots(1)$$

Taking Laplace transformation of both sides of equation (1), we have

$$L(y'') + 3L(y') + 2L(y) = 0$$

$$\Rightarrow \quad p^2 L(y) - py(0) - y'(0) + 3pL(y) - y(0) + 2L(y) = 0$$

$$\Rightarrow \quad (p^2 + 3p + 2)L(y) - (p + 3)y(0) - y'(0) = 0$$

$$\Rightarrow \quad (p^2 + 3p + 2)L(y) - (p + 3)y_0 - y_1 = 0 \qquad [\text{Given } y(0) = y_0, y'(0) = y_1]$$

$$\Rightarrow \quad L(y) = \frac{(p+3)}{(p^2 + 3p + 2)} y_0 + \frac{y_1}{(p^2 + 3p + 2)}$$

$$\Rightarrow \quad L(y) = \frac{(p+3)}{(p+1)(p+2)} y_0 + \frac{y_1}{(p+1)(p+2)}$$

$$\Rightarrow \quad L(y) = \left[\frac{2}{(p+1)} - \frac{1}{(p+2)} \right] y_0 + \left[\frac{1}{(p+1)} - \frac{1}{(p+2)} \right] y_1$$

$$\Rightarrow \quad L(y) = \frac{2y_0 + y_1}{(p+1)} - \frac{y_0 + y_1}{(p+2)}$$

Taking inverse Laplace transformation of both sides, we have

$$y = (2y_0 + y_1) L^{-1} \left(\frac{1}{(p+1)} \right) - (y_0 + y_1) L^{-1} \left[\frac{1}{(p+2)} \right]$$

$$y = (2y_0 + y_1)e^{-t} + (y_0 + y_1)e^{-2t}.$$

EXAMPLE 12. *Solve $(D^2 + 9)y = 18t$, if $y(0) = 0$, $y\left(\dfrac{\pi}{2}\right) = 0$.* (Meerut–1996)

SOLUTION. We can write the given equation in the form of

$$y'' + 9y = 18t$$

Taking Laplace transformation of both siedes of equation (1), we have

$$L(y'') + 9L(y) = 18L(t)$$

$$\Rightarrow \quad p^2 L(y) - py(0) - y'(0) + 9L(y) = \frac{18}{p^2}$$

$$\Rightarrow \quad (p^2 + 9)L(y) = py(0) + y'(0) + \frac{18}{p^2} \quad [\text{Given } y(0) = 0 \text{ and } y'(0) = A]$$

$$\Rightarrow \quad (p^2 + 9)L(y) = A + \frac{18}{p^2}$$

$$\Rightarrow \quad L(y) = \frac{A}{(p^2 + 9)} + \frac{18}{p^2(p^2 + 9)}$$

Taking inverse Laplace transformation, we have

$$y = L^{-1} \left(\frac{A}{p^2 + 9} \right) + 18 L^{-1} \left(\frac{1}{p^2(p^2 + 9)} \right)$$

$$y = L^{-1} \left(\frac{A}{p^2 + 9} \right) + 18 L^{-1} \left[\frac{1}{9} \left(\frac{(p^2 + 9) - p^2}{p^2(p^2 + 9)} \right) \right]$$

$$\Rightarrow \qquad y = L^{-1}\left(\frac{A}{p^2+9}\right) + 2L^{-1}\left[\frac{(p^2+9)-p^2}{p^2(p^2+9)}\right]$$

$$\Rightarrow \qquad y = L^{-1}\left(\frac{A}{p^2+9}\right) + 2L^{-1}\left[\frac{1}{p^2} - \frac{1}{p^2+9}\right]$$

$$\Rightarrow \qquad y = L^{-1}\left(\frac{A-2}{p^2+9}\right) + 2L^{-1}\left(\frac{1}{p^2}\right)$$

$$\Rightarrow \qquad y = (A-2)\frac{1}{3}\sin 3t + 2t$$

Since $\qquad y\left(\dfrac{\pi}{2}\right) = 0 \qquad\qquad$ (given)

$$\Rightarrow \qquad 0 = \frac{1}{3}(A-2)\sin\left(\frac{3\pi}{2}\right) + \pi \text{ where } A = 3\pi + 2$$

Hence, $\qquad y = \pi \sin 3t + 2t.$

EXAMPLE 13. Solve $(D^2 + 2D + 1)y = 3te^{-t}$, $t > 0$ subject to the conditions $y = 4$, $Dy = 2$ when $t = 0$. (Meerut–1996; Kanpur–1982; MDU-2007)

SOLUTION. We can write the given equation in the form of
$$y'' + 2y' + y = 3te^{-t} \qquad\qquad ...(1)$$
Taking Laplace transformation of both sides of equation (1), we have
$$L(y'') + 2L(y') + L(y) = 3L(te^{-t})$$

$$\Rightarrow \quad p^2 L(y) - py(0) - y'(0) + 2pL(y) - 2y(0) + L(y) = 3L(te^{-t})$$

$$\Rightarrow \quad (p^2 + 2p + 1)L(y) - (p+2)y(0) - y'(0) = \frac{3d}{dp}(e^{-t})$$

$$\Rightarrow \quad (p+1)^2 L(y) - (p+2)(4) - 2 = -3\frac{d}{dp}\left(\frac{1}{p+1}\right)$$

$$\Rightarrow \quad (p+1)^2 L(y) = \frac{3}{(p+1)^2} + 4p + 10$$

$$\Rightarrow \quad L(y) = \frac{3}{(p+1)^4} + \frac{4(p+1)+6}{(p+1)^2} \Rightarrow L(y) = \frac{3}{(p+1)^4} + \frac{4}{(p+1)} + \frac{6}{(p+1)^2}$$

Taking inverse Laplace transformation of both sides, we have

$$y = 3L^{-1}\left(\frac{1}{(p+1)^4}\right) + 4L^{-1}\left(\frac{1}{(p+1)}\right) + 6L^{-1}\left(\frac{1}{(p+1)^2}\right)$$

$$\Rightarrow \quad y = 3e^{-t}L^{-1}\left(\frac{1}{p^4}\right) + 4e^{-t} + 6e^{-t}L^{-1}\left(\frac{1}{p^2}\right)$$

$$\Rightarrow \quad y = 3e^{-t}\frac{t^3}{3!} + 4e^{-t} + 6e^{-t}.t \Rightarrow y = \frac{1}{2}e^{-t}.t^3 + 4e^{-t} + 6te^{-t}.$$

EXAMPLE 14. Solve $(D^2 + 1)y = \sin t \sin 2t$, $t > 0$ if $y = 1$, $Dy = 0$, when $t = 0$.

SOLUTION. We can write the given equation as
$$y'' + y = \sin t \sin 2t \qquad\qquad ...(1)$$

Taking Laplace transformation of both sides of equation (1), we have

$$L(y'') + L(y) = L[\sin t \sin 2t]$$

$$\Rightarrow \quad p^2 L(y) - py(0) - y'(0) + L(y) = \frac{p}{2(p^2+1)} - \frac{p}{2(p^2+9)}$$

$$\Rightarrow \quad (p^2+1)L(y) - py(0) - y'(0) = \frac{p}{2(p^2+1)} - \frac{p}{2(p^2+9)}$$

$$\Rightarrow \quad (p^2+1)L(y) - p = \frac{p}{2(p^2+1)} - \frac{p}{2(p^2+9)}$$

$$\Rightarrow \quad L(y) = \frac{p}{(p^2+1)} + \frac{p}{2(p^2+1)^2} - \frac{p}{2(p^2+9)(p^2+1)}$$

$$\Rightarrow \quad L(y) = \frac{p}{(p^2+1)} + \frac{p}{2(p^2+1)^2} - \frac{p}{16}\left[\frac{(p^2+9)-(p^2+1)}{(p^2+9)(p^2+1)}\right]$$

$$\Rightarrow \quad L(y) = \frac{p}{(p^2+1)} - \frac{1}{4}\left[\frac{d}{dp}\left(\frac{1}{p^2+1}\right)\right] - \frac{p}{16(p^2+1)} + \frac{p}{16(p^2+9)}$$

Taking inverse Laplace transformation of both sides, we have

$$y = L^{-1}\left(\frac{p}{p^2+1}\right) - \frac{1}{4}L^{-1}\left[\frac{d}{dp}\left(\frac{1}{p^2+1}\right)\right] - L^{-1}\left(\frac{p}{16(p^2+1)}\right) + L^{-1}\left(\frac{p}{16(p^2+9)}\right)$$

$$\Rightarrow \quad y = \cos t + \frac{1}{4}t L^{-1}\left(\frac{1}{p^2+1}\right) - \frac{1}{16}\cos t + \frac{1}{16}\cos 3t$$

$$\Rightarrow \quad y = \frac{15}{16}\cos t + \frac{1}{4}t\sin t + \frac{1}{16}\cos 3t .$$

EXAMPLE 15. *Solve* $(D^2 + n^2)y = a\sin(nt + \alpha)$, *if* $Dy = 0$ *when* $t = 0$.   ~ (Kanpur–1986, 87)

SOLUTION. We can write the given equation in the form of

$$y'' + n^2 y = a\sin(nt + \alpha) \qquad \qquad ...(1)$$

Taking Laplace transformation both sides of equation (1), we have

$$L(y'') + n^2 L(y) = aL[\sin(nt + \alpha)]$$

$$\Rightarrow \quad p^2 L(y) - py(0) - y'(0) + n^2 L(y) = aL[\sin nt \cos \alpha + \cos nt \sin \alpha]$$

$$\Rightarrow \quad (p^2 + n^2)L(y) - py(0) - y'(0) = a\cos\alpha \cdot \frac{n}{p^2+n^2} + a\sin\alpha \cdot \frac{p}{p^2+n^2}$$

$$\Rightarrow \quad (p^2 + n^2)L(y) = a\cos\alpha \cdot \frac{n}{p^2+n^2} + a\sin\alpha \cdot \frac{p}{p^2+n^2}$$

$$[\because \text{ given } y'(0) = y(0) = 0]$$

$$\Rightarrow \quad L(y) = a\cos\alpha \cdot \frac{n}{(p^2+n^2)^2} + a\sin\alpha \cdot \frac{p}{(p^2+n^2)^2}$$

Taking inverse Laplace transformation of both sides, we have

$$y = an\cos\alpha\, L^{-1}\left(\frac{1}{(p^2+n^2)^2}\right) + a\sin\alpha\, L^{-1}\left(\frac{p}{(p^2+n^2)^2}\right)$$

$$\Rightarrow \quad y = an\cos\alpha \int_0^t \left(\frac{1}{n}\sin nx\right)\frac{1}{n}\sin n(t-x)dx + \frac{a\sin\alpha}{2}\cdot L^{-1}\left\{\frac{d}{dp}\frac{1}{(p^2+n^2)^2}\right\}$$

(by using convolution theorem)

$$\Rightarrow \quad y = \frac{a\cos\alpha}{2n}\int_0^t[\cos n(t-2x)-\cos nt]dx + \frac{a\sin\alpha}{2}tL^{-1}\left(\frac{1}{p^2+n^2}\right)$$

$$\Rightarrow \quad y = \frac{a\cos\alpha}{2n}\left[-\frac{1}{2n}\sin(t-2x)-x\cos nt\right]_0^t + \frac{at\sin\alpha}{2}\sin nt$$

$$\Rightarrow \quad y = \frac{a\cos\alpha\sin nt}{2n^2} - \frac{at}{2n}(\cos\alpha\cos nt - \sin\alpha\sin nt)$$

$$\Rightarrow \quad y = \frac{a\cos\alpha\sin nt}{2n^2} - \frac{at}{2n}\cos(\alpha+nt)$$

Hence, $y = \dfrac{a}{2n^2}[\cos\alpha\sin nt - tn\cos(\alpha+nt)]$.

EXAMPLE 16. *Solve* $(D^3+1)y = 1, t > 0$, *if* $y = Dy = D^2y = 0$ *when* $t = 0$.   (Osmania–1992)

SOLUTION. We can write this equation in the form of
$$y''' + y = 1.$$
Taking Laplace transformation of both sides of equation (1), we have
$$L(y''') + L(y) = 1$$

$$\Rightarrow \quad p^3 L(y) - p^2 y(0) - py'(0) - y''(0) + L(y) = L(1)$$

$$\Rightarrow \quad (p^3+1)L(y) - p^2 y(0) - py'(0) - y''(0) = \frac{1}{p}$$

$$\Rightarrow \quad (p^3+1)L(y) = \frac{1}{p} \qquad\qquad [\text{given: } y(0) = y'(0) = y''(0) = 0]$$

$$\Rightarrow \quad L(y) = \frac{1}{p(p^3+1)}$$

$$\Rightarrow \quad L(y) = \frac{1}{p(p+1)(p^2-p+1)}$$

$$\Rightarrow \quad L(y) = \frac{1}{p} - \frac{1}{3(p+1)} - \frac{(2p-1)}{3(p^2-p+1)}$$

$$\Rightarrow \quad L(y) = \frac{1}{p} - \frac{1}{3(p+1)} - \frac{2\left(p-\dfrac{1}{2}\right)}{3\left[\left(p-\dfrac{1}{2}\right)^2 + \dfrac{3}{4}\right]}$$

Taking inverse Laplace transformation of both sides, we have

$$y = L^{-1}\left(\frac{1}{p}\right) - \frac{1}{3}L^{-1}\left(\frac{1}{p+1}\right) - \frac{2}{3}L^{-1}\left(\frac{p-\dfrac{1}{2}}{\left(p-\dfrac{1}{2}\right)^2 + \left(\dfrac{\sqrt{3}}{2}\right)^2}\right)$$

$$\Rightarrow \qquad y = 1 - \frac{1}{3}e^{-t} - \frac{2}{3}e^{t/2}L^{-1}\left(\frac{p}{p^2 + (\sqrt{3}/2)^2}\right)$$

Hence, $\qquad y = 1 - \frac{1}{3}e^{-t} - \frac{2}{3}e^{t/2}\cos(\sqrt{3}t/2).$

EXAMPLE 17. *Solve* $(D^3 - D)y = 2\cos t, y = 3, Dy = 2, D^2y = 1$, *when* $t = 0$. (Meerut–1981)

SOLUTION. We can write the equation in the form of

$$y''' - y' = 2\cos t \qquad \qquad \ldots(1)$$

Taking Laplace transformation of both sides of equation (1), we have

$$L(y''') - L(y') = 2L(\cos t)$$

$$\Rightarrow \quad p^3 L(y) - p^2 y(0) - py'(0) - y''(0) - pL(y) - y(0) = \frac{2p}{p^2 + 1}$$

$$\Rightarrow \quad (p^3 - p)L(y) - (p^2 + 1)y(0) - py'(0) - y''(0) = \frac{2p}{p^2 + 1}$$

$$\Rightarrow \quad p(p^2 - 1)L(y) - 3(p^2 + 1) - 2p - 1 = \frac{2p}{p^2 + 1}$$

[Since given $y(0) = 3, y'(0) = 2, y''(0) = 1$]

$$\Rightarrow \quad p(p^2 - 1)L(y) - 3p^2 - 2p + 2 = \frac{2p}{p^2 + 1}$$

$$\Rightarrow \quad p(p^2 - 1)L(y) = \frac{2p}{p^2 + 1} + 3p^2 + 2p - 2$$

$$\Rightarrow \quad L(y) = \frac{2p}{p(p^2 + 1)(p^2 - 1)} + \frac{3p^2 + 2p - 2}{p(p-1)(p+1)}$$

$$\Rightarrow \quad L(y) = \frac{2}{(p^2 - 1)(p^2 + 1)} + \frac{2}{p} + \frac{3}{2(p-1)} - \frac{1}{2(p+1)}$$

$$\Rightarrow \quad L(y) = \frac{1}{(p^2 - 1)} - \frac{1}{(p^2 + 1)} + \frac{2}{p} + \frac{3}{2(p-1)} - \frac{1}{2(p+1)}$$

Taking inverse Laplace transformation of both sides, we have

$$y = L^{-1}\left(\frac{1}{p^2 - 1}\right) - L^{-1}\left(\frac{1}{p^2 + 1}\right) + 2L^{-1}\left(\frac{1}{p}\right) + \frac{3}{2}L^{-1}\left(\frac{1}{p-1}\right) - \frac{1}{2}L^{-1}\left(\frac{1}{p+1}\right)$$

$$\Rightarrow y = \sinh t - \sin t + 2 + \frac{3}{2}e^t - \frac{1}{2}e^{-t} = \sinh t - \sin t + 2 + \frac{1}{2}(e^t + e^{-t}) + 2 \cdot \frac{1}{2}(e^t - e^{-t})$$

$$\Rightarrow \qquad y = \sinh t - \sin t + 2 + \cosh t + 2 \sinh t$$

Hence, $\quad y = 3\sinh t - \sin t + \cosh t + 2.$

EXAMPLE 18. *Solve* $(D^3 + D)y = e^{2t}, y(0) = y'(0) = y''(0) = 0$. (Meerut–1999)

SOLUTION. We can write the given equation in the form of

$$y''' + y' = e^{2t} \qquad \qquad \ldots(1)$$

Taking Laplace transformation of both sides of equation (1), we have

$$L(y''') + L(y') = L(e^{2t})$$

$$\Rightarrow \quad p^3 L(y) - p^2 y(0) - py'(0) - y''(0) + pL(y) - y(0) = \frac{1}{(p-2)}$$

$$\Rightarrow \quad (p^3 + p)L(y) - (p^2 + 1) \cdot y(0) - py'(0) - y''(0) = \frac{1}{(p-2)}$$

$$\Rightarrow \quad (p^3 + p)L(y) - (p^2 + 1).0 - 0 - 0 = \frac{1}{(p-2)} \quad [\text{Given } y(0) = y'(0) = y''(0) = 0]$$

$$\Rightarrow \quad (p^3 + p)L(y) = \frac{1}{(p-2)} \quad \Rightarrow \quad L(y) = \frac{1}{p(p^2+1)(p-2)}$$

$$\Rightarrow \quad L(y) = -\frac{1}{2p} + \frac{1}{p(p-2)} + \frac{(2p-1)}{5(p^2+1)}$$

Taking inverse Laplace transformation of both sides, we have

$$y = -\frac{1}{2}L^{-1}\left(\frac{1}{p}\right) + \frac{1}{10}L^{-1}\left(\frac{1}{p-2}\right) + \frac{2}{5}L^{-1}\left(\frac{p}{p^2+1}\right) - \frac{1}{5}L^{-1}\left(\frac{1}{p^2+1}\right)$$

Hence, $\quad y = -\frac{1}{2} + \frac{1}{10}e^{2t} + \frac{2}{5}\cos t - \frac{1}{5}\sin t$.

EXAMPLE 19. Solve $(D^4 - 1)y = 1$ if $y = Dy = D^2y = D^3y = 0$ at $t = 0$.

(Meerut–1987; Rohilkhand–2003)

SOLUTION. We can write the given equation in the form of

$$(y''' - y) = 1 \qquad \qquad ...(1)$$

Taking Laplace transformation of both sides of equation (1), we have

$$L(y''') - L(y) = 1$$

$$\Rightarrow p^4L(y) - p^3y(0) - p^2y'(0) - py''(0) - y'''(0) - L(y) = \frac{1}{p}$$

$$\Rightarrow (p^4 - 1)L(y) - p^3y(0) - p^2y'(0) - py''(0) - y'''(0) = \frac{1}{p}$$

$$\Rightarrow (p^4 - 1)L(y) = \frac{1}{p} \qquad [\text{Given } y(0) = y'(0) = y''(0) = y'''(0) = 0]$$

$$\Rightarrow L(y) = \frac{1}{p(p^4-1)} \quad \Rightarrow \quad L(y) = -\frac{1}{p} + \frac{p}{2(p^2-1)} + \frac{p}{2(p^2+1)}$$

Taking inverse Laplace transformation of both sides, we have

$$y = -L^{-1}\left(\frac{1}{p}\right) + \frac{1}{2}L^{-1}\left(\frac{p}{p^2-1}\right) + \frac{1}{2}L^{-1}\left(\frac{p}{p^2+1}\right)$$

$$\Rightarrow \qquad y = -1 + \frac{1}{2}\cosh t + \frac{1}{2}\cos t.$$

EXAMPLE 20. Solve $(D^4 + 2D^2 + 1)y = 0$, if $y(0) = 0$, $y'(0) = 1$, $y''(0) = 2$ and $y'''(0) = -3$.

(Meerut–1988, 94)

SOLUTION. We can write the given equation in the form of

$$y''' + 2y'' + y = 0 \qquad \qquad ...(1)$$

Taking Laplace transformation of both sides of equation (1), we have

$$L(y''') + 2L(y'') + L(y) = 0$$

$$\Rightarrow p^4L(y) - p^3y(0) - p^2y'(0) - py''(0) - y'''(0) + 2p^2L(y) - 2py(0)$$
$$- 2y'(0) + L(y) = 0$$

$\Rightarrow \quad (p^4 + 2p^2 + 1)L(y) - (p^3 + 2p)y(0) - (p^2 + 2)y'(0) - py''(0) - y'''(0) = 0$

$\Rightarrow \quad (p^4 + 2p^2 + 1)L(y) - (p^2 + 2) - 2p + 3 = 0 \qquad$ [Since given $y(0) = 0$,

$$y'(0) = 1, y''(0) = 2y'''(0) = -3]$$

$\Rightarrow \qquad\qquad (p^4 + 2p^2 + 1)L(y) = p^2 + 2p - 1$

$\Rightarrow \qquad\qquad L(y) = \dfrac{p^2 + 2p - 1}{(p^4 + 2p^2 + 1)} \Rightarrow L(y) = \dfrac{p^2 + 2p - 1}{(p^2 + 1)^2}$

$\Rightarrow \qquad\qquad L(y) = \dfrac{1}{(p^2 + 1)} + \dfrac{2p}{(p^2 + 1)^2} - \dfrac{2}{(p^2 + 1)^2}$

Taking inverse Laplace transformation of both sides, we have

$$y = L^{-1}\left(\frac{1}{p^2 + 1}\right) + 2L^{-1}\left(\frac{p}{(p^2 + 1)^2}\right) - 2L^{-1}\frac{1}{(p^2 + 1)^2}$$

$\Rightarrow \qquad y = L^{-1}\left(\dfrac{1}{p^2 + 1}\right) + 2L^{-1}\left\{-\dfrac{1}{2}\dfrac{d}{dp}\left(\dfrac{1}{p^2 + 1}\right)\right\} - 2\int_0^t \sin x . \sin(t - x)dx$

$$\text{[By convolution theorem]}$$

$\Rightarrow \qquad y = \sin t + 2\left(\dfrac{t}{2}\right)L^{-1}\left(\dfrac{1}{p^2 + 1}\right) - 2 \cdot \dfrac{1}{2}\int_0^t[\cos(2x - t) - \cos t]dx$

$\Rightarrow \qquad y = \sin t + t\sin t - (\sin t - t\cos t)$

Hence, $\qquad y = t(\sin t + \cos t)$.

3.2 SOLUTION OF ORDINARY DIFFERENTIAL EQUATION WITH VARIABLE COEFFICIENTS

The Laplace transform can also be used in solving some ordinary differential equations in which the coefficients are variable. A particular differential equation when the method proves useful is one in which the terms have the form $t^m\, y^n\,(t)$ whose Laplace transform is

$$(-1)^m \frac{d^m}{dp^m}[L\{y^n(t)\}]$$

Solved Examples

EXAMPLE 1.   *Solve* $(tD^2 + D + 4t)y = 0$ *if* $y(0) = 3$, $y'(0) = 0$.   (Andhra–1990; Agra–1987)

SOLUTION.   The given equation can be written as

$$ty'' + y' + 4ty = 0 \qquad\qquad\qquad ...(1)$$

Taking the Laplace transform of both sides of (1), we get

$L\{ty''\} + L\{y'\} + 4L\{ty\} = 0$

$\Rightarrow \quad -\dfrac{d}{dp}L\{y''\} + L\{y'\} + 4(-1)\dfrac{d}{dp}L\{y\} = 0$

$\Rightarrow \quad -\dfrac{d}{dp}[p^2L\{y\} - py(0) - y'(0)] + [pL\{y\} - y(0)] - 4\dfrac{d}{dp}L\{y\} = 0$

$\Rightarrow \quad -\dfrac{d}{dp}[p^2L\{y\} - 3p] + (pL\{y\} - 3) - \dfrac{4d[L\{y\}]}{dp} = 0$

$$\Rightarrow \quad -(p^2+4)\frac{d[L\{y\}]}{dp} - pL\{y\} = 0 \qquad \Rightarrow \quad \frac{d[L\{y\}]}{L\{y\}} + \frac{p}{p^2+4}dp = 0$$

On integrating, we get

$$\log[L\{y\}] + \frac{1}{2}\log(p^2+4) = \log C_1 \quad \Rightarrow \quad L\{y\} = \frac{C_1}{\sqrt{p^2+4}}$$

Therefore, $y = L^{-1}\left\{\dfrac{C_1}{\sqrt{p^2+4}}\right\} = C_1 J_0(2t)$.

EXAMPLE 2. Solve $[tD^2 + (t-1)D - 1]y = 0$ if $y(0) = 5$, $y(\infty) = 0$. (Meerut–1983; Kanpur–1997)

SOLUTION. The given equation can be written as

$$ty'' + ty' - y' - y = 0 \qquad \text{... (1)}$$

Taking the Laplace transforms of both sides of (1), we get

$$L\{ty''\} + L\{ty'\} - L\{y'\} - L\{y\} = 0$$

$$\Rightarrow \quad -\frac{d}{dp}[L\{y''\}] - \frac{d}{dp}[L\{y'\} - [pL\{y\} - y(0) - L\{y\}] = 0$$

$$\Rightarrow \quad -\frac{d}{dp}[p^2 L\{y\} - py(0) - y'(0)] - \frac{d}{dp}[pL\{y\} - y(0)] - pL\{y\} + 5 - L\{y\} = 0$$

$$\Rightarrow -\frac{d}{dp}[p^2 L\{y\} - 5p - A] - \frac{d}{dp}[pL\{y\} - 5] - (p+1)L\{y\} + 5 = 0, \text{ where } A = y'\{0\}$$

$$\Rightarrow \quad \frac{d[L\{y\}]}{dp} + \frac{3p+2}{p^2+p}L\{y\} = \frac{10}{p^2+p} \qquad \text{... (2)}$$

which is a linear differential equation in $L\{y\}$.
Therefore,

$$\text{I.F.} = e^{\int \left\{\frac{3p+2}{(p^2+p)}\right\}dp} = e^{\int \left(\frac{2}{p} + \frac{1}{p+1}\right)dp}$$

$$= e^{[2\log p + \log(p+1)]} = p^2(p+1).$$

Hence, the solution of equation (2) is given by

$$L\{y\}.p^2(p+1) = C_1 + \int \frac{10}{p^2+p}.p^2(p+1)dp$$

$$= C_1 + 10\int p\, dp = C_1 + 5p^2$$

$$\Rightarrow \quad L\{y\} = \frac{C_1}{p^2(p+1)} + \frac{5}{p+1}$$

$$= C_1\left\{\frac{1}{p^2} - \frac{1}{p} + \frac{1}{p+1}\right\} + \frac{5}{p+1}$$

> **AID**
>
> General form of linear differential equation is
>
> $$\frac{dy}{dx} + Py = Q \qquad \text{..(1)}$$
>
> where P and Q are functions of x only or constant
> Now, I.F. $= e^{\int Pdx}$
> and solution of (1) is given by
> $$y(\text{I.F.}) = \int [Q.(\text{I.F.})]dx + C$$

Therefore, $y = C_1 L^{-1}\left\{\dfrac{1}{p^2} - \dfrac{1}{p} + \dfrac{1}{p+1}\right\} + 5L^{-1}\left\{\dfrac{1}{p+1}\right\}$

$$= C_1(t - 1 + e^{-t}) + 5e^{-t}.$$

Now, using the given conditions $y(\infty) = 0$.

We must have $C_1 = 0$. Hence, $y = 5e^{-t}$ is the required solution.

EXAMPLE 3. *Solve* $\dfrac{d^2y}{dx^2} + 2\dfrac{dy}{dx} + 5y = e^{-x}\sin x$, *where* $y(0) = 0,\ \ y'(0) = 1$.

(UPTU–2004, 08; (SUM)–2009; PTU–2010; MTU–2011; Kanpur–2003)

SOLUTION. The given equation is $\dfrac{d^2y}{dx^2} + 2\dfrac{dy}{dx} + 5y = e^{-x}\sin x$

Taking the Laplace transform on both the sides,
we get $[p^2L\{y\} - py(0) - y'(0)] + 2[pL\{y\} - y(0)] + 5L\{y\} = L\{e^{-x}\sin x\}$

$$[p^2L\{y\} - py(0) - y'(0)] + 2[pL\{y\} - y(0)] + 5L\{y\} = \frac{1}{(p+1)^2 + 1} \qquad \text{...(1)}$$

On substituting the values of $y(0)$ and $y'(0)$ in (1), we get

$$(p^2L\{y\} - 1) + 2pL\{y\} + 5L\{y\} = \frac{1}{p^2 + 2p + 2}$$

$$(p^2 + 2p + 5)L\{y\} = 1 + \frac{1}{p^2 + 2p + 2} = \frac{p^2 + 2p + 3}{p^2 + 2p + 2}$$

$$L\{y\} = \frac{p^2 + 2p + 3}{(p^2 + 2p + 5)(p^2 + 2p + 2)}$$

On resolving R.H.S. into partial fractions, we get

$$L\{y\} = \frac{2}{3} \cdot \frac{1}{p^2 + 2p + 5} + \frac{1}{3} \cdot \frac{1}{p^2 + 2p + 2}$$

Hence, $\qquad y = \dfrac{2}{3}L^{-1}\dfrac{1}{p^2 + 2p + 5} + \dfrac{1}{3}L^{-1}\dfrac{1}{p^2 + 2p + 2}$

$$y = \frac{1}{3}L^{-1}\frac{2}{(p+1)^2 + (2)^2} + \frac{1}{3}L^{-1}\frac{1}{(p+1)^2 + (1)^2}$$

$\Rightarrow \qquad y = \dfrac{1}{3}e^{-x}\sin 2x + \dfrac{1}{3}e^{-x}\sin x$

$\Rightarrow \qquad y = \dfrac{1}{3}.e^{-x}(\sin x + \sin 2x)$

EXAMPLE 4. *Solve* $(D^2 + 1)y = t\cos 2t$ *subject to the condition* $y = 0,\ \dfrac{dy}{dt} = 0$ *when* $t = 0$.

(UPTU–2004, 05; UKTU–2012; Raipur–2005; Kanpur–1986, 98; Meerut–1990, 97, 2014;
Agra–1984; Rohilkhand–1999, 2005; Garhwal–1996; Kurukshetra–1981)

SOLUTION. The given equation can be written as

$$y'' + y = t\cos 2t \qquad \text{...(1)}$$

Taking the Laplace transform of both sides of (1), we get

$$L\{y''\} + L\{y\} = L\{t\cos 2t\}$$

$\Rightarrow \ p^2L\{y\} - py(0) - y'(0) + L\{y\} = -\dfrac{d}{dp}[L\{\cos 2t\}]$

$\Rightarrow \ (p^2 + 1)L\{y\} = -\dfrac{d}{dp}\left(\dfrac{p}{p^2 + 4}\right) = -\dfrac{1}{p^2 + 4} + \dfrac{2p^2}{(p^2 + 4)^2}$

$\therefore \qquad L\{y\} = \dfrac{p^2 - 4}{(p^2 + 1)(p^2 + 4)^2} = -\dfrac{5}{9(p^2 + 1)} + \dfrac{5}{9(p^2 + 4)} + \dfrac{8}{3(p^2 + 4)^2}$

[Resolving into partial fractions]

$$\Rightarrow \quad y = -\frac{5}{9}L^{-1}\left\{\frac{1}{p^2+1}\right\} + \frac{5}{9}L^{-1}\left\{\frac{1}{p^2+4}\right\} + \frac{8}{3}L^{-1}\left\{\frac{1}{(p^2+4)^2}\right\}$$

$$= -\frac{5}{9}\sin t + \frac{5}{18}\sin 2t + \frac{8}{3}\int_0^t \frac{1}{2}\sin 2x \cdot \frac{1}{2}\sin 2(t-x)dx$$

[By convolution theorem and using $L^{-1}\left\{\frac{1}{p^2+4}\right\} = \frac{1}{2}\sin 2t$]

$$= -\frac{5}{9}\sin t + \frac{5}{18}\sin 2t + \frac{1}{3}\int_0^t \{\cos(2t-4x) - \cos 2t\}dx$$

$$= -\frac{5}{9}\sin t + \frac{5}{18}\sin 2t + \frac{1}{3}\left[-\frac{1}{4}\sin(2t-4x) - x\cos 2t\right]_0^t$$

$$= -\frac{5}{9}\sin t + \frac{5}{18}\sin 2t + \frac{1}{12}\sin 2t - \frac{1}{3}t\cos 2t + \frac{1}{12}\sin 2t$$

$$= -\frac{5}{9}\sin t + \frac{4}{9}\sin 2t - \frac{1}{3}t\cos 2t$$

EXAMPLE 5. Solve $(D^3 - D^2 - D + 1)y = 8te^{-t}$ if $y = D^2y = 0$, $Dy = 0$ when $t = 0$.

(Meerut–1980, 88, 95; Jhansi–2012)

SOLUTION. The given equation can be written as

$$y''' - y'' - y' + y = 8te^{-t} \qquad \text{... (1)}$$

Taking the Laplace transforms of both sides of (1), we get

$$L\{y'''\} - L\{y''\} - L\{y'\} + L\{y\} = 8L\{te^{-t}\}$$

$$\Rightarrow \quad p^3 L\{y\} - p^2 y(0) - py'(0) - y''(0) - [p^2 L\{y\} - py(0) - y'(0)]$$

$$-[pL\{y\} - y(0)] + L\{y\} = -8\frac{d}{dp}L\{e^{-t}\}$$

or $\quad (p^3 - p^2 - p + 1)L\{y\} - p + 1 = -8\dfrac{d}{dp}\left[\dfrac{1}{p+1}\right]$

$$\Rightarrow \quad (p-1)^2(p+1)L\{y\} = p - 1 + \frac{8}{(p+1)^2}$$

$$\Rightarrow \quad L\{y\} = \frac{1}{(p-1)(p+1)} + \frac{8}{(p-1)^2(p+1)^3}$$

$$= \frac{1}{2}\left(\frac{1}{p-1} - \frac{1}{p+1}\right) - \frac{3}{2(p-1)} + \frac{1}{(p-1)^2}$$

$$+ \frac{3}{2(p+1)} + \frac{2}{(p+1)^2} + \frac{2}{(p+1)^3}$$

$$= -\frac{1}{p-1} + \frac{1}{p+1} + \frac{1}{(p-1)^2} + \frac{2}{(p+1)^2} + \frac{2}{(p+1)^3}$$

Therefore, $\quad y = -L^{-1}\left\{\frac{1}{p-1}\right\} + L^{-1}\left\{\frac{1}{p+1}\right\} + L^{-1}\left\{\frac{1}{(p-1)^2}\right\}$

$$+ 2L^{-1}\left\{\frac{1}{(p+1)^2}\right\} + 2L^{-1}\left\{\frac{1}{(p+1)^3}\right\}$$

$$= -e^t + e^{-t} + e^t L^{-1}\left\{\frac{1}{p^2}\right\} + 2e^{-t} L^{-1}\left\{\frac{1}{p^2}\right\} + 2e^{-t} L^{-1}\left\{\frac{1}{p^3}\right\}$$

$$= -e^t + e^{-t} + e^t.t + 2e^{-t}.t + 2e^{-t}\left(\frac{t^2}{2!}\right)$$

$$= (1 + 2t + t^2)\, e^{-t} - (1 - t)\, e^t.$$

EXAMPLE 6. *Solve* $[tD^2 + (1 - 2t)D - 2]y = 0$, *where* $y(0) = 1$, $y'(0) = 2$.

(UPTU–2002; PTU–2002; Meerut–1996; Rohilkhand–1996; Agra–1998; Raj.–1983)

SOLUTION. Here, $tD^2y + (1 - 2t)Dy - 2y = 0$

$\Rightarrow \qquad ty'' + y' - 2ty' - 2y = 0$

Taking Laplace transform of given differential equation, we get

$\qquad L\{ty''\} + L\{y'\} - 2L\{ty'\} - 2L\{y\} = 0$

$\Rightarrow \quad -\dfrac{d}{dp}L\{y''\} + L\{y'\} + 2\dfrac{d}{dp}L\{y'\} - 2L\{y\} = 0$

$\qquad -\dfrac{d}{dp}\left[p^2L\{y\} - py(0) - y'(0)\right] + [pL\{y\}$

$\qquad\qquad\qquad - y(0)] + 2\left[pL\{y\} - y(0)\right] - 2L\{y\} = 0$

Putting the values of $y(0)$ and $y'(0)$, we get

$\qquad -\dfrac{d}{dp}(p^2L\{y\} - p - 2) + (pL\{y\} - 1) + 2\dfrac{d}{dp}(pL\{y\} - 1) - 2L\{y\} = 0$

$$[\because y(0) = 1, y'(0) = 2]$$

$\Rightarrow \quad -p^2\dfrac{dL\{y\}}{dp} - 2pL\{y\} + 1 + pL\{y\} - 1 + 2\left(p\dfrac{dL\{y\}}{dp} + L\{y\}\right) - 2L\{y\} = 0$

$\Rightarrow \qquad\qquad -(p^2 - 2p)\dfrac{dL\{y\}}{dp} - pL\{y\} = 0$

$\Rightarrow \qquad\qquad -\dfrac{dL\{y\}}{\overline{y}} - \dfrac{1}{p-2}dp = 0 \qquad$ [Separating the variables]

$\Rightarrow \qquad\qquad \int\dfrac{dL\{y\}}{\overline{y}} + \int\dfrac{dp}{p-2} = 0$

$\Rightarrow \qquad\qquad \log L\{y\} + \log(p-2) = \log C$

$\Rightarrow \qquad\qquad \log L\{y\}(p-2) = \log C$

$\Rightarrow \qquad\qquad L\{y\}(p-2) = C$

$\Rightarrow \qquad\qquad L\{y\} = \dfrac{C}{p-2}$

$\Rightarrow \qquad\qquad y = CL^{-1}\left\{\dfrac{1}{p-2}\right\} \Rightarrow y = Ce^{2t} \qquad\qquad$ ...(1)

At $\qquad\qquad x = 0,\ y(0) = Ce^0 \qquad\qquad$ ...(2)

Putting $y(0) = 1$, in (2), we get

$$1 = Ce^0 \Rightarrow C = 1$$

Putting $C = 1$ in (1), we get $y = e^{2t}$. This is the required solution.

EXAMPLE 7. Solve $y'' - ty' + y = 1$ if $y(0) = 1$, $y'(0) = 2$. (Agra–1984)

SOLUTION. Taking the Laplace transforms of both sides of the given equation, we get

$$L\{y''\} - L\{ty'\} + L\{y\} = L\{1\}$$

$$\Rightarrow \quad p^2 L\{y\} - py(0) - y'(0) + \frac{d}{dp}[L\{y'\}] + L\{y\}] = \frac{1}{p}$$

$$\Rightarrow \quad p^2 L\{y\} - p - 2 + \frac{d}{dp}[pL\{y\} - y(0)] + L\{y\} = \frac{1}{p}$$

$$\Rightarrow \quad p^2 L\{y\} - p - 2 + \frac{d}{dp}[pL\{y\} - 1] + L\{y\} = \frac{1}{p}$$

$$\Rightarrow \quad p\frac{d[L\{y\}]}{dp} + (p^2 + 2)L\{y\} = p + 2 + \frac{1}{p}$$

$$\Rightarrow \quad d\frac{[L\{y\}]}{dp} + \left(p + \frac{2}{p}\right)L\{y\} = 1 + \frac{2}{p} + \frac{1}{p^2} \quad \ldots(1)$$

which is a linear differential equation in $L(y)$.

$$\therefore \quad \text{I.F.} = e^{\int\left(p + \frac{2}{p}\right)dp} = e^{\frac{p^2}{2} + 2\log p} = p^2 e^{p^2/2}.$$

Therefore, solution of (1) is given by

$$p^2 e^{p^2/2} L\{y\} = C_1 + \int\left(1 + \frac{2}{p} + \frac{1}{p^2}\right)p^2 e^{p^2/2} dp$$

Hence, the solution of (1) is given by

$$p^2 e^{p^2/2} L\{y\} = C_1 + \int\left(1 + \frac{2}{p} + \frac{1}{p^2}\right)p^2 e^{p^2/2}.dp$$

$$= C_1 + \int (p^2 + 2p + 1) e^{p^2/2} dp = C_1 + \int (p^2 + 1)e^{p^2/2} dp + 2\int pe^{p^2/2} dp$$

$$= C_1 + \int (2v + 1)e^v.\frac{dv}{\sqrt{2v}} + 2\int \sqrt{2v}.e^v.\frac{dv}{\sqrt{2v}}$$

$$\left[\text{where } \frac{p^2}{2} = v \Rightarrow pdp = dv, \text{ i.e., } dp = \frac{dv}{\sqrt{2v}}\right]$$

$$\doteq C_1 + \sqrt{2v}.e^v - \int \frac{e^v}{\sqrt{2v}}dv + \int \frac{e^v}{\sqrt{2v}}dv + 2\int e^v dv$$

$$= C_1 + \sqrt{2v}\,e^v + 2e^v = C_1 + pe^{p^2/2} + 2e^{p^2/2}.$$

Therefore, $L\{y\} = \dfrac{C_1}{p^2}e^{-p^2/2} + \dfrac{1}{p} + \dfrac{2}{p^2}$

$$= \frac{C_1}{p^2}\left(1 - \frac{p^2}{2} + \frac{p^4}{4 \cdot 2!} - \ldots\right) + \frac{1}{p} + \frac{2}{p^2} = \frac{(2 + C_1)}{p^2} - \frac{C_1}{2} + \frac{C_1}{8}p^2 \ldots + \frac{1}{p}$$

[On expanding the exponential function]

Hence, $\quad y = (2 + C_1)L^{-1}\left\{\dfrac{1}{p^2}\right\} - \dfrac{1}{2}C_1 L^{-1}\{1\} + \dfrac{1}{8}C_1 L^{-1}\{p^2\} + \ldots + L^{-1}\left\{\dfrac{1}{p}\right\}$

$$= (2+C_1)t + 1 \qquad [\because L^{-1}\{p^n\} = 0, \text{ for } n = 0,1,2,\ldots]$$

Also, given that
$$y'(0) = 2$$
$$\therefore \qquad 2 = 2 + C_1 \Rightarrow C_1 = 0$$
which gives $y = 2t + 1$ is the required solution.

EXAMPLE 8. *Using Laplace transform, solve the following differential equation*
$$y'' + 2ty' - y = t$$
where, $y(0) = 0$ *and* $y'(0) = 1$. (UPTU–2003)

SOLUTION. We have $\qquad y'' + 2ty' - y = t$ ...(1)
Taking Laplace transform of (1), we get

$$[p^2 L\{y\} - py(0) - y'(0)] - 2\frac{d}{dp}[pL\{y\} - y(0)] - L\{y\} = \frac{1}{p^2} \qquad ...(2)$$

On putting $y(0) = 0$ and $y'(0) = 1$ in (2), we get

$$(p^2 L\{y\} - 1) - 2\frac{d}{dp}(pL\{y\} - 0) - L\{y\} = \frac{1}{p^2}$$

$$\Rightarrow \quad (p^2 L\{y\} - 1) - 2L\{y\} - 2p\frac{dL\{y\}}{dp} - L\{y\} = \frac{1}{p^2}$$

$$\Rightarrow \qquad -2p\frac{dL\{y\}}{dp} + (p^2 - 3)L\{y\} = \frac{1}{p^2} + 1 = \frac{1+p^2}{p^2}$$

$$\Rightarrow \qquad \frac{dL\{y\}}{dp} - \frac{p^2-3}{2p}L\{y\} = \frac{1+p^2}{-2p^3}$$

$$\Rightarrow \qquad \frac{dL\{y\}}{dp} - \left(\frac{p}{2} - \frac{3}{2p}\right)L\{y\} = -\frac{1}{2p^3} - \frac{1}{2p} \qquad ...(3)$$

Thus, (3) is a linear differential equation in $L(y)$

$$\text{I.F.} = e^{\frac{1}{2}\int\left(\frac{3}{p} - p\right)dp} = e^{\frac{1}{2}\left(3\log p - \frac{p^2}{2}\right)} = e^{\frac{p^2}{4}} \cdot p^{3/2}$$

Solution of differential equation (3) is

$$L\{y\}\, e^{-p^2/4} \cdot p^{3/2} = \frac{1}{2}\int\left(\frac{1}{p^3} + \frac{1}{p}\right)p^{3/2}\cdot e^{-p^2/4}\,dp = -\frac{1}{2}\int\left(\sqrt{p} + \frac{1}{p^{3/2}}\right)e^{-p^2/4}\,dp$$

Put $p^2 = ut \Rightarrow p = 2\sqrt{t}$ so that $dp = \dfrac{dt}{\sqrt{t}}$. Then we have

$$L\{y\}p^{3/2}e^{-p^2/4} = -\frac{1}{2}\int\left(\sqrt{2}\,t^{1/4} + \frac{1}{2\sqrt{2}}t^{-3/4}\right)e^{-t}\frac{dt}{\sqrt{t}}$$

$$= -\frac{1}{\sqrt{2}}\int\left(t^{-1/4} + \frac{1}{4}t^{-5/4}\right)e^{-t}\,dt$$

$$= -\frac{1}{\sqrt{2}}\int t^{-1/4}e^{-t}\,dt - \frac{1}{4\sqrt{2}}\int t^{-5/4}e^{-t}\,dt$$

$$= -\frac{1}{\sqrt{2}}\left[t^{-1/4}\frac{e^{-t}}{-1} + \int\left(-\frac{1}{4}\right)t^{-5/4}e^{-t}\,dt\right] + \frac{1}{4\sqrt{2}}\int t^{-5/4}e^{-t}\,dt$$

$$= \frac{1}{\sqrt{2}} e^{-t} \cdot t^{-1/4} = \frac{1}{\sqrt{2}} e^{-p^2/4} \left(\frac{p^2}{4} \right)^{-1/4} = \frac{1}{\sqrt{p}} e^{-p^2/4}$$

$$\Rightarrow \qquad L\{y\} = \frac{1}{p^2}$$

$$\Rightarrow \qquad L\{y\} = \frac{1}{p^2} + C \Rightarrow y = L^{-1} \left\{ \frac{1}{p^2} + C \right\} = t + C.$$

EXAMPLE 9. Solve by using Laplace transform $\dfrac{d^2 y}{dx^2} + y = \cos x$, given $y(0) = 0, y'(0) = 0$.

(Kanpur–2002, 11; Meerut–2013)

SOLUTION. Here, we have $\dfrac{d^2 y}{dx^2} + y = \cos x$ ...(1)

Taking Laplace transform of both the sides of eq^n (1), we get

$$L(y'') + L(y) = L(\cos x)$$

$$\Rightarrow \quad p^2 \bar{y} - p y(0) - y'(0) + \bar{y} = \frac{p}{p^2 + 1}$$

Using $y(0) = 0, y'(0) = 0$, then we get

$$p^2 \bar{y} + \bar{y} = \frac{p}{p^2 + 1} \qquad \Rightarrow \qquad \bar{y} = \frac{p}{(p^2 + 1)^2}$$

$$\Rightarrow \quad L^{-1}(\bar{y}) = L^{-1} \left\{ \frac{p}{(p^2 + 1)^2} \right\} \quad \Rightarrow \quad y(x) = L^{-1} \left\{ \frac{p}{(p^2 + 1)^2} \right\} \quad ...(2)$$

$$L^{-1} \left\{ \frac{p}{p^2 + 1} \right\} = \sin x \qquad \Rightarrow \quad L^{-1} \left\{ \frac{d}{dp} \left(\frac{1}{p^2 + 1} \right) \right\} = (-1)^1 x^1 \sin x$$

$$\Rightarrow \quad L^{-1} \left\{ \frac{-2p}{(p^2 + 1)^2} \right\} = -x \sin x \qquad \Rightarrow \quad L^{-1} \left\{ \frac{p}{(p^2 + 1)^2} \right\} = \frac{x}{2} \sin x$$

3.3 SOLUTION OF SIMULTANEOUS ORDINARY DIFFERENTIAL EQUATIONS

The Laplace transform can be used to solve two or more simultaneous ordinary differential equations. The procedure is essentially the same as that described in previous sections.

Solved Examples

EXAMPLE 1. Solve $(D^2 + 2)x - Dy = 1$, $Dx + (D^2 + 2)y = 0$, if $x = 0 = Dx = y = Dy$, when $t = 0$.

(Meerut–1997; Kumaun–1997; Kanpur–1998; Kurukshetra-2004)

SOLUTION. Taking Laplace transforms of both sides of the given equations, we have

$$L\{x''\} + 2L\{x\} - L\{y'\} = L\{1\}$$

and $\qquad L\{x'\} + 2L\{y''\} + 2L\{y\} = 0$

$$\Rightarrow \quad p^2 L\{x\} - p x(0) - x'(0) + 2L\{x\} - [p L\{y\} - y(0)] = \frac{1}{p}$$

and $\quad p L\{x\} - x(0) + p^2 L\{y\} - p y(0) - y'(0) + 2L\{y\} = 0$

which gives $\qquad (p^2 + 2)L(x) - pL\{y\} = \dfrac{1}{p}$

and $\qquad pL\{x\} + (p^2 + 2)L\{y\} = 0$

Solving for $L\{x\}$ and $L\{y\}$, we have

$$L\{x\} = \frac{p^2 + 2}{p(p^4 + 5p^2 + 4)} = \frac{1}{2p} - \frac{1}{6}\left[\frac{2p}{p^2 + 1} + \frac{p}{p^2 + 4}\right]$$

and $\qquad L\{y\} = \dfrac{-1}{p^4 + 5p^2 + 4} = \dfrac{1}{3}\left[\dfrac{1}{p^2 + 4} - \dfrac{1}{p^2 + 1}\right].$

Therefore, $\qquad x = \dfrac{1}{2}L^{-1}\left\{\dfrac{1}{p}\right\} - \dfrac{1}{6}\left[2L^{-1}\left\{\dfrac{p}{p^2 + 1}\right\} + L^{-1}\left\{\dfrac{p}{p^2 + 4}\right\}\right]$

$$= \frac{1}{2} - \frac{1}{6}[2\cos t + \cos 2t]$$

and $\qquad y = \dfrac{1}{3}\left[\dfrac{1}{2}\sin 2t - \sin t\right] = \dfrac{1}{6}[\sin 2t - 2\sin t]$.

EXAMPLE 2. *Solve the simultaneous equation* $\dfrac{dx}{dt} - y = e^t$, $\dfrac{dy}{dt} + x = \sin t$, *given* $x(0) = 1$, $y(0) = 0$. $\qquad$ (UPTU–2006; Q.Bank–2001; GBTU (SUM)–2010; UKTU–2011; Delhi–2002)

SOLUTION. $\qquad$ Taking Laplace transforms of the given equations, we get

$$[p\bar{x} - x(0)] - \bar{y} = \frac{1}{p - 1}, \text{ where } \bar{x} = L(x), \ \bar{y} = L(y)$$

i.e., $\qquad p\bar{x} - 1 - \bar{y} = \dfrac{1}{p - 1} \qquad\qquad\qquad\qquad [\because x(0) = 1]$

$$p\bar{x} - \bar{y} = \frac{p}{p - 1} \text{ and } [p\bar{y} - y(0)] + \bar{x} = \frac{1}{p^2 + 1}$$

i.e., $\qquad \bar{x} + p\bar{y} = \dfrac{1}{p^2 + 1} \qquad\qquad\qquad\qquad [\because y(0) = 0] \ \ ...(2)$

Solving (1) and (2) for $\bar{x}$ and $\bar{y}$, we have

$$\bar{x} = \frac{p^2}{(p - 1)(p^2 + 1)} + \frac{1}{(p^2 + 1)^2}$$

$$= \frac{1}{2}\left[\frac{1}{p - 1} + \frac{p}{p^2 + 1} + \frac{1}{p^2 + 1}\right] + \frac{1}{(p^2 + 1)^2}$$

$$\bar{y} = \frac{p}{(p^2 + 1)^2} - \frac{p}{(p - 1)(p^2 + 1)}$$

$$= \frac{p}{(p^2 + 1)^2} - \frac{1}{2}\left[\frac{1}{p - 1} - \frac{p}{p^2 + 1} + \frac{1}{p^2 + 1}\right]$$

Taking inverse Laplace transform of both sides, we get

$$x = \frac{1}{2}L^{-1}\left\{\frac{1}{p - 1} + \frac{p}{p^2 + 1} + \frac{1}{p^2 + 1}\right\} + L^{-1}\left\{\frac{1}{(p^2 + 1)^2}\right\}$$

$$= \frac{1}{2}\left[e^t + \cos t + \sin t\right] + \frac{1}{2}(\sin t - t \cos t)$$

$$= \frac{1}{2}\left[e^t + \cos t + 2\sin t - t \cos t\right]$$

$$y = L^{-1}\left\{\frac{p}{(p^2+1)^2}\right\} - \frac{1}{2}L^{-1}\left\{\frac{1}{p-1} - \frac{p}{p^2+1} + \frac{1}{p^2+1}\right\}$$

$$= \frac{1}{2}t \sin t - \frac{1}{2}\left[e^t - \cos t + \sin t\right] = \frac{1}{2}\left[t \sin t - e^t + \cos t - \sin t\right]$$

Hence, $\qquad x = \frac{1}{2}(e^t + \cos t + 2\sin t - t \cos t)$

$$y = \frac{1}{2}(t \sin t - e^t + \cos t - \sin t)$$

EXAMPLE 3.   *Using Laplace transformation, solve*

$$(D-2)x - (D+1)y = 6e^{3t}$$

$$(2D-3)x + (D-3)y = 6e^{3t}$$

Given $x = 3$, $y = 0$ *when* $t = 0$.   (UPTU–2000, 01; Kanpur–1983; Vikram–1996)

SOLUTION.   Taking Laplace transformation of the given equations, we get

$$\left. \begin{array}{l} LDx - 2Lx - LDy - Ly = 6Le^{3t} \\ 2LDx - 3Lx + LDy - 3Ly = 6Le^{3t} \end{array} \right\}$$

$$\Rightarrow \qquad \left. \begin{array}{l} p\bar{x} - x(0) - 2\bar{x} - p\bar{y} + y(0) - \bar{y} = 6\dfrac{1}{p-3} \\ 2p\bar{x} - 2x(0) - 3\bar{x} + p\bar{y} - y(0) - 3\bar{y} = \dfrac{6}{p-3} \end{array} \right\}, \text{ where } \bar{x} = L(x)$$
$$\text{and} \quad \bar{y} = L(y)$$

$$\Rightarrow \qquad \left. \begin{array}{l} (p-2)\bar{x} - (p+1)\bar{y} - 3 = \dfrac{6}{p-3} \\ (2p-3)\bar{x} + (p-3)\bar{y} - 6 = \dfrac{6}{p-3} \end{array} \right\}$$

$$\Rightarrow \qquad \left. \begin{array}{l} (p-2)\bar{x} - (p+1)\bar{y} = \dfrac{3p-3}{p-3} \\ (2p-3)\bar{x} + (p-3)\bar{y} = \dfrac{6p-12}{p-3} \end{array} \right\}$$

$$\Rightarrow \qquad \left. \begin{array}{l} (p-3)(p-2)\bar{x} - (p-3)(p+1)\bar{y} = 3p-3 \\ (p+1)(2p-3)\bar{x} + (p+1)(p-3)\bar{y} = \dfrac{(p+1)(6p-12)}{p-3} \end{array} \right\}$$

On adding, we get

$$(3p^2 - 6p + 3)\bar{x} = 3(p-1) + \frac{6(p^2 - p - 2)}{p-3}$$

$$\Rightarrow \qquad \bar{x} = \frac{3(p-1)}{3(p-1)^2} + \frac{6(p^2 - p - 2)}{3(p-1)^2(p-3)}$$

$$x = L^{-1}\left\{\frac{1}{p-1}+\frac{2}{(p-1)^2}+\frac{2}{p-3}\right\}$$

$$= e^t + 2te^t + 2e^{3t}$$

Putting the value of x in (1), we get

$$(D-2)(e^t + 2te^t + 2e^{3t}) - (D+1)y = 6e^{3t}$$

$$\Rightarrow e^t + 2te^t + 2e^t + 6e^{3t} - 2e^t - 4te^t - 4e^{3t} - (D+1)y = 6e^{3t}$$

$$\Rightarrow \quad (D+1)y = e^t - 2te^t - 4e^{3t} \qquad \qquad \dots (2)$$

Taking Laplace transform of (2), we get

$$p\bar{y} - y(0) + \bar{y} = \frac{1}{p-1} - \frac{2}{(p-1)^2} - \frac{4}{p-3}$$

$$\Rightarrow \quad (p+1)\bar{y} = \frac{1}{p-1} - \frac{2}{(p-1)^2} - \frac{y}{p-3}$$

$$\bar{y} = \frac{1}{p^2-1} - \frac{2}{(p+1)(p-1)^2} - \frac{4}{(p+1)(p-3)}$$

$$\bar{y} = \frac{1}{p^2-1} - \frac{1/2}{p+1} + \frac{1/2}{p-1} - \frac{1}{(p-1)^2} + \frac{1}{p+1} - \frac{1}{p-3}$$

$$\bar{y} = \frac{1}{p^2-1} + \frac{1/2}{p+1} + \frac{1/2}{p-1} - \frac{1}{(p-1)^2} - \frac{1}{p-3}$$

$$\Rightarrow \quad y = L^{-1}\left\{\frac{1}{p^2-1} + \frac{1}{2}\frac{1}{p+1} + \frac{1}{2}\frac{1}{p-1} - \frac{1}{p-3} - \frac{1}{(p-1)^2}\right\}$$

$$\Rightarrow \quad y = \sinh t + \frac{1}{2}e^{-t} + \frac{1}{2}e^t - e^{3t} - te^t$$

$$\Rightarrow \quad y = \sinh t + \cosh t - e^{-3t} - te^t$$

EXAMPLE 4. Solve$(D^2-1)x + 5Dy = t, -Dx + (D^2-4)y = -2$, when $x = 0 = Dx = y = Dy$, $dt = 0$.

(Meerut–1986; Agra–1997; Rohilkhand–1990)

SOLUTION. Taking the Laplace transforms of both sides of the given equations, we have

$$L\{x''\} - L\{x\} + 5L\{y'\} = L\{t\}$$

and

$$-2L\{x'\} + L\{y''\} - 4L\{y\} = -2L\{1\}$$

or

$$p^2 L\{x\} - px(0) - x'(0) - L\{x\} + 5[pL\{y\} - y(0)] = 1/p^2$$

and $-2[pL\{x\} - x(0)] + p^2 L\{y\} - py(0) - y'(0) - 4L\{y\} = -2/p$

which gives

$$(p^2-1)L\{x\} + 5pL\{y\} = 1/p^2 \qquad \qquad \dots(1)$$

and

$$-2pL\{x\} + (p^2-4)L\{y\} = -2/p \qquad \qquad \dots(2)$$

On solving (1) and (2) for $L(x)$ and $L(y)$, we get

$$L(x) = \frac{11p^2-4}{p^2(p^2+1)(p^2+4)} = -\frac{1}{p^2} + \frac{5}{p^2+1} - \frac{4}{p^2+4}$$

and
$$L\{y\} = \frac{-2p^2 + 4}{p(p^2 + 1)(p^2 + 4)} = \frac{1}{p} - \frac{2p}{p^2 + 1} + \frac{p}{p^2 + 4}$$

Therefore, we get
$$x = -L^{-1}\left\{\frac{1}{p^2}\right\} + 5L^{-1}\left\{\frac{1}{p^2 + 1}\right\} - 4L^{-1}\left\{\frac{1}{p^2 + 4}\right\}$$

$$= -t + 5\sin t - 2\sin 2t$$

and
$$y = L^{-1}\left\{\frac{1}{p}\right\} - 2L^{-1}\left\{\frac{p}{p^2 + 1}\right\} + L^{-1}\left\{\frac{p}{p^2 + 4}\right\}$$

$$= 1 - 2\cos t + \cos 2t .$$

EXAMPLE 5.   Solve $Dx + Dy = t$; $D^2x - y = e^{-t}$, when $x(0) = 3$, $x'(0) = -2$, $y(0) = 0$.

(Meerut–1987, 94, 2002; Rohilkhand–1989; Agra–1980; Kanpur–2000, 01; SVTU–2004; MDU-2006)

SOLUTION.   Taking the Laplace transforms of both the sides of the given equations, we get
$$L\{x'\} + L\{y'\} = L\{t\} \quad \text{and} \quad L\{x''\} - L\{y\} = L\{e^{-t}\}$$

which gives $pL\{x\} - x(0) + pL\{y\} - y(0) = 1/p^2$

and
$$p^2L\{x\} - px(0) - x'(0) - L\{y\} = \frac{1}{p+1}$$

or $pL\{x\} + pL\{y\} = 3 + \dfrac{1}{p^2}$ ...(1)

and $p^2L\{x\} - L\{y\} = 3p - 2 + \dfrac{1}{p+1}$ ...(2)

Solving (1) and (2) for $L\{x\}$ and $L\{y\}$, we get
$$L\{x\} = \frac{2}{p} + \frac{1}{p^3} + \frac{1}{2(p+1)} + \frac{p}{2(1+p^2)} - \frac{3}{2(p^2+1)}$$

and
$$L\{y\} = \frac{1}{p(p+1)(p^2+1)} + \frac{2}{p^2+1}$$

$$= \frac{1}{p} - \frac{p}{2(p+1)} - \frac{p}{2(p^2+1)} - \frac{1}{2(p^2+1)} + \frac{2}{p^2+1} .$$

$$= \frac{1}{p} - \frac{1}{2(p+1)} - \frac{p}{2(p^2+1)} + \frac{3}{2(p^2+1)}$$

Therefore,
$$x = 2L^{-1}\left\{\frac{1}{p}\right\} + L^{-1}\left\{\frac{1}{p^3}\right\} + \frac{1}{2}L^{-1}\left\{\frac{1}{p+1}\right\} + \frac{1}{2}L^{-1}\left\{\frac{p}{p^2+1}\right\} - \frac{3}{2}L^{-1}\left\{\frac{1}{p^2+1}\right\}$$

$$= 2 + \frac{1}{2}t^2 + \frac{1}{2}e^{-t} + \frac{1}{2}\cos t - \frac{3}{2}\sin t$$

and $y = L^{-1}\left\{\dfrac{1}{p}\right\} - \dfrac{1}{2}L^{-1}\left\{\dfrac{1}{p+1}\right\} - \dfrac{1}{2}L^{-1}\left\{\dfrac{p}{p^2+1}\right\} + \dfrac{3}{2}L^{-1}\left\{\dfrac{1}{p^2+1}\right\}$

$$= 1 - \frac{1}{2}e^{-t} - \frac{1}{2}\cos t + \frac{3}{2}\sin t .$$

EXAMPLE 6. *Using Laplace transform to solve*

$$Dx + 2D^2y = e^{-t}, \ (D + 2)x - y = 1, \ if \ x(0) = y(0) = y'(0) = 0. \qquad \text{(Agra–2011)}$$

SOLUTION. Here, we have

$$Dx + 2D^2y = e^{-t} \text{ and } (D + 2)x - y = 1$$

$$\Rightarrow \qquad x' + y'' = e^{-t} \text{ and } x' + 2x - y = 1$$

Taking Laplace transform of both the sides of both the equations, we get

$$L\{x'\} + 2L\{y''\} = L\{e^{-t}\} \text{ and } L\{x'\} + 2L\{x\} - L\{y\} = L(1)$$

$$\Rightarrow \qquad p\bar{x} - x(0) + 2[p^2\bar{y} - py(0) - y'(0)] = \frac{1}{p+1} \quad \text{[Here, } \bar{x} = L\{x\} \text{ and } \bar{y} = L\{y\}\text{]}$$

and $\quad p\bar{x} - x(0) + 2\bar{x} - \bar{y} = \dfrac{1}{p}$

$$\Rightarrow \qquad p\bar{x} + 2p^2\bar{y} = \frac{1}{p+1} \quad \text{and} \quad (p+2)\bar{x} - \bar{y} = \frac{1}{p}$$

Now, solving both the equations for $\bar{x}$ and $\bar{y}$, we get

$$\bar{x} = \frac{1}{p(p+1)(2p^2 + 4p + 1)} + \frac{2}{(2p^2 + 4p + 1)}$$

$$= \frac{1}{2p(p+1)\left(p + \dfrac{2-\sqrt{2}}{2}\right)\left(p + \dfrac{2+\sqrt{2}}{2}\right)} + \frac{1}{\left(p + \dfrac{2-\sqrt{2}}{2}\right)\left(p + \dfrac{2+\sqrt{2}}{2}\right)}$$

$$= \frac{1}{p} + \frac{1}{p+1} - \frac{1}{(2-\sqrt{2})\left(p + \dfrac{2-\sqrt{2}}{2}\right)} - \frac{1}{(2+\sqrt{2})\left(p + \dfrac{2+\sqrt{2}}{2}\right)}$$

$$+ \frac{1}{\sqrt{2}\left(p + \dfrac{2-\sqrt{2}}{2}\right)} - \frac{1}{\sqrt{2}\left(p + \dfrac{2+\sqrt{2}}{2}\right)}$$

$$= \frac{1}{p} + \frac{1}{p+1} - \frac{1}{p + \dfrac{2-\sqrt{2}}{2}} - \frac{1}{\left(p + \dfrac{2+\sqrt{2}}{2}\right)}$$

and $\quad \bar{y} = \dfrac{1}{p(p+1)(2p^2 + 4p + 1)}$

$$= \frac{1}{p} + \frac{1}{p+1} - \frac{1}{(2-\sqrt{2})\left(p + \dfrac{2-\sqrt{2}}{2}\right)} - \frac{1}{(2+\sqrt{2})\left(p + \dfrac{2+\sqrt{2}}{2}\right)}$$

$$= \frac{1}{p} + \frac{1}{p+1} - \left(\frac{2+\sqrt{2}}{2}\right)\frac{1}{\left(p + \dfrac{2-\sqrt{2}}{2}\right)} - \left(\frac{2-\sqrt{2}}{2}\right)\frac{1}{\left(p + \dfrac{2+\sqrt{2}}{2}\right)}$$

$$x = L^{-1}\left\{\frac{1}{p}\right\} + L^{-1}\left\{\frac{1}{p+1}\right\} - L^{-1}\left\{\frac{1}{p+a}\right\} - L^{-1}\left\{\frac{1}{p+b}\right\}$$

$$= .1 + e^{-t} - e^{-at} - e^{-bt}$$

and $\quad y = L^{-1}\left\{\dfrac{1}{p}\right\} + L^{-1}\left\{\dfrac{1}{p+1}\right\} - bL^{-1}\left\{\dfrac{1}{p+a}\right\} - aL^{-1}\left\{\dfrac{1}{p+b}\right\}$

$$= 1 + e^{-t} - be^{-at} - ae^{-bt} \qquad \left[\because a = \dfrac{2-\sqrt{2}}{2} \text{ and } b = \dfrac{2+\sqrt{2}}{2}\right]$$

EXAMPLE 7.   *Using Laplace's transform, solve the equations :*

$$\frac{dx}{dt} + \frac{dy}{dt} = t, \quad \frac{d^2x}{dt^2} - y = e^{-t}$$

given that $x(0) = 0$, $y(0) = 0$, $\dfrac{dx}{dt} = 0$ *if* $t = 0$.   (Agra–1996)

SOLUTION.   Taking Laplace transform of both the equations, we get

$$p\bar{x} - x(0) + p\bar{y} - y(0) = \frac{1}{p^2} \qquad \text{[Here } \bar{x} = L(x),\ \bar{y} = L(y)\text{]}$$

and $\quad p^2\bar{x} - px(0) - x'(0) - \bar{y} = \dfrac{1}{(p+1)}$

On substituting the values, we get

$$\bar{x} + \bar{y} = \frac{1}{p^3} \qquad\qquad\qquad\qquad\qquad\qquad …(1)$$

$\therefore \qquad p^2\bar{x} - \bar{y} = \dfrac{1}{p+1} \qquad\qquad\qquad\qquad\qquad …(2)$

Adding eq^n (1) & (2), we get

$$(1+p^2)\bar{x} = \frac{1}{p^3} + \frac{1}{p+1}$$

$$\Rightarrow \qquad \bar{x} = \frac{1}{p^3(1+p^2)} + \frac{1}{(1+p)(1+p^2)}$$

From eq^n (1) $\quad \bar{y} = \dfrac{1}{p^3} - \dfrac{1}{p^3(1+p^2)} - \dfrac{1}{(1+p)(1+p^2)}$

$$\Rightarrow \qquad \bar{y} = \frac{p^2}{p^3(1+p^2)} - \frac{1}{(1+p)(1+p^2)} = \left(\frac{1}{p} - \frac{p}{1+p^2}\right) - \frac{1}{2}\left(\frac{1}{1+p} - \frac{p-1}{1+p^2}\right)$$

$$= \frac{1}{p} - \frac{1}{2(1+p)} - \frac{p}{2(1+p^2)} - \frac{1}{2(1+p^2)}$$

Now, taking inverse Laplace transform, we get

$$y = 1 - \frac{1}{2}e^{-t} - \frac{1}{2}\cos t - \frac{1}{2}\sin t$$

$$\Rightarrow \qquad y = 1 - \frac{1}{2}(e^{-t} + \cos t + \sin t)$$

Taking inverse Laplace transform of eq^n (1), we get

$$x = \frac{t^2}{2} - L^{-1}\{\bar{y}\} = \frac{t^2}{2} - y = \frac{t^2}{2} - \left[1 - \frac{1}{2}(e^{-t} + \cos t + \sin t)\right]$$

$$x = \frac{t^2}{2} - 1 + \frac{1}{2}(e^{-t} + \cos t + \sin t)$$

EXAMPLE 8. *Using Laplace transform method, solve*

$$(D^2 - 2)x - 3y = e^{2t}$$

and $\qquad (D^2 + 2)y + x = 0$

satisfying the conditions $x = 1, y = 1, Dx = Dy = 0$ *when* $t = 0$. (Kanpur–1988)

SOLUTION. Let $\qquad Dx = \dfrac{dx}{dt} = x'$ and $\dfrac{d^2x}{dt^2} = x''$

Then $\qquad x'' - 2x - 3y = e^{2t}$ ...(1)

and $\qquad x + y'' + 2y = 0$ ...(2)

Now, taking Laplace transform of both the equations, we get

$$[p^2\bar{x} - px(0) - x'(0)] - 2\bar{x} - 3\bar{y} = \frac{1}{p-2}$$

and $\qquad \bar{x} + [p^2 y - py(0) - y'(0)] + 2\bar{y} = 0$

Substituting $x(0) = y(0) = 1, x'(0) = 0, y'(0) = 0$, we get

$$(p^2\bar{x} - p) - 2\bar{x} - 3\bar{y} = \frac{1}{p-2}$$

and $\qquad p^2\bar{y} - p + 2\bar{y} + \bar{x} = 0$

$\Rightarrow \qquad (p^2 - 2)\bar{x} - 3\bar{y} = p + \dfrac{1}{p-2}$ ...(3)

and $\qquad (p^2 + 2)\bar{y} + \bar{x} = p$ ...(4)

and $\qquad (p^2 - 2)\bar{x} - 3\bar{y} = p + \dfrac{1}{p-2}$

Now, $\qquad (p^2 - 2)\bar{x} + (p^4 - 4)\bar{y} = p(p^2 - 2)$ ...(5)

$\Rightarrow \qquad (1 - p^4)\bar{y} = 3p - p^3 + \dfrac{1}{p-2}$ (Subtracting the last two)

$\Rightarrow \qquad \bar{y} = \dfrac{3p}{1 - p^4} - \dfrac{p^3}{1 - p^4} + \dfrac{1}{(p-2)(1 - p^4)}$

$\Rightarrow \qquad \bar{y} = \dfrac{-3p}{p^4 - 1} + \dfrac{p^3}{p^4 - 1} - \dfrac{1}{(p-2)(p^4 - 1)}$ ...(6)

From eq^n (3) and (4), we get

$$(p^4 - 4)\bar{x} - 3(p^2 + 2)\bar{y} = p(p^2 + 2) + \frac{(p^2 + 2)}{p-2}$$

and $\qquad 3\bar{x} + 3(p^2 + 2)\bar{y} = 3p$

On adding, we get

$$(p^4 - 1)\bar{x} = p^3 + 5p + \frac{p^2 + 2}{p-2}$$

$\Rightarrow \qquad \bar{x} = \dfrac{p^3}{p^4 - 1} + \dfrac{5p}{p^4 - 1} + \dfrac{p^2 + 2}{(p-2)(p^4 - 1)}$ ...(7)

$\Rightarrow \qquad \dfrac{p}{p^4 - 1} = \dfrac{p}{2}\left[\dfrac{1}{p^2 - 1} - \dfrac{1}{p^2 + 1}\right]$

$$L^{-1}\left\{\frac{p}{p^4-1}\right\} = \frac{1}{2}[\cosh t - \cos t] \qquad \ldots(8)$$

$$\Rightarrow \qquad \frac{p^3}{p^4-1} = \frac{p^3}{(p-1)(p+1)(p^2+1)}$$

$$\Rightarrow \qquad \frac{p^3}{p^4-1} = \frac{A}{p-1} + \frac{B}{p+1} + \frac{Cp+d}{p^2+1} \qquad \ldots(9)$$

$$p = 1, A = \frac{1}{(1+1)(1^2+1)} = \frac{1}{4}$$

$$p = -1, B = \frac{1}{4}$$

From eq^n (9),

$$p^3 = A(p+1)(p^2+1) + B(p-1)(p^2+1) + (Cp+d)(p+1)(p-1) \qquad \ldots(10)$$

Here, coefficient of $p^3 \Rightarrow 1 = A + B + C = \frac{1}{2} + C$

$$\Rightarrow \qquad C = \frac{1}{2}$$

Putting $p = 0$ in eq^n (10), we get

$$0 = A - B - D \text{ or } D = 0$$

$$\therefore \qquad \frac{p^3}{p^4-1} = \frac{1}{4}\left(\frac{1}{p-1} + \frac{1}{p+1}\right) + \frac{1}{2}\cdot\frac{p}{p^2+1} = \frac{1}{2}\left[\frac{p}{p^2-1} + \frac{p}{p^2+1}\right]$$

$$\therefore \quad L^{-1}\left\{\frac{p^3}{p^4-1}\right\} = \frac{1}{2}[\cosh t + \cos t] \qquad \ldots(11)$$

Now, $\dfrac{p^2+2}{(p-2)(p^4-1)} = \dfrac{A}{p-2} + \dfrac{B}{p-1} + \dfrac{C}{p+1} + \dfrac{Dp+E}{p^2+1} \qquad \ldots(12)$

$$p = 2 \Rightarrow A = \frac{6}{1\cdot 3\cdot 5} = \frac{2}{5}$$

$$p = 1 \Rightarrow B = \frac{3}{-1\cdot 2\cdot 2} = \frac{-3}{4}$$

$$p = -1 \Rightarrow C = \frac{3}{-3(-2)\cdot 2} = \frac{1}{4}$$

From eq^n (12), we get

$$p^2 + 2 = A(p^2-1)(p^2+1) + B(p-2)(p^2+1)(p+1)$$
$$+ C(p-2)(p-1)(p^2+1) + (Dp+E)(p^2-1)(p-2)$$

Coeff. of $p^4 \Rightarrow \quad 0 = A + B + C + D$

$$p = 0 \Rightarrow 2 = A - 2B + 2C + 2E$$

On solving, we get

$$D = \frac{1}{10} \text{ and } E = \frac{1}{5}$$

Substituting these values in eq^n (12), we get

$$\frac{p^2+2}{(p-2)(p^4-1)} = \frac{2}{5}\frac{1}{p-2} - \frac{3}{4}\frac{1}{p-1} + \frac{1}{4}\cdot\frac{1}{p+1} + \frac{p+2}{10(p^2+1)}$$

$$\therefore \quad L^{-1}\left\{\frac{p^2+2}{(p-2)(p^4-1)}\right\} = \frac{2}{5}e^{-t} - \frac{3}{4}e^t + \frac{1}{4}e^{-t} + \frac{1}{10}(\cos t + 2\sin t) \qquad \ldots(13)$$

Putting values from (8), (11) and (13) in (7), we get

$$x = \frac{1}{2}(\cosh t + \cos t) + \frac{5}{2}(\cosh t - \cos t) + \frac{2}{5}e^{2t}$$

$$- \frac{3}{4}e^t + \frac{1}{4}e^{-t} + \frac{1}{10}(\cos t + 2\sin t)$$

$$= \frac{3}{2}(e^t + e^{-t}) - 2\cos t + \frac{2}{5}e^{2t} - \frac{3}{4}e^t + \frac{1}{4}e^{-t} + \frac{1}{10}(\cos t + 2\sin t)$$

$$x = \frac{3}{4}e^t + \frac{7}{4}e^{-t} + \frac{2}{5}e^{2t} + \frac{1}{10}(2\sin t - 19\cos t)$$

Now, putting values from (8) and (11) in (6), we get

$$y = -\frac{3}{2}(\cosh t - \cos t) + \frac{1}{2}(\cosh t + \cos t) - \frac{1}{15}e^{2t} + \frac{1}{4}e^t$$

$$- \frac{1}{12}e^{-t} - \frac{1}{10}(\cos t + 2\sin t)$$

$$= -\cosh t + 2\cos t - \frac{1}{15}e^{2t} + \frac{1}{4}e^t - \frac{1}{12}e^{-t} - \frac{1}{10}(\cos t + 2\sin t)$$

$$\Rightarrow \qquad y = \frac{19}{10}\cos t - \frac{1}{5}\sin t - \frac{1}{15}e^{2t} - \frac{1}{4}e^t - \frac{7}{12}e^{-t}$$

EXAMPLE 9. *Solve* $\dfrac{d^2x}{dt^2} + \dfrac{dy}{dt} + 3x = 15e^{-t}$ $\qquad\qquad\qquad\ldots(1)$

$$\frac{d^2y}{dt^2} - 4\frac{dx}{dt} + 3y = 15\sin 2t \qquad\qquad\qquad\ldots(2)$$

if $x(0) = 35,\ x'(0) = -48,\ y(0) = 27,\ y'(0) = -55$

SOLUTION. Taking Laplace transform of eq^n (1) & (2), we get

$$L\{x''\} + L\{y'\} + 3L\{x\} = 15L\{e^{-t}\}$$

and $\qquad L\{y''\} + L\{x'\} + 3L\{y\} = 15L\{\sin 2t\}$

$$\Rightarrow \quad p^2\bar{x} - px(0) - x'(0) + p\bar{y} - y(0) + 3\bar{x} = \frac{15}{(p+1)}$$

and $p^2\bar{y} - py(0) - y'(0) + 4[p\bar{x} - x(0)] + 3\bar{y} = \dfrac{15\times 2}{(p^2+4)}$

Using given conditions, we have

$$(p^2+3)\bar{x} + p\bar{y} = 35p - 21 + \frac{15}{p+1} \qquad\qquad\ldots(3)$$

and $\ (p^2+3)\bar{y} - 4p\bar{x} = 27p + 195 + \dfrac{30}{p^2+4}$

Now, multiplying eq^n (3) by $(p^2 + 3)$ and eq^n (4) by p and then subtracting, we get

$$\bar{x} = \frac{35p^3 - 48p + 300p + 63}{(p^2+1)(p^2+9)} + \frac{15(p^2+3)}{(p+1)(p^2+1)(p^2+9)} - \frac{30p}{(p^2+1)(p^2+4)(p^2+9)}$$

Now, using partial fraction, we have

$$\bar{x} = \frac{30}{p^2+1} - \frac{45}{p^2+9} + \frac{3}{p+1} + \frac{2p}{p^2+4} \qquad ...(5)$$

Now, multiplying eq^n (3) by $4p+2$ and eq^n (4) by (p^2+3) and then adding, we get

$$\bar{y} = \frac{7p^3 - 55p^2 - 3p - 585}{(p^2+1)(p^2+9)} + \frac{60p}{(p+1)(p^2+1)(p^2+9)} + \frac{30(p^2+2)}{(p^2+1)(p^2+4)(p^2+9)}$$

Using partial fraction, we have

$$\bar{y} = \frac{30p}{p^2+9} - \frac{60}{p^2+1} - \frac{3}{p+1} + \frac{2}{p^2+4} \qquad ...(6)$$

Taking inverse Laplace transform of eq^n (5) and (6), we get

$$x = 30\cos t - 15\sin 3t + 3e^{-t} + 2\cos 2t$$

and $\quad y = 30\cos 3t - 60\sin t - 3e^{-t} + \sin 2t$

EXAMPLE 10. *Solve the following simultaneous equations*

$$3\frac{dx}{dt} + \frac{dy}{dt} + 2x = 1$$

$$\frac{dx}{dt} + 4\frac{dy}{dt} + 3y = 0$$

when $x(0) = 3$, $y(0) = 0$. $\qquad$ (MDU-2004)

SOLUTION. Here, we have

$$3\frac{dx}{dt} + \frac{dy}{dt} + 2x = 1$$

and $\quad \dfrac{dx}{dt} + 4\dfrac{dy}{dt} + 3y = 0$

Now, taking Laplace transform of both the equations, we get

$$3[p\bar{x} - x(0)] + [p\bar{y} - y(0)] + 2\bar{x} = \frac{1}{p} \quad \text{where } \bar{x} = L(x) \text{ and } \bar{y} = L(y)$$

and $\quad p\bar{x} - x(0) + 4[p\bar{y} - y(0)] + 3\bar{y} = 0$

Using the given conditions $x(0) = 3$ and $y(0) = 0$, then

$$3p\bar{x} - 9 + p\bar{y} - 0 + 2\bar{x} = \frac{1}{p}$$

and $\quad p\bar{x} - 3 + 4p\bar{y} + 3\bar{y} = 0$

Using the given conditions $x(0) = 3$ and $y(0)=0$, then

$$3p\bar{x} - 9 + p\bar{y} - 0 + 2\bar{x} = \frac{1}{p}$$

and $\quad p\bar{x} - 3 + 4p\bar{y} + 3\bar{y} = 0$

$$(3p+2)\bar{x} + p\bar{y} = 9 + \frac{1}{p} \qquad ...(1)$$

and $\quad p\bar{x} + (4p+3)\bar{y} = 3 \qquad ...(2)$

Now, eq^n (1) $\times$ $(4p+3)$ and (2) $\times$ p, then we get

$$(3p+2)(4p+3)\bar{x} + p(4p+3)\bar{y} = 9(4p+3) + \frac{4p+3}{p}$$

and $\qquad p^2\bar{x} + p(4p+3)\bar{y} = 3p$

Subtracting both the equations, we get

$$[(3p+2)(4p+3) - p^2]\bar{x} = 9.(4p+3) + \frac{4p+3}{p} - 3p$$

$$(11p^2 + 17p + 6)\bar{x} = 9.(4p+3) + \frac{4p+3}{p} - 3p$$

$$\bar{x} = \frac{9.(4p+3)}{(p+1)(11p+6)} + \frac{4p+3}{p(p+1)(11p+6)} - \frac{3p}{(p+1)(11p+6)}$$

$$= \left(\frac{A}{p+1} + \frac{B}{11p+6}\right) + \left(\frac{C}{p} + \frac{D}{(p+1)} + \frac{E}{11p+6}\right) - \left(\frac{F}{p+1} + \frac{G}{11p+6}\right)$$

$$= \left(\frac{9}{5}\cdot\frac{1}{p+1} + \frac{81}{5}\cdot\frac{1}{11p+6}\right) + \left(\frac{1}{2}\cdot\frac{1}{p} - \frac{1}{5}\cdot\frac{1}{p+1} - \frac{33}{10}\cdot\frac{1}{11p+6}\right)$$

$$\qquad\qquad - \left(\frac{3}{5}\cdot\frac{1}{p+1} - \frac{18}{5}\cdot\frac{1}{11p+6}\right)$$

$\Rightarrow \qquad \bar{x} = \frac{1}{2p} + \frac{1}{p+1} + \frac{33}{2}\cdot\frac{1}{11p+6}$

Now, taking inverse Laplace transform, we get

$$x = \frac{1}{2}L^{-1}\left(\frac{1}{p}\right) + L^{-1}\left(\frac{1}{p+1}\right) + \frac{3}{2}L^{-1}\left(\frac{1}{p+\dfrac{6}{11}}\right) = \frac{1}{2} + e^{-t} + \frac{3}{2}e^{-\frac{6}{11}t}$$

Again by applying eq^n (1) $\times$ p and eq^n (2) $\times$ (3p+2), we get

$$p(3p+2)\bar{x} + p^2\bar{y} = 9p+1$$

$$p(3p+2)\bar{x} + (4p+3)(3p+2)\bar{y} = 3(3p+2)$$

Subtracting both the equations, we have

$$[(4p+3)(3p+2) - p^2]\bar{y} = 9p+6-9p-1$$

$\Rightarrow \qquad (p+1)(11p+6)\bar{y} = 5$

$\Rightarrow \qquad \bar{y} = \frac{5}{(p+1)(11p+6)} = \frac{-1}{p+1} + \frac{11}{11p+6}$

Now, taking inverse Laplace transforms, we get

$$\bar{y} = L^{-1}\left(-\frac{1}{p+1}\right) + L^{-1}\left(\frac{1}{p+\dfrac{6}{11}}\right)$$

$$= -e^{-t} + e^{-6/11t}$$

Thus, the required solution is given by

$$x = \frac{1}{2} + e^{-t} + \frac{3}{2}e^{-\frac{6}{11}t}$$

$$y = -e^{-t} + e^{-\frac{6}{11}t}$$

3.4 APPLICATION OF LAPLACE TRANSFORM TO ENGINEERING PROBLEMS

EXAMPLE 1. *A mass m moves along the x-axis under the influence of a force which is proportional to its instantaneous speed and in a direction opposite to the direction of motion. Assuming that at t = 0, the particle is located at x = a and moving to the right with speed V_0. Find the position where the mass comes to rest.*

(Meerut–1983, 88, 90, 96, 2007, 15)

SOLUTION.   The motion of the particle is described as below :

$$P_____\bullet_____$$

Fig. 1

By the Newton's second law of motion, we get the equation of motion of the particle, given by

$$m\frac{d^2x}{dt^2} = -\mu\frac{dx}{dt} \qquad \qquad \text{... (1)}$$

with initial conditions $x(0) = a$ and $x'(0) = V_0$.

Taking the Laplace transform of both sides of (1), we get

$$mL\left\{\frac{d^2x}{dt^2}\right\} = -\mu L\left\{\frac{dx}{dt}\right\}$$

$$\Rightarrow \quad m[p^2 L\{x\} - px(0) - x'(0)] = -\mu[pL\{x\} - x(0)]$$

$$\Rightarrow \quad\quad\quad (mp^2 + \mu p)L\{x\} = m(ap + V_0) + a\mu$$

$$\Rightarrow \quad L\{x\} = \frac{m(ap + V_0) + a\mu}{p(mp + \mu)} = \frac{mV_0 + \mu a}{\mu p} - \frac{mV_0}{\mu\left(p + \dfrac{\mu}{m}\right)}$$

Therefore, $x = \left(\dfrac{mV_0}{\mu} + a\right)L^{-1}\left\{\dfrac{1}{p}\right\} - \dfrac{mV_0}{\mu}L^{-1}\left\{\dfrac{1}{p + \dfrac{\mu}{m}}\right\}$

$$= \left(\frac{mV_0}{\mu} + a\right) - \frac{mV_0}{\mu}.e^{-\mu t/m}. \qquad \qquad \text{...(2)}$$

If $\quad\quad \dfrac{dx}{dt} = V_0 e^{\mu t/m} = 0.$

Then, from (2), we have $x = \dfrac{mV_0}{\mu} + a$.

Hence, the mass m comes to rest at a distance $\dfrac{mV_0}{\mu} + a$ from the centre.

EXAMPLE 2. *A particle moves along a line so that its displacement X from a fixed point at any time x is given by*

$$X''(t) + 4X'(t) + 5X(t) = 80\sin 5t \cdot$$

Find its displacement at any time $x > 0$, if at t = 0, the particle is at rest at X = 0.

SOLUTION.   Here, the displacement of the particle is given by the differential equation

$$X''(t) + 4X'(t) + 5X(t) = 80\sin 5t \qquad \qquad \text{... (1)}$$

where $X(0) = 0$ and $X'(0) = 0$.

Taking Laplace transform of both sides of (1), we get

$$L\{X''(t)\} + 4L\{X'(t)\} + 5L\{X(t)\} = 80L\{\sin 5t\}$$

$$\Rightarrow \quad p^2 L\{X(t)\} - pX(0) - X'(0) + 4[pL\{X(t)\} - X(0)] + 5L\{X(t)\} = 80 \times \frac{5}{p^2 + 25}$$

$$\Rightarrow \quad (p^2 + 4p + 5) L\{X(t)\} = \frac{400}{p^2 + 25}$$

$$\Rightarrow \quad L\{X(t)\} = \frac{400}{(p^2 + 4p + 5)(p^2 + 25)}$$

$$= \frac{-2(p + 5)}{p^2 + 25} + \frac{2(p + 9)}{p^2 + 4p + 5}.$$

Therefore, $\quad X(t) = -2L^{-1}\left\{\dfrac{p}{p^2 + 25} + \dfrac{5}{p^2 + 25}\right\} + 2L^{-1}\left\{\dfrac{(p + 2) + 7}{(p + 2)^2 + 1}\right\}$

$$= -2(\cos 5t + \sin 5t) + 2e^{-2t} L^{-1}\left\{\frac{p + 7}{p^2 + 1}\right\}$$

$$\Rightarrow \quad X(t) = -2(\cos 5t + \sin 5t) + 2e^{-2t}(\cos t + 7 \sin t).$$

EXAMPLE 3. *A resistance R in series with inductance L is connected with emf E(t) . The current i is given by*

$$L\frac{di}{dt} + Ri = E(t)$$

If the switch is connected at $t = 0$ and disconnected at $t = a$, find the current i in terms of t. [UPTU–2001]

SOLUTION. Conditions under which current i flows are $i = 0$ at $t = 0$.

$$E(t) = \begin{cases} E, & 0 < t < a \\ 0, & t > a \end{cases}$$

Given equation is

$$L\frac{di}{dt} + Ri = E(t) \qquad \qquad ...(1)$$

Taking Laplace transform of (1), we get

$$L[pL\{i\} - i(0)] + Ri = \int_0^\infty e^{-pt} E(t)\, dt$$

$$LpL\{i\} + Ri = \int_0^\infty e^{-pt} E(t)\, dt \qquad \qquad [\because i(0) = 0]$$

$$(Lp + R)L\{i\} = \int_0^\infty e^{-pt} E(t)\, dt = \int_0^a e^{-pt} E\, dt + \int_a^\infty e^{-pt} E\, dt$$

$$= E\left[\frac{e^{-pt}}{-p}\right]_0^a + 0 = \frac{E}{p}\left[1 - e^{-ap}\right] = \frac{E}{p} - \frac{E}{p}e^{-ap}$$

$$\Rightarrow \quad L\{i\} = \frac{E}{p(Lp + R)} - \frac{Ee^{-ap}}{p(Lp + R)}$$

Taking inverse Laplace transform, we obtain

$$i = L^{-1}\left\{\frac{E}{p(Lp + R)}\right\} - L^{-1}\left\{\frac{Ee^{-ap}}{p(Lp + R)}\right\} \qquad \qquad ...(2)$$

Now, we have to find the value of $L^{-1}\left\{\dfrac{E}{p(Lp + R)}\right\}$

$$L^{-1}\left\{\frac{E}{p(Lp+R)}\right\} = \frac{E}{L}L^{-1}\left\{\frac{1}{p\left(p+\dfrac{R}{L}\right)}\right\}$$

$$= \frac{E}{L}\cdot\frac{L}{R}L^{-1}\left\{\frac{1}{p} - \frac{1}{p+\dfrac{R}{L}}\right\} = \frac{E}{R}\left[1 - e^{-R/Lt}\right]$$

and $L^{-1}\left\{\dfrac{Ee^{-ap}}{p(Lp+R)}\right\} = \dfrac{E}{R}\left[1 - e^{-\frac{R}{L}(t-a)}\right]u(t-a)$ [By the second shifting theorem]

On substituting the values of the inverse transforms in (2), we get

$$i = \frac{E}{R}\left[1 - e^{-\frac{R}{L}t}\right] - \frac{E}{R}\left[1 - e^{-\frac{R}{L}(t-a)}\right]u(t-a)$$

Hence, $i = \dfrac{E}{R}\left[1 - e^{-\frac{R}{L}t}\right]$, for $0 < t < a$, $[u(t-a) = 0]$

$$i = \frac{E}{R}\left[1 - e^{-\frac{R}{L}t}\right] - \frac{E}{R}\left[1 - e^{-\frac{R}{L}(t-a)}\right], \text{ for } t > a, \ [\because u(t-a) = 1].$$

$$= \frac{E}{R}\left[e^{-\frac{R}{L}(t-a)} - e^{-\frac{R}{L}t}\right] = \frac{E}{R}e^{-\frac{R}{L}t}\left[e^{\frac{Ra}{L}} - 1\right]$$

Additional Solved Examples

EXAMPLE 1.   Solve $y'' + ty' - y = 0$ if $y(0) = 0, y'(0) = 1$. (Agra–1998; Meerut–1988, 98; SVTU–2004)

SOLUTION.   Taking the Laplace transform of the given equation, we get

$$L(y'') + tL(y') - L(y) = 0$$

$$\Rightarrow \quad p^2 L(y) - py(0) - y'(0) - \frac{d}{dp}[L(y')] - L(y) = 0$$

$$\Rightarrow \quad (p^2 - 1)L(y) - 1 - \left[L(y) + p\frac{d}{dp}\{L(y)\}\right] = 0$$

[Since given $y(0) = 0, y'(0) = 1$]

$$\Rightarrow \quad \frac{d}{dp}[L(y)] - \left(p - \frac{2}{p}\right)L(y) = -\frac{1}{p} \quad\quad ...(1)$$

It is a linear differential equation of the first order in $L(y)$, therefore,

$\therefore$ I.F. $= e^{-\int(p - 2/p)dp} = e^{-p^2/2 + 2\log p} = p^2 e^{-p^2/2}$

Thus the solution of equation (1) is given by

$$L(y)p^2 e^{-p^2/2} = C_1 + \int\left(-\frac{1}{p}\right)\cdot p^2 e^{-p^2/2}dp = C_1 - \int pe^{-p^2/2}dp = C_1 + e^{-p^2/2}$$

Hence, $L(y) = \dfrac{C_1}{p^2}e^{p^2/2} + \dfrac{1}{p^2}$

$$\Rightarrow \quad L(y) = \frac{C_1}{p^2}\left[1 + \frac{1}{2}p^2 + \frac{1}{2!}\cdot\frac{1}{4}p^4 + ...\right] + \frac{1}{p^2}$$

$\Rightarrow \qquad L(y) = (C_1 + 1)\dfrac{1}{p^2} + \dfrac{1}{2}C_1 + \dfrac{1}{8}C_1 p^2 + \ldots$

Taking inverse Laplace transformation of both sides, we have

$$y = (C_1 + 1)L^{-1}\left(\dfrac{1}{p^2}\right) + \dfrac{1}{2}C_1 L^{-1}(1) + \dfrac{1}{8}C_1 L^{-1}(p^2) + \ldots$$

$\Rightarrow \qquad y = (C_1 + 1)\cdot t + 0 \qquad$ [Since $L^{-1}(p^n) = 0, n = 0, 1, 2, \ldots$]

Since, given $\qquad y'(0) = 1$

$\Rightarrow \qquad y'(0) = C_1 + 1 = 1 \quad \Rightarrow \; C_1 = 0.$

Hence, $y = t$ is the required solution.

EXAMPLE 2. *Solve* $y''(t) + aty'(t) - 2ay(t) = 1$ *if* $y(0) = y'(0) = 0, a > 0$.

SOLUTION. Taking Laplace transformation of both sides of given equation, we have

$$L(y'') + aL(ty') - 2aL(y) = L(1)$$

$$\Rightarrow p^2 L(y) - py(0) - y'(0) - a\dfrac{d}{dp}[L(y')] - 2aL(y) = \dfrac{1}{p}$$

$$\Rightarrow \qquad p^2 L(y) - a\dfrac{d}{dp}[pL(y) - y(0)] - 2aL(y) = \dfrac{1}{p}$$

$$\Rightarrow \qquad \dfrac{d}{dp}[L(y)] + \left[-\dfrac{p}{a} + \dfrac{3}{p}\right]L(y) = -\dfrac{1}{ap^2} \qquad \ldots(1)$$

It is a linear differential equation in L(y), therefore

$$\text{I.F.} = e^{\int(-p/a + 3/p)dp} = e^{-p^2/2a + 3\log p} = p^3 e^{-p^2/2a}$$

Thus the solution of equation (1) will be given by

$$L(y)p^3 e^{-p^2/2a} = C_1 - \dfrac{1}{a}\int\dfrac{1}{p^2}\cdot p^3 e^{-p^2/2a}\,dp$$

$$\Rightarrow \; L(y)p^3 e^{-p^2/2a} = C_1 - \dfrac{1}{a}\int pe^{-p^2/2a}\,dp$$

$$\Rightarrow \; L(y)p^3 e^{-p^2/2a} = C_1 + e^{-p^2/2a}$$

$$\Rightarrow \qquad L(y) = \dfrac{C_1}{p^3}e^{p^2/2a} + \dfrac{1}{p^3}$$

$$\Rightarrow \qquad L(y) = \dfrac{C_1}{p^3}\left[1 + \dfrac{1}{2a}p^2 + \dfrac{1}{2!}\left(\dfrac{1}{2a}p^2\right)^2 + \ldots\right] + \dfrac{1}{p^3}$$

$$\Rightarrow \qquad y = C_1\left[L^{-1}\left(\dfrac{1}{p^3}\right) + \dfrac{1}{2a}L^{-1}\left(\dfrac{1}{p}\right) + \dfrac{1}{8a^2}L(p) + \ldots\right] + L^{-1}\left(\dfrac{1}{p^3}\right)$$

$$\Rightarrow \qquad y = (C_1 + 1)\dfrac{t^2}{2!} + C_1\cdot\dfrac{1}{2a} \qquad [\because L^{-1}(p^n) = 0, n = 0, 1, 2, \ldots]$$

Now, using $\quad y(0) = 0$, we get $C_1 = 0$

Hence, $\qquad y = \dfrac{1}{2}t^2$

EXAMPLE 3. *Solve* $(D-2)x + 3y = 0$; $2x + (D-1)y = 0$, *if* $x(0) = 8$ *and* $y(0) = 3$.

(Meerut–1983, 2001; Rajasthan–1983)

SOLUTION. We can write the given equation as

$$x' - 2x + 3y = 0$$
$$2x + y' - y = 0$$

Taking Laplace transformation of both sides, we have

$$L(x') - 2L(x) + 3L(y) = 0 \qquad \Rightarrow \quad pL(x) - x(0) - 2L(x) + 3L(y) = 0$$
$$2L(x) + L(y') - L(y) = 0 \qquad \Rightarrow \quad 2L(x) + pL(y) - y(0) - L(y) = 0$$

So the equation becomes as

$$(p-2)L(x) - x(0) + 3L(y) = 0$$

$$\Rightarrow \qquad (p-2)L(x) + 3L(y) = 8 \qquad \qquad \text{...(1)}$$

and $$\qquad 2L(x) + (p-1)L(y) = 3 \qquad \qquad \text{...(2)}$$

On solving equation (1) and (2), we have

$$L(x) = \frac{8p-17}{p^2 - 3p - 4}$$

$$\Rightarrow \qquad L(x) = \frac{8p-17}{(p-4)(p+1)} = \frac{5}{p+1} + \frac{3}{p-4}$$

Taking inverse Laplace transformation of both sides, we get

$$x = L^{-1}\left(\frac{5}{p+1} + \frac{3}{p-4}\right)$$

$$\Rightarrow \qquad x = 5e^{-t} + 3e^{4t}$$

and $$\qquad L(y) = \frac{3p-22}{p^2 - 3p - 4}$$

$$\Rightarrow \qquad L(y) = \frac{3p-22}{(p-4)(p+1)} = \frac{5}{(p+1)} - \frac{2}{(p-4)}$$

Taking inverse Laplace transformation of both sides, we get

$$y = L^{-1}\left(\frac{5}{p+1} - \frac{2}{p-4}\right)$$

$$\Rightarrow \qquad y = 5e^{-t} - 2e^{4t}$$

EXAMPLE 4. *Solve* $(D^2 - 3)x - 4y = 0$

$$x + (D^2 + 1)y = 0, \quad t > 0$$

if $x = y = Dy = 0$, $D(x) = 2$, *when* $t = 0$.

(Meerut–1985)

SOLUTION. We can write the given equation as

$$x'' - 3x - 4y = 0$$

and $$\qquad x + y'' + y = 0$$

Taking Laplace transformation of both sides, we have

$$L(x'') - 3L(x) - 4L(y) = 0$$

$$\Rightarrow \quad p^2 L(x) - py(0) - y'(0) - 3L(x) - 4L(y) = 0$$

$$\Rightarrow \quad (p^2 - 3)L(x) - py(0) - y'(0) - 4L(y) = 0$$

$\Rightarrow$ $\qquad (p^2 - 3)L(x) - 4L(y) = 2 \qquad$ [Given $x = 0, x' = 2$ at $t = 0$]
$$\dots(1)$$

and $\qquad L(x) + L(y'') + L(y) = 0$

$\Rightarrow \qquad L(x) + p^2 L(y) - py(0) - y'(0) + L(y) = 0$

$\Rightarrow \qquad L(x) + (p^2 + 1)L(y) - py(0) - y'(0) = 0$

$\Rightarrow \qquad L(x) + (p^2 + 1)L(y) = 0 \qquad$ [Given $y = y' = 0$]
$$\dots(2)$$

On solving, equation (1) and (2), we have
$$L(x) = 2(p^2 + 1)/(p^2 - 1)^2$$

$\Rightarrow \qquad L(x) = \dfrac{1}{(p-1)^2} + \dfrac{1}{(p+1)^2}$

Taking inverse Laplace transformation of both the sides, we have

$$x = L^{-1}\left(\frac{1}{(p-1)^2}\right) + L^{-1}\left(\frac{1}{(p+1)^2}\right)$$

$\Rightarrow \qquad x = e^t L^{-1}\left(\dfrac{1}{p^2}\right) + e^{-t} L^{-1}\left(\dfrac{1}{p^2}\right)$

$\Rightarrow \qquad x = (e^t + e^{-t})t$

and $\qquad L(y) = \dfrac{-2}{(p+1)^2(p-1)^2}$

$\Rightarrow \qquad L(y) = \dfrac{1}{2}\left[\dfrac{1}{(p+1)} + \dfrac{1}{(p-1)} + \dfrac{1}{(p+1)^2} + \dfrac{1}{(p-1)^2}\right]$

Taking inverse Laplace transformation of both the sides, we have

$$y = \frac{1}{2}\left[L^{-1}\left(\frac{1}{(p+1)}\right) + L^{-1}\left(\frac{1}{(p-1)}\right) + L^{-1}\left(\frac{1}{(p+1)^2}\right) + L^{-1}\left(\frac{1}{(p-1)^2}\right)\right]$$

$\Rightarrow \qquad y = \dfrac{1}{2}[e^{-t} + e^t - te^{-t} + te^t]$

Hence, $\qquad y = \dfrac{1}{2}(1-t)e^{-t} - \dfrac{1}{2}(1+t)e^t$

EXAMPLE 5. Solve $(D-2)x - (D+1)y = 6e^{3t}$, $(2D-3)x + (D-3)y = 6e^{3t}$, if $x = 3, y = 0$ when $t = 0$.

SOLUTION. We can write the given equation in the form of
$$x' - 2x - y' - y = 6e^{3t}$$
and $\qquad 2x' - 3x + y' - 3y = 6e^{3t}$

Taking Laplace transformation of both sides, we have
$$L(x') - 2L(x) - L(y') - L(y) = 6L(e^{3t})$$

$\Rightarrow \quad pL(x) - x(0) - 2L(x) - pL(y) + y(0) - L(y) = \dfrac{6}{(p-3)}$

$\qquad\qquad$ [Given $x = 3$ and $y = 0$ at $t = 0$]

$$\Rightarrow \qquad (p-2)L(x) - 3 - (p+1)L(y) = \frac{6}{(p-3)}$$

$$\Rightarrow \qquad (p-2)L(x) - (p+1)L(y) = 3 + \frac{6}{(p-3)} = \frac{3p-3}{(p-3)} \qquad \ldots(1)$$

and $\qquad 2L(x') - 3L(x) + L(y') - 3L(y) = 6L(e^{3t})$

$$\Rightarrow \quad 2pL(x) - 2x(0) - 3L(x) + pL(y) - y(0) - 3L(y) = \frac{6}{(p-3)}$$

$$\Rightarrow \qquad (2p-3)L(x) - 6 - (p-3)L(y) = \frac{6}{(p-3)} \qquad \text{[Given } x(0) = 3, y(0) = 0]$$

$$\Rightarrow \qquad (2p-3)L(x) - (p-3)L(y) = 6 + \frac{6}{p-3} = \frac{6p-12}{p-3} \qquad \ldots(2)$$

On solving, equation (1) and (2), we have

$$L(x) = \frac{1}{(p-1)} + \frac{2}{(p-1)^2} + \frac{2}{(p-3)}$$

Taking inverse Laplace transformation of both sides, we have

$$x = L^{-1}\left(\frac{1}{(p-1)}\right) + L^{-1}\left(\frac{2}{(p-1)^2}\right) + L^{-1}\left(\frac{2}{(p-3)}\right)$$

$$\Rightarrow \qquad x = e^t + 2te^t + 2e^{3t}$$

and $\qquad L(y) = \dfrac{-3p+5}{(p-3)(p-1)^2}$

$$L(y) = \frac{1}{(p-1)} - \frac{1}{(p-1)^2} - \frac{1}{(p-3)}$$

Taking inverse Laplace transformation of both sides, we have

$$y = L^{-1}\left(\frac{1}{(p-1)}\right) - L^{-1}\left(\frac{1}{(p-1)^2}\right) - L^{-1}\left(\frac{1}{p-3}\right)$$

Hence, $y = e^t - te^t - e^{3t}$ is the required solution.

EXAMPLE 6. *Solve* $(D-2)x - (D-2)y = \sin t$, $(D^2 + 1)x + 2Dy = 0$ *if* $x = 0 = x'(0) = y(0)$.

SOLUTION. We can write the given equation in the form of

$$x' - 2x - y' + 2y = \sin t$$
$$x'' + x + 2y' = 0$$

Taking Laplace transformation of both sides, we have

$$L(x') - 2L(x) - L(y') + 2L(y) = L(\sin t)$$

$$\Rightarrow \quad pL(x) - x(0) - 2L(x) - pL(y) + y(0) + 2L(y) = \frac{1}{(p^2+1)}$$

$$\Rightarrow \qquad (p-2)L(x) - (p-2)L(y) = \frac{1}{(p^2+1)} \quad \text{[Since } x = 0 = x'(0) = y(0)]$$

$$\ldots(1)$$

and $\qquad L(x'') + L(x) + 2L(y') = 0$

$$\Rightarrow \quad p^2 L(x) - py(0) - y'(0) + L(x) + 2pL(y) - 2y(0) = 0$$

$$\Rightarrow \quad (p^2 + 1)L(x) + 2pL(y) = 0 \qquad [\text{Since } x(0) = 0 = x'(0) = y(0)]$$
$$\hspace{10cm} ...(2)$$

On solving, equation (1) and (2), we have

$$L(x) = \frac{2p}{(p^2 + 1)(p^3 - 3p - 2)} = \frac{2p}{(p^2 + 1)(p + 1)^2 (p - 2)}$$

$$\Rightarrow \quad L(x) = \frac{1}{9(p + 1)} + \frac{1}{3(p + 1)^2} + \frac{4}{45(p - 2)} - \frac{p + 2}{5(p^2 + 1)}$$

Taking inverse Laplace transformation of both sides, we have

$$x = \frac{1}{9} L^{-1} \left(\frac{1}{(p + 1)} \right) + \frac{1}{3} L^{-1} \left(\frac{1}{(p + 1)^2} \right) + \frac{4}{45} L^{-1} \left(\frac{1}{p - 2} \right) - \frac{1}{5} L^{-1} \left(\frac{p + 2}{p^2 + 1} \right)$$

$$\Rightarrow \quad x = \frac{1}{9} e^{-t} + \frac{1}{3} e^{-t} L^{-1} \left(\frac{1}{p^2} \right) + \frac{4}{45} e^{2t} - \frac{1}{5} \cos t - \frac{2}{5} \sin t$$

$$\Rightarrow \quad x = \frac{1}{9} e^{-t} + \frac{1}{3} t e^{-t} + \frac{4}{45} e^{2t} - \frac{1}{5} \cos t - \frac{2}{5} \sin t$$

and $\quad L(y) = \dfrac{1}{p^3 - 3p - 2} = \dfrac{1}{(p + 1)^2 (p - 2)}$

$$\Rightarrow \quad L(y) = \frac{1}{9(p + 1)} + \frac{1}{3(p + 1)^2} - \frac{1}{9(p - 2)}$$

Taking inverse Laplace transformation of both sides, we have

$$y = \frac{1}{9} L^{-1} \left(\frac{1}{p + 1} \right) + \frac{1}{3} L^{-1} \left(\frac{1}{(p + 1)^2} \right) - \frac{1}{9} L^{-1} \left(\frac{1}{p - 2} \right)$$

$$\Rightarrow \quad y = \frac{1}{9} e^{-t} - \frac{1}{9} e^{2t} + \frac{1}{3} e^{-t} L^{-1} \left(\frac{1}{p^2} \right)$$

Hence, $\quad y = \dfrac{1}{9} e^{-t} - \dfrac{1}{9} e^{2t} + \dfrac{1}{3} t e^{-t}$

EXAMPLE 7. *Solve* $\dfrac{\partial y}{\partial x} = 2 \dfrac{\partial y}{\partial t} + y, y(x, 0) = 6e^{-3x}$ *which is bounded for* $x > 0, t > 0$.

SOLUTION. Taking Laplace transformation of both side of the given equation, we have

$$L \left(\frac{\partial y}{\partial x} \right) = 2L \left(\frac{\partial y}{\partial t} \right) + L(y)$$

$$\Rightarrow \quad \frac{d}{dx} L(y) = 2[pL(y) - y(x, 0)] + L(y)$$

$$\Rightarrow \quad \frac{d}{dx} L(y) - (2p + 1)L(y) = -12e^{-3x} \hspace{4cm} ...(1)$$

which is a linear equation, therefore,

$$\text{I.F.} = e^{-\int (2p+1)dx} = e^{-(2p+1)x}$$

Thus the solution is given by

$$e^{-(2p+1)x} L(y) = C - 12 \int e^{-3x} \cdot e^{-(2p+1)x} dx$$

$$= C - 12\int e^{-(2p+4)x}dx = C + \frac{6}{(p+2)}e^{-(2p+4)x}$$

$$\Rightarrow \qquad L(y) = Ce^{(2p+1)x} = \frac{6}{(p+2)}e^{-3x}$$

Since, $y(x, t)$ is bounded as $x \to \infty$ therefore, $L[Y(x, p)]$ also bounded as $x \to \infty$.

$$\Rightarrow \qquad C = 0$$

So, $\qquad L(y) = \frac{6}{(p+2)}e^{-3x}$

Taking inverse Laplace transformation of both sides, we have

$$y = L^{-1}\left(\frac{6}{(p+2)}e^{-3x}\right)$$

Hence, $\qquad y(x, t) = 6e^{-2t-3x}$

EXAMPLE 8. Solve $\dfrac{\partial y}{\partial t} = 3\dfrac{\partial^2 y}{\partial x^2}$ where $y\left(\dfrac{\pi}{2}, t\right) = 0, \left(\dfrac{\partial y}{\partial x}\right)_{x=0} = 0$ and $y(x,0) = 30\cos 5x$.

SOLUTION. Taking Laplace transformation of both sides of given equation, we have

$$L\left(\frac{\partial y}{\partial t}\right) = 3L\left(\frac{\partial^2 y}{\partial x^2}\right)$$

$$\Rightarrow \qquad PLy(x,p) - y(x,0) = 3\frac{d^2}{dx^2}[L(y)]$$

$$\Rightarrow \qquad \frac{d^2}{dx^2}L(y) - \frac{P}{3}L(y) = -10\cos 5x$$

whose solution is $\qquad L(y) = C_1 e^{x\sqrt{p/3}} + C_2 e^{-x\sqrt{p/3}} + \dfrac{10}{-5^2 - \left(\dfrac{p}{3}\right)}\cos 5x$

$$\Rightarrow \qquad L(y) = C_1 e^{x\sqrt{p/3}} + C_2 e^{-x\sqrt{p/3}} + \frac{30}{75+p}\cos 5x \qquad \dots(1)$$

Since given conditions are $y\left(\dfrac{\pi}{2}, t\right) = 0, \left(\dfrac{\partial y}{\partial x}\right)_{x=0} = 0$

Taking Laplace transformation of both sides, we have

$$Ly\left(\frac{\pi}{2}, t\right) = 0, \quad L\left(\frac{\partial y}{\partial x}\right)_{x=0} = 0$$

$$\Rightarrow Ly\left(\frac{\pi}{2}, t\right) = 0, \qquad \frac{d}{dx}L(y) = 0$$

$$\Rightarrow Ly\left(\frac{\pi}{2}, p\right) = 0, \qquad \frac{d}{dx}L(y) = 0$$

After putting these value in equation (1), we have

$$Ly\left(\frac{\pi}{2}, p\right) = 0 \qquad \Rightarrow \qquad C_1 e^{\pi/2\sqrt{p/3}} + C_2 e^{-\pi/2\sqrt{p/3}} = 0$$

and when $x = 0$,

$$\frac{dL(y)}{dx} = 0 \qquad \Rightarrow \qquad C_1\sqrt{\frac{p}{3}} - C_2\sqrt{\frac{p}{3}} = 0$$

On solving, we get

$$C_1 = C_2 = 0$$

Thus, $L(y) = \dfrac{30}{75+p}\cos 5x$

Taking inverse Laplace transformation of both sides, we have

$$\Rightarrow \qquad y = L^{-1}\left(\frac{30}{75+p}\cos 5x\right)$$

Hence, $\quad y = 30e^{-75t}\cos 5x$

EXAMPLE 9. Solve $\dfrac{\partial y}{\partial t} = \dfrac{\partial^2 y}{\partial x^2}, y(x,0) = 3\sin 2\pi x, y(0,t) = 0 = y(1,t)\ 0 < x < 1, t > 0$.

(Meerut–1981, 83)

SOLUTION. After taking Laplace transformation of both sides of given equation, we get

$$L\left(\frac{\partial y}{\partial t}\right) = L\left(\frac{\partial^2 y}{\partial x^2}\right)$$

$$\Rightarrow \qquad pLy(x,p) - y(x,0) = \frac{d^2}{dx^2}L(y)$$

$$\Rightarrow \qquad \frac{d^2}{dx^2}L(y) - pL(y) = -3\sin 2\pi x$$

$$L(y) = Ly(x,p) = C_1 e^{x\sqrt{p}} + C_2 e^{-x\sqrt{p}} - \frac{3}{-(2\pi)^2 - p}\sin 2\pi x$$

$$Ly(x,p) = C_1 e^{x\sqrt{p}} + C_2 e^{-x\sqrt{p}} + \frac{3}{4\pi^2 + p}\sin 2\pi x \qquad \ldots(1)$$

Since given conditions are $y(x, 0) = 3\sin 2\pi x, y(0, t) = y(1, t) = 0$

Taking Laplace transformation of both sides, we get

$$Ly(x, 0) = 3L(\sin 2\pi x), \qquad Ly(0, t) = Ly(1, t) = 0$$
$$Ly(p, 0) = 3L(\sin 2\pi p), \qquad Ly(0, p) = C_1 + C_2 = 0$$
$$Ly(1, p) = C_1 e^{\sqrt{p}} + C_2 e^{-\sqrt{p}} = 0$$

On solving, we have $C_1 = C_2 = 0$

$$Ly(x,p) = \frac{3}{4\pi^2 + p}\sin 2\pi x$$

$$\Rightarrow \qquad y(x,t) = 3L^{-1}\left(\frac{1}{4\pi^2 + p}\right)\sin 2\pi x$$

Hence, $\quad y(x,t) = 3e^{-4\pi^2 t}\sin 2\pi x$

EXAMPLE 10. Solve $\dfrac{\partial y}{\partial t} = 2\dfrac{\partial^2 y}{\partial x^2}$ with $y(0, t) = 0, y(5, t) = 0$

$$y(x, 0) = 10\sin 4\pi x - 5\sin 6\pi x$$

(Meerut–1982, 83, 84, 87, 88, 92; Agra–1981; Raj.–1983)

SOLUTION. Taking Laplace transformation of both sides of the given equation, we have

$$L\left(\frac{\partial y}{\partial t}\right) = 2L\left(\frac{\partial^2 y}{\partial x^2}\right)$$

$$\Rightarrow \qquad pLy(x,p) - y(x,0) = 2\frac{d^2}{dx^2}L(y)$$

$$\Rightarrow \qquad \frac{d^2}{dx^2}L(y) - \frac{p}{2}L(y) = \frac{1}{2}(5\sin 6\pi x - 10\sin 4\pi x)$$

$$\Rightarrow \qquad L(y) = Ly(x,p)$$

$$L(y) = C_1 e^{x\sqrt{p/2}} + C_2 e^{-x\sqrt{p/2}} - \frac{5\sin 6\pi x}{72\pi^2 + p} + \frac{10\sin 4\pi x}{32\pi^2 + p} \qquad \dots(1)$$

Since given condition, we have
$$y(0, t) = y(5, t) = 0$$
Taking Laplace transformation both sides, we have
$$Ly(0, p) = Ly(5, p) = 0$$

$$\Rightarrow \qquad Ly(0, p) = Ly(5, p) = 0$$

$$\Rightarrow \qquad C_1 + C_2 = 0 = C_1 e^{5\sqrt{p/2}} + C_2 e^{-5\sqrt{p/2}}$$

On solving, we have $C_1 = C_2 = 0$

$$L\{y(x,p)\} = -\frac{5\sin 6\pi x}{72\pi^2 + p} + \frac{10\sin 4\pi x}{32\pi^2 + p}$$

Taking inverse Laplace transformation of both sides, we have

$$y(x,t) = -5e^{-72\pi^2 t}\sin 6\pi x + 10e^{-32\pi^2 t}\sin 4\pi x$$

EXAMPLE 11. Solve $\dfrac{\partial y}{\partial t} = 3\dfrac{\partial^2 y}{\partial x^2}, y_x(0,t) = 0, y\left(\dfrac{\pi}{2}, t\right) = 0$

and $y(x,0) = 20\cos 3x - 5\cos 9x$

SOLUTION. Taking Laplace transformation of both sides of the given equation, we have

$$L\left(\frac{\partial y}{\partial t}\right) = 3L\left(\frac{\partial^2 y}{\partial x^2}\right)$$

$$\Rightarrow \qquad pLy(x,p) - y(x,0) = 3\frac{d^2}{dx^2}L(y)$$

$$\Rightarrow \qquad 3\frac{d^2}{dx^2}L(y) - pL(y) = -20\cos 3x + 5\cos 9x$$

Thus solution is

$$L(y) = Ly(x,p) = C_1 e^{x\sqrt{p/3}} + C_2 e^{-x\sqrt{p/3}} + \frac{20}{27+p}\cos 3x - \frac{5}{243+p}\cos 9x \quad \dots(1)$$

The given conditions are

$$y\left(\frac{\pi}{2}, t\right) = 0 \qquad \text{and} \qquad \frac{\partial y}{\partial x} = 0, x = 0$$

$$Ly\left(\frac{\pi}{2}, p\right) = 0 \qquad \text{and} \qquad \frac{d}{dx}L(y) = 0, x = 0$$

$$Ly\left(\frac{\pi}{2},p\right)=0 \quad \Rightarrow \quad C_1 e^{\frac{\pi}{2}\sqrt{p/3}} + C_2 e^{-\frac{\pi}{2}\sqrt{p/3}} = 0 \qquad \ldots(2)$$

$$\frac{d}{dx}L(y) = C_1\sqrt{p/3}\,e^{x\sqrt{p/3}} - C_2\sqrt{p/3}\,e^{-x\sqrt{p/3}}$$

$$-\frac{60}{27+\pi}\sin 3x + \frac{45}{243+p}\sin 9x$$

$$\frac{d}{dx}L(y) = 0 = C_1\sqrt{p/3} - C_2\sqrt{p/3} = 0 \qquad \ldots(3)$$

On solving, equation (2) and (3), we have $C_1 = C_2 = 0$

$$Ly(x,p) = \frac{20}{27+p}\cos 3x - \frac{5}{243+p}\cos 9x$$

Taking inverse Laplace transformation of both sides, we have

$$y(x,t) = L^{-1}\left(\frac{20}{27+p}\right)\cos 3x - L^{-1}\left(\frac{5}{243+p}\right)\cos 9x$$

$$y(x,t) = 20e^{-27t}\cos 3x - 5e^{-243t}\cos 9x$$

EXAMPLE 12. *Solve* $y^{iv}(t) + y'''(t) = \cos t$, *with* $y(0) = y'(0) = y'''(0) = 0$, $y''(0)$ *is arbitrary.*

SOLUTION. Taking the Laplace transformation of both sides of the given euqation, we have

$$L(y^{iv}) + L(y''') = L(\cos t)$$

Therefore,

$$p^4 L(y) - p^3 y(0) - p^2 y'(0) - py''(0) - y'''(0)$$

$$+ p^3 L(y) - p^2 y(0) - py'(0) - y''(0) = \frac{p}{(p^2+1)}$$

$$\Rightarrow (p^4 + p^3)L(y) - (p+1)A = \frac{p}{(p^2+1)}$$

where $y''(0) = A$ and given $y(0) = y'(0) = y''(0) = 0$

$$\Rightarrow \qquad L(y) = \frac{A}{p^3} + \frac{p}{p^2(p+1)(p^2+1)}$$

$$\Rightarrow \qquad L(y) = \frac{A}{p^3} + \frac{1}{2(p+1)} - \frac{1}{p} + \frac{1}{p^2} + \frac{(p-1)}{2(p^2+1)}$$

$$\Rightarrow \qquad L(y) = \frac{A}{p^3} + \frac{1}{2(p+1)} - \frac{1}{p} + \frac{1}{p^2} + \frac{p}{2(p^2+1)} - \frac{1}{2(p^2+1)}$$

Taking inverse Laplace transformation of both sides, we have

$$y = AL^{-1}\left(\frac{1}{p^3}\right) + \frac{1}{2}L^{-1}\left(\frac{1}{p+1}\right) - L^{-1}\left(\frac{1}{p}\right) + L^{-1}\left(\frac{1}{p^2}\right)$$

$$+ \frac{1}{2}L^{-1}\frac{p}{(p^2+1)} - \frac{1}{2}L^{-1}\cdot\frac{1}{(p^2+1)}$$

$$\Rightarrow \qquad y = \frac{1}{2}At^2 + \frac{1}{2}e^{-t} - 1 + t + \frac{1}{2}\cos t - \frac{1}{2}\sin t$$

$$\Rightarrow \qquad y = -1 + t + ct^2 + \frac{1}{2}(e^{-t} + \cos t - \sin t)$$

EXAMPLE 13.   *Solve* $y'''(t) + y''(t) - 4y'(t) - 4y(t) = F(t)$ *if* $y(0) = y''(0) = 0$ *and* $y'(0) = 2$.

SOLUTION.   Taking the Laplace transformation of both sides of the given equation, we get

$$L(y''') + L(y'') - 4L(y') - 4L(y) = L[F(t)]$$

$$\Rightarrow \quad p^3 L(y) - p^2 y(0) - py'(0) - y''(0) + p^2 L(y) - py(0)$$
$$- y'(0) - 4pL(y) + 4y(0) - 4L(y) = L[F(t)]$$

Let $LF(t) = f(p)$

$$\Rightarrow \quad (p^3 + p^2 - 4p - 4)L(y) - (p^2 + p - 4)y(0) - (p + 1)y'(0) - y''(0) = f(p)$$

$$\Rightarrow \quad (p^3 + p^2 - 4p - 4)L(y) - 2(p + 1) = f(p)$$

$$[\text{Given } y'(0) = 2, y(0) = y''(0) = 0]$$

$$\Rightarrow \quad L(y) = \frac{2(p + 1)}{(p^3 + p^2 - 4p - 4)} + \frac{f(p)}{(p^3 + p^2 - 4p - 4)}$$

$$\Rightarrow \quad L(y) = \frac{2}{(p^2 - 4)} + \frac{f(p)}{(p + 1)(p - 2)(p + 2)}$$

Taking inverse Laplace transformation of both sides, we have

$$y = L^{-1}\left(\frac{2}{(p^2 - 4)}\right) + L^{-1}\frac{f(p)}{(p + 1)(p - 2)(p + 2)}$$

$$\Rightarrow \quad y = \sinh 2t + \frac{1}{12}\left[L^{-1}f(p)\left\{\frac{-4}{p + 1} + \frac{1}{p - 2} + \frac{3}{p + 2}\right\}\right]$$

Hence, $y = \sinh 2t + \dfrac{1}{12}f(t)^{*}(-4e^{-t} + e^{2t} + 3e^{-2t})$

EXAMPLE 14.   *Solve* $(D^3 - D^2 + 4D - 4)y = 68e^t \sin 2t, y = 1, Dy = -19, D^2y = -37 \text{ at } t = 0$.

(Meerut–1987, 89, 2008)

SOLUTION.   We can write the given equation in the form as

$$y''' - y'' + 4y' - 4y = 68e^t \sin 2t \qquad \qquad \dots(1)$$

Taking Laplace transformation of equation (1) of both sides, we have

$$L(y''') - L(y'') + 4L(y') - 4L(y) = 68L(e^t \sin 2t)$$

$$\Rightarrow \quad p^3 L(y) - p^2 y(0) - py'(0) - y''(0) - p^2 L(y) + py(0) + y'(0)$$
$$+ 4pL(y) - 4y(0) - 4L(y) = \frac{68 \times 2}{(p - 1)^2 + 4}$$

$$\Rightarrow (p^3 - p^2 + 4p - 4)L(y) - p^2 + 19p + 37 + p - 19 - 4 = \frac{136}{p^2 - 2p + 5}$$

$$[\text{Since given } y(0) = 1, y'(0) = -19, y''(0) = -37]$$

$$\Rightarrow \quad (p - 1)(p^2 + 4)L(y) - p^2 + 20p + 14 = \frac{136}{p^2 - 2p + 5}$$

$$\Rightarrow \quad L(y) = \frac{p^2 - 20p - 14}{(p - 1)(p^2 + 4)} + \frac{136}{(p - 1)(p^2 + 4)(p^2 - 2p + 5)}$$

$$L(y) = \frac{33}{5(p - 1)} + \frac{38p - 62}{5(p^2 + 4)} + \frac{34}{5(p - 1)} - \frac{136(3p - 7)}{85(p^2 + 4)} - \frac{(2p + 14)}{(p^2 - 2p + 5)}$$

$\Rightarrow \qquad L(y) = \dfrac{1}{5(p-1)} + \dfrac{14p}{5(p^2+4)} - \dfrac{6}{5(p^2+4)} - \dfrac{2(p-1)}{(p-1)^2+4} - \dfrac{16}{(p-1)^2+4}$

Taking inverse Laplace transformation of both sides, we have

$$y = \frac{1}{5}L^{-1}\left(\frac{1}{p-1}\right) + \frac{14}{5}L^{-1}\left(\frac{p}{p^2+4}\right) - \frac{6}{5}L^{-1}\left(\frac{1}{p^2+4}\right)$$

$$- 2L^{-1}\left(\frac{p-1}{(p-1)^2+4}\right) - 16L^{-1}\left(\frac{1}{(p-1)^2+4}\right)$$

$$= \frac{1}{5}e^t + \frac{14}{5}\cos 2t - \frac{6}{5}\cdot\frac{1}{2}\sin 2t - 2e^t\cos 2t - 16\cdot\frac{1}{2}e^t\sin 2t$$

$$= \frac{1}{5}[e^t + 14\cos 2t - 3\sin 2t] - 2e^t(\cos 2t + 4\sin 2t)$$

EXAMPLE 15. Solve $y'' - 4y' + 3y = F(t)$ if $y(0) = 1, y'(0) = 0$.

SOLUTION. Taking Laplace transformation of both sides of the given equation, we have

$$L(y'') - 4L(y') + 3L(y) = L[F(t)]$$

$\Rightarrow \quad p^2 L(y) - py(0) - y'(0) - 4pL(y) + 4y(0) + 3L(y) = LF(t)$

Let $\qquad L\{F(t)\} = f(p)$

$\Rightarrow \qquad (p^2 - 4p + 3)L(y) - (p-4) = f(p)$ [Given $y(0) = 1, y'(0) = 0$]

$\Rightarrow \qquad L(y) = \dfrac{(p-4)}{(p^2-4p+3)} + \dfrac{f(p)}{(p^2-4p+3)}$

$\Rightarrow \qquad L(y) = \dfrac{(p-4)}{(p-1)(p-3)} + \dfrac{f(p)}{(p-1)(p-3)}$

$\Rightarrow \qquad L(y) = \dfrac{3}{2(p-1)} - \dfrac{1}{2(p-3)} + \dfrac{1}{2}\left(\dfrac{1}{(p-3)} - \dfrac{1}{(p-1)}\right)f(p)$

Taking inverse Laplace transformation of both sides, we have

$$y = \frac{3}{2}L^{-1}\left(\frac{1}{p-1}\right) - \frac{1}{2}L^{-1}\left(\frac{1}{p-3}\right) + \frac{1}{2}L^{-1}\left(\frac{1}{(p-3)} - \frac{1}{(p-1)}f(p)\right)$$

$\Rightarrow \qquad y = \dfrac{3}{2}e^t - \dfrac{1}{2}e^{3t} + \dfrac{1}{2}(e^{3t} - e^t)F^*(t)$

$\Rightarrow \qquad y = \dfrac{3}{2}e^t - \dfrac{1}{2}e^{3t} + \dfrac{1}{2}\int_0^t (e^{3x} - e^x)F(t-x)dx$

EXAMPLE 16. A particle P of mass 2 grams moves on the X-axis and is attached towards origin 'O' with a force numerically about to 8X. If it is initially at rest at X = 10, find its position at any subsequent time assuming :
  (a) no other force acts
  (b) a damping force numerically equal to 8 times the instantaneous velocity acts.

SOLUTION. Since given that mass of particle = 2 gm,
It attracts towards origin O with a force equal to 8X.
Initially X = 10 at rest.
(a) From Newton's second law of motion, the equation of motion,

$$2\frac{d^2X}{dt^2} = -8X \qquad \Rightarrow \qquad \frac{d^2X}{dt^2} = -4X$$

$$\Rightarrow \qquad \frac{d^2X}{dt^2} + 4X = 0 \qquad\qquad\qquad …(1)$$

Initially $X = 10$ and $X' = 0$ at $t = 0$

Taking Laplace transformation of both sides of equation (1), we have

$$L\left(\frac{d^2X}{dt^2}\right) + 4L(X) = 0$$

$$\Rightarrow \qquad p^2L(X) - pX(0) - X'(0) + 4L(X) = 0$$

$$\Rightarrow \qquad (p^2 + 4)L(X) - pX(0) - X'(0) = 0$$

$$\Rightarrow \qquad (p^2 + 4)L(X) = 10p$$

$$[\text{On applying given condition } X(0) = 10, X'(0) = 0]$$

$$\Rightarrow \qquad L(X) = \frac{10p}{(p^2 + 4)}$$

Taking inverse Laplace transformation of both sides of equation, we get

$$X = L^{-1}\left(\frac{10p}{p^2 + 4}\right)$$

$$X = 10 \cos 2t$$

(b) Since there is damping force equal to 8 times of the instantaneous velocity acts, thus the equation of motion of the particle is

$$2\frac{d^2X}{dt^2} = -8X - 8\frac{dX}{dt}$$

$$\Rightarrow \qquad \frac{d^2X}{dt^2} = -4X - 4\frac{dX}{dt}$$

$$\Rightarrow \qquad \frac{d^2X}{dt^2} + 4\frac{dX}{dt} + 4X = 0 \qquad\qquad …(2)$$

Initially $\qquad\qquad X(0) = 10, X'(0) = 0$

Taking Laplace transformation of equation (2) of both sides, we have

$$L\left(\frac{d^2X}{dt^2}\right) + 4L\left(\frac{dX}{dt}\right) + 4L(X) = 0$$

$$\Rightarrow \qquad p^2L(X) - pX(0) - X'(0) + 4pL(X) - 4X(0) + 4L(X) = 0$$

$$\Rightarrow \qquad (p^2 + 4p + 4)L(X) = 10p + 40$$

$$\Rightarrow \qquad L(X) = \frac{10p + 40}{(p^2 + 4p + 4)} = 10\left(\frac{p + 4}{(p^2 + 4p + 4)}\right)$$

$$\Rightarrow \qquad L(X) = 10\left[\frac{(p + 2) + 2}{(p + 2)^2}\right]$$

$$\Rightarrow \qquad L(X) = 10\left[\frac{1}{(p+2)} + \frac{2}{(p+2)^2}\right]$$

Taking inverse Laplace transformation of both sides, we have

$$X = 10L^{-1}\left[\frac{1}{(p+2)} + \frac{2}{(p+2)^2}\right]$$

$$X = 10e^{-2t} + 20te^{-2t}$$

3.5 APPLICATIONS OF LAPLACE TRANSFORM TO SOLUTION OF INITIAL AND BOUNDARY VALUE PROBLEMS

Many problems in science and engineering, when formulated mathematically, lead to partial differential equations involving one or more unknown functions together with certain prescribed conditions on the functions which arise from the physical situations.

3.5.1 SOME IMPORTANT PARTIAL DIFFERENTIAL EQUATIONS

(A) One dimensional heat conduction equation

$$\frac{\partial U}{\partial t} = k\frac{\partial^2 U}{\partial x^2}$$

where $U(x, t)$ is the temperature in a solid at position x at time t.

The constant k is called diffusivity.

(B) One dimensional wave equation

$$\frac{\partial^2 Y}{\partial t^2} = a^2\frac{\partial^2 Y}{\partial x^2}$$

where $Y(x, t)$ is the displacement of any point x of the string at time t.

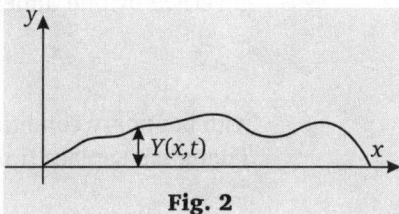

Fig. 2

(C) Longitudinal vibrations of a beam

$$\frac{\partial^2 Y}{\partial t^2} = C^2\frac{\partial^2 Y}{\partial x^2}$$

where $Y(x, t)$ is the longitudinal displacement from the equilibrium position of the cross section at x.

Fig. 3

(D) Transverse vibrations of a beam

$$\frac{\partial^2 Y}{\partial t^2} + b^2\frac{\partial^4 Y}{\partial x^4} = 0$$

The above equation describes the motion of a beam, initially located on the x-axis which is vibrating transversely.

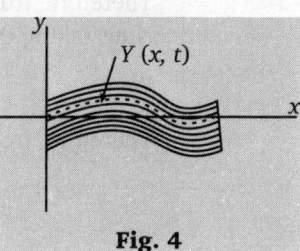

Fig. 4

(E) Heat conduction in a cylinder

$$\frac{\partial U}{\partial t} = k\left(\frac{\partial^2 U}{\partial r^2} + \frac{1}{r}\frac{\partial U}{\partial r}\right)$$

Here U(r, t) is the temperature at any time t at a distance r from the axis of a cylindrical solid. It is assumed that heat flow can take place only in the radial direction.

(F) Transmission lines

$$\frac{\partial E}{\partial x} = - RI - L\frac{\partial I}{\partial t}$$

$$\frac{\partial I}{\partial x} = -GE - C\frac{\partial E}{\partial t}$$

These are simultaneous equations for the current I and voltage E in a transmission line at any position x and at any time t. The constants R, L, G and C are respectively the resistance, inductance, conductance and capacitance per unit length.

Fig. 5

Solved Examples

EXAMPLE 1. *A semi-infinite solid x > 0 is initially at temperature zero. At time t > 0, a constant temperature V_0 > 0 is applied and maintained at the force x = 0. Find the temperature at any point of the solid at any time t > 0.* (Agra–1992; Meerut–1996, 90)

SOLUTION. We know that the temperature $u(x, t)$ at any point of the solid at any time $t > 0$ is governed by one dimensional heat equation

$$\frac{\partial u}{\partial t} = C^2 \frac{\partial^2 u}{\partial x^2}, (x > 0, t > 0) \qquad ...(1)$$

with boundary conditions $u(0, t) = V_0$, $u(x, 0) = 0$

Taking the Laplace transforms of both sides of (1), we get

$$L\left\{\frac{\partial u}{\partial t}\right\} = C^2 L\left\{\frac{\partial^2 u}{\partial x^2}\right\}$$

$$\Rightarrow \quad p\bar{u}(x,p) - u(x,0) = C^2 \frac{d^2\bar{u}}{dx^2} \Rightarrow \frac{d^2\bar{u}}{dx^2} - \frac{p}{C^2}\bar{u} = 0 \qquad ...(2)$$

The solution of (2) is given by

$$\bar{u}(x,p) = Ae^{\{\sqrt{p/c^2}\}\cdot x} + Be^{-\{\sqrt{p/c^2}\}\cdot x} \qquad ...(3)$$

Since u is finite when $x \to \infty$.

Therefore, $\bar{u}$ is also finite when $x \to \infty$.

Therefore, from (3) $A = 0$, otherwise $\bar{u} \to \infty$ as $x \to \infty$.

Now, taking the Laplace transform of the condition $u(0, t) = V_0$

We have $\bar{u}(0,p) = \int_{t=0}^{\infty} V_0 e^{-pt} dt = \frac{V_0}{p}$

Therefore, from (3), we have

$$\bar{u}(0,p) = B = \frac{V_0}{p} \qquad \text{(Using } A = 0\text{)}$$

Hence, $$\bar{u}(x,p) = \frac{V_0}{p}e^{-\{\sqrt{p/c^2}\}\cdot x}$$

$$\Rightarrow \quad u(x,t) = L^{-1}\left\{\frac{V_0}{p}e^{-\{\sqrt{p/c^2}\}\cdot x}\right\}$$

$$= V_0 erfc\left\{\frac{x}{2c\sqrt{t}}\right\} \qquad \left[\because L^{-1}\left\{\frac{e^{-\lambda/p}}{p}\right\} = erfc\left(\frac{\lambda}{2\sqrt{t}}\right)\right].$$

EXAMPLE 2. *A semi-infinite solid $x > 0$ has its initial temperature equal to zero. A constant heat flux A is applied at the face $x = 0$. Find the temperature at any point $x > 0$ of the solid, given that*

$$L^{-1}\left\{\frac{e^{-x\sqrt{p}}}{p^{3/2}}\right\} = \sqrt{\left(\frac{t}{\pi}\right)}e^{-x^2/4t^2} - \frac{x}{\pi}erfc\left(\frac{x}{2\sqrt{t}}\right)$$

SOLUTION. The temperature $u(x, t)$ at any point in the solid is governed by the heat equation

$$\frac{\partial u}{\partial t} = C^2 \frac{\partial^2 u}{\partial x^2}, x > 0, t > 0 \qquad \qquad \text{...(1)}$$

with boundary conditions

$$-ku_x(0,t) = A, \lim_{x \to \infty} u(x,t) = 0, t > 0 \qquad \qquad \text{...(2)}$$

and initial condition $u(x, 0) = 0$.

Taking the Laplace transform of both sides of (1) and (2), we get

$$p\bar{u}(x,p) - u(x,0) = C^2 \frac{d^2u}{dx^2}$$

$$\Rightarrow \qquad \frac{d^2\bar{u}}{dx^2} - \frac{p}{C^2}\bar{u} = 0 \qquad \qquad \text{...(3)}$$

$$L\{u_x(0,t)\} = -\frac{A}{k}L\{1\} \quad \text{or} \quad \left(\frac{d\bar{u}}{dx}\right)_{x=0} = -\frac{A}{kp}$$

and $\bar{u}(x,p) \to 0$ as $x \to \infty$

Now, the solution of (3) is given by

$$\bar{u} = C_1 e^{\sqrt{(p/c^2)}\cdot x} + C_2 e^{-\sqrt{(p/c^2)}\cdot x}$$

Since $\bar{u} \to 0$ as $x \to \infty$, therefore, $C_1 = 0$

$$\Rightarrow \qquad \bar{u} = C_2 e^{-\sqrt{(p/c^2)}\cdot x}$$

Also, then

$$\left(\frac{d\bar{u}}{dx}\right)_{x=0} = -C_2 \sqrt{\left(\frac{p}{C^2}\right)}$$

$$\Rightarrow \qquad \frac{-A}{kp} = -C_2 \sqrt{\left(\frac{p}{C^2}\right)} \Rightarrow C_2 = \frac{-AC}{kp^{3/2}}$$

Hence,

$$\bar{u}(x,p) = \frac{AC}{kp^{3/2}} e^{-\{\sqrt{p/c^2}\}\cdot x}$$

$$\Rightarrow \qquad u(x,p) = \frac{AC}{k} \cdot L^{-1}\left\{\frac{e^{-\{\sqrt{p/c^2}\}\cdot x}}{p^{3/2}}\right\}$$

$$= \frac{AC}{k}\left[\sqrt{\left(\frac{t}{\pi}\right)}e^{-x^2/4c^2t^2} - \frac{x}{\pi c}erfc\left(\frac{x}{2\sqrt{(tc)^2}}\right)\right].$$

EXAMPLE 3. *Solve the boundary value problem $\dfrac{\partial^2 u}{\partial t^2} = a^2 \dfrac{\partial^2 u}{\partial x^2}$ $x > 0, t > 0$*

with boundary conditions $u(x,0) = 0, u_t(x,0) = 0, t > 0$

$$u(0,t) = F(t), \lim_{x\to\infty} u(x,t) = 0, t > 0 \quad \text{(SVTU–2005; Agra–1993)}$$

SOLUTION. Taking the Laplace transforms of both the sides of the given equation and boundary conditions, we get

$$L\left\{\frac{\partial^2 u}{\partial t^2}\right\} = a^2 L\left\{\frac{\partial^2 u}{\partial x^2}\right\}$$

$$\Rightarrow \quad p^2\bar{u}(x,p) - pu(x,0) - u_t(x,0) = a^2\frac{d^2\bar{u}}{dx^2}$$

$$\Rightarrow \quad \frac{d^2\bar{u}}{dx^2} - \frac{p^2}{a^2}\bar{u} = 0 \qquad\qquad \text{...(1)}$$

Also, $\bar{u}(0,p) = \int_0^\infty F(t)e^{-pt}dt = \bar{F}(p)$ and $\bar{u}(x,p) = 0$ as $x \to \infty$

Solution of (1), is given by

$$\bar{u}(x,p) = Ae^{px/a} + Be^{-px/a}$$

Since, $\bar{u}(x,p) = 0$ as $x \to \infty \Rightarrow A = 0$

and $\bar{u}(0,p) = \bar{F}(p) = B$

Therefore, $\bar{u}(x,p) = \bar{F}(p)e^{-px/a}$

Hence, $\bar{u}(x,t) = L^{-1}\{e^{-px/a} \cdot \bar{F}(p)\} = \begin{cases} F\left(t - \dfrac{x}{a}\right), & t > \dfrac{x}{a} \\[2mm] 0, & t < \dfrac{x}{a} \end{cases}$

$$= F\left(t - \frac{x}{a}\right)H\left(t - \frac{x}{a}\right).$$

EXAMPLE 4. *Solve the following boundary value problem* $\dfrac{\partial^2 u}{\partial t^2} = a^2\dfrac{\partial^2 u}{\partial x^2}, x > 0, t > 0$ *with boundary conditions* $u_x(0, t) = A \sin \omega t, u(x, 0) = 0, u_t(x, 0) = 0, |u(x, t)| < M.$

(Meerut–1987)

SOLUTION. Taking the Laplace transform of the given equation and boundary conditions, we have

$$L\left\{\frac{\partial^2 u}{\partial t^2}\right\} = a^2 L\left\{\frac{\partial^2 u}{\partial x^2}\right\}$$

$$\Rightarrow \quad p^2\bar{u}(x,p) - pu(x,0) - u_t(x,0) = a^2\frac{d^2\bar{u}(x,p)}{dx^2}$$

$$\Rightarrow \quad a^2\frac{d^2\bar{u}}{dx^2} - p^2\bar{u} = 0 \qquad\qquad \text{...(1)}$$

and $L\{u_x(0,t)\} = AL\{\sin wt\}$

$$\Rightarrow \quad \left(\frac{d\bar{u}}{dx}\right)_{x=0} = \frac{Aw}{p^2 + w^2} \qquad\qquad \text{...(2)}$$

Now, solution of (1) is given by

$$\bar{u}(x,p) = C_1 e^{px/a} + C_2 e^{-px/a} \qquad\qquad \text{...(3)}$$

Here, it is given that $|u(x,t)| < M$, *i.e.*, $u(x, t)$ is bounded from which it follows that $L\{u(x,t)\} = \bar{u}(x,p)$ is also bounded, *i.e.*, $\bar{u}(x,p) \to a$ finite value as $x \to \infty$.

Therefore, from (3), we must have $C_1 = 0$, otherwise $\bar{u}(x,p) \to \infty$ as $x \to \infty$.

$\Rightarrow \qquad \bar{u}(x,p) = C_2 e^{-px/a}$ and therefore $\dfrac{d}{dx}\bar{u}(x,p) = C_2\left(-\dfrac{p}{a}\right)e^{-px/a}$

But from (2),

$$\left(\frac{d\bar{u}}{dx}\right)_{x=0} = \frac{Aw}{p^2 + w^2} \qquad\qquad \frac{Aw}{p^2 + w^2} = -\frac{C_2 p}{a}$$

$\Rightarrow \qquad C_2 = -\dfrac{Aaw}{p(p^2 + w^2)}$

$\Rightarrow \qquad \bar{u}(x,p) = -\dfrac{Aaw}{p(p^2 + w^2)} e^{-px/a}$

$\therefore \qquad u(x,t) = -L^{-1}\left\{\dfrac{Aaw}{p(p^2 + w^2)} e^{-px/a}\right\} = -F\left(t - \dfrac{x}{a}\right)H\left(t - \dfrac{x}{a}\right)$

where $\qquad F(t) = L^{-1}\left\{\dfrac{Aaw}{p(p^2 + w^2)}\right\} = \dfrac{Aa}{w}L^{-1}\left\{\dfrac{1}{p} - \dfrac{p}{p^2 + w^2}\right\} = \dfrac{Aa}{w}(1 - \cos wt)$

Hence, $\qquad u(x,t) = -\left(\dfrac{Aa}{w}\right)\left[1 - \cos\left(t - \dfrac{x}{a}\right)\right]H\left(t - \dfrac{x}{a}\right)$

EXAMPLE 5. *A string is stretched between two fixed points (0, 0) and (c, 0). If it is displaced into the curve* $y = b\sin\left(\dfrac{\pi x}{c}\right)$ *and released from rest in that position at time t = 0. Find its displacement at any time t > 0 and at any point 0 < x < c.* (Meerut–1983)

SOLUTION. We have, the displacement $y(x, t)$ of the string is governed by the differential equation

$$\frac{\partial^2 y}{\partial t^2} = a^2\frac{\partial^2 y}{\partial x^2} \qquad\qquad\qquad ...(1)$$

with boundary condition
$$y(0, t) = 0, \, y(c, t) = 0 \qquad\qquad\qquad ...(2)$$
and initial conditions

$$y(x,0) = b\sin\frac{\pi x}{c} \text{ and } y_t(x,0) = 0$$

Taking the Laplace transforms of (1) and (2), we have

$$L\left\{\frac{\partial^2 y}{\partial t^2}\right\} = a^2 L\left\{\frac{\partial^2 y}{\partial x^2}\right\}$$

$\Rightarrow \quad p^2\bar{y}(x,p) - py(x,0) - y_t(x,0) = a^2\dfrac{d^2\bar{y}}{dx^2}.$

$\Rightarrow \quad \dfrac{d^2\bar{y}}{dx^2} - \dfrac{p^2\bar{y}}{a^2} = -\dfrac{bp}{a^2}\sin\dfrac{\pi x}{C} \qquad\qquad ...(3)$

and $\qquad\qquad \bar{y}(0,p) = 0$ and $\bar{y}(c,p) = 0.$

Solution of (3) is given by

$$\bar{y}(x,p) = Ae^{(p/a)\cdot x} + Be^{-(p/a)\cdot x} + \frac{b\sin\dfrac{\pi x}{c}}{p^2 + (\pi^2 a^2 / c^2)}$$

$\Rightarrow \qquad\qquad \overline{y}(0,p) = 0 \qquad \Rightarrow \qquad A + B = 0$

and $\qquad\qquad \overline{y}(c,p) = 0$

$\Rightarrow \quad Ae^{pc/a} + Be^{-pc/a} = 0$

On solving, we get $A = 0, B = 0$

Hence, $\qquad\qquad \overline{y}(x,p) = \dfrac{bp\sin\dfrac{\pi x}{c}}{p^2 + (\pi^2 a^2 / c^2)}$

$\Rightarrow \qquad\qquad y(x,t) = b\sin\dfrac{\pi x}{c} L^{-1}\left\{\dfrac{p}{p^2 + (\pi^2 a^2 / c^2)}\right\} = b\cos\dfrac{\pi a t}{c}\sin\dfrac{\pi x}{c}$

EXAMPLE 6. *An infinite long string having one end $x = 0$ is initially at rest on the x-axis. The end $x = 0$ under goes a periodic transverse displacement given by $A_0 \sin nt$, $t > 0$, find the displacement of any point on the string at $t > 0$.* (Meerut–1992; Agra–1998)

SOLUTION. The displacement of any point of the string is governed by the equation

$$\frac{\partial^2 u}{\partial t^2} = a^2 \frac{\partial^2 u}{\partial x^2}, x > 0, t > 0 \qquad\qquad \dots(1)$$

with the boundary and initial conditions

$$u(0,t) = A_0 \sin nt, t > 0 \qquad\qquad \dots(2)$$

$$u(x,0) = 0, u_t(x,0) = 0, t > 0$$

and the displacement is finite.

Taking the Laplace transform of (1) and (2), we have

$$L\left\{\frac{\partial^2 u}{\partial t^2}\right\} = a^2 L\left\{\frac{\partial^2 u}{\partial x^2}\right\}$$

$$\Rightarrow \quad p^2\overline{u}(x,p) - pu(x,0) - u_t(x,0) = a^2 \frac{d^2\overline{u}}{dx^2}$$

$$\Rightarrow \quad \frac{d^2\overline{u}}{dx^2} - \frac{p^2}{a^2}\overline{u} = 0 \qquad\qquad \dots(3)$$

Also, $\overline{u}(0,p) = A_0 L(\sin nt) = \dfrac{A_0 n}{p^2 + n^2}$ and $\overline{u}(x,p)$ is finite for $x \geq 0$.

The solution of (3) is given by

$$\overline{u}(x,p) = Ae^{(p/a)\cdot x} + Be^{-(p/a)\cdot x}$$

Since $\overline{u}(x,p)$ is finite, therefore, $A = 0$ otherwise $\overline{u}(x,p) \to \infty$ as $x \to \infty$.

$\therefore \qquad\qquad \overline{u}(x,p) = Be^{-(p/a)\cdot x}$

Now, $\quad \overline{u}(0,p) = A_0 n L^{-1}\left\{\dfrac{1}{p^2 + n^2} e^{-(p/a)\cdot x}\right\}$

$$= \begin{cases} A_0 \sin n\left(t - \dfrac{x}{a}\right), & t > \dfrac{x}{a} \\[3mm] 0, & t < \dfrac{x}{a} \end{cases} \qquad \left[\because L^{-1}\left\{\dfrac{1}{p^2 + n^2}\right\} = \dfrac{1}{n}\sin nt\right]$$

EXAMPLE 7. *Solve the boundary value problem* $\dfrac{\partial^2 u}{\partial x^2} = \dfrac{1}{k}\dfrac{\partial u}{\partial t}$ *subject to the conditions*

$u = 0$, *when* $x = \infty$: $u = 0$, *when* $t = 0$. (Meerut–1989, 91)

SOLUTION. Taking the Laplace transforms of the given equation and the boundary conditions, we have

$$L\left\{\frac{\partial^2 u}{\partial x^2}\right\} = \frac{1}{k}L\left\{\frac{\partial u}{\partial t}\right\}$$

$$\Rightarrow \quad \frac{d^2}{dx^2}\bar{u}(x,p) = \frac{1}{k}\{p\bar{u}(x,p) - u(x,0)\}$$

$$\Rightarrow \quad \frac{d^2\bar{u}}{dx^2} - \frac{1}{k}p\bar{u} = 0 \qquad\qquad …(1)$$

and $\lim\limits_{x\to\infty} \bar{u}(x,p) = 0$ $\qquad\qquad …(2)$

Here, solution of (1) is given by

$$\bar{u}(x,p) = C_1 e^{x\sqrt{(p/k)}} + C_2 e^{-x\sqrt{(p/k)}}$$

But $\bar{u}(x,p) = 0$ when $x \to \infty$. Therefore, we must have $C_1 = 0$ otherwise

$\bar{u}(x,p) \to \infty$ as $x \to \infty$

$$\Rightarrow \quad \bar{u}(x,p) = C_2 e^{-x\sqrt{p/k}}$$

$$\Rightarrow \quad u(x,p) = C_2 L^{-1}\{e^{-x\sqrt{(p/k)}}\}$$

$$= C_2 L^{-1}\left[1 - \left(\frac{x}{\sqrt{k}}\right)p^{1/2} + \left(\frac{x^2}{2!\cdot k}\right)p - \left(\frac{x^3}{3!\cdot k^{3/2}}\right)p^{3/2}\right.$$

$$\left. + \left(\frac{x^4}{4!\cdot k^2}\right)p^2 - \left(\frac{x^5}{5!\cdot k^{5/2}}\right)p^{5/2} + …\right]$$

$$= C_2\left[0 - \frac{x}{\sqrt{k}}\frac{(-1)^1}{\sqrt{\pi}}\cdot\frac{1}{2}t^{-3/2} + 0 - \frac{x^3}{3!\,k^{3/2}}\frac{(-1)^2}{\sqrt{\pi}}\cdot\frac{1}{2}\frac{3}{2}\cdot t^{-5/2}\right.$$

$$\left. + 0 - \frac{x^5}{5!\,k^{5/2}}\frac{(-1)^3}{\sqrt{\pi}}\cdot\frac{1}{2}\frac{3}{2}\frac{5}{2}\cdot t^{-7/2} + …\right]$$

$$\left[\because L^{-1}\{p^{n+(1/2)}\} = \frac{(-1)^{n+1}}{\sqrt{\pi}}\cdot\frac{1}{2}\frac{3}{2}\frac{5}{2}…\left(\frac{2n+1}{2}\right)t^{-n-(3/2)}\right.$$

$$\left. \text{and } L^{-1}\{p^n\} = 0 \text{ if } n \text{ is zero or a positive integer}\right]$$

$$= \frac{Cx}{\sqrt{k}}\cdot\frac{1}{2\sqrt{(\pi t^3)}}\left[1 - \frac{1}{1!}\left(\frac{x^2}{4kt}\right) + \frac{1}{2!}\left(\frac{x^2}{4kt}\right)^2 - …\right] \text{ (By taking } C_2 = Cx)$$

$$\Rightarrow \quad u(x,t) = \frac{Cx}{\sqrt{k}}\cdot\frac{1}{2\sqrt{(\pi t^3)}}e^{-x^2/4kt}$$

EXAMPLE 8. *A beam is hinged at its ends $x = 0$ and $x = l$ carries a uniform load w_0 per unit length. Find the deflection at any point.* (Meerut–1982, 86)

SOLUTION. We have, the deflection $y(x)$ at any point x satisfies the differential equation

$$\frac{d^4 y}{dx^4} = \frac{w_0}{E \cdot I} \quad t < x < l \qquad \ldots(1)$$

with boundary conditions

$$y(0) = 0 \qquad y''(0) = 0 \qquad \ldots(2)$$
$$y(l) = 0 \qquad y''(l) = 0 \qquad \ldots(3)$$

Taking the Laplace transform of both sides of (1), we have

$$p^4 \bar{y}(p) - p^3 y(0) - p^2 y'(0) - p y''(0) - y'''(0) = \frac{w_0}{EIp}$$

$$\Rightarrow \quad p^4 \bar{y}(p) - Ap^2 - B = \frac{w_0}{EIp}, \text{ where } A = y'(0) \text{ and } B = y''(0)$$

$$\Rightarrow \quad \bar{y}(p) = \frac{A}{p^2} + \frac{B}{p^4} + \frac{w_0}{EIp^5}$$

$$\Rightarrow \quad y(x) = Ax + \frac{Bx^3}{3!} + \frac{w_0}{EI} \cdot \frac{x^4}{4!} \qquad \text{(By taking inverse Laplace transform)}$$

$$= Ax + \frac{Bx^3}{6} + \frac{w_0 x^4}{24 \cdot E \cdot I}$$

Using condition (3), we get

$$y(l) = 0 = Al + \frac{Bl^3}{6} + \frac{w_0 l^4}{24 \cdot E \cdot I} \text{ and } \quad y''(l) = 0 = B \cdot l + \frac{w_0 l^2}{2 \cdot E \cdot I}$$

On solving, we get $A = \dfrac{w_0 l^3}{24EI}$ and $B = -\dfrac{w_0 l}{2EI}$

Therefore, the required deflection is

$$y(x) = \frac{w_0 \cdot l^3 \cdot x}{24EI} - \frac{w_0 l}{2EI} \cdot \frac{x^3}{6} + \frac{w_0 x^4}{24EI}$$

$$= \frac{w_0}{24EI}(l^3 x - 2lx^3 + x^4) = \frac{w_0}{24EI} x(l - x)(l^2 + lx - x^2)$$

EXAMPLE 9. *A beam which is clamped at its ends $x = 0$ and $x = l$ carries a uniform load w_0 per unit length. Show that the deflection at any point is*

$$y(x) = \frac{w_0 x^2 (l - x)^2}{24EI} \qquad \text{(Meerut-1985, 86, 97)}$$

SOLUTION. Here, the deflection $y(x)$ at any point x, satisfies the differential equation

$$\frac{d^4 y}{dx^4} = \frac{w_0}{E \cdot I}, 0 < x < l \qquad \ldots(1)$$

with boundary conditions

$$y(0) = 0 \qquad y'(0) = 0 \qquad \ldots(2)$$
$$y(l) = 0 \qquad y'(l) = 0 \qquad \ldots(3)$$

Taking the Laplace transform of both sides of (1), we have

$$p^4 \bar{y}(p) - p^3 y(0) - p^2 y'(0) - p y''(0) - y'''(0) = \frac{w_0}{EIp}$$

$$\Rightarrow \quad p^4 \bar{y}(p) - Ap - B = \frac{w_0}{E \cdot I \cdot p}, \text{ where } A = y''(0) \text{ and } B = y'''(0)$$

$$\Rightarrow \quad \bar{y}(p) = \frac{A}{p^3} + \frac{B}{p^4} + \frac{w_0}{E \cdot I \cdot p^4}$$

Now, taking the inverse Laplace transform, we get

$$y(x) = A \cdot \frac{x^2}{2!} + \frac{Bx^3}{3!} + \frac{w_0 x^4}{4! \cdot E \cdot I} = \frac{Ax^2}{2} + \frac{Bx^3}{6} + \frac{w_0 x^4}{24 \cdot E \cdot I}$$

Now, using (3), we get

$$y(l) = 0 = \frac{Al^2}{2} + \frac{Bl^3}{6} + \frac{w_0 l^4}{24 E \cdot I}$$

and

$$y'(l) = 0 = Al + \frac{Bl^2}{2} + \frac{w_0 l^3}{6EI}$$

On solving, we get $A = \dfrac{w_0 l^2}{12 E \cdot I}, B = -\dfrac{w_0 l}{2EI}$

Therefore, the required deflection is

$$y(x) = \frac{w_0 l^2}{24 EI} x^2 - \frac{w_0 l}{12 EI} x^3 + \frac{w_0 x^4}{24 EI} = \frac{w_0 x^2}{24 EI}(l-x)^2$$

EXAMPLE 10. *The faces $x = 0$ and $x = 1$ of a slab of material for which $k = 1$ are kept at temperature 0 and 1 respectively until the temperature distribution becomes $u = x$. After time $t = 0$ both faces are held at temperature 0. Determine the temperature formula, given that*

$$L^{-1}\left\{ \frac{\sinh x\sqrt{p}}{p \sinh a\sqrt{p}} \right\} = \frac{x}{a} + \frac{2}{\pi} \sum_{n=1}^{\infty} \frac{(-1)^n}{n} e^{-n^2\pi^2 t/a^2} \sin\left(\frac{n\pi x}{a}\right)$$

SOLUTION. The temperature $u(x, t)$ at any point in the slab is governed by the heat equation

$$\frac{\partial u}{\partial t} = \frac{\partial^2 u}{\partial x^2} \qquad \qquad ...(1)$$

with boundary conditions

$$u(0, t) = 0, \quad u(1, t) = 0, u(x, 0) = x \qquad \qquad ...(2)$$

Taking the Laplace transform of both sides of (1) & (2), we get

$$p\bar{u} - u(x,0) = \frac{d^2\bar{u}}{dx^2}$$

$$\Rightarrow \quad \frac{d^2\bar{u}}{dx^2} - p\bar{u} = -x \qquad \qquad ...(3)$$

Now, the solution of (3) is given by

$$\bar{u} = ae^{-x\sqrt{p}} + be^{x\sqrt{p}} + \frac{1}{D^2 - p}(-x)$$

$$= ae^{-x\sqrt{p}} + be^{x\sqrt{p}} + \frac{1}{p}\left(1 - \frac{D^2}{p}\right)^{-1} x$$

$$\Rightarrow \quad \bar{u} = ae^{-x\sqrt{p}} + be^{x\sqrt{p}} + \frac{x}{p}$$

which can be written as

$$\bar{u} = a \cosh x\sqrt{p} + b \sinh x\sqrt{p} + \frac{x}{p}$$

From eq^n (2), we have

$$L\{u(0,t)\} = 0 \qquad \Rightarrow \qquad \bar{u}(0,p) = 0$$
$$\Rightarrow \qquad a = 0$$

Now, eq^n (1) is given by

$$\bar{u} = b \sinh x\sqrt{p} + \frac{x}{p} \qquad \qquad ...(4)$$

Using $u(1, t) = 0$, we get

$$L\{u(1, t)\} = 0 \qquad \Rightarrow \qquad \bar{u}(1,p) = 0$$
$$\Rightarrow \qquad 0 = b \sinh\sqrt{p} + \frac{1}{p} \qquad \Rightarrow \qquad b = \frac{-1}{p\sinh\sqrt{p}}$$

Putting this value in eq^n (4), we get

$$\bar{u} = \frac{x}{p} - \frac{\sinh x\sqrt{p}}{p\sinh\sqrt{p}}$$

$$\Rightarrow \qquad u = L^{-1}\left\{\frac{x}{p}\right\} - L^{-1}\left\{\frac{\sinh x\sqrt{p}}{p\sinh\sqrt{p}}\right\}$$

$$\Rightarrow \qquad u = x - L^{-1}\left\{\frac{\sinh x\sqrt{p}}{p\sinh\sqrt{p}}\right\}$$

$$= x - \left[x + \frac{2}{\pi}\sum_{n=1}^{\infty}\frac{(-1)^n}{n}\cdot e^{-n^2\pi^2 t}\sin(n\pi x)\right]$$

Hence,

$$u = -\frac{2}{\pi}\sum_{n=1}^{\infty}\frac{(-1)^n}{n}\cdot e^{-n^2\pi^2 t}\sin(n\pi x)$$

Additional Solved Examples

EXAMPLE 1. *Solve the boundary value problem* $\dfrac{\partial u}{\partial x} + 4\dfrac{\partial u}{\partial t} = -8t, t > 0, x > 0$ *subject to the conditions* $u = 0$, *when* $t = 0$, $x > 0$, $u = 2t^2$ *when* $x = 0$, $t > 0$.   (Meerut–1990)

SOLUTION. Taking the Laplace transformation of both sides of the given equation, we have

$$L\left(\frac{\partial u}{\partial x}\right) + 4L\left(\frac{\partial u}{\partial t}\right) = -8L(t)$$

$$\Rightarrow \frac{d}{dx}L\{u(x,p)\} + 4pL\{u(x,p)\} - 4u(x,0) = -\frac{8}{p^2}$$

$$\Rightarrow \qquad \frac{d}{dx}L(u) + 4pL(u) = -\frac{8}{p^2} \qquad \qquad ...(1)$$

It is a linear equation and $L[u(0, t)] = 2L(t^2)$

[Since given conditions are $u(x, 0) = 0$, $x > 0$ and $u(0, t) = 2t^2$, $t > 0$]

$$\Rightarrow \qquad Lu(0,p) = \frac{4}{p^3} \qquad \qquad ...(2)$$

The solution of differential equation (1) is

$$L(y)(I.F.) = C - 8\int \left\{ \left(\frac{1}{p^2}\right) I.F. \right\} dx \qquad \qquad \text{...(3)}$$

Since, $I.F. = e^{\int 4p dx} = e^{4px}$

On substituting this value in equation (3), we have

$$L(u)e^{4px} = C - 8\int \left(\frac{1}{p^2}\right) e^{4px} dx$$

$$\Rightarrow \qquad L(u)e^{4px} = C - \frac{2}{p^3} e^{4px}$$

$$\Rightarrow \qquad L(u) = L\{u(x,p)\} = Ce^{-4px} - \frac{2}{p^3} \qquad \qquad \text{...(4)}$$

From equation (2) and (3), we get

$$Lu(0,p) = C - \left(\frac{2}{p^3}\right) = 0 \quad \Rightarrow \quad C = \frac{6}{p^3}$$

Hence, $$Lu(x,p) = \frac{6}{p^3} e^{-4px} - \frac{2}{p^3}$$

Taking inverse Laplace transformation of both sides, we get

$$u(x,t) = 6L^{-1}\left(\frac{1}{p^3} e^{-4px}\right) - 2L^{-1}\left(\frac{1}{p^3}\right)$$

$$\Rightarrow \qquad u(x,t) = 3(t-4x)^2 H(t-4x) - t^2$$

EXAMPLE 2. *A semi-infinite solid x > 0 is initially at temperature zero. At time t = 0, a constant temperature u_0 > 0 is applied and maintained at the face x = 0. Show that the temperature at any point of the solid at $t = \dfrac{1}{4}$ is given by $u_0 erfC\left(\dfrac{x}{\sqrt{k}}\right)$.*

SOLUTION. The temperature $u(x, t)$ at any point of semi-infinite solid at any time t greater than zero is defined by one dimensional heat equation

$$\frac{\partial u}{\partial t} = C^2 \frac{\partial^2 u}{\partial x^2} \qquad \qquad \text{...(1)}$$

when $x > 0, t > 0$ and the boundary conditions are $u(0, t) = u_0, C^2 = k, u(x, 0) = 0$

Taking Laplace transformation of both sides of equation (1), we have

$$L\left(\frac{\partial u}{\partial t}\right) = kL\left(\frac{\partial^2 u}{\partial x^2}\right)$$

$$\Rightarrow \qquad pL\{u(x,p)\} - u(x,0) = k\frac{d^2}{dx^2}\{L(u)\}$$

$$\Rightarrow \qquad \frac{d^2}{dx^2}L(u) - \frac{p}{k}L(u) = 0$$

and its solution is

$$Lu(x,p) = Ae^{(\sqrt{p/k})x} + Be^{-(\sqrt{p/k})x} \qquad \qquad \text{...(2)}$$

Since u is finite as $x \to \infty$. $L(u)$ will be finite as $x \to \infty$.
Hence, $\qquad\qquad\qquad A = 0$.
Taking the Laplace transformation of the condition, $u(0, t) = u_0$, we get

$$L\{u(0,p)\}\int_{t=0}^{\infty}u_0 e^{-pt}dt = \frac{u_0}{p} \quad\Rightarrow\quad L\{u(0,p)\} = B = \frac{u_0}{p}$$

Hence, $\qquad\qquad L\{u(x,p)\} = \frac{u_0}{p}e^{(-\sqrt{p/k})x}$

Taking inverse Laplace transformation of both sides, we have

$$u(x,t) = L^{-1}\left(\frac{u_0}{p}e^{(-\sqrt{p/k})x}\right)$$

$$\Rightarrow \qquad\qquad u(x,t) = u_0 \operatorname{erf} C\left(\frac{x}{2\sqrt{kt}}\right)$$

Since given temperature at $t = \frac{1}{4}$ is $u\left(x,\frac{1}{4}\right) = u_0 \operatorname{erf} C\left(\dfrac{x}{2\sqrt{\dfrac{k}{u}}}\right)$

$$\Rightarrow \qquad\qquad u\left(x,\frac{1}{4}\right) = u_0 \operatorname{erf} C\left(\frac{x}{\sqrt{k}}\right)$$

EXAMPLE 3. *The initial temperature of a slab of homogeneous bounded by the planes $x = 0$ and $x = l$ is u_0. Find the temperature in this solid after the face $x = 0$ is insulated and temperature of face $x = l$ is reduced to zero.*

SOLUTION. The temperature $u(x, t)$ in the solid is governed by the heat equation

$$\frac{\partial u}{\partial t} = C^2 \frac{\partial^2 u}{\partial x^2}(0 < x < l, t > 0) \qquad\qquad ...(1)$$

and the boundary condition $u_x(0, t) = 0$
and $\qquad\qquad\qquad u(l, t) = 0$
and initially $\qquad\qquad u(x, 0) = u_0$
Now taking the Laplace transformation of both sides of equation (1), we get

$$L\left(\frac{\partial u}{\partial t}\right) = C^2 L\left(\frac{\partial^2 u}{\partial x^2}\right)$$

$$\Rightarrow \qquad pL\{u(x,p)\} - u(x,0) = C^2 \frac{d^2}{dx^2}L(u)$$

$$\Rightarrow \qquad \frac{d^2}{dx^2}L(u) - \frac{p}{C^2}L(y) = -\frac{u_0}{C^2} \qquad \text{[Given } u(x, 0) = u_0] \qquad ...(2)$$

$$\Rightarrow \qquad L\{u_x(0, t)\} = 0$$

$$\Rightarrow \qquad \left[\frac{d}{dx}L(u)\right]_{x=0} = 0 \qquad\qquad ...(3)$$

$$L[u, (l, t)] = 0 \Rightarrow Lu(l, p) = 0 \qquad\qquad ...(4)$$

The solution of equation (2) is given by

$$Lu(x,p) = \frac{u_0}{p} + A\cosh x\sqrt{\frac{p}{C^2}} + B\sinh x\sqrt{\frac{p}{C^2}} \qquad\qquad ...(5)$$

From equation (3), (4) and (5), we have
$$B = 0$$
$$0 = \frac{u_0}{p} + A \cosh l \sqrt{\frac{p}{C^2}} \qquad \Rightarrow A = -\frac{u_0}{p \cosh l(\sqrt{p/C^2})}$$

Put the values of A and B in equation (5), we have

$$L\{u(x,p)\} = \frac{u_0}{p} - \frac{u_0}{p} \frac{\cosh x \sqrt{(p/C^2)}}{\cosh l \sqrt{p/C^2}}$$

$$\Rightarrow \qquad L\{u(x,p)\} = \frac{u_0}{p} - \frac{u_0}{p} \frac{e^{x\sqrt{p/C^2}} + e^{-x\sqrt{p/C^2}}}{2} \cdot \frac{2}{e^{l\sqrt{p/C^2}} + e^{-l\sqrt{p/C^2}}}$$

$$\Rightarrow \qquad L\{u(x,p)\} = \frac{u_0}{p} - \frac{u_0}{p} e^{-l\sqrt{p/C^2}} \{e^{x\sqrt{p/C^2}} + e^{-x\sqrt{p/C^2}}\}(1 + e^{-2l\sqrt{p/C^2}})^{-1}$$

$$\Rightarrow \qquad Lu(x,p) = \frac{u_0}{p} - \frac{u_0}{p} \{e^{-(x+l)\sqrt{p/C^2}} + e^{-\sqrt{p/C^2}(-x+l)}\} \sum_{n=0}^{\infty} (-1)^n e^{-2nl\sqrt{p/C^2}}$$

$$\Rightarrow \qquad Lu(x,p) = \frac{u_0}{p} - \frac{u_0}{p} \sum_{n=0}^{\infty} (-1)^n [e^{-\sqrt{p/C^2}(ml+x)} + e^{-\sqrt{p/C^2}(ml-x)}]$$

[where $m = 2n + 1$]

Taking inverse Laplace transformation of both sides, we have

$$u(x,t) = u_0 L^{-}\left(\frac{1}{p}\right) - u_0 \sum_{n=0}^{\infty} (-1)^n \left[L^{-1}\left(\frac{e^{-\sqrt{p/C^2}(ml+x)}}{p} \right) + L^{-}\left(\frac{e^{-\sqrt{p/C^2}(ml-x)}}{p} \right) \right]$$

$$\Rightarrow \quad u(x,t) = u_0 - u_0 \sum_{n=0}^{\infty} (-1)^n \left[\text{erf } C\left(\frac{ml+x}{2\sqrt{C^2 t}} \right) + \text{erf } C\left(\frac{ml-x}{2\sqrt{C^2 t}} \right) \right]$$

Hence, $u(x,t) = u_0 - u_0 \sum_{n=0}^{\infty} (-1)^n \left[\text{erf } C\left(\frac{(2n+1)l+x}{2\sqrt{C^2 t}} \right) + \text{erf } C\left(\frac{(2n+1)l-x}{2\sqrt{C^2 t}} \right) \right]$

EXAMPLE 4. *Solve the boundary value problem $\dfrac{\partial^2 u}{\partial t^2} = a^2 \dfrac{\partial^2 u}{\partial x^2} (x > 0, t > 0)$ subject to the conditions*

$$u(x,0) = 0, x \geq 0, u_t(x,0) = 0, x > 0$$
$$u(0,t) = t, \lim_{x \to \infty} u(x,t) = 0, t \geq 0 \qquad \text{(SVTU–2005)}$$

SOLUTION.   Taking the Laplace transformation of both the sides of the given equation, we get

$$L\left\{ \frac{\partial^2 u}{\partial t^2} \right\} = a^2 L\left\{ \frac{\partial^2 u}{\partial x^2} \right\}$$

$$\Rightarrow \qquad p^2 L\{u(x,p)\} - pu(x,0) - u_t(x,0) = a^2 \frac{d^2}{dx^2} L(u)$$

$$\Rightarrow \qquad \frac{d^2}{dx^2} L(u) - \frac{p^2}{a^2} L(u) = 0 \qquad \text{[Given } u(x, 0) = 0, x \geq 0, u_t(x, 0) = 0]$$

...(1)

Also $\qquad L\{u(0,p)\} = L(t) = \dfrac{1}{p^2}$ and $L\{u(x,p)\} = 0$ as $x \to \infty$

The solution of (1) is given by

$$L\{u(x,p)\} = Ae^{(p/a)x} + Be^{-(p/a)x} \qquad \qquad ...(2)$$

Since, $\qquad L\{u(x,p)\} = 0$ as $x \to \infty$

$\Rightarrow \qquad \qquad \qquad A = 0$

and $\qquad L\{u(0,p)\} = \dfrac{1}{p^2} = B$

Putting these value in equation (2), we have

$$L\{u(x,p)\} = \dfrac{1}{p^2}e^{-(p/a)x}$$

Taking inverse Laplace transformation of both sides, we have

$$u(x,t) = L^{-1}\left(\dfrac{1}{p^2}e^{-(p/a)x}\right) = \left(t - \dfrac{x}{a}\right)H\left(t - \dfrac{x}{a}\right).$$

EXAMPLE 5. Solve the boundary value problem $\dfrac{\partial^2 u}{\partial x^2} = \dfrac{1}{C^2}\dfrac{\partial^2 u}{\partial t^2}$ $t > 0, x > 0$ subject to the conditions

(a) $u = E \sin at$ when $x = 0, t > 0$ (b) u is finite when $x \to \infty$

(c) $u = \dfrac{\partial u}{\partial t} = 0$ when $t = 0, x > 0$

SOLUTION. Taking Laplace transformation of both sides of the given equation and apply the given conditions, we get

$$L\left(\dfrac{\partial^2 u}{\partial x^2}\right) = \dfrac{1}{C^2}L\left(\dfrac{\partial^2 u}{\partial t^2}\right)$$

$\Rightarrow \qquad \dfrac{d^2}{dx^2}Lu(x,p) = \dfrac{1}{C^2}[p^2 L\{u(x,p)\} - pu(x,0) - u_t(x,0)]$

$\Rightarrow \qquad \dfrac{d^2}{dx^2}L(u) - \dfrac{p^2}{C^2}L(u) = 0 \qquad$ [Since $u(x, 0) = 0$ and $u_t(x, 0) = 0, x > 0$]

$$...(1)$$

and $\qquad \qquad Lu(0, p) = EL(\sin at) \qquad \Rightarrow \qquad Lu(0,p) = \dfrac{Ea}{(a^2 + p^2)}$

$L\{u(x, p)\}$ is finite when $x \to \infty$

Solution of equation (1) is given by

$$Lu(x, p) = C_1 e^{px/C} + C_2 e^{-px/C} \qquad \qquad ...(2)$$

as $Lu(x, p) \to \infty$ as $x \to \infty \Rightarrow C_1 = 0$

Thus equation (2) becomes

$$Lu(x, p) = C_2 e^{-px/C}$$

$\Rightarrow \qquad \qquad Lu(0,p) = \dfrac{Ea}{(a^2 + p^2)} = C_2$

$\Rightarrow \qquad \qquad Lu(x,p) = \left(\dfrac{Ea}{a^2 + p^2}\right)e^{-px/C}$

Taking inverse Laplace transformation of both sides, we have

$$\Rightarrow \qquad u(x,p) = EL^{-1}\left(\frac{e^{-px/C}a}{a^2 + p^2}\right)$$

$$\Rightarrow \qquad u(x,t) = E\sin\{a(t - x / C) \cdot H(t - x / C)\}$$

EXAMPLE 6. *A string is stretched between two fixed points* (0, 0) *and* (C, 0). *If it is displaced into the curve* $y = b\sin\left(\dfrac{n\pi x}{C}\right)$ *and released from rest in that position at time* $t = 0$, *solve the boundary value problem for the displacement* $y(x, t)$.

SOLUTION. The displacement $y(x, t)$ of the string is given by the differential equation,

$$\frac{\partial^2 y}{\partial t^2} = a^2 \frac{\partial^2 y}{\partial x^2} \qquad \qquad ...(1)$$

with initial conditions

$$y(x,0) = b\sin(n\pi x / C)\ 0 \le x \le C$$
$$y_t(x,0) = 0$$

and boundary condition are given by $y(0, t) = 0, y(C, t) = 0$.

Taking the Laplace transformation of both sides of equation (1), we have

$$L\left\{\frac{\partial^2 y}{\partial t^2}\right\} = a^2 L\left\{\frac{\partial^2 y}{\partial x^2}\right\}$$

$$\Rightarrow \qquad p^2 Ly(x,p) - py(x,0) - y_t(x,0) = a^2 \frac{d^2}{dx^2}Ly(x,p)$$

$$\Rightarrow \qquad a^2 \frac{d^2}{dx^2}L(y) - p^2 L(y) - pb\sin(n\pi x / C) = 0 \qquad ...(2)$$

$$[\because y_t(x, 0) = 0, y(x, 0) = b \sin (n\pi x/C)]$$

and also $\qquad Ly(0,p) = 0,\ \ Ly(C,p) = 0$. $\qquad \qquad ...(3)$

Thus the solution of equation (2) is

$$Ly(x,p) = C_1 e^{xp/a} + C_2 e^{-xp/a} + \{pb / (a^2 n^2 \pi^2 / c^2 + p^2)\}\sin\frac{n\pi x}{C}$$

$$...(4)$$

Now we have $Ly(0, p) = 0 \qquad \Rightarrow \qquad C_1 + C_2 = 0$

and $\qquad \qquad Ly(C, p) = 0 \qquad \Rightarrow \qquad C_1 e^{pc/a} + C_2 e^{-pc/a} = 0$

On solving, we get $C_1 = C_2 = 0$

After substituting these value in equation (4), we have

$$Ly(x,p) = b\{p / n^2\pi^2 a^2 / c^2 + p^2\}\sin\left(\frac{n\pi x}{C}\right)$$

Taking inverse Laplace transformation of both sides, we have

$$y(x,t) = bL^{-1}\frac{p}{(n^2\pi^2 a^2 / c^2 + p^2)}\sin\frac{n\pi x}{c}$$

$$\Rightarrow \qquad y(x,t) = b\cos\left(\frac{n\pi at}{C}\right)\sin\left(\frac{n\pi x}{C}\right)$$

EXAMPLE 7. *Solve the boundary value problem* $\dfrac{\partial^2 u}{\partial t^2} = a^2 \dfrac{\partial^2 u}{\partial x^2} - g, x > 0, t \le 0$ *with the boundary conditions*

$$u(x, 0) = 0 = u_t(x, 0) \; x \ge 0$$

$$u(0, t) = 0, \lim_{x \to \infty} u_x(x,t) = 0, t \ge 0$$

SOLUTION.   Taking Laplace transformation of both sides of the given equation, we have

$$L\left(\frac{\partial^2 u}{\partial t^2}\right) = a^2 L\left(\frac{\partial^2 u}{\partial x^2}\right) - gL(1)$$

$$\Rightarrow \quad p^2 Lu(x,p) - pu(x,0) - u_t(x,0) = a^2 \frac{d^2}{dx^2} L(u) - \frac{g}{p}$$

$$\Rightarrow \quad \frac{d^2}{dx^2} L(u) - \frac{p^2}{a^2} L(u) = \frac{g}{a^2 p} \qquad [\text{Given } u(x, 0) = u_t(x, 0) = 0] \qquad \dots(1)$$

and also $\qquad Lu(0, p) = 0$ and given $\lim\limits_{x \to \infty} \dfrac{d}{dx} L(u) = 0 \qquad \dots(2)$

The solution of equation (1) is

$$Lu(x,p) = Ae^{(p/a)x} + Be^{-(p/a)x} - \frac{g}{p^3} \qquad \dots(3)$$

From equation (2), we have

$$\lim_{x \to \infty} \frac{d}{dx} L(u) = 0$$

$$\Rightarrow \qquad A = 0, L\{u(x, p)\} = Be^{-(p/a)x} - \frac{g}{p^3}$$

$$\Rightarrow \qquad Lu(0, p) = B - \frac{g}{p^3} = 0 \Rightarrow B = \frac{g}{p^3}$$

Thus equation (3) becomes as

$$Lu(x,p) = -\frac{g}{p^3} + \frac{g}{p^3} e^{-(p/a)x}$$

Taking inverse Laplace transformation of both sides, we have

$$u(x,t) = -gL^{-1}\left(\frac{1}{p^3}\right) + gL^{-1}\left(\frac{e^{-(p/a)x}}{p^3}\right)$$

$$= -\frac{gt^2}{2} + g \begin{cases} \dfrac{1}{2}\left(t - \dfrac{x}{a}\right)^2, & \text{if } \; t \ge \dfrac{x}{a} \\[2mm] 0, & \text{if } \; t \le \dfrac{x}{a} \end{cases}$$

$$= \begin{cases} -\dfrac{g}{2a^2}(2axt - x^2), & \text{if } \; x \le at \\[2mm] -\dfrac{gt^2}{2}, & \text{if } \; x \ge at \end{cases}$$

EXAMPLE 8. *A tightly stretched flexible string has its ends fixed at x = 0 and x = l. At time t = 0, the string is given a shape defined by u = μx(l – x), when μ is constant and then released. Find the displacement of any point x of the string at any time t > 0 given that*

$$L^{-1}\left\{\frac{\cosh px}{p^3 \cosh pa}\right\} = \frac{1}{2}(t^2 + x^2 - a^2) - \frac{16a^2}{\pi^3}\frac{(-1)^n}{(2n-1)^3}$$

$$\left(\frac{\cos(2n-1)\pi x}{2a} \cdot \frac{\cos(2n-1)\pi t}{2a}\right)$$

SOLUTION. The displacement $u(x, t)$ of the string is given by

$$\frac{\partial^2 u}{\partial t^2} = a^2 \frac{\partial^2 u}{\partial x^2} \qquad\qquad …(1)$$

Initial condition are given by
$$u(x, 0) = \mu x(l - x) \quad \text{and} \quad u_t(x, 0) = 0$$
and the boundary conditions are
$$u(0, t) = 0 \qquad \text{and} \quad u(l, t) = 0, t > 0$$
Taking Laplace transformation of both sides of equation (1), we have

$$L\left(\frac{\partial^2 u}{\partial t^2}\right) = a^2 L\left(\frac{\partial^2 u}{\partial x^2}\right)$$

$$\Rightarrow \quad p^2 Lu(x, p) - pu(x, 0) - u_t(x, 0) = a^2 \frac{d^2}{dx^2} Lu(x, p)$$

$$\Rightarrow \quad a^2 \frac{d^2}{dx^2} L(u) - p^2 L(u) = -\mu p(l - x)x \quad \text{[Since } u(x, 0) = \mu x(l - x), u_t(x, 0) = 0]$$

and $Lu(0, p) = 0, Lu(l, p) = 0$ \qquad\qquad …(2)

The solution of equation (2), we have

$$Lu(x, p) = C_1 \cosh(px/a) + C_2 \sinh(px/a) + \left(\frac{\mu}{p}\right)x(l - x) - \frac{2a^2\mu}{p^3} \qquad …(3)$$

$$Lu(0, p) = 0 \quad \Rightarrow \quad C_1 - \frac{2a^2\mu}{p^3}$$

$$\Rightarrow \quad Lu(l, p) = 0 \quad \Rightarrow \quad C_1 = \cosh\left(\frac{pl}{a}\right) + C_2 \sinh\left(\frac{pl}{a}\right) - \frac{2a^2\mu}{p^3} = 0$$

On solving, we have

$$C_1 = \frac{2a^2\mu}{p^3} \quad \text{and} \quad C_2 = \frac{2a^2\mu}{p^3}\left(\frac{1 - \cosh pl/a}{\sinh pl/a}\right)$$

$$\Rightarrow \quad C_2 = -\frac{2a^2\mu}{p^3} \tanh\left(\frac{pl}{2a}\right)$$

Substituting in equation (3), we have

$$Lu(x, p) = \frac{2a^2\mu}{p^3}\left[\cosh\left(\frac{px}{a}\right) - \tanh\left(\frac{pl}{2a}\right)\sin\left(\frac{px}{a}\right)\right] + \frac{\mu}{p}x(l - x) - \frac{2a^2\mu}{p^3}$$

$$\Rightarrow \quad Lu(x, p) = \mu x(l - x)\frac{1}{p} - \frac{2a^2\mu}{p^3} + \frac{2a^2\mu}{p^3}\frac{\cosh(p(2x - l)/2a)}{\cosh(pl/2a)}$$

Taking inverse Laplace transformation of both sides, we have

$$L^{-1}\{L(u(x,p))\} = \mu x(l-x)L^{-1}\left(\frac{1}{p}\right) - 2a^2\mu L^{-1}\left(\frac{1}{p^3}\right)$$

$$+ 2a^2\mu L^{-1}\left(\frac{1}{p^3} \frac{\cosh\{p(2x-l)/2a\}}{\cosh(pl/2a)}\right)$$

$$\Rightarrow \quad u(x,t) = \mu x(l-x) - a^2\mu t^2 + 2a^2\mu \cdot \frac{1}{2}(t^2 + (2x-l)^2/4a^2 - l^2/4a^2)$$

$$- \frac{16}{\pi^3}\cdot\frac{l^2}{4a^2}\sum_{n=1}^{\infty}\frac{(-1)^n}{(2n-1)^3}\left[\frac{\cos\{(2n-1)\pi(2x-l)/2a\}}{2\left(\frac{l}{2a}\right)}\frac{\cos(2n-1)\pi t}{2(l/2a)}\right]$$

Hence, $u(x,t) = \dfrac{8\mu l^2}{\pi^3}\displaystyle\sum_{n=1}^{\infty}\frac{1}{(2n-1)^3}\sin\frac{(2n-1)\pi x}{l}\cdot\frac{\cos(2n-1)\pi a t}{l}$

EXAMPLE 9. *A cantilinear beam clampled at $x = 0$ and free at $x = l$ carries a uniform load ω_0 per unit length. Show that the deflection is $y(x) = \dfrac{\omega_0 x^2}{24EI}(x^2 - 4lx + 6l^2)$.*

SOLUTION. The length $y(x)$ at any point x satisfies the differential equation

$$\frac{d^4 y}{dx^4} = \frac{W_0}{EI}, 0 < x < l \qquad \qquad \dots(1)$$

and the boundary conditions are

$$y(0) = 0, y'(0) = 0 \text{ at } x = 0$$

$$y''(l) = 0, y'''(l) = 0 \text{ at end } x = l \text{ is free}$$

Taking Laplace transformation of both sides of equation (1), we get

$$p^4 L\{y(p)\} - p^3 y(0) - p^2 y'(0) - p y''(0) - y'''(0) - \frac{W_0}{(EIp)}$$

$$\Rightarrow \qquad p^4 Ly(p) = Ap + B + \frac{W_0}{EIp}.$$

[Since $y(0) = 0, y'(0) = 0$ and let $A = y''(0), y'''(0) = B$]

$$\Rightarrow \qquad Ly(p) = \frac{A}{p^3} + \frac{B}{p^4} + \frac{W_0}{EIp^5} \qquad \qquad \dots(2)$$

Taking the inverse Laplace transform, we have

$$\Rightarrow \qquad y(x) = \frac{Ax^2}{2} + \frac{Bx^3}{6} + \frac{W_0 x^4}{24EI}$$

$$\Rightarrow \qquad y'(x) = Ax + \frac{Bx^2}{2} + \frac{W_0 x^3}{6EI}, y'' = A + Bx + \frac{W_0 x^2}{2EI}$$

$$\Rightarrow \qquad y'''(x) = B + \frac{W_0 x}{EI}$$

Since given $\quad y''(l) = 0, y'''(l) = 0$

$$\Rightarrow \qquad y''(l) = 0 = A + Bl + \frac{W_0 l^2}{2EI} \qquad \text{and} \qquad y'''(l) = 0 = B + \frac{W_0 l}{EI}$$

On solving these, we have

$$A = \frac{W_0 l^2}{2EI}, B = -\frac{W_0 l}{EI}$$

On substituting these value in equation (2), we have

$$y(x) = \frac{W_0 l^2 x^2}{4EI} - \frac{W_0 l x^3}{6EI} + \frac{W_0 x^4}{24EI}$$

$$\Rightarrow \qquad y(x) = \frac{W_0 x^2}{24EI}(6l^2 - 4lx + x^2)$$

$$\Rightarrow \qquad y(x) = \frac{W_0 x^2}{24EI}(x^2 - 4lx + 6l^2)$$

EXAMPLE 10. *An inductor of 2 henrys, a resistance of 16 ohm and a capacitor of 0.02 farads are connected in series with an E.M.F. of 100 sin 3t volts. At t = 0 the charge on the capacitor and the current in the circuit are zero. Find the charge and current at any time t > 0.*

SOLUTION. It is given that $R = 16$ ohms, $C = 0.02$ farads, $L = 2$ henrys, $E = 100 \sin 3t$,

$$\frac{dQ}{dt} = I = 0 \text{ at } t = 0$$

Here, we have

$$L\frac{d^2Q}{dt^2} + \frac{RdQ}{dt} + \frac{Q}{C} = E$$

$$\Rightarrow \qquad 2\frac{d^2Q}{dt^2} + 16\frac{dQ}{dt} + \frac{Q}{0.02} = 100 \sin 3t$$

$$\Rightarrow \qquad 2\frac{d^2Q}{dt^2} + 16\frac{dQ}{dt} + \frac{100}{2}Q = 100 \sin 3t$$

$$\Rightarrow \qquad \frac{d^2Q}{dt^2} + 8\frac{dQ}{dt} + 25Q = 50 \sin 3t$$

$$\Rightarrow \qquad Q'' + 8Q' + 25Q = 50 \sin 3t$$

$$\Rightarrow \qquad L\{Q''\} + 8L\{Q'\} + 25L\{Q\} = 50L\{\sin 3t\}$$

$$\Rightarrow \qquad \{p^2\bar{Q} - pQ(0) - Q'(0)\} + 8\{p\bar{Q} - Q(0)\} + 25\bar{Q} = \frac{50 \times 3}{p^2 + 9}$$

Now, substituting $Q(0) = 0 = Q'(0)$, then

$$p^2\bar{Q} + 8p\bar{Q} + 25\bar{Q} = \frac{150}{p^2 + 9}$$

$$\Rightarrow \qquad \bar{Q} = \frac{150}{(p^2 + 9)(p^2 + 8p + 25)} = \frac{75}{52}\left[\frac{-p+2}{p^2+9} + \frac{p+6}{p^2+8p+25}\right]$$

$$= \frac{75}{52}\left[\frac{-p}{p^2+9} + \frac{2}{p^2+9} + \frac{p+4}{(p+4)^2+9} + \frac{2}{(p+4)^2+9}\right]$$

$$Q = L^{-1}\{\bar{Q}\} = \frac{75}{52}\left[-\cos 3t + \frac{2}{3}\sin 3t + e^{-4t}\cos 3t + \frac{2}{4}e^{-4t}\sin 3t\right]$$

$$= \frac{75}{52 \times 3}[(2\sin 3t - 3\cos 3t) + e^{-4t}(2\sin 3t + 3\cos 3t)]$$

$$\therefore \quad Q = \frac{25}{52}[(2\sin 3t - 3\cos 3t) + e^{-4t}(2\sin 3t + 3\cos 3t)]$$

$$I = \frac{dQ}{dt} = \frac{25}{52}[3(2\cos 3t + 3\sin 3t) + e^{-4t}\{-4(2\sin 3t + 3\cos 3t)$$

$$+ 3(2\cos 3t - 3\sin 3t)\}]$$

$$= \frac{25}{52}[3(2\cos 3t + 3\sin 3t) - e^{-4t}(17\sin 3t + 6\cos 3t)]$$

EXAMPLE 11. *An alternating E.M.F. E sin ωt is applied to an inductance and a capacitance C in series.*
Show that the current in the circuit is $\dfrac{E\omega}{(n^2 - \omega^2)L}(\cos\omega t - \cos nt)$ *where* $n^2 = 1/EC$.

SOLUTION.   Consider,

$$L\frac{dI}{dt} + \frac{Q}{C} = E\sin\omega t \qquad \qquad ...(1)$$

where,

$$I = \frac{dQ}{dt} \qquad \qquad ...(2)$$

Now, taking Laplace transform of eq^n (1), we get

$$L[p\bar{I} - I(0)] + \frac{1}{C}\bar{Q} = \frac{E\omega}{p^2 + \omega^2}$$

But

$$I(0) = 0$$

Since,

$$Lp\bar{I} + \frac{1}{C}\bar{Q} = \frac{E\omega}{p^2 + \omega^2} \qquad \qquad ...(3)$$

Here,

$$L\{I\} = L\left\{\frac{dQ}{dt}\right\}$$

$$\Rightarrow \qquad \bar{I} = p\bar{Q} - Q(0), \text{But } Q(0) = 0$$

Thus,

$$\bar{I} = p\bar{Q} \qquad \qquad [\text{From eq}^n (3)]$$

$$Lp\bar{I} + \frac{1}{Cp}\bar{I} = \frac{E\omega}{p^2 + \omega^2}$$

$$\Rightarrow \qquad \bar{I} = \frac{E\omega}{(p^2 + \omega^2)\left(Lp + \dfrac{1}{Cp}\right)}$$

$$Lp + \frac{1}{Cp} = (LCp^2 + 1)\cdot\frac{1}{Cp}$$

$$= \frac{1}{Cp}\left(\frac{p^2}{n^2} + 1\right) = \frac{(p^2 + n^2)}{n^2 pC} = \frac{(p^2 + n^2)L}{p}$$

$$\bar{I} = \frac{E\omega p}{(p^2 + \omega^2)(p^2 + n^2)L}$$

$$= \frac{E\omega}{(n^2 - \omega^2)L}\left\{\frac{p}{p^2 + \omega^2} - \frac{p}{p^2 + n^2}\right\}$$

Now, taking inverse Laplace transform

$$I = \frac{E\omega}{L(n^2 - \omega^2)}\{\cos \omega t - \cos nt\}$$

EXAMPLE 12. An inductor of 3 henrys is in series with a resistance of 30 ohms and an e.m.f. of 150 volts. Assuming that t = 0, the current is zero, find the current at time t > 0.

SOLUTION. It is given that, L = 3 henrys, R = 30 ohms, E = 150 volts and I = 0 at t = 0. Then

$$L\frac{d^2Q}{dt^2} + R\frac{dQ}{dt} = E \qquad \qquad ...(1)$$

Putting all the values in eq^n (1), we get

$$3\frac{d^2Q}{dt^2} + 30\frac{dQ}{dt} = 150$$

$$\Rightarrow \qquad 3Q'' + 30Q' = 150$$

$$\Rightarrow \qquad 3L\{Q''\} + 30L\{Q'\} = 150L\{1\}$$

$$\Rightarrow \quad 3\{p^2\bar{Q} - pQ(0) - Q'(0)\} + 30\{p\bar{Q} - Q(0)\} = \frac{150}{p} \qquad \qquad ...(2)$$

Given that at t = 0 $\Rightarrow$ I = 0, then

$$\frac{dQ}{dt} = Q' = I = 0 \text{ at } t = 0, \, Q(0) = 0$$

Now, eq^n (2), becomes

$$3\{p^2\bar{Q}\} + 30p\bar{Q} = \frac{150}{p} \Rightarrow \bar{Q} = \frac{50}{p^2(p+10)}$$

$$\Rightarrow \qquad Q = 50L^{-1}\left\{\frac{1}{p^2(p+10)}\right\} \qquad \qquad ...(3)$$

$$\Rightarrow \qquad L^{-1}\left\{\frac{1}{p+10}\right\} = e^{-10t}$$

$$L^{-1}\left\{\frac{1}{p(p+10)}\right\} = \int_0^t e^{-10u}du = \frac{1}{10}(1 - e^{-10t})$$

Again $\qquad L^{-1}\left\{\frac{1}{p^2(p+10)}\right\} = \int_0^t \frac{1}{10}(1 - e^{-10u})du = \frac{1}{10}\left(t + \frac{e^{-10t} - 1}{10}\right)$

Now, eq^n (3) becomes

$$Q = 50\left(t + \frac{e^{-10t} - 1}{10}\right)$$

$$I = \frac{dQ}{dt} = 50\left[1 + \frac{-10e^{-10t} - 0}{10}\right]$$

$$= 50(1 - e^{-10t})$$

EXAMPLE 13. *A mass m moves along the x-axis under the influence of a force which is proportional to its instantaneous speed and in a direction opposite to the direction of motion. Assuming that t = 0, the particle is located at x = a and moving to the right with speed V_0, find the position where the mass comes to rest.*

SOLUTION. The equation of motion is

$$m\frac{d^2x}{dt^2} = -\mu\frac{dx}{dt}$$

$$\Rightarrow \qquad mx'' = -\mu x' \qquad \qquad ...(1)$$

where boundary conditions are
$$x(0) = a, x'(0) = V_0 \qquad \qquad ...(2)$$

From eq^n(1), we get

$$mL\{x''\} + \mu L\{x'\} = 0$$

$$\Rightarrow \quad m(p^2\bar{x} - px(0) - x'(0)) + \mu\{p\bar{x} - x(0)\} = 0$$

$$\Rightarrow \qquad m(p^2\bar{x} - pa - V_0) + \mu\{p\bar{x} - a\} = 0$$

$$\Rightarrow \qquad \bar{x} = \frac{mV_0 + \mu a + mpa}{mp^2 + \mu p} = \frac{mV_0 + \mu a + mpa}{mp(p + \mu / m)}$$

$$= \frac{mV_0 + \mu a}{mp(p + \mu / m)} + \frac{a}{p + \mu / m}$$

$$= \left(\frac{mV_0 + \mu a}{m}\right)\left\{\frac{m}{\mu} - e^{-\mu t/m}\right\} + ae^{-\mu t/m}$$

Now, taking inverse Laplace transform, we get

$$x = \left(\frac{mV_0 + \mu a}{m}\right)\left\{\frac{m}{\mu} - e^{-\mu t/m}\right\} + ae^{-\mu t/m} \qquad ...(3)$$

$$\Rightarrow \qquad \frac{dx}{dt} = \left[\frac{\mu(mV_0 + \mu a)}{m^2} - \frac{a\mu}{m}\right]e^{-\mu t/m}$$

$$\Rightarrow \quad \text{If } \frac{dx}{dt} = 0 \text{, then}$$

$$\left[\frac{\mu(mV_0 + \mu a)}{m^2} - \frac{a\mu}{m}\right]e^{-\mu t/m} = 0$$

$$\Rightarrow \qquad \frac{mV_0 + a\mu}{m} - a = 0$$

Eq^n (3) becomes

$$x = a\left\{\frac{m}{\mu} - e^{-\mu t/m}\right\} + ae^{-\mu t/m} = \frac{am}{\mu} = \frac{mV_0 + a\mu}{\mu} \quad \text{[From eq}^n \text{ (4)]}$$

If $\frac{dx}{dt} = 0$ then $x = \frac{mV_0}{\mu} + a$

Thus, the mass comes to rest at a distance $\frac{mV_0}{\mu} + a$ from the centre O.

EXAMPLE 14. *A mass of 2 grams moves on the x-axis and is attracted towards origin O with a force numerically equal to 8x. If it is initially at rest at x = 10, find its position at any time t. Also a damping force numerically equal to 8 times the instantaneous velocity acts.*

SOLUTION. It is given that $m = 2$, Resistance $= 8x$, $x(0) = 10$, $x'(0) = 0$

and damping force $= 8\dfrac{dx}{dt}$

Now, the equation of motion is $2\dfrac{d^2x}{dt^2} = -8x - 8\dfrac{dx}{dt}$

$\Rightarrow$ $\qquad x'' + 4x + 4x' = 0$

$\Rightarrow$ $\qquad L\{x''\} + 4L\{x\} + 4L\{x'\} = 0$

$\Rightarrow$ $\quad p^2\overline{x} - px(0) - x'(0) + 4\overline{x} + 4\{p\overline{x} - x(0)\} = 0$

Now, putting $x'(0) = 0$, $x(0) = 10$, then we get

$\qquad p^2\overline{x} - 10p + 4\overline{x} + 4p\overline{x} - 40 = 0$

$\Rightarrow$ $\qquad\qquad (p+2)^2\overline{x} = 40 + 10p$

$$\overline{x} = \frac{40}{(p+2)^2} + \frac{10p}{(p+2)^2} = \frac{40}{(p+2)^2} + \frac{10(p+2) - 20}{(p+2)^2}$$

$$x = L^{-1}\left\{\frac{40}{(p+2)^2} + \frac{10}{(p+2)} - \frac{20}{(p+2)^2}\right\} = L^{-1}\left\{\frac{20}{(p+2)^2} + \frac{10}{(p+2)}\right\}$$

$$= 20e^{-2t}L^{-1}\left\{\frac{1}{p^2}\right\} + 10e^{-2t} = (20t + 10)e^{-2t}$$

EXAMPLE 15. *Find the solution of the equation*

$$\frac{\partial u}{\partial t} = k\frac{\partial^2 u}{\partial x^2} \qquad\qquad\qquad ...(1)$$

which tends to zero as $x \to \infty$ and which satisfies the conditions

$\qquad u = f(t)$, *when* $x = 0$, $t > 0$

$\qquad u = 0$, *when* $x > 0$, $t = 0$

SOLUTION. Taking Laplace transform of eq^n (1), we get

$$p\overline{u} - u(x,0) = k\frac{d^2\overline{u}}{dx^2}$$

Here, $\qquad\qquad u(x, 0) = 0$

$\therefore$ $\qquad\qquad \dfrac{d^2\overline{u}}{dx^2} - a^2\overline{u} = 0$

$\Rightarrow$ $\qquad (D^2 - a^2)\overline{u} = 0$ $\quad$ where $\quad a^2 = \dfrac{p}{k}$

The solution of the equation is

$$\overline{u} = Ae^{-ax} + Be^{ax}$$

But given that $u \to 0$ as $x \to \infty$. Therefore $\overline{u} \to 0$ as $x \to \infty$

Which gives $B = 0$. Hence, $\overline{u} = Ae^{-ax}$

$\qquad u = f(t), x = 0$ $\quad\Rightarrow\quad$ $\overline{u} = \overline{f}(p), x = 0$

Now, subjecting (1) to condition (2), we get

$$\overline{f}(p) = A$$

$$\overline{u} = \overline{f}(p)e^{-ax} = \overline{f}(p)e^{-x(p/k)^{1/2}} dp$$

Now, applying inversion formula of Laplace, we get

$$u(x,t) = \frac{1}{2\pi i} \int_{\gamma-i\infty}^{\gamma+i\infty} e^{pt} \overline{f}(p) e^{-x(p/k)^{1/2}} dp$$

EXAMPLE 16. *Solve the boundary value problem*

$$\frac{\partial^2 u}{\partial t^2} = a^2 \frac{\partial^2 u}{\partial x^2} - g, (x > 0, t > 0) \qquad \ldots(1)$$

with the boundary conditions :

$$u(x, 0) = 0 = u_t(x, 0), x \geq 0$$

$$u(0,t) = 0, \lim_{x \to \infty} u_x(x,t) = 0, t \geq 0$$

SOLUTION.   Taking Laplace transform of eq^n (1), we get

$$L\left\{\frac{\partial^2 u}{\partial t^2}\right\} = a^2 L\left\{\frac{\partial^2 u}{\partial x^2}\right\} - gL\{1\}$$

$$\Rightarrow \quad p^2 u(x,p) - pu(x,0) - u_t(x,0) = a^2 \frac{d^2 u}{dx^2} - \frac{g}{p}$$

Using boundary conditions, we get

$$\Rightarrow \quad p^2 u(x,p) = a^2 \frac{d^2 u}{dx^2} - \frac{g}{p}$$

$$\Rightarrow \quad (D^2 - \frac{p^2}{a^2})\overline{u}(x,p) = \frac{g}{pa^2}$$

$$\Rightarrow \quad \overline{u} = Ae^{-px/a} + Be^{px/a} + \frac{1}{D^2 - \frac{p^2}{a^2}} \cdot \frac{g}{pa^2} \qquad \ldots(2)$$

But $\frac{1}{D^2 - \frac{p^2}{a^2}} \cdot \frac{g}{pa^2} = \frac{g}{pa^2} \cdot \frac{1}{-(p^2/a^2)} \left[1 - \frac{a^2}{p^2}D^2\right]^{-1} \cdot 1 = \frac{-g}{p^3}\left[1 + \frac{a^2}{p^2}D \ldots\right], 1 = \frac{-g}{p^3}$

Equation (1) becomes

$$\overline{u}(x,p) = Ae^{-p(x/a)} + Be^{(px/a)} - \frac{g}{p^3} \qquad \ldots(3)$$

$$u(0,t) = 0 \rightarrow L\{u(0,t)\} = L\{0\} \Rightarrow \overline{u}(0,p) = 0 \qquad \ldots(4)$$

$$\lim_{x \to \infty} u_x(x,t) = 0 \Rightarrow \overline{u}_x(x,p) = 0 \text{ if } x \to \infty \qquad \ldots(5)$$

From eq^n (3) and (4), we get $0 = A + B - \frac{g}{p^3}$

Eq^n (3) gives $\overline{u}(x,p) = \frac{g}{p^3}e^{-px/a} - \frac{g}{p^3}$

$$\Rightarrow \qquad u = L^{-1}\left\{\frac{g}{p^3}e^{-px/a}\right\} - gL^{-1}\left\{\frac{1}{p^3}\right\}$$

$$u = -gL^{-1}\left\{\frac{1}{p^{2+1}}\right\} + gL^{-1}\left\{\frac{e^{-px/a}}{p^{2+1}}\right\} = -g\frac{t^2}{2} + g\cdot\frac{1}{2}\left(t-\frac{x}{a}\right)^2 \cdot H\left(t-\frac{x}{a}\right)$$

$$\text{where} \quad H\left(t-\frac{x}{a}\right) = \begin{cases} 1 & \text{if} \quad t > \dfrac{x}{a} \\ 0 & \text{if} \quad t < \dfrac{x}{a} \end{cases}$$

MISCELLANEOUS EXERCISE

1. Solve $ty'' + y' + ty = 0, y(0) = 1, y'(0) = 0$
(Agra–2001, 05)

2. Solve $y''(t) + y(t) = 4e^t, y(0) = 0 = y'(0)$
(Kanpur–2010)

3. Solve $(D^3 + D)x = 2, x_0 = 3, x_1 = 1, x_2 = -2$

4. Solve $(D^2 - D - 2)y = 20 \sin 2t, y = -1,$ $Dy = 2$ at $t = 0$.

5. $\dfrac{d^2y}{dt^2} + y = 0, y = 1, \dfrac{dy}{dt} = 0$ when $t = 0$.
(Meerut–1985)

6. Solve $(D^2 + D)y = t^2 + 2t$ if $y(0) = 4, y'(0) = -2$.
(Agra–1989, 95, 2005, Rohilkhand–2001
MDU-2007)

7. Solve $\dfrac{d^3y}{dt^3} - \dfrac{3d^2y}{dt^2} + \dfrac{3dy}{dt} - y = t^2e^t$ where
$y(0) = 1, \left(\dfrac{dy}{dt}\right)_{t=0} = 0, \left(\dfrac{d^2y}{dt^2}\right)_{t=0} = -2$
(UPTU–2008; SVTU–2009)

8. Voltage Ee^{-at} is applied at $t = 0$ to a circuit of inductance L and resistance R. Show that the current at time t is $\dfrac{E}{R-aL}(e^{-at} - e^{-Rt/L})$
(UPTU–2007; VTU–2000)

9. Solve $y'' + 4y' + 3y = e^{-t}, y(0) = y'(0) = 1$.
(VTU–2008; Kurukshetra–2005)

10. Solve $y'' + y = t, y(0) = 1, y'(0) = 0$.
(Mumbai–2009)

11. Solve $y'' - 3y' + 2y = e^{3t}$ when $y(0) = 1$ and $y'(0) = 0$.
(VTU–2010)

12. Solve $(D^2 - 3D + 2)y = 4e^{2t}$ with $y(0) = 3,$ $y(0) = 5$.
(Mumbai–2008)

13. Solve $y'' + 25y = 10 \cos 5t$ given that $y(0) = 2,$ $y''(0) = 0$.
(Kanpur–2004, 06; SVTU–2008)

14. Solve $\dfrac{d^2y}{dt^2} + \dfrac{2dy}{dt} - 3y = \sin t, y = \dfrac{dy}{dt} = 0,$
when $t = 0$.
(Kurukshetra–2005; Madras–2003; MDU-2007)

15. Solve $y'' + 2y' + 5y = 5(t - 2), y(0) = 0,$ $y'(0) = 0$.
(PTU–2005)

16. Solve $\dfrac{d^2y}{dt^2} + 9x = \sin 2t, x(0) = 1, x'(0) = 0$.
(GBTU–2011)

17. Solve $\dfrac{d^2y}{dt^2} + 9x = \sin 3t,$ given $y = 0, \dfrac{dy}{dt} = 0,$ $x'(0) = 0$.
(MTU–2012)

18. Solve $\dfrac{d^2y}{dx^2} + 6\dfrac{dx}{dt} + 8x = e^{-3t} - e^{-5t}, x(0) = 0,$ $x'(0) = 0$.
(UPTU–2009)

19. Solve $y'' + 3y' + 2y = te^{-t}; y(0) = 1, y'(0) = 0$.
(GBTU–2012)

20. Solve $y'' + 2y' + y = te^{-t}; y(0) = 1, y'(0) = 2$.
(MTU–2011)

21. Solve $\dfrac{d^2x}{dt^2} + 3\dfrac{dx}{dt} + 2x = r(t)$
where $r(t) = \begin{cases} e^t, & 0 < t < 2 \\ 0, & t > 2 \end{cases}$ and $x(0) = 1,$ $x'(0) = -2$.
(GBTU–2010)

ANSWERS

1. $y = J_0(t)$

2. $y(t) = 2(e^t - \cos t - \sin t)$

3. $x = 1 + 2t + 2\cos t - \sin t$

4. $y = 2e^{2t} - 4e^{-t} + \cos 2t - \dfrac{t}{3}\cos 2t$

5. $y = \cos t$

6. $y = \dfrac{1}{3}t^3 + 2 + 2e^{-t}$

7. $y = \left(1 - t - \dfrac{t^2}{2} + \dfrac{t^5}{60}\right)e^t$

9. $y = \dfrac{7}{4}e^{-t} - \dfrac{3}{4}e^{-3t} - \dfrac{1}{2}te^{-t}$

10. $y = t - 3\sin t + \cos t$

11. $y = 2t + 3 + \dfrac{1}{2}(e^{3t} - e^t) - 2e^{2t}$

12. $y = 4e^{2t}(1 + t) - 7e^t$

13. $y = 2\cos 5t + t\sin 5t$

14. $y = \dfrac{1}{8}e^t - \dfrac{1}{40}e^{-3t} - \dfrac{1}{10}(2\sin t + \cos t)$

15. $y = -\dfrac{12}{5} + \dfrac{12}{5}e^{-t}\cos 2t + \dfrac{7}{10}e^{-t}\sin 2t$

16. $x = \cos 3t + \dfrac{1}{5}\sin 2t - \dfrac{2}{15}\sin 3t$

17. $y = \dfrac{1}{18}(\sin 3t - 3t\cos 3t)$

18. $x = \dfrac{1}{3}(e^{-2t} - e^{-5t}) - e^{-3t} + e^{-4t}$

19. $y = 3e^{-t} - 2e^{-2t} + e^{-t}\left(\dfrac{t^2}{2} - t\right)$

20. $y = e^{-t}\left(1 + 3t + \dfrac{t^3}{6}\right)$

21. $x = \dfrac{4}{3}e^{-2t} + \dfrac{1}{6}e^t[1 - u(t-2)] - \dfrac{1}{2}e^{-t} + \dfrac{1}{2}e^{4-t}u(t-2) - \dfrac{1}{3}e^{6-2t}u(t-2)]$

Self Assessment Test

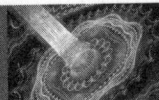

1. Solve $\dfrac{d^2y}{dt^2} + 9y = r(t)$ with initial conditions

   $y(0) = 0$ and $y'(0) = 4$ where

   $$r(t) = \begin{cases} 8\sin t, & 0 < t < \pi \\ 0, & t > \pi \end{cases} \qquad \text{(GBTU–2011)}$$

2. A particle moves in a line so that its displacement x from a fixed point O at any time t, is given by

   $$\dfrac{d^2x}{dt^2} + 4\dfrac{dx}{dt} + 5x = 80\sin 5t$$

   Using Laplace transform, find its displacement at any time t if initially particle is at rest at $x = 0$.
   (UPTU–2009)

3. An alternating E.M.F. $E\sin\omega t$ is applied to circuit with an inductance L and a capacitance C in series. Show that the current in the circuit is

   $$\dfrac{E\omega}{(n^2 - \omega^2)L}(\cos\omega t - \cos nt) \text{ where } n^2 = \dfrac{1}{LC}.$$

   (GBTU–2010)

4. Using $\dfrac{dy}{dt} + 2x = \sin 2t$, $\dfrac{dx}{dt} - 2y = \cos 2t$ $(t > 0)$

   such that $t = 0, x = 1$ and $y = 0$. Show that the particle moves along the curve $4x^2 + 4xy + 5y^2 = 4$.
   (UPTU–2003; MTU–2011)

5. Solve $\dfrac{dx}{dt} + y = \sin t$, $\dfrac{dy}{dt} + x = \cos t$, using $x = 2$, $y = 0$ at $t = 0$.

   (UPTU–2004; GBTU–2012; Kerala–2005)

6. Solve $\dfrac{dx}{dt} + 4\dfrac{dy}{dt} - y = 0$; $\dfrac{dx}{dt} + 2y = e^{-t}$ with

   $x(0) = y(0) = 0$. (UPTU–2008)

7. A body falls from rest in a liquid whose density is one fourth that of the body. If the liquid offers resistance proportional to the velocity and the velocity approaches a limiting value of 9 meter/sec, find the distance falled in 5 seconds. (UPTU–2001)

8. A beam of length l is clamped horizontally at its ends $x = 0$ and is free at the end $x = l$. A point load ω is applied at the end $x = l$ in addition to a uniform load ω per unit length

from $x = 0$ to $x = l/2$. Find the deflection of the beam. (UPTU–2001)

9. An electric circuit consists of an inductor of L henrys in series with a capacitor C farads. At $t = 0$, an E.M.F. given by

   $$E(t) = \begin{cases} tE_0 / T_0, & 0 < t < T_0 \\ 0, & t > 0 \end{cases}$$

   is applied. Assuming that the current and charge on the capacitor are zero at $t = 0$, find the charge at any time $t > 0$.

10. An iron ball of mass m falls from rest under gravity in a liquid which resists the motion with a force mk times the velocity. Determine the motion.

11. Constant voltage E is applied at $t = 0$ to a circuit with an inductance L, capacitance C and resistance R. Find the current I at time t, if the initial current and charge are zero.

12. A bar of length l is at constant temperature u_0. At $t = 0$ the end $x = l$ is suddenly given the constant temperature u_1 and the end $x = 0$ is insulated. Assuming that the surface of the bar is insulated, find the temperature at any point x of the bar at any time $t > 0$, given that

   $$L^{-1}\left\{\dfrac{\cosh(x\sqrt{p})}{p\cosh(a\sqrt{p})}\right\} = 1 + \dfrac{4}{\pi}\sum_{n=1}^{\infty}\dfrac{(-1)^n}{2n-1}$$

   $$e^{-(2n-1)^2\pi^2 t/4a^2} \times \cos\left(\dfrac{2n-1}{2a}\right)\pi x$$

13. Solve the boundary value problem

    $$\dfrac{\partial^2 u}{\partial t^2} = a^2\dfrac{\partial^2 u}{\partial x^2} + g, t > 0, 0 < x < l$$

    with $u = \dfrac{\partial u}{\partial t} = 0$ when $t = 0, 0 < x < l$

    $u = 0$ when $x = 0, t > 0$

    given that $L^{-1}\left\{\dfrac{\cosh px}{p^3\cosh pl}\right\} = \dfrac{1}{2}(t^2 + x^2 - l^2)$

    $$-\dfrac{16l^2}{\pi^3}\sum_{n=1}^{\infty}\dfrac{(-1)^n}{(2n-1)^3}\cos\left\{\dfrac{(2n-1)\pi x}{2l}\right\}$$

    $$\cos\left\{\dfrac{(2l-1)\pi t}{2l}\right\}$$

14. Find the solution of $\dfrac{\partial u}{\partial t} = \dfrac{\partial^2 u}{\partial x^2}$, which tends to zero as $x \to \infty$ and which satisfies the conditions

  (i) $u = f(t)$ when $x = 0, t > 0$

  (ii) $u = 0$ when $x > 0, t = 0$

15. An inductor of 3 henrys is in series with a resistance of 30 ohms and an E.M.F. of 150 sin 20t volts. Assuming that $t = 0$, the current is zero, find the current at time $t > 0$.

16. A string is stretched between two fixed points $(0, 0)$ and $(C, 0)$. If it is displaced into the curve $y = b \sin(\pi x/C)$ and released from rest in the position at time $t = 0$, sut up and solve the boundary value problem for the displacement $y(x, t)$.

17. A centilever beam, clamped at $x = 0$ and free at $x = l$, carries a uniform load ω_0 per unit length. Show that the deflection is

$$y(x) = \frac{W_0 x^2}{24EI}(x^2 - 4lx + 6l^2).$$

18. Solve $\dfrac{\partial^2 y}{\partial t^2} = a^2 \dfrac{\partial^2 y}{\partial x^2}$, with conditions $y = 0$ when $t = 0$ for $x \geq 0$, $\dfrac{dy}{dt} = 0$ when $t = 0$ for $x > 0$, $y = 0$ when $x = 0$ for $t \geq 0$, $y(x, t) = 0$, when $x \to \infty$ for $t \geq 0$.

--- ANSWERS ---

1. $y = \sin 3t + \sin t + [\sin(t - \pi) - \frac{1}{3}\sin 3(t - \pi)]u(t - \pi)$

2. $x = e^{-2t}(2\cos t + 14 \sin t) - 2\cos 5t - 2\sin 5t$

5. $x = e^{-t} + e^t, y = \sin t + e^{-t} - e^{-t}$

6. $x = -\dfrac{5}{7}e^{-t} + \dfrac{8}{21}e^{3t/4} + \dfrac{1}{4}; y = \dfrac{1}{7}e^{-t} - \dfrac{1}{7}e^{3/4 \cdot t}$

7. $x = 34.17$ meters

8. $y = \dfrac{1}{24EI}\left[\omega x^2\left(x^2 - 2lx + \dfrac{3l^2}{2}\right) - \omega\left(x - \dfrac{l}{2}\right)^4 H\left(x - \dfrac{l}{2}\right) - 4\omega x^2(3l - x)\right]$

9. $Q = \dfrac{CE_0}{T_0}\left[T_0 \cos\left(\dfrac{t - T_0}{\sqrt{CL}}\right) + (CL)^{1/2}\sin\left(\dfrac{t - T_0}{\sqrt{LC}}\right) - (CL)^{1/2}\sin\left(\dfrac{t}{\sqrt{CL}}\right)\right]$ if $t > T_0$

10. $\dfrac{dx}{dt} = \dfrac{ag}{k}(1 - e^{-kt})$

11. $I = \begin{cases} \dfrac{Et}{L}e^{-Rt/2L} & \text{if } R^2 = \dfrac{4L}{C} \\[2mm] \dfrac{E}{L}e^{-Rt/2} \cdot \dfrac{1}{n}\sin nt & \text{if } R^2 < \dfrac{4L}{C} \\[2mm] \dfrac{E}{Lk}e^{-Rt/2L} \cdot \sinh(kt) & \text{if } R^2 > \dfrac{4L}{C} \end{cases}$

12. $u = u_1 + (u_1 - u_0)\dfrac{4}{\pi}\sum_{n=1}^{\infty}\dfrac{(-1)^n}{2n - 1}e^{-(2n-1)^2\pi^2 kt/4t^2}\cos\dfrac{(2n - 1)\pi x}{2l}$

13. $u = \dfrac{g}{2}\left[\dfrac{l^2}{a^2} - \dfrac{(l - x)^2}{a^2}\right] + \dfrac{16l^2 g}{\pi^3}\sum_{n=1}^{\infty}\dfrac{(-1)^n}{(2n - 1)^3}\cos\left\{\dfrac{(2n - 1)\pi(l + x)}{2l}\right\}\cos\left\{\dfrac{(2n - 1)\pi ta}{2l}\right\}$

14. $u = \dfrac{1}{2\pi i}\int_{\gamma - i\infty}^{\gamma + i\infty} e^{pt}\overline{f}(p)e^{-x\sqrt{p}}\,dp$

15. $\sin 20t - 2\cos 20t + 2e^{-10t}$

16. $y(x, t) = b\cos\left(\dfrac{\pi at}{C}\right) \cdot \sin\left(\dfrac{\pi x}{C}\right)$

18. $\left(t - \dfrac{x}{a}\right)H\left(t - \dfrac{x}{a}\right)$

Objective Evaluations

❖ Fill in the Blanks

1. The solution of $y''(t) + y(t) = t$ if $y'(0) = 1$ and $y(\pi) = 0$ is _____.

2. If $y = y(x, t)$, then $L\left\{\dfrac{\partial y}{\partial t}\right\} =$ _____

3. $L\{y(t)\} =$ _____.

4. The solution of $\dfrac{d^2 y}{dt^2} + y = 0$ under the condition

that $y = 1, \dfrac{dy}{dt} = 0$, when $t = 0$ is _____.

5. The solution of $(tD^2 + D + 4t)y = 0$, if $y(0) = 3, y'(0) = 0$ is _____.

6. The solution of $(D - 2)x + 3y = 0$; $2x + (D - 1)y = 0$, if $x(0) = 8$ and $y(0) = 3$ is _____.

7. $\int Q e^{\int P dx} dx = y$ _____.

❖ True or False

Write 'T' for True and 'F' for False statement.

1. The solution of $\dfrac{\partial^2 u}{\partial x^2} - \dfrac{\partial^2 u}{\partial t^2} = xt$ if $u = \dfrac{\partial u}{\partial t} = 0$

when $t = 0$ is $u = \dfrac{-xt^3}{6}$. **(T/F)**

2. If $y = y(x, t)$ then

$L = \left\{\dfrac{\partial^2 y}{\partial t^2}\right\} = p\bar{y}(x, p) - py(x, 0) - y_t(x, 0).$

(T/F)

3. $\int_0^\infty e^{-pt} \dfrac{\partial y}{\partial x} dt = L\left\{\dfrac{\partial^2 y}{\partial x^2}\right\}$ **(T/F)**

4. $L\left\{\dfrac{\partial^2 y}{\partial x^2}\right\} = \dfrac{d^2 \bar{y}}{dx^2}$ **(T/F)**

5. The solution of $ty'' + y' + 4ty = 0, y(0) = 3,$ $y'(0) = 0$ is $Y = 3J_0(3t).$ **(T/F)**

❖ Multiple Choice Questions

Choose the most appropriate one.

1. If $ty'' + y' + 4ty = 0, y(0) = 3, y'(0) = 0,$ then $y =$
(a) $2J_0(3t)$ (b) $3J_0(3t)$
(c) $3J_0(2t)$ (d) none of these

2. The solution of the differential equation $ty'''(t) + y'(t) + ty(t) = 0$ under the conditions that $y(0) = 1$ and $y(t)$ and its derivatives have transforms :
(a) $y(t) = J_0(t)$ (b) $y(t) = 1$
(c) $y(t) = J_1(t)$ (d) none of these

3. If $y = y(x, t)$, then $L\left(\dfrac{\partial y}{\partial x}\right) =$

(a) $p\bar{y}(x, 0) + y(x, 0)$ (b) $p\bar{y}(x, p) - y(x, 0)$
(c) $p\bar{y}(x, p) - y(x, p)$ (d) none of these

4. If $\dfrac{d^2 y}{dt^2} + t\dfrac{dy}{dt} - y = 0$ and $y(0) = 0, \dfrac{dy}{dt} = 1$ at $t = 0$ then $y =$
(a) 1 (b) t
(c) $t + 1$ (d) none of these

5. If $ty'' + (1 - 2y)y' - 2y = 0, y(0) = 1$ and $y'(0) = -2$, then $y =$

(a) e^t (b) e^{3t}
(c) e^{2t} (d) e^{4t}

6. If $\dfrac{d^2 y}{dt^2} + y = 0$ under the conditions

$y = 1, \dfrac{dy}{dt} = 0$ when $t = 0$, then $y =$
(a) $\sin t$ (b) $\cos t$
(c) $\tan t$ (d) none of these

7. If $(D^2 - 2D + 2)y = 0$ and $y = 1 = Dy$ when $t = 0$, then $y =$
(a) $e^t \cos t$ (b) $e^t \sin t$
(c) $e^t \tan t$ (d) none of these

8. If $y''(t) + 6y'(t) + 9y(t) = 6t^2 e^{-3t}$ and $y(0) = 0 = y_1$, and $y =$
(a) $\dfrac{t}{2} e^{-3t}$ (b) $\dfrac{t^4}{3} e^{-3t}$

(c) $\dfrac{t^4}{2} e^{-3t}$ (d) none of these

9. If $y''(t) + y(t) = t$, if $y'(0) = 1$ and $y(\pi) = 0$, then $y =$
(a) $\pi \cos t$ (b) $\pi \cos t + t$
(c) $t \cos t$ (d) $t \cos t + 1$

10. $L\left\{\dfrac{\partial^2 y}{\partial x^2}\right\} =$

   (a) 0   (b) 1

   (c) 2   (d) $\dfrac{d^2\bar{y}}{dx^2}$

11. If $(D^2 - 2D + 2)y = 0$ and $y = 1 = Dy$ when $t = 0$, then $y =$

   (a) $e^{-t}\sin t$ (b) $e^{-t}\cos t$

   (c) $e^t \sin t$ (d) $e^t \cos t$

12. If $(D^2 + 6D + 9)y = 6t^2 e^{-3t}$ and $y(0) = y'(0) = 0$, then $y =$

   (a) $\dfrac{t}{2}e^{-3t}$ (b) $\dfrac{t^4}{2}e^{-3t}$

   (c) $\dfrac{t^2}{2}e^{-3t}$ (d) none of these

13. If $\dfrac{d^2 y}{dt^2} + y = 0$, if $y = 1, \dfrac{dy}{dt} = 0$, when $t = 0$, then $y =$

   (a) $\sin t$ (b) $\cos t$

   (c) $\tan^2 t$ (d) none of these

14. If $\dfrac{d^2 y}{dt^2} + t\dfrac{dy}{dt} - y = 0,$ if $y(0) = 0, \dfrac{dy}{dt} = 1$, when $t = 0$, then $y =$

   (a) t (b) $t + 1$

   (c) $t + 2$ (d) $t + 3$

15. Consider the differential equation $\dfrac{d^2 y(t)}{dt^2} + \dfrac{2 dy(t)}{dt} + y(t) = \delta(t)$ with $y(t)\big|_{t=0} = -2$ and $\dfrac{dy}{dt}\bigg|_{t=0} = 0$. The numerical value of $\dfrac{dy}{dt}\bigg|_{t=0}$ is :   [GATE(EC, IN)–2012]

   (a) 1   (b) –2

   (c) –1   (d) 0

16. A solution for the differential equation $\dot{x}(t) + 2x(t) = \delta(t)$ with initial condition $x(0) = 0$ is :   [GATE(EC)–2006]

   (a) $e^{2t}u(t)$ (b) $e^{-2t}u(t)$

   (c) $e^{-t}u(t)$ (d) $e^t u(t)$

ANSWERS

❖ Fill in the Blanks

1. $y = \pi\cos t + t$ **2.** $p\bar{y}(x, p) - y(x, 0)$ **3.** $\bar{y}(p)$ **4.** $\cos t$ **5.** $C_1 J_0(2t)$

6. $y = 5e^{-t} - 2e^{4t}$ **7.** $e^{\int P dx}$

❖ True or False

1. T   **2.** T   **3.** F   **4.** T   **5.** F

❖ Multiple Choice Questions

1. (b)   **2.** (a)   **3.** (b)   **4.** (b)   **5.** (c)   **6.** (b)   **7.** (a)   **8.** (c)   **9.** (b)

10. (d)   **11.** (d)   **12.** (c)   **13.** (b)   **14.** (a)   **15.** (a)   **16.** (b)

ⅤⅤⅤⅤⅤⅤ

Chapter 4

Fourier Transform and Its Applications

4.1 INTRODUCTION

If a function $f(x)$ defined on the interval $]-\infty,\infty[$, and piecewise continuous in each finite partial interval and absolutely integrable in $]-\infty,\infty[$, then

$$F(f(x)) = \tilde{f}(p) = \int_{-\infty}^{\infty} e^{ipx} f(x)dx$$

is defined as Fourier transform of $f(x)$.

The inverse formula for Fourier transform is given by

$$f^{-1}[\tilde{f}(p)] = f(x) = \frac{1}{2\pi}\int_{-\infty}^{\infty} \tilde{f}(p)e^{-ipx}dp .$$

REMARK

- We can also define $\tilde{f}(p) = F(f(x)) = \frac{1}{\sqrt{2\pi}}\int_{-\infty}^{\infty} e^{-ipx} f(x)dx$

  and $F^{-1}[\tilde{f}(p)] = f(x) = \frac{1}{\sqrt{2\pi}}\int_{-\infty}^{\infty} e^{-ipx} \tilde{f}(p)dp .$

4.2 FOURIER SINE AND COSINE TRANSFORMS

Definition (1): *The infinite Fourier sine transform of the function $f(x)$, $0 < x < \infty$ is defined by $F_s[f(x)]$ or $\tilde{f}_s(p)$ and is defined by* (Meerut–1983)

$$\tilde{f}_s(p) = F_s[f(x)] = \sqrt{\frac{2}{\pi}}\int_0^{\infty} f(x)\sin pxdx$$

The inverse formula for infinite Fourier sine transform is given by

$$f(x) = F_s^{-1}[\tilde{f}_s(p)] = \sqrt{\frac{2}{\pi}}\int_0^{\infty} \tilde{f}_s(p)\sin pxdx .$$

Definition (2): *The infinite Fourier cosine transform of $f(x)$, $0 < x < \infty$ is denoted by $F_c[f(x)]$ or $\tilde{f}_c(p)$ and is defined by*

$$\tilde{f}_c[f(x)] = \tilde{f}_c(p) = \sqrt{\frac{2}{\pi}}\int_0^{\infty} f(x)\cos pxdx$$

The inversion formula for Fourier cosine transform is given by

$$f(x) = \sqrt{\frac{2}{\pi}}\int_0^{\infty} \tilde{f}_c(p)\cos pxdx .$$

4.2.1 LINEARITY PROPERTY OF FOURIER TRANSFORMS

Let $\tilde{f}(p)$ and $\tilde{g}(p)$ be Fourier transform of $f(x)$ and $g(x)$ respectively. Then

$F\{af(x)+bg(x)\} = a\tilde{f}(p)+b\tilde{g}(p)$ where a and b constants.

4.2.2 CHANGE OF SCALE PROPERTY

THEOREM 1. **(For Complex Fourier Transform).** *If $\tilde{f}(p)$ is the complex Fourier transform of $f(x)$, the complex Fourier transform of $f(ax)$ is given by*

$$\frac{1}{a}\tilde{f}\left(\frac{p}{a}\right).$$ (Meerut–1986, 2001, 05; Rohilkhand–1989; Purvanchal–1994)

PROOF. By definition, we have

$$\int_{-\infty}^{\infty} e^{ipx} f(x)dx = \tilde{f}(p).$$ ...(1)

Consider $\tilde{f}(ap) = \int_{-\infty}^{\infty} e^{ipx} f(x)dx$.

Putting $ax = t \Rightarrow dx = dt$, we get

$$\tilde{f}(ap) = \frac{1}{a}\int_{-\infty}^{\infty} e^{ip(t/a)} f(t)dt = \frac{1}{a}\int_{-\infty}^{\infty} e^{i(p/a)x} f(x)dx = \frac{1}{a}\tilde{f}\left(\frac{p}{a}\right).$$

REMARK

- In a similar way, we can prove that :

  (a) If $\tilde{f}_s(p)$ is the Fourier sine transform of $f(x)$, then Fourier sine transform of $f(ax)$ is given by

  $$\frac{1}{a}\tilde{f}_s\left(\frac{p}{a}\right).$$

  (b) If $\tilde{f}_c(p)$ is the Fourier cosine transform of $f(x)$, then Fourier cosine transform of $f(ax)$ is given

  by $\dfrac{1}{a}\tilde{f}_c\left(\dfrac{p}{a}\right)$.

THEOREM 2. **(Shifting Property).** *If $\tilde{f}(p)$ is the complex Fourier transform of $f(x)$, then complex Fourier transform of $f(x - a)$ is $e^{ipa}\,\tilde{f}(p)$.*

PROOF. By definition, we have

$$\int_{-\infty}^{\infty} e^{ipx} f(x)dx = \tilde{f}(p)$$ ...(1)

Consider $\tilde{f}(x - a) = \int_{-\infty}^{\infty} e^{ipx} f(x - a)dx = \int_{-\infty}^{\infty} e^{ip(t+a)} f(t)dt$

$$= e^{ipx}\int_{-\infty}^{\infty} e^{ipt} f(t)dt = e^{ipa}\tilde{f}(p) \qquad \text{[Putting } x - a = t\text{]}$$

4.2.3 SOME IMPORTANT INTEGRALS (TO BE USED DIRECTLY)

1. $\displaystyle\int e^{ax}\sin bx\,dx = \frac{e^{ax}}{a^2 + b^2}(a\sin bx - b\cos bx)$

2. $\displaystyle\int e^{ax}\cos bx\,dx = \frac{e^{ax}}{a^2 + b^2}(a\cos bx + b\sin bx)$

3. $\displaystyle\int_0^{\infty} e^{-ax}\sin bx\,dx = \frac{b}{a^2 + b^2}$ **4.** $\displaystyle\int_0^{\infty} e^{-ax}\cos bx\,dx = \frac{a}{a^2 + b^2}$

5. $\displaystyle\frac{d^n}{dx^n}\left(\frac{x}{a^2 + x^2}\right) = \frac{(-1)^n . n!}{(a^2 + x^2)^{(n+1)/2}}\cos\left[(n+1)\tan^{-1}\left(\frac{a}{x}\right)\right]$

6. $\displaystyle\frac{d^n}{dx^n}\left(\frac{a}{a^2 + x^2}\right) = \frac{(-1)^n . n!}{(a^2 + x^2)^{(n+1)/2}}\sin\left[(n+1)\tan^{-1}\left(\frac{a}{x}\right)\right]$

7. $\int_0^\infty \dfrac{\sin px}{x} dx = \begin{bmatrix} \pi/2 & ; & \text{if } p > 0 \\ -\pi/2 & ; & \text{if } p < 0 \end{bmatrix}$ **8.** $\int_0^\infty e^{-x^2} dx = \sqrt{\pi}, \int_0^\infty e^{-x^2} dx = \dfrac{\sqrt{\pi}}{2}$

9. $\int_0^\infty \dfrac{e^{ax} - e^{-ax}}{e^{\pi x} - e^{-\pi x}} dx = \dfrac{1}{2} \tan \dfrac{a}{2}, \int_0^\infty \dfrac{e^{ax} + e^{-ax}}{e^{\pi x} - e^{-\pi x}} dx = \dfrac{1}{2} \sec \dfrac{a}{2}$

10. $\int_0^\infty \dfrac{\cos px}{1 + p^2} dp = \dfrac{\pi e^{-x}}{2}, \int_0^\infty \dfrac{p \sin px}{1 + p^2} dp = \dfrac{\pi}{2} e^{-x}$

<u>THEOREM 3</u>   **(Modulation Theorem). If $\tilde{f}(p)$ is the complex Fourier transform of $f(x)$ then, the Fourier transform of $f(x) \cos ax$ is $\dfrac{1}{2}[\tilde{f}(p-a) + \tilde{f}(p+a)]$.**

(Meerut–1992; Osmania–2004; Purvanchal–1995)

<u>PROOF.</u>   By definition, we have

$$\tilde{f}(p) = \int_{-\infty}^{\infty} e^{ipx} f(x) dx$$

Now, $F[f(x)\cos ax] = \int_{-\infty}^{\infty} e^{ipx} f(x) \cos ax \, dx$

$$= \int_{-\infty}^{\infty} e^{ipx} f(x) \left[\frac{e^{iax} + e^{-iax}}{2} \right] dx$$

$$= \frac{1}{2} \int_{-\infty}^{\infty} e^{i(p+a)x} f(x) dx + \frac{1}{2} \int_{-\infty}^{\infty} e^{i(p-a)x} f(x) dx$$

$$= \frac{1}{2} f(p+a) + \frac{1}{2} f(p-a)$$

<u>REMARKS</u>

- In a similar way, we can prove that

If $\tilde{f}_s(p)$ is the complex Fourier transform of $f(x)$, then

1. Fourier transform of $f(x) \cos ax$ is $\dfrac{1}{2}[f_s(p+a) + f_s(p-a)]$

2. Fourier transform of $f(x) \sin ax$ is $\dfrac{1}{2}[f_s(p+a) - f_s(p-a)]$

3. Fourier transform of $f(x) \sin ax$ is $\dfrac{1}{2}[f_c(p-a) - f_c(p+a)]$

4.2.4 SOME MORE IMPORTANT RESULTS

1. Multiple Fourier Transforms

Let $f(x, y)$ be a function of two variables x and y. Temporarily, as a function of x, its Fourier transform is given by

$$\tilde{f}(p, y) = \frac{1}{\sqrt{2\pi}} \int_{-\infty}^{\infty} f(x, y) e^{ipx} dx$$

Also, regarding $\tilde{f}(p, y)$ as a function of y, its Fourier transform is given by

$$\tilde{f}(p, q) = \frac{1}{\sqrt{2\pi}} \int_{-\infty}^{\infty} \tilde{f}(p, y) e^{iqy} dy = \frac{1}{2\pi} \int_{-\infty}^{\infty} \int_{-\infty}^{\infty} f(x, y) e^{i(px + qy)} dx \, dy$$

2. Inversion Formula : If

$$f(x, y) = \frac{1}{\sqrt{2\pi}} \int_{-\infty}^{\infty} \tilde{f}(p, y) e^{-ipx} dp \text{ and } \tilde{f}(p, y) = \frac{1}{\sqrt{2\pi}} \int_{-\infty}^{\infty} \tilde{f}(p, q) e^{-iqy} dq$$

Then $f(x, y) = \dfrac{1}{2\pi} \int_{-\infty}^{\infty} \tilde{f}(p, q) e^{-i(px + qy)} dp \, dq$

3. Convolution

The function $H(x) = F * G = \dfrac{1}{\sqrt{2\pi}} \int_{-\infty}^{\infty} F(u)G(x-u)du$ is called the convolution or falting

of two integrable functions F and G over the interval $]-\infty, \infty [$.

Sometime it can also be defined as $F * G = \int_{-\infty}^{\infty} F(u)G(x-u)du$.

THEOREM 1.   **(The convolution or Falting Theorem for Fourier Transforms)**

If F[f(x)] and F[g(x)] are the Fourier transforms of the functions f(x) and g(x) respectively, then the Fourier transform of the convolution of f(x) and g(x) is the product of their Fourier transforms, i.e.,

$$F\{f(x) * g(x)\} = F\{f(x)\}.F\{g(x)\}$$

(Meerut–1982, 86; Rajasthan–1983; MKU–2008; Delhi–1998, 99)

PROOF.   Consider   LHS $= F\{f(x) * g(x)\}$

$$= F\int_{-\infty}^{\infty} f(u)g(x-u)du \qquad\qquad \text{(By definition of convolution)}$$

$$= \int_{-\infty}^{\infty} e^{ipx}\left\{\int_{-\infty}^{\infty} f(u)g(x-u)du\right\}dx$$

$$= \int_{-\infty}^{\infty} f(u)\left\{\int_{-\infty}^{\infty} e^{ipx}g(x-u)dx\right\}du$$

$$= \int_{-\infty}^{\infty} f(u)\left\{\int_{-\infty}^{\infty} e^{-ip(u+v)}g(v)dv\right\}du$$

$$\text{[Putting } x - u = v \Rightarrow dx = dv]$$

$$= \int_{-\infty}^{\infty} e^{ipu} f(u)\left\{\int_{-\infty}^{\infty} e^{ipv}g(v)dv\right\}du$$

$$= \int_{-\infty}^{\infty} e^{ipu} f(u)F\{g(x)\}du = F\{g(x)\}\int_{-\infty}^{\infty} e^{ipu} f(u)du$$

$$= F\{g(x)\}.F\{f(x)\}.$$

THEOREM 2.   **(Parseval's identity for Fourier Transform: Rayleigh Theorem or Plancharel's Theroem)**   (MKU–2008)

If f(p) and g(p) are the complex Fourier transform of F(x) and G(x) respectively, then

(i) $\dfrac{1}{2\pi}\int_{-\infty}^{\infty} f(p)\bar{g}(p)dp = \int_{-\infty}^{\infty} F(x)\overline{G(x)}dx$ **(ii)** $\dfrac{1}{2\pi}\int_{-\infty}^{\infty} |f(p)|^2 dp = \int_{-\infty}^{\infty} |F(x)|^2 dx$

where bar denotes the complex conjugate.

PROOF.   (i)   By inversion formula, we have

$$G(x) = \frac{1}{2\pi}\int_{-\infty}^{\infty} g(p)e^{-ipx} dp \qquad\qquad\qquad \text{...(1)}$$

$$\Rightarrow \qquad \overline{G(x)} = \frac{1}{2\pi}\int_{-\infty}^{\infty} \overline{g(p)}e^{ipx} dp \qquad\qquad\qquad \text{...(2)}$$

Thus, $\int_{-\infty}^{\infty} F(x)\overline{G(x)}dx = \int_{-\infty}^{\infty} F(x)\left\{\dfrac{1}{2\pi}\int_{-\infty}^{\infty} \overline{g(p)}e^{ipx} dp\right\}dx$

$$= \frac{1}{2\pi}\int_{-\infty}^{\infty} \overline{g(p)}\left\{\int_{-\infty}^{\infty} F(x)e^{ipx}dx\right\}dp = \frac{1}{2\pi}\int_{-\infty}^{\infty} \overline{g(p)}f(p)dp$$

(ii) Taking $G(x) = F(x)$ in part (i), we get

$$\frac{1}{2\pi}\int_{-\infty}^{\infty} f(p)\overline{f(p)}dp = \int_{-\infty}^{\infty} F(x)\overline{F(x)}dx$$

Hence, $\quad \frac{1}{2\pi}\int_{-\infty}^{\infty} |f(p)|^2\, dp = \int_{-\infty}^{\infty} |F(x)|^2\, dx$

REMARKS
- In a similar way, we can prove that

(a) $\frac{2}{\pi}\int_0^{\infty} f_c(p)g_c(p)dp = \int_0^{\infty} F(x)G(x)dx$

(b) $\frac{2}{\pi}\int_0^{\infty} f_s(p)g_s(p)dp = \int_0^{\infty} F(x)G(x)dx$

(c) $\frac{2}{\pi}\int_0^{\infty} [f_c(p)]^2 dp = \int_0^{\infty} [F(x)]^2 dx$

(d) $\frac{2}{\pi}\int_0^{\infty} [f_s(p)]^2 dp = \int_0^{\infty} [F(x)]^2 dx$

4.3 RELATION BETWEEN FOURIER AND LAPLACE TRANSFORM

Consider the function $\quad f(t) = \begin{cases} e^{-xt}g(t) & ; \quad t < 0 \\ 0 & ; \quad t > 0 \end{cases}$ ... (1)

Then the Fourier transform of $f(t)$ is given by

$$F[f(t)] = \int_{-\infty}^{\infty} e^{ipt} f(t)dt$$

$$= \int_{-\infty}^{0} 0 e^{ipt} dt + \int_0^{\infty} e^{-xt}g(t)e^{ipt} dt = \int_0^{\infty} e^{(ip-x)t}g(t)dt$$

$$= \int_0^{\infty} e^{-xst}g(t)dt = L\{g(t)\} \qquad \text{[Putting } x - ip = s]$$

Hence, Fourier transformation of the function $f(t)$ defined by (1) is the Laplace transform of function $g(t)$.

4.4 FOURIER TRANSFORMS OF THE DERIVATIVE OF A FUNCTION

THEOREM-1. *The Fourier transform of the function* $\dfrac{d^n F}{dx^n}$ *is* $(-ip)^n$ *times the Fourier transform of the function* $f(x)$ *provided that the first* $(n-1)$ *derivatives of* $F(x)$ *vanish as* $x \to \pm\infty$, *i.e,* $F\{f^n(x)\} = (-ip)^n F\{f(x)\}$ (Rohilkhand–1992, 94; Kanpur–2004)

Proof. By definition, we have

$$F\{f^n(x)\} = \int_{-\infty}^{\infty} f^n(x)e^{ipx}dx = \left[f^{n-1}(x)e^{ipx} \right]_{-\infty}^{\infty} - \int_{-\infty}^{\infty} f^{n-1}(x)ipe^{ipx}dx$$

$$= -ip\int_{-\infty}^{\infty} f^{n-1}(x)e^{ipx}dx \qquad \left[\because \lim_{x\to\pm\infty} f^{n-1}(x) = 0 \right]$$

Repeating this process of integration by parts $(n-1)$ times more and assuming that

$$\lim_{x\to\pm\infty} f^r(x) = 0 \text{ for } r = 1, 2, 3, ..., n-1$$

Thus, we have $\quad F\{f^n(x)\} = (-ip)^n \int_{-\infty}^{\infty} f(x)e^{ipx}dx$

Hence, $\quad F[f^n(x)] = (-ip)^n F[f(x)].$

REMARK

- In a similar way, we can prove the following results:

(a) $\tilde{f}_c^{2n}(p) = -\sum_{r=0}^{n-1} (-1)^r \alpha_{2n-2r-1} p^{2r} + (-1)^n p^{2n} \tilde{f}_c(p)$

(b) $\tilde{f}_c^{2n+1}(p) = -\sum_{r=0}^{n-1} (-1)^r \alpha_{2n-2r} p^{2r} + (-1)^n p^{2n+1} \tilde{f}_s(p)$

(c) $\tilde{f}_c^{2n}(p) = -\sum_{r=1}^{n} (-1)^r \alpha_{2n-2r} p^{2r-1} + (-1)^{n+1} p^{2n} \tilde{f}_s(p)$

(d) $\tilde{f}_s^{2n+1}(p) = -\sum_{r=1}^{n} (-1)^r \alpha_{2n-2r+1} p^{2r-1} + (-1)^{n+1} p^{2n+1} \tilde{f}_c(p)$

Provided that first $(n-1)$ derivatives of $f(x)$ vanish as $x \to \infty$ and $\dfrac{d^{n+1}f}{dx^{n-1}} \to \alpha_{n-1}$, etc. $x \to 0$.

✍ RECAPITULATIONS

S. No.	Operation	Result
1.	Linearity property	$F\{af(x) + bg(x)\} = a\tilde{f}(p) + b\tilde{g}(p)$
2.	Change of scale property	$\tilde{f}(ap) = \dfrac{1}{a}\int_{-\infty}^{\infty} e^{ip(t/a)} f(t)dt = \dfrac{1}{a}\tilde{f}\left(\dfrac{p}{a}\right)$
3	Shifting property	$F[f(x-a)] = e^{ipa}\tilde{f}(p)$ where $F[f(x)] = \tilde{f}(p)$
4.	Modulation Theorem	$F[f(x)\cos ax] = \dfrac{1}{2}f(p+a) + \dfrac{1}{2}f(p-a)$
5.	Convolution theorem	$F\{f(x) * g(x)\} = F\{f(x)\}.F\{g(x)\}$
6.	Fourier transform of derivatives	$F[f^n(x)] = (-ip)^n F[f(x)]$

Solved Examples

EXAMPLE 1. *Find the complex Fourier transform of $f(x) = e^{-a|x|}$, where $a > 0$ and x belongs to $]-\infty, \infty[$.*
(VTU–2010; SVTU–2008; Kottayam–2005; Meerut–1990, 2014)

SOLUTION. We have $\tilde{f}(p) = \int_{-\infty}^{\infty} f(x)e^{ipx}dx = \int_{-\infty}^{\infty} e^{ipx} e^{-a|x|}dx$

$= \int_{-\infty}^{0} e^{ipx} e^{ax}dx + \int_{0}^{\infty} e^{ipx} e^{-ax}dx = \int_{-\infty}^{0} e^{(a+ip)x}dx + \int_{0}^{\infty} e^{-(a-ip)x}dx$

$= \left[\dfrac{e^{(a+ip)x}}{a+ip}\right]_{-\infty}^{0} + \left[\dfrac{e^{-(a-ip)x}}{-(a-ip)}\right]_{0}^{\infty} = \dfrac{1}{a+ip} + \dfrac{1}{a-ip} = \dfrac{2a}{a^2 + p^2}$

EXAMPLE 2. *Find the Fourier transform of $f(x)$ defined by*
$f(x) = \begin{cases} 1 & ; \ |x| < a \\ 0 & ; \ |x| > a \end{cases}$ *and hence evaluate*

(i) $\int_{-\infty}^{\infty} \dfrac{\sin pa \cos px}{p}dx$ and (ii) $\int_{0}^{\infty} \dfrac{\sin p}{p}dp$

(WBTU–2005; Madras–2003; PTU–2003; Kottayam–2005; Meerut–1982, 83, 91, 92, 97, 2004; Bundelkhand–2006, 07; Agra–1981, 2003; Purvanchal–1995, 96; Rohilkhand–1989, 1992; Rajasthan–1984; Kanpur–2006; Avadh–2013; SNU(AP)–1997; Mysore–1997; Osmania–2004; Delhi–2005)

SOLUTION. We have $|x| < a \Rightarrow -a < x < a$

and $\quad |x| > a \Rightarrow x < -a \text{ or } x > a$

Thus, $\quad f(x) = \begin{cases} 1 & ; \quad \text{if } -a < x < a \\ 0 & ; \quad \text{if } x < -a \text{ or } x > a \end{cases}$ ...(1)

So, $\quad \tilde{f}(p) = F[f(x)]$

$$= \int_{-\infty}^{\infty} e^{ipx} f(x) dx$$

$$= \int_{-\infty}^{-a} e^{ipx} f(x) dx + \int_{-a}^{a} e^{ipx} f(x) dx + \int_{a}^{\infty} e^{ipx} f(x) dx$$

$$= 0 + \int_{-a}^{a} e^{ipx} dx + 0 \qquad \text{[Using (1)]}$$

$$= \left[\frac{e^{ipx}}{ip} \right]_{-a}^{a} = \frac{1}{ip}(e^{ipa} - e^{-ipa}) = \frac{1}{ip}(2i \sin pa)$$

$$\Rightarrow \quad \tilde{f}(p) = [Ff(x)] = \frac{2 \sin pa}{p} \qquad \text{...(2)}$$

Using corresponding inversion formula, we get

$$\frac{1}{2\pi} \int_{-\infty}^{\infty} \tilde{f}(p) e^{-ipx} dp = f(x)$$

$$\Rightarrow \quad \int_{-\infty}^{\infty} \frac{2 \sin pa}{p} e^{-ipx} dp = 2\pi f(x)$$

$$\Rightarrow \quad 2\int_{-\infty}^{\infty} \frac{\sin pa}{p}(\cos px - i \sin px) dp = 2\pi f(x)$$

$$\Rightarrow \quad \int_{-\infty}^{\infty} \frac{1}{p}(\sin pa \cos px - i \sin pa \sin px) dx = \pi f(x).$$

Equating real parts of both the sides, we get

$$\int_{-\infty}^{\infty} \frac{\sin pa \cos px}{p} dx = \pi f(x)$$

$$\Rightarrow \quad \int_{-\infty}^{\infty} \frac{\sin pa \cos px}{p} dx = \begin{cases} \pi & ; \quad |x| < a \\ 0 & ; \quad |x| > a \end{cases}$$

Putting $x = 0$ and $a = 1$, we get

$$\int_{-\infty}^{\infty} \frac{\sin p}{p} dp = \pi \text{ or } 2\int_{-\infty}^{\infty} \frac{\sin p}{p} dp = \pi$$

Hence, $\int_{0}^{\infty} \frac{\sin p}{p} dp = \frac{\pi}{2}$.

EXAMPLE 3. *Find Fourier sine and cosine transforms of $f(x) = x$.*

SOLUTION. By definition

$$\tilde{f}_c(p) = \int_{0}^{\infty} x \cos px \, dx \text{ and } \tilde{f}_s(p) = \int_{0}^{\infty} x \sin px \, dx$$

$$\Rightarrow \tilde{f}_c(p) - i\tilde{f}_s(p) = \int_{0}^{\infty} x(\cos px - i \sin px) dx$$

$$= \int_{0}^{\infty} x e^{-ipx} dx = \int_{0}^{\infty} \frac{y}{ip} e^{-y} \frac{dy}{ip} \qquad \text{[By putting } ipx = y\text{]}$$

$$= -\frac{1}{p^2} \int_{0}^{\infty} y^{2-1} . e^{-y} dy = -\frac{\Gamma(2)}{p^2}$$

$$\Rightarrow \tilde{f}_c(p) - i\tilde{f}_s(p) = -\frac{1}{p^2} + i.0$$

Equating real and imaginary parts of both the sides, we get

$$\tilde{f}_c(p) = -\frac{1}{p^2} \text{ and } \tilde{f}_s(p) = 0$$

EXAMPLE 4.   *Find f(x), if its Fourier cosine transform is* $\dfrac{1}{(1+p^2)}$.

(Kanpur–1997, 2001; Purvanchal–1995)

SOLUTION.   We have to evaluate $f(x) = F_c^{-1}\left\{\dfrac{1}{1+p^2}\right\}$. By inversion formula, we have

$$F(x) = \frac{2}{\pi}\int_0^\infty \frac{1}{1+p^2}\cos px\, dp \qquad \text{...(1)}$$

$$\Rightarrow \qquad \frac{dF}{dx} = \frac{2}{\pi}\int_0^\infty \frac{-p\sin px}{1+p^2}\, dp$$

$$= -\frac{2}{\pi}\int_0^\infty \frac{-p\sin px}{1+p^2}\, dp = -\frac{2}{\pi}\int_0^\infty \frac{(1+p^2)-1}{1+p^2}\cdot\frac{\sin px}{p}\, dp$$

$$= -\frac{2}{\pi}\int_0^\infty \left(1 - \frac{1}{1+p^2}\right)\cdot\frac{\sin px}{p}\, dp + \frac{2}{\pi}\int_0^\infty \frac{\sin px}{p(1+p^2)}\, dp$$

$$= -\frac{2}{\pi}\cdot\frac{\pi}{2} + \frac{2}{\pi}\int_0^\infty \frac{\sin px}{p(1+p^2)}\, dx \qquad \left[\because \int_0^\infty \frac{\sin px}{p}\, dp = \frac{\pi}{2}\right]$$

$$= -1 + \frac{2}{\pi}\int_0^\infty \frac{\sin px}{p(1+p^2)}\, dx \qquad \text{... (2)}$$

Also, $\qquad \dfrac{d^2F}{dx^2} = \dfrac{2}{\pi}\displaystyle\int_0^\infty \dfrac{p\cos px}{p(1+p^2)}\, dp = \dfrac{2}{\pi}\displaystyle\int_0^\infty \dfrac{\cos px}{(1+p^2)}\, dx$

$$\Rightarrow \qquad \frac{d^2F}{dx^2} = F(x) \qquad\qquad\qquad\qquad \text{[Using (1)]}$$

$$\Rightarrow \qquad (D^2 - 1)F = 0 \qquad\qquad\qquad\qquad\qquad \text{...(3)}$$

$$\Rightarrow \qquad F(x) = Ae^{-x} + Be^{x} \qquad\qquad\qquad\qquad \text{...(4)}$$

Now, $\qquad F(0) = \dfrac{2}{\pi}\displaystyle\int_0^\infty \dfrac{dp}{1+p^2} = \dfrac{2}{\pi}\Big[\tan^{-1}p\Big]_0^\infty = \dfrac{2}{\pi}\cdot\dfrac{\pi}{2} = 1$

and $\qquad F(0) = A + B \qquad\qquad\qquad\qquad\qquad\qquad \text{...(5)}$

$\therefore \qquad A + B = 1 \qquad\qquad\qquad\qquad\qquad\qquad\qquad \text{...(6)}$

From (4) $\qquad \dfrac{dF}{dx} = -Ae^{-x} + Be^{x}.$

Putting $x = 0$ in (2) and (6), we get

$$\frac{dF}{dx} = -1 \text{ and } \frac{dF}{dx} = -A + B$$

$$\Rightarrow \qquad -A + B = 1 \qquad\qquad\qquad\qquad\qquad\qquad \text{...(7)}$$

On solving (5) and (7), we get $A = 1$, $B = 0$. Hence from (4) $F(x) = e^{-x}$

Example 5. If $f(x) = \begin{cases} 1 & ; \ |x| < a \\ 0 & ; \ |x| > a \end{cases}$ and $\tilde{f}(p) = \dfrac{2\sin pa}{p}$, where $p \neq 0$ then show that

$\displaystyle\int_0^\infty \dfrac{\sin^2 ax}{x}\,dx = \dfrac{\pi a}{2}$. (VTU–2010; SVTU–2009; UPTU–2008; Kanpur–1996; Purvanchal–1994)

Solution. Using Parseval's identity for Fourier transform, we get

$$\int_{-\infty}^{\infty} [f(x)]^2\,dx = \frac{1}{2\pi}\int_{-\infty}^{\infty} [\tilde{f}(p)]^2\,dp$$

$$\Rightarrow \quad \int_{-\infty}^{-a} [f(x)]^2\,dx + \int_{-a}^{a} [f(x)]^2\,dx + \int_{a}^{\infty} [f(x)]^2\,dx = \frac{1}{2\pi}\int_{-\infty}^{\infty} [\tilde{f}(p)]^2\,dp$$

$$\Rightarrow \quad 0 + \int_{-a}^{a} 1^2\,dx + 0 = \frac{1}{2\pi}\int_{-\infty}^{\infty} \frac{4\sin^2 pa}{p^2}$$

$$\Rightarrow \quad 2a = \frac{2}{\pi}\int_{-\infty}^{\infty} \frac{\sin^2 ax}{x^2}\,dx \ \text{ or } \ \int_{-\infty}^{\infty} \frac{\sin^2 ax}{x^2}\,dx = \pi a$$

or $\quad 2\displaystyle\int_0^\infty \frac{\sin^2 ax}{x^2}\,dx = \pi a \ \text{ or } \ \int_0^\infty \frac{\sin^2 ax}{x^2}\,dx = \frac{\pi a}{2}$.

Example 6. Find the Fourier transform of $f(x)$, if

$$f(x) = \begin{cases} \dfrac{\sqrt{2\pi}}{2\varepsilon} & , \ |x| \le \varepsilon \\ 0 & , \ |x| > \varepsilon \end{cases}$$

Solution. We have,

$$F\{f(x)\} = \frac{1}{2\pi}\int_{-\infty}^{\infty} e^{ipx} f(x)\,dx = \frac{1}{\sqrt{2\pi}}\int_{-\varepsilon}^{\varepsilon} e^{ipx}\cdot\frac{\sqrt{2\pi}}{2\varepsilon}\,dx = \frac{1}{2\varepsilon}\left[\frac{e^{ipx}}{ip}\right]_{-\varepsilon}^{\varepsilon}$$

$$= \frac{e^{ipt} - e^{-ipt}}{2ip\varepsilon} = \frac{\sin p\varepsilon}{p\varepsilon}.$$

Example 7. Find the Fourier transform of $f(x)$, if

$$f(x) = \begin{cases} x^2 & , \ |x| < a \\ 0 & , \ |x| > a \end{cases}$$ (SVTU–2008; Meerut–2000)

Solution. We have

$$\tilde{f}(p) = \frac{1}{\sqrt{2\pi}}\int_{-\infty}^{\infty} f(x)e^{ipx}\,dx = \frac{1}{2\pi}\int_{-a}^{a} x^2 e^{ipx}\,dx$$

$$= \frac{1}{\sqrt{2\pi}}\cdot\left[x^2\cdot\frac{e^{ipx}}{ip} - 2x\cdot\frac{e^{ipx}}{(ip)^2} + 2\cdot\frac{e^{ipx}}{(ip)^3}\right]_{-a}^{a}$$

$$= \frac{1}{\sqrt{2\pi}}\left[\frac{a^2}{ip}(e^{ipa} - e^{-ipa}) + \frac{2a}{p^2}(e^{ipa} + e^{-ipa}) - \frac{2}{ip^3}(e^{ipa} - e^{-ipa})\right]$$

$$= \frac{1}{p^3\sqrt{2\pi}}\left[\frac{a^2 p^2}{i}\cdot 2i\sin pa + 2ap\cdot 2\cos pa - \frac{2}{i}\cdot 2i\sin pa\right]$$

$$= \frac{1}{p^3}\sqrt{\frac{2}{\pi}}\left[(a^2 p^2 - 2)\sin pa + 2ap\cdot\cos pa\right]$$

EXAMPLE 8. *Find the Fourier transform of*

$$F(x) = \begin{bmatrix} 1 - x^2 & , & |x| \le 1 \\ 0 & , & |x| > 1 \end{bmatrix}$$

and hence evaluate $\int_0^\infty \left(\dfrac{x \cos x - \sin x}{x^3} \right) \cos \dfrac{x}{2} dx$

(VTU–2011; Anna–2005, 12;
Mumbai –2005; Meerut–1982; Agra–2000, 02, 10;
Rajasthan–1983; Rohilkhand–1990, 91, 2013; Kanpur– 2002, 03;
Purvanchal–2003, 04; Jhansi–2012; Delhi–2001; Banglore–2004)
(Madras–1997; Mysore–1997; Rohilkhand–1990, 91, 99; Warangal–1996, SVTU-2006)

SOLUTION. We have

$$\tilde{F}(p) = \frac{1}{\sqrt{2\pi}} \int_{-\infty}^\infty e^{ipx} F(x) dx$$

$$= \frac{1}{\sqrt{2\pi}} \int_{-1}^1 (1 - x^2) e^{ipx} dx = \frac{1}{\sqrt{2\pi}} \left(\frac{1 - x^2}{ip} e^{ipx} \right)_{-1}^1 + \frac{2}{\sqrt{2\pi}} \int_{-1}^1 x \cdot \frac{e^{ipx}}{ip} dx$$

$$= \frac{\sqrt{2}}{ip\sqrt{\pi}} \left[\left(\frac{x e^{ipx}}{ip} \right)_{-1}^1 - \int_{-1}^1 1 \cdot \frac{e^{ipx}}{ip} dx \right] = \frac{\sqrt{2}}{ip\sqrt{\pi}} \left\{ \frac{e^{ip} + e^{-ip}}{ip} - \frac{(e^{ipx}) - 1}{(ip)^2} \right\}$$

$$= \frac{\sqrt{2}}{ip\sqrt{\pi}} \left[\frac{2 \cos p}{ip} + \frac{e^{ip} - e^{-ip}}{p^2} \right] = \frac{\sqrt{2}}{\sqrt{\pi}} \left[-\frac{2 \cos p}{p^2} + \frac{2i \sin p}{ip^3} \right]$$

$$= -2 \sqrt{\frac{2}{\pi}} \left[\frac{p \cos p - \sin p}{p^3} \right]$$

If $\qquad \tilde{F}(p) = \dfrac{1}{\sqrt{2\pi}} \int_{-\infty}^\infty F(x) e^{ipx} dx$

then $\qquad F(x) = \dfrac{1}{\sqrt{2\pi}} \int_{-\infty}^\infty \tilde{F}(p) e^{-ipx} dp$

$$\therefore \quad -\frac{1}{\sqrt{2\pi}} \int_{-\infty}^\infty \frac{2\sqrt{2/\pi}(p \cos p - \sin p)}{p^3} \cdot e^{-ipx} dp = \begin{bmatrix} 1 - x^2 & , & |x| < 1 \\ 0 & , & |x| > 1 \end{bmatrix}$$

or $\quad -\int_{-\infty}^\infty \left(\dfrac{p \cos p - \sin p}{p^3} \right) \cos px dp$

or $\quad +i \int_{-\infty}^\infty \left(\dfrac{p \cos p - \sin p}{p^3} \right) \sin px dp = \begin{bmatrix} \dfrac{\pi}{2}(1 - x^2) & , & |x| < 1 \\ 0 & , & |x| > 1 \end{bmatrix}$

or $\quad -\int_{-\infty}^\infty \dfrac{p \cos p - \sin p}{p^3} \cos px dp = \begin{bmatrix} \dfrac{\pi}{2}(1 - x^2) & , & |x| < 1 \\ 0 & , & |x| > 1 \end{bmatrix}$

Now taking $x = \dfrac{1}{2}$, we have

$$-\int_{-\infty}^\infty \frac{p \cos p - \sin p}{p^3} \cdot \cos \frac{p}{2} dp = \frac{\pi}{2} \left(1 - \frac{1}{4} \right) = \frac{3\pi}{8}$$

or $\quad 2\int_0^\infty \dfrac{p\cos p - \sin p}{p^3}\cdot\cos\dfrac{p}{2}dp = -\dfrac{3\pi}{8}$

or $\quad \int_0^\infty \left(\dfrac{x\cos x - \sin x}{x^3}\right)\cdot\cos\dfrac{x}{2}dx = -\dfrac{3\pi}{16}$.

EXAMPLE 9. *Find the cosine transform of a function of x which is unity for $0 < x < a$ and zero for $x \ge a$. What is the function whose cosine transform is $\sqrt{\dfrac{2}{\pi}}\dfrac{\sin ap}{p}$?*

(VTU–2010, SVTU–2009; UPTU–2008; Garhwal–2000; Meerut–1998)

SOLUTION. It is given that $f(x) = \begin{cases} 1 & , \ 0 < x < a \\ 0 & , \ x \ge a \end{cases}$

We have $\quad \tilde{f}_c(p) = \sqrt{\dfrac{2}{\pi}}\int_0^\infty f(x)\cos px\,dx = \sqrt{\dfrac{2}{\pi}}\int_0^a \cos px\,dx = \sqrt{\dfrac{2}{\pi}}\dfrac{\sin pa}{p}$

Again , $\quad f(x) = \sqrt{\dfrac{2}{\pi}}\int_0^\infty \tilde{f}_c(p)\cos px\,dp$

$$= \sqrt{\dfrac{2}{\pi}}\int_0^\infty \sqrt{\dfrac{2}{\pi}}\dfrac{\sin pa}{p}\cos px\,dp = \dfrac{1}{\pi}\int_0^\infty \dfrac{\sin(a+x)p + \sin(a-x)p}{p}dp$$

$$= \dfrac{1}{\pi}\int_0^\infty \dfrac{\sin(a+x)p}{p}dp + \dfrac{1}{\pi}\int_0^\infty \dfrac{\sin(a-x)p}{p}dp$$

$$= \dfrac{1}{\pi}\left(\dfrac{\pi}{2} + \dfrac{\pi}{2}\right) = 1 \text{ if } x < a$$

and $\quad f(x) = \dfrac{1}{\pi}\left(\dfrac{\pi}{2} - \dfrac{\pi}{2}\right) = 0 \text{ if } x > a \qquad \left[\because \int_0^\infty \dfrac{\sin ax}{x}dx = \dfrac{\pi}{2}, \text{ if } a > 0\right]$

EXAMPLE 10. *Find the Fourier sine and cosine transform of f(x), if*

$$f(x) = \begin{cases} x & , \ 0 < x < 1 \\ 2-x & , \ 1 < x < 2 \\ 0 & , \ x > 2 \end{cases}$$

(JNTU–2006; Meerut–1991; Rohilkhand–1989, 97; Kanpur–1988, 2002, 07, 10, 14; Mysore–1997)

SOLUTION. We have

$$\tilde{f}_s(p) = \sqrt{\dfrac{2}{\pi}}\int_0^\infty f(x)\sin px\,dx$$

$$= \sqrt{\dfrac{2}{\pi}}\cdot\left[\int_0^1 x\sin px\,dx + \int_1^2 (2-x)\sin px\,dx\right]$$

$$= \sqrt{\dfrac{2}{\pi}}\cdot\left[\left(-\dfrac{x}{p}\cos px + \dfrac{1}{p^2}\sin px\right)_0^1 + \left\{-\left(\dfrac{2-x}{p}\right)\cos px - \dfrac{1}{p^2}\sin px\right\}_1^2\right]$$

$$= \sqrt{\dfrac{2}{\pi}}\cdot\left[\dfrac{2}{p^2}\sin p - \dfrac{1}{p^2}\sin 2p\right] = 2\sqrt{\dfrac{2}{\pi}}\cdot\dfrac{\sin p}{p^2}(1-\cos p)$$

$$\tilde{f}_c(p) = \sqrt{\dfrac{2}{\pi}}\int_0^\infty f(x)\cos px\,dx$$

$$= \sqrt{\dfrac{2}{\pi}}\left[\int_0^1 x\cos px\,dx + \int_1^2 (2-x)\cos px\,dx\right]$$

$$= \sqrt{\frac{2}{\pi}} \left[\left(\frac{x}{p} \sin px + \frac{1}{p^2} \cos px \right)_0^1 + \left\{ \left(\frac{2-x}{p} \right) \sin px - \frac{\cos px}{p^2} \right\}_1^2 \right]$$

$$= \sqrt{\frac{2}{\pi}} \frac{1}{p^2} [2 \cos p - 1 - \cos 2p] = 2\sqrt{\frac{2}{\pi}} \cdot \frac{\cos p}{p^2} (1 - \cos p)$$

EXAMPLE 11.   *Find Fourier sine and cosine transforms of e^{-x} and using the inversion formulae recover the original functions, in both the cases.*

(Meerut–1983, 86; Agra–1987; Kanpur–2001, 2009; Osmania–1992)

SOLUTION.   Let $f(x) = e^{-x}$

Then

$$\tilde{f}_s(p) = \sqrt{\frac{2}{\pi}} \int_0^\infty f(x) \sin px \, dx = \sqrt{\frac{2}{\pi}} \int_0^\infty e^{-x} \sin px \, dx$$

$$= \sqrt{\frac{2}{\pi}} \left[\frac{e^{-x}}{1+p^2} (-\sin px - p \cos px) \right]_0^\infty = \frac{p}{1+p^2} \sqrt{\frac{2}{\pi}}$$

and

$$\tilde{f}_c(p) = \sqrt{\frac{2}{\pi}} \int_0^\infty f(x) \cos px \, dx = \sqrt{\frac{2}{\pi}} \int_0^\infty e^{-x} \cos px \, dx$$

$$= \sqrt{\frac{2}{\pi}} \left[\frac{e^{-x}}{1+p^2} (-\cos px + p \sin px) \right]_0^\infty = \frac{1}{1+p^2} \sqrt{\frac{2}{\pi}}.$$

Applying inversion to the sine transform, we have

$$f(x) = \sqrt{\frac{2}{\pi}} \int_0^\infty \tilde{f}_s(p) \cdot \sin px \, dp = \frac{2}{\pi} \int_0^\infty \frac{p \sin px}{1+p^2} dp$$

and applying inversion to the cosine transform, we have

$$f(x) = \sqrt{\frac{2}{\pi}} \int_0^\infty \tilde{f}_c(p) \cos px \, dp = \frac{2}{\pi} \int_0^\infty \frac{\cos px}{1+p^2} dp$$

Therefore, from Fourier integral theorem, we have

$$f(x) = \frac{1}{\pi} \int_0^\infty dp \int_{-\infty}^\infty f(v) \cos p(x-v) dv$$

or

$$f(x) = \frac{1}{\pi} \int_0^\infty \cos px \, dp \int_{-\infty}^\infty f(v) \cos pv \, dv + \frac{1}{\pi} \int_0^\infty \sin px \, dp \int_0^\infty f(v) \sin pv \, dv$$

...(3)

Case I: Define $f(x)$ in $[-\infty, 0]$ such that $f(x)$ is an even function of x, from (3), we have

$$f(x) = \frac{2}{\pi} \int_0^\infty \cos px \, dp \int_0^\infty f(v) \cos pv \, dv$$

Taking $f(x) = e^{-x}$, we have

$$e^{-x} = \frac{2}{\pi} \int_0^\infty \cos px \, dp \int_0^\infty e^{-v} \cos pv \, dv$$

$$= \frac{2}{\pi} \int_0^\infty \cos px \left[\frac{e^{-x}}{1+p^2} (-\cos pv + p \sin pv) \right]_0^\infty dp$$

$$= \frac{2}{\pi} \int_0^\infty \frac{\cos px}{1+p^2} dp$$

$$\therefore \qquad \int_0^\infty \frac{\cos px}{1+p^2}\,dp = \frac{\pi}{2}e^{-x}$$

$\therefore$ From (2), we have $f(x) = \frac{2}{\pi}\cdot\frac{\pi}{2}e^{-x} = e^{-x}$

Case II. Now, we define $f(x)$ in $(-\infty, 0)$ such that $f(x)$ is an odd function of x.
From (2), we have

$$f(x) = \frac{2}{\pi}\int_0^\infty \sin px\,dp\int_0^\infty f(v)\sin pv\,dv$$

Taking $f(x) = e^{-x}$ and simplifying, we have

$$\int_0^\infty \frac{p\sin px}{1+p^2}\,dp = \frac{\pi}{2}e^{-x}$$

$\therefore$ From (1), $\qquad\qquad f(x) = \frac{2}{\pi}\cdot\frac{\pi}{2}e^{-x} = e^{-x}.$

EXAMPLE 12. *Find Fourier cosine transform of* $f(x) = \dfrac{1}{1+x^2}$ *and hence find Fourier sine transform*
of $F(x) = \dfrac{x}{1+x^2}.$ (Meerut–1987, 88, 90, 98, 2013; Agra–1989, 2000, 04;
Bundelkhand–2005, 06, Agra–2004; Kanpur–1993, 97, 2007; Rohilkhand–2014)

SOLUTION. We have

$$\tilde{f}_c(p) = \sqrt{\frac{2}{\pi}}\int_0^\infty f(x)\cos px\,dx = \sqrt{\frac{2}{\pi}}\int_0^\infty \frac{\cos px}{1+x^2}\,dx$$

Differentiating both sides w.r.t p, we get

$$\frac{d}{dp}\tilde{f}_c(p) = -\sqrt{\frac{2}{\pi}}\int_0^\infty \frac{x\sin px}{1+x^2}\,dx = -\sqrt{\frac{2}{\pi}}\int_0^\infty \frac{(x^2+1-1)\sin px}{x(1+x^2)}\,dx$$

$$= -\sqrt{\frac{2}{\pi}}\int_0^\infty \frac{\sin px}{x}\,dx + \sqrt{\frac{2}{\pi}}\int_0^\infty \frac{\sin px}{x(1+x^2)}\,dx$$

$$= -\sqrt{\frac{2}{\pi}}\cdot\frac{\pi}{2} + \sqrt{\frac{2}{\pi}}\int_0^\infty \frac{\sin px}{x(1+x^2)}\,dx$$

Differentiating again w.r.t p, we get

$$\frac{d^2}{dp^2}\tilde{f}_c(p) = \sqrt{\frac{2}{\pi}}\int_0^\infty \frac{\cos px}{1+x^2}\,dx = \tilde{f}_c(p)$$

or $\quad (D^2-1)\tilde{f}_c(p) = 0$

whose general solution is $\tilde{f}_c(p) = Ae^p + Be^{-p}$ ... (1)

Now when $p = 0$,

$$\tilde{f}_c(p) = \sqrt{\frac{2}{\pi}}\int_0^\infty \frac{dx}{1+x^2} = \sqrt{\frac{2}{\pi}}\left[\tan^{-1}x\right]_0^\infty = \frac{\pi}{2}\sqrt{\frac{2}{\pi}} = \sqrt{\frac{\pi}{2}}$$

and $\quad \dfrac{d}{dp}\tilde{f}_c(p) = -\sqrt{\dfrac{\pi}{2}}$

$\therefore$ From (1), we have

$$\sqrt{\frac{\pi}{2}} = A+B \quad \text{and} \quad -\sqrt{\frac{\pi}{2}} = A-B$$

Solving, $\qquad A = 0, B = \sqrt{\dfrac{\pi}{2}}$

∴ From (1), we have $\tilde{f}_c(p) = \sqrt{\dfrac{\pi}{2}} e^{-p}$

Now, we have

$$\tilde{f}_c(p) = \sqrt{\frac{2}{\pi}} \int_0^\infty \frac{\cos px}{1 + x^2} dx = \sqrt{\frac{\pi}{2}} . e^{-p}$$

Now differentiating both sides w.r.t. p, we have

$$-\int_0^\infty \frac{x \sin px}{1 + x^2} dx = -\frac{\pi}{2} . e^{-p}$$

or $\qquad \tilde{F}_s(p) = \sqrt{\dfrac{2}{\pi}} \int_0^\infty \dfrac{x}{1 + x^2} \sin px\, dx = \sqrt{\dfrac{\pi}{2}} . e^{-p}$

EXAMPLE 13. *Find the sine and cosine transform of* $x^n e^{-ax}$ (Meerut–1989, 91; Kanpur–1981; Agra–2001)

SOLUTION. Let $\quad f(x) = x^n e^{-ax}$

$$\therefore \quad \tilde{f}_s(p) = \sqrt{\frac{2}{\pi}} \int_0^\infty f(x) \sin px\, dx$$

$$= \sqrt{\frac{2}{\pi}} \int_0^\infty x^n e^{-ax} \sin px\, dx \qquad \qquad \ldots(1)$$

We have

$$\int_0^\infty e^{-ax} \sin px\, dx$$

$$= \left[\frac{e^{-ax}}{a^2 + p^2} (-a \sin px - p \cos px) \right]_0^\infty$$

$$= \frac{p}{a^2 + p^2} = \frac{1}{2i} \left(\frac{1}{a - ip} - \frac{1}{a + ip} \right)$$

Differentiating both sides w.r.t. a, n times, we get

$$(-1)^n \int_0^\infty x^n e^{-ax} \sin px\, dx$$

$$= \frac{1}{2i} \left[\frac{d^n}{da^n} (a - ip)^{-1} - \frac{d^n}{da^n} (a + ip)^{-1} \right]$$

$$= \frac{1}{2i} (-1)^n n! \left[(a - ip)^{-(n+1)} - (a + ip)^{-(n+1)} \right]$$

$$= \frac{1}{2i} (-1)^n n! \left[2i r^{-(n+1)} \sin(n+1)\theta \right] \qquad \text{Putting } a = r \cos\theta,\ p = r \sin\theta$$

$$= (-1)^n n! (1/r)^{n+1} \sin(n+1)\theta$$

$$\therefore \int_0^\infty x^n e^{-ax} \sin px\, dx$$

$$= n! . [1 / (a^2 + p^2)^{(n+1)/2}] \sin\left\{ (n+1) \tan^{-1}(p/a) \right\}$$

$$[\because\ r = (a^2 + p^2)^{1/2} \text{ and } \theta = \tan^{-1}(p/a)]$$

Hence, from (1)

$$\tilde{f}_s(p) = \sqrt{\frac{2}{\pi}} \cdot \frac{n!\sin\{(n+1)\tan^{-1}(p/a)\}}{(a^2+p^2)^{(n+1)/2}}$$

Also, $\tilde{f}_c(p) = \sqrt{\frac{2}{\pi}} \int_0^\infty f(x)\cos px\, dx$

$$\tilde{f}_c(p) = \sqrt{\frac{2}{\pi}} \int_0^\infty x^n e^{-ax}\cos px\, dx \qquad \ldots(2)$$

We have

$$\int_0^\infty e^{-ax}\cos px\, dx$$

$$= \left[\frac{e^{-ax}}{a^2+p^2}(-a\cos px + p\sin px)\right]_0^\infty$$

$$= \frac{a}{a^2+p^2} = \frac{1}{2}\left(\frac{1}{a-ip}+\frac{1}{a+ip}\right)$$

Differentiating both sides w.r.t. a, n times, we have

$$(-1)^n \int_0^\infty x^n e^{-ax}\cos px\, dx = \frac{1}{2}(-1)^n n!\left[(a-ip)^{-(n+1)} + (a+ip)^{-(n+1)}\right]$$

or $\int_0^\infty x^n e^{-ax}\cos px\, dx = (-1)^n n!(1/r)^{n+1}\cos(n+1)\theta$

Putting $a = r\cos\theta$, $p = r\sin\theta$ and on simplification

$$= (-1)^n n!\frac{\cos\{(n+1)\tan^{-1}(p/a)\}}{(a^2+p^2)^{(n+1)/2}}$$

Hence from (2), we get

$$\tilde{f}_c(p) = \sqrt{\frac{2}{\pi}}\frac{n!\cos\{(n+1)\tan^{-1}(p/a)\}}{(a^2+p^2)^{(n+1)/2}}.$$

EXAMPLE 14. *Find Fourier sine transform of* $e^{-ax/x}$. (PTU–2006; VTU–2010; Rohtak–2005; Agra–1981; Meerut–1988, 89, 2000; Kanpur–2005; Rohilkhand–1990)

SOLUTION. If $f(x) = \dfrac{e^{-ax}}{x}$, then we have

$$\tilde{f}_s(p) = \sqrt{\frac{2}{\pi}} \int_0^\infty \frac{e^{-ax}}{x}\sin px\, dx$$

Differentiating both sides w.r.t. p, we have

$$\frac{d}{dp}\tilde{f}_s(p) = \sqrt{\frac{2}{\pi}} \int_0^\infty e^{-ax}\cos px\, dx$$

$$= \sqrt{\frac{2}{\pi}}\left[\frac{e^{-ax}}{a^2+p^2}(-a\cos px + p\sin px)\right]_0^\infty = \frac{a}{a^2+p^2}\cdot\sqrt{\frac{2}{\pi}}$$

$$\therefore \quad \tilde{f}_s(p) = a\sqrt{\frac{2}{\pi}}\int\frac{dp}{a^2+p^2} + c = \sqrt{2/\pi}\,\tan^{-1}(p/a) + c$$

But when $p = 0, \tilde{f}_s(p) = 0$

$\therefore \qquad c = 0$

Hence, $\tilde{f}_s(p) = \sqrt{2/\pi} \cdot \tan^{-1}(p/a)$.

EXAMPLE 15. *Find the Fourier sine transform of*

$$f(x) = \frac{1}{x(a^2 + x^2)}$$

(UPTU–2008; Meerut–1984)

SOLUTION. We have

$$\tilde{f}_s(p) = \sqrt{\frac{2}{\pi}} \int_0^\infty \frac{1}{x(a^2 + x^2)} \sin px \, dx \qquad \qquad ...(1)$$

Let

$$I = \int_0^\infty \frac{1}{x(a^2 + x^2)} \sin px \, dx \qquad \qquad ...(2)$$

Then

$$\frac{dI}{dp} = \frac{d}{dp} \int_0^\infty \frac{\sin px}{x(a^2 + x^2)} dx = \int_0^\infty \left[\frac{\partial}{\partial p} \left\{ \frac{\sin px}{x(a^2 + x^2)} \right\} \right] dx$$

$$= \int_0^\infty \frac{\cos px}{a^2 + x^2} dx \qquad \qquad ...(3)$$

$\therefore \qquad \dfrac{d^2 I}{dp^2} = -\int_0^\infty \dfrac{x \sin px}{a^2 + x^2} dx = -\int_0^\infty \dfrac{x^2 \sin px}{x(a^2 + x^2)} dx$

$$= -\int_0^\infty \frac{(x^2 + a^2) - a^2}{x(a^2 + x^2)} \sin px \, dx$$

$$= -\int_0^\infty \frac{\sin px}{x} dx + a^2 \int_0^\infty \frac{\sin px}{x(a^2 + x^2)} dx$$

$$= -\frac{\pi}{2} + a^2 I. \qquad \qquad \left[\because \int_0^\infty \frac{\sin px}{x} dx = \frac{\pi}{2} \right]$$

$\therefore \qquad \dfrac{d^2 I}{dp^2} - a^2 I = -\dfrac{\pi}{2}$

or $\quad (D^2 - a^2)I = -\dfrac{\pi}{2}$, where $D \equiv \dfrac{d}{dp}$

The solution of the above differential equation is

$$I = Ae^{-ap} + Be^{ap} + \frac{\pi}{2a} \qquad \qquad ...(4)$$

$\therefore \qquad \dfrac{dI}{dp} = -Aae^{-ap} + Bae^{-ap} \qquad \qquad ...(5)$

Now from (1), when $p = 0$, we have $I = 0$ and from (2), when $p = 0$, we have

$$\frac{dI}{dp} = \int_0^\infty \frac{1}{a^2 + x^2} dx = \frac{1}{a} \left[\tan^{-1} \frac{x}{a} \right]_0^\infty = \frac{\pi}{2a}$$

So putting $p = 0$ in (4) and (5), we get

$$A + B = -\frac{\pi}{2a^2} \qquad \qquad ...(6)$$

and $a(-A + B) = \dfrac{\pi}{2a^2}, i.e., -A + B = \dfrac{\pi}{2a^2}$ ...(7)

Solving (6) and (7), we get $B = 0, A = -\dfrac{\pi}{2a^2}$ ...(8)

Putting the values of A and B in (4), we get

$$I = \int_0^\infty \frac{\sin px}{x(a^2 + x^2)} dx = -\frac{\pi}{2a^2} e^{-ap} + \frac{\pi}{2a^2} = \frac{\pi}{2a^2}(1 - e^{-ap}).$$

Now putting the value of I in (1), we get

$$\tilde{f}_s(p) = \sqrt{\frac{2}{\pi}} \cdot \frac{\pi}{2a^2}(1 - e^{-ap}) = \frac{1}{a^2}\sqrt{\frac{\pi}{2}} \cdot (1 - e^{-ap})$$

EXAMPLE 16. *Find the Fourier cosine transform of e^{-x^2}.* (VTU–2010; Rajasthan–2006; Anna–2009; Meerut–1989, 91, 92, 95; Agra–1982, 85; Kanpur–1994, 96; Rohilkhand–1994, 96; Calicut–1994; Purvanchal–1990)

SOLUTION. We have

$$\tilde{F}_c\{e^{-x^2}\} = \sqrt{\frac{2}{\pi}} \int_0^\infty e^{-x^2} \cos px\, dx = I$$

Differentiating w.r.t. 'p', we have

$$\frac{dI}{dp} = -\sqrt{\frac{2}{\pi}} \int_0^\infty x e^{-x^2} \sin px\, dx = \frac{1}{2}\sqrt{\frac{2}{\pi}} \int_0^\infty (-2x e^{-x^2}) \cdot \sin px\, dx$$

$$= \frac{1}{2}\sqrt{\frac{2}{\pi}}\left[(e^{-x^2} \sin px)_0^\infty - p\int_0^\infty e^{-x^2} \cos px\, dx \right]$$

(Integrating by parts taking $\sin px$ as first function.)

$$= -\frac{p}{2} I$$

$$\therefore \qquad \frac{dI}{I} = -\frac{p}{2} dp$$

Integrating, we have

$$\log I = -\frac{p^2}{4} + \log A \text{ or } I = Ae^{-p^2/4} \qquad \text{...(2)}$$

But when $p = 0$, from (1),

$$I = \sqrt{\frac{2}{\pi}} \int_0^\infty e^{-x^2} dx = \frac{1}{\sqrt{2}}$$

$\therefore$ From (2), $A = 1/\sqrt{2}$

Hence, $\quad I = F_c\{e^{-x^2}\} = (1/\sqrt{2}) e^{-p^2/4}$

EXAMPLE 17. *Find $f(x)$ if its sine transform is $\dfrac{\pi}{2}$.*

SOLUTION. We have $\quad \tilde{f}_s(p) = \dfrac{\pi}{2}$

$\therefore$ Applying Fourier sine inversion formula, we have

$$f(x) = \sqrt{\frac{2}{\pi}} \int_0^\infty \tilde{f}_s(p) \sin px\, dp$$

$$f(x) = \sqrt{\frac{2}{\pi}} \int_0^\infty \frac{\pi}{2} . \sin pxdp = \sqrt{\frac{\pi}{2}} \int_0^\infty \sin pxdp \qquad \ldots(1)$$

Now, we have

$$\int_0^\infty e^{-ipx} dp = \left[\frac{e^{-ipx}}{(-ix)}\right]_{p=0}^\infty = 1/(ix) = -i/x$$

or $\int_0^\infty (\cos px - i \sin px) dp = -i/x$.

Equating imaginary parts on both sides, we have

$$\int_0^\infty \sin pxdp = 1/x$$

Hence from (1) $f(x) = \sqrt{(\pi/2)}.(1/x)$

EXAMPLE 18. *Find the inverse Fourier transform of $\tilde{f}(p) = e^{-|p|y}$.* (Kanpur–1998, 99)

SOLUTION. We have $|p| = \begin{cases} -p & , & p \leq 0 \\ p & , & p \geq 0 \end{cases}$

$$\therefore \quad f(x) = \frac{1}{\sqrt{2\pi}} \int_{-\infty}^\infty \tilde{f}(p).e^{-ipx} dp = \frac{1}{\sqrt{2\pi}} \int_{-\infty}^\infty e^{-|p|y}.e^{-ipx} dp$$

$$= \frac{1}{\sqrt{2\pi}} \int_{-\infty}^0 e^{py}.e^{-ipx} dp + \frac{1}{\sqrt{2\pi}} \int_0^\infty e^{-py}.e^{-ipx} dp$$

$$= \frac{1}{\sqrt{2\pi}} \int_{-\infty}^0 e^{(y-ix)p} dp + \frac{1}{\sqrt{2\pi}} \int_0^\infty e^{-p(y+ix)} dp$$

$$= \frac{1}{\sqrt{2\pi}} \left[\frac{e^{(y-ix)p} x}{(y-ix)}\right]_{-\infty}^0 + \frac{1}{\sqrt{2\pi}} \left[\frac{e^{-p(y+ix)}}{-(y+ix)}\right]_0^\infty$$

$$= \frac{1}{\sqrt{2\pi}} \left[\frac{1}{(y-ix)} + \frac{1}{(y+ix)}\right] = \frac{y\sqrt{2}}{\sqrt{\pi}(y^2 + x^2)}.$$

EXAMPLE 19. *Find $f(x)$ if $\tilde{f}_c(p) = p^n e^{-ap}$.*

SOLUTION. Using Fourier cosine inversion formula, we have

$$f(x) = \sqrt{\frac{2}{\pi}} \int_0^\infty p^n e^{-ap} \cos pxdp \qquad \ldots(1)$$

We have $\int_0^\infty e^{-ap} \cos pxdp = \dfrac{a}{a^2 + x^2}$.

Differentiating both sides w.r.t. a, n times, we have

$$(-1)^n \int_0^\infty p^n e^{-ap} \cos pxdp = \frac{d^n}{da^n}\left(\frac{a}{a^2 + x^2}\right) = \frac{1}{2}\frac{d^n}{da^n}\left(\frac{1}{a - ix} + \frac{1}{a + ix}\right)$$

$$= \frac{1}{2}[(-1)^n n!(a - ix)^{-n-1} + (-1)^n n!(a + ix)^{-n-1}]$$

$$= \frac{(-1)^n n!}{x^{n+1}} \cos(n+1)\theta \sin^{n+1}\theta$$

Where $\theta = \tan^{-1}\left(\dfrac{x}{a}\right)$

$$\therefore \qquad \int_0^\infty p^n e^{-ap} \cos px\, dp$$

$$= \frac{n!\cos(n+1)\theta}{x^{r+1}} \cdot \frac{x^{n+1}}{(a^2+x^2)^{(n+1)/2}}$$

$$= \frac{n!\cos(n+1)\theta}{(a^2+x^2)^{(n+1)/2}}$$

$\therefore$ From (1) $\qquad f(x) = \sqrt{\dfrac{2}{\pi}} \dfrac{n!\cos(n+1)\theta}{(a^2+x^2)^{(n+1)/2}}$

EXAMPLE 20. *Use the sine inversion formula to obtain f(x) if*

$$\tilde{f}_s(p) = \frac{p}{1+p^2}. \qquad \text{(Meerut–2006; Rohilkhand–1993, 95; Kanpur–2006; SVTU–2005)}$$

SOLUTION. Using Fourier sine inversion formula, we have

$$f(x) = \sqrt{\frac{2}{\pi}} \int_0^\infty \frac{p}{1+p^2} \cdot \sin px\, dp$$

$$= \sqrt{\frac{2}{\pi}} \int_0^\infty \frac{p^2}{p(1+p^2)} \sin px\, dp = \sqrt{\frac{2}{\pi}} \int_0^\infty \frac{(p^2+1)-1}{p(1+p^2)} \sin px\, dp$$

$$= \sqrt{\frac{2}{\pi}} \int_0^\infty \frac{\sin px}{p} dp - \sqrt{\frac{2}{\pi}} \int_0^\infty \frac{\sin px}{p(1+p^2)} dp$$

or $\qquad f(x) = \sqrt{\dfrac{\pi}{2}} - \sqrt{\dfrac{2}{\pi}} \int_0^\infty \dfrac{\sin px}{p(1+p^2)} dp \qquad \left[\because \int_0^\infty \dfrac{\sin px}{p} dp = \dfrac{\pi}{2}\right] \qquad \ldots(1)$

$\therefore \qquad \dfrac{df}{dx} = -\sqrt{\dfrac{2}{\pi}} \int_0^\infty \dfrac{\cos px}{1+p^2} dp \qquad \qquad \ldots(2)$

and $\qquad \dfrac{d^2 f}{dx^2} = \sqrt{\dfrac{2}{\pi}} \int_0^\infty \dfrac{p\sin px}{1+p^2} dp \qquad$ or $\qquad \dfrac{d^2 f}{dx^2} - f = 0$

Whose solution is $f = Ae^x + Be^{-x}$ $\qquad\qquad\qquad \ldots (3)$

$\therefore \qquad \dfrac{df}{dx} = Ae^x - Be^{-x}$ $\qquad\qquad\qquad \ldots (4)$

Now when $x=0, f = \sqrt{\dfrac{\pi}{2}}$, $\qquad\qquad\qquad$ [From (1)]

and $\qquad \dfrac{df}{dx} = -\sqrt{\dfrac{2}{\pi}} \int_0^\infty \dfrac{dp}{1+p^2} = -\sqrt{\dfrac{\pi}{2}} \qquad$ [From (2)]

$\therefore$ From (3) and (4)

$$\sqrt{\frac{\pi}{2}} = A+B \text{ and } -\sqrt{\frac{\pi}{2}} = A-B$$

Solving, $\qquad A = 0, B = \sqrt{\dfrac{\pi}{2}}$

Hence, $\qquad f(x) = \sqrt{\dfrac{\pi}{2}} e^{-x}.$

EXAMPLE 21. *Find the Fourier transform of*

$$f(x) = \begin{cases} x, & |x| \le a \\ 0 & |x| > a \end{cases}$$ (Meerut–1991; Kanpur–1980, 97, 2008; Haridwar–2001; Rohilkhand–1997)

SOLUTION. Here we have

$$\tilde{f}(p) = \frac{1}{\sqrt{2\pi}} \int_{-\infty}^{\infty} f(x) e^{ipx} dx = \frac{1}{\sqrt{2\pi}} \int_{-a}^{a} x e^{ipx} dx$$

$$= \frac{1}{\sqrt{2\pi}} \left[x \frac{e^{ipx}}{ip} - \frac{e^{ipx}}{(ip)^2} \right]_{-a}^{a} = \frac{1}{\sqrt{2\pi}} \cdot \frac{1}{ip^2} [pa(e^{ipa} + e^{-ipa}) + i(e^{ipa} - e^{-ipa})]$$

$$= -\frac{i}{p^2} \cdot \frac{1}{\sqrt{2\pi}} [pa \cdot 2\cos pa + i \cdot 2i \sin pa] = -\frac{i}{p^2} \sqrt{\frac{2}{\pi}} (ap \cos ap - \sin ap)$$

EXAMPLE 22. *What is the function whose cosine transform is* $\dfrac{\sin ap}{p}$ *?*

(Bundelkhand–2005; Kanpur–1979; Meerut–1999; Agra–2000)

SOLUTION. Here, we have

$$\tilde{f}_c(p) = \int_0^{\infty} f(x) \cos px dx$$

Then $$f(x) = \frac{2}{\pi} \int_0^{\infty} \tilde{f}_c(p) \cos px dp = \frac{2}{\pi} \int_0^{\infty} \frac{\sin ap}{p} \cos px dp$$

EXAMPLE 23. *Find the sine and cosine transform of* $2e^{-5x} + 5e^{-2x}$. (Kanpur–1998, 2003)

SOLUTION. Consider $f_c\{2e^{-5x} + 5e^{-2x}\} = 2f_c\{e^{-5x}\} + 5f_c\{e^{-2x}\}$

$$= 2\int_0^{\infty} e^{-5x} \cos px dx + 5\int_0^{\infty} e^{-2x} \cos px dx$$

$$= 2 \cdot \frac{5}{5^2 + p^2} + 5 \cdot \frac{2}{2^2 + p^2} = 10 \left(\frac{1}{p^2 + 4} + \frac{1}{p^2 + 25} \right)$$

Now, $f_s\{2e^{-5x} + 5e^{-2x}\} = 2f_s\{e^{-5x}\} + 5f_s\{e^{-2x}\}$

$$= 2\int_0^{\infty} e^{-5x} \sin px dx + 5\int_0^{\infty} e^{-2x} \sin px dx = 2f_s\{e^{-5x}\} + 5f_s\{e^{-2x}\}$$

$$= 2 \cdot \frac{p}{5^2 + p^2} + 5 \cdot \frac{p}{2^2 + p^2} = p \left(\frac{2}{p^2 + 25} + \frac{5}{p^2 + 4} \right)$$

EXAMPLE 24. *Find* $f_s^{-1}\{e^{-\pi p}\}$ (Agra–2013)

SOLUTION. Taking inverse Fourier sine transform, we get

$$f_s^{-1}\{e^{-\pi p}\} = \frac{2}{\pi} \int_0^{\infty} e^{-\pi p} \sin px dp = \frac{2}{\pi} \cdot \frac{x}{\pi^2 + x^2} = \frac{2x}{\pi(\pi^2 + x^2)}$$

EXAMPLE 25. *Find the sine transform of the function*

$$f(x) = \begin{cases} \sin x & when & 0 < x < a \\ 0 & when & x > 0 \end{cases}$$ (Kanpur–2003, 04)

SOLUTION. Here, we have

$$f_s\{f(x)\} = \int_0^{\infty} f(x) \sin px dx = \int_0^a f(x) \sin px dx + \int_a^{\infty} f(x) \sin px dx$$

$$= \int_0^a \sin x \cdot \sin px dx + \int_a^{\infty} 0 \cdot \sin px dx = \int_0^a \sin x \cdot \sin px dx + 0$$

$$= \frac{1}{2} \int_0^a [\cos(1-p)x - \cos(1+p)x] dx$$

$$= \frac{1}{2} \left[\frac{\sin(1-p)x}{1-p} - \frac{\sin(1+p)x}{1+p} \right]_0^a = \frac{1}{2} \left[\frac{\sin(1-p)a}{1-p} - \frac{\sin(1+p)a}{1+p} \right]$$

EXAMPLE 26. *Find the Fourier sine transform of $e^{-|x|}$. Hence, evaluate $\int_0^\infty \frac{x\sin(mx)}{1+x^2}dx$.*

(Kanpur–2001; Purvanchal–1994)

SOLUTION. Here, Fourier sine transform of $e^{-|x|} = F(x)$ is

$$F_p\{f(x)\} = f_p(p) = \int_0^\infty F(x)\cdot\sin(px)dx$$

$$= \int_0^\infty e^{-|x|}\sin(px)dx = \int_0^\infty e^{-x}\sin(px)dx = \frac{p}{1+p^2}$$

Now, $\qquad F(x) = F_p^{-1}\{f_s(p)\} = \frac{2}{\pi}\int_0^\infty f_s(p)\sin(px)dp$

$\Rightarrow \qquad F(x) = \frac{2}{\pi}\int_0^\infty \frac{p}{1+p^2}\sin(px)dp = \frac{\pi}{2}e^{-x}$

$$= \int_0^\infty \frac{p}{1+p^2}\sin(px)dp$$

Therefore replacing x by m

$$\frac{\pi}{2}e^{-m} = \int_0^\infty \frac{p}{1+p^2}\sin(mp)dp$$

$\Rightarrow \int_0^\infty \left(\frac{x}{1+x^2}\right)\sin(mx)dx = \frac{\pi}{2}e^{-m}$

EXAMPLE 27. *Show that $e^{-x}\cos x = \frac{2}{\pi}\int_0^\infty \frac{(u^2+2)\cos ux}{u^2+4}du, x > 0$.* (Kanpur–2007)

SOLUTION. We know that

$$f(x) = \frac{2}{\pi}\int_0^\infty \cos ux\left\{\int_0^\infty f(v)\cos uvdv\right\}du \qquad \text{...(1)}$$

Now, $\quad f(x) = e^{-x}\cos x$ and $f(v) = e^{-v}\cos v$

$\Rightarrow e^{-x}\cos x = \frac{2}{\pi}\int_0^\infty \cos ux\left\{\int_0^\infty e^{-v}\cos v\cos uvdv\right\}du$

$$= \frac{1}{\pi}\int_0^\infty \cos ux\left\{\int_0^\infty e^{-v}(2\cos uv\cos v)dv\right\}du$$

$$= \frac{1}{\pi}\int_0^\infty \cos ux\left[\int_0^\infty e^{-v}\{\cos(u+1)v+\cos(u-1)v\}dv\right]du$$

$$= \frac{1}{\pi}\int_0^\infty \cos ux\left[\int_0^\infty e^{-v}\cos(u+1)vdv + \int_0^\infty e^{-v}\cos(u-1)vdv\right]du$$

$$= \frac{1}{\pi}\int_0^\infty \cos ux\left[\frac{1}{1+(u+1)^2}+\frac{1}{1+(u-1)^2}\right]du$$

where $\int_0^\infty e^{-ax}\cos bxdx = \frac{a}{a^2+b^2}$

$$= \frac{1}{\pi}\int_0^\infty \frac{(u^2+2-2u)+(u^2+2+2u)}{(u^2+2+2u)(u^2+2-2u)}\cos uxdu$$

$$= \frac{1}{\pi}\int_0^\infty \frac{2(u^2+2)\cos ux}{(u^2+2)^2-(2u)^2}du = \frac{2}{\pi}\int_0^\infty \frac{(u^2+2)\cos ux}{u^4+4}du$$

EXAMPLE 28. *Find Fourier sine transform of e^{-ax}/x,*

(Meerut–2000; Kanpur–2005; Purvanchal–1995, 96, Banglore–1994; Delhi–2005)

SOLUTION.   From definition of Fourier transform

$$F_p\left\{\frac{e^{-ax}}{x}\right\} = \int_0^\infty \frac{e^{-ax}}{x} \sin px\, dx = 1, \text{ say} \qquad \ldots(1)$$

$$\frac{dI}{dp} = \int_0^\infty \frac{e^{-ax}}{x}(x\cos px)dx$$

$$= \int_0^\infty e^{-ax}\cos px\, dx = \frac{a}{a^2 + p^2}$$

$$\Rightarrow \qquad dI = \{a/a^2 + p^2\}dp \qquad \ldots(2)$$

On integrating eq^n (2), we get

$$I = \tan^{-1}(p/a) + C,$$

where C is any arbitrary constant

Now, if $p = 0$, then from (1) and (2)

$$C + \tan^{-1}0 = 0 \text{ such that } C = 0$$

Now, from eq^n (1) and (3), we get

$$I = \tan^{-1}(p/a)$$

$$\Rightarrow \qquad F_p\{e^{-ax}/x\} = \tan^{-1}(p/a)$$

Additional Solved Examples

EXAMPLE 1.   *Find the Fourier complex transform of $f(x)$ if*

$$f(x) = \begin{cases} e^{i\omega x}, & a < x < b \\ 0, & x < a, x > b \end{cases}.$$ (Meerut–1991; Avadh–2014)

SOLUTION.   Here, we have

$$F\{f(x)\} = \frac{1}{\sqrt{2\pi}} \int_{-\infty}^{\infty} e^{ipx} f(x)dx$$

$$= \frac{1}{\sqrt{2\pi}}\left[\int_{-\infty}^{a} 0 \cdot e^{ipx}dx + \int_a^b e^{ipx} \cdot e^{i\omega x}dx + \int_b^\infty 0 \cdot e^{ipx}dx\right]$$

$$= \frac{1}{\sqrt{2\pi}}\int_a^b e^{i(p+\omega)}dx = \frac{1}{\sqrt{2\pi}}\left[\frac{e^{i(p+\omega)x}}{i(p+\omega)}\right]_a^b$$

$$= \frac{i}{\sqrt{2\pi}}\left[\frac{e^{i(p+\omega)a} - e^{i(p+\omega)b}}{(p+\omega)}\right].$$

EXAMPLE 2.   *Find the sine transform of a function of x which is equal to $\sin x$ for $0 < x < a$ and 0 for $x > a$.*   (Meerut–1980; Rohilkhand–1995; Kanpur–2004)

SOLUTION.   Here, we have

$$\overline{f}_s(p) = \sqrt{\frac{2}{\pi}}\int_0^\infty f(x)\sin px\, dx = \sqrt{\frac{2}{\pi}}\left[\int_0^a \sin x \cdot \sin px\, dx + 0\right]$$

$$= \frac{1}{2}\sqrt{\frac{2}{\pi}}\int_0^a [\cos(p-1)x - \cos(p+1)x]dx$$

$$= \frac{1}{2}\sqrt{\frac{2}{\pi}}\left[\frac{\sin(p-1)x}{p-1} - \frac{\sin(p+1)x}{p+1}\right]_0^a$$

$$= \frac{1}{\sqrt{2\pi}}\left[\frac{\sin(p-1)a}{p-1} - \frac{\sin(p+1)a}{p+1}\right].$$

EXAMPLE 3. *Find the cosine transform of $f(x)$ if $f(x) = \begin{cases} \cos x, & 0 < x < a \\ 0, & x > a. \end{cases}$*

(Meerut–1983, 88; Kanpur–2001)

SOLUTION. Here, we have

$$\tilde{f}_c(p) = \sqrt{\frac{2}{\pi}} \int_0^\infty f(x) \cdot \cos px\, dx = \sqrt{\frac{2}{\pi}} \int_0^a \cos x \cdot \cos px\, dx$$

$$= \frac{1}{\sqrt{2\pi}} \int_0^a [\cos(1+p)x + \cos(1-p)x]\, dx$$

$$= \frac{1}{\sqrt{2\pi}} \left[\frac{\sin(1+p)x}{1+p} + \frac{\sin(1-p)x}{1-p} \right]_0^a$$

$$= \frac{1}{\sqrt{2\pi}} \left[\frac{\sin(1+p)a}{1+p} + \frac{\sin(1-p)a}{1-p} \right].$$

EXAMPLE 4. *Find the Fourier sine and cosine transform of $f(x)$ if*

$$f(x) = \begin{cases} 1 & ; \quad 0 \le x < 1 \\ 0 & ; \quad x > 1 \end{cases}$$

(Meerut–1990, 92; Raj.–1983; Agra–2013; Avadh–2013; Kanpur–2004; Guwahati–2000)

SOLUTION. Here, we have

$$\tilde{f}_s(p) = \sqrt{\frac{2}{\pi}} \int_0^\infty f(x) \sin px\, dx$$

$$= \sqrt{\frac{2}{\pi}} \int_0^1 \sin px\, dx = \sqrt{\frac{2}{\pi}} \left[-\frac{1}{p} \cos px \right]_0^1 = \sqrt{\frac{2}{\pi}} \cdot \frac{(1 - \cos p)}{p}$$

$$\tilde{f}_c(p) = \sqrt{\frac{2}{\pi}} \int_0^\infty f(x) \cos px\, dx$$

$$= \sqrt{\frac{2}{\pi}} \int_0^1 1 \cdot \cos px\, dx = \sqrt{\frac{2}{\pi}} \left[\frac{1}{p} \sin px \right]_0^1 = \sqrt{\frac{2}{\pi}} \cdot \frac{\sin p}{p}.$$

EXAMPLE 5. *Find the Fourier sine transform of $\dfrac{x}{1+x^2}$.*

(UPTU–2004; Kanpur–2007, 10; Meerut–1990; Agra–2001)

SOLUTION. Here, we have

$$\tilde{f}_s(p) = \sqrt{\frac{2}{\pi}} \int_0^\infty f(x) \sin px\, dx = \sqrt{\frac{2}{\pi}} \int_0^\infty \frac{x \sin px}{1+x^2}\, dx$$

$$= \sqrt{\frac{2}{\pi}} \int_0^\infty \frac{(x^2 + 1 - 1)}{x(1+x^2)} \sin px\, dx$$

$$= \sqrt{\frac{2}{\pi}} \int_0^\infty \frac{\sin px}{p}\, dx - \sqrt{\frac{2}{\pi}} \int_0^\infty \frac{\sin px}{x(1+x^2)}\, dx$$

$$\tilde{f}_s(p) = \sqrt{\frac{2}{\pi}} \cdot \frac{\pi}{2} - \sqrt{\frac{2}{\pi}} \int_0^\infty \frac{\sin px}{x(1+x^2)}\, dx \qquad \text{...(1)}$$

Now, differentiating w.r.to p, we get

$$\frac{d}{dp} \tilde{f}_s(p) = -\sqrt{\frac{2}{\pi}} \int_0^\infty \frac{\cos px}{1+x^2}\, dx \qquad \text{...(2)}$$

Again differentiating

$$\frac{d^2}{dp^2}\tilde{f}_s(p) = \sqrt{\frac{2}{\pi}}\int_0^\infty \frac{x\sin px}{1+x^2}dx = \tilde{f}_s(p) \qquad \text{...(3)}$$

$$= (D^2 - 1)\tilde{f}_s(p) = 0$$

$$\tilde{f}_s(p) = Ae^p + Be^{-p} \qquad \text{...(4)}$$

which is the general solution.

Let $p = 1$, then from eq^n (1), we get

$$\tilde{f}_s(p) = \sqrt{\frac{\pi}{2}}\int_0^\infty \frac{dx}{1+x^2}dx = -\sqrt{\frac{\pi}{2}}\left(\tan^{-1}x\right)_0^\infty = -\sqrt{\frac{\pi}{2}}$$

Now from eq^n (4), we get

$$\frac{d}{dp}\tilde{f}_s(p) = Ae^p - Be^{-p}$$

If $p = 0 \Rightarrow \tilde{f}_s(p) = \sqrt{\pi/2} \Rightarrow A + B = \sqrt{\pi/2}$

and $\qquad \dfrac{d}{dp}\tilde{f}_s(p) = -\sqrt{\pi/2} \Rightarrow A - B = -\sqrt{\pi/2}$

On solving, we get $A = 0, B = \sqrt{\pi/2}$

$$\tilde{f}_s(p) = \sqrt{\pi/2}\cdot e^{-p}. \qquad\qquad \text{[From (3)]}$$

EXAMPLE 6.   *Find the sine and cosine transform of* $\dfrac{e^{ax}+e^{-ax}}{e^{\pi x}-e^{-\pi x}}$

(Meerut–1987, 90; Agra–1986; Bundelkhand–2008; Kanpur–2014)

SOLUTION.   Let $\quad f(x) = \dfrac{e^{ax}+e^{-ax}}{e^{\pi x}-e^{-\pi x}}$

Now, $\tilde{f}_s(p) = \sqrt{\dfrac{2}{\pi}}\int_0^\infty f(x)\sin px\,dx = \sqrt{\dfrac{2}{\pi}}\int_0^\infty \dfrac{e^{ax}+e^{-ax}}{e^{\pi x}-e^{-\pi x}}\sin px\,dx$

$$= \sqrt{\frac{2}{\pi}}\int_0^\infty \frac{e^{ax}+e^{-ax}}{e^{\pi x}+e^{-\pi x}}\cdot\frac{e^{ipx}-e^{-ipx}}{2i}dx$$

$$= \sqrt{\frac{2}{\pi}}\left[\frac{1}{2i}\int_0^\infty \frac{e^{(a+ip)x}-e^{-(a+ip)x}}{e^{\pi x}+e^{-\pi x}}dx - \frac{1}{2i}\int_0^\infty \frac{e^{(a-ip)x}-e^{-(a-ip)x}}{e^{\pi x}-e^{-\pi x}}dx\right]$$

$$= \sqrt{\frac{2}{\pi}}\left[\frac{1}{2i}\cdot\frac{1}{2}\tan\frac{a+ip}{2} - \frac{1}{2i}\cdot\frac{1}{2}\tan\frac{a-ip}{2}\right]$$

$$\left[\because \int_0^\infty \frac{e^{ax}-e^{-ax}}{e^{\pi x}-e^{-\pi x}}dx = \frac{1}{2}\tan\frac{a}{2}\right]$$

$$= \sqrt{\frac{2}{\pi}}\left[\frac{1}{4i}\frac{\sin\dfrac{a+ip}{2}}{\cos\dfrac{a+ip}{2}} - \frac{1}{4i}\frac{\sin\dfrac{a-ip}{2}}{\cos\dfrac{a-ip}{2}}\right]$$

$$= \sqrt{\frac{2}{\pi}}\frac{\sin\dfrac{a+ip}{2}\cos\dfrac{a-ip}{2} - \sin\dfrac{a-ip}{2}\cos\dfrac{a+ip}{2}}{4i\cos\dfrac{a+ip}{2}\cos\dfrac{a-ip}{2}}$$

$$= \sqrt{\frac{2}{\pi}} \frac{\sin a + \sin ip - (\sin a - \sin ip)}{2 \cdot 2i[\cos ip + \cos a]}$$

$$= \frac{\sinh p}{\sqrt{2\pi}(\cosh p + \cos a)} = \frac{e^p - e^{-p}}{(\sqrt{2\pi})(e^p + e^{-p} + 2\cos a)}$$

and, $\quad \tilde{f}_c(p) = \sqrt{\frac{2}{\pi}} \int_0^\infty \frac{e^{ax} + e^{-ax}}{e^{\pi x} - e^{-\pi x}} dx = \sqrt{\frac{2}{\pi}} \int_0^\infty \frac{e^{ax} + e^{-ax}}{e^{ax} - e^{-ax}} \cdot \frac{e^{ipx} + e^{-ipx}}{2} dx$

$$= \frac{1}{\sqrt{2\pi}} \int_0^\infty \frac{e^{ax} + e^{-(a+ip)x}}{e^{\pi x} - e^{-\pi x}} dx + \frac{1}{\sqrt{2\pi}} \int_0^\infty \frac{e^{(a-ip)x} - e^{-(a-ip)x}}{e^{\pi x} - e^{-\pi x}} dx$$

$$= \frac{1}{\sqrt{2\pi}} \cdot \frac{1}{2} \sec \frac{a+ip}{2} + \frac{1}{\sqrt{2\pi}} \cdot \frac{1}{2} \sec \frac{a-ip}{2}$$

$$= \frac{\cos \dfrac{a-ip}{2} + \cos \dfrac{a+ip}{2}}{2\sqrt{2\pi} \cos \dfrac{a+ip}{2} \cos \dfrac{a-ip}{2}} = \sqrt{\frac{2}{\pi}} \frac{\left(\cos \dfrac{a}{2}\right)\left(\cos \dfrac{ip}{2}\right)}{\cos a + \cos ip}$$

$$= \sqrt{\frac{2}{\pi}} \frac{\cos(a/2)\cosh(p/2)}{\cos a + \cosh p} = \sqrt{\frac{2}{\pi}} \frac{\cos(a/2) \cdot (e^{p/2} + e^{-p/2})}{\cos a + e^p + e^{-p}} \cdot$$

EXAMPLE 7. *Find the Fourier sine transform of $f(x)$ if*

$$f(x) = \begin{cases} 0 & , \quad 0 < x < a \\ x & , \quad a \le x \le b \\ 0 & , \quad x > b \end{cases}$$

(Rohilkhand–1989)

SOLUTION. Here, we have

$$\tilde{f}_s(p) = \sqrt{\frac{2}{\pi}} \int_0^\infty f(x) \sin px\, dx$$

$$= \sqrt{\frac{2}{\pi}} \left[\int_0^a 0 \sin px\, dx + \int_a^b x \sin px\, dx + \int_b^\infty 0 \sin px\, dx \right]$$

$$= \sqrt{\frac{2}{\pi}} \int_a^b x \sin px\, dx = \sqrt{\frac{2}{\pi}} \left[\left(\frac{-x \cos px}{p} \right)_a^b + \left(\frac{\sin px}{p^2} \right)_a^b \right]$$

$$= \sqrt{\frac{2}{\pi}} \left[\frac{-b \cos pb + a \cos pa}{p} + \frac{\sin pb - \sin pa}{p^2} \right].$$

EXAMPLE 8. *Find the cosine transform of $\dfrac{1}{x^2 + a^2}$.* (VTU–2011, Anna–2009)

SOLUTION. Let $\quad f(x) = \dfrac{1}{x^2 + a^2}$. Then we have

$$\tilde{f}_c(p) = \sqrt{\frac{2}{\pi}} \int_0^\infty \frac{\cos px}{x^2 + a^2} dx \qquad\qquad\qquad ...(1)$$

If $\quad I = \int_0^\infty \dfrac{\cos px}{x^2 + a^2} dx$, then $\qquad\qquad\qquad\qquad\qquad ...(2)$

$$\frac{dI}{dp} = -\int_0^\infty \frac{x \sin px}{x^2 + a^2} dx = -\int_0^\infty \frac{x^2 \sin px}{x(x^2 + a^2)} dx$$

$$= -\int_0^\infty \frac{(x^2 + a^2) - a^2}{x(x^2 + a^2)} \sin px \, dx$$

$$= -\int_0^\infty \frac{\sin px}{x} dx + a^2 \int_0^\infty \frac{\sin px}{x(x^2 + a^2)} dx$$

$$= -\frac{\pi}{2} + a^2 \int_0^\infty \frac{\sin px}{x(x^2 + a^2)} dx \qquad \qquad ...(3)$$

Now, $\qquad \dfrac{d^2 I}{dp^2} = 0 + a^2 \int_0^\infty \dfrac{\cos px}{x^2 + a^2} dx = a^2 I$

$$\frac{d^2 I}{dp^2} - a^2 I = 0 \qquad \Rightarrow \qquad (D^2 - a^2)I = 0 \qquad \left(\because D \equiv \frac{d}{dp} \right)$$

The solution of the equation

$$I = A e^{ap} + B e^{-ap} \qquad \qquad ...(4)$$

$\therefore \qquad \dfrac{dI}{dp} = aA e^{ap} - aB e^{-ap} \qquad \qquad ...(5)$

If $p = 0$, then (2) becomes

$$I = \int_0^\infty \frac{dx}{a^2 + x^2} = \frac{1}{a} \left[\tan^{-1} \frac{x}{a} \right]_0^\infty = \frac{\pi}{2a}$$

and if $p = 0$, then (3) becomes

$$\frac{dI}{dp} = -\frac{\pi}{2}$$

Now, substituting $p = 0$ in eq^n (4) and (3), we get

$$A + B = \pi/2a \qquad \qquad ...(6)$$

and $\qquad a(A - B) = -\pi/2, \text{ i.e., } A - B = -\pi/2a \qquad \qquad ...(7)$

Now, from eq^n (6) and (7), we get

$$A = 0, B = \pi/2a$$

Putting these values in eq^n (4), we get

$$I = \int_0^\infty \frac{\cos px}{x^2 + a^2} dx = \frac{\pi}{2a} e^{-ap}.$$

EXAMPLE 9.   *Show that the Fourier transform of* $f(x) = e^{-x^2/2}$ *is* $e^{-p^2/2}$.

(Meerut–1985, 2001; Bundelkhand–2008; Agra–1982, 2000; Kanpur–2011)

SOLUTION.   Here, we have

$$F\{f(x)\} = \frac{1}{\sqrt{2\pi}} \int_{-\infty}^\infty f(x) e^{ipx} dx = \frac{1}{\sqrt{2\pi}} \int_{-\infty}^\infty e^{-x^2/2} e^{ipx} dx$$

$$= \frac{1}{\sqrt{2\pi}} \int_{-\infty}^\infty e^{-1/2 x^2 + ipx} dx = \frac{1}{\sqrt{2\pi}} \int_{-\infty}^\infty e^{-1/2(x^2 - 2ipx)} dx$$

$$= \frac{1}{\sqrt{2\pi}} \int_{-\infty}^\infty e^{-1/2(x - ip)^2 - 1/2 p^2} dx = \frac{1}{\sqrt{2\pi}} \int_{-\infty}^\infty e^{-1/2(x - ip)^2} e^{-p^2/2} dx$$

Putting $(x - ip)/\sqrt{2} = y \qquad \Rightarrow \qquad dx = \sqrt{2} dy,$

then, $\qquad F\{f(x)\}] = \dfrac{e^{-p^2/2}}{\sqrt{\pi}} \int_{-\infty}^\infty e^{-y^2} dy = \dfrac{e^{-p^2/2}}{\sqrt{\pi}} \cdot \sqrt{\pi} \qquad \left(\because \int_{-\infty}^\infty e^{-y^2} dy = \sqrt{\pi} \right)$

$$= e^{-p^2/2}.$$

EXAMPLE 10. *Show that* $\int_0^\infty \dfrac{\cos xt}{1+t^2}\,dt = \dfrac{\pi}{2}e^{-x}, x \geq 0$. (Rohilkhand–1992; Roorkee–1967; Purvanchal–1995)

SOLUTION. Here, we know that

$$\int_0^\infty e^{-ax}\cos bx\,dx = \frac{a}{a^2+b^2}, \text{ then}$$

$$\Rightarrow \int_0^\infty e^{-x}\cos txdx = \frac{1}{1+t^2}$$

Comparing this equation with

$$F_c\{f(x)\} = \tilde{f}_c(t) = \int_0^\infty f(x)\cos txdx \qquad \ldots(1)$$

we get $\tilde{f}_c(t) = \dfrac{1}{1+t^2}, f(x) = e^{-x}$

Now, inverse formula corresponding to eq^n (1) is

$$F_c^{-1}\{\tilde{f}_c(t)\} = f(x) = \frac{2}{\pi}\int_0^\infty \tilde{f}_c(t)\cos txdt$$

Substituting the values, we get

$$e^{-x} = \frac{2}{\pi}\int_0^\infty \frac{\cos tx}{1+t^2}\,dt$$

$$\Rightarrow \qquad \frac{\pi}{2}e^{-x} = \int_0^\infty \frac{\cos tx}{1+t^2}\,dt$$

EXAMPLE 11. *Find $f(x)$ if its cosine transform is* $\dfrac{1}{1+p^2}$. (Kanpur–1997; Meerut–1997; Purvanchal–1995)

SOLUTION. Consider $\tilde{f}_c(p) = \dfrac{1}{(1+p^2)}$

Using Fourier cosine inversion formula, we have

$$f(x) = \sqrt{\frac{2}{\pi}}\int_0^\infty \tilde{f}_c(p)\cos pxdx = \sqrt{\frac{2}{\pi}}\int_0^\infty \frac{\cos px}{1+p^2}\,dp = \sqrt{(\pi/2)}e^{-x}.$$

EXAMPLE 12. *Find $f(x)$ if*
(i) its sine transform is e^{-ap} (ii) its cosine transform is e^{-ap}.

(Agra–2003, 04; Bundelkhand–2005)

SOLUTION. (i) Consider $\tilde{f}_s(p) = e^{-ap}$

$$\therefore \qquad f(x) = \sqrt{\frac{2}{\pi}}\int_0^\infty \tilde{f}_s(p)\sin pxdp = \sqrt{\frac{2}{\pi}}\int_0^\infty e^{-ap}\sin pxdp \cdot$$

$$= \sqrt{\frac{2}{\pi}}\left[\frac{e^{-ap}}{a^2+x^2}(-a\sin px - x\cos px)\right]_0^\infty = \sqrt{\frac{2}{\pi}}\cdot\frac{x}{a^2+x^2}$$

(ii) Now $\tilde{f}_c(p) = e^{-ap}$

$$\therefore \qquad f(x) = \sqrt{\frac{2}{\pi}}\int_0^\infty \tilde{f}_c(p)\cos pxdx = \sqrt{\frac{2}{\pi}}\int_0^\infty e^{-ap}\cos pxdx$$

$$= \sqrt{\frac{2}{\pi}}\left[\frac{e^{-ap}}{a^2+x^2}(-a\cos px - x\sin px)\right]_0^\infty = \sqrt{\left(\frac{2}{\pi}\right)}\cdot\frac{a}{a^2+x^2}\cdot$$

EXAMPLE 13. *Find $f(x)$ if its cosine transform is*

$$\tilde{f}_c(p) = \begin{cases} \dfrac{1}{\sqrt{2\pi}}\left(a - \dfrac{p}{2}\right) & , \text{ if } p < 2a \\ 0 & , \text{ if } p \geq 2a \end{cases}$$

(Meerut–1984; Rohilkhand–1990, 96)

SOLUTION.   Using Fourier cosine inversion formula, we get

$$f(x) = \sqrt{\frac{2}{\pi}} \int_0^\infty \tilde{f}_c(p) \cdot \cos pxdx$$

$$= \sqrt{\frac{2}{\pi}} \int_0^{2a} \frac{1}{\sqrt{2\pi}}\left(a - \frac{p}{2}\right)\cos pxdx + \sqrt{\frac{2}{\pi}}\int_{2a}^\infty 0 \cdot \cos pxdx$$

$$= \left[\frac{1}{\pi x}\left(a - \frac{p}{2}\right)\sin px\right]_0^{2a} - \left[\frac{1}{2\pi x^2}\cos px\right]_0^{2a}$$

$$= \frac{1 - \cos 2ax}{2\pi x^2} = \pi^{-1}x^{-2}\sin^2 ax \cdot$$

EXAMPLE 14.   *Find $f(x)$ if $\tilde{f}_s(p) = \dfrac{e^{-ap}}{p}$. Hence deduce that $F_s^{-1}\left\{\dfrac{1}{p}\right\}$.*

(Rohilkhand–1983; Agra–2003; Kanpur–2006)

SOLUTION.   Consider

$$F_s^{-1}\{e^{-ap/p}\} = f(x) = \sqrt{\frac{2}{\pi}}\int_0^\infty \frac{e^{-ap}}{p}\sin pxdx$$

$$\therefore \qquad \frac{df}{dx} = \sqrt{\frac{2}{\pi}}\int_0^\infty e^{-ap}\cos pxdx = \sqrt{\frac{2}{\pi}}\cdot\frac{a}{a^2 + x^2}$$

$$= \sqrt{\frac{2}{\pi}}\int \frac{adx}{a^2 + x^2} + A$$

$$f = \sqrt{\frac{2}{\pi}}\tan^{-1}(x/a) + A$$

If $x = 0$ and $f = 0$ then $A = 0$

Thus, $f(x) = F_s^{-1}\{e^{-ap}/p\} = \sqrt{(2/\pi)}\tan^{-1}(x/a)$ ...(1)

Now, putting $a = 0$ in eq[n] (1), we get

$$F_s^{-1}\{1/p\} = \sqrt{\frac{2}{\pi}}\tan^{-1}\infty = \sqrt{\frac{\pi}{2}}\cdot$$

EXAMPLE 15.   *Find Fourier sine and cosine transform of x^{n-1}, where $0 < n < 1$.*

(Madras–2006; Agra–2000; Meerut–1991)

SOLUTION.   Here, we have $f(x) = x^{n-1}$

Now, sine transform is $\tilde{f}_s(p) = \int_0^\infty x^{n-1}\sin pxdx$

and cosine transform is $\tilde{f}_c(p) = x$.

EXAMPLE 16.   *Solve the integral equation*

$$\int_0^\infty f(\theta)\cos \alpha\theta d\theta = \begin{cases} 1 - \alpha & ; \ 0 \leq \alpha \leq 1 \\ 0 & ; \ \alpha > 1 \end{cases}$$

(Purvanchal–1996; VTU–2011; Kurukshetra–2005; Kanpur–2003)

SOLUTION. Here, we have

$$\int_0^\infty f(\theta)\cos\alpha\theta d\theta = f_c(\alpha)$$

$$f_c(\alpha) = \begin{cases} 1-\alpha, & 0 \le \alpha \le 1 \\ 0, & \alpha > 1 \end{cases} \qquad \qquad ...(1)$$

Using Fourier cosine inversion formula, we have

$$f(\theta) = \frac{2}{\pi}\int_0^\infty f_c(\alpha)\cos\alpha\theta d\theta = \frac{2}{\pi}\int_0^1 (1-\alpha)\cos\alpha\theta d\alpha + \frac{2}{\pi}\int_1^\infty (0)\cos\alpha\theta d\theta$$

$$= \frac{2}{\pi}\int_0^1 (1-\alpha)\cos\alpha\theta d\alpha$$

On integrating by parts, we get

$$f(\theta) = \frac{2}{\pi}\left[\left|(1-\alpha)\frac{\sin\alpha\theta}{\theta}\right|_0^1 - \int_0^1 (-1)\frac{\sin\alpha\theta}{\theta} d\alpha \right]$$

$$= \frac{2}{\pi\theta}\left|\frac{-\cos\alpha\theta}{\theta}\right|_0^1 = \frac{2(1-\cos\theta)}{\pi\theta^2}$$

Since,

$$f_c(\alpha) = \int_0^\infty f(\theta)\cos\alpha\theta d\theta = \int_0^\infty \frac{2(1-\cos\theta)}{\pi\theta^2}\cos\alpha\theta d\theta \qquad ...(2)$$

From eq^n (1) and (2), we get

$$\frac{2}{\pi}\int_0^\infty \frac{1-\cos\theta}{\theta^2}\cos\alpha\theta d\theta = \begin{cases} 1-\alpha, & 0 \le \alpha \le 1 \\ 0, & \alpha > 1 \end{cases}.$$

EXAMPLE 17. *If the Fourier sine transform of f(x) is* $\dfrac{1-\cos n\pi}{n^2\pi^2}, 0 \le x \le \pi$, *find f(x).*   (Delhi–2002)

SOLUTION. Consider $\quad f(x) = \dfrac{1-\cos n\pi}{n^2\pi^2}$

$$f_s\{f(x)\} = \frac{1-\cos n\pi}{n^2\pi^2}, 0 < x < \pi = f_s(n), n = 1, 2, ...$$

Now, taking Fourier sine inversion formula, we get

$$f(x) = \frac{2}{\pi}\sum_{n=1}^\infty f_s(n)\sin nx = \frac{2}{\pi}\sum_{n=1}^\infty \left(\frac{1-\cos n\pi}{n^2\pi^2}\right)\sin nx = \frac{2}{\pi^3}\sum_{n=1}^\infty \left(\frac{1-\cos n\pi}{n^2}\right)\sin nx.$$

EXAMPLE 18. *Find the Fourier cosine transform of* $\left(1-\dfrac{x}{\pi}\right)^2$.   (Kanpur–1981, 95, 98; PTU–2006)

SOLUTION. Consider $\quad f(x) = \left(1-\dfrac{x}{\pi}\right)^2$

Now, $\quad \tilde{f}_c(p) = \int_0^\pi f(x)\cos\left(\dfrac{p\pi x}{\pi}\right)dx = \int_0^\pi \left(1-\dfrac{x}{\pi}\right)^2\cos px\, dx$

$$= \left[\frac{\sin px}{p}\left(1-\frac{x}{\pi}\right)^2\right]_0^\pi - \int_0^\pi -2\left(1-\frac{x}{\pi}\right)\cdot\frac{1}{\pi}\cdot\frac{\sin px}{p} dx$$

$$= 0 + \frac{2}{2\pi}\left[\left\{\frac{-\cos px}{p}\left(1-\frac{x}{\pi}\right)\right\}_{x=0}^\pi - \int_0^\pi \frac{\cos px}{p}\cdot\frac{1}{\pi} dx\right]$$

$$= \frac{2}{p^2\pi}\left[1 - \frac{1}{\pi}\left(\frac{\sin px}{p}\right)^{\pi}_{x=0}\right] = \frac{2}{p^2\pi}[1-0]$$

$$\tilde{f}_c(p) = \frac{2}{\pi p^2}, p > 0$$

When $p = 0$, then we get

$$\tilde{f}_c(0) = \int_0^{\pi}\left(1 - \frac{x}{\pi}\right)^2 (\cos 0x)dx = \int_0^{\pi}\left(1 - \frac{x}{\pi}\right)^2 dx$$

$$= \left[\frac{1}{3}\left(1 - \frac{x}{\pi}\right)^3 \cdot \left(\frac{1}{-1/\pi}\right)\right]_0^{\pi} = -\frac{\pi}{3}\left[\left(1 - \frac{x}{\pi}\right)^3\right]_0^{\pi} = \frac{\pi}{3}$$

Hence,
$$\tilde{f}_c(p) = \begin{cases} \pi/3, & \text{if} \quad p = 0 \\ 2/\pi p^2, & \text{if} \quad p = 1,2,3,... \end{cases}.$$

4.5 FINITE FOURIER TRANSFORMS

4.5.1 FINITE FOURIER SINE TRANSFORMS

Let $f(x)$ be a function, which is sectionally continuous over some finite interval]0, 1[for the variable x. Then finite Fourier sine transforms of $f(x)$ on this interval is given by $\tilde{f}(s) = \int_0^l f(x)\sin\frac{p\pi x}{l}dx$, where p is any integer.

If the end points of the interval become $x = 0$ and $x = \pi$, then $\tilde{f}_s(p) = \int_0^{\pi} f(x)\sin pxdx$.

REMARKS

- $\tilde{f}_s(p)$ is always zero when $p = 0$.
- The function $f(x)$ is called the inverse finite Fourier sine transform of $\tilde{f}_s(p)$, *i.e.*, $f(x) = F_s^{-1}\{\tilde{f}_s(p)\}$.

4.5.2 INVERSION FORMULA FOR SINE TRANSFORM

If $\tilde{f}_s(p)$ is the finite Fourier sine transform of $f(x)$ over the interval]0, l[, then the inversion formula for sine transform is given by

$$f(x) = \frac{2}{l}\sum_{p=1}^{\infty}\tilde{f}_s(p)\sin\frac{p\pi x}{l}$$

or in the interval]0, π[, we have

$$f(x) = \frac{2}{\pi}\sum_{p=1}^{\infty}\tilde{f}_s(p)\sin px.$$

4.5.3 FINITE FOURIER COSINE TRANSFORM

Let $f(x)$ be a function which is sectionally continuous over some finite interval]0, l[of the variable x. Then the finite Fourier cosine transform of $f(x)$ on this interval is defined as

$$\tilde{f}_c(p) = \int_0^l f(x)\cos\frac{p\pi x}{l}dx, \text{ where } p \text{ is any integer.}$$

4.5.4 INVERSION FORMULA FOR COSINE TRANSFORM

If $\tilde{f}_c(p)$ is the finite Fourier cosine transform of $f(x)$ over the interval]0, l[then the inversion formula for cosine transform is given by

$$f(x) = \frac{1}{l}\tilde{f}_c(0) + \frac{2}{l}\sum_{p=1}^{\infty}\tilde{f}_c(p)\cos\frac{p\pi x}{l}, \text{ where } \tilde{f}_c(0) = \int_0^l f(x)dx.$$

Also, if π is taken as the upper limit for the finite Fourier cosine transform, then inversion formula is given by

$$f(x) = \frac{1}{\pi}\tilde{f}_c(0) + \frac{2}{\pi}\sum_{p=1}^{\infty}\tilde{f}_c(p)\cos px, \text{ where } \tilde{f}_c(0) = \int_0^{\pi} f(x)dx.$$

4.5.5 MULTIPLE FINITE FOURIER TRANSFORMS

Let $f(x,y)$ be a function of two variables x and y defined in the square $0 \le x \le \pi$ and $0 \le y \le \pi$. Let us consider $f(x, y)$ temporarily as a function of x, then finite sine transform is given by

$$\tilde{f}_s(p,y) = \int_0^{\pi} f(x,y)\sin px\,dx$$

and now the finite sine transform of $\tilde{f}_s(p,y)$ which is a function of y is given by

$$\tilde{F}_s(p,q) = \int_0^{\pi}\tilde{f}_s(p,y)\sin qy\,dy$$

Thus,
$$\tilde{F}_s(p,q) = \int_0^{\pi}\int_0^{\pi} f(x,y)\sin px\,\sin qy\,dx\,dy$$

4.5.6 SOME OPERATIONAL PROPERTIES OF FINITE FOURIER TRANSFORM

THEOREM 1. *The finite Fourier sine transforms resolves the differential form $f''(x)$ into a linear algebraic form in the transform $\tilde{f}_s(p)$ and the boundary value $f(0)$ and $f(\pi)$ such that $F_s[f''(x)] = -p^2\tilde{f}_s(p) + p\{f(0) - (-1)^p f(\pi)\}$ whenever $f(x)$ and $f'(x)$ are continuous and $f''(x)$ is sectionally continuous on the interval $0 \le x \le \pi$.*

PROOF. Consider $F_s\{f''(x)\} = \int_0^{\pi} f''(x)\sin px\,dx$

Let $f'(x)$ be second function, then integrating by parts, we get

$$= \{f'(x)\sin px\}_0^{\pi} - p\int_0^{\pi} f'(x)\cos px\,dx$$

$$= 0[-pf(x)\cos px]_0^{\pi} - p^2\int_0^{\pi} f(x)\sin px\,dx$$

$$= -p(\pi)\cos p\pi + pf(0) - p^2\tilde{f}_s(p)$$

Hence, $F_s\{f''(x)\} = -p^2\tilde{f}_s(p) + p[f(0) - (-1)^p f(\pi)]$

THEOREM 2. *If $f(x)$ and $f'(x)$ are continuous and if $f''(x)$ is sectionally continuous, the finite Fourier transformation resolved the differential form $f''(x)$ into an algebraic form is $\tilde{f}_c(p)$ and the boundary value $f'(0)$ and $f'(\pi)$ such that*
$$F_c[f''(x)] = -p^2\tilde{f}_c(p) - f'(0) + (-1)^p f'(\pi)$$

PROOF. Consider $F_c\{f''(x)\} = \int_0^{\pi} f''(x)\cos px\,dx$

Let $f'(x)$ be second function then integrating by parts, we get

$$= \{f'(x)\cos px\}_0^{\pi} + p\int_0^{\pi} f'(x)\sin px\,dx$$

$$= f'(\pi)\cos p\pi - f'(0) + p\left[f(x)\sin px\right]_0^{\pi} - p^2\int_0^{\pi} f(x)\cos px\,dx$$

$$= (-1)^p f'(\pi) - f'(0) - p^2\tilde{f}_c(p)$$

$$F_c\{f''(x)\} = -p^2\tilde{f}_c(p) - f'(0) + (-1)^p f'(\pi)$$

REMARK

- If $\tilde{f}_c(p)$ is the cosine transform of a sectionally continuous function $f(x)$, $0 \le x \le \pi$ then
$$F_c^{-1}\left\{\frac{\tilde{f}_c(p)}{p^2}\right\} = \int_0^x\int_t^{\pi} f(r)dr\,dt = \frac{\tilde{f}_c(0)}{2\pi}(x - \pi^2) + A$$

where A is any arbitrary constant.

THEOREM 3.   **If $f(x)$ is continuous and $f'(x)$ is sectionally continuous, then**

(i) $F_s[f'(x)] = -pF_c\{f(x)\}; p = 1, 2, 3...$

and   (ii) $F_c[f'(x)] = pF_s\{f(x) - f(0) + (-1)^p f(\pi)\}; p = 0, 1, 2, 3...$

PROOF.   (i) Consider

$$F_s\{f'(x)\} = \int_0^\pi f'(x) \sin px\, dx$$

On integrating by parts, we get

$$= \left[f(x) \sin px \right]_0^\pi - p\int_0^\pi f(x) \cos px\, dx$$

$$= -pF_c\{f(x)\} \text{ where } p = 1, 2, 3, ...$$

(ii) Now, consider

$$F_c\{f'(x)\} = \int_0^\pi f'(x) \cos px\, dx$$

On integrating by parts, we get

$$= \left[f(x) \cos px \right]_0^\pi + p\int_0^\pi f(x) \sin px\, dx$$

$$= f(\pi) \cos p\pi - f(0) + pF_s\{f(x)\}, \text{ where } p = 0, 1, 2, ...$$

$$= pF_s\{f(x)\} - f(0) + (-1)^p f(\pi)$$

REMARK

- If $H(x)$ is sectionally continuous function, then

$$F_s[H(x)] = -pF_c\left[\int_0^x H(r)dr \right], p = 1, 2...$$

and $$F_c\left[H(x) - \frac{1}{\pi} \tilde{H}_c(0) \right] = pF_s\left[\int_0^x H(r)dr - \frac{x}{\pi} \tilde{H}_c(0) \right], p = 0, 1, 2...$$

4.5.7 CONVOLUTION

Let $F(x)$ and $G(x)$ be two functions defined on the interval $-2\pi < x < 2\pi$, then the function $F(x) * G(x) = \int_{-\pi}^{\pi} F(x - y)G(y)dy$ is called the convolution of $F(x)$ and $G(x)$ on the interval $-\pi < x < \pi$.

THEOREM 1.   **(Fourier Integral Theorem). If $f(x)$ satisfies the following conditions**

(i) **$f(x)$ satisfies Dirichlet conditions in every interval $-l \leq x \leq l$.**

(ii) **$\int_{-\infty}^{\infty} |f(x)| dx$ converges, i.e., $f(x)$ is absolutely integrable in the interval $-\infty < x < \infty$,**

then $f(x) = \dfrac{1}{2\pi} \int_{p=-\infty}^{\infty} \int_{t=-\infty}^{\infty} f(t) \cos p(x - t)dp\, dt$

The integral on R.H.S. is called Fourier integral or Fourier integral expansion of $f(x)$.   (Gurukul Kangri Haridwar–2001)

PROOF.   Let $f(x)$ be a function which satisfy Dirichlet's condition in the interval $(-C, C)$, then $\int_{-\infty}^{\infty} |f(x)| dx$ is convergent.

$$\Rightarrow \quad f(x) = \frac{a_0}{2} + \sum_{n=1}^{\infty} \left(a_n \cos \frac{n\pi x}{C} + b_n \sin \frac{n\pi x}{C} \right) \qquad ...(1)$$

Here, $a_n = \dfrac{1}{C} \int_{-C}^{C} f(t) \cos \dfrac{n\pi t}{C} dt, n = 0, 1, 2, ...$

and $b_n = \dfrac{1}{C} \int_{-C}^{C} f(t) \sin \dfrac{n\pi t}{C} dt, n = 1, 2, 3, ...$

Putting these values in eq^n (1), we get

$$f(x) = \frac{1}{2C}\int_{-C}^{C} f(t)dt + \frac{1}{C}\int_{-C}^{C} f(t)\left[\sum_{n=1}^{\infty} \cos\frac{n\pi(t-x)}{C}\right]dt$$

$$= \frac{1}{2C}\int_{-C}^{C} f(t)\left[1 + \lim_{n\to\infty}\sum_{r=1}^{n} 2\cos\frac{r\pi(t-x)}{C}\right]dt$$

$$= \frac{1}{2C}\int_{-C}^{C} f(t)\left[1 + \lim_{n\to\infty}\sum_{r=1}^{n} \cos\frac{r\pi(t-x)}{C} + \cos\frac{-r\pi(t-x)}{C}\right]dt$$

$$= \frac{1}{2C}\int_{-C}^{C} f(t)\left[1 + \lim_{n\to\infty}\sum_{r=-n}^{n} \frac{\cos r\pi(t-x)}{C}\right]dt$$

$$= \frac{1}{2C}\int_{-C}^{C} f(t)dt + \frac{1}{2\pi}\int_{-C}^{C} f(t)\left[\lim_{n\to\infty}\sum_{r=-n}^{n} \frac{1}{(C/\pi)}\cos\frac{r}{C/\pi}(t-x)\right]dt$$

Using definition of integral as a limit of sum, we get

$$f(x) = \frac{1}{2C}\int_{-C}^{C} f(t)dt + \frac{1}{2\pi}\int_{-C}^{C} f(t)\left[\int_{-\infty}^{\infty}\cos u(t-x)du\right]dt$$

$$\Rightarrow \quad f(x) = \frac{1}{2C}\int_{-C}^{C} f(t)dt + \frac{1}{2\pi}\int_{-C}^{C} f(t)dt\int_{-\infty}^{\infty}\cos u(t-x)du$$

$$\Rightarrow \quad f(x) = 0 + \frac{1}{2\pi}\int_{-\infty}^{\infty} f(t)dt\int_{-\infty}^{\infty}\cos u(t-x)du$$

$$\Rightarrow \quad f(x) = \frac{1}{2\pi}\int_{-\infty}^{\infty}\int_{-\infty}^{\infty} f(t)\cos u(t-x)dt\,du$$

If $-\infty < x < \infty$, then

$$f(x) = \frac{1}{2\pi}\int_{p=-\infty}^{\infty}\int_{t=-\infty}^{\infty} f(t)\cos p(t-x)dp\,dt$$

4.5.8 Different Form of Fourier Integral Formula

(1) $f(x) = \frac{1}{\pi}\int_{p=-\infty}^{\infty}\int_{t=-\infty}^{\infty} f(t)\cos p(t-x)dp\,dt$

(2) Cosine Form : $f(x) = \frac{2}{\pi}\int_{0}^{\infty}\int_{0}^{\infty} f(t)\cos pt\cos px\,dp\,dt$

(3) Sine Form : $f(x) = \frac{2}{\pi}\int_{0}^{\infty}\int_{0}^{\infty} f(t)\sin pt\sin px\,dp\,dt$

(4) Exponential Form : $f(x) = \frac{1}{2\pi}\int_{-\infty}^{\infty}\int_{-\infty}^{\infty} f(t)e^{-ipt}e^{ipx}dp\,dt$

4.5.9 Some Important Results

(1) $\frac{2}{\pi}\int_{0}^{\infty} f_c(p)g_c(p)dp = \int_{0}^{\infty} F(x)G(x)dx$

(2) $\frac{2}{\pi}\int_{0}^{\infty} f_s(p)g_s(p)dp = \int_{0}^{\infty} F(x)G(x)dx$

(3) $\frac{2}{\pi}\int_{0}^{\infty}|f_c(p)|^2 dp = \int_{0}^{\infty}|F(x)|^2 dx$

(4) $\frac{2}{\pi}\int_{0}^{\infty}|f_s(p)|^2 dp = \int_{0}^{\infty}|F(x)|^2 dx$

Solved Examples

EXAMPLE 1. *Find the finite Fourier sine and cosine transforms of $f(x) = x$.*

(Meerut–1987; Rohilkhand–1983, 91; Kanpur–2006)

SOLUTION. We have

$$\tilde{f}_s(p) = \int_0^\pi f(x) \sin px \, dx = \int_0^\pi x \sin px \, dx = \left(\frac{-x \cos px}{p} \right)_0^\pi + \frac{1}{p} \int_0^\pi \cos px \, dx$$

$$= \frac{\pi(-1)^{p+1}}{p} + \frac{1}{p} \left[\frac{\sin px}{p} \right]_0^\pi = \frac{\pi(-1)^{p+1}}{p}$$

Similarly,

$$\tilde{f}_c(p) = \int_0^\pi f(x) \cos px \, dx = \int_0^\pi x \cos px \, dx$$

$$= \left(\frac{x \sin px}{p} \right)_0^\pi - \frac{1}{p} \int_0^\pi \sin px \, dx = \frac{1}{p} \left[\frac{\cos px}{p} \right]_0^\pi = \frac{(-1)^p - 1}{p^2}, p = 1, 2, 3 \ldots$$

Also, if $p = 0$, then $\tilde{f}_c(p) = \int_0^\pi x . 1 \, dx = \frac{\pi^2}{2}$.

EXAMPLE 2. *Find the finite sine transform of $f(x)$ if*

(i) $f(x) = \begin{cases} x & ; \quad 0 \le x \le \pi/2 \\ \pi - x & ; \quad \pi/2 \le x \le \pi \end{cases}$

(Agra–1982, 89; Meerut–1991)

(ii) $f(x) = \begin{cases} -x & ; \quad x < c \\ \pi - c & ; \quad x > c, \end{cases}$ *where $0 \le c \le \pi$*

(Agra–1982, 2001; Meerut–1991)

SOLUTION. (i) We have

$$\tilde{f}_s(p) = \int_0^\pi f(x) \sin px \, dx = \int_0^{\pi/2} x \sin px \, dx + \int_{\pi/2}^\pi (\pi - x) \sin px \, dx$$

$$= \left[x \frac{(-\cos px)}{p} + \frac{\sin px}{p^2} \right]_0^{\pi/2} + \left[(\pi - x) \left(\frac{-\cos px}{p} \right) - \frac{\sin px}{p^2} \right]_{\pi/2}^\pi$$

$$= \frac{2}{p^2} \sin \left(\frac{p\pi}{2} \right).$$

(ii) We have

$$\tilde{f}_s(p) = \int_0^\pi f(x) \sin px \, dx = \int_0^C -x \sin px \, dx + \int_C^\pi (\pi - x) \sin px \, dx$$

$$= \left[x \frac{\cos px}{p} \right]_0^C - \int_0^C 1 . \frac{\cos px}{p} \, dx + \left[(\pi - x) \left(-\frac{\cos px}{p} \right) \right]_C^\pi - \frac{1}{p} \int_C^\pi \cos px \, dx$$

$$= \frac{C}{p} \cos pC - \frac{1}{p^2} [\sin px]_0^C + \frac{\pi - C}{p} \cos pC - \frac{1}{p^2} [\sin px]_C^\pi$$

$$= \frac{C}{p} \cos pC - \frac{1}{p^2} \sin pC + \frac{\pi - C}{p} \cos pC + \frac{1}{p^2} \sin pC$$

EXAMPLE 3. *Find the finite Fourier sine and cosine transforms of the function $f(x) = 2x, 0 < x < 4$.*

(Agra–1985, 87, 2014; Meerut–1991, 99, 2004; Rohilkhand–1989, 2007; SVTU–2005)

SOLUTION. We have $\tilde{f}_s(p) = \int_0^l f(x)\sin(p\pi x/l)dx$

$$= \int_0^4 2x\sin(p\pi x/4)dx, \; l = 4 \text{ (Given)}$$

$$= \left[\frac{-2x\cos(p\pi x/4)}{p\pi/4}\right]_0^4 + 2\int_0^4 \frac{\cos(px\pi/4)}{p\pi/4}dx$$

$$= -\frac{32}{p\pi}\cos p\pi + \frac{8}{p\pi}\left[\frac{\sin(p\pi x/4)}{p\pi/4}\right]_0^4 = -\frac{32}{p\pi}\cos p\pi$$

Also, $\tilde{f}_c(p) = \int_0^l f(x)\cos(p\pi x/l)dx$

$$= \int_0^4 2x\cos(p\pi x/4)dx, \quad \text{as } l = 4$$

$$= \left[\frac{2x\sin(p\pi x/4)}{p\pi/4}\right]_0^4 - 2\int_0^4 \frac{\sin(p\pi x/4)}{p\pi/4}dx$$

$$= \frac{8}{p\pi}\left[\frac{\cos(p\pi x/4)}{p\pi/4}\right]_0^4 = \frac{32}{p^2\pi^2}(\cos p\pi - 1), \text{ if } p > 0$$

and if $p = 0$, then $\tilde{f}_c(p) = \int_0^4 2x.1 dx = 16$

EXAMPLE 4. *Find the finite Fourier sine transforms of $f(x)$ if*

(i) $x(\pi - x)$ and (ii) $x(\pi^2 - x^2)$ (Meerut–1990)

SOLUTION. (i) $\tilde{f}_s\{x(\pi - x)\} = \int_0^\pi x(\pi - x)\sin px\,dx$

$$= \left[x(\pi - x)\left(-\frac{\cos px}{p}\right)\right]_0^\pi + \int_0^\pi (\pi - 2x)\left(\frac{\cos px}{p}\right)dx$$

$$= \left[(\pi - 2x)\left(\frac{\sin px}{p^2}\right) - 2.\left(\frac{\cos px}{p^3}\right)\right]_0^\pi = \frac{2}{p^3}[1 - (-1)^p].$$

(ii) $\tilde{f}_s\{x(\pi^2 - x^2)\} = \int_0^\pi x(\pi^2 - x^2)\sin px\,dx$

$$= \left[x(\pi^2 - x^2)\left(-\frac{\cos px}{p}\right)\right]_0^\pi + \int_0^\pi (\pi^2 - 3x^2)\left(\frac{\cos px}{p}\right)dx$$

$$= \left[(\pi^2 - 3x^2)\left(\frac{\sin px}{p^2}\right)\right]_0^\pi - \int_0^\pi (-6x)\left(\frac{\sin px}{p^2}\right)dx$$

$$= 6\left[x\left(\frac{\cos px}{p^3}\right) + \left(\frac{\sin px}{p^4}\right)\right]_0^\pi = \frac{6\pi}{p^3}(-1)^{p+1}.$$

EXAMPLE 5. *Find the finite Fourier sine transform of $f(x) = \sin nx$.* (Agra–1982; Kanpur–1983)

SOLUTION. We have $\tilde{f}_s(p) = \int_0^\pi \sin nx \sin px\,dx$

$$= \frac{1}{2}\int_0^\pi [\cos(p-n)x - \cos(p+n)x]dx$$

$$= \frac{1}{2}\left[\frac{\sin(p-n)x}{p-n} - \frac{\sin(p+n)x}{p+n}\right]_0^\pi = 0, \text{ if } p \neq n.$$

If $p = n$, then

$$\tilde{f}_s(p) = \int_0^\pi \sin nx \sin px\, dx = \int_0^\pi \sin^2 nx\, dx$$

$$= \frac{1}{2}\int_0^\pi (1 - \cos 2nx)dx = \frac{1}{2}\left[x - \frac{\sin 2nx}{2n}\right]_0^\pi = \frac{\pi}{2}.$$

Hence, $\tilde{f}_s(p) = 0$ if $p \neq n$, and $\tilde{f}_s(p) = \pi/2$ if $p = n$

EXAMPLE 6. *Find the finite cosine transform of $f(x) = -\dfrac{\cos k(\pi - x)}{k \sin k\pi}$.* (Meerut–1991)

SOLUTION. We have $\tilde{f}_c(p) = -\int_0^\pi \dfrac{\cos\{k(\pi - x)\}}{k \sin k\pi} \cos px\, dx$

$$= -\frac{1}{2k \sin k\pi} \int_0^\pi [\cos\{k(\pi - x) + px\} + \cos\{k(\pi - x) - px\}]dx$$

$$= -\frac{1}{2k \sin k\pi}\left[\frac{\sin(k\pi - kx + px)}{p - k} - \frac{\sin(k\pi - kx - px)}{p + k}\right]_0^\pi$$

$$= -\frac{1}{2k \sin k\pi}\left[\frac{\sin p\pi}{p - k} - \frac{\sin(-p\pi)}{p + k} - \frac{\sin k\pi}{p - k} + \frac{\sin k\pi}{p + k}\right]$$

$$= \frac{1}{2k}\left(\frac{1}{p - k} - \frac{1}{p + k}\right) = \frac{1}{p^2 - k^2}, k \neq 0,1,2,3...$$

EXAMPLE 7. *Find finite Fourier sine transform of $f(x) = \dfrac{\sin k(\pi - x)}{\sin(k\pi)}$.* (Meerut–1991, 92)

SOLUTION. We have $\tilde{f}_s(p) = \int_0^\pi f(x) \sin px\, dx = \int_0^\pi \dfrac{\sin\{k(\pi - x)\}}{\sin(k\pi)} \cdot \sin px\, dx$

$$= \frac{1}{2\sin(k\pi)} \int_0^\pi [\cos\{px - k(\pi - x)\} - \cos\{px + k(\pi - x)\}]dx$$

$$= \frac{1}{2\sin(k\pi)} \cdot \left[\frac{\sin\{px - k(\pi - x)\}}{p + k} - \frac{\sin\{px + k(\pi - x)\}}{p - k}\right]_0^\pi$$

$$= \frac{1}{2\sin(k\pi)} \cdot \left[\frac{\sin p\pi + \sin k\pi}{p + k} - \frac{\sin p\pi - \sin k\pi}{p - k}\right]$$

$$= \frac{1}{2\sin(k\pi)} \cdot \left(\frac{1}{p + k} + \frac{1}{p - k}\right) \sin(k\pi) = \frac{p}{p^2 - k^2}, k \neq 0,1,2...$$

EXAMPLE 8. *Find $f(x)$ if its finite sine transform is given by*

$$\tilde{f}_s(p) = \frac{1 - \cos p\pi}{p^2 \pi^2}, \text{ where } 0 < x < \pi.$$ (Bundelkhand–2007; Kanpur–2008)

SOLUTION. We have $f(x) = \dfrac{2}{\pi} \displaystyle\sum_{p=1}^\infty \tilde{f}_s(p) \sin px = \dfrac{2}{\pi} \displaystyle\sum_{p=1}^\infty \left(\dfrac{1 - \cos p\pi}{p^2 \pi^2}\right) \sin px$

$$= \frac{2}{\pi^3} \sum_{p=1}^\infty \left(\frac{1 - \cos p\pi}{p^2}\right) \sin px$$

EXAMPLE 9. *When $f(x) = \sin mx$, where m is a positive integer, show that*
$$\tilde{f}_s(p) = 0 \text{ if } p \neq m \text{ and } \tilde{f}_s(p) = \pi/2 \text{ if } p = m$$

SOLUTION. We have $\tilde{f}_s(p) = \int_0^\pi f(x) \sin px \, dx$

$$= \int_0^\pi \sin mx \sin px \, dx$$

$$= \frac{1}{2} \int_0^\pi [\cos(m-p)x - \cos(m+p)x] dx$$

$$= \frac{1}{2} \left[\frac{\sin(m-p)x}{m-p} - \frac{\sin(m+p)x}{m+p} \right]_0^\pi = 0, \text{ if } m \neq p$$

If $m = p$, then

$$\tilde{f}_s(p) = \int_0^\pi \sin^2 px \, dx = \frac{1}{2} \int_0^\pi (1 - \cos 2px) dx$$

$$= \frac{1}{2} \left[x - \frac{\sin 2px}{2p} \right]_0^\pi = \frac{\pi}{2}.$$

EXAMPLE 10. *Find the finite Fourier sine transform of $f(x)$ if*

(i) $f(x) = \dfrac{2}{\pi} \tan^{-1} \dfrac{b \sin x}{1 - b \cos x}$ \qquad (ii) $f(x) = \dfrac{2}{\pi} \tan^{-1} \dfrac{2b \sin x}{1 - b^2}$

SOLUTION. (i) If $f(x) = \dfrac{2}{\pi} \tan^{-1} \dfrac{b \sin x}{1 - b \cos x}$, then

$$\tilde{f}_s(p) = \int_0^\pi \frac{2}{\pi} \cdot \left[\tan^{-1} \frac{b \sin x}{1 - b \cos x} \right] \sin px \, dx$$

Now let $\tan \theta = \dfrac{b \sin x}{1 - b \cos x}$

Then $\dfrac{i \sin \theta}{\cos \theta} = \dfrac{ib \sin x}{1 - b \cos x}$

Applying componendo and dividendo, we have

$$\frac{\cos \theta + i \sin \theta}{\cos \theta - i \sin \theta} = \frac{1 - b \cos x + ib \sin x}{1 - b \cos x - ib \sin x}$$

or $\qquad \dfrac{e^{i\theta}}{e^{-i\theta}} = \dfrac{1 - b(\cos x - i \sin x)}{1 - b(\cos x + i \sin x)} = \dfrac{1 - be^{-ix}}{1 - be^{ix}}$

$\therefore \qquad e^{2i\theta} = \dfrac{1 - be^{-ix}}{1 - be^{ix}}$

$\therefore \qquad 2i\theta = \log(1 - be^{-ix}) - \log(1 - be^{ix})$

$$= -\left\{ be^{-ix} + \frac{b^2 e^{-2ix}}{2} + \frac{b^3 e^{-3ix}}{3} + \dots \right\}$$

$$+ \left\{ be^{ix} + \frac{b^2 e^{2ix}}{2} + \frac{b^3 e^{3ix}}{3} + \dots \right\}, \text{ if } |b| \leq 1$$

$$= b(e^{ix} - e^{-ix}) + \frac{b^2}{2}(e^{2ix} - e^{-2ix}) + \frac{b^3}{3}(e^{3ix} - e^{-3ix}) + \dots$$

$$= 2ib\sin x + \frac{b^2}{2}.(2i\sin 2x) + \frac{b^3}{3}.(2i\sin 3x) + \dots$$

$$\therefore \qquad \theta = \tan^{-1}\frac{b\sin x}{1 - b\cos x} - b\sin x + \frac{b^2}{2}\sin 2x + \frac{b^3}{3}\sin 3x + \dots$$

∴ From (1),

$$\tilde{f}_s(p) = \int_0^\pi \frac{2}{\pi}.\left[b\sin x + \frac{b^2}{2}\sin 2x + \frac{b^3}{3}\sin 3x + \dots\right]\sin px\,dx$$

$$= \int_0^\pi \frac{2}{\pi}.\frac{b^p}{p}\sin^2 px\,dx$$

since all other integrals vanish as

$$\int_0^\pi \sin mx \sin nx\,dx = 0 \text{ if } m \text{ and } n \text{ are integers and } m \neq n$$

$$= \frac{b^p}{\pi p}\int_0^\pi(1 - \cos 2px\,dx) = \frac{b^p}{\pi p}\left[x - \frac{\sin 2px}{2p}\right]_0^\pi$$

$$= \frac{b^p}{\pi p}.\pi = \frac{b^p}{p}, |b| \leq 1$$

(ii) Let $\qquad \tan\theta = \dfrac{2b\sin x}{1 - b^2}$ then $\dfrac{i\sin\theta}{\cos\theta} = \dfrac{2ib\sin x}{1 - b^2}.$

Applying componendo and dividendo, we have

$$\frac{\cos\theta + i\sin\theta}{\cos\theta - i\sin\theta} = \frac{1 - b^2 + 2ib\sin x}{1 - b^2 - 2ib\sin x} = \frac{1 - b^2 + b(e^{ix} - e^{-ix})}{1 - b^2 - b(e^{ix} - e^{-ix})}$$

$$\therefore \qquad \frac{e^{i\theta}}{e^{-i\theta}} = \frac{(1 + be^{ix}) - be^{-ix}(1 + be^{ix})}{(1 - be^{ix}) + be^{-ix}(1 - be^{ix})}$$

or $\qquad e^{2i\theta} = \dfrac{(1 + be^{ix})(1 - be^{-ix})}{(1 - be^{ix})(1 + be^{-ix})}$

$$\therefore \qquad 2i\theta = \log(1 + be^{ix}) + \log(1 - be^{-ix}) - \log(1 - be^{ix}) - \log(1 + be^{-ix})$$

$$= \{\log(1 + be^{ix}) - \log(1 - be^{ix})\} - \{\log(1 + be^{-ix}) - \log(1 - be^{-ix})\}$$

$$= 2\left\{be^{ix} + \frac{b^3}{3}e^{3ix} + \frac{b^5}{5}e^{5ix}\dots\right\} - 2\left\{be^{-ix} + \frac{b^3}{3}e^{-3ix} + \frac{b^5}{5}e^{-5ix} + \dots\right\}$$

$$\text{if } |b| \leq 1$$

$$= 2\left[b(e^{ix} - e^{-ix}) + \frac{b^3}{3}(e^{3ix} - e^{-3ix}) + \frac{b^5}{5}(e^{5ix} - e^{-5ix}) + \dots\right]$$

$$= 2\left[b(2i\sin x) + \frac{b^3}{3}(2i\sin 3x) + \frac{b^5}{5}(2i\sin 5x) + \dots\right]$$

$$\therefore \qquad \theta = \tan^{-1}\frac{2b\sin x}{1-b^2}$$

$$= 2\left[b\sin x + \frac{b^3}{3}\sin 3x + \frac{b^5}{5}\sin 5x + ...\right] \qquad ...(1)$$

Now if $f(x) = \frac{2}{\pi}\tan^{-1}\frac{2b\sin x}{1-b^2}$, then

$$\tilde{f}_s(p) = \int_0^\pi \frac{2}{\pi}\left[\tan^{-1}\frac{2b\sin x}{1-b^2}\right]\sin pxdx$$

$$= \int_0^\pi \frac{2}{\pi}.2\left[b\sin x + \frac{b^3}{3}\sin 3x + \frac{b^5}{5}\sin 5x...\right]\sin pxdx \qquad \text{[From (1)]}$$

Now if m and n are integers and $m \neq n$ then $\int_0^\pi \sin mx\sin nxdx = 0$

If p is even, then $\tilde{f}_s(p) = 0$

Again if p is odd, then

$$\tilde{f}_s(p) = \frac{2}{\pi}\int_0^\pi \frac{b^p}{p}.2\sin^2 pxdx = \frac{2}{\pi}\frac{b^p}{p}\int_0^\pi (1-\cos 2px)dx$$

$$= \frac{2}{\pi}\frac{b^p}{p}\left[x - \frac{\sin 2px}{2p}\right]_0^\pi = \frac{2}{\pi}\frac{b^p}{p}.\pi = 2\frac{b^p}{p}.$$

Hence, $\tilde{f}_s(p) = \frac{1-(-1)^p}{p}b^p, |b| \leq 1$.

EXAMPLE 11. *Show that* $\tilde{f}_c(x) = 0, n \neq m$

$$\tilde{f}_c(m) = \frac{\pi}{2}, n = m$$

where $f(x) = \cos mx$ *and* m *is positive integer.* (Agra–2000)

SOLUTION. Here, we have

$$\tilde{f}_c(p) = \int_0^\pi f(x)\cos pxdx = \frac{1}{2}\int_0^\pi [\cos(m-n)x + \cos(m+n)x]dx$$

$$= \frac{1}{2}\left[\frac{-\sin(m-n)x}{m-n} + \frac{\sin(m+n)x}{m+n}\right]_0^\pi = 0 \text{ if } m \neq n \qquad ...(1)$$

If $m = n$, then

$$\tilde{f}_c m = \int_0^\pi \cos mx\cos mxdx = \int_0^\pi \cos^2 mxdx = \frac{1}{2}\int_0^\pi (1+\cos 2mx)dx$$

$$= \frac{1}{2}\left[x + \frac{\sin 2mx}{2m}\right]_0^\pi = \frac{1}{2}(\pi+0) = \frac{\pi}{2}. \qquad ...(2)$$

EXAMPLE 12. *Find the finite cosine transform of* $f(x)$ *if*

$$f(x) = \begin{cases} 1, & 0 < x < \pi/2 \\ -1, & \pi/2 < x < \pi \end{cases} \qquad \text{(Meerut–1988, 91, 2000; Kanpur–2012)}$$

SOLUTION. Here, we have

$$\tilde{f}_c(p) = \int_0^\pi f(x)\cos pxdx$$

$$= \int_0^{\pi/2} 1 \cdot \cos px \, dx + \int_{\pi/2}^{\pi} (-1) \cos px \, dx$$

$$= \left[(\cos px) / p \right]_0^{\pi/2} - \left[(\sin px) / p \right]_{\pi/2}^{\pi}$$

$$= (2/p) \sin(p\pi / 2), p > 0$$

If $p = 0$, then

$$\tilde{f}_c(p) = \int_0^{\pi} f(x) \cdot 1 \, dx = \int_0^{\pi/2} 1 \cdot dx + \int_{\pi/2}^{\pi} (-1) \, dx = 0.$$

EXAMPLE 13. *Find $f(x)$ if*

$$\tilde{f}_c(p) = \left(6 \sin \frac{p\pi}{2} - \cos p\pi \right) / (2p+1)\pi \ \text{ for } p = 1, 2, 3, \dots$$

and $\qquad \tilde{f}_c(p) = \dfrac{2}{\pi} \ \text{ for } p = 0$

where $0 < x < 4$. (Kanpur–1999)

SOLUTION. We know that

$$f(x) = F_C^{-1}\{\tilde{f}_c(p)\} = \frac{1}{l}\tilde{f}_c(0) + \frac{2}{l} \sum_{p=1}^{\infty} \tilde{f}_c(p) \cos \frac{p\pi x}{l}$$

$$= \frac{1}{4} \cdot \frac{2}{\pi} + \frac{2}{4} \sum_{p=1}^{\infty} \frac{(6 \sin p\pi / 2 - \cos p\pi)}{(2p+1)\pi} \cos \left(\frac{p\pi x}{4} \right)$$

$$= \frac{1}{2\pi} + \frac{1}{2\pi} \sum_{p=1}^{\infty} \frac{\left(6 \sin \dfrac{p\pi}{2} - \cos p\pi \right)}{(2p+1)} \cos \left(\frac{p\pi x}{4} \right).$$

EXAMPLE 14. *Find finite Fourier cosine transform of e^{ax} in the interval $(0, l)$.* (Kanpur–2013)

SOLUTION. Here, $\qquad F_c(e^{ax}) = \int_0^l e^{ax} \cos\left(\dfrac{p\pi x}{l} \right) dx$

$$= \frac{\left[e^{ax} \left\{ a \cos\left(\dfrac{p\pi x}{l} \right) + \left(\dfrac{p\pi}{l} \right) \sin\left(\dfrac{p\pi x}{l} \right) \right\} \right]_{x=0}^{l}}{a^2 + \left(\dfrac{p\pi}{l} \right)^2}$$

$$= \frac{a}{a^2 + \left(\dfrac{p\pi}{l} \right)^2} [e^{al} \cos(p\pi - 1)].$$

EXAMPLE 15. *Find the Fourier sine integral for $f(x) = e^{-\beta x} (\beta > 0)$. Hence, show that*

$$\frac{\pi}{2} e^{-\beta x} = \int_0^{\infty} \frac{\lambda \sin(\lambda x) d\lambda}{\lambda^2 + \beta^2}$$ (Kanpur–2004; UPTU–2004)

SOLUTION. We have $\qquad f(x) = \dfrac{2}{\pi} \int_0^{\infty} \int_0^{\infty} f(t) \sin(ut) \cdot \sin(ux) \, du \, dt$

$$\Rightarrow \qquad \frac{\pi}{2} e^{-\beta x} = \int_0^{\infty} \int_0^{\infty} e^{-\beta t} \sin(ut) \cdot \sin(ux) \, du \, dt$$

Now, $\qquad \dfrac{\pi}{2} e^{-\beta x} = \int_0^{\infty} \sin(ux) \, du \left[\int_0^{\infty} e^{-\beta t} \sin(ut) \, dt \right]$

$$= \int_0^{\infty} \sin(ux) \, du \left[\frac{u}{u^2 + \beta^2} \right] = \int_0^{\infty} \frac{\lambda \sin(\lambda x) d\lambda}{\lambda^2 + \beta^2}.$$

EXAMPLE 16. *Find the finite sine transform of x^3.*

SOLUTION. We have $F_p\{x^3\} = \int_0^\pi x^3 \sin px\, dx = \left(\dfrac{-x^3 \cos px}{p}\right)_{x=0}^{\pi} + \dfrac{3}{p}\int_0^\pi x^2 \cos px\, dx$

$$= -\frac{\pi^3}{p}\cos p\pi + \frac{3}{p}\left[-\left(\frac{x^2 \sin px}{p}\right)_{x=0}^{\pi} - \frac{2}{p}\int_0^\pi x \sin px\, dx\right]$$

$$= -\frac{\pi^3}{p}\cos p\pi - \frac{6}{p^2}\left[-\left(\frac{x\cos px}{p}\right)_{x=0}^{\pi} + \frac{1}{p}\int_0^\pi \cos(px)\, dx\right]$$

$$= -\frac{\pi^3}{p}\cos \pi p + \frac{6\pi}{p^3}\cos \pi p - \frac{6}{p^4}(\sin \pi x)_{x=0}^{\pi}$$

$$= -\frac{\pi^3}{p}\cos p\pi + \frac{6\pi}{p^3}\cos p\pi = \frac{\pi}{p}(-1)^p\left\{\frac{6}{p^2} - \pi^2\right\}.$$

if $p = 1, 2, 3, 4, \ldots$

EXAMPLE 17. *Find finite Fourier sine and cosine transform of $f(x) = x^2$, $0 < x < 4$.*

(Agra–1986; Kanpur–2003, 04; Purvanchal–2002; Meerut–1991)

SOLUTION. We have $F_p\{f(x)\} = \int_0^c f(x)\sin\dfrac{n\pi x}{c}dx = \int_0^4 x^2 \sin\dfrac{n\pi x}{4}dx$

$$= \left[-\frac{4}{n\pi}x^2 \cos\frac{n\pi x}{4}\right]_{x=0}^{4} + \int_0^4 2x \cdot \frac{4}{n\pi}\cos\frac{n\pi x}{4}dx$$

$$= -\frac{4^3}{n\pi}\cos n\pi + \frac{8}{n\pi}\left[\frac{4x}{n\pi}\sin\frac{n\pi x}{4} + \frac{4^2}{n^2\pi^2}\cos\frac{n\pi x}{4}\right]_{x=0}^{4}$$

$$= -\frac{4^3}{n\pi}\cos n\pi + \frac{8\cdot 4^2}{n\pi\cdot n^2\pi^2}(\cos n\pi - 1)$$

$$\tilde{f}_s(n) = -\frac{64}{n\pi}\cos n\pi + \frac{128}{n^3\pi^3}(\cos n\pi - 1)$$

Now, $F_c\{f(x)\} = \tilde{f}_c(n) = \int_0^4 f(x)\cos\dfrac{n\pi x}{4}dx = \int_0^4 x^2 \cos\dfrac{n\pi x}{4}dx$

$$= \left[\frac{4x^2}{n\pi}\sin\frac{n\pi x}{4}\right]_{x=0}^{4} - \int_0^4 \frac{4}{n\pi}2x\sin\frac{n\pi x}{4}dx$$

$$= 0 - \frac{8}{n\pi}\left[-\frac{4x}{n\pi}\cos\frac{n\pi x}{4} + \frac{4^2}{n^2\pi^2}\sin\frac{n\pi x}{4}\right]_{x=0}^{4} = \frac{128}{n^2\pi^2}\cos n\pi.$$

EXAMPLE 18. *Using Fourier integral formula, show that*

$$e^{-x}\cos x = \frac{2}{\pi}\int_0^\infty \frac{(u^2 + 2)\cos ux}{u^4 + 4}du$$

(Kanpur–2004, 07)

SOLUTION. Here, $f(x) = \dfrac{2}{\pi}\int_0^\infty \int_0^\infty f(t)\cos ut \cos ux\, du\, dt$

Let $f(t) = e^{-t} \cos t$, then we have

$$e^{-x} \cos x = \frac{2}{\pi} \int_0^\infty \cos ux \, du \int_0^\infty e^{-t} \cos t \cos ut \, dt$$

$$= \frac{1}{\pi} \int_0^\infty \cos ux \, du \cdot \int_0^\infty e^{-t} [\cos(u+1)t + \cos(u-1)t] dt$$

$$= \frac{1}{\pi} \int_0^\infty \cos ux \, du \cdot \left[\frac{1}{(u+1)^2 + 1} + \frac{1}{(u-1)^2 + 1} \right]$$

$$= \frac{1}{\pi} \int_0^\infty \frac{2(u^2 + 2) \cos ux \, du}{(u^2 + 2u + 2)(u^2 + 2 - 2u)}$$

$$= \frac{1}{\pi} \int_0^\infty \frac{2(u^2 + 2) \cos ux \, du}{[(u^2 + 2)^2 - (2u)^2]} = \frac{2}{\pi} \int_0^\infty \frac{(u^2 + 2) \cos ux \, du}{u^4 + 4}.$$

EXAMPLE 19. *Express the function*

$$f(x) = \begin{cases} 1 & \text{for } |x| \leq 1 \\ 0 & \text{for } |x| > 1 \end{cases}$$

as a Fourier integral. Hence, evaluate

$$\int_0^\infty \frac{\sin \lambda \cos \lambda x}{\lambda} d\lambda \qquad \text{(Kanpur–2005; UPTU–2008)}$$

SOLUTION. By Fourier integral formula, we have

$$f(x) = \frac{1}{\pi} \int_{p=0}^\infty \int_{t=-\infty}^\infty f(t) \cos p(t - x) dp \, dt$$

$$= \frac{1}{\pi} \int_{p=0}^\infty \int_{t=-1}^1 \cos p(t - x) dp \, dt = \frac{1}{\pi} \int_{p=0}^\infty \left[\frac{\sin p(t - x)}{p} \right]_{t=-1}^1 dp$$

$$= \frac{1}{\pi} \int_{p=0}^\infty \left[\frac{\sin p(1 - x) - \sin p(-1 - x)}{p} \right] dp$$

$$= \frac{1}{\pi} \int_0^\infty \left[\frac{\sin p(1 - x) + \sin p(1 + x)}{p} \right] dp$$

$$= \frac{2}{\pi} \int_0^\infty \frac{\sin p \cos px}{p} dp = \frac{2}{\pi} \int_0^\infty \frac{\sin \lambda \cos \lambda x}{\lambda} d\lambda$$

$$\Rightarrow \qquad f(x) = \frac{2}{\pi} \int_0^\infty \frac{\sin \lambda \cos \lambda x}{\lambda} d\lambda$$

Now, $\qquad \int_0^\infty \frac{\sin \lambda \cos \lambda x}{\lambda} d\lambda = \frac{\pi}{2} f(x) = \begin{cases} \pi/2, & |x| \leq 1 \\ 0, & |x| > 1 \end{cases}$

$$\Rightarrow \qquad \int_0^\infty \frac{\sin \lambda \cos \lambda x}{\lambda} d\lambda = \begin{cases} \pi/2 & \text{for } |x| \leq 1 \\ 0 & \text{for } |x| > 1 \end{cases}$$

EXAMPLE 20. *Find Fourier transform of f(x) defined by*

$$f(x) = \begin{cases} 1, & |x| < a \\ 0, & |x| > a \end{cases}$$

and hence prove that $\int_0^\infty \dfrac{\sin^2 ax}{x^2}dx = \dfrac{\pi a}{2}$ (SVTU–2004; Kanpur–2008)

SOLUTION. Cosider $F\{t\} = \int_{-\infty}^\infty e^{-ipx}f(x)dx$

$$= \int_{-\infty}^{-a} e^{-ipx}f(x)dx + \int_{-a}^a e^{-ipx}f(x)dx + \int_a^\infty e^{-ipx}f(x)dx$$

$$= \int_\infty^a e^{ipy}f(-y)(-dy) + \int_{-a}^a e^{-ipx}dx + \int_a^\infty e^{-ipx}\cdot 0 dx$$

$$= \int_\infty^a e^{ipy}\cdot 0\cdot dy + \frac{1}{ip}(e^{ipx})_{-a}^a + 0$$

$$= \frac{e^{ipa} - e^{-ipa}}{ip} = \frac{2}{p}\sin pa = \tilde{f}(p)$$

Now, using Parseval's identity for Fourier integral, we get

$$\int_{-\infty}^\infty |f(x)|^2 dx = \frac{1}{2\pi}\int_{-\infty}^\infty |\tilde{f}(p)|^2 dp$$

$$\Rightarrow \qquad \int_{-a}^a 1^2 dx = \frac{1}{2\pi}\int_{-\infty}^\infty \frac{4}{p^2}\sin^2 padp$$

$$\Rightarrow \qquad 2a = \frac{2}{2\pi}\int_0^\infty \frac{4}{p^2}\sin^2 padp$$

$$\Rightarrow \int_0^\infty \frac{\sin^2(ax)dx}{x^2} = \frac{\pi a}{2}$$

EXAMPLE 21. *Using Fourier integral, show that*

$$e^{-ax} = \frac{2a}{\pi}\int_0^\infty \frac{\cos\lambda x}{\lambda^2 + a^2}d\lambda, a > 0, x \geq 0$$

Or, Using Fourier integral representation of an appropriate function, show that

$$\int_0^\infty \frac{\cos(\omega x)d\omega}{k^2 + \omega^2} = \frac{\pi}{2k}e^{-kx}, \text{where } x > 0, k > 0. \qquad \text{(Kanpur–2003)}$$

SOLUTION. From Fourier cosine integral formula

$$f(x) = \frac{2}{\pi}\int_0^\infty \int_0^\infty f(t)\cos pt \cos pxdpdt$$

Let $f(t) = e^{-at}$, then we get

$$e^{-ax} = \frac{2}{\pi}\int_0^\infty \cos pxdx \cdot \int_0^\infty e^{-at}\cos ptdt$$

$$\Rightarrow \qquad e^{-ax} = \frac{2}{\pi}\int_0^\infty \cos pxdp \cdot \frac{a}{p^2 + a^2}$$

$$= \frac{2a}{\pi}\int_0^\infty \frac{\cos pxdx}{p^2 + a^2} = \frac{2a}{\pi}\int_0^\infty \frac{\cos\lambda xd\lambda}{\lambda^2 + a^2}$$

$$\Rightarrow \qquad e^{-ax} = \frac{2a}{\pi}\int_0^\infty \frac{\cos\lambda x}{\lambda^2 + a^2}d\lambda$$

Additional Solved Examples

EXAMPLE 1. *Find the finite Fourier sine and cosine transform of $f(x) = 1$.*

SOLUTION. Here, we have

$$\tilde{f}_s(p) = \int_0^\pi f(x) \cdot \sin px\, dx = \int_0^\pi 1 \cdot \sin px\, dx = \int_0^\pi \sin px\, dx$$

$$= \left[-\frac{\cos px}{p} \right]_0^\pi = -\frac{\cos p\pi}{p} + \frac{1}{p} = \frac{1}{p}[1 - (-1)^p]$$

Now, $\tilde{f}_c(p) = \int_0^\pi f(x) \cos px\, dx = \int_0^\pi 1 \cdot \cos px\, dx$

$$= [(1/p)\sin px]_0^\pi = 0.$$ [if $p = 1, 2, 3, ...$]

EXAMPLE 2. *Show that the finite Fourier sine transform of $\dfrac{x}{\pi}$ is $\dfrac{1}{p}(-1)^{p+1}$.* (Meerut–1987, 99)

SOLUTION. Here, we have

$$\tilde{f}_s\left(\frac{x}{\pi}\right) = \int_0^\pi \frac{x}{\pi}\sin px\, dx = \frac{1}{\pi}\int_0^\pi x \sin px\, dx$$

$$= \frac{1}{\pi}\left[\left(-\frac{\cos px}{p}\right)x + \frac{\sin px}{p^2}\right]_0^\pi = \frac{1}{\pi}\left[-\frac{\pi}{p}\cos p\pi\right]$$

$$= -\frac{1}{p}(-1)^p = \frac{(-1)^{p+1}}{p}.$$

EXAMPLE 3. *Find the finite Fourier sine transform of $\left(1 - \dfrac{x}{\pi}\right)$ and $\dfrac{x}{4\pi}$.* (Meerut–1998, 99)

SOLUTION. Here, finite Fourier sine transform of $\left(1 - \dfrac{x}{\pi}\right)$

$$\tilde{f}_s(p) = \int_0^\pi \left(1 - \frac{x}{\pi}\right)\sin px\, dx$$

$$= \left[\left(1 - \frac{x}{\pi}\right)\left(-\frac{\cos px}{p}\right)\right]_0^\pi - \int_0^\pi \left(-\frac{1}{\pi}\right)\left(-\frac{\cos px}{p}\right)dx$$

$$= \frac{1}{p} - \frac{1}{\pi}\left[\frac{\sin px}{p^2}\right]_0^\pi = \frac{1}{p}$$

Now, Finite Fourier sine transform of $x/4\pi$

$$\tilde{f}_s(p) = \int_0^\pi \left(\frac{x}{4\pi}\right)\sin px\, dx$$

$$= \left[\left(\frac{x}{4\pi}\right)\left(-\frac{\cos px}{p}\right)\right]_0^\pi - \int_0^\pi \frac{1}{4\pi}\left(-\frac{\cos px}{p}\right)dx$$

$$= \frac{-1}{4p}\cos p\pi + \frac{1}{4\pi}\left[\frac{\sin px}{p^2}\right]_0^\pi = \frac{(-1)^{p+1}}{4p}.$$

EXAMPLE 4. *Find the finite sine and cosine transforms of $f(x) = x^2, 0 < x < \pi$.*

(Agra–1986, 2001, 03; Meerut–1991)

SOLUTION. Here, we have

$$\tilde{f}_s(p) = \int_0^\pi x^2 \sin px \, dx$$

$$= \left[x^2 \cdot \left(-\frac{\cos px}{p} \right) \right]_0^\pi + 2\int_0^\pi x \cdot \frac{\cos px}{p} \, dx$$

$$= \frac{-\pi^2 \cos p\pi}{p} + 2\left[x \cdot \frac{\sin px}{p^2} + \frac{\cos px}{p^3} \right]_0^\pi$$

$$= -(\pi^2 / p)\cos p\pi + 2(\cos p\pi - 1) / p^3$$

$$= \pi^2(-1)^{p+1} + 2\{(-1)^p - 1\} / p^3$$

Now, $\tilde{f}_c(p) = \int_0^\pi x^2 \cos px \, dx$

$$= \left[x^2 \cdot \frac{\sin px}{p} \right]_0^\pi - 2\int_0^\pi x \cdot \frac{\sin px}{p} \, dx$$

$$= 0 - 2\left[x \cdot \frac{(-\cos px)}{p^2} + \frac{\sin px}{p^3} \right]_0^\pi$$

$$= \frac{2\pi \cos p\pi}{p^2} = \frac{2\pi(-1)^p}{p^2}, \text{ if } p > 0$$

Let $p = 0$, then $\tilde{f}_c(p) = \int_0^\pi x^2 \cdot 1 \, dx = \pi^3 / 3$.

EXAMPLE 5. *Find the finite cosine transform of $\left(1 - \frac{x}{\pi} \right)^2$.*

(Meerut–1989, 94; Rohilkhand–1990; Agra–1981; Avadh–2013; Kanpur–1995, 98; SVTU–2006)

SOLUTION. Here, we have

$$\tilde{f}_c(p) = \int_0^\pi \left(1 - \frac{x}{\pi} \right)^2 \cos px \, dx$$

$$= \left[\left(1 + \frac{x}{\pi} \right)^2 \cdot \frac{\sin px}{p} \right]_0^\pi + \frac{2}{\pi p} \int_0^\pi \left(1 - \frac{x}{\pi} \right) \sin px \, dx \cdot$$

$$= 0 + \frac{2}{\pi p} \left[\left(1 - \frac{x}{\pi} \right) \left(-\frac{\cos px}{p} \right) - \frac{1}{p\pi} \cdot \frac{\sin px}{p} \right]_0^\pi = \frac{2}{\pi p^2}, p > 0$$

If $p = 0$, then

$$\tilde{f}_c(p) = \int_0^\pi \left(1 - \frac{x}{\pi} \right)^2 \cdot 1 \, dx = \left[-\frac{\pi}{3} \left(1 - \frac{x}{\pi} \right)^3 \right]_0^\pi = \frac{\pi}{3}$$

EXAMPLE 6. *Find the finite Fourier sine transform of $f(x) = \frac{\pi}{3} - x + \frac{x^2}{2\pi}$.* (Meerut–1980, 2001, 02)

SOLUTION. Here, we have

$$\tilde{f}_s(p) = \int_0^\pi \left(\frac{\pi}{3} - x - \frac{x^2}{2\pi} \right) \sin px \, dx$$

$$= \left[\left(\frac{\pi}{3} - x - \frac{x^2}{2\pi}\right)\left(-\frac{\cos px}{p}\right)\right]_0^\pi + \int_0^\pi \left(-1 + \frac{x}{\pi}\right)\frac{\cos px}{p}dx$$

$$= \frac{\pi}{6p}\{(-1)^p + 2\} + \left[\left(-1 + \frac{x}{\pi}\right)\left(\frac{\sin px}{p^2}\right)\right]_0^\pi - \int_0^\pi \left(\frac{1}{\pi}\right)\left(\frac{\sin px}{p^2}\right)dx$$

$$= \frac{\pi}{6p}\{(-1)^p + 2\} + \frac{1}{\pi p^3}[\cos px]_0^\pi$$

$$= \frac{\pi}{6p}\{(-1)^p + 2\} + \frac{1}{\pi p^3}\{(-1)^p - 1\}.$$

EXAMPLE 7.   *Find the finite sine transforms of $f(x) = \cos kx$.*

(Kanpur–1997, 2006; Agra–1980, 82; Meerut–1991)

SOLUTION.   Here, we have

$$\tilde{f}_s(p) = \int_0^\pi \cos kx \sin px\,dx$$

$$= \frac{1}{2}\int_0^\pi [\sin(k+p)x + \sin(p-k)x]dx$$

$$= \frac{1}{2}\left[-\frac{\cos(k+p)x}{(k+p)} - \frac{\cos(p-k)x}{(p-k)}\right]_0^\pi$$

$$= \frac{1}{2}\left[-\frac{\cos(k+p)\pi}{k+p} - \frac{\cos(p-k)\pi}{p-k} + \frac{1}{k+p} + \frac{1}{p-k}\right]$$

$$= \frac{1}{2}\left[-\frac{(p-k)\cos(k+p)\pi + (k+p)\cos(p-k)\pi}{p^2 - k^2} + \frac{p-k+k+p}{p^2 - k^2}\right]$$

$$= \frac{1}{2(p^2 - k^2)}[-p\{\cos(k+p)\pi + \cos(p-k)\pi\}$$
$$+ k\{\cos(k+p)\pi - \cos(p-k)\pi\} + 2p]$$

$$= \frac{1}{(p^2 - k^2)}[-p\cos k\pi - k\sin k\pi \sin p\pi + p]$$

$$= \frac{p}{p^2 - k^2}[1 - \cos k\pi \cos p\pi] = \frac{p}{p^2 - k^2}[1 - (-1)^p \cos k\pi]$$

EXAMPLE 8.   *Find the finite Fourier cosine transforms of $f(x) = \sin nx$.*

(Meerut–1991, 2001, 02; Kanpur–2001; Agra–1982, 2002)

SOLUTION.   Here, we have

$$\tilde{f}_c(p) = \int_0^\pi \sin nx \cos px\,dx$$

$$= \frac{1}{2}\int_0^\pi [\sin(n+p)x + \sin(n-p)x]dx$$

$$= \frac{1}{2}\left[-\frac{\cos(n+p)x}{n+p} - \frac{\cos(n-p)x}{n-p}\right]_0^\pi, \text{ if } p \neq n, \text{ then}$$

$$\tilde{f}_c(p) = \frac{1}{2}\left[-\frac{\cos(n+p)\pi}{n+p} - \frac{\cos(n-p)\pi}{n-p} + \frac{1}{n+p} + \frac{1}{n-p}\right]$$

If $n - p$ is even, then

$$\tilde{f}_c(p) = \frac{1}{2}\left[-\frac{1}{n+p} - \frac{1}{n-p} + \frac{1}{n+p} + \frac{1}{n-p}\right] = 0$$

If $n - p$ is odd, then

$$\tilde{f}_c(p) = \frac{1}{2}\left[\frac{2}{n+p} + \frac{2}{n-p}\right] = \frac{2n}{n^2 - p^2}$$

Now, if $p = n$, then

$$\tilde{f}_c(p) = \int_0^\infty \sin nx \cos nx\, dx = \frac{1}{2}\int_0^\pi \sin 2nx\, dx$$

$$= \frac{1}{2}\left[-\frac{\cos 2nx}{2n}\right]_0^\pi = 0$$

Hence, $\qquad \tilde{f}_c(p) = 0 \ \text{ or } \ \dfrac{2n}{n^2 - p^2}$.

EXAMPLE 9. *Find finite Fourier cosine transform of* $f(x) = \dfrac{\cosh[c(\pi - x)]}{\sinh(\pi c)}$. $\qquad$ (Meerut–1991)

SOLUTION. Here, we have

$$\tilde{f}_c(p) = \int_0^\pi f(x)\cos px\, dx = \int_0^\pi \frac{\cosh\{c(\pi - x)\}}{\sinh(\pi c)} \cdot \cos px\, dx$$

$$= \frac{1}{\sinh(\pi c)} \cdot \frac{1}{2}\int_0^\pi \{e^{c(\pi-x)} + e^{-c(\pi-x)}\}\cos px\, dx$$

$$= \frac{1}{2\sinh(\pi c)}\left[e^{c\pi}\int_0^\pi e^{-cx}\cos px\, dx + e^{-c\pi}\int_0^\pi e^{cx}\cos px\, dx\right]$$

$$= \frac{1}{2\sinh(\pi c)}\left[\frac{e^{c\pi} \cdot e^{-cx}}{c^2 + p^2}(-c\cos px + p\sin px)\right.$$

$$\left. + \frac{e^{-c\pi} \cdot e^{cx}}{c^2 + p^2}(c\cos px + p\sin px)\right]_0^\pi$$

$$= \frac{1}{2\sinh(\pi c)}\left[\frac{\{-c\cos(\pi p) + ce^{-\pi}\} + \{c\cos(\pi p) - ce^{-c\pi}\}}{c^2 + p^2}\right]$$

$$= \frac{c(e^{c\pi} - e^{-c\pi})}{2(c^2 + p^2)\sinh(\pi c)} = \frac{c \cdot 2\sinh(\pi c)}{2(c^2 + p^2)\sinh(\pi c)} = \frac{c}{c^2 + p^2}.$$

EXAMPLE 10. *Find* $f(x)$ *if its finite sine transforms is given by* $\tilde{f}_s(p) = \dfrac{2\pi(-1)^{p-1}}{p^3}, p = 1,2,3...$

where $0 < x < \pi$. $\qquad$ (Kanpur–1999)

SOLUTION. Here, we have

$$f(x) = \frac{2}{\pi}\sum_{p=1}^\infty \tilde{f}_s(p)\sin px = \frac{2}{\pi}\sum_{p=1}^\infty \frac{2\pi(-1)^{p-1}}{p^3}\sin px$$

$$= 4\sum_{p=1}^\infty \frac{(-1)^{p-1}}{p^3}\sin px .$$

4.6 APPLICATIONS OF FOURIER TRANSFORM

The finite sine and cosine transforms can be applied when the range of the variable selected for exclusion is 0 to ∞. The choice of sine and cosine transform is decided by the form of the boundary conditions at the lower limit of the variable selected for exclusion.

Hence, we have

$$F_s\left\{\frac{\partial^2 u}{\partial x^2}\right\} = \int_0^\infty \frac{\partial^2 u}{\partial x^2} \sin px\, dx = \left[\frac{\partial u}{\partial x}\sin px\right]_0^\infty - p\int_0^\infty \frac{\partial u}{\partial x}\cos px\, dx$$

$$= -p\int_0^\infty \frac{\partial u}{\partial x}\cos px\, dx \text{ if } \frac{\partial u}{\partial x} \to 0 \text{ as } x \to \infty$$

$$= -p\left\{[u\cos px]_0^\infty + p\int_0^\infty u\sin px\, dx\right\} = p(u)_{x=0} - p^2\bar{u}_s$$

[By assuming $u \to 0$ as $x \to \infty$]

Therefore, $\qquad F_s\left\{\frac{\partial^2 u}{\partial x^2}\right\} = pu(0,t) - p^2\bar{u}_s(p,t)$ $\qquad\qquad$ ...(1)

where $u(x, t)$ is a function of two variable x and t and $\bar{u}_s(p,t)$ is the Fourier sine transform of $u(x, t)$ with respect to x.

Further, $\qquad F_c\left\{\frac{\partial^2 u}{\partial x^2}\right\} = \int_0^\infty \frac{\partial^2 u}{\partial x^2}\cos px\, dx \overset{\bullet}{=} \left[\frac{\partial u}{\partial x}\cos px\right]_0^\infty + p\int_0^\infty \frac{\partial u}{\partial x}\sin px\, dx$

$$= -\left(\frac{\partial u}{\partial x}\right)_{x=0} + p\int_0^\infty \frac{\partial u}{\partial x}\sin px\, dx \quad \left[\text{Assuming } \frac{\partial u}{\partial x} \to 0 \text{ as } x \to \infty\right]$$

$$= -\left(\frac{\partial u}{\partial x}\right)_{x=0} + p\left\{[u\sin px]_0^\infty - p\int_0^\infty u\cos px\, dx\right\}$$

$$= -\left(\frac{\partial u}{\partial x}\right)_{x=0} - p^2\int_0^\infty u(x,t)\cos px\, dx\,.$$

Then, $\qquad F_s\left\{\frac{\partial^2 u}{\partial x^2}\right\} = -\left(\frac{\partial u}{\partial x}\right)_{x=0} - p^2\bar{u}_c(p,t)$ $\qquad\qquad$ ... (2)

Where, $\bar{u}_c(p,t)$ is the Fourier cosine transform of $u(x, t)$ with respect to x.

REMARKS

- It must be noted that the successful use of a sine transform in removing a term $\frac{\partial^2 u}{\partial x^2}$ required $u(0, t)$, i.e., u at $x = 0$ while the use of a cosine transform for the same purpose requires, $u_x(0,t)$, i.e, $\frac{\partial u}{\partial x}$ at $x = 0$.

- The terms $\frac{\partial u}{\partial x}$ or any partial derivative of odd order cannot be removed with the help of sine or cosine transforms.

- When one of the variables in a differential equation ranges form $-\infty$ to $+\infty$ then that variable can be excluded with the help of complex Fourier transforms.

Solved Examples

EXAMPLE 1. *Solve* $\dfrac{\partial u}{\partial t} = 2\dfrac{\partial^2 u}{\partial x^2}$ *if* $u(0,t) = 0, u(x,0) = e^{-x}$, $x > 0$, $u(x, t)$ *is bounded where*

$x > 0, t > 0.$ (Rohtak–2006; Meerut–1986, 2004, 14; Kanpur–2004, 13; Patna–2003, 04)

SOLUTION.   As per given $\dfrac{\partial u}{\partial t} = 2\dfrac{\partial^2 u}{\partial x^2}$ ... (1)

Subject to the boundary conditions

$$u(0,t) = 0, u(x,t) \text{ is bounded} \qquad \text{... (2)}$$

and initial condition

$$u(x,0) = e^{-x}, x > 0 \qquad \text{... (3)}$$

Since, $u(0, t)$ is given, taking the Fourier sine transform of both sides of (1), we get

$$\int_0^\infty \frac{\partial u}{\partial t}\sin px\, dx = 2\int_0^\infty \frac{\partial^2 u}{\partial x^2}\sin px\, dx$$

$$\Rightarrow \frac{d}{dt}\int_0^\infty u(x,t)\sin px\, dx$$

$$= 2\left\{\left(\frac{\partial u}{\partial x}\sin px\right)_0^\infty - \int_0^\infty \frac{\partial u}{\partial x}p\cos px\, dx\right\}$$

$$\Rightarrow \qquad \frac{d\bar{u}_s}{dt} = -2p\int_0^\infty \frac{\partial u}{\partial x}\cos px\, dx \text{ if } \frac{du}{dx} \to 0 \text{ as } x \to \infty$$

$$\left[\text{Assume } \bar{u}_s(p,t) = \int_0^\infty u(x,t)\sin px\, dx\right]$$

$$= -2p\left\{[u(x,t)\cos px]_0^\infty - \int_0^\infty u(x,t)(-p\sin px)dx\right\}$$

$$= -2p\left\{0 - u(0,t) + p\int_0^\infty u(x,t)\sin px\, dx\right\}$$

$$[\because u(x,t) \to 0, \cos x \to \infty]$$

$$= 2pu(0,t) - 2p^2\bar{u}_s$$

$$\Rightarrow \qquad \frac{d\bar{u}_s}{dt} = -2p^2\bar{u}_s \qquad\qquad [\because u(0, t) = 0]$$

On separating the variables, we get

$$\frac{d\bar{u}_s}{\bar{u}_s} = -2p^2 dt \Rightarrow \log \bar{u}_s - \log C = -2p^2 t$$

$$\Rightarrow \qquad \log\left(\frac{\bar{u}_s}{C}\right) = -2p^2 t \Rightarrow \bar{u}_s(p,t) = Ce^{-2p^2 t} \qquad \text{...(4)}$$

Now, taking the Fourier sine transform of both sides of (3), we get

$$\int_0^\infty u(x,0)\sin px\, dx = \int_0^\infty e^{-x}\sin px\, dx$$

$$\Rightarrow \qquad \bar{u}_s(p,0) = \left[\frac{e^{-x}}{1+p^2}(-\sin px - p\cos px)\right]_0^\infty = \frac{p}{1+p^2} \qquad \text{...(5)}$$

Putting $t = 0$ in (4) and (5), we get

$$\frac{p}{1+p^2} = C$$

$\therefore \qquad \bar{u}_s(p,t) = \frac{p}{1+p^2}e^{-2p^2t}.$

Taking the inverse Fourier sine transform, we get

$$u(x,t) = \frac{2}{\pi}\int_0^\infty \frac{p}{1+p^2}e^{-p^2t}\sin px\,dx$$

EXAMPLE 2. Solve $\dfrac{\partial u}{\partial t} = \dfrac{\partial^2 u}{\partial x^2}, x > 0, t > 0$ subject   to   the   conditions $u(0, \quad t) \quad = \quad 0,$

$u(x,0) = \begin{cases} 1 & ; \quad 0 < x < 1 \\ 0 & ; \qquad x > 1 \end{cases},\ u(x, t)$ *is bounded.*

(UPTU–2003; Rohilkhand–1991; Meerut–1995; Kanpur–1999, 2000)

SOLUTION. Taking the Fourier sine transform of both side of given PDE, we get

$$\int_0^\infty \frac{\partial u}{\partial t}\sin px\,dx = \int_0^\infty \frac{\partial^2 u}{\partial x^2}\sin px\,dx$$

or $\dfrac{d}{dt}\displaystyle\int_0^\infty u\sin px\,dx$

$$= \left[\frac{\partial u}{\partial x}\sin px\right]_0^\infty - p\int_0^\infty \frac{\partial u}{\partial x}\cos px\,dx$$

$\therefore \qquad \dfrac{d\bar{u}_s}{dt} = -p\displaystyle\int_0^\infty \frac{\partial u}{\partial x}\cos px\,dx \qquad\qquad$ if $\dfrac{\partial u}{\partial x} \to 0$ as $x \to \infty$

$$= -p\left\{[u\cos px]_0^\infty + p\int_0^\infty u\sin px\,dx\right\} = -pu(0,t) - p^2\bar{u}_s;$$

$$= -p^2\bar{u}_s. \qquad\qquad\qquad \text{if } u \to 0 \text{ as } x \to \infty$$

On separating the variables, we get

$$\frac{d\bar{u}_s}{\bar{u}_s} = -p^2dt$$

whose solution is given by

$$\bar{u}_s(p,t) = Ce^{-p^2t} \qquad\qquad\qquad\qquad\qquad ...(1)$$

Putting $t = 0$, we get $C = \bar{u}_s(p,0)$ $\qquad\qquad\qquad\qquad ...(2)$

Now, $\bar{u}_s(p,0) = \displaystyle\int_0^\infty u(x,0)\sin px\,dx$

$$= \int_0^1 u(x,0)\sin px\,dx + \int_1^\infty u(x,0)\sin px\,dx = \int_0^1 \sin px\,dx$$

Now, from (2), $C = \displaystyle\int_0^1 \sin px\,dx = \left[\frac{\cos px}{-p}\right]_0^1 = \frac{1-\cos p}{p}.$

Thus, (1) gives $\bar{u}_s(p,t) = \left[\dfrac{(1-\cos p)}{p}\right]e^{-p^2t}.$

Finally, taking the inverse Fourier sine transform, we get

$$u(x,t) = \frac{2}{\pi}\int_0^\infty \frac{1-\cos p}{p}e^{-p^2t}\sin px\,dp$$

which is the required solution.

EXAMPLE 3. *Using the Fourier sine transform, solve the partial differential equation* $\dfrac{\partial V}{\partial t} = k\dfrac{\partial^2 V}{\partial x^2}$

for $x > 0$, $t > 0$ *under the boundary conditions* $V = V_0$ *when* $x = 0$, $t > 0$ *and the initial condition* $v = 0$ *when* $t = 0$, $x > 0$.

(Meerut–1988, 89, 91, 93; Rohilkhand–1988, 2001; Kanpur–1991; Himachal–2001, 02, 05)

SOLUTION. Taking the Fourier sine transform of both the sides of the given equation, we get

$$\sqrt{\frac{2}{\pi}}\int_0^\infty \frac{\partial V}{\partial t}\sin px\,dx = k\sqrt{\frac{2}{\pi}}\frac{\partial^2 V}{\partial x^2}.\sin px\,dx$$

$$\Rightarrow \quad \frac{d}{dt}\sqrt{\frac{2}{\pi}}\int_0^\infty V\sin px\,dx = -kp^2\tilde{V}_s + k_p\sqrt{\frac{2}{\pi}}V(0,t)$$

$$\Rightarrow \quad \frac{d\tilde{V}_s}{dt} + kp^2\tilde{V}_s + k_p\sqrt{\frac{2}{\pi}}V_s \qquad\qquad [\because V(0,t) = V_0]$$

which is a linear differential equation of first order

$$\text{I.F.} = e^{\int kp^2 dt} = e^{kp^2 t}$$

Its solution is given by

$$\tilde{V}_s e^{kp^2 t} = C + \sqrt{\frac{2}{\pi}}\int kpV_0.e^{kp^2 t}dt = C + \sqrt{\frac{2}{\pi}}\frac{V_0}{p}e^{kp^2 t} \qquad ...(1)$$

But when $\quad t = 0, \tilde{V}_s = 0 \qquad\qquad\qquad (\because V = 0 \text{ when } t = 0)$

Using (1), we have $\quad 0 = C + \dfrac{V_0}{p}\sqrt{\dfrac{2}{\pi}} \Rightarrow C = -\dfrac{V_0}{p}\sqrt{\dfrac{2}{\pi}}.$

Putting this value in (1), we get

$$\tilde{V}_s e^{kp^2 t} = \frac{V_0}{p}\sqrt{\frac{2}{\pi}}(e^{kp^2 t} - 1) \quad\Rightarrow\quad \tilde{V}_s = \sqrt{\frac{2}{\pi}}\frac{V_0}{p}(1 - e^{-kp^2 t}).$$

Applying the inverse Fourier transform, we have

$$V = \frac{2}{\pi}V_0\int_0^\infty \frac{(1-e^{-kp^2 t})}{p}\sin px\,dp$$

$$= \frac{2V_0}{\pi}\left[\int_0^\infty \frac{\sin px\,dx}{p}dp - \int_0^\infty \frac{e^{-kp^2 t}}{p}.\sin px\,dp\right]$$

$$= \frac{2V_0}{\pi}\left[\frac{\pi}{2} - \int_0^\infty \frac{e^{-kp^2 t}}{p}.\sin px\,dp\right]$$

Hence, $\quad V(x,t) = V_0\left[1 - \dfrac{2}{\pi}\int_0^\infty \dfrac{e^{-kp^2 t}}{p}.\sin px\,dp\right].$

EXAMPLE 4. *Use the method of Fourier transform to determine the displacement $y(x,t)$ of an infinite string, given that the string is initially at rest and that the initial displacement is $f(x)$, $-\infty < x < \infty$ show that $y(x,t) = \frac{1}{2}[f(x + Ct) + f(x - Ct)]$.*

(Rohtak–2000; Meerut–1992, 2006; Kurukshetra–1997; Allahabad–2003; Kolkata–2003, 05, 06)

SOLUTION. It is known that the displacement of a string is governed by one dimensional wave equation

$$\frac{\partial^2 y}{\partial t^2} = C^2 \frac{\partial^2 y}{\partial x^2} \qquad ...(1)$$

where $y(x, t)$ is the displacement at any time, $t, -\infty < x < \infty, t > 0$, and $C^2 = \frac{T}{\rho}$

Taking the Fourier transform of both sides of (1), we get

$$\frac{1}{\sqrt{2\pi}} \int_{-\infty}^{\infty} \frac{\partial^2 y}{\partial t^2} e^{ipx} dx = C^2 \frac{1}{\sqrt{2\pi}} \int_{-\infty}^{\infty} \frac{\partial^2 y}{\partial x^2} e^{ipx} dx$$

$$\Rightarrow \qquad \frac{d^2}{dt^2} \frac{1}{\sqrt{2\pi}} \int_{-\infty}^{\infty} y e^{ipx} dx = C^2 (-ip)^2 \tilde{y}(p,t)$$

$$\Rightarrow \qquad \frac{d^2 \tilde{y}(p,t)}{dt^2} + C^2 p^2 \tilde{y}(p,t) = 0$$

whose solution is given by

$$\tilde{y}(p,t) = A \cos Cpt + B \sin Cpt \qquad ...(2)$$

As per given, the string is initially at rest, *i.e.*, $\frac{\partial y}{\partial t} = 0$ at $t = 0$.

Thus, $\frac{1}{\sqrt{2\pi}} \int_{-\infty}^{\infty} \frac{\partial y}{\partial t} e^{ipx} dx = \frac{d}{dt} \frac{1}{\sqrt{2\pi}} \int_{-\infty}^{\infty} y e^{ipx} dx$

$$= \frac{d\tilde{y}(p,t)}{dt} = 0 \text{ at } t = 0 \cdot$$

Now from (2), we have $0 = BC_p \Rightarrow B = 0$

Also, at $t = 0$, $y = f(x)$.

So, at $t = 0$,

$$\tilde{y}(p,0) = \frac{1}{\sqrt{2\pi}} \int_{-\infty}^{\infty} f(u) e^{ipu} du = \tilde{f}(p)$$

$$\therefore \qquad \tilde{y}(p,t) = \tilde{f}(p) \cos Cpt$$

Now, taking the inverse Fourier transform, we have

$$y(x,t) = \frac{1}{\sqrt{2\pi}} \int_{-\infty}^{\infty} \tilde{f}(p) \cos Cpt \, e^{-ipx} dp$$

$$= \frac{1}{2\pi} \int_{-\infty}^{\infty} \left[\int_{-\infty}^{\infty} f(u) e^{ipu} du \right] \cos Cpt \, e^{-ipx} dp$$

$$= \frac{1}{4\pi} \int_{-\infty}^{\infty} \left[\int_{-\infty}^{\infty} f(u) e^{ipu} du \right] (e^{iCpt} + e^{-iCpt}) e^{-ipx} dp$$

$$= \frac{1}{2} \left[\frac{1}{2\pi} \int_{-\infty}^{\infty} \left\{ \int_{-\infty}^{\infty} f(u) e^{-i\alpha u} du \right\} (e^{-iC\alpha t} + e^{iC\alpha t}) e^{i\alpha x} d\alpha \right]$$

Putting $\quad p = -\alpha \;\Rightarrow\; dp = -\,d\alpha$

We get

$$y(x,t) = \frac{1}{2}\left[\frac{1}{2\pi}\int_{-\infty}^{\infty} f(u)e^{-i\alpha u}\left\{\int_{-\infty}^{\infty} e^{i\alpha(x+Ct)}dx\right\}du + \frac{1}{2\pi}\int_{-\infty}^{\infty} f(u)e^{-i\alpha u}\right.$$
$$\left.\left\{\int_{-\infty}^{\infty} e^{i\alpha(x-Ct)}d\alpha\right\}du\right].$$

Finally, using Fourier integral formula, we get

$$y(x,t) = \frac{1}{2}[f(x+Ct) + f(x-Ct)].$$

EXAMPLE 5. *A thin membrane of great extent is release from rest in the position $z = f(x,y)$. Show that the displacement at any subsequent time is given by*

$$z(x,y) = \frac{1}{2\pi}\int_{-\infty}^{\infty}\int_{-\infty}^{\infty} F(p,q)\,\cos\{Ct\sqrt{p^2+q^2}\}e^{-i(px+qy)}dpdq$$

where $F(p,t)$ is double Fourier's transform of $f(x,y)$. (UPTU–2005; Lucknow–2005)

SOLUTION. It is known that the displacement of the membrane is governed by two dimensional wave equation

$$\frac{\partial^2 z}{\partial t^2} = C^2\left(\frac{\partial^2 z}{\partial x^2} + \frac{\partial^2 z}{\partial y^2}\right), \text{ when } C^2 = \frac{T}{\rho}$$

Taking the Fourier transforms of both the sides, we get

$$\frac{1}{2\pi}\int_{-\infty}^{\infty}\int_{-\infty}^{\infty}\frac{\partial^2 z}{\partial t^2}e^{i(px+qy)}dxdy = \frac{C^2}{2\pi}\int_{-\infty}^{\infty}\int_{-\infty}^{\infty}\left(\frac{\partial^2 z}{\partial x^2} + \frac{\partial^2 z}{\partial y^2}\right)e^{i(px+qy)}dxdy$$

$\Rightarrow \qquad \dfrac{d^2 z}{dt^2} = C^2(-p^2 - q^2)\tilde{z} = 0$, where $\tilde{z} = \dfrac{1}{2\pi}\int_{-\infty}^{\infty}\int_{-\infty}^{\infty} z.e^{i(px+qy)}.dxdy$

So, $\qquad \dfrac{d^2\tilde{z}}{dt^2} + C^2(p^2 + z^2)\tilde{z} = 0$ whose solution is given by

$$\tilde{z} = A\cos\{C\sqrt{p^2+q^2}.t\} + B\sin[C\sqrt{p^2+q^2}.t]$$

As per given, the initial conditions are $z = f(x,y), \dfrac{\partial z}{\partial t} = 0$ at $t = 0$.

Taking the Fourier transforms of these conditions, we get

$$\tilde{z} = \frac{1}{2\pi}\int_{-\infty}^{\infty}\int_{-\infty}^{\infty} f(x,y)e^{i(px+qy)}dxdy = F(p,q) \text{ and } \frac{d\tilde{z}}{dt} = 0 \text{ at } t = 0.$$

$\therefore \qquad 0 = \left(\dfrac{d\tilde{z}}{dt}\right)_{t=0} = BC\sqrt{p^2+q^2}$

$\Rightarrow \qquad B = 0$

Thus, the solution is given by $z = F(p,q)\cos\{C\sqrt{p^2+q^2}t\}$.

Applying the inversion Formula for double Fourier transform, we get

$$z(x,y,t) = \frac{1}{2\pi}\int_{-\infty}^{\infty}\int_{-\infty}^{\infty} F(p,q)\,\cos\{Ct\sqrt{p^2+q^2}\}e^{-i(px+qy)}dpdq.$$

EXAMPLE 6. *Use a cosine transform to show that the steady temperature in the semi-finite solid $y > 0$ when a temperature on the surface $y = 0$ is kept at unity over the strip $|x| < a$ and at zero outside the strip is*

$$\frac{1}{\pi}\left\{\tan^{-1}\left(\frac{a+x}{y}\right) + \tan^{-1}\left(\frac{a-x}{y}\right)\right\}$$

The result $\int_{-\infty}^{\infty} e^{-sx} x^{-1} \sin rx du = \tan^{-1}\frac{r}{s}, r > 0, s > 0$ may be assumed.

(Meerut–1980, 82, 91, 92, 93, 95, 2005; Himachal–2003; Nagpur–2001; Allahabad–2003)

SOLUTION.   It is known that the steady temperature $u(x, y)$ in the semi-finite solid is governed by two dimensional Laplace equation

$$\frac{\partial^2 u}{\partial x^2} + \frac{\partial^2 u}{\partial y^2} = 0, 0 < y < \infty, -\infty < x < \infty \qquad \text{...(1)}$$

subject to the conditions
$u = 1, y = 0, -a < x < a$ and $u = 0, y = 0$ and $|x| < |a|$.
Taking Fourier cosine transform of both sides of (1), we get

$$\sqrt{\frac{2}{\pi}}\int_0^\infty \frac{\partial^2 u}{\partial x^2}\cos px\,dx + \sqrt{\frac{2}{\pi}}\int_0^\infty \frac{\partial^2 u}{\partial y^2}\cos px\,dx = 0$$

$$\Rightarrow \sqrt{\frac{2}{\pi}}\left[\frac{\partial u}{\partial x}.\cos px\right]_0^\infty + p\sqrt{\frac{2}{\pi}}\int_0^\infty \sin px\,dx + \frac{d^2\tilde{u}_c}{dy^2} = 0$$

$$\left[\frac{\partial u}{\partial x} \to 0 \text{ as } x \to \infty \text{ and } \frac{\partial u}{\partial x} \to 0 \text{ as } x \to 0 \text{ by symmetry}\right]$$

$$\Rightarrow p\sqrt{\frac{2}{\pi}}[u\sin px]_0^\infty - p^2\sqrt{\frac{2}{\pi}}\int_0^\infty u\cos px\,dx + \frac{d^2\tilde{u}}{dy^2} = 0$$

$$\Rightarrow \qquad\qquad \frac{d^2\tilde{u}_c}{dy^2} - p^2\tilde{u}_c = 0$$

whose solution is given by

$$\tilde{u} = Ae^{py} + Be^{-py} \qquad \text{...(2)}$$

As $y \to \infty, \tilde{u}_c \to 0$ therefore $A = 0$

$$\Rightarrow \qquad \tilde{u}_c = Be^{-by} \qquad \text{...(3)}$$

When $y = 0$,

$$\tilde{u}_c = \sqrt{\frac{2}{\pi}}\int_r^a 1.\cos px\,dx + \sqrt{\frac{2}{\pi}}\int_0^\infty 0.\cos px\,dx$$

$$= \sqrt{\frac{2}{\pi}}\left(\frac{\sin px}{p}\right)_0^\infty = \sqrt{\frac{2}{\pi}}\frac{\sin pa}{p}.$$

Also, from (3),

$$\sqrt{\frac{2}{\pi}}\frac{\sin pa}{p} = B \Rightarrow \tilde{u}_c = \sqrt{\frac{2}{\pi}}\frac{\sin pa}{p}e^{-py}$$

Taking the inverse Fourier cosine transform, we get

$$u(x, y) = \sqrt{\frac{2}{\pi}}\int_0^\infty \sqrt{\frac{2}{\pi}}\frac{\sin pa}{p}e^{-py}\cos px\,dp$$

$$= \frac{1}{\pi} \int_0^\infty \frac{e^{-py}}{p} [\sin(a+x)p + \sin(a-x)p] dp$$

$$= \frac{1}{\pi} \int_0^\infty e^{-py} p^{-1} \sin(a+x)p dp + \frac{1}{\pi} \int_0^\infty e^{-py} p^{-1} \sin(a-x)p dp$$

$$= \frac{1}{\pi} \left[\tan^{-1}\left(\frac{a+x}{y}\right) + \tan^{-1}\left(\frac{a-x}{y}\right) \right]$$

EXAMPLE 7. *Solve* $\dfrac{\partial^4 V}{\partial x^4} + \dfrac{\partial^2 V}{\partial y^2} = 0$, $-\infty < x < \infty, y \geq 0$ *subject to the conditions*

(i) *V and its partial derivative tends to zero as* $x \to \pm\infty$.

(ii) $V = f(x), \dfrac{\partial V}{\partial y} = 0$ *on* $y = 0$.

(Meerut–1989, 93; Raj.–1981, 2003; Banglore–2001; Kanpur–2003; Rewa–2006)

SOLUTION. Taking the Fourier transform of both the sides of the given equation, we get

$$\frac{1}{\sqrt{2\pi}} \int_{-\infty}^\infty \frac{\partial^4 V}{\partial x^4} e^{ipx} dx + \frac{1}{\sqrt{2\pi}} \int_{-\infty}^\infty \frac{\partial^2 V}{\partial y^2} e^{ipx} dx = 0$$

$$\Rightarrow \quad \frac{1}{\sqrt{2\pi}} \left(\frac{\partial^3 V}{\partial x^3} e^{ipx} \right)_{-\infty}^\infty - \frac{ip}{\sqrt{2\pi}} \int_{-\infty}^\infty \frac{\partial^3 V}{\partial x^3} e^{ipx} dx + \frac{d^2}{dy^2} \frac{1}{\sqrt{2\pi}} \int_{-\infty}^\infty V . e^{ipx} dx = 0$$

$$\Rightarrow \quad \frac{-ip}{\sqrt{2\pi}} \left[\left(\frac{\partial^2 V}{\partial x^2} e^{ipx} \right)_{-\infty}^\infty - ip \int_{-\infty}^\infty \frac{\partial^2 V}{\partial x^2} e^{ipx} dx \right] + \frac{d^2 \tilde{V}}{dy^2} = 0 \quad \left[\because \frac{\partial^3 V}{\partial x^3} \to 0 \text{ as } x \to \pm\infty \right]$$

$$\Rightarrow \quad \frac{(ip)^2}{\sqrt{2\pi}} \left[\left(\frac{\partial V}{\partial x} e^{ipx} \right)_{-\infty}^\infty - ip \int_{-\infty}^\infty \frac{\partial V}{\partial x} e^{ipx} dx \right] + \frac{d^2 \tilde{V}}{dy^2} = 0$$

$$\Rightarrow \quad -\frac{(ip)^3}{\sqrt{2\pi}} \left[\left(V e^{ipx} . dx \right)_{-\infty}^\infty - ip \int_{-\infty}^\infty V e^{ipx} dx \right] + \frac{d^2 \tilde{V}}{dy^2} = 0$$

$$\Rightarrow \quad \frac{(ip)^4}{\sqrt{2\pi}} \int_{-\infty}^\infty V e^{ipx} . dx + \frac{d^2 \tilde{V}}{dy^2} = 0$$

$$\Rightarrow \quad \frac{d^2 \tilde{V}}{dy^2} + p^4 \tilde{V} = 0$$

whose general solution is given by

$$\tilde{V} = C_1 \cos p^2 y + C_2 \sin p^2 y \qquad \qquad \dots(1)$$

Since, on $y = 0$, $V = f(x)$ and $\dfrac{\partial V}{\partial y} = 0$

$$\therefore \qquad \tilde{V} = \frac{1}{\sqrt{2\pi}} \int_{-\infty}^\infty f(x) e^{ipx} dx = \tilde{f}(p)$$

and $\qquad \dfrac{1}{\sqrt{2\pi}} \int_{-\infty}^\infty \dfrac{\partial V}{\partial y} e^{ipx} dx = 0 \Rightarrow \dfrac{d\tilde{V}}{dy} = 0$

Also, from (1), $\tilde{f}(p) = C_1$ and $d\tilde{V} = 0 = C_2$

$\Rightarrow \qquad C_2 = 0$

Then , from (1), we have $\tilde{V} = \tilde{f}(p)\cos p^2 y$.

Finally, applying inversion theorem for Fourier transform, we get

$$V = \frac{1}{\sqrt{2\pi}} \int_{-\infty}^{\infty} \tilde{f}(p)\cos(p^2 y)e^{-ipx}dp .$$

EXAMPLE 8. *Use the finite cosine transform to solve* $\dfrac{\partial V}{\partial t} = k\dfrac{\partial^2 V}{\partial x^2}$ *with boundary conditions* $\dfrac{\partial V}{\partial x} = 0$, *when* $x = 0$ *and* $x = \pi, t > 0$ *the initial conditions* $V = f(x)$, *when* $t = 0$, $0 < x < \pi$.

(Agra–1993)

SOLUTION. Taking the finite Fourier cosine transform (with $l = p$) of both the sides of the given equation, we get

$$\int_0^\pi \frac{\partial V}{\partial t}\cos px\,dx = k\int_0^\pi \frac{\partial^2 V}{\partial x^2}\cos px\,dx$$

$$\Rightarrow \quad \frac{d\tilde{V}_c}{\partial t} = k[-p^2\tilde{V}_c - \{V_x(0,t) - V_x(\pi,t)\cos p\pi\}]$$

where $\tilde{V}_c$ is finite Fourier cosine transform of V.

$$\therefore \qquad\qquad \frac{d\tilde{V}_c}{dt} = -kp^2\tilde{V}_c \Rightarrow \tilde{V}_c = Ae^{-kp^2 t} \qquad\qquad \dots (1)$$

Taking finite Fourier cosine transform of initial conditions, we have at $t = 0$

$$\tilde{V}_c = \int_0^\pi f(y)\cos py\,dy$$

From (1), we have

$$A = \int_0^\pi f(x)\cos px\,dx$$

$$\Rightarrow \qquad\qquad \tilde{V}_c = e^{-kp^2 t}\int_0^\pi f(y)\cos py\,dy$$

On taking the inverse finite Fourier cosine transform, we have

$$V(x,t) = \frac{1}{\pi}\tilde{V}_c(0) + \frac{2}{\pi}\sum_{p=1}^{\infty} \tilde{V}_c(p)\cos px$$

$$= \frac{1}{\pi}\int_0^\pi f(y)dy + \frac{2}{\pi}\sum_{p=1}^{\infty}\left[e^{-kp^2 t}\cos px.\int_0^\pi f(y)\cos py\,dy\right]$$

EXAMPLE 9. *Show that the solution of Laplace's equation for r inside the semi-infinite strip* $x = 0$, $0 < y < b$, *such that*

$$\begin{aligned} r &= f(x) &&; \quad \text{when } y = 0, 0 < x < \infty \\ r &= 0 &&; \quad \text{when } y = b, 0 < x < \infty \\ r &= 0 &&; \quad \text{when } x = 0, 0 < y < b \end{aligned}$$

is given by

$$V = \frac{2}{\pi}\int_0^\infty f(u)du\int_0^\infty \frac{\sinh(b-y)p}{\sinh bp}\sin xp.up.dp .$$

(Meerut–1984, 94)

SOLUTION. Two dimensional Laplace's equation is given by

$$\frac{\partial^2 V}{\partial x^2} + \frac{\partial^2 V}{\partial y^2} = 0,\ 0 < x < \infty, 0 < y < b$$

$$\dots(1)$$

Taking Fourier sine transform of (1) of both the sides, we get

$$\sqrt{\frac{2}{\pi}} \int_0^\infty \frac{\partial^2 V}{\partial x^2} \sin px\, dx + \sqrt{\frac{2}{\pi}} \int_0^\infty \frac{\partial^2 V}{\partial y^2} \sin px\, dx = 0$$

$$\sqrt{\frac{2}{\pi}} \left[\frac{\partial V}{\partial x} \sin px \right]_0^\infty - p\sqrt{\frac{2}{\pi}} \int_0^\infty \frac{\partial V}{\partial x} \cos px\, dx + \frac{d^2 \tilde{V}_s}{dy^2} = 0$$

$$\Rightarrow \quad -p\sqrt{\frac{2}{\pi}} \left[V \cdot \cos px \right]_0^\infty - p^2 \sqrt{\frac{2}{\pi}} \int_0^\infty V \sin px\, dx + \frac{d^2 \tilde{V}_s}{dy^2} = 0 \qquad \left[\frac{\partial V}{\partial x} \to 0 \text{ as } x \to \infty \right]$$

$$\Rightarrow \quad \frac{d^2 \tilde{V}_s}{dy^2} - p^2 \tilde{V}_s = 0 \qquad\qquad\qquad [V \to 0 \text{ as } x \to \infty]$$

whose solution is given by

$$\tilde{V}_s = A \cosh py + B \sinh py \qquad\qquad\qquad ...(2)$$

Further, taking Fourier sine transform of given conditions, we have

When $y = 0$, $\tilde{V}_s = \sqrt{\frac{2}{\pi}} \int_0^\infty f(u) \sin pu\, du$

and when $y = b$, $\tilde{V}_s = 0$ So, from (2), we have

$$\sqrt{\frac{2}{\pi}} \int_0^\infty f(u) \sin pu\, du = A$$

and $\qquad\qquad\qquad 0 = A \cosh pb + B \sinh pb$

$$\Rightarrow \qquad\qquad\qquad B = -\frac{\cosh pb}{\sinh pb} \cdot \sqrt{\frac{2}{\pi}} \int_0^\infty f(u) \sin pu\, du$$

Therefore,

$$\tilde{V}_s = \sqrt{\frac{2}{\pi}} \cosh py \int_0^\infty f(u) \sin pu\, du - \sqrt{\frac{2}{\pi}} \frac{\cosh pb}{\sinh pb} \sinh py \int_0^\infty f(u) \sin pu\, du$$

$$= \sqrt{\frac{2}{\pi}} \int_0^\infty f(u) \sin pu \left[\cosh py - \frac{\cosh pb \sinh py}{\sinh pb} \right] du$$

$$= \sqrt{\frac{2}{\pi}} \int_0^\infty f(u) \sin pu \left[\frac{\sinh pb \cosh py - \cosh pb \sinh py}{\sinh pb} \right] du$$

$$= \sqrt{\frac{2}{\pi}} \int_0^\infty f(u) \sin pu \cdot \frac{\sinh(b-y)p}{\sinh pb} du$$

Finally, taking the inverse Fourier sine transform, we get

$$V = \sqrt{\frac{2}{\pi}} \int_0^\infty \tilde{V}_s \sin px\, dp$$

$$= \frac{2}{\pi} \int_0^\infty f(u)\, du \int_0^\infty \frac{\sinh(b-y)p}{\sinh pb} \cdot \sin pu \cdot \sin px\, dp \cdot$$

EXAMPLE 10. *Use finite Fourier transform to solve*

$$\frac{\partial u}{\partial t} = \frac{\partial^2 u}{\partial x^2}, \quad u(0,t) = 0, u(4,t) = 0, u(x,0) = 2x$$

(Agra–1988, 2003; Kanpur–2009, 12; Meerut–2013; Rohilkhand–2014)

SOLUTION. Taking the finite Fourier sine transform of both sides of the given partial differential equation, we get

$$\int_0^4 \frac{\partial u}{\partial t} \sin \frac{p\pi x}{4} dx = \int_0^4 \frac{\partial^2 u}{\partial x^2} \sin \frac{p\pi x}{4} dx$$

$$\Rightarrow \qquad \frac{d\tilde{u}_s}{dt} = -\frac{p^2\pi^2}{4}\tilde{u}_s + \frac{p\pi}{4}[u(0,t) - u(4,t)\cos p\pi]$$

where $\tilde{u}_s$ is finite Fourier sine transform of u.

$$\therefore \qquad \frac{d\tilde{u}_s}{dt} = \frac{-p^2\pi^2}{15}\tilde{u}_s \Rightarrow \tilde{u}_s = Ae^{-p^2\pi^2 t/16}$$

Now, since $u(x, 0) = 2x$, where $0 < x < 4$, taking finite Fourier sine transform, we have

at $t = 0$, $\qquad \tilde{u}_s = \int_0^4 2x.\sin\frac{px\pi}{4}dx$

$$= \left[-2x\frac{4}{p\pi}\cos\frac{p\pi x}{4} + 2.\frac{4}{2\pi}.\frac{4}{p\pi}\sin\frac{p\pi x}{4}\right]_0^4$$

$$= \frac{32(-\cos p\pi)}{p\pi} = \frac{32[-(-1)]^p}{p\pi}$$

$$\therefore \qquad \frac{32(-1)^{p+1}}{p\pi} = A,$$

Hence, $\qquad \tilde{u}_s = \frac{32(-1)^{p+1}}{p\pi}e^{-tp^2\pi^2/16}$

Finally, taking the inverse finite Fourier sine transform, we get

$$u(x,t) = \frac{2}{4}\sum_{p=1}^{\infty}\frac{32(-1)^{p+1}}{p\pi}e^{-p^2\pi^2 t/16}.\sin\frac{p\pi x}{4}$$

$$= \frac{16}{\pi}\sum_{p=1}^{\infty}\frac{(-1)^{p+1}}{p}e^{-p^2\pi^2 t/16}.\sin\frac{p\pi x}{4}$$

EXAMPLE 11. *Using finite Fourier transform. Find the solution of the wave equation* $\dfrac{\partial^2 u}{\partial t^2} = 4\dfrac{\partial^2 u}{\partial x^2}$
subject to the conditions $u(0, t) = 0$. $u(\pi, t) = 0, u(x, 0) = 0.1\sin x + 0.1\sin 4x$ *and*
$u_t(x, 0) = 0$ *for* $0 < x < \pi, t > 0$.

SOLUTION. Taking the finite Fourier transform on both the sides of the given equation, we get

$$\int_0^\pi\frac{\partial^2 u}{\partial t^2}\sin pxdx = 4\int_0^\pi\frac{\partial^2 u}{\partial x^2}\sin pxdx$$

$$\Rightarrow \qquad \frac{d^2\tilde{u}_s}{dt^2} = -4p^2\tilde{u}_s + 4p[u(0,t) - u(\pi,t)\cos p\pi]$$

$$\Rightarrow \qquad \frac{d^2\tilde{u}_s}{dt^2} + 4p^2\tilde{u}_s = 0$$

$$\Rightarrow \qquad \tilde{u}_s = A\cos 2pt + B\sin 2pt$$

Now, at $t = 0, u = (0.1)\sin x + (0.01)\sin 4x$ and $\dfrac{\partial u}{\partial t} = 0$

Taking finite sine transforms, we get
At $t = 0$,

$$\tilde{u}_s = \int_0^\pi\{(0.1)\sin x + (0.01)\sin 4x\}\sin pxdx$$

$$= (0.1)\int_0^\pi\sin x\, px\, dx + (0.01)\int_0^\pi\sin 4x\sin px\, dx$$

and $\dfrac{d\tilde{u}_s}{dt} = 0$.

Also, from (1), we have

$$(0.1)\int_0^\pi \sin x \sin px\, dx + (0.01)\int_0^\pi \sin 4x \sin px\, dx = A$$

$\Rightarrow \quad \dfrac{d\tilde{u}_s}{dt} = 0 = 2B_p \Rightarrow B = 0.$

Now, $\quad \tilde{u}_s = \left[(0.1)\int_0^\pi \sin px\, dx + (0.01)\int_0^\pi \sin 4x \sin px\, dx \right]\cos 2pt$

Finally, taking the inverse finite Fourier sine transform, we get

$$u(x,t) = \dfrac{2}{\pi} \sum_{p=1}^\infty \tilde{u}_s \sin px$$

$$= \dfrac{2}{\pi}(0.1) \sum_{p=1}^\infty \left\{ \int_0^\pi \sin x \sin px\, dx \right\} \cos 2pt \sin px$$

$$+ \dfrac{2}{\pi}(0.01) \sum_{p=1}^\infty \left\{ \int_0^\pi \sin 4x px\, dx \right\} \cos 2pt \sin px$$

$$= \dfrac{2}{\pi}(0.1) \left[\int_0^\pi \sin x \sin x\, dx \right] \cos 2t \sin x$$

$$+ \dfrac{2}{\pi}(0.01) \left[\int_0^\pi \sin 4x . \sin 4x\, dx \right] \cos 8t \sin 4x$$

$\Rightarrow \quad u(x,t) = (0.1)\cos 2t \sin x + (0.01)\cos 8t \sin 4t.$

EXAMPLE 12. *Find the steady temperature $V(x, y)$ in a long square bar of side π when one face is kept at constant and the other faces at zero temperature. Also $V(x, y)$ is bounded.*

Or

Find a function $V(x, y)$ which is harmonic in the open square $0 < x < \pi$, $0 < y < \pi$ takes a constant value V_0 on the edges $y = \pi$ and vanishes on the other edges of the square. (Meerut–1985, 95; IAS–2003; Delhi–2004)

Solution. It is known that the steady temperature is governed by two dimensional Laplace equation

$$\dfrac{\partial^2 V}{\partial x^2} + \dfrac{\partial^2 V}{\partial y^2} = 0 \qquad \qquad \dots(1)$$

subject to the boundary conditions a $V(0, y) = 0 = V(\pi, y), V(x, 0) = 0$ and $V(x, \pi) = V_0$, a constant.

Taking the Fourier sine transform of both sides of (1), we get

$$\int_0^\pi \dfrac{\partial^2 V}{\partial x^2} \sin px\, dx + \int_0^\pi \dfrac{\partial^2 V}{\partial y^2} \sin px\, dx = 0$$

$$\Rightarrow \left[\dfrac{\partial V}{\partial x} \sin px \right]_0^\pi - p\int_0^\pi \dfrac{\partial V}{\partial x} \cos px\, dx + \dfrac{\partial^2}{\partial y^2} \int_0^\pi V \sin px\, dx = 0$$

$$\left[\because \dfrac{\partial V}{\partial x} \text{ is finite at } x = 0 \text{ and } x = \pi \right]$$

$$\Rightarrow -p\left[V \cos px \right]_0^\pi - p^2 \int_0^\pi V . \sin px\, dx + \dfrac{d^2 \tilde{V}_s}{dy^2} = 0$$

$$\Rightarrow \dfrac{d^2 \tilde{V}_s}{dy^2} - p^2 \tilde{V}_s = 0 \qquad \qquad \left[\because V = 0 \text{ at } x = 0 \text{ and } x = \pi \right]$$

$$\Rightarrow \qquad \tilde{V}_s = A \cosh py + B \sinh py \qquad \qquad ...(2)$$

As per given when $y = 0, V = 0$ and when $y = \pi, V = V_0$

Taking finite Fourier sine transform of above boundary conditions, we get when

$y = 0, \tilde{V}_s = 0$ and when $y = \pi, \tilde{V}_s = \int_0^\pi V_0 \sin px dx$

$$= V_0 \left(-\frac{\cos px}{p} \right)_0^\pi = V_0 \left(\frac{1 - \cos p\pi}{p} \right)$$

Thus, from (2), we have $0 = A$

and $\qquad V_0 \left(\dfrac{1 - \cos p\pi}{p} \right) = B \sinh p\pi$

$$\Rightarrow \qquad B = \frac{V_0}{\sinh p\pi} \left(\frac{1 - \cos p\pi}{p} \right).$$

Putting these values in (2), we get

$$\tilde{V}_s = \frac{V_0}{\sinh p\pi} \left(\frac{1 - \cos p\pi}{p} \right) \sinh py$$

Finally, taking the inverse finite Fourier sine transform, we get

$$V(x, y) = \frac{2}{\pi} \sum_{p=1}^{\infty} \frac{V_0}{\sinh p\pi} \left(\frac{1 - \cos p\pi}{p} \right) \sinh py \sin px$$

$$= \frac{4V_0}{\pi} \sum_{n=0}^{\infty} \frac{\sinh(2n+1)y \sin(2n+1)x}{(2n+1) \sinh(2n+1)\pi}$$

EXAMPLE 13.   *The cross section of a long bar of diffusivity k is the square $0 < x < \pi, 0 < y < \pi$. If the four faces of the bar are maintained at zero temperature and the initial temperature is unity, use respective finite transforms to show that the temperature at time t is $\phi(x)\,\phi(y)$ where*

$$\phi(x) = \frac{4}{\pi} \sum_{n=0}^{\infty} (2n+1)^{-1} \sin(2n+1)x e^{[-k(2n+1)^2 t]}$$

SOLUTION.   It is known that the temperature in the bar is governed by two dimensional heat equation

$$\frac{\partial^2 V}{\partial x^2} + \frac{\partial^2 V}{\partial y^2} = \frac{1}{k} \frac{\partial V}{\partial t} \qquad \qquad ...(1)$$

where k is the diffusivity of the bar and $V(x, y, t)$ is the temperature at any point and at any time t.

Here, boundary conditions are

$$V(0, y, t) = 0, \ V(\pi, y, t) = 0, \qquad \qquad ...(2)$$
$$V(x, 0, t) = 0, \ V(x, \pi, t) = 0, \qquad \qquad ...(3)$$

and initial condition is given by

$$V(x, y, 0) = 1 \qquad \qquad ...(4)$$

Taking the finite Fourier sine transform of both sides of (1), we get

$$\int_0^\pi \frac{\partial^2 V}{\partial x^2} \sin px dx + \int_0^\pi \frac{\partial^2 V}{\partial y^2} \sin px dx = \frac{1}{k} \int_0^\pi \frac{\partial V}{\partial t} \sin px dx$$

$$\Rightarrow \qquad -p^2 \tilde{V}_s + \frac{\partial^2 \tilde{V}_s}{\partial y^2} = \frac{1}{k} \frac{\partial \tilde{V}_s}{\partial t} \qquad \qquad ...(5)$$

Further, the finite Fourier sine transform of boundary and initial conditions are on $x = 0, x = \pi, \tilde{V}_s = 0$ on $y = 0, y = \pi, \tilde{V}_s = 0$ and

at $t = 0$ $\quad \tilde{V}_s = \int_0^\pi 1 . \sin px \, dx = \left(-\dfrac{\cos px}{p} \right)_0^\pi$

$$= \dfrac{1 - \cos p\pi}{p} = 0 \text{ or } \dfrac{2}{p} \qquad\qquad \text{[As } p \text{ is even or odd.]}$$

Taking again the finite Fourier sine transform of (5), we get

$$-p^2 \int_0^\pi \tilde{V}_s \sin qy \, dy + \int_0^\pi \dfrac{\partial^2 \tilde{V}_s}{\partial y^2} \sin qy \, dy = \dfrac{1}{k} \int_0^\pi \dfrac{\partial \tilde{V}_s}{\partial t} \sin qy \, dy$$

$\Rightarrow$ $\qquad\qquad -p^2 \tilde{V}_s' - q^2 \tilde{V}_s' = \dfrac{1}{k} \dfrac{d\tilde{V}_s'}{dt}$

subject to the boundary condition (3).

Here $\tilde{V}_s'$ is finite Fourier sine transform of $\tilde{V}_s$

or $\qquad\qquad \dfrac{d\tilde{V}_s'}{dt} = -k(p^2 + q^2)\tilde{V}_s'$

$\Rightarrow$ $\qquad\qquad V\tilde{V}_s' = Ae^{-k(p^2 + q^2)t}$ $\qquad\qquad$ ...(6)

Taking finite Fourier sine transform of $\tilde{V}_s$, we get

at $\qquad\qquad t = 0, \tilde{V}_s' = 0, \text{ if } p \text{ is even}$

and $\qquad\qquad \tilde{V}_s' = \int_0^\pi \dfrac{2}{p} \sin qy \, dy \text{ if } p \text{ is odd}$

$$= \dfrac{2}{pq}(1 - \cos q\pi) = 0 \text{ or } \dfrac{4}{pq}$$

$$\text{[According as } q \text{ is even or odd]}$$

Thus, when p and q are both even, we have at $t = 0, \tilde{V}_s' = 0$

So, from (6), $A = 0 \Rightarrow \tilde{V}_s' = 0 \Rightarrow V = 0$

Further, when p, q are both odd, we have at $t = 0$, $\tilde{V}_s' = \dfrac{4}{pq}$ therefore, from (6), we

have $\dfrac{4}{pq} = A$

$\Rightarrow$ $\qquad\qquad \tilde{V}_s' = \dfrac{4}{pq} e^{-k(p^2 + q^2)t}$

$\Rightarrow$ $\qquad\qquad \tilde{V}_s' = \dfrac{4}{(2n+1)(2m+1)} e^{-k\{(2n+1)^2 + (2m+1)^2\}t}$

Where $p = 2n + 1, q = 2m + 1$.

Taking again inverse finite sine transform, we get

$$V(x, y, t) = \dfrac{2}{\pi} \dfrac{8}{\pi} \sum_{n=0}^{\infty} \sum_{m=0}^{\infty} \dfrac{e^{-k\{(2n+1)^2 + (2m+1)^2\}t}}{(2n+1)(2m+1)} \sin(2m+1)y \sin(2n+1)x$$

$\Rightarrow$ $\qquad\qquad \phi(x) = \dfrac{4}{\pi} \sum_{n=0}^{\infty} (2n+1)^{-1} e^{-k(2n+1)^2 t} \sin(2n+1)x$

EXAMPLE 14. *Determine the displacement $u(x, t)$ in a horizontal string stretched from origin to the point $(\pi, 0)$ when the motion is due to the weight of the string alone. The string may be taken to be initially at rest in the position $u = 0$.* (Meerut–1991, 2009)

SOLUTION. Let the displacement $u(x, t)$ be taken positive along the vertical downward line. Suppose that string is released from the position $u = 0$ along horizontal line. Then the displacement $u(x,t)$ of the string is governed by the differential equation.

$$\frac{\partial^2 u}{\partial t^2} = C^2 \frac{\partial^2 u}{\partial x^2} + g, \ 0 < x < \pi, t > 0,$$

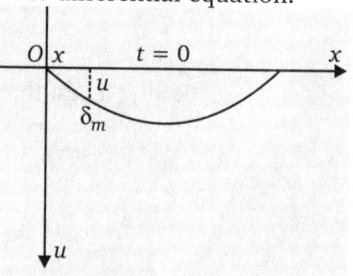

Fig. 1

where $C^2 = T/\rho$. ...(1)

The boundary conditions are

$u(0,t) = 0 = u(\pi,t), t > 0$ initial conditions are

$u(x,0) = 0 = u_t(x,0), 0 < x < \pi$.

Taking the finite Fourier sine transform of both sides of (1), we have

$$\int_0^\pi \frac{\partial^2 u}{\partial t^2} \sin pxdx = C^2 \int_0^\pi \frac{\partial^2 u}{\partial x^2} \sin pxdx + g \int_0^\pi \sin pxdx$$

$$\Rightarrow \frac{d^2}{dt^2} \tilde{u}_s = -C^2 p^2 \tilde{u}_s + C^2 p[u(0,t) - u(\pi,t)\cos p\pi] + \frac{q}{p}[1-(-1)^p]$$

$$\Rightarrow \frac{d^2}{dt^2} \tilde{u}_s + C^2 p^2 \tilde{u}_s = \frac{q}{p}[1-(-1)^p]$$

Whose solution is given by

$$\tilde{u}_s(p,t) = A\cos pat + B\sin pat + \left(\frac{q}{C^2 p^3}\right)[1-(-1)^p]. \quad ...(2)$$

Now, taking the finite Fourier sine transform of the given initial conditions, we have

When $t = 0, \tilde{u}_s = 0$ and $\frac{d^2}{dt^2} \tilde{u}_s = 0$.

Now, $\frac{d}{dt}\tilde{u}_s = -paA\sin pat + paB\cos pat$

When $t = 0, \tilde{u}_s = 0 \Rightarrow A = \frac{q}{C^2 p^3}[1-(-1)^p][1-\cos pat]$

$$\Rightarrow \tilde{q}_s(p,t) = \frac{q}{C^2 p^3}[1-(-1)^p][1-\cos pat].$$

Taking the inverse Fourier sine transform, we get

$$u(x,t) = \frac{2}{\pi} \sum_{p=1}^\infty \tilde{u}_s(p,t)\sin px$$

$$= \frac{2q}{\pi C^2} \sum_{p=1}^\infty \frac{1}{p^3}[1-(-1)^p](1-\cos pat)\sin px.$$

EXAMPLE 15. *Solve the boundary value problem*

$$\frac{\partial U}{\partial t} = \frac{\partial^2 U}{\partial x^2}, U(0,t) = 1, U(\pi,t) = 3, U(x,0) = 1 \text{ where } 0 < x < \pi, t > 0.$$

(Meerut–1991, 2011)

SOLUTION.   Here, we have

$$\frac{\partial U}{\partial t} = \frac{\partial^2 U}{\partial x^2} \qquad \qquad \dots(1)$$

Taking the finite Fourier sine transform of both the side, we get

$$\int_0^\pi \frac{\partial U}{\partial t} \sin px\, dx = \int_0^\pi \frac{\partial^2 U}{\partial x^2} \sin px\, dx$$

$$\Rightarrow \qquad \frac{d\tilde{U}_s}{dt} = -p^2 \tilde{U}_s + p[U(0,t) - U(\pi,t)\cos p\pi]$$

$$\Rightarrow \qquad \frac{d\tilde{U}_s}{dt} = -p^2 \tilde{U}_s + p[1 - 3\cos p\pi]$$

$$\Rightarrow \qquad \frac{d\tilde{U}_s}{dt} + p^2 \tilde{U}_s = p(1 - 3\cos p\pi) \qquad \qquad \dots(2)$$

Now,   I.F. $= e^{\int p^2 dt} = e^{p^2 t}$

Solution of eq^n (2), is

$$\tilde{U}_s \cdot e^{p^2 t} = A + \int p(1 - 3\cos p\pi) e^{p^2 t} dt$$

$$= A + \frac{(1 - 3\cos p\pi)}{p} e^{p^2 t}$$

$$\Rightarrow \qquad \tilde{U}_s = A e^{-p^2 t} + \frac{(1 - 3\cos p\pi)}{p} \qquad \qquad \dots(3)$$

If $t = 0$, $U = 1$

Again taking finite Fourier sine transform

$$\tilde{U}_s = \int_0^\pi 1 \cdot \sin px\, dx = \frac{1 \cdot (1 - \cos p\pi)}{p}$$

If $t = 0$ then from (3),

$$\frac{1 \cdot (1 - \cos p\pi)}{p} = A + \frac{1 - 3\cos p\pi}{p}$$

$$\Rightarrow \qquad A = \frac{2\cos p\pi}{p}$$

Now, from eq^n (2), we get

$$\tilde{U}_s = \frac{2\cos p\pi}{p} e^{-p^2 t} + \frac{(1 - 3\cos p\pi)}{p}$$

Now, taking inverse infinite Fourier sine transform, we get

$$U(x,t) = \frac{2}{\pi} \sum_{p=1}^{\infty} \frac{2\cos p\pi}{p} \cdot e^{-p^2 t} \sin px + \frac{2}{\pi} \sum_{p=1}^{\infty} \frac{(1 - 3\cos p\pi)}{p} \sin px$$

$$\qquad \qquad \dots(4)$$

Therefore, $F_s\{1\} = \int_0^\pi 1 \cdot \sin px\, dx = \dfrac{1 - \cos p\pi}{p}$

$$= \frac{1 - (-1)^p}{p}, (p = 1, 2, \dots)$$

and $F_s\{x\} = \int_0^\pi x \cdot \sin px\, dx$

$$= \left(-\frac{x}{p} \cos px \right)_0^\pi + \int_0^\pi \frac{\cos px}{p} dx = \frac{-\pi}{p} \cos p\pi = \frac{\pi}{p}(-1)^{p+1}$$

$$1 = \frac{2}{\pi} \sum_{p=1}^{\infty} \frac{1 - (-1)^p}{p} \sin px$$

Hence
$$x = \frac{2}{\pi} \sum_{p=1}^{\infty} \frac{\pi(-1)^{n+1}}{p} \sin px = 2 \sum_{p=1}^{\infty} \frac{(-1)^{p+1}}{p} \sin px$$

$$= \frac{2}{\pi} \sum_{p=1}^{\infty} \frac{(1 - 3\cos p\pi)}{p} \sin px = \frac{2}{\pi} \sum_{p=1}^{\infty} \frac{1 - 3(-1)^p}{p} \sin px$$

$$= \frac{2}{\pi} \sum_{p=1}^{\infty} \frac{1 - (-1)^p}{p} \sin px + \frac{4}{\pi} \sum_{p=1}^{\infty} \frac{(-1)^{p+1}}{p} \sin px$$

$$= 1 + \frac{2}{\pi} x$$

From eq^n (4), we have

$$U(x,t) = \frac{4}{\pi} \sum_{p=1}^{\infty} \frac{\cos p\pi}{p} e^{-p^2 t} \sin px + 1 + \frac{2x}{\pi}$$

EXAMPLE 16. *Solve the boundary value problem by using Fourier transform*

$$\frac{\partial u}{\partial t} = 2 \frac{\partial^2 u}{\partial x^2}, (0 < x < 4, t > 0)$$

with boundary conditions

$$u(0,t) = u(4,t), u(x,0) = 3\sin(\pi x) - 2\sin(5\pi x) \qquad \text{(Kanpur–2009)}$$

SOLUTION. Here, it is given $u(0, t)$ and $u(4, t)$, so we take the interval $(0, 4)$, then we consider

$$\int_0^4 \frac{\partial u}{\partial t} \sin\left(\frac{p\pi x}{4}\right) dx = 2 \int_0^4 \frac{\partial^2 u}{\partial x^2} \sin\left(\frac{p\pi x}{4}\right) dx$$

On integrating R.H.S. by parts and $\sin(p\pi) = 0 = \sin(0)$

$$\frac{d}{dt} \int_0^4 u \sin\left(\frac{p\pi x}{4}\right) dx = 2\left[\left(\frac{\partial u}{\partial x} \sin\frac{p\pi x}{4} \right)^4 - \frac{p\pi}{4} \int_0^4 \frac{\partial u}{\partial x} \cos\left(\frac{p\pi x}{4}\right) dx \right]$$

$$\Rightarrow \qquad \frac{d}{dt} \bar{u}_s = 0 - \frac{p\pi}{2} \left[\left\{ u(x,t) \cos\left(\frac{p\pi x}{4}\right) \right\}^4_{x=0} + \frac{p\pi}{4} \int_0^4 \frac{\partial u}{\partial x} \cdot u \sin\left(\frac{p\pi x}{4}\right) dx \right]$$

$$= 0 - \frac{(p\pi)^2}{8} \bar{u}_s \text{ as } u(0, t) = 0 = u(4, t)$$

$$\Rightarrow \quad \frac{d}{dt} \bar{u}_s + \frac{p^2\pi^2}{8} \bar{u}_s = 0 \quad \Rightarrow \quad \int \frac{d\bar{u}_s}{\bar{u}_s} + \frac{p^2\pi^2}{8} \int dt = 0$$

$$\Rightarrow \quad \log \bar{u}_s + \frac{p^2\pi^2}{8} t = \log A_p \quad \Rightarrow \quad \bar{u}_s(p,t) = A_p e^{-p^2\pi^2 t/8} \qquad \text{...(1)}$$

where A_p being constant of integration.

Let $t = 0$, then (1) $\Rightarrow A_p = \bar{u}_s(p,0) = \int u(x,0) \sin\left(\frac{p\pi x}{4}\right) dx$

$$\Rightarrow \qquad A_p = \int_0^4 [3\sin(\pi x) - 2\sin(5\pi x)] \sin\left(\frac{p\pi x}{4}\right) dx$$

Putting $\frac{\pi x}{4} = \theta$, then we get

$$A_p = \frac{4}{\pi} \int_0^\pi [3\sin(4\theta) - 2\sin(20\theta)] \sin(p\theta) d\theta$$

But $\qquad \int_0^\pi \sin(m\theta) \cdot \sin(n\theta) d\theta = \begin{cases} 0 & \text{if} \quad m \neq n \\ \pi/2 & \text{if} \quad m = n \end{cases}$

Then $A_p = 0$, $p = 4$ and $p = 20$

From eq^n (2),

$$A_4 = \frac{4}{\pi} \int_0^\pi [3\sin(4\theta) - 2\sin(20\theta)]\sin(4\theta)d\theta$$

$$= \frac{4}{\pi}\left[3 \cdot \frac{\pi}{2} \cdot 2(0)\right] = 6$$

and $\qquad A_{20} = \frac{4}{\pi} \int_0^\pi [3\sin(4\theta) - 2\sin(20\theta)]\sin(20\theta)d\theta$

$$= \frac{4}{\pi}\left[3(0) - 2\left(\frac{\pi}{2}\right)\right] = -4$$

Hence, $A_4 = 6$ and $A_{20} = -4$

Now taking inverse sine transform, we get

$$u(x,t) = \frac{2}{4} \sum_{p=1}^\infty \tilde{u}_s(p,t)\sin\left(\frac{p\pi x}{4}\right)$$

Using eq^n (1), we get

$$u(x,t) = \frac{1}{2} \sum_{p=1}^\infty A_p e^{-\frac{p\pi x}{8}} \sin\left(\frac{p\pi x}{4}\right)$$

$$= \frac{1}{2}\left[A_4 e^{-2p\pi^2 t} \sin(\pi x) + e^{-50\pi^2 t} \sin(5\pi x)\right]$$

$$= \frac{1}{2}\left[6 \cdot e^{-2p\pi^2 t} \sin(\pi x) - 4 - e^{-50\pi^2 t} \sin(5\pi x)\right]$$

Hence, $\qquad u(x,t) = 3e^{-2p\pi^2 t} \sin(\pi x) - 2e^{-50\pi^2 t} \sin(5\pi x)$

EXAMPLE 17. *Use the complex form of the Fourier transform to show that*

$$u = \frac{1}{2\sqrt{pt}} \int_{-\infty}^\infty f(u)\exp.\left\{\frac{-(x-u)^2}{4t}\right\} du$$

is the solution of the boundary value problem $\dfrac{\partial u}{\partial t} = \dfrac{\partial^2 u}{\partial x^2}, -\infty < x < \infty, t > 0$

$$u = f(x), \text{ when } t = 0.$$

SOLUTION. Here, we have $\qquad \dfrac{\partial u}{\partial t} = \dfrac{\partial^2 u}{\partial x^2}$ $\qquad\qquad$ ...(1)

Now, taking Fourier transform of both the sides, we get

$$\int_{-\infty}^\infty \frac{\partial u}{\partial t} e^{-ipx} dx = \int_{-\infty}^\infty \frac{\partial^2 u}{\partial x^2} e^{-ipx} dx$$

$\Rightarrow \qquad \dfrac{\partial}{\partial t}\int_{-\infty}^\infty u e^{-ipx} dx = (ip)^2 F\{u\}$ $\qquad\qquad \Rightarrow \qquad \dfrac{\partial \bar{u}}{\partial t} = -p^2 \bar{u}$

$\Rightarrow \qquad \dfrac{\partial \bar{u}}{\partial t} + p^2 \bar{u} = 0$ $\qquad\qquad\qquad\qquad \Rightarrow \qquad \bar{u} = Ae^{-p^2 t}$ $\qquad$ ...(2)

$\Rightarrow \qquad \bar{u}(p,t) = Ae^{-p^2 t}$ $\qquad\qquad\qquad\qquad \Rightarrow \quad \bar{u}(p,0) = A$

$\Rightarrow \qquad A = \int_{-\infty}^\infty u(x,0)e^{-ipx} dx$ $\qquad\qquad \Rightarrow \qquad A = \int_{-\infty}^\infty f(x)e^{-ipx} dx$

Putting this value in eq^n (2), we get

$$\bar{u} = e^{-p^2t} \int_{-\infty}^{\infty} f(x)e^{-ipx}dx$$

$$\bar{u} = e^{-p^2t} \int_{-\infty}^{\infty} f(u)e^{-ipu}du$$

Now, taking inverse Fourier transform, we get

$$u(x,t) = \frac{1}{2\pi} \int_{-\infty}^{\infty} e^{-p^2t} \left[\int_{-\infty}^{\infty} f(u)e^{-ipu}du \right] e^{ipx}dp$$

$$= \frac{1}{2\pi} \int_{-\infty}^{\infty} f(u) \left[\int_{-\infty}^{\infty} e^{-p^2t+ip(x-u)}dp \right] du$$

$$= \frac{1}{2\pi} \int_{-\infty}^{\infty} f(u) \left[\int_{-\infty}^{\infty} \exp\left\{ -t\left(p^2 - ip\left(\frac{x-u}{2t}\right) \right) \right\} dp \right] du$$

$$= \frac{1}{2\pi} \int_{-\infty}^{\infty} f(u) \left[\int_{-\infty}^{\infty} \exp\left\{ -t\left(p - i\left(\frac{x-u}{2t}\right) \right)^2 + \left(\frac{x-u}{2t}\right)^2 \right\} dp \right] du$$

$$= \frac{1}{2\pi} \int_{-\infty}^{\infty} f(u) \left[\int_{-\infty}^{\infty} \exp\left\{ -t\left(p - i\left(\frac{x-u}{2t}\right) \right)^2 \right\} \times \exp\left\{ -\frac{(x-u)^2}{4t} \right\} dp \right] du$$

Substituting $\sqrt{t}\left\{ p - i\left(\frac{x-u}{2t}\right) \right\} = y \Rightarrow dp\sqrt{t} = dy$, we get

$$u(x,t) = \frac{1}{2\pi} \int_{-\infty}^{\infty} f(u) \left[\int_{-\infty}^{\infty} e^{-y^2} e^{-(x-u)^2/4t} \frac{dy}{\sqrt{t}} \right] du$$

$$= \frac{1}{2\pi\sqrt{t}} \int_{-\infty}^{\infty} f(u) \cdot e^{-x(x-u)^2/4t} \left[\int_{-\infty}^{\infty} e^{-y^2} dy \right] du$$

$$= \frac{1}{2\pi\sqrt{t}} \int_{-\infty}^{\infty} f(u) \sqrt{\pi} \exp\left\{ -\frac{(x-u)^2}{4t} \right\} du$$

$$= \frac{1}{2\sqrt{\pi t}} \int_{-\infty}^{\infty} f(u) \exp\left\{ -\frac{(x-u)^2}{4t} \right\} du$$

EXAMPLE 18. *Suppose $V(x, y)$ denotes electrostatic potential in region bounded by the planes $x = 0$, $x = \pi$ and $y = 0$ in which there is a uniform distribution of space charge of density $h/4\pi$. If the planes $x = 0$ and $y = 0$ are kept at potential zero, the plane $x = \pi$ at another fixed potential $V = 1$ and V is finite as $y = \infty$ determine V.*

SOLUTION. Here, Poisson's equation is

$$\nabla^2 V = -4\pi\rho \qquad \qquad ...(1)$$

where ρ is the density and $\rho = \dfrac{h}{4\pi}$

with boundary conditions

(a) $V(0, y) = 0, y > 0$ (b) $V(x, 0) = 0, 0 < x < \pi$

(c) $V(x, y) = 1, y > 0$ (d) $\lim\limits_{y \to \infty} V(x,y) = $ finite, $0 < x < \pi$

Here, eq^n (1) is equivalent to $\dfrac{\partial^2 V}{\partial x^2} + \dfrac{\partial^2 V}{\partial y^2} = -h$ ...(2)

Applying finite sine transform on eq^n (2), we get

$$\int_0^\pi \frac{\partial^2 V}{\partial x^2} \sin px dx + \int_0^\pi \frac{\partial^2 V}{\partial y^2} \sin px dx = -h\int_0^\pi \sin px dx$$

$$\Rightarrow \quad \left(\frac{\partial V}{\partial x} \sin px\right)_0^\pi - p\int_0^\pi \frac{\partial V}{\partial x} \cos px dx + \frac{\partial^2}{\partial y^2} \int_0^\pi V \sin px dx = -h\left(\frac{-\cos px}{p}\right)_0^\pi$$

$$\Rightarrow \quad 0 - p\left\{(V \cos px)_0^\pi + p\int_0^\pi V \sin px dx\right\} + \frac{\partial^2 \bar{V}_s}{\partial y^2} = \frac{h(\cos p\pi - 1)}{p}$$

$$\Rightarrow \quad \frac{\partial^2 \bar{V}_s}{\partial y^2} - p^2 \bar{V}_s - p[V(\pi, y)\cos p\pi - V(0, y)] = \frac{h}{p}(\cos p\pi - 1)$$

Using boundary conditions (a) and (b), we get

$$\frac{\partial^2 \bar{V}_s}{\partial y^2} - p^2 \bar{V}_s - p\cos p\pi = \frac{h}{p}(\cos p\pi - 1)$$

$$\Rightarrow \quad \frac{\partial^2 \bar{V}_s}{\partial y^2} - p^2 \bar{V}_s = p\cos p\pi - hF_p\{1\} \qquad \left[\because F_p\{1\} = \frac{1}{p}\{\cos(p\pi) - 1\}\right]$$

$$\Rightarrow \quad \frac{\partial^2 \bar{V}_s}{\partial y^2} - p^2 \bar{V}_s = p(-1)^p - hF_p\{1\}$$

Therefore, the solution is

$$\bar{V}_s = Ae^{-py} + Be^{py} + [p(-1)^p - hF_p\{1\}] \cdot \frac{1}{D^2 - p^2} \cdot e^{0 \cdot y}$$

$$\Rightarrow \quad \bar{V}_s = Ae^{-py} + Be^{py} + \frac{[p(-1)^p - hF_p\{1\}]}{-p^2} \qquad \qquad \dots(3)$$

Here, V is finite $\Rightarrow \bar{V}$ is finite as $y \to \infty$, $B = 0$.

$$\bar{V}_s = Ae^{-py} + \frac{hF_p\{1\} - p(-1)^p}{p^2} \qquad \qquad \dots(4)$$

$$\Rightarrow \quad \bar{V}_s(p, 0) = A + \frac{hF_p\{1\} - p(-1)^p}{p^2} \qquad \qquad \dots(5)$$

Now $F_p\{V(x, 0)\} = 0 \Rightarrow V_s(p, 0) = 0$ $\qquad \qquad \dots(6)$

From eq^n (5) and (6), we get

$$A + \frac{hF_p\{1\} - p(-1)^p}{p^2} = 0 \quad \Rightarrow \quad A = \frac{p(-1)^p - hF_p\{1\}}{p^2}$$

Putting this value in eq^n (4), we get

$$\bar{V}_s = \left[\frac{hF_p\{1\} - p(-1)^p}{p^2}\right](1 - e^{-py})$$

Taking inverse finite sine transform, we get

$$V(x, y) = \frac{2}{\pi} \sum_{p=1}^{\infty} \frac{[hF_p\{1\} - p(-1)^p](1 - e^{-py})\sin px}{p^2}$$

EXAMPLE 19. *If the flow of heat is linear so that the variation of θ (temperature) with z and y may be neglected and if it is assumed that no heat is generated in the medium, then solve the differential equation $\partial\theta/\partial t = k(\partial^2\theta/\partial x^2)$ where $-\infty < x < \infty$ and $\theta = f(x)$ when $t = 0$, $f(x)$ being a given function of x.* (CSIR–2002; IFS–2005)

SOLUTION. It is given that

$$\partial\theta/\partial t = k(\partial^2\theta/\partial x^2), \quad -\infty < x < \infty \qquad \text{...(1)}$$

where initial conditions are

$$\theta(x, 0) = f(x) \qquad \text{...(A)}$$

Now, taking Fourier transform of eq^n (1), we get

$$\int_{-\infty}^{\infty} \frac{\partial\theta}{\partial t} e^{ipx} dx = k\int_{-\infty}^{\infty} \frac{\partial^2\theta}{\partial x^2} e^{ipx} dx \qquad \text{...(2)}$$

$$\Rightarrow \quad \frac{d}{dt}\int_{-\infty}^{\infty} \theta(x,t) e^{ipx} dx = k(-ip)^2 F\{\theta(x,t)\}$$

$$\Rightarrow \quad \frac{d\bar\theta}{dt} = -kp^2\bar\theta$$

where

$$\bar\theta(p,t) = F\{\theta(x,t)\} = \int_{-\infty}^{\infty} \theta(x,t) e^{ipx} dx \qquad \text{...(3)}$$

$$\bar\theta(p,t) = Ae^{-kp^2 t} \qquad \text{[General form of eq}^n \text{ (3)]}$$

where A is any arbitrary constant

Taking fourier transform of eq^n (A), we get

$$\int_{-\infty}^{\infty} \theta(x,0) e^{ipx} dx = \int_{-\infty}^{\infty} f(x) e^{ipx} dx$$

$$\Rightarrow \quad \bar\theta(p,0) = \bar f(p) \qquad \text{...(4)}$$

Now, putting $t = 0$ in eq^n (2) and from eq^n (4), we get

$$\bar f(p) = A$$

$$\Rightarrow \quad \bar\theta(p,t) = \bar f(p) e^{-kp^2 t} \qquad \text{[From (3)]}$$

Now, taking inverse Fourier transform, we get

$$\theta(x,t) = \frac{1}{2\pi}\int_{-\infty}^{\infty} \bar f(p) e^{-kp^2 t} e^{-ipx} dp$$

$$= \frac{1}{2\pi}\int_{-\infty}^{\infty} \bar f(p) e^{-kp^2 t - ipx} dp$$

where $\bar f(p) = \int_{-\infty}^{\infty} f(x) e^{ipx} dx$.

Additional Solved Examples

EXAMPLE 1. *Solve $\dfrac{\partial u}{\partial t} = \dfrac{\partial^2 u}{\partial x^2}$ subject to the conditions*

(i) $u = 0$ when $x = 0, t > 0$ (ii) $u = \begin{cases} 1 & ; & 0 < x < 1 \\ 0 & ; & x > 1 \end{cases}$ *and*

(iii) $u(x, t)$ is bounded. (Rohilkhand–2013; SVTU–2006; Kerala–1999, 2001, 05)

SOLUTION. Here, it is given that $(u)_x = 0$. Now, apply sine transform to the given equation, we get

$$\int_0^\infty \frac{\partial u}{\partial t} \sin px\, dx = \int_0^\infty \frac{\partial^2 u}{\partial x^2} \sin px\, dx$$

$$\frac{\partial}{\partial t}\int_0^\infty u \sin px\, dx = \left[\frac{\partial u}{\partial x} \sin px\right]_0^\infty - p\int_0^\infty \frac{\partial u}{\partial x} \cos px\, dx$$

$$\frac{\partial \tilde{u}_s}{\partial t} = -p\int_0^\infty \frac{\partial u}{\partial x} \cos px\, dx$$

$$= -p\left[(u\cos px)_0^\infty + p\int_0^\infty u\sin px\, dx \right]$$

$$= -pu(0,t) - p^2\tilde{u}_s \text{, when } u \to 0 \text{ as } x \to \infty$$

$$= -p^2\tilde{u}_s \qquad \text{[From boundary conditions]}$$

$$\Rightarrow \qquad \frac{d\tilde{u}_s}{dt} + p^2\tilde{u}_s = 0$$

$$\Rightarrow \qquad \tilde{u}_s = Ae^{-p^2 t}$$

$$\Rightarrow \qquad \tilde{u}_s(p,0) = A$$

$$\Rightarrow \qquad A = \int_0^\infty u(x,0)\sin px\, dx$$

$$\Rightarrow \qquad A = \int_0^1 1\cdot\sin px\, dx + \int_1^\infty 0\sin px\, dx = -\left[\frac{\cos px}{p} \right]_{x=0}^1 + 0$$

$$\Rightarrow \qquad A = \frac{1-\cos p}{p}$$

Now (1) $\qquad \Rightarrow \qquad \tilde{u}_s = \left(\frac{1-\cos p}{p} \right)e^{-p^2 t}$

Now, applying Fourier sine transform, we get

$$u(x,t) = \frac{2}{\pi}\int_0^\infty \left(\frac{1-\cos p}{p} \right)e^{-p^2 t}\cdot\sin px\, dp .$$

EXAMPLE 2. *Solve the boundary value problem* $\dfrac{\partial^2 u}{\partial t^2} = 9\dfrac{\partial^2 u}{\partial x^2}$ *subject to conditions*

$$u(0,t) = 0, u(2,t) = (0.05)\times(2-x); u_t(x,0) = 0, where\ 0 < x < 2, t > 0$$

(Nagpur–1991, 97, 2003)

SOLUTION. Here, the given equation is

$$\frac{\partial^2 u}{\partial t^2} = 9\frac{\partial^2 u}{\partial x^2} \qquad \qquad ...(1)$$

with boundary conditions

$$u(0, t) = 0 = u(2, t) \qquad \qquad ...(2)$$

$$u(x, 0) = 0.05x(2-x) \qquad \qquad ...(3)$$

$$u_t(x, 0) = 0,\ 0 < x < 2 \qquad \qquad ...(4)$$

Now taking finite Fourier sine transform of given eq[n], we get

$$\int_0^2 \frac{\partial^2 u}{\partial t^2}\cdot\sin\left(\frac{p\pi x}{2} \right)dx = 9\int_0^2 \frac{\partial^2 u}{\partial x^2}\cdot\sin\left(\frac{p\pi x}{2} \right)dx$$

$$\Rightarrow \quad \frac{\partial^2}{\partial t^2}\int_0^2 u\sin\left(\frac{p\pi x}{2} \right)dx = 9\left[\left\{ \frac{\partial u}{\partial x}\cdot\sin\left(\frac{p\pi x}{2} \right) \right\}_{x=0}^2 - \frac{p\pi}{2}\int_0^2 \frac{\partial u}{\partial x}\cdot\cos\left(\frac{p\pi x}{2} \right)dx \right]$$

$$\Rightarrow \quad \frac{d^2}{dt^2}\tilde{u}_s = -9\frac{p\pi}{2}\left[\left\{ u(x,t)\cos\left(\frac{p\pi x}{2} \right) \right\}_{x=0}^2 + \frac{p\pi}{2}\int_0^2 u(x,t)\sin\left(\frac{p\pi x}{2} \right)dx \right]$$

$$= -\frac{9p\pi}{2}\left[u(2,t)\cos(p\pi) - u(0,t) + \frac{p\pi}{2}\tilde{u}_s \right]$$

Now, using eq^n (2), we get

$$\frac{d^2\tilde{u}_s}{dt^2} = -\frac{9}{4}p^2\pi^2\tilde{u}_s$$

$$\Rightarrow \quad (D^2 + \frac{9}{4}p^2\pi^2)\tilde{u}_s = 0 \qquad \qquad \dots(5)$$

Here, auxiliary equation is

$$D^2 + \frac{9}{4}p^2\pi^2 = 0$$

$$\Rightarrow \qquad \qquad D = \pm i\frac{3}{2}p\pi, \frac{d}{dt} = D$$

$$\Rightarrow \qquad \tilde{u}_s(p,t) = C_1\cos\left(\frac{3\pi pt}{2}\right) + C_2\sin\left(\frac{3\pi pt}{2}\right) \qquad \dots(6)$$

Now, taking finite sine transform of eq^n (4), we get

$$\tilde{u}_t(p,0) = 0 \qquad \Rightarrow \frac{d\tilde{u}_s}{dt} = 0, \text{ when } t = 0. \qquad \dots(7)$$

Diff. eq^n (6) w.r.to t, we get

$$\frac{d\tilde{u}_s}{dt} = \frac{3\pi p}{2}\left[-C_1\sin\left(\frac{3\pi pt}{2}\right) + C_2\cos\left(\frac{3\pi pt}{2}\right)\right] \qquad \dots(8)$$

Substituting $t = 0$ in eq^n (7), we get

$$0 = \frac{3\pi p}{2}[0 + C_2(1)] \quad \Rightarrow \quad C_2 = 0$$

From eq^n (6), we get

$$\Rightarrow \qquad \tilde{u}_s(p,t) = C_1\cos\left(\frac{3\pi pt}{2}\right) \qquad \qquad \dots(9)$$

$$\Rightarrow \qquad \tilde{u}_s(p,0) = C_1$$

From eq^n (3), $\tilde{u}_s(p,0) = \int_0^2 u(x,0)\sin\left(\frac{\pi px}{2}\right)dx \qquad \dots(10)$

From eq^n (10) and (3), we get

$$C_1 = 0.05\int_0^2(2x - x^2)\sin\left(\frac{p\pi x}{2}\right)dx$$

$$= 0.05\left[-\frac{2}{\pi}(2x - x^2)\cos\left(\frac{p\pi x}{2}\right)\right]_{x=0}^{2}$$

$$+ \frac{0.1}{\pi p}\int_0^2(2 - 2x)\cos\left(\frac{\pi px}{2}\right)dx$$

$$\Rightarrow \qquad C_1 = (0 - 0) + \frac{0.1}{\pi p}\left\{\left(\frac{2}{\pi p}\right)(2 - 2x)\sin\left(\frac{\pi px}{2}\right)\right\}_{x=0}^{2}$$

$$- \frac{0.1}{\pi p}\int_0^2(2 - 2x)\cos\left(\frac{\pi px}{2}\right)dx$$

$$= (0 - 0)\left(\frac{0.1}{\pi p}\right) - \left\{\frac{4}{\pi p}\left(\frac{0.1}{\pi p}\right)\right\}\frac{2}{\pi p}\cos\left(\frac{p\pi x}{2}\right)_{x=0}^{2}$$

$$\Rightarrow \qquad C_1 = \frac{-0.8}{(\pi p)^3}[\cos(\pi p) - 1]$$

Putting this value in eq^n (8), we get

$$\tilde{u}_s(p,t) = \frac{0.8}{(\pi p)^3}[1-(-1)^p]\cos\left(\frac{3\pi pt}{2}\right)$$

Now, taking inverse finite fourier transform, we get

$$u(x,t) = \frac{2}{2}\sum_{p=1}^{\infty}\tilde{u}_s(p,t)\sin\left(\frac{p\pi x}{2}\right)$$

$$= \sum_{p=1}^{\infty}\frac{0.8}{(\pi p)^3}[1-(-1)^p]\cos\left(\frac{3\pi}{2}pt\right)\cdot\sin\left(\frac{p\pi x}{2}\right)$$

$$\Rightarrow \qquad u(x,t) = \frac{0.8}{\pi^3}\sum_{p=1}^{\infty}\frac{2}{(2n+1)^3}\cos\left[\frac{3}{2}\pi t(2n+1)\right]\cdot\sin\left(\frac{\pi x}{2}(2n+1)\right)$$

where $\qquad p = 2n + 1$.

EXAMPLE 3. *Use finite Fourier transform to solve* $\dfrac{\partial v}{\partial t} = \dfrac{\partial^2 v}{\partial x^2}, 0 < x < 6, t > 0$

and $v_x(0,t) = 0, v_x(6,t) = 0, v(x,0) = 2x$. (Meerut–1986)

SOLUTION. Taking the finite Fourier sine transform of both sides of the given partial differential equation, we get

$$\int_0^6 \frac{\partial v}{\partial t}\cos\frac{p\pi x}{6}dx = \int_0^6 \frac{\partial^2 v}{\partial x^2}\cos\frac{p\pi x}{6}dx$$

$$\Rightarrow \qquad \frac{d\tilde{v}_c}{dt} = -\frac{p^2\pi^2}{6^2}\tilde{v}_c - \{v_x(0,t) - v_s(6,t)\cos p\pi\}$$

$$\Rightarrow \qquad \frac{d\tilde{v}_c}{dt} = -\frac{\pi^2 p^2}{36}\tilde{v}_c$$

$$\Rightarrow \qquad \tilde{v}_c = Ae^{-p^2\pi^2 t/36} \qquad\qquad ...(1)$$

At $t = 0$, $V = 2x$ and

at $t = 0, \tilde{v}_c = \int_0^6 2x\cos\frac{p\pi x}{6}dx = \frac{72}{p^2\pi^2}(\cos p\pi - 1)$

Now from eq^n (1), we get

$$\frac{72}{p^2\pi^2}(\cos p\pi - 1) = A$$

Thus, $\qquad \tilde{v}_c = \frac{72}{p^2\pi^2}(\cos p\pi - 1)\cdot e^{-p^2\pi^2 t/36}$

Finally, taking the inverse finite Fourier cosine transform, we get

$$v(x,t) = \frac{1}{6}\tilde{v}_c(0) + \frac{2}{6}\sum_{p=1}^{\infty}\tilde{v}_c(p)\cos\frac{p\pi x}{6}$$

$$= \frac{1}{6}\int_0^6 2xdx + \frac{24}{\pi^2}\sum_{p=1}^{\infty}\left(\frac{\cos p\pi - 1}{p^2}\right)e^{-p^2\pi^2 t/36}\cos\frac{p\pi x}{6}$$

$$\Rightarrow \qquad v(x,t) = 6 + \frac{24}{\pi^2}\sum_{p=1}^{\infty}\left(\frac{\cos p\pi - 1}{p^2}\right)e^{-p^2\pi^2 t/36}\cos\frac{p\pi x}{6}.$$

EXAMPLE 4. *A string of density ρ and length π is stretched to a tension ρC^2. At time t = 0, one end (x = 0) is given a small oscillation a sin ωt. If the other end remains fixed, use a finite sine transform to show that the displacement of the point x at time t is*

$$a \sin \omega t \sin \frac{\omega(\pi - x)}{c} \cosec \frac{\pi \omega}{c} + \left(\frac{2ac\omega}{\pi}\right) \sum_{p=0}^{\infty} (\omega^2 - p^2 c^2)^{-1} \sin px \sin pct$$

(Meerut–1998, 2001, 02)

SOLUTION. It is known that the displacement is governed by one dimensional wave equation

$$\frac{\partial^2 U}{\partial t^2} = c^2 \frac{\partial^2 U}{\partial x^2} \qquad \ldots(1)$$

where $\qquad c^2 = \dfrac{T}{\rho}, 0 < x < \pi, t > 0$

Here, boundary conditions are

$$U(0, t) = a \sin \omega t, U(\pi, t) = 0, t > 0$$
$$U_t(x, 0) = 0, U(x, 0) = 0, 0 < x < \pi$$

Taking the finite fourier sine transform of both the sides of (1), we get

$$\int_0^\pi \frac{\partial^2 U}{\partial t^2} \sin px\, dx = c^2 \int_0^\pi \frac{\partial^2 U}{\partial x^2} \sin px\, dx$$

$$\Rightarrow \qquad \frac{d^2 \tilde{U}_s}{dt^2} = -c^2 p^2 \tilde{U}_s + c^2 \cdot p[U(0,t) - U(\pi,t)\cos p\pi]$$

$$\Rightarrow \qquad \frac{d^2 \tilde{U}_s}{dt^2} + c^2 p^2 \tilde{U}_s = ac^2 p \sin \omega t$$

$$\tilde{U}_s = A \cos cpt + B \sin cpt + \frac{ac^2 p}{-\omega^2 + c^2 p^2} \sin \omega t \qquad \ldots(2)$$

Taking again the finite Fourier sine transform of both the sides of (2), we get

at $t = 0, \tilde{U}_s = 0$ and $\dfrac{d\tilde{U}_s}{dt} = 0$

$$A = 0 \text{ and } Bcp + \frac{ac^2 p\omega}{c^2 p^2 - \omega^2} \qquad \text{[Form eq}^n \text{ (2)]}$$

Now, taking the inverse finite Fourier sine transform

$$U(x,t) = \frac{2}{\pi} ac\omega \sum_{p=1}^{\infty} \frac{\sin cpt}{\omega^2 - c^2 p^2} \sin px + \frac{2}{\pi} a \sin \omega t \sum_{p=1}^{\infty} \frac{c^2 p \cdot \sin px}{c^2 p^2 - \omega^2}$$

$$= \frac{2ac\omega}{\pi} \sum_{p=1}^{\infty} (\omega^2 - c^2 p^2)^{-1} \sin cpt \sin px + \frac{2a \sin \omega t}{\pi} \sum_{p=1}^{\infty} \frac{p \sin px}{p^2 - \frac{\omega^2}{c^2}}$$

$$\ldots(3)$$

$$F_s\left(\frac{\sin\{\omega(\pi - x)/c\}}{\sin(\pi\omega/c)}\right) = \int_0^\pi \frac{\sin\{\omega(\pi - x)/c\}}{\sin(\pi\omega/c)} \cdot \sin px\, dx$$

$$= \frac{1}{2\sin(\pi\omega/c)} \left[\frac{\cos\left\{\dfrac{\omega(\pi - x)}{c} - px\right\}}{\left(-\dfrac{\omega}{c} - p\right)} - \frac{\sin\left\{\dfrac{\omega(\pi - x)}{c} + px\right\}}{\left(-\dfrac{\omega}{c} + p\right)}\right]_0^\pi$$

$$= \frac{1}{2\sin\frac{\pi\omega}{c}}\left[\frac{\sin\frac{\omega\pi}{c}}{\frac{\omega}{c}+p}+\frac{\sin\frac{\omega\pi}{c}}{p-\frac{\omega}{c}}\right] = \frac{p}{p^2-\frac{\omega^2}{c^2}}$$

$$\therefore \quad \frac{\sin\frac{\omega(\pi-x)}{c}}{\sin(\pi\omega/c)} = \frac{2}{\pi}\sum_{p=1}^{\infty}\frac{p}{p^2-\omega^2/c^2}\sin px$$

Now, from eq^n (2), we get

$$U(x,t) = \frac{2ac\omega}{\pi}\sum_{p=1}^{\infty}(\omega^2-c^2p^2)^{-1}\sin cpt\sin px$$

$$+ a\sin\omega t\sin\frac{\omega(\pi-x)}{c}\operatorname{cosec}\frac{\omega\pi}{c}$$

EXAMPLE 5. *Solve* $\dfrac{\partial u}{\partial t}=\dfrac{\partial^2 u}{\partial x^2}$ *if* $u_x(0,t)=0, u(x,0)=\begin{cases}x & ; & 0\le x\le 1\\ 0 & ; & x>1\end{cases}$ *and* $u(x,\,t)$ *is bounded*

where $x>0,\ t>0$. (SVTU–2005; Meerut–1992)

SOLUTION. Taking Fourier cosine transform of the given equation, we get

$$\int_0^\infty\frac{\partial u}{\partial t}\cos px\,dx = \int_0^\infty\frac{\partial^2 u}{\partial x^2}\cos px\,dx$$

$$\Rightarrow \quad \frac{d\tilde{u}_c}{dt} = \left[\left(\frac{\partial u}{\partial x}\right)_{x=0}-p^2\tilde{u}_c\right]$$

But $\dfrac{\partial u}{\partial x}=0$ at $x=0$, then

$$\frac{d\tilde{u}_c}{dt} = -p^2\tilde{u}_c$$

Thus the solution is $\tilde{u}_c = Ae^{-p^2 t}$ ...(1)

$$F_c\{u(x,0)=\tilde{u}_c(p,0)=\int_0^\infty u(x,0)\cos px\,dx$$

$$= \int_0^1 x\cos px\,dx + \int_1^\infty 0\cdot\cos px\,dx = \frac{\sin p}{p}+\frac{1}{p^2}(\cos p-1)$$

By eq^n (1)

$$A = \frac{\sin p}{p}+(\cos p-1)\cdot\frac{1}{p^2}$$

$$\tilde{u}_c(p,t) = \left[\frac{\sin p}{p}+\frac{\cos(p-1)}{p^2}\right]e^{-p^2 t}$$

Now, taking inverse cosine transformation, we get

$$u(x,t) = \frac{2}{\pi}\int_0^\infty\left[\frac{\sin p}{p}+\frac{\cos(p-1)}{p^2}\right]e^{-p^2 t}\cos px\,dp$$

EXAMPLE 6. *Use the finite cosine transform to solve* $\dfrac{\partial V}{\partial t}=k\dfrac{\partial^2 V}{\partial x^2}$, *one dimensional heat equation with the boundary conditions* $\dfrac{\partial V}{\partial x}=0$ *when* $x=0$ *and* $x=\pi, t>0$ *and the initial condition* $V=f(x)$ *when* $t=0, 0<x<\pi$. (Kanpur–2014)

SOLUTION. Here, the given equation is

$$\frac{\partial V}{\partial t} = k\frac{\partial^2 V}{\partial x^2}, 0<x<\pi, t>0$$ ...(1)

Taking finite Fourier cosine transform of (1), we get

$$\frac{1}{k}\int_0^\pi \frac{\partial V}{\partial t}\cos px\,dx = \int_0^\pi \frac{\partial^2 V}{\partial x^2}\cos px\,dx$$

$$\Rightarrow \qquad \frac{1}{k}\frac{d\tilde{V}_c}{dt} = \left\{\frac{\partial V}{\partial x}\cos px\right\}_{x=0}^\pi + p\int_0^\pi \frac{\partial V}{\partial x}\sin px\,dx$$

$$= 0 + p\left[\{V(x,t)\sin px\}_{x=0}^\pi - p\int_0^\pi V\cos px\,dx\right] = -p^2\tilde{V}_c$$

$$\Rightarrow \qquad \frac{d\tilde{V}_c}{dt} = -kp^2\tilde{V}_c$$

Thus, the solution is

$$\tilde{V}_c(p,t) = A e^{-p^2 kt}$$

$$V = f(x) = \tilde{V}_c(p,0) = \int_0^\pi f(x)\cos px\,dx$$

$$A = \int_0^\pi f(x)\cos px\,dx = A(p)$$

Now, by definition of inverse cosine finite Fourier transform

$$V(x,t) = \frac{1}{\pi}\tilde{V}_c(0,t) + \frac{2}{\pi}\sum_{p=1}^\infty \tilde{V}_c \cos px$$

Hence, $\qquad V(x,t) = \dfrac{A}{\pi} + \dfrac{2}{\pi}\sum_{p=1}^\infty A(p)e^{-p^2 kt}\cos px$

EXAMPLE 7. *If the initial temperature of an infinite bar is given by*

$$\theta(x) = \begin{cases} \theta_0 & \text{for } |x| < a \\ 0 & \text{for } |x| > a \end{cases}$$

Show that the temperature at any point x at any instant t is given by

$$\theta(x,t) = \frac{\theta_0}{\pi}\left[erf\frac{(a+x)}{2c\sqrt{t}} + erf\frac{(a-x)}{2c\sqrt{t}} \right] \qquad \text{(Meerut–1986, 89, 93)}$$

SOLUTION. Assume that $\theta(x, t)$ be the temperature at any point x and at time t, then we have

$$\frac{\partial\theta}{\partial t} = C^2\frac{\partial^2\theta}{\partial x^2}(t > 0) \qquad \qquad \qquad ...(1)$$

with initial conditions

$$\theta(x) = \begin{cases} \theta_0 & \text{for } |x| < a \\ 0 & \text{for } |x| > a \end{cases}$$

Using Fourier transform, we get

$$\frac{d\tilde{\theta}}{dt} = C^2 f\left(\frac{\partial^2\theta}{\partial x^2}\right) = -C^2 p^2 f[\theta(x,t)] \qquad (\because f[\theta(x,t)] = \tilde{\theta})$$

$$\Rightarrow \qquad \frac{d\tilde{\theta}}{dt} = -C^2 p^2\tilde{\theta} \qquad \qquad \qquad ...(2)$$

Again taking Fourier transform, we get

$$\tilde{\theta}(p,0) = \int_{-\infty}^\infty \theta(x,0)e^{ipx}\,dx = \int_{-a}^a \theta_0 e^{ipx}\,dx$$

$$= \left[\theta_0\frac{e^{ipx}}{ip}\right]_{-a}^a = \theta_0\frac{e^{ipa} - e^{-ipa}}{ip} = 2\theta_0\frac{\sin ap}{p} \qquad ...(3)$$

Eq^n (2) becomes

$$\frac{d\tilde{\theta}}{\tilde{\theta}} = -C^2 p^2\,dt$$

Separating the variable and on integrating, we get

$$\int \frac{d\tilde{\theta}}{\tilde{\theta}} - \log C = -C^2 p^2 \int dt$$

$$\Rightarrow \quad \log \tilde{\theta} - \log C = -C^2 p^2 t \qquad \Rightarrow \qquad \tilde{\theta} = Ce^{-C^2 p^2 t}$$

From eq^n (3), we get

$$2\theta_0 \frac{\sin ap}{p} = \tilde{\theta}(p, 0) = C \qquad \Rightarrow \qquad C = 2\theta_0 \frac{\sin ap}{p}$$

Since, $$\tilde{\theta} = \frac{2\theta_0 \sin ap}{p} e^{-C^2 p^2 t}$$

Now, taking inverse Fourier transform, we get

$$\theta(x, t) = \frac{1}{2\pi} \int_{-\infty}^{\infty} \frac{2\theta_0 \sin ap}{p} e^{-C^2 p^2 t} e^{-ipx} dp$$

$$= \frac{\theta_0}{\pi} \int_{-\infty}^{\infty} \frac{\sin ap}{p} e^{-C^2 p^2 t} (\cos xp - i \sin xp) dp$$

$$= \frac{2\theta_0}{\pi} \int_{0}^{\infty} \frac{\sin ap}{p} e^{-C^2 p^2 t} \cos xp \, dp$$

$$= \frac{\theta_0}{\pi} \int_{0}^{\infty} e^{-C^2 p^2 t} \frac{\sin(a+x)p + \sin(a-x)p}{p} dp$$

$$= \frac{\theta_0}{\pi} \int_{0}^{\infty} e^{-v^2} \left\{ \frac{\sin(a+x)v}{c\sqrt{t}} + \frac{\sin(a-x)v}{c\sqrt{t}} \right\} \frac{dv}{v}$$

$$\left[\because v^2 = c^2 p^2 t \Rightarrow v \, dv = c^2 pt \Rightarrow \frac{dv}{v} = \frac{dp}{p} \right]$$

$$= \frac{\theta_0}{\pi} \left\{ erf \frac{(a+x)}{2c\sqrt{t}} + erf \frac{(a-x)}{2c\sqrt{t}} \right\}$$

MISCELLANEOUS EXERCISE

1. Find the Fourier transform of $e^{-a^2 x^2}, a < 0$. Hence, deduce that $e^{-x^2/2}$ is self reciprocal in respect to Fourier transform.

   (Kottayam–2005; Madras–2006)

2. Find the Fourier transform of

$$f(x) = \begin{cases} a^2 - x^2; & |x| \le a \\ 0; & |x| > a \end{cases} \qquad \text{(VTU–2007)}$$

3. Find the Fourier transform of

$$f(x) = \begin{cases} 4x & \text{for} \quad 0 < x < 1 \\ 4 - x & \text{for} \quad 1 < x < 4 \\ 0 & \text{for} \quad x > 4 \end{cases}$$

4. Show that the inverse finite Fourier sine transform of $f(x) = \frac{1}{\pi}\left[1 + \cos x\pi - 2\cos \frac{x\pi}{2}\right]$ is

$$f(x) = \begin{cases} 1; & 0 < x < \pi/2 \\ -1; & \pi/2 < x < \pi \end{cases} \qquad \text{(VTU–2008)}$$

5. A string is stretched between the two fixed points $(0, 0)$ and $(C, 0)$. If it is displaced into the curve $y = b\sin\left(\frac{n\pi x}{C}\right)$ and released from rest in that position at time $t = 0$. Solve the boundary value problem for the displacement $y(x, t)$. (Aligarh–1997, 2003; Lucknow–2005)

6. A tightly stretched flexible string has its ends fixed at $x = 0$ and $x = l$. At time $t = 0$, the string is given a shape defined by $P(x) = \mu x(l - x)$ where μ is a constant and then released. Find the displacement of any point x of the string at any point $t = 0$ (VTU(ME) – 2006)

ANSWERS

1. $\dfrac{\sqrt{\pi}}{a} e^{-p^2/4a^2}$

2. $\dfrac{4}{p^3}[\sin p\alpha - p\alpha \cos p\alpha]$

3. $\dfrac{(2\cos p - \cos 4p - 1)}{p^2} - \dfrac{2\sin p}{p}$

Self Assessment Test

1. Find the Fourier sine transform of

$$f(x) = \begin{cases} \sin x, & 0 < x < a \\ 0, & x > a \end{cases}$$ (Kanpur–2003, 04)

2. Find Fourier sine transform of $f(x)$, if

$$f(x) = \begin{cases} 0, & 0 < x < a \\ x, & a \le x < b \\ 0, & x > b \end{cases}$$ (Meerut–1991)

3. Find the finite Fourier cosine transform of

$$\tilde{f_c}(p) = \begin{cases} 0, & n + p \text{ is even} \\ \dfrac{2n}{n^2 - p^2}, & n + p \text{ is odd} \end{cases}$$ (Agra–2003)

4. If $f(x) = x$, for $0 < x < \pi$, deduce for $0 < x < 4$, then find the finite Fourier sine and cosine transforms. (Bundelkhand–2006)

5. Find $f(x)$, when $0 < x < 1$, if its finite cosine transform is given by

$$\tilde{f_c}(p) = \frac{\cos(2p\pi/3)}{(2p+1)^2}$$ (Bundelkhand–2008)

6. Find the Fourier cosine transform of the function $\dfrac{1}{1 + x^2}$.

7. Find $F(x)$, if its Fourier transform is

$$\frac{\sqrt{\pi}}{a} e^{-p^2/4a}.$$ (Agra–2004)

8. Find $f(x)$, if its Fourier cosine transform is

$$\tilde{f_c}(p) = \begin{cases} \dfrac{1}{2n}\left(a - \dfrac{p}{2}\right), & p < 2a \\ 0, & p \ge 2a \end{cases}$$

9. Using the Fourier cosine inversion theorem, show that

$$\int_0^\infty \frac{(p^2 + 2)\cos px}{p^4 + 4} dp = \frac{\pi}{2} e^{-x} \cos px > 0$$ (Agra–2002)

10. Find the complex Fourier transform of $e^{-|x|}$. Hence show that

$$\int_0^\infty \frac{x \sin mx}{1 + x^2} dx = \frac{\pi}{2} e^{-m}, m > 0$$ (Agra–2002)

ANSWERS

1. $\dfrac{1}{2}\left\{\dfrac{\sin[a(1-p)]}{1-p} - \dfrac{\sin[a(1+p)]}{1+p}\right\}$

2. $\dfrac{1}{p}(-b\cos pb + a\cos pa) + \dfrac{1}{p^2}(\sin pb - \sin pa)$

4. $\tilde{f_s}(p) = \dfrac{\pi}{p}(-1)^{p+1}, \tilde{f_c}(p) = \begin{cases} \pi^2/2, & p = 0 \\ \dfrac{1}{P^2}[(-1)^p - 1], & p = 1, 2, 3, \dots \end{cases}$

5. $1 + 2\sum\limits_{p=1}^{\infty} \dfrac{\cos(2p\pi/3)}{(2p+1)^2} \cos p\pi x$

6. $\tilde{f_c}(p) = \dfrac{\pi}{2}e^{-p}$ **7.** e^{-ax^2} **8.** $\dfrac{1}{x}e^{-x}\sin x$ **10.** $\dfrac{2}{1+p^2}$

Objective Evaluations

❖ Fill in the Blanks

1. If $f(x)$ is an odd function, then
$$f(x) = \frac{2}{\pi} \int_0^\infty \sin \lambda x \cdot \int_0^\infty f(t) \sin \lambda t \, dt \, d\lambda \text{ is called}$$
_____.

2. The infinite fourier sine transform of e^{-x} is
$$\frac{p}{(\underline{})}.$$

3. If $f(x)$ is an even function, then
$$f(x) = \frac{2}{\pi} \int_0^\infty \cos \lambda x \cdot \int_0^\infty f(t) \cos \lambda t \, dt \, d\lambda \text{ is called}$$
_____.

4. The finite Fourier cosine transform of $f(x)$ is defined as $\int_0^\infty (\underline{}) \sin pxdx$, where p is an integer.

5. The finite Fourier cosine transform of $\frac{\partial u}{\partial x}$, where u is a function of x and t for
$$0 < x < l, t > 0 \text{ is } \frac{p\pi}{l} F_p\{u\} - \{u(0,t) - u(l,t)$$
_____}.

❖ True or False

Write 'T' for True and 'F' for False statement.

1. $(f*g)*h = f*(g*h)$ **(T/F)**

2. Every Fourier transform is a linear transformation. **(T/F)**

3. The function $f(x)$ whose (infinite) cosine transform is $\frac{\sin ap}{p}$, will be $\frac{2}{\pi} \int_0^\infty \frac{\sin ap}{p} \cos pxdx$ **(T/F)**

4. The infinite Fourier cosine transform of $f(x) = e^{-x}$ is $\frac{1}{1+p^2}$. **(T/F)**

5. The complex form of Fourier integral is
$$f(x) = \frac{1}{2\pi} \int_{-\infty}^\infty \int_{-\infty}^\infty f(x)e^{-i\lambda(t-x)} dt d\lambda \quad \textbf{(T/F)}$$

6. The finite Fourier cosine transform of $f(x)$ is π, if $p = 0$ and 0, if $p = 1, 2, 3, \dots$. **(T/F)**

❖ Multiple Choice Questions

Choose the most appropriate one.

1. The infinite Fourier sine transform of e^{-x} is :
(Bundelkhand–2003)

  (a) $\dfrac{p}{1+p^2}$ (b) $\dfrac{1}{1+p^2}$

  (c) $\dfrac{p}{1-p^2}$ (d) $1+p^2$

2. The finite Fourier cosine transform of $f(x) = x$ is : (Bundelkhand–2006)

  (a) $\dfrac{\pi}{p}(-1)^{p+1}$ (b) $\dfrac{\pi}{p}(-1)^p$

  (c) $\dfrac{\pi}{p^2}(-1)^{p+1}$ (d) $\dfrac{(-1)^p - 1}{p^2}$

3. The finite Fourier sine transform of $\dfrac{x}{\pi}$ is :

(Bundelkhand–2007, 08)

  (a) $\dfrac{(-1)^{p-1}}{p}$ (b) $\dfrac{(-1)^{p+1}}{p^2}$

  (c) $\dfrac{(-1)^p}{p^2}$ (d) $\dfrac{(-1)^{p+1}}{p}$

4. If $f_s(p) = p/(1+p^2)$, then $F(x)$ is :

(Agra–2004)

  (a) e^{-x^2} (b) e^x

  (c) e^{x^2} (d) e^{-x}

5. The finite Fourier sine transform of $f(x) = x$ is : (Bundelkhand–2005)

  (a) $\dfrac{\pi}{p^2}(-1)^{p+1}$ (b) $\dfrac{\pi}{p}(-1)^p$

  (c) $\dfrac{\pi}{p^2}(-1)^p$ (d) $\dfrac{\pi}{p}(-1)^{p+1}$

6. The Fourier cosine transform of $f(x)$ where
$$f(x) = \begin{cases} 1, & 0 \le x \le 1 \\ 0, & x > 1 \end{cases} :$$

  (a) 1   (b) $\dfrac{\cos p}{p}$

  (c) $\dfrac{\sin p}{p}$ (d) none of these

7. The Fourier cosine tranform of e^{-x} is :

(Agra–2004)

(a) $\dfrac{1}{1+p^2}$ (b) $\dfrac{p}{1+p^2}$

(c) $\dfrac{1}{1-p^2}$ (d) $\dfrac{p}{1-p^2}$

8. If $F(p)$ is the complex Fourier transform of $f(x)$, then $F\{f(ax)\} =$

(a) $\dfrac{1}{a}F\left(\dfrac{a}{p}\right), a \neq 0$ (b) $\dfrac{1}{a}F\left(\dfrac{p}{a}\right), a \neq 0$

(c) $aF\left(\dfrac{p}{a}\right), a \neq 0$ (d) $\dfrac{1}{a}F(ap), a \neq 0$

9. If $F\{f(x)\} = F(p)$, then $F\{f(x)\cos ax\}$ is equal

to :

(a) $\dfrac{1}{2}\{F(p-a)-F(p+a)\}$

(b) $\dfrac{1}{2}\{F(p+a)-F(p-a)\}$

(c) $\dfrac{1}{2}\{F(p+a)+F(p-a)\}$

(d) none of these

10. The complex Fourier transform of $e^{-|x|}$ is :

(a) $\dfrac{1}{1-p^2}$ (b) $\dfrac{1}{1+p^2}$

(c) $\dfrac{2}{1-p^2}$ (d) $\dfrac{2}{1+p^2}$

ANSWERS

✧ Fill in the Blanks

1. Fourier sine integral   **2.** $1 + p^2$ **3.** Fourier cosine integral   **4.** $f(x)$ **5.** $\cos p\pi$

✧ True or False

1. T   **2.** T   **3.** F   **4.** T   **5.** T   **6.** T

✧ Multiple Choice Questions

1. (a)   **2.** (d)   **3.** (d)   **4.** (a)   **5.** (d)   **6.** (c)   **7.** (a)   **8.** (b)   **9.** (c)

10. (d)

Comprehensive Fourier Cosine Transform Tables

(1) GENERAL FORMULAS

S. No.	$F(x)$	$f_c(u) = \int_0^\infty f(x)\cos ux\,dx$
1.	$\begin{cases} 1 & \text{if } 0 < x < a \\ 0 & \text{if } a < x \end{cases}$	$\dfrac{1}{u}\sin(au)$
2.	$\begin{cases} 2 & \text{if } 0 < x < 1 \\ 2-x & \text{if } 1 < x < 2 \\ 0 & \text{if } 2 < x \end{cases}$	$\dfrac{u}{u^2}\cos u \sin^2\dfrac{u}{2}$
3.	$\dfrac{a}{a^2+(b+x)^2}+\dfrac{a}{a^2+(b-x)^2}$	$\pi e^{-au}\cos(bu)$
4.	$\dfrac{b+x}{a^2+(b+x)^2}+\dfrac{b-x}{a^2+(b-x)^2}$	$\pi e^{-au}\sin(bu)$
5.	$\dfrac{1}{a^4+x^4}, a>0$	$\dfrac{1}{2}\pi a^{-3}e^{\left(-\frac{au}{\sqrt{2}}\right)}\sin\left(\dfrac{\pi}{4}+\dfrac{au}{\sqrt{2}}\right)$
6.	$\dfrac{1}{(a^2+x^2)(b^2+x^2)}, a,b>0$	$\dfrac{\pi}{2}\dfrac{ae^{-bu}-be^{-au}}{ab(a^2-b^2)}$
7.	$\dfrac{1}{\sqrt{x}}$	$\sqrt{\dfrac{\pi}{2u}}$
8.	$\begin{cases} 0 & \text{if } 0 < x < a \\ \dfrac{1}{\sqrt{x-a}} & \text{if } a < x \end{cases}$	$\sqrt{\dfrac{\pi}{2u}}[\cos(au)-\sin(au)]$
9.	$\begin{cases} \dfrac{1}{\sqrt{a^2-x^2}} & \text{if } 0 < x < a \\ 0 & \text{if } a < x \end{cases}$	$\dfrac{\pi}{2}J_0(au)$

(2) FOR EXPONENTIAL FUNCTIONS

S. No.	$F(x)$	$f_c(u) = \int_0^\infty f(x)\cos ux\,dx$
1.	e^{-ax}	$\dfrac{a}{a^2+u^2}$
2.	$\dfrac{1}{x}(e^{-ax}-e^{-bx})$	$\dfrac{1}{2}\log\dfrac{b^2+u^2}{a^2+u^2}$
3.	$\sqrt{x}e^{-ax}$	$\dfrac{1}{2}\sqrt{\pi}(a^2+u^2)^{-3/4}\cos\left(\dfrac{3}{2}\tan^{-1}\dfrac{u}{a}\right)$
4.	$\dfrac{1}{\sqrt{x}}e^{-ax}$	$\sqrt{\dfrac{\pi}{2}}\left[\dfrac{a+(a^2+u^2)^{1/2}}{a^2+u^2}\right]^{1/2}$

5.	$x^n e^{-ax}, n = 1, 2, \ldots$	$\dfrac{a^{n+1}n!}{(a^2+u^2)^{n+1}} \sum\limits_{0 \le 2K \le n+1} (-1)^K C_{n+1}^{2K}\left(\dfrac{u}{a}\right)^{2K}$
6.	$\dfrac{x}{e^{ax}-1}$	$\dfrac{1}{2u^2} - \dfrac{\pi^2}{2a^2 \sinh^2(\pi a^{-1}u)}$
7.	$\dfrac{1}{x}\left(\dfrac{1}{2} - \dfrac{1}{x} + \dfrac{1}{e^x-1}\right)$	$-\dfrac{1}{2}\log(1 - e^{-2\pi u})$
8.	$e^{(-ax^2)}$	$\dfrac{1}{2}\sqrt{\dfrac{\pi}{a}}e^{(-u^2/4a)}$
9.	$\dfrac{1}{\sqrt{x}}e^{(-a/x)}$	$\sqrt{\dfrac{\pi}{2u}}e^{-\sqrt{2au}}[\cos(\sqrt{2au}) - \sin(\sqrt{2au})]$
10.	$\dfrac{1}{x\sqrt{x}}e^{(-a/x)}$	$\sqrt{\dfrac{\pi}{a}}e^{-\sqrt{2au}}\cos(\sqrt{2au})$

(3) FOR HYPERBOLIC FUNCTIONS

S. No.	$F(x)$	$f_c(u) = \int_0^\infty f(x)\cos ux\,dx$
1.	$\dfrac{1}{\cosh(ax)}, a > 0$	$\dfrac{\pi}{2a\cosh\left(\dfrac{1}{2}\pi a^{-1}u\right)}$
2.	$\dfrac{1}{\cosh^2(ax)}, a > 0$	$\dfrac{\pi u}{2a^2\sinh\left(\dfrac{1}{2}\pi a^{-1}u\right)}$
3.	$\dfrac{\cosh(ax)}{\cosh(bx)}, |a| < b$	$\dfrac{\pi}{b}\left[\dfrac{\cos\left(\dfrac{1}{2}\pi ab^{-1}\right)\cosh\left(\dfrac{1}{2}\pi b^{-1}u\right)}{\cos(\pi ab^{-1}) + \cosh(\pi b^{-1}u)}\right]$
4.	$\dfrac{1}{\cosh(ax) + \cos b}$	$\dfrac{\pi\sinh(a^{-1}bu)}{a\sin b\sinh(\pi a^{-1}u)}$
5.	$e^{-ax^2}\cosh(bx), a > 0$	$\dfrac{1}{2}\sqrt{\dfrac{\pi}{a}}e^{\frac{(b^2-u^2)}{4a}} \cdot \cos\left(\dfrac{abu}{2}\right)$
6.	$\dfrac{x}{\sinh(ax)}$	$\dfrac{\pi^2}{4a^2\cosh^2\left(\dfrac{1}{2}\pi a^{-1}u\right)}$
7.	$\dfrac{\sinh(ax)}{\cosh(ax)}, |a| < b$	$\dfrac{\pi}{2b}\dfrac{\sin(\pi ab^{-1})}{\cos(\pi ab^{-1}) + \cosh(\pi b^{-1}u)}$
8.	$\dfrac{1}{x}\tanh(ax), a > 0$	$\log\left[\coth\left(\dfrac{1}{4}\pi a^{-1}u\right)\right]$

(4) FOR LOGARITHMIC FUNCTIONS

S. No.	$F(x)$	$f_c(u) = \int_0^\infty f(x)\cos ux\,dx$
1.	$\log\left(1 + \dfrac{a^2}{x^2}\right), a > 0$	$\dfrac{\pi}{u}(1 - e^{-au})$
2.	$e^{-ax}\log x, a > 0$	$-\dfrac{aC + \dfrac{1}{2}\log(u^2 + a^2) + u\tan^{-1}\left(\dfrac{u}{a}\right)}{u^2 + a^2}$
3.	$\log(1 + e^{-ax}), a > 0$	$\dfrac{a}{2u^2} - \dfrac{\pi}{2u\sinh(\pi a^{-1}u)}$
4.	$\log(1 - e^{-ax}), a > 0$	$\dfrac{a}{2u^2} - \dfrac{\pi}{2u}\coth(\pi a^{-1}u)$

(5) FOR TRIGONOMETRIC FUNCTIONS

S. No.	$F(x)$	$f_c(u) = \int_0^\infty f(x)\cos ux\,dx$
1.	$\dfrac{\sin ax}{x}, a > 0$	$\begin{cases} \pi/2, & u < a \\ \pi/4, & u = a \\ 0, & u > a \end{cases}$
2.	$\dfrac{x\sin(ax)}{x^2 + b^2}, a, b > 0$	$\begin{cases} \dfrac{1}{2}\pi e^{-ab}\cosh(bu) & \text{if } u < a \\ -\dfrac{1}{2}\pi e^{-bu}\sinh(ab) & \text{if } u > a \end{cases}$
3.	$\dfrac{1 - \cos ax}{x}, a > 0$	$\dfrac{1}{2}\log\left|1 - \dfrac{a^2}{u^2}\right|$
4.	$\dfrac{1 - \cos ax}{x^2}, a > 0$	$\begin{cases} \dfrac{\pi}{2}(a - u) & \text{if } u < a \\ 0 & \text{if } u > a \end{cases}$
5.	$\dfrac{1}{\sqrt{x}}\cos a\sqrt{x}$	$\sqrt{\dfrac{\pi}{u}}\sin\left(\dfrac{a^2}{4u} + \dfrac{\pi}{4}\right)$
6.	$\cos ax^2, a > 0$	$\sqrt{\dfrac{\pi}{8a}}\left[\cos\left(\dfrac{1}{4}a^{-1}u^2\right) + \sin\left(\dfrac{1}{4}a^{-1}u^2\right)\right]$

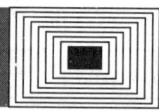

Comprehensive Fourier Sine Transform Tables

(1) FOR POWER LAW FUNCTIONS

S. No.	$F(x)$	$f_s(u) = \int_0^\infty f(x)\sin(ux)dx$
1.	$\begin{cases}1; & 0 < x < a \\ 0; & a < x\end{cases}$	$\dfrac{1}{u}[1 - \cos(au)]$
2.	$\dfrac{1}{x}$	$\dfrac{\pi}{2}$
3.	$\dfrac{x}{a^2 + x^2}, a > 0$	$\dfrac{\pi}{2}e^{-au}$
4.	$\dfrac{a}{a^2 + (x-b)^2} - \dfrac{a}{x^2 + (x+b)^2}$	$\pi e^{-au}(\sin bu)$
5.	$\dfrac{1}{\sqrt{x}}$	$\sqrt{\dfrac{\pi}{2u}}$
6.	$\dfrac{1}{x\sqrt{x}}$	$\sqrt{2\pi u}$
7.	$\dfrac{(\sqrt{a^2 + x^2} - a)^{1/2}}{\sqrt{a^2 + x^2}}$	$\sqrt{\dfrac{\pi}{2u}}e^{-au}$
8.	$x^{-n}, 0 < n < 2$	$\cos\left(\dfrac{1}{2}\pi n\right)\Gamma(1-n)u^{n-1}$

(2) FOR EXPONENTIAL FUNCTIONS

S. No.	$F(x)$	$f_s(u) = \int_0^\infty f(x)\sin(ux)dx$
1.	$e^{-ax}, a > 0$	$\dfrac{u}{a^2 + u^2}$
2.	$x^n e^{-ax}, a > 0, n \in N$	$n!\left(\dfrac{a}{a^2+u^2}\right)^{n+1} \cdot \sum_{K=0}^{\left[\frac{n}{2}\right]}(-1)^K C_{n+1}^{2K+1}\left(\dfrac{u}{a}\right)^{2K+1}$
3.	$\dfrac{1}{x}e^{-ax}, a > 0$	$\tan^{-1}\dfrac{x}{a}$
4.	$\sqrt{x}e^{-ax}, a > 0$	$\dfrac{\sqrt{\pi}}{2}(a^2+u^2)^{-3/4}\sin\left(\dfrac{3}{2}\tan^{-1}\dfrac{u}{a}\right)$
5.	$\dfrac{1}{\sqrt{x}}e^{-ax}, a > 0$	$\sqrt{\dfrac{\pi}{2}}\dfrac{(\sqrt{a^2+u^2}-a)^{1/2}}{\sqrt{a^2+u^2}}$
6.	$\dfrac{1}{e^{ax}+1}, a > 0$	$\dfrac{1}{2u} - \dfrac{\pi}{2a\sinh\left(\dfrac{\pi u}{a}\right)}$
7.	$\dfrac{1}{e^{ax}-1}, a > 0$	$\dfrac{\pi}{2a}\coth\left(\dfrac{\pi u}{a}\right) - \dfrac{1}{2u}$

8.	$\dfrac{e^{x/2}}{e^x - 1}$	$-\dfrac{1}{2}\tanh(\pi u)$
9.	xe^{-ax^2}	$\dfrac{\sqrt{\pi}}{4a^{3/2}}ue^{(-u^2/4a)}$
10.	$\dfrac{1}{x}e^{-ax^2}$	$\dfrac{\sqrt{\pi}}{2}\mathrm{erf}\left(\dfrac{u}{2\sqrt{a}}\right)$

(3) FOR HYPERBOLIC FUNCTIONS

S. No.	$F(x)$	$f_s(u) = \int_0^\infty f(x)\sin(ux)dx$
1.	$\dfrac{1}{\sinh(ax)}, a > 0$	$\dfrac{\pi}{2a}\tanh\left(\dfrac{1}{2}\pi a^{-1}u\right)$
2.	$\dfrac{x}{\sinh(ax)}, a > 0$	$\dfrac{\pi^2 \sinh\left(\dfrac{1}{2}\pi a^{-1}u\right)}{4a^2 \cosh^2\left(\dfrac{1}{2}\pi a^{-1}u\right)}$
3.	$\dfrac{1}{x\cosh(ax)}, a > 0$	$\tan^{-1}\left[\sinh\left(\dfrac{1}{2}\pi a^{-1}u\right)\right]$
4.	$1 - \tanh\left(\dfrac{1}{2}ax\right), a > 0$	$\dfrac{1}{u} - \dfrac{\pi}{a\sinh(\pi a^{-1}u)}$
5.	$\coth\left(\dfrac{1}{2}ax\right) - 1, a > 0$	$\dfrac{\pi}{a}\coth(\pi a^{-1}u) - \dfrac{1}{u}$
6.	$\dfrac{\cosh(ax)}{\sinh(bx)}, |a| < b$	$\dfrac{\pi}{2b}\dfrac{\sinh(\pi b^{-1}u)}{\cos(\pi ab^{-1}) + \cosh(\pi b^{-1}u)}$
7.	$\dfrac{\sinh(ax)}{\cosh(bx)}, |a| < b$	$\dfrac{\pi}{b}\dfrac{\sin\left(\dfrac{1}{2}\pi ab^{-1}\right)\sinh\left(\dfrac{1}{2}\pi b^{-1}u\right)}{\cos(\pi ab^{-1}) + \cosh(\pi b^{-1}u)}$

(4) FOR LOGARITHMIC FUNCTIONS

S. No.	$F(x)$	$f_s(u) = \int_0^\infty f(x)\sin(ux)dx$
1.	$\dfrac{\log x}{x}$	$-\dfrac{1}{2}\pi(\log u + C)$
2.	$\dfrac{\log x}{\sqrt{x}}$	$-\sqrt{\dfrac{\pi}{2u}}\left[\log 4u + C - \dfrac{\pi}{2}\right]$
3.	$\log\left|\dfrac{a+x}{a-x}\right|, a, b > 0$	$\dfrac{2\pi}{u}e^{-au}\sin au$
4.	$\log\dfrac{(x+b)^2 + a^2}{(x-b)^2 + a^2}, a, b > 0$	$\dfrac{2\pi}{u}e^{-au}\sin bu$

(5) FOR TRIGONOMETRIC FUNCTIONS

S. No.	$F(x)$	$f_s(u) = \int_0^\infty f(x)\sin(ux)dx$
1.	$\dfrac{\sin ax}{x}, a > 0$	$\dfrac{1}{2}\log\left\|\dfrac{u+a}{u-a}\right\|$
2.	$\dfrac{\sin ax}{x^2}, a > 0$	$\begin{cases} \dfrac{1}{2}\pi u, & 0 < u < a \\ \dfrac{1}{2}\pi a, & u > a \end{cases}$
3.	$\dfrac{\sin \pi x}{1-x^2}$	$\begin{cases} \sin u, & 0 < u < \pi \\ 0, & u > \pi \end{cases}$
4.	$e^{-ax}\sin bx, a > 0$	$\dfrac{a}{2}\left[\dfrac{1}{a^2+(b-u)^2} - \dfrac{1}{a^2+(b+u)^2}\right]$
5.	$\dfrac{1}{x}e^{-ax}\sin bx, a > 0$	$\dfrac{1}{4}\log\dfrac{(u+b)^2+a^2}{(u-b)^2+a^2}$
6.	$\dfrac{1}{x}\sin^2 ax, a > 0$	$\begin{cases} \dfrac{\pi}{4}; & 0 < u < 2a \\ \dfrac{\pi}{8}; & u = 2a \\ 0; & u > 2a \end{cases}$
7.	$\dfrac{1}{x}\sin ax \sin bx, a, b > 0$	$\begin{cases} 0; & 0 < u < a-b \\ \dfrac{\pi}{4}; & a-b < u < a+b \\ 0; & a+b < u \end{cases}$
8.	$\dfrac{\cos ax}{x}, a > 0$	$\begin{cases} 0; & 0 < u < a \\ \dfrac{\pi}{4}; & u = a \\ \dfrac{\pi}{2}; & a < u \end{cases}$
9.	$\dfrac{1-\cos ax}{x^2}, a > 0$	$\dfrac{u}{2}\log\left\|\dfrac{u^2-a^2}{u^2}\right\| + \dfrac{a}{2}\log\left\|\dfrac{u+a}{u-a}\right\|$
10.	$\dfrac{1}{\sqrt{x}}\cos a\sqrt{x}$	$\sqrt{\dfrac{\pi}{u}}\cos\left(\dfrac{a^2}{4u}+\dfrac{\pi}{4}\right)$

VVVVVV

5.1 INTRODUCTION

In this chapter we shall discuss about a very important concept of integral transform, *i.e.*, Hankel and Mellin Transforms. Before discussing these topics, firstly we give some basic idea of Bessel's function, which is frequently used in the present chapter.

5.2 CONCEPTS OF BESSEL'S FUNCTION

The homogeneous linear differential equation of the form

$$x^2 \frac{d^2y}{dx^2} + x\frac{dy}{dx} + (x^2 - n^2)y = 0 \qquad \dots(1)$$

is known as Bessel's differential equation, where n is a non-negative real number.

5.2.1 SOLUTION OF THE BESSEL'S EQUATION

Change the differential equation (1) into standard form by dividing (1) by x^2.

$$\frac{d^2y}{dx^2} + \frac{1}{x}\frac{dy}{dx} + \left(1 - \frac{n^2}{x^2}\right)y = 0 \qquad \dots(2)$$

Now compare this differential equation with following equation

$$\frac{d^2y}{dx^2} + P(x)\frac{dy}{dx} + Q(x)y = 0$$

We get $\qquad P(x) = \frac{1}{x}, Q(x) = \left(1 - \frac{n^2}{x^2}\right)$

It is obvious from $P(x)$ and $Q(x)$ that $x = 0$ is a singular point which is located at the origin. Therefore, we assume the solution of (1) in the form of a power series of the following type :

$$y = \sum_{m=0}^{\infty} a_m x^{m+r} (a_0 \neq 0) \qquad \dots(3)$$

Differentiating (3) w.r.to x, we get

$$\frac{dy}{dx} = \sum_{m=0}^{\infty} a_m (m+r)x^{m+r-1} \qquad \dots(4)$$

Again differentiating (4) w.r.to x, we get

$$\frac{d^2y}{dx^2} = \sum_{m=0}^{\infty} a_m (m+r)(m+r-1)x^{m+r-2} \qquad \dots(5)$$

Now, substitute the values of $y, \dfrac{dy}{dx}, \dfrac{d^2y}{dx^2}$ from (3), (4) and (5) into (1), we have

$$x^2 \sum_{m=0}^{\infty} a_m(m+r)(m+r-1)x^{m+r-2} + x \sum_{m=0}^{\infty} a_m(m+r)x^{m+r-1}$$

$$(x^2 - n^2) \sum_{m=0}^{\infty} a_m x^{m+r} = 0$$

$$\sum_{m=0}^{\infty} a_m(m+r)(m+r-1)x^{m+r} + \sum_{m=0}^{\infty} a_m(m+r)x^{m+r}$$

$$- \sum_{m=0}^{\infty} a_m n^2 x^{m+r} + \sum_{m=0}^{\infty} a_m x^{m+r+2} = 0 \qquad \ldots(6)$$

Equation (6) will be an identity if the equation (3) is a solution of (1), the coefficient of each terms in (6) will be zero. Thus, taking the coefficient of x^r, x^{r+1}.

$$a_0 r(r-1) + a_0 r - n^2 a_0 = 0 \qquad \ldots(7)$$

$$a_1(r+1)r + a_1(r+1) - n^2 a_1 = 0 \qquad \ldots(8)$$

In general taking the coefficients of x^{s+r}

$$a_s(s+r)(s+r-1) + a_s(s+r) - n^2 a_s + a_{s-2} = 0 \qquad \ldots(9)$$

For $s = 2, 3, 4, \ldots$

From (7), we have

$$r(r-1) + r - n^2 = 0 \qquad\qquad (\because a_0 \neq 0)$$

$$r^2 - n^2 = 0 \qquad \Rightarrow \qquad r = n, -n$$

From (8), we have

$$[(r+1)r + (r+1) - n^2]a_1 = 0$$

For any value of $r = n, -n$, we get $a_1 = 0$

From (9), we have

$$a_s[(s+r)(s+r-1) + s + r - n^2] + a_{s-2} = 0$$

or

$$a_s[(s+r)^2 - n^2] + a_{s-2} = 0$$

or

$$a_s(s+r-n)(s+r+n) + a_{s-2} = 0 \qquad \ldots(10)$$

For case if $r = n$, then (1) becomes

$$a_s(s)(s+2n) + a_{s-2} = 0$$

or

$$a_s = \frac{-1}{s(s+2n)} \cdot a_{s-2}$$

Putting $s = 2, 3, 4, 5, \ldots$

$$a_2 = \frac{-1}{2(2+2n)} \cdot a_0$$

$$a_3 = \frac{-1}{3(3+2n)} \cdot a_1 = 0 \qquad (\because a_1 = 0)$$

$$a_4 = \frac{-1}{4(4+2n)} \cdot a_2 = \frac{-1}{4(4+2n)} - \frac{1}{2(2+2n)} a_0$$

$$= (-1)^2 \frac{1}{2 \cdot 4(2+2n)(4+2n)} \cdot a_0$$

$$\vdots$$

etc.

We observed that $a_1 = a_3 = a_5 = ... = 0$. Since a_0 is arbitrary. Let us choose

$$a_0 = \frac{1}{2^n \, \Gamma(n+1)}$$

where $\Gamma(n + 1)$ is a Gamma function, therefore we know that $\Gamma(n + 1) = n\Gamma(n)$ and if n is positive integer. $\Gamma(n + 1) = n!$. Thus,

$$a_2 = -\frac{1}{2(2+2n)} a_0$$

$$= -\frac{1}{2^2(1+n)} \cdot \frac{1}{2^n \Gamma(n+1)} \qquad \left(\because a_0 = \frac{1}{2^n \, \Gamma(n+1)} \right)$$

$$= \frac{-1}{2^{n+2} \Gamma(n+2)}$$

$$a_4 = (-1)^2 \frac{1}{2^4 (2)!(1+n)(2+n)} \cdot \frac{1}{2^n \Gamma(n+1)}$$

$$= (-1)^2 \frac{1}{2^{n+4} (2)! \Gamma(n+3)}$$

and so on. Now from (3), we have

$$y = \sum_{m=0}^{\infty} a_m x^{m+r}$$

$$= a_0 x^r + a_1 x^{1+r} + a_2 x^{2+r} + a_3 x^{3+r} + a_4 x^{4+r} + ...$$

$$= \frac{1}{2^n \, \Gamma(n+1)} x^n - \frac{1}{2^{n+2} (1)! \Gamma(n+2)} x^{n+2} + \frac{1}{2^{n+4} (2)! \Gamma(n+3)} x^{n+4} + ...$$

$$= \sum_{m=0}^{\infty} \frac{(-1)^m x^{n+2m}}{2^{n+2m} (m)! \Gamma(n+m+1)}$$

This solution is known as Bessel's function, which is denoted by $J_n(x)$. This function is also known as Bessel's function of first kind.

$$\therefore \qquad J_n(x) = \sum_{m=0}^{\infty} (-1)^m \frac{x^{n+2m}}{2^{n+2m} (m)! \Gamma(m+n+1)} \qquad ...(11)$$

For case if $r = -n$, we have

$$J_{-n}(x) = \sum_{m=0}^{\infty} \frac{(-1)^m x^{-n+2m}}{2^{-n+2m} (m)! \Gamma(-n+m+1)} \qquad ...(12)$$

5.2.2 GENERAL SOLUTIONS

The solution of the Bessel's differential equation of the type

$$y(x) = A J_n(x) + B J_{-n}(x)$$

where A and B are arbitrary constants, is called general solution.

5.2.3 LINEAR DEPENDENCE

THEOREM-1. **For an integer $r = n$, the Bessel's function $J_n(x)$ and $J_{-n}(x)$ are linearly dependent, because**

$$J_{-n}(x) = (-1)^n J_n(x), \text{ for } n = 1, 2, ...$$

PROOF. SINCE

$$J_{-n}(x) = \sum_{m=0}^{\infty} \frac{(-1)^m x^{-n+2m}}{2^{-n+2m} (m)! \Gamma(-n+m+1)} \qquad ...(1)$$

if n is a positive integer, then the gamma function in the coefficients of first n terms becomes infinite and coefficients of (1) becomes zero. Thus, the summation will start at $m = n$ and in this case $\Gamma(-n + m + 1) = (m - n)!$.

From (1), we have

$$J_{-n}(x) = \sum_{m=n}^{\infty} \frac{(-1)^m x^{-n+2m}}{2^{-n+2m}(m)!(m-n)!}$$

$$= \sum_{k=0}^{\infty} \frac{(-1)^{n+k} x^{n+2k}}{2^{n+2k}(k)!(n+k)!} \qquad (\because m = n + k)$$

$$= (-1)^n \sum_{k=0}^{\infty} \frac{(-1)^k x^{n+2k}}{2^{n+2k}(k)!\Gamma(n+k+1)}$$

$$J_{-n}(x) = (-1)^n J_n(x)$$

5.2.4 DEFINITION OF $J_n(x)$, WHEN $n = 0$

Putting $n = 0$ in the Bessel's differential equation, we get

$$x\frac{d^2 y}{dx^2} + \frac{dy}{dx} + xy = 0 \qquad \qquad ...(1)$$

Let us assume the solution

$$y = \sum_{m=0}^{\infty} a_m x^{m+r} \quad (a_0 \neq 0) \qquad \qquad ...(2)$$

$$\frac{dy}{dx} = \sum_{m=0}^{\infty} a_m (m+r)x^{m+r-1}$$

and

$$\frac{d^2 y}{dx^2} = \sum_{m=0}^{\infty} a_m (m+r)(m+r-1)x^{m+r-2}$$

Substituting these values in (1), we get

$$x \sum_{m=0}^{\infty} a_m (m+r)(m+r-1)x^{m+r-2} + \sum_{m=0}^{\infty} a_m (m+r)x^{m+r-1} + x \sum_{m=0}^{\infty} a_m x^{m+r} = 0$$

or $\quad \sum_{m=0}^{\infty} a_m (m+r)(m+r-1)x^{m+r-1} + \sum_{m=0}^{\infty} a_m (m+r)x^{m+r-1} + \sum_{m=0}^{\infty} a_m x^{m+r-1} = 0 \qquad ...(3)$

If (2) is the solution of (1), then (3) will be an identity. Thus coefficients of each terms will be zero. So that taking the coefficients of x^{r-1}, we get

$$a_0 r(r-1) + a_0 r = 0$$

or $\qquad\qquad r^2 a_0 = 0 \qquad \Rightarrow \qquad r = 0 \qquad (\because a_0 \neq 0)$

Now taking the coefficient of x^r, we have

$$a_1(1+r)r + a_1(1+r) = 0$$

$\Rightarrow \qquad\qquad a_1(1+r)^2 = 0$

$\Rightarrow \qquad\qquad\qquad a_1 = 0 \qquad\qquad\qquad\qquad (\because r = 0)$

In general, taking the coefficient of x^{m+r}

$$a_{m+1}(m+r+1)(m+r) + a_{m+1}(m+r+1) + a_{m-1} = 0$$

or $\qquad a_{m+1}(m+r+1)^2 + a_{m-1} = 0$

or

$$a_{m+1} = \frac{a_{m-1}}{(m+r+1)^2}$$

For the case $r = 0$, $a_{m+1} = -\dfrac{a_{m-1}}{(m+1)^2}$

Putting $m = 1, 2, 3, 4, 5, \ldots$

$$a_3 = -\frac{a_1}{9} = 0 \qquad\qquad (\because a_1 = 0)$$

$$a_2 = -\frac{a_0}{2^2}$$

$$a_4 = -\frac{a_2}{4^2} = \frac{(-1)^2 a_0}{2^2 \cdot 4^2}$$

$$a_5 = 0 \text{ etc.}$$

Thus, we obtained $a_1 = a_3 = a_5 = \ldots = 0$. Hence,

$$y = a_0 \left(1 - \frac{x^2}{2^2} + \frac{x^4}{2^2 \cdot 4^2} - \frac{x^6}{2^2 \cdot 4^2 \cdot 6^2} + \ldots \right)$$

If $a_0 = 1$, then $y = J_0(x)$.

$$J_0(x) = 1 - \frac{x^2}{2^2} + \frac{x^4}{2^2 \cdot 4^2} - \frac{x^6}{2^2 \cdot 4^2 \cdot 6^2} + \ldots$$

J_0 is also known as Bessel's function of order zero.

5.2.5 GENERATING FUNCTION FOR $J_n(x)$

The function of the form $e^{\left[\frac{1}{2} x \left(t - \frac{1}{t} \right) \right]}$ *generates $J_n(x)$, if taking coefficients of t^n. Thus this function is known as Generating function for $J_n(x)$.*

Proof. Expand $e^{\left[\frac{1}{2} x \left(t - \frac{1}{t} \right) \right]}$

$$= e^{\frac{xt}{2}} \cdot e^{-\frac{x}{2t}}$$

$$= \left[1 + \frac{xt}{2} + \frac{1}{(2)!} \left(\frac{xt}{2} \right)^2 + \ldots + \frac{1}{(n)!} \left(\frac{xt}{2} \right)^n + \frac{1}{(n+1)!} \left(\frac{xt}{2} \right)^{n+1} \right.$$

$$\left. + \frac{1}{(n+2)!} \left(\frac{xt}{2} \right)^{n+2} + \ldots \right] \cdot \left[1 - \frac{x}{2t} + \frac{1}{(2)!} \left(\frac{x}{2t} \right)^2 + \ldots \right.$$

$$\left. + \frac{(-1)^n}{(n)!} \left(\frac{x}{2t} \right)^n + \frac{(-1)^{n+1}}{(n+1)!} \left(\frac{x}{2t} \right)^{n+1} + \frac{(-1)^{n+2}}{(n+2)!} \left(\frac{x}{2t} \right)^{n+2} + \ldots \right]$$

Now, collecting the coefficient of t^n, in above expression obtained after multiplication

$$= \frac{1}{(n)!} \left(\frac{x}{2} \right)^n - \frac{1}{(n+1)!} \left(\frac{x}{2} \right)^{n+2} + \frac{1}{(n+2)!} \cdot \frac{1}{(2)!} \left(\frac{x}{2} \right)^{n+4} + \ldots$$

$$= \sum_{m=0}^{\infty} (-1)^m \cdot \frac{1}{(m)!(m+n)!} \left(\frac{x}{2} \right)^{n+2m}$$

$$= \sum_{m=0}^{\infty} \frac{(-1)^m x^{n+2m}}{2^{n+2m}(m)!\Gamma(m+n+1)} \qquad (\because \Gamma(m+n+1) = (m+n)!)$$

$$= J_n(x)$$

$$\therefore \qquad e^{\left[\frac{1}{2}x\left(t-\frac{1}{t}\right)\right]} = \sum_{m=0}^{\infty} t^n J_n(x)$$

If taking the coefficient of t^{-n}, we get

$$= \frac{(-1)^n}{(n)!}\left(\frac{x}{2}\right)^n + \frac{(-1)^{n+1}}{(n+1)!}\left(\frac{x}{2}\right)^{n+2} + \frac{(-1)^{n+2}}{(n+2)!}\cdot\frac{1}{(2)!}\left(\frac{x}{2}\right)^{n+4} + \dots$$

$$= (-1)^n\left[\frac{1}{(n)!}\left(\frac{x}{2}\right)^n - \frac{1}{(n+1)!}\left(\frac{x}{2}\right)^{n+2} + \frac{1}{(n+2)!}\cdot\frac{1}{2!}\left(\frac{x}{2}\right)^{n+4} - \dots\right]$$

$$= (-1)^n \sum_{m=0}^{\infty} \frac{(-1)^m x^{n+2m}}{2^{n+2m}(m)!\Gamma(n+m+1)} = (-1)^n J_n(x)$$

$$= J_{-n}(x) \qquad\qquad [\because J_{-n}(x) = (-1)^n J_n(x)]$$

Hence, we obtained

$$e^{\left[\frac{1}{2}x\left(t-\frac{1}{t}\right)\right]} = \sum_{n=-\infty}^{\infty} t^n J_n(x)$$

5.3 SOME IMPORTANT RECURRENCE RELATIONS

1. $xJ_n'(x) = nJ_n(x) - xJ_{n+1}(x)$ where $J_n'(x) = \dfrac{dJ_n(x)}{dx}$

2. $xJ_n'(x) = -nJ_n(x) + xJ_{n-1}(x)$ 3. $2J_n'(x) = J_{n-1}(x) - J_{n+1}(x)$

4. $2nJ_n(x) = x[J_{n-1}(x) + J_{n+1}(x)]$ 5. $\dfrac{d}{dx}[x^{-n}J_n(x)] = -x^{-n}J_{n+1}(x)$

6. $\dfrac{d}{dx}[x^n J_n(x)] = x^n J_{n-1}(x)$

5.4 SOME IMPORTANT RESULTS

1. $J_n(x)$ is even and odd function for even n and for odd n respectively.

2. $J_0'(x) = -J_1(x)$

3. $\dfrac{d}{dx}[J_n^2 + J_{n+1}^2] = 2\left(\dfrac{n}{x}J_n^2 - \dfrac{n+1}{x}J_{n+1}^2\right)$ 4. $\dfrac{d}{dx}(xJ_nJ_{n+1}) = x(J_n^2 - J_{n+1}^2)$

5. $J_{1/2}(x) = \sqrt{\dfrac{2}{\pi x}}\cdot\sin x$ 6. $J_{-1/2}(x) = \sqrt{\dfrac{2}{\pi x}}\cdot\cos x$

7. $[J_{1/2}(x)]^2 + [J_{-1/2}(x)]^2 = \dfrac{2}{\pi x}$ 8. $J_{-3/2}(x) = -\sqrt{\dfrac{2}{\pi x}}\left(\dfrac{1}{x}\cos x + \sin x\right)$

9. $\lim\limits_{x\to 0} \dfrac{J_n(x)}{x^n} = \dfrac{1}{2^2\Gamma(n+1)}, n > -1$ 10. $J_0^2 + 2(J_1^2 + J_2^2 + J_3^2 + \dots) = 1$

11. $\int_0^\infty e^{-ax}J_0(bx)dx = \dfrac{1}{\sqrt{a^2+b^2}}, a > 0$ 12. $J_n(x) = \dfrac{1}{\pi}\int_0^\infty (\cos n\theta - x\sin\theta)d\theta$

13. $J_{n+3} + J_{n+5} = \dfrac{2}{x}(n+4)J_{n+4}$

14. $J_n' = \dfrac{2}{x}\left[\dfrac{n}{2}J_n - (n+2)J_{n-2} + (n+4)J_{n+4} - \cdots\right]$

15. $\int J_{n+1}(x)dx = \int J_{n-1}(x)dx - 2J_n(x) + A$ **16.** $J_{3/2}(x) = \sqrt{\dfrac{2}{\pi x}}\left[\dfrac{1}{x}\sin x - \cos x\right]$

17. $J_{5/2}(x) = \sqrt{\dfrac{2}{\pi x}}\left[\left(\dfrac{3-x^2}{x^2}\right)\sin x - \dfrac{3}{x}\cos x\right]$

18. $J_{-5/2}(x) = \sqrt{\dfrac{2}{\pi x}}\left[\left(\dfrac{3-x^2}{x^2}\right)\cos x + \dfrac{3}{x}\sin x\right]$

19. $4J_n''(x) = J_{n-2}(x) - 2J_n(x) + J_{n+2}(x)$ **20.** $J_n J_{-n}' - J_{-n}J_n' = -\dfrac{2\sin n\pi}{\pi x}$

21. $J_2 = J_0'' - \dfrac{1}{x}J_0'$ **22.** $J_2 - J_0 = 2J_0''$

23. $\int_0^x x^n J_{n-1}(x)dx = x^n J_n(x)$ **24.** $\int_0^x x^{n+1}J_n(x)dx = x^{n+1}J_{n+1}(x)$

25. $\int_0^{\pi/2}\sqrt{\pi x}J_{1/2}(2x)dx = 1$ **26.** $J_0(x) = \dfrac{1}{\pi}\int_0^\pi \cos(x\sin\phi)d\phi$

5.5 HANKEL TRANSFORMS

5.5.1 INFINITE HANKEL TRANSFORMS

Definition 1. *Let $f(x)$, $0 < x < \infty$ be a function. Then infinite Hankel Transform of $f(x)$ be given by*

$$H\{f(x)\} = \overline{f}(p) = \int_0^\infty f(x)\cdot xJ_n(px)dx$$

where $J_n(px)$ is the Bessel's function of the first kind of order n and $\overline{f}(p)$ is the Hankel transform of order n of the function $f(x)$.

NOTATION

If $\overline{f}(p)$ is the Hankel transform of order n of the function $f(x)$, then $\overline{f}(p)$ is denoted by $\overline{f}_n(p)$. Therefore,

$$\int_0^\infty xf(x)\cdot J_n(px)dx = \overline{f}_n(p) = H[f(x), n]$$

Definition 2. *Let $\overline{f}(p)$ be the infinite Hankel transform of order n of $f(x)$, i.e.,*

$$H\{f(x)\} = \overline{f}(p) = \int_0^\infty f(x)xJ_n(px)dx$$

Then $f(x)$ is known as the inverse transform of the function $\overline{f}(p)$ and is written as

$$f(x) = H^{-1}\{\overline{f}(p)\}$$

i.e., $f(x) = H^{-1}\{\overline{f}(p)\} = \int_0^\infty \overline{f}(p)\cdot pJ_n(px)dp$

THEOREM 1.   **(Linearty property) If $f(x)$ and $g(x)$ are two functions and a, b two constants, then**

$$H\{af(x) + bg(x)\} = aH\{f(x)\} + bH\{g(x)\} \qquad \text{(Kanpur–2004)}$$

PROOF.   Consider,

$$H\{af(x) + bg(x)\} = \int_0^\infty x[af(x) + bg(x)]J_n(px)dx$$

$$= a\int_0^\infty xf(x)\cdot J_n(px)dx + b\int_0^\infty xg(x)\cdot J_n(px)dx$$

$$= aH\{f(x)\} + bH\{g(x)\}$$

THEOREM 2.   **(Change of scale property) If** $H\{f(x)\} = \bar{f}(p)$, **then** $H\{f(ax)\} = \dfrac{1}{a^2}\bar{f}\left(\dfrac{p}{a}\right)$,

a being a constant.

(Kanpur–1998)

PROOF.   Let us suppose $H\{f(x)\} = \bar{f}(p)$, then

$$H\{f(ax)\} = \int_0^\infty xf(ax) \cdot J_n(px)dx$$

$$= \int_0^\infty \frac{y}{a} \cdot f(y) \cdot J_n\left(\frac{py}{a}\right)\frac{dy}{a}, \text{ where } ax = y$$

$$= \frac{1}{a^2}\int_0^\infty yf(y) \cdot J_n\left(\frac{p}{a} \cdot y\right)dy$$

$$= \frac{1}{a^2}\bar{f}\left(\frac{p}{a}\right)$$

THEOREM 3.   **(Hankel Transform of the derivatives of a functions)**

Let $\bar{f}_n(p)$ be the Hankel transform of order n of the function f(x) and $\bar{f}'_n(p)$ is the transform of f'(x). Then

$$\bar{f}'_n(p) = -\frac{p}{2n}\Big[(n+1)\bar{f}_{n-1}(p) - (n-1)\bar{f}_{n+1}(p)\Big]$$

PROOF.   Consider $\bar{f}_n(p) = \int_0^\infty xf(x)J_n(px)dx$

$$\bar{f}'_n(p) = \int_0^\infty x\frac{df}{dx} \cdot J_n(px)dx = \int_0^\infty \frac{df}{dx}[xJ_n(px)]dx$$

On integrating by parts and assume that $xf(x) \to 0$ as $x \to 0$ and $xf(x) \to 0$ as $x \to \infty$, we get

$$\bar{f}'_n(p) = \Big[f(x) \cdot xJ_n(px)\Big]_0^\infty - \int_0^\infty f(x) \cdot \frac{d}{dx}[xJ_n(px)]dx$$

$$= 0 - \int_0^\infty f(x) \cdot [J_n(px) + pxJ'_n(px)]dx \hspace{2cm} \dots(1)$$

We know that

$$xJ'_n(x) = -nJ_n(x) + xJ_{n-1}(x)$$

Now, replacing x by px, then we get

$$pxJ'_n(px) = -nJ_n(px) + pxJ_{n-1}(px)$$

$$\Rightarrow \quad pxJ'_n(px) + J_n(px) = (1-n)J_n(px) + pxJ_{n-1}(px)$$

Using in eq^n (1), we get

$$\bar{f}'_n(p) = -\int_0^\infty f(x)[(1-n)J_n(px) + pxJ_{n-1}(px)]dx$$

$$= -(1-n)\int_0^\infty f(x) \cdot J_n(px)dx - p\int_0^\infty xf(x) \cdot J_{n-1}(px)dx$$

$$\bar{f}'_n(p) = -(1-n)\int_0^\infty f(x) \cdot J_n(px)dx - p\bar{f}_{n-1}(p) \hspace{1.5cm} \dots(2)$$

Now, we also know that

$$2nJ_n(x) = x[J_{n-1}(x) + J_{n+1}(x)]$$

Replacing x by px, then we get

$$2nJ_n(px) = px[J_{n+1}(px) + J_{n-1}(px)]$$

Multiplying eq^n (3) by $f(x)$ and then integrating

$$2n\int_0^\infty f(x) \cdot J_n(px)dx$$

$$= p\left[\int_0^\infty xf(x)J_{n-1}(px)dx + \int_0^\infty xf(x)J_{n+1}(px)dx\right]$$

$$= p[\bar{f}_{n-1}(p) + \bar{f}_{n+1}(p)]$$

$$\Rightarrow \int_0^\infty f(x) J_n(px) dx = \frac{p}{2n}[\bar{f}_{n-1}(p) + \bar{f}_{n+1}(p)]$$

Putting in eq^n (2), we get

$$\bar{f}_n'(p) = -\frac{(1-n)p}{2n}[\bar{f}_{n-1}(p) + \bar{f}_{n+1}(p)] - p\bar{f}_{n-1}(p)$$

$$= \frac{p}{2n}[(n-1)\bar{f}_{n-1}(p) + (n-1)\bar{f}_{n+1}(p) - 2n\bar{f}_{n-1}(p)]$$

$$= \frac{p}{2n}[-(n+1)\bar{f}_{n-1}(p) + (n-1)\bar{f}_{n+1}(p)]$$

$$\Rightarrow \qquad \bar{f}_n'(p) = -\frac{p}{2n}[(n+1)\bar{f}_{n-1}(p) - (n-1)\bar{f}_{n+1}(p)] \qquad \text{...(3)}$$

DEDUCTIONS

1. When putting $n = 1$ in eq^n (3), then $H\{f'(x), n = 1\} = -pH\{f(x), n = 0\}$

2. Putting $n = 2$ in eq^n (3), then we get $\bar{f}_2'(p) = -\frac{p}{4}[3\bar{f}_1(p) - \bar{f}_3(p)]$

3. Putting $n = 3$ in eq^n (3), then $\bar{f}_3'(p) = -\frac{p}{6}[4\bar{f}_2(p) - 2\bar{f}_4(p)]$

4. $\bar{f}_n''(p) = -\frac{p^2}{4}\left[\left(\frac{n+1}{n-1}\right)\bar{f}_{n-2}(p) - 2\left(\frac{n^2-3}{n^2-1}\right)\bar{f}_n(p) + \left(\frac{n-1}{n+1}\right)\bar{f}_{n+2}(p)\right]$

THEOREM 4. **(Parseval's Theorem)** *If $\bar{f}(p)$ and $\bar{g}(p)$ be the Hankel transforms of the function $f(x)$ and $g(x)$ respectively, then*

$$\int_0^\infty xf(x) \cdot g(x) dx = \int_0^\infty p\bar{f}(p)\bar{g}(p) dp$$

PROOF. Consider

$$\bar{f}(p) = \int_0^\infty f(x) \cdot xJ_n(px) dx$$

$$\bar{g}(p) = \int_0^\infty g(x) \cdot xJ_n(px) dx$$

$$\int_0^\infty p\bar{f}(p)\bar{g}(p) dp = \int_0^\infty p\bar{f}(p)[\int_0^\infty g(x) \cdot xJ_n(px) dx] dp$$

$$= \int_0^\infty p\bar{f}(p) dp \cdot \int_0^\infty g(x) \cdot xJ_n(px) dx$$

Now, changing the order of integration, then we get

$$= \int_0^\infty xg(x) dx \int_0^\infty p\bar{f}(p)J_n(px) dp$$

$$= \int_0^\infty xg(x) dx \cdot f(x) \qquad [\because f(x) = \int_0^\infty \bar{f}(p) \cdot pJ_n(px) dx]$$

$$= \int_0^\infty xf(x) \cdot g(x) dx$$

5.6 THE HANKEL TRANSFORM OF $\dfrac{d^2f}{dx^2} + \dfrac{1}{x} \times \dfrac{df}{dx} - \dfrac{n^2}{x^2}f$

We have $\qquad H\left(\dfrac{d^2f}{dx^2}\right) = \int_0^\infty \dfrac{d^2f}{dx^2} \cdot xJ_n(px) dx$

$$= \left[\frac{df}{dx} \cdot xJ_n(px)\right]_0^\infty - \int_0^\infty \frac{df}{dx}[J_n(px) + (px)J_n'(px)] dx$$

Let $xf'(x) \to 0$ as $x \to 0$ or $x \to \infty$, we get

$$H\left(\frac{d^2f}{dx^2}\right) + \int_0^\infty \frac{f'(x)}{x} xJ_n(px) dx = -p\int_0^\infty xf'(x)J_n'(px) dx$$

$\Rightarrow \qquad H\left(\dfrac{d^2 f}{dx^2} + \dfrac{f'}{x}\right) = -p\left[xf(x)J_n'(px)\right]_0^\infty + p\int_0^\infty f(x)\dfrac{d}{dx}[xJ_n'(px)]dx$

Now, taking $xf(x) \to 0$ as $x \to 0$ or $x \to \infty$, then we get

$$H\left(\dfrac{d^2 f}{dx^2} + \dfrac{f'}{x}\right) = p\int_0^\infty f\dfrac{d}{dx}[xJ_n'(px)]dx$$

Here, Bessel's function $J_n(x)$ satisfies the differential equation, then

$$\dfrac{d}{dx}\left\{x\dfrac{dy}{dx}\right\} + \left(1 - \dfrac{n^2}{x^2}\right)xy = 0$$

$\Rightarrow \qquad \dfrac{d}{dx}\{xJ_n'(x)\} = -\left(1 - \dfrac{n^2}{x^2}\right)xJ_n(x)$

Now, replacing x by px, then we get

$$\dfrac{d}{pdx}\{pxJ_n'(px)\} = \left(\dfrac{n^2}{p^2 x^2} - 1\right)pxJ_n(px)$$

$\Rightarrow \qquad \dfrac{d}{dx}\{xJ_n'(px)\} = \dfrac{n^2}{px}J_n(px) - pxJ_n(px)$

$\Rightarrow \qquad H\left(\dfrac{d^2 f}{dx^2} + \dfrac{f'}{x} - \dfrac{n^2}{x^2}f\right) = -p^2 \overline{f}_n(p)$

$\Rightarrow \qquad \int_0^\infty \left[\dfrac{d^2 f}{dx^2} + \dfrac{1}{x}\dfrac{df}{dx} - \dfrac{n^2}{x^2}f\right]xJ_n(px)dx = -p^2 \overline{f}_n(p)$

Deductions

    1. If $n = 0$, then $\int_0^\infty \left(\dfrac{d^2 f}{dx^2} + \dfrac{1}{x}\dfrac{df}{dx}\right)xJ_0(px)dx = -p^2 \overline{f}_0(p)$

    where $\overline{f}_0(p)$ is the Hankel transform of $f(x)$ of order zero.

$$H\{f(x)\}, n = 0\} = \overline{f}_0(p)$$

    2. If $n = 1$, then

$$\int_0^\infty \left(\dfrac{d^2 f}{dx^2} + \dfrac{1}{x}\dfrac{df}{dx} - \dfrac{1}{x^2}f\right)xJ_1(px)dx = -p^2 \overline{f}_1(p)$$

✍ RECAPITULATIONS

S. No.	Operation	$f(x)$	$H\{f(x)\}$
1.	Linearity property	$\{af(x) + bg(x)\}$	$aH[f(x)] + bH[g(x)]$
2.	Change of scale property	$f(ax)$	$\dfrac{1}{a^2}\overline{f}\left(\dfrac{p}{a}\right)$
3	Hankel Transform of derivative	$f'(x)$	$-\dfrac{p}{2n}[(n+1)\overline{f}_{n-1}(p) - (n-1)\overline{f}_{n+1}(p)]$
4.	Parseval's theorem	$\int_0^\infty xf(x)g(x)dx$	$\int_0^\infty p\overline{f}(p)\overline{g}(p)dp$
5.	—	$\dfrac{d^2 f}{dx^2} + \dfrac{f'}{x} - \dfrac{n^2}{x^2}f$	$-p^2 \overline{f}_n(p)$

Solved Examples

EXAMPLE 1.   *Evaluate* $H\left\{\dfrac{e^{-ax}}{x}, n = 1\right\}$

SOLUTION.   Here, we have

$$H\left\{\frac{e^{-ax}}{x}, n = 1\right\} = \int_0^\infty \frac{e^{-ax}}{x} \cdot xJ_1(px)dx$$

$$= \int_0^\infty e^{-ax} J_1(px)dx = \frac{1}{p} - \frac{a}{p(a^2 + p^2)^{1/2}}.$$

EXAMPLE 2.   *Find inverse Hankel transform of* $p^{-2}e^{-ap}$ *of order one.*   (Kanpur–1996, 99)

SOLUTION.   Here, we have

$$H^{-1}\{p^{-2}e^{-ap}, n = 1\} = \int_0^\infty p^{-2} \cdot e^{-ap} \cdot pJ_1(px)dp = \int_0^\infty \frac{e^{-ap}}{p} J_1(px)dp$$

$$= \frac{(a^2 + x^2)^{1/2} - a}{x}.$$

EXAMPLE 3.   *Find the Hankel transform of* e^{-ax} *taking* $xJ_1(px)$ *as the kernal of the transform.*

(Agra–1987; Kanpur–1997)

SOLUTION.   Here, we have

$$H\{e^{-ax}, n = 0\} = \int_0^\infty e^{-ax} \cdot xJ_0(px)dx$$

$$= a(a^2 + p^2)^{-3/2}.$$

EXAMPLE 4.   *Find the Hankel transform of*

$$f(x) = \begin{cases} 1, & 0 < x < a, \quad n = 0 \\ 0, & x > 0, \quad\quad n = 0 \end{cases}$$ (Kanpur–1997, 2001)

SOLUTION.   Consider $H\{f(x), n = 0\} = \int_0^\infty f(x) \cdot xJ_0(px)dx$

$$= \int_0^a f(x) \cdot xJ_0(px)dx + \int_a^\infty f(x) \cdot xJ_0(px)dx$$

$$= \int_0^a 1 \cdot xJ_0(px)dx + \int_a^\infty 0 \cdot xJ_0(px)dx$$

$$= \int_0^a xJ_0(px)dx + 0 \qquad\qquad ...(1)$$

From Recurrence formula for Bessel's function is given by

$$\frac{d}{dx}\{x^n J_n(x)\} = x^n J_{n-1}(x)$$

Now, replacing n and x by 1 and px respectively, then we get

$$\frac{d}{pdx}\{pxJ_1(px)\} = pxJ_0(px)$$

$$\Rightarrow \qquad \frac{d}{dx}\{xJ_1(px)\} = pxJ_0(px)$$

On integrating from $x = 0$ to $x = a$,

$$[xJ_1(px)]_0^a = p\int_0^a xJ_0(px)dx$$

$$\int_0^a xJ_0(px)dx = \frac{a}{p}J_1(pa)$$

Now, eq^n (1) $\Rightarrow$

$$H\{f(x), n = 0\} = \frac{a}{p}J_1(pa).$$

EXAMPLE 5. *Find the Hankel transform of*

$$f(x) = \begin{cases} a^2 - x^2, & 0 < x < a, \quad n = 0 \\ 0, & x > a, \quad n = 0 \end{cases}$$

(Kanpur–1995, 98)

SOLUTION. Consider

$$H\{f(x), n = 0\} = \int_0^\infty f(x) \cdot xJ_0(px)dx \qquad \dots(1)$$

$$= \int_0^a f(x) \cdot xJ_0(px)dx + \int_a^\infty f(x) \cdot xJ_0(px)dx$$

$$= \int_0^a (a^2 - x^2)xJ_0(px)dx + \int_a^\infty 0 \cdot xJ_0(px)dx$$

$$= \int_0^a (a^2 x - x^3)J_0(px)dx + 0$$

$$= a^2 \int_0^a xJ_0(px)dx - \int_0^a x^3 J_0(px)dx$$

$$= a^2 I_1 - I_2 \qquad \dots(2)$$

where

$$I_1 = \int_0^a xJ_0(px)dx$$

$$I_2 = \int_0^a x^3 J_0(px)dx$$

From Recurrence formula for Bessel's function, we have

$$\frac{d}{dx}\{x^n J_n(x)\} = x^n J_{n-1}(x)$$

Replacing x by px, we get

$$\frac{1}{p}\frac{d}{dx}\{(px)^n J_n(px)\} = (px)^n J_{n-1}(px)$$

$$\Rightarrow \qquad \frac{1}{p}\frac{d}{dx}\{x^n J_n(px)\} = x^n J_{n-1}(px) \qquad \dots(3)$$

When $n = 1$, then

$$\frac{1}{p}\frac{d}{dx}\{xJ_1(px)\} = xJ_0(px) \qquad \dots(4)$$

On integrating, we get

$$\int_0^a xJ_0(px)dx = \frac{1}{p}[xJ_1(px)]_0^a = \frac{a}{p}J_1(pa)$$

$$I_1 = \int_0^a xJ_0(px)dx = \frac{a}{p}J_1(pa) \qquad \dots(5)$$

From (4), we get

$$I_2 = \int_0^a x^3 J_0(px)dx = \int_0^a \frac{x^2}{p} \cdot \frac{d}{dx}\{xJ_1(px)\}dx$$

$$= \frac{1}{p}\left[\left\{x^2 \cdot xJ_1(px)\right\}_0^a - \int_0^a 2x \cdot xJ_1(px)dx\right]$$

$$= \frac{1}{p}\left[a^3 J_1(pa) - 2\int_0^a x^2 J_1(px)dx\right]$$

Now, putting $n = 2$ in eq^n (3), we get

$$I_2 = \frac{1}{p}\left[a^3 J_1(pa) - 2\int_0^a \frac{1}{p}\frac{d}{dx}\{x^2 J_2(px)\}dx\right]$$

$$= \frac{a^3}{p}J_1(pa) - \frac{2}{p^2}\left\{x^2 \cdot J_2(px)\right\}_0^a$$

$$= \frac{a^3}{p}J_1(pa) - \frac{2a^2}{p^2}J_2(pa) \qquad \dots(6)$$

Using eq^n (5) & (6) in eq^n (2), we get

$$H\{f(x), n = 0\} = a^2 \cdot \frac{a}{p} J_1(pa) - \left[\frac{a^3}{p} J_1(pa) - \frac{2a^2}{p^2} J_2(pa)\right] = \frac{2a^2}{p^2} J_2(pa).$$

EXAMPLE 6. *Show that the Hankel transform of order zero of*

$$f(x) = \begin{cases} a^2 - x^2, & 0 < x < a \\ 0, & x > a \end{cases}$$

is $\dfrac{4a}{p^3} J_1(pa) - \dfrac{2a^2}{p^2} J_0(pa)$

SOLUTION. From previous example, we have

$$H\{f(x), n = 0\} = \frac{2a^2}{p^2} J_2(pa) \qquad \ldots(1)$$

Also, we know that

$$2n J_n(x) = x[J_{n-1}(x) + J_{n+1}(x)]$$

Substituting $n = 1$, we get

$$2J_1(x) = x[J_0(x) + J_2(x)]$$

Now, replacing x by pa, we get

$$2J_1(pa) = pa[J_0(pa) + J_2(pa)]$$

$$\Rightarrow \quad \frac{2}{pa} J_1(pa) - J_0(pa) = J_2(pa)$$

Now, multiplying both the sides by $\dfrac{2a^2}{p^2}$, we get

$$\frac{4a}{p^3} J_1(pa) - \frac{2a^2}{p^2} J_0(pa) = \frac{2a^2}{p^2} J_2(pa)$$

From eq^n (1), we get

$$H\{f(x), n = 0\} = \frac{4a}{p^3} J_1(pa) - \frac{2a^2}{p^2} J_0(pa).$$

EXAMPLE 7. *Find the Hankel transform of order zero of the function* $\dfrac{d^2 f}{dx^2} + \dfrac{1}{x} \cdot \dfrac{df}{dx}$, *where*

$$f(x) = \frac{e^{-ax}}{x}.$$

SOLUTION. We know that

$$\int_0^\infty \left(\frac{d^2 f}{dx^2} + \frac{1}{x} \cdot \frac{df}{dx} - \frac{n^2 f}{x^2}\right) x J_n(px) dx = -p^2 \bar{f}_n(p)$$

If $n = 0$, then

$$\int_0^\infty \left(\frac{d^2 f}{dx^2} + \frac{1}{x} \cdot \frac{df}{dx}\right) x J_0(px) dx = -p^2 \bar{f}_0(p)$$

$$\Rightarrow \qquad H\left\{\frac{d^2 f}{dx^2} + \frac{1}{x} \cdot \frac{df}{dx}; n = 0\right\} = -p^2 \bar{f}_0(p)$$

$$= -p^2 \int_0^\infty f(x) \cdot x J_0(px) dx = -p^2 \int_0^\infty \frac{e^{-ax}}{x} \cdot x J_0(px) dx$$

$$= -p^2 \int_0^\infty e^{-ax} J_0(px) dx = -p^2 (a^2 + p^2)^{-1/2}.$$

EXAMPLE 8.   *Prove that*

$$\int_0^a r(a^2 - r^2)J_0(pr)dr = \frac{4a}{p^3}J_1(pa) - \frac{2a^2}{p^2}J_0(pa)$$

SOLUTION.   Consider,

$$I = \int_0^a r(a^2 - r^2)J_0(pr)dr$$

$$\Rightarrow \qquad I = a^2 \int_0^a rJ_0(pr)dr - \int_0^a r^3 J_0(pr)dr$$

$$I = a^2 I_1 - I_2 \qquad\qquad \text{...(1)}$$

$$I_1 = \int_0^a rJ_0(pr)dr$$

$$I_2 = \int_0^a r^3 J_0(pr)dr$$

Since,

$$I_1 = \frac{a}{p}J_1(pa)$$

$$I_2 = \frac{a^3}{p}J_1(pa) - \frac{2a^2}{p^2}J_2(pa) \text{ [Equation(6) of example-5]}$$

Putting these values in eq^{n} (1), we get

$$I = \frac{a^3}{p}J_1(pa) - \frac{a^3}{p}J_1(pa) + \frac{2a^2}{p^2}J_2(pa)$$

$$I = \frac{2a^2}{p^2}J_2(pa) = \frac{4a}{p^3}J_1(pa) - \frac{2a^2}{p^2}J_0(pa).$$

EXAMPLE 9.   *Find the Hankel transform of order zero of $1/x$ and then apply the inversion formula to get the original formula.*   (Kanpur–1999)

SOLUTION.   Consider

$$f(x) = \frac{1}{x}$$

$$H\{f(x), n = 0\} = \int_0^\infty \frac{1}{x} \cdot xJ_0(px)dx = \int_0^\infty J_0(px)dx$$

$$= \int_0^\infty e^{-0x}J_0(px)dx = (0^2 + p^2)^{-1/2} = \frac{1}{p}$$

$$\bar{f}_0(p) = \frac{1}{p}$$

Now, applying inversion formula, we get

$$f(x) = \int_0^\infty \tilde{f}_0(p) \cdot pJ_0(px)dp$$

$$= \int_0^\infty \frac{1}{p}pJ_0(px)dp = \int_0^\infty J_0(px)dp$$

$$= \int_0^\infty e^{-0 \cdot p}J_0(px)dp = (0^2 + x^2)^{-1/2} = \frac{1}{x}.$$

EXAMPLE 10.   *Find Hankel transform of order zero of the function*

$$f(x) = \begin{cases} 4 - x^2, & 0 < x < 2 \\ 0, & x > 2 \end{cases}$$
     (Kanpur–1988)

SOLUTION.   Here, we have

$$f(x) = \begin{cases} 4 - x^2, & 0 < x < 2 \\ 0, & x > 2 \end{cases}$$

Let $a = 2$, then

$$f(x) = \begin{cases} a^2 - x^2, & 0 < x < a \\ 0, & x > a \end{cases}$$

Now, $\qquad H\{f(x), n = 0\} = \dfrac{4a}{p^3} J_1(pa) - \dfrac{2a^2}{p^2} J_0(pa)$

$$= \dfrac{8}{p^3} J_1(2p) - \dfrac{8}{p^2} J_0(2p).$$

EXAMPLE 11. *Find the Hankel transform of* $(\cos ax)/x$ *taking* $xJ_0(px)$ *as the kernal.* (Kanpur–1995)

SOLUTION. Consider $\qquad H\left\{\dfrac{\cos ax}{x}\right\} = \int_0^\infty \dfrac{\cos ax}{x} \cdot xJ_0(px)dx$

$$= \int_0^\infty \cos ax \cdot J_0(px)dx$$

$$= \text{R.P. of } \int_0^\infty e^{-iax} J_0(px)dx$$

$$= \text{R.P. of } (i^2a^2 + p^2)^{-1/2} = \text{R.P. of } (p^2 - a^2)^{-1/2}$$

$$= \begin{cases} (p^2 - a^2)^{-1/2} & \text{if} \quad p > 0 \\ 0 & \text{if} \quad 0 < p < a \end{cases}.$$

5.7 APPLICATIONS OF HANKEL TRANSFORM TO BOUNDARY VALUE PROBLEMS

EXAMPLE 1. *The magnetic potential V for a circular disc of radius a and strength* ω, *magnetised parallel to its axis, satisfies Laplace's equations is equal to* $2\pi\omega$ *on the disc itself and vanishes at exterior points in the plane of the disc, show that at the point* (r, z), $z > 0$.

$$V = 2\pi\omega\int_0^\infty e^{-pz} J_0(rp)J_1(ap)dp$$

SOLUTION. We know that, the magnetic potential V satisfies the Laplace's equation

$$\dfrac{\partial^2 V}{\partial x^2} + \dfrac{1}{r} \cdot \dfrac{\partial V}{\partial r} + \dfrac{\partial^2 V}{\partial z^2} = 0 \qquad \qquad ...(1)$$

with boundary conditions

(i) $V = 2\pi\omega$, $0 \le r \le a$, $z = 0$

(ii) $V = 0$, $r > a$, $z = 0$

Now, Hankel transform of order zero of eq^n (1) is

$$\int_0^\infty \left(\dfrac{\partial^2 V}{\partial r^2} + \dfrac{1}{r}\dfrac{\partial V}{\partial r}\right) r \cdot J_0(pr)dr + \int_0^\infty \dfrac{\partial^2 V}{\partial z^2} rJ_0(pr)dr = 0$$

$$\Rightarrow \qquad -p^2\bar{V} + \dfrac{\partial^2 V}{\partial z^2} = 0 \qquad \qquad ...(2)$$

where $\qquad \bar{V} = H\{V, n = 0\} = \int_0^\infty V \cdot rJ_0(pr)dr$

Now, solution of eq^n (2) is given by

$$\bar{V} = Ae^{pz} + Be^{-pz}$$

If V is bounded then $\bar{V}$ is also bounded.

Either $A = 0$ or $\bar{V} \to \infty$ as $z \to \infty$

$$\bar{V} = Be^{-pz} \qquad \qquad ...(3)$$

Now, let $z = 0$, then we get

$$B = [\bar{V}]z = 0 \qquad \qquad ...(4)$$

and
$$\overline{V} = \int_0^\infty (V)_{z=0} r J_0(pr) dr$$

$$= \int_0^a (V)_{z=0} r J_0(pr) dr + \int_a^\infty (V)_{z=0} r J_0(pr) dr$$

$$= \int_0^a 2\pi\omega r J_0(pr) dr + \int_a^\infty 0 r J_0(pr) dr$$

$$= 2\pi\omega \int_0^a r J_0(pr) dr + 0 = 2\pi\omega \int_0^a \frac{1}{p} \cdot \frac{d}{dr} [r J_1(pr) dr]$$

$$B = \frac{2\pi\omega}{p} a J_1(pa)$$

Now, (3) $\Rightarrow$ $\overline{V} = \dfrac{2\pi\omega a}{p} J_1(pa) e^{-pz}$

Now, applying inversion formula of Hankel transform of zero order, then we get

$$V = \int_0^\infty \overline{V} \cdot p J_0(pr) dp = \int_0^\infty \frac{2\pi\omega a}{p} J_1(pa) e^{-pz} p J_0(pr) dp$$

$$V(r,z) = 2\pi\omega a \int_0^\infty e^{-pz} J_1(pa) \cdot J_0(pr) dp .$$

EXAMPLE 2.   *Find Hankel transform of*

$$f(x) = \begin{cases} x^n, & 0 < x < a \\ 0, & x > a \end{cases}$$

taking $x J_n(px)$ as kernal.

SOLUTION.   We know that

$$H\{f(x)\} = \int_0^\infty f(x) \cdot x J_n(px) dx \qquad \ldots(1)$$

$$= \int_0^a f(x) \cdot x J_n(px) dx + \int_a^\infty f(x) \cdot x J_n(px) dx$$

$$= \int_0^a x^n \cdot x J_n(px) dx + \int_a^\infty 0 \cdot x J_n(px) dx$$

$$\overline{f}_n(p) = \int_0^a x^{n+1} J_n(px) dx \qquad \ldots(2)$$

From Recurrence formula, we get

$$\frac{d}{dx}\{x^n J_n(x)\} = x^n J_{n-1}(x)$$

Now, replacing x by px, we get

$$\frac{d}{pdx}\{(px)^n J_n(px)\} = (px)^n J_{n-1}(px)$$

$$\Rightarrow \quad \frac{1}{p}\frac{d}{dx}\{x^n J_n(px)\} = x^n J_{n-1}(px)$$

Now, replacing n by $n + 1$, then we get

$$\frac{d}{pdx}\{x^{n+1} J_{n+1}(px)\} = x^{n+1} J_n(px)$$

From eqⁿ (2)

$$\overline{f}_n(p) = \int_0^a \frac{1}{p} \cdot \frac{d}{dx}\{x^{n+1} J_{n+1}(px) dx\}$$

$$= \frac{1}{p}\left[x^{n+1} J_{n+1}(px)\right]_0^a = \frac{a^{n+1}}{p} J_{n+1}(pa) .$$

EXAMPLE 3. *Find the potential V(r, z) of a field due to a flat circular disc with its centre at the origin and radius equal to one. The axis of the disc is along z-axis and it satisfies the differential equation*

$$\frac{\partial^2 V}{\partial x^2} + \frac{1}{r} \cdot \frac{\partial V}{\partial r} + \frac{\partial^2 V}{\partial z^2} = 0, 0 \le r < \infty, z \ge 0$$

and the boundary conditions :

$$V = V_0 \text{ when } z = 0, 0 \le r < 1$$

$$\frac{\partial V}{\partial z} = 0 \text{ when } z = 0, r > 1$$

SOLUTION. The Hankel transform of the given differential equation is

$$\int_0^\infty \left(\frac{\partial^2 V}{\partial r^2} + \frac{1}{r} \cdot \frac{\partial V}{\partial r} \right) r J_0(pr) dr + \int_0^\infty \frac{\partial^2 V}{\partial z^2} r J_0(pr) dr = 0$$

$$\Rightarrow \qquad -p^2 \overline{V} + \frac{d^2 \overline{V}}{dz^2} = 0, \text{ where } \overline{V} = \int_0^\infty V r J_0(pr) dr$$

The solution of the equation is

$$\overline{V} = A e^{pz} + B e^{-pz}$$

If V is finite then $\overline{V}$ is also finite as $z \to \infty$

Either $A = 0$ or $\overline{V}$ becomes infinite as $z \to \infty$

Now, $\qquad\qquad \overline{V} = B e^{-pz}$

From inversion formula, we have

$$V(r, z) = \int_0^\infty B e^{-pz} p J_0(pr) dp \qquad\qquad \text{...(1)}$$

Here, $B = B(p) \quad \Rightarrow \quad \frac{\partial V}{\partial z} = \int_0^\infty B e^{-pz} (-p^2) J_0(pr) dp \qquad\qquad \text{...(2)}$

Substituting $z = 0$ in eq^n (1) and (2), we get

$$\int_0^\infty p B J_0(pr) dp = (V)_{z=0} = V_0, 0 \le r < 1$$

and $\quad \int_0^\infty -p^2 B J_0(pr) dp = \left(\frac{\partial V}{\partial z} \right)_{z=0} = 0, r > 1$

$$\Rightarrow \qquad \left. \begin{array}{l} \int_0^\infty p B J_0(pr) dp = V_0, 0 \le r < 1 \\ \text{and } \int_0^\infty -p^2 B J_0(pr) dp = 0, r > 1 \end{array} \right\} \qquad\qquad \text{...(3)}$$

Now, comparing eq^n (3) with

$$\left. \begin{array}{l} \int_0^\infty \frac{\sin p}{p} \cdot J_0(pr) dp = \frac{\pi}{2}, 0 \le r < 1 \\ \int_0^\infty \sin p J_0(pr) dp = 0, r > 1 \end{array} \right\}$$

then, we get $\qquad\qquad B = \frac{2}{\pi} V_0 \frac{\sin p}{p^2}$

Putting this value in eq^n (1), we get

$$V(r, z) = \frac{2V_0}{\pi} \int_0^\infty e^{-pz} \frac{\sin p}{p} \cdot J_0(pr) dp .$$

EXAMPLE 4. *Heat is supplied at a constant rate Q per unit area time over a circular area of radius a in the plane z = 0 to an infinite solid of conductivity k. Show that the steady temperature at a distant r from the axis of circular area and distance z from the* *plane z = 0 is given by*

$$\frac{Qa}{2k}\int_0^\infty e^{-pz}J_0(pr)J_1(pa)\cdot p^{-1}dp$$

(Kanpur–2001)

SOLUTION.   Let $V(r, z)$ be the temperature at a point (r, z), then

$$\frac{\partial V}{\partial t} = k\left[\frac{\partial^2 V}{\partial r^2} + \frac{1}{r}\cdot\frac{\partial V}{\partial r} + \frac{\partial^2 V}{\partial z^2}\right]$$

Here, $\dfrac{\partial V}{\partial t} = 0$, because temperature is steady. Therefore, we have

$$\frac{\partial^2 V}{\partial r^2} + \frac{1}{r}\cdot\frac{\partial V}{\partial r} + \frac{\partial^2 V}{\partial z^2} = 0$$

Now, Hankel transform of order zero of the equation, is

$$\int_0^\infty\left(\frac{\partial^2 V}{\partial r^2} + \frac{1}{r}\cdot\frac{\partial V}{\partial r}\right)rJ_0(pr)dr + \int_0^\infty\frac{\partial^2 V}{\partial z^2}rJ_0(pr)dr = 0$$

It can be written as

$$H(V, n = 0) = \int_0^\infty VrJ_0(pr)dr = \overline{V}$$

$$\Rightarrow \qquad -p^2\overline{V} + \frac{d^2V}{dz^2} = 0 \quad\Rightarrow\quad (D^2 - p^2)\overline{V} = 0$$

The solution of the equation is

$$\overline{V} = Ae^{pz} + Be^{-pz} \qquad\qquad\qquad\qquad\dots(1)$$

But $\overline{V} \to 0$ as $z \to \infty \Rightarrow A = 0$

$$\overline{V} = Be^{-pz} \qquad\qquad\qquad\qquad\dots(2)$$

This gives $\dfrac{d\overline{V}}{dz} = -Bpe^{-pz} \qquad\Rightarrow\qquad \dfrac{d\overline{V}}{dz} = -Bp$ at $z = 0$ ...(3)

Here, the boundary conditions are

(i) $2\left(-k\dfrac{\partial V}{\partial z}\right) = Q, 0 \le r < a, z = 0$

(ii) $2\left(-k\dfrac{\partial V}{\partial z}\right) = 0, r > a, z = 0$

Now, condition (1) gives

$$\frac{\partial V}{\partial z} = -\frac{Q}{2k}, 0 \le r < a, z = 0$$

Since, $\dfrac{\partial V}{\partial z} = 0, r > a$ ...(4)

Consider $z = 0$

$$\frac{\partial \overline{V}}{\partial z} = H\left\{\frac{\partial V}{\partial z}\right\} = \int_0^\infty \frac{\partial V}{\partial z}rJ_0(pr)dr = \int_0^a \frac{\partial V}{\partial z}rJ_0(pr)dr + \int_a^\infty \frac{\partial V}{\partial z}rJ_0(pr)dr$$

Now, using eq^n (4), we get

$$\frac{d\overline{V}}{dz} = \int_0^a\left(-\frac{Q}{2k}\right)rJ_0(pr)dr + \int_a^\infty 0\cdot rJ_0(pr)dr$$

$$\Rightarrow \qquad \frac{d\overline{V}}{dz} = -\frac{Q}{2k}\int_0^a rJ_0(pr)dr \qquad\qquad\qquad\dots(5)$$

Therefore, $\quad \dfrac{d}{dx}[x^n J_n(x)] = x^n J_{n-1}(x)$

Putting $n = 1$ and replace x by pr, we get

$$\frac{d}{pdr}[(pr)^1 J_1(pr)] = (pr)^1 J_0(pr)$$

$\Rightarrow \qquad\qquad \dfrac{d}{dr}[rJ_1(pr)] = prJ_0(pr)$

Using in eq^n (5), we get

$$\frac{d\overline{V}}{dz} = -\frac{Q}{2k}\int_0^a \frac{1}{p}\frac{d}{dr}[rJ_1(pr)]dr$$

$$\frac{d\overline{V}}{dz} = -\frac{Q}{2pk}[rJ_1(pr)]_{r=0}^a = -\frac{Qa}{2pk}J_1(pa)$$

Equating this to (3), we get

$$-Bp = -\frac{Qa}{2kp}J_1(pa) \qquad \Rightarrow \qquad B = \frac{Qa}{2kp^2}J_1(pa)$$

Putting this value in eq^n (2), we get

$$\overline{V} = \frac{Qa}{2kp^2}J_1(pa)$$

Finally, applying inversion formula of Hankel transform, we get

$$V(r,z) = \int_0^\infty \left[\frac{Qa}{2kp^2}J_1(pa)e^{-pz}\right]pJ_0(pr)dp$$

$\Rightarrow \qquad V(r,z) = \dfrac{Qa}{2k}\int_0^\infty J_1(pa)J_0(pr)e^{-pz}dp$.

EXAMPLE 5. *The vibrations of very large membrane are governed by the equation*

$$\frac{\partial^2 U}{\partial r^2} + \frac{1}{r}\cdot\frac{\partial U}{\partial r} = \frac{1}{C^2}\cdot\frac{\partial^2 U}{\partial t^2}, r \ge 0, t \ge 0$$

with $\qquad U = f(r), \dfrac{\partial U}{\partial t} = g(r),$ *when* $t = 0$

Show that for $t > 0$

$$U(r,t) = \int_0^\infty p, \overline{f}(p)\cos(pct)f_0(pr)dp + \frac{1}{C}\int_0^\infty \overline{g}(p)\sin(pct)J_0(pr)dp$$

where $\overline{f}(p)$ and $\overline{g}(p)$ are the zero order Hankel transform of $f(r)$, $g(r)$ respectively.

SOLUTION. Taking the Hankel transform of zero order of given equation, we get

$$\int_0^\infty \left(\frac{\partial^2 U}{\partial r^2} + \frac{1}{r}\frac{\partial U}{\partial r}\right)rJ_0(pr)dr = \frac{1}{C^2}\int_0^\infty \frac{\partial^2 U}{\partial t^2}rJ_0(pr)dr$$

$$-p^2\overline{U} = \frac{1}{C^2}\frac{d^2\overline{U}}{dt^2} \qquad \text{or} \qquad [D^2 + (pC)^2]\overline{U} = 0$$

Solution of the equation is

$$\overline{U} = A\cos(pct) + B\sin(pct) \qquad\qquad\qquad \dots(1)$$

$\therefore \qquad \dfrac{d\overline{U}}{dt} = cp[-A\sin cpt + B\cos cpt]$

At $t = 0, \dfrac{d\bar{U}}{dt} = Bcp$ ...(2)

$U = f(r), t = 0 \quad \Rightarrow \quad \bar{U} = \int_0^\infty f(r)rJ_0(pr)dr = \bar{f}(p)$

$\bar{U} = \bar{f}(p)$ at $t = 0$ ...(3)

This gives $\dfrac{\partial U}{\partial t} = g(r)$ when $t = 0$

$H\left(\dfrac{\partial U}{\partial t}\right) = H\{g(r)\} \quad \Rightarrow \quad \dfrac{d\bar{U}}{dt} = \bar{g}(p)$ at $t = 0$ ...(4)

From eq^n (1) and (3), we get $\bar{f}(p) = 4$

and From eq^n (2) and (4), we get $Bcp = \bar{g}(p)$

Using these values in eq^n (1), we get

$$\bar{U} = \bar{f}(p)\cos pct + \dfrac{1}{Cp}\bar{g}(p)\sin pct$$

Finally, applying inversion formula, we get

$$U(r,t) = \int_0^\infty [\bar{f}(p)\cos(pct)]pJ_0(pr)dp + \int_0^\infty \left[\dfrac{1}{cp}\bar{g}(p)\sin pct\right]pJ_0(pr)dp .$$

EXAMPLE 6. *Applying Hankel transform to solve the differential equation* $\dfrac{\partial^2 V}{\partial r^2} + \dfrac{1}{r}\dfrac{\partial V}{\partial r} + \dfrac{\partial^2 V}{\partial z^2} = 0$ *for the region* $r \geq 0$, $z \geq 0$ *satisfying the following conditions :*

$$V \to 0 \text{ as } z \to \infty \text{ and } r \to \infty; V = f(r) \text{ on } z = 0, r \geq 0$$

SOLUTION. We know that

$$\bar{V} = Be^{-pz}$$ ...(1)

at $z = 0$, $V = f(r)$

$\Rightarrow \qquad \bar{V} = \bar{f}(p) = \int_0^\infty f(r)rJ_0(pr)dr$

$\Rightarrow \qquad \bar{V} = \bar{f}(p)e^{-pz}$

Now, using inversion formula, we get

$$V = \int_0^\infty \bar{V}pJ_0(pr)dp = \int_0^\infty \bar{f}(p)e^{-pz}pJ_0(pr)dp .$$

5.8 THE FINITE HANKEL TRANSFORM

Let $f(x)$ satisfies the Dirichlet's condition in the interval [0, a], then finite Hankel transform $\tilde{f}(p_i)$ of order n be given by

$$\bar{f}(p_i) = \int_0^a f(x)a \cdot J_n(xp_i)dx$$

where a is a positive root of the transcendental equation $J_n(ap_i) = 0$.

5.8.1 INVERSION FORMULA

Let $f(x)$ be continuous at any point of the interval [0, a] then inversion formula for $\tilde{f}(p_i)$ be given by

$$f(x) = \dfrac{2}{a^2}\sum_i \bar{f}(p_i)\dfrac{J_n(xp_i)}{[J_n'(ap_i)]^2}$$

PROOF. Let $f(x)$ be represented by

$$f(x) = \sum_i C_0 J_i(xp_i) \quad 0 \leq x \leq a$$

where $$C_i = \frac{2}{a^2 J_{n+1}^2(ap_i)} \int_0^a f(x) x J_n(xp_i) dx$$

$$= \frac{2\bar{f}}{a^2[J_{n+1}(ap_i)^2]} = \frac{2\bar{f}(p_i)}{a^2[J_n'(ap_i)]^2}$$

Now, put $x = ap_i$ in the recurrence relation given by

$$xJ_n'(x) = nJ_n(x) - xJ_{n+1}(x)$$

We get $$ap_iJ_n'(ap_i) = nJ_n(ap_i) - ap_iJ_{n+1}(ap_i)$$

$\Rightarrow$ $$ap_iJ_n'(ap_i) = -ap_iJ_{n+1}(ap_i)$$ $$(\because J_n(ap_i) = 0)$$

Hence, $$f(x) = \frac{2}{a^2}\sum_i \bar{f}(p_i)\frac{J_n(xp_i)}{[J_{n+1}'(ap_i)]^2}$$

5.8.2 Hankel Transform of $\dfrac{df}{dx}$

To prove

(1) $H_n\left(\dfrac{df}{dx}\right) = \int_0^a \dfrac{df}{dx}xJ_n(px)dx$ **(2)** $\bar{f}_n' = \dfrac{p}{2n}\Big[(n-1)\bar{f}_{n+1} - (n+1)\bar{f}_{n-1}\Big]$

(3) $H_n\left(\dfrac{df}{dx}\right) = \dfrac{p}{2n}[(n-1)H_{n+1}\{f(x)\} - (n+1)H_{n+1}\{f(x)\}]$

PROOF.   Let $H_n\left\{\dfrac{df}{dx}\right\}$ be the nth order finite Hankel transform. Then

$$H_n\left\{\frac{df}{dx}\right\} = \int_0^a \frac{df}{dx}\cdot xJ_n(px)dx \qquad \text{...(1)}$$

$$= \Big[f(x)\cdot xJ_n(px)\Big]_{x=0}^a - \int_0^a f(x)\cdot\frac{d}{dx}\{xJ_n(px)\}dx$$

On integrating by parts, we get

$$= -\int_0^a f(x)\frac{d}{dx}\{xJ_n(px)\}dx \qquad \text{...(2)}$$

Replacing x by px in Recurrence relations

$$2J_n'(x) = J_{n-1}(x) - J_{n+1}(x)$$

and $2nJ_n(x) = x[J_{n-1}(x) + J_{n+1}(x)]$

then we get

$$2J_n'(px) = J_{n-1}(px) - J_{n+1}(px)$$

and $2nJ_n(px) = px[J_{n-1}(px) + J_{n+1}(px)]$

$$\frac{d}{dx}\{xJ_n(px)\} = J_n(px) + pxJ_n'(px)$$

$$= \frac{px}{2n}[J_{n-1}(px) + J_{n+1}(px)] + \frac{px}{2}[J_{n-1}(px) - J_{n+1}(px)]$$

$$= \frac{px}{2n}[J_{n-1}(px) + (1-n)J_{n+1}(px)]$$

Putting this value in eq^n (2), we get

$$H_n\left\{\frac{df}{dx}\right\} = -\int_0^a f(x)\cdot\frac{px}{2n}[J_{n-1}(px)\cdot(1+n)+(1-n)J_{n+1}(px)]dx$$

$$= -\frac{p}{2n}\int_0^a [f(x)\cdot xJ_{n-1}(px)\cdot(1+n)-(n-1)f(x)\cdot xJ_{n+1}(px)]dx$$

$$= -\frac{p}{2n}[(1+n)H_{n-1}\{f\}-(n-1)H_{n+1}\{f\}]$$

$$= \frac{p}{2n}[(n-1)H_{n+1}\{f(x)\}-(1+n)H_{n-1}\{f\}] \quad\quad ...(3)$$

DEDUCTION.   Putting $n = 1$ in eq^n (3), we get

$$H_1\left\{\frac{df}{dx}\right\} = \frac{p}{2}[0-2H_0\{f(x)\}] = -pH_0\{f(x)\}$$

5.8.3 HANKEL TRANSFORM OF $\dfrac{d^2f}{dx^2}+\dfrac{1}{x}\dfrac{df}{dx}$

To prove $H_n\left\{\dfrac{d^2f}{dx^2}+\dfrac{1}{x}\dfrac{df}{dx}\right\} = \dfrac{p}{2}\left[-H_{n-1}\left\{\dfrac{df}{dx}\right\}+H_{n+1}\left\{\dfrac{df}{dx}\right\}\right]$

PROOF.   We have already prove that

$$H_n\left\{\frac{df}{dx}\right\} = \frac{p}{2n}[-(n+1)H_{n-1}(f)+(n-1)H_{n-1}(f)] \quad\quad ...(1)$$

Replacing f by $\dfrac{df}{dx}$ in eq^n (1), we get

$$H_n\left\{\frac{d^2f}{dx^2}\right\} = \frac{p}{2n}\left[-(n+1)H_{n-1}\left(\frac{df}{dx}\right)+(n-1)H_{n-1}\left(\frac{df}{dx}\right)\right] \quad\quad ...(2)$$

and $H_n\left\{\dfrac{I}{x}\dfrac{df}{dx}\right\} = \int_0^a \dfrac{1}{x}\dfrac{df}{dx}\cdot xJ_n(px)dx = \int_0^a \dfrac{df}{dx}J_n(px)dx$ ...(3)

Now, replacing x by px in the recurrence relation

$$2nJ_n(x) = x[J_{n-1}(x)+J_{n+1}(x)]$$

We get $2nJ_n(px) = px[J_{n-1}(px)+J_{n+1}(px)]$

$\Rightarrow$ $J_n(px) = \dfrac{px}{2n}[J_{n-1}(px)+J_{n+1}(px)]$ ...(4)

Multiplying eq^n (4) by $\dfrac{df}{dx}$ and then integrating, we get

$$\int_0^a \frac{df}{dx}\cdot J_n(px)dx = \frac{p}{2n}\int_0^a\left[\frac{df}{dx}\cdot xJ_{n-1}(px)+\frac{df}{dx}xJ_{n+1}(px)\right]dx$$

$$= \frac{p}{2n}\left[H_{n-1}\left\{\frac{df}{dx}\right\}+H_{n+1}\left\{\frac{df}{dx}\right\}\right]$$

Substituting this value in eq^n (3), we get

$$H_n\left\{\frac{1}{x}\cdot\frac{df}{dx}\right\} = \frac{p}{2n}\left[H_{n-1}\left\{\frac{df}{dx}\right\}+H_{n+1}\left\{\frac{df}{dx}\right\}\right] \quad\quad ...(5)$$

On, adding eq^n (2) and (5), we get

$$H_n\left\{\frac{d^2f}{dx^2}+\frac{1}{x}\cdot\frac{df}{dx}\right\} = \frac{p}{2n}\left[(-n-1+1)H_{n-1}\left\{\frac{df}{dx}\right\}+(n-1+1)H_{n+1}\left\{\frac{df}{dx}\right\}\right]$$

$$= \frac{p}{2n}\left[-nH_{n-1}\left\{\frac{df}{dx}\right\} + nH_{n+1}\left\{\frac{df}{dx}\right\}\right]$$

$$= \frac{p}{2}\left[-H_{n-1}\left\{\frac{df}{dx}\right\} + H_{n+1}\left\{\frac{df}{dx}\right\}\right]$$

5.8.4 HANKEL TRANSFORM OF $\dfrac{d^2f}{dx^2} + \dfrac{1}{x}\dfrac{df}{dx} - \dfrac{n^2}{x^2}f(x)$

To prove $H_n\left\{\dfrac{d^2f}{dx^2} + \dfrac{1}{x}\dfrac{df}{dx} - \dfrac{n^2}{x^2}f\right\} = -paf(a)J_n'(pa) - p^2H_n\{f\}$

PROOF. Consider

$$H_n\left\{\frac{d^2f}{dx^2} + \frac{1}{x}\frac{df}{dx} - \frac{n^2}{x^2}f\right\} = \int_0^a\left\{\frac{d^2f}{dx^2} + \frac{1}{x}\frac{df}{dx} - \frac{n^2}{x^2}f\right\}x\cdot J_n(px)dx$$

$$= \int_0^a \frac{d^2f}{dx^2}\cdot xJ_n(px)dx + \int_0^a\frac{df}{dx}\cdot J_n(px)dx - n^2\int_0^a\frac{1}{x}\cdot J_n(px)dx$$

On integrating by parts, we get

$$= \left[\frac{df}{dx}\cdot xJ_n(px)\right]_0^a - \int_0^a\frac{df}{dx}\frac{d}{dx}\{xJ_n(px)\}dx + \int_0^a\frac{df}{dx}J_n(px)dx$$

$$\quad - n^2\int_0^a\frac{f}{x}\cdot J_n(px)dx$$

$$= 0 - \int_0^a\frac{df}{dx}[J_n(px) + pxJ_n'(px)]dx + \int_0^a\frac{df}{dx}J_n(px)dx$$

$$\quad - n^2\int_0^a\frac{f}{x}\cdot J_n(px)dx \qquad\qquad [\because J_n(pa) = 0]$$

$$= -p\int_0^a\frac{df}{dx}\cdot xJ_n'(px)dx - n^2\int_0^a\frac{f}{x}J_n(px)dx$$

On integrating first integral by parts, we get

$$= -p\left[f(x)\cdot xJ_n'(px)\right]_0^a + p\int_0^a f(x)\cdot\frac{d}{dx}[xJ_n'(px)]dx - n^2\int_0^a\frac{f}{x}J_n(px)dx$$

$$= -pf(a)\cdot aJ_n'(pa) + p\int_0^a f(x)\cdot[J_n'(px) + pxJ_n''(px)]dx - n^2\int_0^a\frac{f}{x}J_n(px)dx$$

$$\qquad\qquad\qquad\qquad\qquad\qquad\qquad\qquad\qquad\qquad\qquad ...(1)$$

We know that $J_n(x)$ is the solution of Bessel's differential equation

$$x^2\frac{d^2y}{dx^2} + x\frac{dy}{dx} + (x^2 - n^2)y = 0$$

Therefore, we can write

$$x^2\frac{d^2J_n(x)}{dx^2} + x\frac{dJ_n(x)}{dx} + (x^2 - n^2)J_n(x) = 0$$

$$\Rightarrow \qquad x^2J_n''(x) + xJ_n'(x) + (x^2 - n^2)J_n(x) = 0$$

$$\Rightarrow \quad p^2x^2J_n''(px) + pxJ_n'(px) + (x^2 - n^2)J_n(px) = 0 \qquad [\text{Replacing } x \text{ by } px]$$

$$\Rightarrow \qquad pxJ_n''(px) + J_n'(px) = \left(-px + \frac{n^2}{px}\right)$$

Now, eq^n (1) becomes

$$H_n\left\{\frac{d^2f}{dx^2}+\frac{1}{x}\frac{df}{dx}-\frac{n^2}{x^2}f\right\}=-paf(a)\cdot J_n'(pa)+p\int_0^a f(x)\cdot\left[-px+\frac{n^2}{px}\right]J_n(px)dx$$

$$-n^2\int_0^a\frac{f}{x}J_n(px)dx$$

$$=-paf(a)\cdot J_n'(pa)+p\int_0^a -pxf(x)J_n(px)dx$$

$$=-paf(a)\cdot J_n'(pa)-p^2H_n\{f\} \qquad \ldots(2)$$

DEDUCTION.   Putting $a=1$ in eq^n (2), we get

$$H_n\left\{\frac{d^2f}{dx^2}+\frac{1}{x}\frac{df}{dx}-\frac{n^2}{x^2}f\right\}=-pf(1)J_n'(p)-p^2H_n\{f\}$$

and $a=0$, then

$$H_n\left\{\frac{d^2f}{dx^2}+\frac{1}{x}\frac{df}{dx}-\frac{n^2}{x^2}f\right\}=-pf(1)J_0'(p)-p^2H_0\{f\}$$

$$=-pf(1)J_1(p)-p^2H_0\{f\}$$

where $\dfrac{d}{dx}\{x^{-n}J_n(x)\}=-x^{-n}J_{n+1}(x)$

✎ RECAPITULATIONS

1.	$f(x)=\dfrac{2}{a^2}\sum_i \bar{f}(p_i)\dfrac{J_n(xp_i)}{[J_n')ap_i)]^2}$
2.	$H_n\left\{\dfrac{df}{dx}\right\}=\dfrac{P}{2n}[(n-1)H_{n+1}\{f(x)-(n+1)H_{n-1}\{f(x)\}]$
3.	$H_n\left\{\dfrac{d^2f}{dx^2}+\dfrac{1}{x}\dfrac{df}{dx}\right\}=\dfrac{p}{2}\left[-H_{n-1}\left\{\dfrac{df}{dx}\right\}+H_{n+1}\left\{\dfrac{df}{dx}\right\}\right]$
4.	$H_n\left\{\dfrac{d^2f}{dx^2}+\dfrac{1}{x.dx}\dfrac{df}{}-\dfrac{n^2}{x^2}f\right\}=-paf(a)J_n'(pa)-p^2H_n\{f\}$
5.	$H_0\left\{\dfrac{d^2f}{dx^2}+\dfrac{1}{x}\dfrac{df}{dx}\right\}=pf(1)J_1(p)-p^2H_0(f)$
6.	$H_1\left\{\dfrac{df}{dx}\right\}=-pH_0\{f(x)\}$

▌ Solved Examples ▐

EXAMPLE 1.   *Find the finite Hankel transform of x^n, $(n>-1)$ if $xJ_n(px)$ is the Kernal of the transform.*

or, Find Hankel transform of $f(x)=\begin{cases}x^n, & 0<x<a\\ 0, & x>a\end{cases}$

where $n>0$, taking $xJ_n(px)$ as kernal.   (Kanpur–2004)

SOLUTION.   We know that

$$H_n\{x^n\}=\int_0^a x^n\cdot xJ_n(px)dx$$

$$\Rightarrow \qquad H_n\{x^n\} = \int_0^a x^{n+1} J_n(px)dx \qquad \qquad \dots(1)$$

By recurrence formula, we get

$$\frac{d}{dx}\{x^n J_n(x)\} = x^n J_{n-1}(px)$$

Now, replacing x by px, we get

$$\frac{d}{pdx}\{x^n J_n(px)\} = x^n J_{n-1}(px)$$

Again, replacing n by $n + 1$, we get

$$\frac{d}{pdx}\{x^{n+1} J_{n+1}(px)\} = x^{n+1} J_n(px)$$

Finally, from eq^n (1), we get

$$H_n\{x^n\} = \int_0^a \frac{1}{p} \cdot \frac{d}{dx}\{x^{n+1} J_{n+1}(px)\}dx$$

$$= \frac{1}{p}\left[x^{n+1} J_{n+1}(px)\right]_0^a = \frac{a^{n+1}}{p} J_{n+1}(pa) \cdot$$

EXAMPLE 2. *Show that*

$$H_0(c) = \frac{ca}{p} J_1(ap), \text{ c being a constant.}$$

or To show that $\int_0^a cxJ_0(px)dx = \frac{ca}{p}J_1(pa)$

SOLUTION. Consider $\quad H_0\{c\} = \int_0^a cxJ_0(px)dx = c\int_0^a xJ_0(px)dx \qquad \dots(1)$

From recurrence formula, we get

$$\frac{d}{dx}[x^n J_n(x)] = x^n J_{n-1}(x)$$

Putting $n = 1$, $\qquad \dfrac{d}{dx}[xJ_1(x)] = xJ_0(x)$

Now, replacing x by px

$$\frac{d}{pdx}[pxJ_1(px)] = pxJ_0(px)$$

$$\Rightarrow \qquad \frac{d}{dx}[xJ_1(px)] = pxJ_0(px)$$

Using in eq^n (1), we get

$$H_0\{c\} = \frac{c}{p}\int_0^a \frac{d}{dx}\{xJ_1(px)\}dx = \frac{c}{p}[xJ_1(px)]_0^a = \frac{ca}{p}J_1(pa) \cdot$$

EXAMPLE 3. *Find the finite Hankel transform of* $(1 - x^2)$, *taking* $xJ_0(px)$ *as the kernel.*

(Kanpur–2002)

SOLUTION. Consider,

$$H_0\{1 - x^2\} = \int_0^a (1 - x^2)xJ_0(px)dx$$

$$= \int_0^a [xJ_0(px) - x^3 J_0(px)]dx = I_1 - I_2 \qquad \dots(1)$$

where $\qquad I_1 = \int_0^a xJ_0(px)dx = \int_0^a \frac{1}{p}\frac{d}{dx}[xJ_1(px)]dx$

$$\left[\because \frac{d}{dx}[x^n J_n(x)] = x^n J_{n-1}(x)\right]$$

$$= \frac{1}{p}\left[xJ_n(px)\right]_0^a = \frac{a}{p}J_1(ap) \qquad \dots(2)$$

$$I_2 = \int_0^a x^3 J_0(px)dx = H_0\{x^2\}$$

$$= \frac{a^2}{p^2}\left[\left(ap - \frac{4}{ap}\right)J_1(pa) + 2J_0(pa)\right] \qquad \ldots(3)$$

Using eq^n (2) and (3), eq^n (1) can be written as

$$H_0\{1 - x^2\} = \frac{a}{p}J_1(ap) - \frac{a^2}{p^2}\left[\left(ap - \frac{4}{ap}\right)J_1(ap) + 2J_0(pa)\right].$$

EXAMPLE 4. *Find finite tranform of* x^2 *if* $xJ_0(px)$ *is the kernel of the transform.*   (Kanpur–2002)

SOLUTION. From recurrence formula, we get

$$2nJ_n(x) = x[J_{n-1}(x) + J_{n+1}(x)]$$

$$\frac{d}{dx}[x^n J_n(x)] = x^n J_{n-1}(x)$$

Now replacing x by px, we get

$$2nJ_n(x) = x[J_{n-1}(x) + J_{n+1}(x)]$$

$$\frac{d}{dx}[x^n J_n(x)] = x^n J_{n-1}(x)$$

$$H_0\{x^2\} = \int_0^a x^2 \cdot xJ_0(px)dx = \int_0^a x^2 \cdot \frac{d}{pdx}\{xJ_1(px)\}dx$$

On integrating by parts, we get

$$= \frac{1}{p}\left[x^2 \cdot xJ_1(px)\right]_0^a - \frac{1}{p}\int_0^a 2x \cdot xJ_1(px)dx$$

$$= \frac{a^3}{p}J_1(pa) - \frac{2}{p}\int_0^a x^2 J_1(px)dx$$

$$= \frac{a^3}{p}J_1(pa) - \frac{2}{p}\int_0^a \frac{d}{pdx}[x^2 J_2(px)]dx$$

$$= \frac{a^3}{p}J_1(pa) - \frac{2}{p^2}[x^2 J_2(px)]_0^a$$

$$\Rightarrow \qquad H_0\{x^2\} = \frac{a^3}{p}J_1(pa) - \frac{2a^2}{p^2}J_2(pa) \qquad \ldots(3)$$

Now, substituting $n = 1$ and $x = a$ in eq^n (1), we get

$$2J_1(pa) = pa[J_0(pa) + J_2(pa)]$$

$$\Rightarrow \qquad \frac{2}{pa}J_1(pa) - J_0(pa) = J_2(pa)$$

Substituting this value in eq^n (3), we get

$$H_0\{x^2\} = \frac{a^3}{p}J_1(pa) - \frac{2a^2}{p^2}\left[\frac{2}{pa}J_1(pa) - J_0(pa)\right]$$

$$= \frac{a^3}{p}J_1(pa) - \frac{4a^2}{p^3 a}J_1(pa) + \frac{2a^2}{p^2}J_0(pa)$$

$$= \frac{a^2}{p^2}\left[\left(ap - \frac{4}{ap}\right)J_1(pa) + 2J_0(pa)\right].$$

EXAMPLE 5. *Find the Hankel transform of $(a^2 - x^2)$ if $xJ_0(px)$ is the kernel of the transform.*

(Kanpur–1995)

SOLUTION. Consider

$$H_0\{a^2 - x^2\} = \int_0^a (a^2 - x^2) \cdot xJ_0(px)dx, \text{ where } p \text{ is a root of } J_0(pa) = 0$$

$$= a^2 \int_0^a xJ_0(px)dx - \int_0^a x^2 \cdot xJ_0(px)dx$$

$$= a^2 H\{1\} - H_0\{x^2\}$$

$$= a^2 \frac{a}{p} J_1(pa) - \frac{a^2}{p^2}\left[\left(ap - \frac{4}{ap}\right)J_1(pa) + 2J_0(pa)\right]$$

$$= \left(\frac{a^3}{p} - \frac{a^3}{p} + \frac{4a}{p^3}\right)J_1(pa) - \frac{2a^2}{p^3}J_0(pa)$$

$$= \frac{4a}{p^3}J_1(pa) - \frac{2a^2}{p^3}J_0(pa).$$

EXAMPLE 6. *Use the Hankel transform to solve the differential equation*

$$\frac{\partial^2 V}{\partial r^2} + \frac{1}{r}\frac{\partial V}{\partial r} = \frac{1}{k} \cdot \frac{\partial V}{\partial t}, 0 \le r < 1, t > 0$$

where $\frac{\partial V}{\partial r} + hV = 0,$ *when* $r = 1, t > 0;$ $V = 1,$ *when* $t = 0, 0 \le r < 1.$ (Kanpur–2004)

SOLUTION. Here, $\bar{V} = \int_0^a VrJ_0(pr)dr$

$$pJ_0'(p) + hJ_0(p) = 0 \qquad \qquad \ldots(1)$$

Now, taking finite Hankel transform of

$$\frac{\partial^2 V}{\partial r^2} + \frac{1}{r}\frac{\partial V}{\partial r} = \frac{1}{k}\frac{\partial V}{\partial t}$$

We get

$$\int_0^1 \frac{\partial^2 V}{\partial r^2} \cdot rJ_0(pr)dr + \int_0^1 \frac{1}{r}\frac{\partial V}{\partial r}rJ_0(pr)dr = \int_0^1 \frac{1}{k}\frac{\partial V}{\partial t}rJ_0(pr)dr \qquad \ldots(2)$$

L.H.S. of (2)

$$= \left[\frac{\partial V}{\partial r}rJ_0(pr)\right]_{r=0}^1 - \int_0^1 \frac{\partial V}{\partial r}[J_0(pr) + rpJ_0'(pr)]dr + \int_0^1 \frac{\partial V}{\partial r}J_0(pr)dr$$

$$= \left[r\frac{\partial V}{\partial r}J_0(pr)\right]_{r=0}^1 - \int_0^1 \frac{\partial V}{\partial r} \cdot rpJ_0'(pr)]dr \qquad \left[\because \frac{\partial V}{\partial r} + hV = 0 \text{ if } r = 1\right]$$

$$= -hJ_0(p)(V)_{r=1} - p\left[\{VrJ_0'(pr)\}_0^1 - \int_0^1 V[J_0'(pr) - prJ_0''(pr)]dr\right]$$

$$= pJ_0'(p)(V)_{r=1} - pJ_0'(p)(V)_{r=1} + p\int_0^1 V[J_0'(pr) + prJ_0''(pr)]dr$$

[From eq^n (1)]

$$= 0 + p\int_0^1 V[J_0'(pr) + prJ_0''(pr)]dr$$

$$= p\int_0^1 V[J_0'(pr) + prJ_0''(pr)]dr \qquad \ldots(3)$$

Let $J_n(x)$ is the solution of Bessel's equation

$$x^2 \frac{d^2y}{dx^2} + x\frac{dy}{dx} + (x^2 - n^2)y = 0$$

Hence, $J_0(x)$ is the solution of

$$x^2 \frac{d^2y}{dx^2} + x\frac{dy}{dx} + x^2 y = 0 \qquad \Rightarrow \qquad x\frac{d^2y}{dx^2} + \frac{dy}{dx} + xy = 0$$

This gives

$$\frac{xd^2 J_0(x)}{dx^2} + \frac{d}{dx}J_0(x) + xJ_0(x) = 0 \qquad \Rightarrow \qquad xJ_0''(x) + J_0'(x) + xJ_0(x) = 0$$

Now, replacing x by px, then we get

$$pxJ_0''(px) + J_0'(px) + pxJ_0(px) = 0 \qquad \Rightarrow \quad pxJ_0''(pr) + J_0'(pr) + prJ_0(pr) = 0$$

$$\Rightarrow \qquad prJ_0''(pr) + J_0'(pr) = -prJ_0(pr)$$

Then, eq^n (3) becomes

$$\text{L.H.S. of eq}^n \text{ (2)} = p\int_0^1 V[-prJ_0(pr)]dr$$

$$= -p^2\int_0^1 VrJ_0(pr)]dr = -p^2\bar{V}$$

Now, eq^n (2) becomes

$$-p^2\bar{V} = \int_0^1 \frac{1}{k}\frac{\partial V}{\partial t} \cdot r \cdot J_0(pr)dr = \frac{1}{k}\frac{d\bar{V}}{dt}$$

$$\frac{d\bar{V}}{dt} = -kp^2\bar{V}$$

Solution of this equation is given by

$$\bar{V} = Ae^{-kp^2 t} \qquad \qquad \dots(4)$$

$$\Rightarrow \qquad \bar{V}(p,t) = Ae^{-kp^2 t}$$

$$\Rightarrow \qquad (\bar{V})_{t=0} = A$$

$$= \int_0^1 (V)_{t=0} rJ_0(pr)dr = A$$

$$\Rightarrow \int_0^1 rJ_0(pr) = A \qquad \qquad \Rightarrow \qquad A = \frac{1}{p}J_1(p)$$

Eq^n (4), becomes

$$\bar{V} = \frac{1}{p}J_1(p)e^{-kp^2 t}$$

Now, applying inversion formula, we get

$$V = 2\sum_p e^{-kp^2 t} \cdot \frac{p^2}{h^2} \cdot \frac{J_1(p)}{p} \cdot \frac{J_0(pr)}{J_0^2(p)} \qquad \left[\because J_0'(p) = -\frac{h}{p}J_0(p)\right]$$

$$\Rightarrow \qquad V = 2\sum_p e^{-kp^2 t} \cdot \frac{pJ_1(p)}{h^2} \cdot \frac{J_0(pr)}{J_0^2(p)}.$$

EXERCISE 5.1

1. Find the Hankel transform of $\frac{df}{dx}$, when $f = \frac{e^{-ax}}{x}$ and $n = 1$. (Kanpur–1984)

2. Find the Hankel transform of $x^{-2}e^{-x}$ of order one.

3. Evaluate $H^{-1}\left\{\frac{e^{-ap}}{p}, n = 0\right\}$. (Kanpur–1994, 96)

4. Obtain the inverse Hankel transform of e^{-ap} of order zero. (Kanpur–1995, 98)

5. Find $H\left\{\frac{e^{-7x}}{x}, n = 0\right\}$.

6. Find Hankel transform of $\frac{\sin ax}{x}$ taking $xJ_0(px)$ as the kernel.

7. Show that

$$\int_0^1 \frac{J_n(ax)}{J_n(a)} \cdot x J_n(px) dx = \frac{p}{a^2 - p^2} J_n'(p)$$

Show also that the Hankel transform of $\frac{J_n(ax)}{J_n(a)}$ of order n is $\frac{p}{a^2 - p^2} J_n'(p)$, where $J_n(p) = 0$.

8. Find finite Hankel transform of $x_{n-1}(n > 1)$ if $x J_{n-1}(p_i x)$ is the kernel of the transform.

$\mathcal{A}NSWERS$

1. $-p(a^2 + p^2)^{-1/2}$ **2.** $\dfrac{(1 + p^2)^{1/2} - 1}{p}$ **3.** $(a^2 + x^2)^{-1/2}$ **4.** $a(a^2 + x^2)^{-3/2}$

5. $(49 + p^2)^{-1/2}$ **6.** $\begin{cases} 0, & \text{if} \quad p > a \\ -(p^2 - a^2)^{-1/2}, & \text{if} \quad 0 < p < a \end{cases}$ **8.** $\dfrac{a^n}{p_i} \cdot J_n(ap_i)$

5.9 MELLIN TRANSFORMS

Let $f(x)$ be a function defined over the interval $]0, \infty[$. Then, Mellin's transform of $f(x)$ denoted by $M[f(x)]$ is given by

$$M[f(x)] = \overline{f(p)} = \int_0^\infty x^{p-1} f(x) dx, p > 0$$

5.10 INVERSE MELLIN TRANSFORM

The inverse Mellin's transform is given by

$$f(x) = M^{-1}[\overline{f(p)}] = \frac{1}{2\pi i} \int_{k-i\infty}^{k+i\infty} x^{-p} \overline{f(p)} dp$$

5.10.1 LINEARITY PROPERTY OF MELLIN'S TRANSFORMS

We have $\qquad M[af(x) + bg(x)] = aMf(x) + bMg(x), a, b \in \mathbf{R}$

5.10.2 SOME MORE PROPERTIES OF MELLIN'S TRANSFORMS

THEOREM 1. **(Change of scale property)** $M[f(ax)] = \dfrac{1}{a^p} \overline{f(p)}$

PROOF. Consider

$$M[f(ax)] = \int_0^\infty x^{p-1} f(ax) dx$$

Let $ax = t$, then we get

$$= \int_0^\infty \left(\frac{t}{a}\right)^{p-1} f(t) \frac{dt}{a} = \frac{1}{a^p} \int_0^\infty t^{p-1} f(t) dt$$

$$= \frac{1}{a^p} \int_0^\infty x^{p-1} f(x) dx = \frac{1}{a^p} \overline{f(p)}$$

THEOREM 2. $M[f(x^a)] = \dfrac{1}{a} \overline{f\left(\dfrac{p}{a}\right)}$

PROOF. Consider

$$M[f(x^a)] = \int_0^\infty x^{p-1} f(x^a) dx$$

Let $x^a = t, x = t^{1/a}$, then

$$= \int_0^\infty t^{p-1/a} f(t) \left\{\frac{1}{a} t^{1/a-1} dt\right\}$$

$$= \frac{1}{a} \int_0^\infty t^{\frac{p}{a}-1} f(t) dt = \frac{1}{a} \int_0^\infty x^{\frac{p}{a}-1} f(x) dx = \frac{1}{a} \overline{f\left(\frac{p}{a}\right)}$$

THEOREM 3. $\quad M[x^a f(x)] = \overline{f(p+a)}$

PROOF. Consider

$$M[x^a f(x)] = \int_0^\infty x^{p-1}[x^a f(x)]dx$$

$$= \int_0^\infty x^{(p+a)-1}f(x)dx = \int_0^\infty x^{q-1}f(x)dx , \qquad \text{where } q = p + a$$

$$= \overline{f(q)} = \overline{f(p+a)}$$

THEOREM 4. $\quad M\left[\dfrac{1}{x}f\left(\dfrac{1}{x}\right)\right] = \overline{f(1-p)}$

PROOF. We have

$$M\left[\frac{1}{x}f\left(\frac{1}{x}\right)\right] = \int_0^\infty x^{p-1}\left[\frac{1}{x}f\left(\frac{1}{x}\right)\right]dx = \int_0^\infty x^{p-2}f\left(\frac{1}{x}\right)dx$$

Let $\dfrac{1}{x} = t \implies x = \dfrac{1}{t}$, then

$$= \int_0^\infty (t^{-1})^{p-2} f(t)\left(-\frac{dt}{t^2}\right) = \int_0^\infty t^{-p} f(t)dt = \int_0^\infty x^{-p} f(x)dx$$

$$= \int_0^\infty x^{(1-p)-1} f(x)dx = \overline{f(1-p)}$$

THEOREM 5. $\quad M[(\log x)f(x)] = \dfrac{d}{dp}\overline{f(p)}$

PROOF. We have

$$M[(\log x)f(x)] = \int_0^\infty x^{p-1}[(\log x)f(x)]dx \qquad \qquad \ldots(1)$$

and $\quad M[f(x)] = \overline{f(p)} = \int_0^\infty x^{p-1} f(x)dx \qquad \qquad \ldots(2)$

Now, differentiating eq^n (2), w.r.to p, we get

$$\frac{d\overline{f(p)}}{dp} = \int_0^\infty x^{p-1}(\log x)f(x)dx = M[(\log x)f(x)] \qquad \text{[From eq}^n\text{ (1)]}$$

Solved Examples

Example 1. *Evaluate* $M[(1+x^a)^{-b}]$

SOLUTION. Consider

$$M[(1+x^a)^{-b}] = \int_0^\infty x^{p-1}(1+x^a)^{-b} dx = \int_0^\infty \frac{x^{p-1}}{(1+x^a)^b}dx$$

$$\text{[Putting } x^a = t, ax^{a-1}dx = dt]$$

$$= \int_0^\infty \frac{x^{p-1}}{ax^{a-1}}\frac{dt}{(1+t)^b} = \frac{1}{a}\int_0^\infty \frac{x^{p-a}dt}{(1+t)^b}$$

$$= \frac{1}{a}\int_0^\infty \frac{t^{\left(\frac{p-a}{a}\right)^{p-a}}}{(1+t)^b} dt = \frac{1}{a}\int_0^\infty \frac{t^{p/a-1}}{\left(b-\dfrac{p}{a}\right)+\dfrac{p}{a}} dt$$

Using $\quad B(m, n) = \int_0^\infty \dfrac{x^{m-n}dx}{(1+x)^{m+n}}$, then we have

$$M[(1+x^a)^{-b}] = \frac{1}{a}B\left[b - \frac{p}{a}, \frac{p}{a}\right] = \frac{1}{a}\frac{\Gamma\left(b - \dfrac{p}{a}\right)\cdot\Gamma\left(\dfrac{p}{a}\right)}{\Gamma\left(b - \dfrac{p}{a} + \dfrac{p}{a}\right)} = \frac{1}{a}\frac{\Gamma\left(b - \dfrac{p}{a}\right)\cdot\Gamma\left(\dfrac{p}{a}\right)}{\Gamma(b)} .$$

EXAMPLE 2.   *Evaluate $M[(1 + x)^{-1}]$.*

SOLUTION.   We have

$$M[(1+x)^{-1}] = \int_0^\infty (x)^{p-1}(1+x)^{-1}dx = \int_0^\infty \frac{x^{p-1}}{1+x}dx = \frac{\pi}{\sin(p\pi)} .$$

EXAMPLE 3.   *Evaluate $M\left[x^2\dfrac{d^2F}{dx^2} + x\dfrac{dF}{dx}\right]$.*

SOLUTION.   We have

$$M\left[x^2\frac{d^2F}{dx^2} + x\frac{dF}{dx}\right] = M\left[\frac{xd}{dx}\left(x\frac{dF}{dx}\right)\right] = M\left[\left(\frac{xd}{dx}\right)^2 F\right] = p^2 f(p) .$$

✎	**RECAPITULATIONS**		
S. No.	**Function $f(x)$**	**Mellin Transform $[Mf(x)]$**	**Property**
1.	$f(x)$	$\overline{f(p)} = \int_0^\infty x^{p-1}f(x)dx, p > 0$	
2.	$af(x) + bg(x)$	$aMf(x) + bMg(x), a, b \in \mathbf{R}$	Linearity property
3	$f(a - x)$	$\dfrac{1}{a^p}\overline{f(p)}$	Change of scale property
4.	$f(x^a)$	$\dfrac{1}{a}\overline{f}\left(\dfrac{p}{a}\right)$	Change of scale property
5.	$x^a f(x)$	$\overline{f(p+a)}$	
6.	$\dfrac{1}{x}f\left(\dfrac{1}{x}\right)$	$\overline{f(1-p)}$	
7.	$\log x\, f(x)$	$\dfrac{d}{dp}\overline{f(p)}$	

Self Assessment Test

1. Find the finite Hankel transform of

$$\frac{1}{r} \cdot \frac{\partial}{\partial r}\left(r \frac{\partial V}{\partial r}\right) - \frac{n^2 V}{r^2},$$

where $V = \begin{cases} 0, & \text{when} \quad r = 0 \\ V_1, & \text{when} \quad r = 1 \end{cases}$.

2. Use the Hankel transform to solve the differential equation

$$\frac{\partial^2 V}{\partial r^2} + \frac{1}{r}\frac{\partial V}{\partial r} = \frac{1}{k} \cdot \frac{\partial V}{\partial t}, 0 \leq r < 1, t > 0$$

Here $V = V_0$ (constant) when $r = 1, t > 0$ and $V = 0$ when $t = 0, 0 \leq r < 1$.

3. Show that $H_0(c) = \dfrac{ac}{p} J_1(ap), c$ being a constant.

4. Show that

$$\int_0^a r^2 J_0(pr)\,dr$$

$$= \frac{a^2}{p^2}\left[2J_0(pa) + \left(ap - \frac{4}{ap}\right)J_1(pa)\right]$$

5. Show that $M(e^{-x}) = \Gamma(p), p > 0$.

6. Show that $M[(1 + x^a)^{-1}] = \dfrac{\pi}{a \sin\left(\dfrac{p\pi}{a}\right)}$.

―――――*ANSWERS*―――――

1. $p^{n-m} J_m(p)$ 2. $2\sum \dfrac{V_0}{p}\dfrac{}{p}(e^{-ktp^2} - 1)\dfrac{J_0(pr)}{J_1(p)}$

Objective Evaluations

❖ Fill in the Blanks

1. The homogeneous linear differential equation of the form $x^2 \dfrac{d^2y}{dx^2} + x \dfrac{dy}{dx} + (x^2 - n^2)y = 0$ is known as _____ differential equation.

2. General solution of Bessel's differential equation is _____.

3. $\dfrac{d}{dx}[x^n J_n(x)] = $ _____.

4. $J_{1/2}(x) = $ _____.

5. $2 J_n'(x) = $ _____.

6. $M[x^a f(x)] = $ _____.

7. $M[f(x^a)] = $ _____.

❖ True or False

Write 'T' for True and 'F' for False statement.

1. $J_n(x)$ is even and odd function for even n and for odd n respectively. **(T/F)**

2. $J_0'(x) = J_1(x)$ **(T/F)**

3. $x J_n'(x) = -J_n(x) - x J_{n-1}(x)$ **(T/F)**

4. $x J_n'(x) = n J_n(x) - x J_{n+1}(x)$ where $J_n'(x) = \dfrac{dJ_n(x)}{dx}$ **(T/F)**

5. $\dfrac{d}{dx}[J_n^2 + J_{n+1}^2] = 2\left[\dfrac{n}{x} J_n^2 - \dfrac{n+1}{x} J_{n+1}^2\right]$ **(T/F)**

6. $M[f(ax)] = \dfrac{1}{a^p} \overline{f}(p)$ **(T/F)**

7. The Mellin's transform of $f(x)$ is $M[f(x)] = \overline{f}(p) = \int_0^\infty x^{p-1} f(x)\,dx, p > 0$. **(T/F)**

❖ Multiple Choice Questions

Choose the most appropriate one.

1. $M\left[\dfrac{1}{x} f\left(\dfrac{1}{x}\right)\right]$ is equal to :

   (a) $\overline{f}(p+a)$
   (b) $\overline{f}(p-a)$
   (c) $f(p+a)$
   (d) $f(p-a)$

2. The homogeneous linear differential equation $x^2 \dfrac{d^2y}{dx^2} + x \dfrac{dy}{dx} + (x^2 - n^2)y = 0$ is known as :

   (a) Hankel transform
   (b) Mellin's transform
   (c) Bessel's differential equation
   (d) none of these

3. The Hankel transform of

   $f(x) = \begin{cases} a^2 - x^2, & 0 < x < a, n = 0 \\ 0, & x > a, n = 0 \end{cases}$:

   (a) $\dfrac{2a^2}{p^2} J_2(pa)$
   (b) $\dfrac{2a^2}{p^2} J_1(pa)$

   (c) $\dfrac{4a^2}{p^2} J_2(pa)$
   (d) $\dfrac{4a^2}{p^2} J_1(pa)$

4. $J_n J_{-n}' - J_{-n} J_n'$ is equal to :

   (a) $-\dfrac{2\sin n\pi}{\pi x}$
   (b) $-\dfrac{2\cos n\pi}{\pi x}$
   (c) $\dfrac{2\sin n\pi}{\pi x}$
   (d) $\dfrac{2\cos n\pi}{\pi x}$

5. Hankel transform of e^{-ax} taking $x J_1(px)$ as the kernal of the transform is :

   (a) $a(a^2 + p^2)^{-1/2}$
   (b) $a(a^2 - p^2)^{3/2}$
   (c) $a(a^2 + p^2)^{-3/2}$
   (d) none of these

6. The value of $M\left[x^2 \dfrac{d^2F}{dx^2} + x \dfrac{dF}{dx}\right]$ is :

   (a) $-p^2 f(p)$
   (b) $-p^{-2} f(p)$
   (c) $p^{-2} f(p)$
   (d) $p^2 f(p)$

7. $M[(\log x)f(x)]$ is equal to :

   (a) $-\dfrac{d}{dp} \overline{f}(p)$
   (b) $\dfrac{d}{dp} \overline{f}(p)$
   (c) $\dfrac{d^2}{dp^2} \overline{f}(p)$
   (d) none of these

8. Mellin's transform $M[f(x)]$ is given by :

   (a) $\int_0^\infty e^{p-1} f(x)\,dx, p > 0$
   (b) $\int_{-\infty}^\infty e^{p-1} f(x)\,dx, p > 0$
   (c) $\int_0^\infty e^{p+1} f(x)\,dx, p > 0$
   (d) none of these

9. The value of $H^{-1}\left\{\dfrac{e^{-ap}}{p}, n = 0\right\}$ is :

  (a) $(a^2 - x^2)^{1/2}$ (b) $(a^2 - x^2)^{-1/2}$

  (c) $(a^2 + x^2)^{1/2}$ (d) $(a^2 + x^2)^{-1/2}$

10. $\dfrac{d}{dx}[x^{-n}J_n(x)]$ is equal to :

  (a) $-x^{-n}J_{n+1}(x)$ (b) $-x^n J_{n+1}(x)$

  (c) $-x^{-n}J_{n+2}(x)$ (d) $-x^n J_{n+2}(x)$

ANSWERS

❖ Fill in the Blanks

1. Bessel's   **2.** $y(x) = AJ_n(x) + BJ_{-n}(x)$ **3.** $x^n J_{n-1}(x)$ **4.** $\sqrt{\dfrac{2}{\pi x}}\sin x$

5. $J_{n-1}(x) - J_{n+1}(x)$ **6.** $\overline{f(p+a)}$ **7.** $\dfrac{1}{a}f\left(\dfrac{p}{a}\right)$

❖ True or False

1. T   **2.** F   **3.** F   **4.** T   **5.** T   **6..** T   **7.** T

❖ Multiple Choice Questions

1. (b)   **2.** (c)   **3.** (a)   **4.** (a)   **5.** (c)   **6.** (d)   **7.** (b)   **8.** (a)   **9.** (d)
10. (a)

Comprehensive Mellin Transform Tables

(1) GENERAL FORMULAS

S.No.	Original Function $F(x)$	Mellin Transform $f(p) = \int_0^\infty F(x) x^{p-1} dx$
1.	$F(ax), a > 0$	$a^{-p} f(p)$
2.	$x^a \cdot F(x)$	$f(p + a)$
3.	$F\left(\dfrac{1}{x}\right)$	$f(-p)$
4.	$\left(x\dfrac{d}{dx}\right)^n F(x)$	$(-1)^n p^n f(p)$
5.	$\left(\dfrac{d}{dx} x\right)^n F(x)$	$(-1)^n (p-1)^n f(p)$

(2) FOR POWER LAW FUNCTIONS

S.No.	Original Function $F(x)$	Mellin Transform $f(p) = \int_0^\infty F(x) x^{p-1} dx$
1.	$\dfrac{1}{x+a}, a > 0$	$\dfrac{\pi a^{p-1}}{\sin \pi p}, 0 < \mathrm{Re}(p) < 1$
2.	$\dfrac{1}{(x+a)(x+b)}, a, b > 0$	$\dfrac{\pi(a^{p-1} - b^{p-1})}{(b-a)\sin(\pi p)}, 0 < \mathrm{Re}(p) < 2$
3.	$\dfrac{1}{x^2 + a^2}, a > 0$	$\dfrac{\pi a^{p-2}}{2\sin\left(\dfrac{\pi p}{2}\right)}, 0 < \mathrm{Re}(p) < 2$
4.	$\dfrac{1}{x^n + a^n}, a > 0, n \in \mathbf{N}$	$\dfrac{\pi a^{p-n}}{n \sin\left(\dfrac{\pi p}{n}\right)}, 0 < \mathrm{Re}(p) < n$
5.	$\dfrac{1}{(x^2 + a^2)(x^2 + b^2)}, a, b > 0$	$\dfrac{\pi(a^{p-2} - b^{p-2})}{2(b^2 - a^2)\sin\left(\dfrac{\pi p}{2}\right)}, 0 < \mathrm{Re}(p) < 4$

(3) FOR EXPONENTIAL FUNCTIONS

S.No.	Original Function $F(x)$	Mellin Transform $f(p) = \int_0^\infty F(x) x^{p-1} dx$
1.	$e^{-ax}, a > 0$	$a^{-b} \Gamma(p), \mathrm{Re}(p) > 0$
2.	$\dfrac{e^{-ax}}{x+b}, a, b > 0$	$e^{ab} b^{p-1} \Gamma(p) \Gamma(1 - p, ab), \mathrm{Re}(p) > 0$
3.	$e^{-\alpha x^\beta}, \alpha, \beta > 0$	$\beta^{-1} \alpha^{-p/\beta} \Gamma(p/\beta), \mathrm{Re}(p) > 0$

(4) FOR LOGARITHMIC FUNCTIONS

S.No.	Original Function $F(x)$	Mellin Transform $f(p) = \int_0^\infty F(x)x^{p-1}dx$
1.	$\begin{cases} \log x; & 0 < x < a \\ 0; & a < x \end{cases}$	$\dfrac{p\log a - 1}{p^2 a^p}, \mathrm{Re}(p) > 0$
2.	$\log(1+ax), a > 0$	$\dfrac{\pi}{pa^p \sin(\pi p)}, -1 < \mathrm{Re}(p) < 0$
3.	$\dfrac{\log x}{x+a}, a > 0$	$\dfrac{\pi a^{p-1}[\log a - \pi \cot(\pi p)]}{\sin(\pi p)}, 0 < \mathrm{Re}(p) < 1$
4.	$\dfrac{\log x}{(x+a)(x+b)}, a, b > 0$	$\dfrac{\pi[a^{p-1}\log a - b^{p-1}\log b - \pi \cot(\pi p)(a^{p-1} - b^{p-1})]}{(b-a)\sin \pi p}$
5.	$\log\left\|\dfrac{1+x}{1-x}\right\|$	$\dfrac{\pi}{p}\tan\left(\dfrac{\pi p}{2}\right), -1 < \mathrm{Re}(p) < 1$
6.	$e^{-x}\log^n x, n \in \mathbf{N}$	$\dfrac{d^n}{dp^n}\Gamma(p), \mathrm{Re}(p) > 0$

(5) FOR TRIGONOMETRIC FUNCTIONS

S.No.	Original Function $F(x)$	Mellin Transform $f(p) = \int_0^\infty F(x)x^{p-1}dx$
1.	$\sin ax, a > 0$	$a^{-p}\Gamma(p)\sin\left(\dfrac{\pi p}{2}\right), -1 < \mathrm{Re}(p) < 1$
2.	$\cos ax, a > 0$	$a^{-p}\Gamma(p)\cos\left(\dfrac{\pi p}{2}\right), 0 < \mathrm{Re}(p) < 1$
3.	$e^{-ax}\sin bx, a > 0$	$\dfrac{\Gamma(p)\sin\left(p\tan^{-1}\dfrac{b}{a}\right)}{(a^2+b^2)^{p/2}}, -1 < \mathrm{Re}(p)$
4.	$e^{-ax}\cos bx, a > 0$	$\dfrac{\Gamma(p)\cos\left(p\tan^{-1}\dfrac{b}{a}\right)}{(a^2+b^2)^{p/2}}, 0 < \mathrm{Re}(p)$
5.	$\tan^{-1}x$	$-\dfrac{\pi}{2p\cos\left(\dfrac{\pi p}{2}\right)}, -1 < \mathrm{Re}(p) < 0$
6.	$\cot^{-1}x$	$\dfrac{\pi}{2p\cos\left(\dfrac{\pi p}{2}\right)}, 0 < \mathrm{Re}(p) < 1$

 # Comprehensive Inverse Mellin's Transform Tables

(1) For Power law Functions

S.No.	Direct Function $F(p)$	Inverse Transform $F(x) = \dfrac{1}{2\pi i} \int_{a-i\infty}^{a+i\infty} f(p)x^p dp$
1.	$\dfrac{1}{p}, \mathrm{Re}(p) > 0$	$\begin{cases} 1 & 0 < x < 1 \\ 0 & 1 < x \end{cases}$
2.	$\dfrac{1}{p}, \mathrm{Re}(p) < 0$	$\begin{cases} 0 & 0 < x < 1 \\ -1 & 1 < x \end{cases}$
3.	$\dfrac{1}{p+a}, \mathrm{Re}(p) > -a$	$\begin{cases} x^a & 0 < x < 1 \\ 0 & 1 < x \end{cases}$
4.	$\dfrac{1}{p+a}, \mathrm{Re}(p) < -a$	$\begin{cases} 0 & 0 < x < 1 \\ -x^a & 1 < x \end{cases}$
5.	$\dfrac{1}{(p+a)^2}, \mathrm{Re}(p) > -a$	$\begin{cases} -x^a \log x & 0 < x < 1 \\ 0 & 1 < x \end{cases}$
6.	$\dfrac{1}{(p+a)^2}, \mathrm{Re}(p) < -a$	$\begin{cases} 0 & 0 < x < 1 \\ x^a \log x & 1 < x \end{cases}$
7.	$\dfrac{1}{(p+a)(p+b)}, -a < \mathrm{Re}(p) < -b$	$\begin{cases} \dfrac{x^a}{b-a}; & 0 < x < 1 \\ \dfrac{x^b}{b-a}; & 1 < x \end{cases}$
8.	$\dfrac{1}{(p+a)(p+b)}, \mathrm{Re}(p) > -a, -b$	$\begin{cases} \dfrac{x^a - x^b}{b-a} & 0 < x < 1 \\ 0 & 1 < x \end{cases}$
9.	$\dfrac{1}{(p+a)(p+b)}, \mathrm{Re}(p) < -a, -b$	$\begin{cases} 0 & 0 < x < 1 \\ \dfrac{x^b - x^a}{b-a} & 1 < x \end{cases}$

(2) For Exponential and Logarithmic Functions

S.No.	Direct Function $F(p)$	Inverse Transform $F(x) = \dfrac{1}{2\pi i} \int_{a-i\infty}^{a+i\infty} f(p)x^p dp$
1.	$e^{ap^2}, a > 0$	$\dfrac{1}{2\sqrt{\pi a}} e^{\left(-\dfrac{\log^2 x}{4a}\right)}$
2.	$\log\left(\dfrac{p+a}{p+b}\right), \mathrm{Re}(p) > -a, -b$	$\begin{cases} \dfrac{x^a - x^b}{\log x} & 0 < x < 1 \\ 0 & 1 < x \end{cases}$

(3) FOR TRIGONOMETRIC FUNCTIONS

S.No.	Direct Function $F(p)$	Inverse Transform $F(x) = \frac{1}{2\pi i}\int_{a-i\infty}^{a+i\infty} f(p)x^p dp$
1.	$\dfrac{\pi}{\sin \pi p}, 0 < \text{Re}(p) < 1$	$\dfrac{1}{x+1}$
2.	$\dfrac{\pi^2}{\sin^2(\pi p)}, 0 < \text{Re}(p) < 1$	$\dfrac{\log x}{x-1}$
3.	$\sin\left(\dfrac{p^2}{a}\right), a > 0$	$\dfrac{1}{2}\sqrt{\dfrac{a}{\pi}}\sin\left(\dfrac{1}{4}\left|\log x\right|^2 - \dfrac{\pi}{4}\right)$
4.	$\dfrac{\pi}{\cos(\pi p)}, -\dfrac{1}{2} < \text{Re}(p) < \dfrac{1}{2}$	$\dfrac{\sqrt{x}}{x+1}$
5.	$\cos\left(\dfrac{p^2}{a}\right), a > 0$	$\dfrac{1}{2}\sqrt{\dfrac{a}{\pi}}\cos\left(\dfrac{1}{4}a\left|\log x\right|^2 - \dfrac{\pi}{4}\right)$
6.	$\tan^{-1}\left(\dfrac{a}{p+b}\right), \text{Re}(p) > -b$	$\begin{cases} \dfrac{x^b}{\left|\log x\right|}\sin(a\left|\log x\right|) & \text{if } 0 < x < 1 \\ 0 & \text{if } 1 < x \end{cases}$

VVVVVV

Chapter 6

Application of Integral Transforms to integral Equations

6.1 INTRODUCTION

Before discussing the application of Laplace and Fourier transform to integral equation, we shall give some basic introduction of integral equations.

6.1.1 INTEGRAL EQUATION

An integral equation is an equation in which the unknown function occurs under the integral sign. The name 'Integral equation' for any equation involving the unknown function $u(x)$ under the integral sign was introduced by Du Bois-Reymond in 1888. In 1782, Laplace used the integral transform

$$f(x) = \int_0^\infty e^{-xt} f(t)dt$$

to solve the linear differential equations and differential equations. In 1826, Abel solved the integral equation named after him having the form.

$$f(x) = \int_0^x (x-t)^{-\alpha} u(t)dt$$

where $f(x)$ is a continuous function satisfying $f(a) = 0$ and $0 < \alpha < 1$. Huygens solved the Abel's integral equation for $\alpha = \dfrac{1}{2}$. In 1826, Poisson obtained an integral equation of the type

$$u(x) = f(x) + \lambda \int_0^x k(x,t)u(t)dt$$

in which the unknown function $u(t)$ occurs outside as well as before the integral sign and the variable x appears as one of the limits of the integral. Dirichlet's problem which is the determination of a function ψ having prescribed values over a certain boundary surface S and satisfying Laplace's equation $\nabla^2 \psi = 0$ within the region enclosed by S, was shown by Heumann in 1870 to be equivalent to the solution of an integral equation. He solved the integral equation by an expansion in powers of a certain parameter λ. In 1896, Volterra gave the first general treatment of the solution of the class of linear integral equation bearing his name and characterized by the variable x appearing as the upper limit of the integral. In 1900, Fredholms have discussed a more general class of linear integral equation having the form

$$u(x) = f(x) + \int_a^b k(x,t)u(t)dt$$

6.1.2 SOME IMPORTANT DEFINITIONS

Definition 1. *An integral equation is an equation in which an unknown function appears under one or more integral signs.*

For example: For $a \le x \le b$, $a \le t \le b$, the equations

$$f(x) = \int_a^x k(x,t)u(t)dt \qquad \qquad \text{...(1)}$$

$$u(x) = f(x) + \lambda \int_a^x k(x,t)u(t)dt \qquad \qquad \text{...(2)}$$

where the function $u(x)$ is the unknown function, while the functions $f(x)$ and $k(x, t)$ are known functions and λ, a and b are constants, are the integral equations.

REMARKS

- If the derivative of the function are involved in the equation, then it is called an integro-differential equations.
- The function $f(x)$ and $k(x, t)$ may be complex valued functions of the real variables x and t.
- The function $u(x)$ is the unknown function, while the other functions are known.

6.1.3 LINEAR AND NON-LINEAR INTEGRAL EQUATIONS

An integral equation is called linear, if, only linear operations are performed in it upon the unknown function. On the other hand, an integral equation, which is not linear is known as a non-linear integral equations.

6.1.4 GENERAL FORM

The most general type of linear equation is of the form

$$v(x) \cdot u(x) = f(x) + \lambda \int_a k(x,t)u(t)dt \qquad \ldots(1)$$

where the upper limit may be either variable x or fixed. The functions f, v and k are known functions, while u is unknown, which is to be determined. λ is a non-zero real or complex parameter. The function $k(x, t)$ is known as kernel of the integral equation.

6.1.5 TYPES OF THE LINEAR INTEGRAL EQUATION

(i) If $v(x) \neq 0$, then (1) is known as linear integral equation of the third kind.

(ii) When $v(x) = 0$, then (1) reduces to

$$f(x) + \lambda \int_a k(x,t)u(t)dt = 0$$

which is known as linear integral equation of the first kind.

(iii) When $v(x) = 1$, then (1) reduces to

$$u(x) = f(x) + \lambda \int k(x,t)u(t)dt$$

which is known as linear integral equation of the second kind.

6.1.6 FREDHOLM INTEGRAL EQUATION

(Amritsar–2004)

A linear integral equation of the form

$$v(x) \cdot u(x) = f(x) + \lambda \int_a^b k(x,t)u(t)dt \qquad \ldots(1)$$

where a, b both are constants, $f(x)$, $v(x)$ and $k(x, t)$ are known functions while $u(x)$ is unknown functions and λ is a non-zero real or complex parameter, is called Fredholm integral equation of third kind.

6.1.7 SPECIAL CASES

(a) **Fredholm Intgral Equation of the First Kind :** Put $v(x) = 0$ in (1), then integral equation of the form   (Garhwal–2000, 04)

$$f(x) + \lambda \int_a^b k(x,t)u(t)dt = 0 \qquad \ldots(2)$$

is known as Fredholm integral equation of the first kind.

(b) **Fredholm Integral Equation of the Second Kind :** Put $v(x) = 1$ in (1), then integral equation of the form

$$u(x) = f(x) + \lambda \int_a^b k(x,t)u(t)dt \qquad \ldots(3)$$

is known as Fredholm integral equation of the second kind.   (Garhwal–2000, 04)

(c) Homogeneous Fredholm Integral Equation of the Second Kind : Put $f(x) = 0$ in (3), then a linear integral equation of the form

$$u(x) = \lambda \int_a^b k(x,t)u(t)dt \qquad \ldots(4)$$

is known as the homogeneous Fredholm integral equation of the second kind.

6.1.8 VOLTERRA INTEGRAL EQUATION

(Kanpur–1990)

A linear integral equation of the form

$$v(x) \cdot u(x) = f(x) + \lambda \int_a^x k(x,t)u(t)dt \qquad \ldots(1)$$

where a is a constant, $f(x)$, $v(x)$ and $k(x, t)$ are known functions, while $u(x)$ is unknown function, λ is a non-zero real or complex parameter is called volterra integral equation of third kind.

SPECIAL CASES

(a) Volterra Integral Equation of the First Kind : Put $v(x) = 0$ in (1), then a linear integral equation of the form

$$f(x) + \lambda \int_a^x k(x,t)u(t)dt = 0 \qquad \ldots(2)$$

is known as Volterra integral equation of the first kind. (Garhwal–2004)

(b) Volterra Integral Equation of the Second Kind : Put $v(x) = 1$ in (1), then a linear integral equation of the form

$$u(x) = f(x) + \lambda \int_a^x k(x,t)u(t)dt \qquad \ldots(3)$$

is known as Volterra integral equation of the second kind. (Garhwal–2004)

(c) Homogeneous Volterra Integral Equation of the Second Kind : A linear equation of the form

$$u(x) = \lambda \int_a^x k(x,t)u(t)dt \qquad \ldots(4)$$

is known as the homogeneous Volterra integral equation of the second kind.

6.1.9 SINGULAR INTEGRAL EQUATIONS

(Meerut–2006; Garhwal–2004)

When one or both limits of integration becomes infinite or when the kernel becomes infinite at one or more points within the range of integration, the integral equation is known as singular integral equation.

For example :

(1) $u(x) = f(x) + \lambda \int_{-\infty}^{\infty} e^{-|x-t|} u(t)dt$ $\qquad \ldots(1)$

(2) $f(x) = \int_0^x \dfrac{1}{(x-t)^\alpha} u(t)dt, 0 < \alpha < 1$ $\qquad \ldots(2)$

6.1.10 CONVOLUTION TYPE INTEGRAL EQUATION

If the kernel $k(x, t)$ of the integral equation is a function of the difference $(x - t)$, i.e.,

$$k(x, t) = k(x - t) \qquad \ldots(1)$$

where k is a certain function of one variable, then the integral equation

$$u(x) = F(x) + \lambda \int_a^x k(x - t)u(t)dt \qquad \ldots(2)$$

are called integral equation of the convolution type.

REMARKS

- $k(x - t)$ is called the difference kernel.
- The function defined by the integral

$$\int_0^x k(x - t)u(t)dt = \int_0^x k(t)u(x - t)dt$$

is called the convolution or the Falting of the two functions k and u and are known as convolution integral.

6.2 APPLICATION OF LAPLACE TRANSFORM TO DETERMINE THE SOLUTIONS OF VOLTERRA INTEGRAL EQUATION WITH CONVOLUTION TYPE KERNELS

Consider the Volterra integral equation of first kind

$$f(x) = \int_0^x k(x-t)\,u(t)\,dt \qquad \ldots(1)$$

where $k(x-t)$ depends only on the difference $(x-t)$.

Taking the Laplace transform to both sides of the equation (1), we have

$$L[f(x)] = L\left[\int_0^x k(x-t)\,u(t)\,dt\right]$$

$$\Rightarrow \qquad F(p) = k(p, u(p)) \quad \Rightarrow \quad u(p) = \frac{F(p)}{k(p)} \qquad \ldots(2)$$

Similarly, the transform method is also applicable to the Volterra integral equation of the second kind. Consider a non-homogeneous integral equation of second kind as

$$u(x) = f(x) + \int_0^x k(x-t)u(t)dt \qquad \ldots(3)$$

Taking Laplace transform of both sides of equation (3) and adding the convolution formula, we have

$$L[u(x)] = L[f(x)] + L\left[\int_0^x k(x-t)u(t)dt\right]$$

$$\Rightarrow \qquad u(p) = F(p) + k(p)\,u(p)$$

$$u(p) = \frac{F(p)}{1 - k(p)} \qquad \ldots(4)$$

The resolvent kernel of equation (3) can be determined by the method of integral transform. Let the kernel $k(x, t)$ be defined as a difference kernel, then so is the resolvent kernel. Since the resolvent kernel $R(x, t)$ is the sum of the iterated kernels and they all depend on the difference $(x - t)$ then

$$R_1(x,t) = k(x,t) = k(x-t)$$

and

$$k_2(x,t) = \int_t^x k(x-z)\,k(z-t)\,dz$$

Let

$$z - t = \sigma \;\Rightarrow\; z = t + \sigma \;\Rightarrow\; dz = d\sigma$$

or

$$k_2(x,t) = \int_0^{x-t} k(x-t-\sigma)\,k(\sigma)\,d\sigma \qquad \ldots(5)$$

Similarly the other integration can be determined. Thus the solution of the integral equation will be

$$u(x) = f(x) + \int_0^x R(x-t)\,f(t)\,dt \qquad \ldots(6)$$

Taking Laplace transform of both sides of equation (6), we have

$$L[u(x)] = L[f(x)] + L\left[\int_0^x R(x-t)\,f(t)\,dt\right]$$

$$u(p) = F(p) + R(p)\,F(p) \quad \text{where} \quad R(p) = L[R(x-t)]$$

$$\frac{F(p)}{1 - k(p)} = [1 + R(p)]\,F(p) \qquad \text{[From equation (4)]}$$

$$R(p) = \frac{k(p)}{1 - k(p)} \qquad \ldots(7)$$

By taking inverse Laplace transform of the equation (7), we have

$$R(x-t) = L^{-1}\left[\frac{k(p)}{1 - k(p)}\right]$$

Solved Examples

EXAMPLE 1. *Solve the Abel integral equation*

(i) $f(x) = \int_0^x \dfrac{u(t)}{(x-t)^\alpha} dt, 0 < \alpha < 1$ (Meerut–2005 BP)

(ii) $f(x) = \int_0^x \dfrac{u(t)}{\sqrt{x-t}} dt$

SOLUTION.

(i) The integral equation may be written as

$$f(x) = u(x) * x^{-\alpha} \qquad \ldots(1)$$

Taking Laplace transform of both sides of equation (1), we have

$$L[f(x)] = Lu(x) L(x^{-\alpha})$$

$$F(p) = L[u(x)] \frac{\overline{|1-\alpha|}}{p^{1-\alpha}}$$

$$L[u(x)] = \frac{p^{1-\alpha} F(p)}{\overline{|1-\alpha|}} = \frac{p}{\overline{|\alpha|}\,\overline{|1-\alpha|}} [\overline{|\alpha|} p^{-\alpha} F(p)]$$

$$L[u(x)] = \frac{p}{\pi/\sin\pi\alpha} [\overline{|\alpha|} p^{-\alpha} F(p)] = \frac{p\sin\pi\alpha}{\pi} L\left[x^{\alpha-1} * F(p) \right]$$

$$\left[\because \overline{|\alpha|}\,\overline{|1-\alpha|} = \frac{\pi}{\sin\pi\alpha} \right]$$

or $L[u(x)] = \dfrac{\sin\pi\alpha}{\pi} p L\left[\int_0^x (x-t)^{\alpha-1} F(t)\, dt \right]$

or $L[u(x)] - \dfrac{\sin\pi\alpha}{\pi} p L[G(x)]$

where $G(x) = \int_0^x (x-t)^{\alpha-1} F(t)\, dt, \qquad G(0) = 0$

We know that

$$L[G'(x)] = p L[G(x)] - G(0) = p L[G(x)]$$

or $L[u(x)] = \dfrac{\sin\pi\alpha}{\pi} L(G'(x))$

$$u(x) = \frac{\sin\pi\alpha}{\pi} \frac{d}{dx} G(x)$$

or $u(x) = \dfrac{\sin\pi\alpha}{\pi} \dfrac{d}{dx}\left[\int_0^x (x-t)^{\alpha-1} F(t)\, dt \right]$

(ii) The integral equation is given by

$$f(x) = \int_0^x \frac{u(t)\, dt}{\sqrt{x-t}} \qquad \ldots(1)$$

Here $k(x) = x^{-1/2}$ is not piecewise continuous but it does have a Laplace transform.

$$k(z) = \int_0^\infty x^{-1/2} e^{-zx}\, dx = \overline{\left|\frac{1}{2}\right|} z^{-1/2} = \sqrt{\pi}\, z^{-1/2}$$

Hence, $T(z) = L(u) = \dfrac{F(z)}{k(z)} = z^{1/2} \dfrac{F(z)}{\sqrt{\pi}} = \dfrac{z}{\pi}\left[\sqrt{\pi}\, z^{-1/2} F(z) \right]$

$$= \frac{z}{\pi} k(z) \, F(z)$$

$$\therefore \quad u(x) = \frac{1}{\pi} \frac{d}{dx} \int_0^x \frac{f(t)}{\sqrt{x-t}} \, dt$$

EXAMPLE 2.   Solve the Abel's equations

(i) $\int_0^x \frac{u(t)}{(x-t)^{1/3}} \, dt = x(1+x)$ (ii) $\int_0^x \frac{u(t)}{(x-t)^{1/2}} \, dt = 1+x+x^2$

(Kanpur-1996)

SOLUTION.   (i) The integral equation is given as

$$\int_0^x \frac{u(t)}{(x-t)^{1/3}} \, dt = x(1+x) \qquad \qquad \dots(1)$$

Equation (1) can be expressed as

$$u(x) \, x^{-1/3} = x(1+x)$$

Taking Laplace transform, we have

$$L\left[u(x) * x^{-1/3} \right] = L\,[x+x^2]$$

$$\Rightarrow \quad L\,[u(x)] \, L[x^{-1/3}] = L\,[x+x^2]$$

$$L\,[u(x)] \frac{\overline{|2/3|}}{p^{2/3}} = \frac{1}{p^2} + \frac{2}{p^3}$$

$$L\,[u(x)] = \frac{1}{\overline{|2/3|}} \left[\frac{1}{p^{4/3}} + \frac{1}{p^{7/3}} \right]$$

Taking inverse Laplace transform, we have

$$u(x) = \frac{1}{\overline{|2/3|}} \left[L^{-1}\left[\frac{1}{p^{4/3}} \right] + 2L^{-1}\left[\frac{1}{p^{7/3}} \right] \right]$$

$$= \frac{1}{\overline{|2/3|}} \left[\frac{x^{1/3}}{\frac{1}{3}\overline{|1/3|}} + 2 \frac{2^{4/3}}{\left(\frac{4}{3}\right)\left(\frac{1}{3}\right)\overline{|1/3|}} \right] = \frac{3x^{1/3}}{\overline{|2/3|}\,\overline{|1/3|}} \left(1 + \frac{3}{2}x\right)$$

$$= \frac{3x^{1/3}}{\pi/\sin(\pi/3)} \left(1 + \frac{3}{2}x\right) \qquad \left[\because \overline{|n|}\,\overline{|n-1|} = \frac{\pi}{\sin n\pi}\right]$$

$$\Rightarrow \quad u(x) = \frac{3\sqrt{3}}{4\pi} x^{1/3} (2+3x)$$

(ii) The integral equation is given as

$$\int_0^x \frac{u(t)}{(x-t)^{1/2}} \, dt = 1+x+x^2 \qquad \qquad \dots(1)$$

Equation (1) can be written as

$$u(x) * x^{-1/2} = 1+x+x^2$$

Taking Laplace transform, we have

$$L\,[u(x) * x^{-1/2}] = L\,[1+x+x^2]$$

$$L\,[u(x)] \, L\,[x^{-1/2}] = L\,[1+x+x^2]$$

$$L[u(x)] \frac{\overline{1/2}}{p^{1/2}} = \frac{1}{p} + \frac{1}{p^2} + \frac{2!}{p^3}$$

$$L[u(x)] = \frac{1}{\sqrt{\pi}} \left(\frac{1}{p^{1/2}} + \frac{1}{p^{3/2}} + \frac{2}{p^{5/2}} \right)$$

Taking inverse Laplace to both sides, we have

$$u(x) = \frac{1}{\sqrt{\pi}} \left[L^{-1} \left(\frac{1}{p^{1/2}} \right) + L^{-1} \left(\frac{1}{p^{3/2}} \right) + 2 L^{-1} \left(\frac{1}{p^{5/2}} \right) \right]$$

$$u(x) = \frac{1}{\sqrt{\pi}} \left[\frac{x^{-1/2}}{\overline{1/2}} + \frac{x^{1/2}}{\overline{3/2}} + \frac{2 x^{3/2}}{\overline{5/2}} \right]$$

$$= \frac{1}{\sqrt{\pi}} \left[\frac{x^{-1/2}}{\sqrt{\pi}} + \frac{x^{1/2}}{\left(\frac{1}{2} \right) \sqrt{\pi}} + \frac{2 x^{3/2}}{\left(\frac{3}{2} \right) \left(\frac{1}{2} \right) \sqrt{\pi}} \right]$$

$$\Rightarrow \quad u(x) = \frac{1}{\pi} \left[x^{-1/2} + 2 x^{1/2} + \frac{8}{3} x^{3/2} \right]$$

EXAMPLE 3. *Solve the integral equation*

$$x = \int_0^x e^{x-t} u(t) dt \qquad \text{(Meerut–2006)}$$

SOLUTION. The integral equation can be written as

$$x = u(x) * e^x \qquad \dots(1)$$

Taking Laplace transform of both sides of equation (1), we have

$$L[x] = L[u(x)] * L[e^x]$$

$$\frac{1}{p^2} = L[u(x)] * L[e^x]$$

$$\frac{1}{p^2} = L[u(x)] \frac{1}{p-1} \cdot$$

Therefore

$$L[u(x)] = \frac{p-1}{p^2} = \frac{1}{p} - \frac{1}{p^2} \qquad \dots(2)$$

Taking inverse Laplace transform of (2), we have

$$u(x) = L^{-1} \left(\frac{1}{p} \right) - L^{-1} \left(\frac{1}{p^2} \right) = 1 - x, \; i.e., u(x) = 1 - x$$

EXAMPLE 4. *Solve the integral equation* $\sin x = \int_0^x J_0 (x-t) u(t) dt$ \qquad (Kanpur–2005)

SOLUTION. The given equation can be written as

$$\sin x = u(x) * J_0(x) \qquad \dots(1)$$

Taking Laplace transform of equation (1), we have

$$L(\sin x) = L[u(x)] L[J_0(x)]$$

$$\frac{1}{p^2 + 1} = L[u(x)] \frac{1}{\sqrt{p^2 + 1}}$$

$$L[u(x)] = \frac{1}{\sqrt{p^2+1}}$$

Taking the inverse Laplace transform, we have

$$u(x) = L^{-1}\frac{1}{\sqrt{p^2+1}} = J_0(x)$$

Hence, $\int_0^x J_0(x-t)J_0(t)\,dt = \sin x$

EXAMPLE 5. *Solve the integral equation of the first kind*

$$\int_0^x u(t)\,u(x-t)\,dt = 16\sin 4x \quad \text{(Meerut–2005, Kanpur–1995, 2000, 01, SVTU-1997)}$$

SOLUTION. The given equation can be written as

$$u(x)*u(x) = 16\sin 4x \qquad \ldots(1)$$

Taking Laplace transform of (1), we have

$$[L[u(x)]]^2 = 16\frac{4}{p^2+16}$$

$$L\,u(x) = \pm\frac{8}{\sqrt{p^2+16}} \qquad \ldots(2)$$

Taking inverse Laplace transform of equation (2), we have

$$u(x) = L^{-1}\left[\pm\frac{8}{\sqrt{p^2+16}}\right] = \pm 8\,J_0[4x]$$

EXAMPLE 6. *Solve the inhomogeneous integral equation*

$$u(x) = 1 + \int_0^x \sin(x-t)\,u(t)\,dt$$

verify your result. (Kanpur–1995, 2000, 01, 02; Kurukshetra–2007)

SOLUTION. The integral equation can be written as

$$u(x) = 1 + u(x)*\sin x \qquad \text{[Using convolution theorem]} \quad \ldots(1)$$

Taking the Laplace transform of equation (1), we have

$$L[u(x)] = L(1) + L[u(x)]*L[\sin x]$$

$$= \frac{1}{p} + L[u(x)].\frac{1}{p^2+1}$$

$$\left(1 - \frac{1}{p^2+1}\right)L[u(x)] = \frac{1}{p}$$

$$L[u(x)] = \frac{p^2+1}{p^3} = \frac{1}{p} + \frac{1}{p^3}$$

Taking inverse Laplace transform, we have

$$u(x) = L^{-1}\left(\frac{1}{p}\right) + L^{-1}\left(\frac{1}{p^3}\right)$$

$$\Rightarrow \qquad u(x) = 1 + \frac{1}{2}x^2$$

We shall show that $u(x) = 1 + \dfrac{x^2}{2}$ is the solution of the given integral equation.

$$u(x) = 1 + \int_0^x \left(1 + \frac{t^2}{2}\right) \sin(x - t)\, dt$$

$$u(x) = 1 + \left[\left(1 + \frac{t^2}{2}\right) \cos(x - t)\right]_0^x - \int_0^x t \cos(x - t)\, dt$$

Hence, $\quad u(x) = 1 + \left(1 + \frac{1}{2}x^2 - \cos x\right) - \left[\left[-t \sin(x - t)\right]_0^x + \left[\cos(x - t)\right]_0^x\right]$

$$= 1 + 1 + \frac{x^2}{2} - \cos x - (1 - \cos x) = 1 + \frac{x^2}{2}$$

EXAMPLE 7.
SOLUTION.

Solve the inhomogeneous integral equation $u(x) = 1 - \int_0^x (x - t)\, u(t)\, dt$.
The integral equation can be written as

$$u(x) \doteq 1 - u(x) * x \qquad\qquad \text{[Using convolution theorem]}$$

Taking Laplace transform, we have

$$L[u(x)] = L(1) - L[u(x) * x] = L(1) - L[u(x)] . \frac{1}{p^2}$$

$$\left(1 + \frac{1}{p^2}\right) L[u(x)] = \frac{1}{p}$$

$$L\, u(x) = \frac{p}{p^2 + 1}$$

Taking inverse Laplace transform, we have

$$u(x) = L^{-1} \frac{p}{p^2 + 1} = \cos x$$

Hence, $\quad u(x) = \cos x$

EXAMPLE 8. Solve the integral equation

$$u(x) = x^2 + \int_0^x \sin(x - t)\, u(t)\, dt$$

Also verify your result.

SOLUTION. The given equation can be written as

$$u(x) = x^2 + u(x) * \sin x \qquad \text{[By convolution theorem]} \qquad ...(1)$$

Taking Laplace transform of both sides of equation (1), we have

$$L[u(x)] = L(x^2) + L[u(x) * \sin x]$$

$$L[u(x)] = \frac{2}{p^3} + L[u(x)] . \frac{1}{p^2 + 1}$$

$$\left[1 - \frac{1}{p^2 + 1}\right] L[u(x)] = \frac{2}{p^3}$$

$$L[u(x)] = \frac{2(p^2 + 1)}{p^5} = \frac{2}{p^3} + \frac{2}{p^5}$$

Taking inverse Laplace transform, we have

$$u(x) = L^{-1}\left(\frac{2}{p^3}\right) + L^{-1}\left(\frac{2}{p^5}\right)$$

$$u(x) = 2\frac{x^2}{2!} + 2\frac{x^4}{4!} = x^2 + \frac{1}{12}x^4$$

Now, we shall show that $u(x) = x^2 + \frac{1}{12}x^4$ is the solution of the given integral equation.

$$u(x) = x^2 + \int_0^x \left(t^2 + \frac{1}{12} t^4 \right) \sin(x - t) \, dt$$

$$u(x) = x^2 + \int_0^x t^2 \sin(x - t) \, dt + \frac{1}{12} \int_0^x t^4 \sin(x - t) \, dt$$

$$u(x) = x^2 + \int_0^x t^2 \sin(x - t) dt$$
$$+ \frac{1}{12} \left[\left[t^4 \cos(x - t) \right]_0^x - 4 \int_0^x t^3 \cos(x - t) dt \right]$$

$$u(x) = x^2 + \int_0^x t^2 \sin(x - t) \, dt + \frac{1}{12} x^4$$
$$- \frac{1}{3} \left[\left[-t^3 \sin(x - t) \right]_0^x + 3 \int_0^x t^2 \sin(x - t) dt \right]$$

$$u(x) = x^2 + \int_0^x t^2 \sin(x - t) dt + \frac{1}{12} x^4 - \int_0^x t^2 \sin(x - t) \, dt$$

$$u(x) = x^2 + \frac{1}{12} x^4$$

EXAMPLE 9. Solve the integral equation

(i) $u(x) = x + 2 \int_0^x \cos(x - t) \, u(t) \, dt$ (Kanpur-1994)

(ii) $u(x) = e^{-x} - 2 \int_0^1 \cos(x - t) \, u(t) \, dt$ (SVTU-1997)

(iii) $u(x) = a \sin x - 2 \int_0^1 \cos(x - t) \, u(t) \, dt$ (Ravishankar-2004)

SOLUTION. (i) The integral equation is

$$u(x) = x + 2 \int_0^x \cos(x - t) \, u(t) \, dt \qquad \ldots(1)$$

Using convolution theorem, the equation (1) can be written as

$$u(x) = x + 2u(x) * \cos x$$

Taking Laplace transform of both sides, we have

$$L[u(x)] = L(x) + L[u(x) * \cos x]$$

$$L[u(x)] = \frac{1}{p^2} + \frac{2p}{p^2 + 1} L\, u(x)$$

$$L\, u(x) = \frac{p^2 + 1}{p^2 (p - 1)^2} = \frac{1}{(p - 1)^2} + \frac{1}{p^2 (p - 1)^2}$$

Taking inverse Laplace transform, we have

$$u(x) = L^{-1} \frac{1}{(p - 1)^2} + L^{-1} \frac{1}{p^2 (p - 1)^2}$$

$$u(x) = x e^x + (x - 2) e^x + x + 2 = 2e^x (x - 1) + x + 2$$

(ii) The given integral equation is

$$u(x) = e^{-x} - 2 \int_0^1 \cos(x - t) \, u(t) \, dt \qquad \ldots(1)$$

Equation (1) can be written as

$$u(x) = e^{-x} - 2u(x) * \cos x \qquad \ldots(2)$$

Taking Laplace transform of equation (2), we have

$$L[u(x)] = L[e^{-x}] - 2L[u(x) * \cos x]$$

$$L[u(x)] = \frac{1}{p + 1} - 2L[u(x)] \frac{p}{p^2 + 1}$$

$$\left(1 + \frac{2p}{p^2 + 1}\right) L[u(x)] = \frac{1}{p+1}$$

$$L[u(x)] = \frac{p^2 + 1}{(p+1)^3} = \frac{[(p+1)-1]^2 + 1}{(p+1)^3}$$

Taking inverse Laplace of both sides, we have

$$u(x) = L^{-1}\left[\frac{[(p+1)-1]^2 + 1}{(p+1)^3}\right]$$

$$= e^{-x} L^{-1}\left[\frac{(p-1)^2 + 1}{p^3}\right] \qquad \text{[Using first shifting theorem]}$$

$$= e^{-x} L^{-1}\left[\frac{p^2 - 2p + 2}{p^3}\right] = e^{-x} L^{-1}\left[\frac{1}{p} - \frac{2}{p^2} + \frac{2}{p^3}\right]$$

$$= e^{-x}\left[1 - 2x + 2\frac{x^2}{2!}\right]$$

$\Rightarrow \qquad u(x) = e^{-x}(1 - 2x + x^2)$

(iii) The given integral equation is

$$u(x) = a\sin x - 2\int_0^1 \cos(x-t)\, u(t)\, dt \qquad \qquad \text{...(1)}$$

The integral equation (1) can be written as

$$u(x) = a\sin x - 2u(x) * \cos x$$

Taking Laplace transform of equation (2), we have

$$L[u(x)] = a\, L[\sin x] - 2L[u(x)]\, L(\cos x)$$

$$= \frac{a}{p^2 + 1} - 2L[u(x)]\frac{p}{p^2 + 1}$$

$$\left[1 + \frac{2p}{p^2 + 1}\right] L[u(x)] = \frac{a}{p^2 + 1}$$

$$L[u(x)] = \frac{a}{(p+1)^2}$$

Taking the inverse Laplace transform, we have

$$u(x) = a\, L^{-1}\frac{1}{(p+1)^2}$$

$\Rightarrow \qquad u(x) = a e^{-x} L^{-1}\left[\frac{1}{p^2}\right] \qquad \qquad \text{[Using first shifting theorem]}$

Hence, $\qquad u(x) = a e^{-x}\, x = ax\, e^{-x}$

EXAMPLE 10. *Solve the integral equation* $u'(x) = x + \int_0^x \cos t\, u(x-t)\, dt$, $\quad u(0) = 4$

(Meerut–2005 BP, Kanpur–1997, 2002, 05, Kurukshetra-2006)

SOLUTION. The integral equation is given by

$$u'(x) = x + \int_0^x \cos t\, u(x-t)\, dt$$

It can be written as

$$u'(x) = x + u(x) * \cos x, \qquad u(0) = 4$$

Taking Laplace transform of both sides, we have

$$L[u'(x)] = L(x) + L[u(x)]\, L[\cos(x)]$$

$$p L[u(x)] - u(0) = \frac{1}{p^2} + L[u(x)] \frac{p}{p^2+1} \qquad [\because \text{ Given } u(0) = 4]$$

$$\left(1 - \frac{1}{p^2+1}\right) p L[u(x)] = \frac{1}{p^2} + 4$$

$$L[u(x)] = \frac{p^2+1}{p^3}\left(\frac{1}{p^2}+4\right) = \frac{4}{p} + \frac{5}{p^3} + \frac{1}{p^5}$$

Taking inverse Laplace transform, we have

$$u(x) = 4 + \frac{5}{2!}x^2 + \frac{1}{4!}x^4 = 4 + \frac{5}{2}x^2 + \frac{1}{24}x^4$$

EXAMPLE 11.   *Solve the integral equation* $u'(x) = \sin x + \int_0^x \cos t \, u(x-t) \, dt$, $u(0) = 0$

SOLUTION.   The given integral equation can be written as

$$u'(x) = \sin x + u(x) * \cos x, \qquad u(0) = 0 \qquad \text{...(1)}$$

Taking Laplace transform of equation (1), we have

$$L[u'(x)] = L[\sin(x)] + L[u(x) * \cos x]$$

$$\Rightarrow \quad p L[u(x)] - u(0) = \frac{1}{p^2+1} + L[u(x)].\frac{p}{p^2+1}$$

$$\Rightarrow \quad \left(1 - \frac{1}{p^2+1}\right) p L[u(x)] = \frac{1}{p^2+1}$$

$$\Rightarrow \quad \frac{p^3}{p^2+1} L[u(x)] = \frac{1}{p^2+1}$$

$$\Rightarrow \quad L[u(x)] = \frac{1}{p^3}$$

Taking inverse Laplace transform, we have

$$u(x) = L^{-1}\left(\frac{1}{p^3}\right) = \frac{x^2}{2}$$

EXAMPLE 12.   *Find the resolvent of the Volterra integral equation and hence its solution*

$$u(x) = F(t) + \int_0^x (x-t) \, u(t) \, dt$$

SOLUTION.   The given integral equation is

$$u(x) = F(t) + u(x) * x \qquad \text{...(1)}$$

Taking Laplace transform and using the convolution theorem, we have

$$L[u(x)] = L[F(t)] + L[u(x) * x] = L[F(t)] + Lu(x).\frac{1}{p^2}$$

$$\left(1 - \frac{1}{p^2}\right) L[u(x)] = L[F(t)]$$

$$L[u(x)] = \frac{p^2}{p^2-1} L[F(t)] \qquad \text{...(2)}$$

Let $R(x-t)$ be the resolvent kernel of the given integral equation, then solution of the integral equation becomes

$$u(x) = F(x) + \int_0^x R(x-t) \, F(x) \, dx \qquad \text{...(3)}$$

$$u(x) = F(x) + R(x) * F(x) \qquad \text{...(4)}$$

Taking the Laplace transform, we have

$$L[u(x)] = L[F(x)] + L[R(x) * F(x)]$$

$$\frac{p^2}{p^2-1} L[F(x)] = L[F(x)] + L[R(x)] \cdot L[F(x)]$$

$$\frac{p^2}{p^2-1} L[F(x)] = [1 + L[R(x)]] L[F(x)]$$

$$1 + L[R(x)] = \frac{p^2}{p^2-1}$$

$$L[R(x)] = \frac{p^2}{p^2-1} - 1 = \frac{1}{p^2-1}$$

Taking inverse Laplace transform, we have $R(x) = \sinh x$

So that $R(x-t) = \sinh(x-t)$

Thus the solution will be

$$u(x) = F(x) + \int_0^x \sinh(x-t) F(x) \, dx$$

EXAMPLE 13. *Determine the resolvent kernel and hence solve the integral equation*

$$u(x) = F(x) + \int_0^x e^{x-t} u(t) \, dt$$

SOLUTION. The integral equation is given by

$$u(x) = F(x) + \int_0^x e^{x-t} u(t) \, dt \qquad \ldots(1)$$

The given equation (1) can be written as

$$u(x) = F(x) + u(x) * e^x$$

Taking Laplace transform and using the convolution theorem, we have

$$L[u(x)] = L[F(x)] + L[u(x) * e^x]$$

$$L[u(x)] = L[F(x)] + L[u(x)] L[e^x]$$

$$L[u(x)] = L[F(x)] + \frac{1}{p-1} L[u(x)]$$

$$\left(1 - \frac{1}{p-1}\right) L[u(x)] = L[F(x)]$$

$$L[u(x)] = \frac{p-1}{p-2} L[F(x)]$$

Let $R(x-t)$ be the resolvent kernel of the given integral equation. Then we know that the required solution is

$$u(x) = F(x) + \int_0^x R(x-t) F(x) \, dx$$

or $\qquad u(x) = F(x) + R(x) * F(x)$

Taking Laplace transform of both sides of this equation, we have

$$L u(x) = L[F(x)] + L[R(x)] \cdot L[F(x)]$$

$$\frac{p-1}{p-2} L[F(x)] = [1 + L R(x)] L[F(x)]$$

$$L R(x) = \frac{p-1}{p-2} - 1 = \frac{1}{p-2}$$

Taking the inverse Laplace transform, we have

$$R(x) = L^{-1} \frac{1}{p-2} = e^{2x}$$

So that $\qquad R(x-t) = e^{2(x-t)}$ giving the required resolvent kernel.

Thus the required solution is

$$u(x) = F(x) + \int_0^x e^{2(x-t)} F(x) \, dx$$

EXAMPLE 14. *Solve the integral equation*

$$u(x) = f(x) + \lambda \int_0^x J_0 (x-t) u(t) dt \qquad \text{(Meerut–2005BP)}$$

SOLUTION. The integral equation is given by

$$u(x) = f(x) + \lambda \int_0^x J_0 (x-t) u(t) dt \qquad \qquad ...(1)$$

The given equation (1) can be written as

$$u(x) = F(x) + \lambda u(x) * J_0 (x)$$

Taking Laplace transform and using the convolution theorem, we have

$$L[u(x)] = L[f(x)] + L[u(x)].L[\lambda J_0(x)]$$

$$= L[f(x)] + L[u(x)] \frac{\lambda}{\sqrt{p^2+1}}$$

$$\left[1 - \frac{\lambda}{\sqrt{p^2+1}} \right] Lu(x) = L[f(x)]$$

$$L[u(x)] = \frac{1}{\left[1 - \frac{\lambda}{\sqrt{p^2+1}} \right]} L f(x) \qquad \qquad ...(2)$$

Let $R(x-t)$ be the resolvent kernel of the given integral equation, then the solution of the integral equation becomes

$$u(x) = f(x) + \int_0^x R(x-t) f(x) dx \qquad \qquad ...(3)$$

or $\qquad\qquad u(x) = f(x) + R(x) * f(x) \qquad \qquad ...(4)$

Taking Laplace transform, we have

$$L[u(x)] = L[f(x)] + L[R(x) * f(x)]$$
$$L[u(x)] = [1 + L R(x)] L[f(x)]$$

From the relation (2), we have

$$\left[1 \bigg/ \left(1 - \frac{\lambda}{\sqrt{p^2+1}} \right) \right] L[F(x)] = [1 + L.R(x)] L f(x)$$

$$L R(x) = \frac{1}{1 - \frac{\lambda}{\sqrt{p^2+1}}} - 1 = \frac{\lambda}{\sqrt{p^2+1} - \lambda}$$

Taking the inverse Laplace transform, we have

$$R(x) = L^{-1} \left[\frac{\lambda}{\sqrt{p^2+1} - \lambda} \right] \qquad \qquad ...(5)$$

The value of the resolvent kernel follows by setting $(x-t)$ for x and the solution of the integral equation is obtained from (3) and (5).

EXAMPLE 15. *Solve the integral equation*

$$f(x) = \int_0^x k(x^2 - t^2) u(t) dt, \ x > 0$$

SOLUTION. The integral equation is given by

$$f(x) = \int_0^x k(x^2 - t^2) u(t) dt, \ x > 0 \qquad \qquad ...(1)$$

Let $\qquad\qquad x^2 = \alpha, \ t^2 = \beta, \ f_1(\alpha) = f(\alpha^{1/2}) = f(x)$

$$\psi(\beta) = \frac{1}{2}\beta^{-1/2}\,u(\beta^{1/2}) \qquad \qquad \ldots(2)$$

The integral equation (1) reduces to

$$f_1(\alpha) = \int_0^\alpha k(\alpha-\beta)\,u(\beta^{1/2}).\frac{1}{2}\beta^{-1/2}\,d\beta$$

$$f_1(\alpha) = \int_0^\alpha k(\alpha-\beta)\,\psi(\beta)\,d\beta, \quad \alpha > 0 \qquad \qquad \ldots(3)$$

Taking Laplace transform and using the convolution theorem, we have

$$L[f_1(\alpha)] = L\left[\int_0^\alpha k(\alpha-\beta)\,\psi(\beta)\,d\beta\right]$$

where the kernel $k(\alpha-\beta)$ depends only on the difference $(\alpha-\beta)$.

$$L[f_1(\alpha)] = L[k(\alpha)*\psi(\alpha)] = L[k(\alpha)].L[\psi(\alpha)]$$

Since

$$L[f_1] = F(p); \quad L(k) = k(p); \quad L(\psi) = \psi(p)$$

$$F(p) = k(p)\,\psi(p) \qquad \qquad \ldots(4)$$

$$\psi(p) = \frac{F(p)}{k(p)} \mp \frac{p\,F(p)}{p\,k(p)} = p\,F(p)\,G(p) \qquad \qquad \ldots(5)$$

where

$$\frac{1}{p\,k(p)} = G(p)$$

or

$$\psi(p) = p\,G(p)\,F(p)$$

$$\psi(p) = L\left[\frac{d}{d\alpha}\int_0^\alpha g(\alpha-\beta)\,f_1(\beta)\,d\beta\right]$$

or

$$u(\alpha) = \frac{d}{d\alpha}\int_0^\alpha g(\alpha-\beta)\,f_1(\beta)\,d\beta \qquad \qquad \ldots(6)$$

where $g(\alpha)$ represents the inverse of the function $G(p)$.

From the relation (2), the integral equation (6) becomes

$$u(x) = 2\frac{d}{dx}\int_0^x t\,f(t)\,g(x^2-t^2)\,dt$$

6.3 APPLICATION OF FOURIER TRANSFORM TO DETERMINE THE SOLUTION OF SINGULAR INTEGRAL EQUATIONS

The procedure will be clear from the following examples.

Solved Examples

EXAMPLE 1. *Solve the integral equation*

$$\int_0^\infty F(t)\cos pt\,dt = \begin{cases} 1-p, & 0 \le p \le 1 \\ 0, & p > 1 \end{cases} \qquad \text{(Meerut–2005BP, 06)}$$

SOLUTION. Let

$$F_c(p) = \begin{cases} 1-p, & 0 \le p \le 1 \\ 0, & p > 1 \end{cases} \qquad \qquad \ldots(1)$$

Then the given integral equation can be written as

$$F_c(p) = \int_0^\infty F(t)\cos pt\,dt \qquad \qquad \ldots(2)$$

By definition of Fourier cosine transform, we see that

$$F(t) = \frac{2}{\pi}\int_0^\infty F_c(p)\cos pt\,dp$$

$$= \frac{2}{\pi}\left[\int_0^1 F_c(p)\cos pt\,dp + \int_1^\infty F_c(p)\cos pt\,dp\right]$$

$$= \frac{2}{\pi} \left[\int_0^1 (1-p) \cos pt \, dp + \int_1^\infty 0. \cos pt \, dp \right]$$

$$= \frac{2}{\pi} \left[(1-p) \frac{\sin pt}{t} \right]_0^1 - \int_0^1 (-1) \frac{\sin pt}{t} \, dp$$

$$= \frac{2}{\pi t} \int_0^1 \sin pt \, dp = \frac{2}{\pi t} \left[\frac{-\cos pt}{t} \right]_0^1$$

$$= \frac{2}{\pi t^2} (-\cos t + 1) = \frac{2(1-\cos t)}{\pi t^2}$$

EXAMPLE 2. *Solve the integral equation*

$$\int_0^\infty F(t) \sin pt \, dt = \begin{cases} 1, & 0 \le p < 1 \\ 2, & 1 \le p < 2 \\ 0, & p \ge 2 \end{cases} \qquad \text{(Meerut–2005; Kottayam-2005)}$$

SOLUTION. Let $F_s(p) = \begin{cases} 1, & 0 \le p < 1 \\ 2, & 1 \le p < 2 \\ 0, & p \ge 2 \end{cases}$ ...(1)

Then the given integral equation can be written as

$$F_s(p) = \int_0^\infty F(t) \sin pt \, dt \qquad \text{...(2)}$$

We have

$$F(t) = \frac{2}{\pi} \int_0^\infty F_s(p) \sin pt \, dp$$

$$= \frac{2}{\pi} \left[\int_0^1 F_s(p) \sin pt \, dp + \int_1^2 F_s(p) \sin pt \, dp + \int_2^\infty F_s(p) \sin pt \, dp \right]$$

$$= \frac{2}{\pi} \left[\int_0^1 \sin pt \, dp + \int_1^2 (2) \sin pt \, dp + \int_2^\infty 0. \sin pt \, dp \right]$$

$$= \frac{2}{\pi} \left\{ \left[\frac{-\cos pt}{t} \right]_0^1 + 2 \left[\frac{-\cos pt}{t} \right]_1^2 \right\}$$

$$= \frac{2}{\pi} \left[\frac{-\cos t + 1}{t} + 2 \frac{-\cos 2t + \cos t}{t} \right] = \frac{2}{\pi t} (1 + \cos t - 2\cos 2t)$$

EXAMPLE 3. *Solve* $\int_0^\infty F(t) \cos pt \, dt = e^{-p}$ (Kanpur-1990; SVTU-2009; Rohtak-2004)

SOLUTION. We have

$$F(t) = \frac{2}{\pi} \int_0^\infty F_c(p) \cos pt \, dp = \frac{2}{\pi} \int_0^\infty e^{-p} \cos pt \, dp \qquad [\because e^{-p} = F_c(p)]$$

$$= \frac{2}{\pi} \left[\frac{e^{-p}}{1+t^2} (-\cos pt + t \sin pt) \right]_0^\infty$$

$$= \frac{2}{\pi(1+t^2)} \qquad \left[\because \int e^{at} \cos bt \, dt = \frac{e^{at}}{a^2 + b^2} (a \cos bt + b \sin bt) \right]$$

$$= \frac{2}{\pi(1+t^2)}$$

6.4 APPLICATION OF MELLIN'S TRANSFORM TO INTEGRAL EQUATIONS

The Mellin transform is given by

$$u(p) = \int_0^\infty x^{p-1} u(x) \, dx \qquad \text{...(1)}$$

and the corresponding convolution theorem states that $k(p)u(1-p)$ is the Mellin transform of
$$\int_0^\infty k(x,t)\,u(t)\,dt \qquad \ldots(2)$$
where $k(p)$ is the Mellin transform of $k(x)$.
$$\int_0^\infty x^{p-1}\,dx \int_0^\infty k(x,t)\,u(t)\,dt = \int_0^\infty y^{p-1}\,k(y)\,dy \int_0^\infty t^{-p}\,u(t)\,dt$$

Let
$$y = xt, \qquad dy = t\,dx$$
$$\int_0^\infty x^{p-1}\,dx \int_0^\infty k(xt)\,u(t)\,dt = k(p)\,u(1-p) \qquad \ldots(3)$$

6.5 FOX'S INTEGRAL EQUATION

Fredholm integral equation of second kind is
$$u(x) = f(x) + \int_0^\infty k(x,t)\,u(t)\,dt, \qquad 0 < x < \infty \qquad \ldots(4)$$
Since $k(p)\,u(1-p)$ is the Mellin transform
$$\int_0^\infty k(xt)\,u(t)\,dt$$
Taking the Mellin transform of equation (4), we have
$$u(p) = F(p) + k(p)\,u(1-p) \qquad \ldots(5)$$
where $u(p)$, $F(p)$ and $k(p)$ is the Mellin transform $u(x)$, $f(x)$ and $k(x)$ respectively. Substituting $(1-p)$ in place of p, we have
$$u(1-p) = F(1-p) + k(1-p)\,u\,p \qquad \ldots(6)$$

Equation (5) gives
$$u(1-p) = \frac{u(p) - F(p)}{k(p)}$$

Substituting this value in equation (6), we have
$$u(p) = \frac{F(p) + k(p)\,F(1-p)}{1 - k(p)\,k(1-p)}$$

Solved Example

EXAMPLE. *Fox integral equation is*
$$u(x) = f(x) + \int_0^\infty k(xt)\,u(t)\,dt, \qquad 0 < x < \infty$$

with kernel $k(x) = \lambda\sqrt{\dfrac{2}{\pi}}\,\sin x$. *Determine its solution.* (Kanpur–2005)

SOLUTION. The Mellin transform of the kernel $k(x)$ is
$$k(p) = \lambda\sqrt{\frac{2}{\pi}} \int_0^\infty x^{p-1}\,\sin x\,dx$$

$$k(p) = \lambda\sqrt{\frac{2}{\pi}}\,\lceil p\,\sin\frac{\pi p}{2}$$

$$k(p)\,k(1-p) = \lambda\sqrt{\frac{2}{\pi}}\,\lceil p\,\sin\frac{\pi p}{2}\cdot\lambda\sqrt{\frac{2}{\pi}}\,\lceil 1-p\,\sin\frac{(1-p)\pi}{2}$$

$$k(p)\,k(1-p) = \frac{2\lambda^2}{\pi}\,\lceil p\,\lceil 1-p\,\sin\frac{\pi p}{2}\,\cos\frac{\pi p}{2}$$

$$k(p)\,k(1-p) = \frac{\lambda^2}{\pi}\,\frac{\pi}{\sin\pi p}\,\sin\pi p = \lambda^2$$

Hence, $$u(p) = \frac{F(p)}{1-\lambda^2} + \frac{\lambda}{1-\lambda^2}\sqrt{\frac{2}{\pi}}\,\sqrt{p}\,\sin\frac{\pi p}{2}\,F(1-p)$$

$\sqrt{p}\,\sin\dfrac{\pi p}{2}\,F(1-p)$ is the Mellin transform of $\int_0^\infty \sin(xt)\,f(t)\,dt$.

Thus, the solution of Fox's integral equation is

$$u(x) = \frac{f(x)}{1-\lambda^2} + \frac{\lambda}{1-\lambda^2}\sqrt{\frac{2}{\pi}}\int_0^\infty \sin(xt)\,f(t)\,dt$$

EXERCISE 6.1

1. Solve the integral equation

$$F(x) = x\int_x^\infty \frac{u'(x)}{(x^2-t^2)^{1/2}}\,dx$$

2. Solve the integral equation

$$u(x) = F(x) + \lambda\int_0^x J_1(x-t)\,u(t)\,dt$$

3. Show that the solution of the integral equation

$$F(x) = 2\int_x^1 \frac{t\,u'(t)}{(x^2-t^2)^{1/2}}\,dt$$

is $u(t) = -\dfrac{1}{\pi t}\dfrac{d}{dt}\int_x^1 \dfrac{t\,F(t)\,dt}{(x^2-t^2)^{1/2}}$

Find the solution for

(i) $F(x) = 2x^2/(1-x^2)^{1/2}$,

(ii) $F(x) = x^2$

4. Solve the integral equation

$$u'(x) = x + \int_0^x u(x-t)\cos t\,dt \quad \phi(0) = 2$$

5. Solve the integral equation

$$u(x) = F(x) + \lambda\int_0^x J_0(x-t)\,u(t)\,dt$$

6. Solve the integral equations

(i) $u(x) = x + 2\int_0^x \sin(x-t)\,u(t)\,dt$

(ii) $u(x) = a\sin x - 2\int_0^1 \cos(x-t)u(t)\,dt$

7. Solve the inhomogeneous Abel integral equation

$$u(x) = F(x) + \lambda\int_0^x \frac{u(t)}{(x-t)^\alpha}\,dt, \quad 0<\alpha<1$$

8. Solve the Abel integral equation of the second kind

$$u(x) = x^{-1/2}\,e^{-a/4} + \frac{i}{\sqrt{\pi}}\int_0^x \frac{u(t)}{(x-t)^{1/2}}\,dt$$

9. Show that the solution of the integral equation

$$e^{-p} = \int_0^\infty F(x)\cos px\,dx \text{ is } \frac{2}{\pi(1+x^2)}$$

10. Solve the Abel's integral equation

$$\int_0^x \frac{u(t)\,dt}{\sqrt{x-t}} = 3x^2 + x^3 + 5x^4$$

11. Find the resolvent kernel of the integral equations

(i) $u(x) = F(x) + \int_0^x k(x^2-t^2)\,u(t)\,dt$

(ii) $u(x) = F(x) + \int_0^x e^{x-t}u(t)\,dt$

Hint to Selected Problems

2. Taking Laplace transform on both the sides, we get

$$L[u(x)] = L[F(x)] + \lambda L[u(x)] * L[J_1(x)]$$

Using $J_n(at) = \dfrac{\left\{\sqrt{p^2+a^2}-p\right\}^n}{a^n\sqrt{p^2+a^2}}$

We get

$$L[u(x)] = L[F(x)] + \lambda L[u(x)]\cdot\frac{\sqrt{p^2-p+1}}{\sqrt{p^2+1}}$$

$$\Rightarrow L[u(x)]$$

$$= \left\{\frac{\sqrt{p^2+1}}{\sqrt{p^2+1}-\lambda\sqrt{p^2-p+1}}\right\}L[F(x)]$$

$$\Rightarrow u(x)$$

$$= L^{-1}\left\{\frac{\sqrt{p^2+1}}{\sqrt{p^2+1}-\lambda\sqrt{p^2-p+1}}\right\}F(x)$$

4. $u' = x + u(x) * \cos x$

Taking Laplace transform on both the sides, we get

$$L[u'(x)] = L(x) + L[u(x)]\,L[\cos x]$$

$$\Rightarrow pL[u(x)] - u(0) = \frac{1}{p^2} + L[u(x)]\cdot\frac{p}{p^2+1}$$

After simplification, we get

$$L[u(x)] = \frac{(2p^2+1)(p^2+1)}{2p^3}$$

Now taking inverse Laplace transform.

5. $u(x) = F(x) + \lambda[u(x) * J_0(x)]$

$$\Rightarrow L[u(x)] = L[F(x)] + \lambda L[u(x)]\cdot L[J_0(x)]$$

$$= L[F(x)] + \lambda . L[u(x)] . \frac{1}{\sqrt{p^2 + 1}}$$

$$\Rightarrow \left[1 - \frac{\lambda}{\sqrt{p^2 + 1}}\right] L[u(x)] = L[F(x)]$$

$$\Rightarrow L[u(x)] = \frac{1}{\left\{1 - \frac{\lambda}{\sqrt{p^2 + 1}}\right\}} L[F(x)]$$

$$\Rightarrow u(x) = L^{-1}\left\{\frac{\sqrt{p^2 + 1}}{\sqrt{p^2 + 1} - \lambda}\right\} F(x)$$

7. $\qquad u(x) = F(x) + \lambda[u(x) * x^{-\alpha}]$

$$\Rightarrow L[u(x)] = L[F(x)] + \lambda . L[u(x)] . L[x^{-\alpha}]$$

$$= L[F(x) + \lambda . L[u(x)] . \frac{\overline{1 - \alpha}}{p^{1-\alpha}}$$

$$\Rightarrow \quad L[(x)] = \frac{1}{\left\{1 - \frac{\lambda\overline{1-\alpha}}{p^{1-\alpha}}\right\}}$$

Now taking inverse Laplace transform.

9. Using the result

$$F(x) = \frac{2}{\pi} \int_0^\infty F_c(p) \cos px \, dp$$

We get $\quad F(x) = \frac{2}{\pi} \int_0^\infty e^{-p} \cos px \, dp$

$$= \frac{2}{\pi} \left[\frac{e^{-p}}{1 + x^2}(-\cos px + x \sin px)\right]_0^\infty$$

$$= \frac{2}{\pi(1 + x^2)}[e^{-\infty}(-\cos\infty + \infty)$$

$$- 1(-1 + 0)]$$

$$\therefore \qquad F(x) = \frac{2}{\pi(1 + x^2)}$$

ANSWERS

---

7. $\quad u(x) = F(x) + \int_0^x \sum\limits_{n=1}^\infty \dfrac{\lambda\sqrt{1-\alpha}(x-t)^{1-\alpha}}{(x-t)[n(1-\alpha)]} \cdot F(t)dt$

Self Assessment Test

1. Solve $\int_0^t \frac{f(u)du}{\sqrt{t-u}} = 1 + 2t - t^2$

(MDU–2004, Kurukshetra-2004)

2. Solve $F(t) = t + \int_0^t F(t-u)\cos u\, du$.

(Kuvempa-2005)

3. Solve $y'(t) + 4y(t) + 5\int_0^t y(u)du = e^{-1}, y(0) = 0$.

(MDU–2005)

4. Solve $\frac{dy}{dt} + 3y + 2\int_0^t y\, dt = t$ given $y(0) = 1$.

(MDU–2006; Madras-1997)

5. Solve $\int_0^t \frac{y(u)\, du}{(t-u)^{1/2}} = \sqrt{t}.$ given $y(0) = 1$.

(Sagar–2004)

6. Solve $y(t) = t + \int_0^t y(u)\sin(t-u)\, du$.

(Kurukshetra-2005)

7. Solve $F'(t) = t - \int_0^t F(t-u)\cos u\, du, F(0) = 1$

(Punjab-2005, Kurukshetra-2006)

8. Solve $f(t) = t - \int_0^t f(t-u)\cos hu\, du,\ f(0) = 1$

(Punjab-2005)

9. Solve $3\sin 2x = y(x) + \int_0^x (x-t)\, y(t)\, dt$.

(Nagpur-2005, Lucknow-1994)

ANSWERS

1. $f(t) = \frac{1}{\pi}\left[t^{-1/2} + 4t^{1/2} - \frac{8}{3}t^{3/2}\right]$

2. $f(t) = 1 + t - e^{t/2}\left\{\cos(t\sqrt{3}/2) - (1/\sqrt{3}) \times \sin\left(\frac{t\sqrt{3}}{2}\right)\right\}$

3. $y(t) = -\frac{1}{2}e^{-t} + \frac{1}{2}e^{-2t}(\cos t + 3\sin t)$

4. $f(t) = \frac{1}{2} - 2e^{-t} + \frac{5}{2}e^{-2t}$

5. $y(t) = 1/2$

6. $y(t) = t + t^3/6$

7. $F(t) = 1 + t^2 + t^4/24$

8. $f(t) = 1 + t - e^{-t/2}\{\cosh(t\sqrt{5}/2) - (1/\sqrt{5}) \times \sinh(t\sqrt{5}/2)\}$

9. $y(x) = 2\sin t - \sin 2t$

vvvvvv

Chapter 7
Z-Transforms

7.1 INTRODUCTION

Z-transforms plays an important role in discrete analysis as Laplace and Fourier transforms in continuous system. It has many properties similar to those of the Laplace transforms. The z-transforms operate on sequences of the discrete integer valued arguments, not on function of continuous arguments. For every operational rule of Laplace transforms, there is a corresponding operational rule of z-transforms and for every application of the Laplace transforms, there is a corresponding application of z-transforms.

7.2 Z-TRANSFORMS

If the function f_n is defined for discrete values $(n = 0, 1, 2, ...)$ and $f_n = 0$, for $n < 0$ then its z-transform is defined by

$$Z(f^n) = F(z) = \sum_{n=-\infty}^{\infty} f(n) z^{-n} = \sum_{n=-\infty}^{\infty} \frac{f(n)}{z^n}$$

whenever the infinite series converges. Here z is a complex number. Z is an operator and $F(z)$ is the Z-transforms of $\{f(n)\}$.

7.2.1 SOME STANDARD Z-TRANSFORMS

(Kotayam–2005)

(1) $Z(a^n) = \dfrac{z}{z-a}$

(2) $Z(n^p) = -z\dfrac{d}{dz} Z(n^{p-1})$, where p is a positive integer.

(3) $Z(1) = \dfrac{z}{z-1}$

(4) $Z(1) = \dfrac{z}{(z-1)^2}$

(5) $Z(n^2) = \dfrac{z^2 + z}{(z-1)^3}$

(6) $Z(n^3) = \dfrac{z^3 + 4z^2 + z}{(z-1)^4}$

(7) $Z(n^4) = \dfrac{z^4 + 11z^3 + 11z^2 + z}{(z-1)^5}$

7.3 PROPERTIES OF Z-TRANSFORM

7.3.1 LINEARITY PROPERTY

If <f(n)> and <g(n)> are two sequences such that they can be added and a, b are any two scalars, then

$$Z[< a f (n) + b g (n) >] = aZ [< f(n) >] + bZ [< g(n) >]$$

Proof. Consider

$$Z[<af(n)+bg(n)>] = \sum_{n=-\infty}^{\infty} [a f(n) + b g(n)] z^{-n}$$ [By definition of z-transform]

$$= \sum_{n=-\infty}^{\infty} [af(n)z^{-n} + bg(n)z^{-n}] = a \sum_{n=-\infty}^{\infty} f(n)z^{-n} + b \sum_{n=-\infty}^{\infty} g(n)z^{-n}$$

$$= aZ[< f(n) >] + b Z[< g(n) >]$$

7.3.2 CHANGE OF SCALE PROPERTY

If Z[<f(n)>] = F(z), then

(1) $z[< a^n f(n) >] = F\left(\dfrac{z}{a}\right)$ **(2)** $Z[<a^{-n} f(n)>] = F(az)$

PROOF.   (1) We have

$$F\left(\frac{z}{a}\right) = \sum_{n=-\infty}^{\infty} f(n)\left(\frac{z}{a}\right)^{-n} \qquad \text{[By definition of z-transform]} \quad ...(1)$$

Also we have

$$Z[< a^n f(n) >] = \sum_{n=-\infty}^{\infty} a^n f(n) z^{-n} = \sum_{n=-\infty}^{\infty} f(n)\left(\frac{z}{a}\right)^{-n} = F\left(\frac{Z}{a}\right)$$

(2) We have

$$F(az) = \sum_{n=-\infty}^{\infty} f(n)(az)^{-n}$$

Also, we have

$$Z[\{a^{-n} f(n)\}] = \sum_{n=-\infty}^{\infty} a^{-n} f(n) z^{-n} = \sum_{n=-\infty}^{\infty} f(n)(az)^{-n} = F(az)$$

[Using equation (2)]

REMARK

- Here, the geometric factor a^{-n} when $|a| < 1$ damps the function u_n, hence, above rule is also known as 'damping rule'.

(Madras–2006)

7.3.3 APPLICATION OF DAMPING RULE

The application of damping rule gives the following important results :

(1) $Z(na^n) = \dfrac{az}{(z-a)^2}$ (Madras–2000)   **(2)** $Z(n^2 a^n) = \dfrac{az^2 + a^2 z}{(z-a)^3}$

(3) $Z(\cos n\theta) = \dfrac{z(z - \cos\theta)}{z^2 - 2z\cos\theta + 1}$ **(4)** $Z(\sin n\theta) = \dfrac{z\sin\theta}{z^2 - 2z\cos\theta + 1}$

(5) $Z(a^n \cos n\theta) = \dfrac{z(z - a\cos\theta)}{z^2 - 2az\cos\theta + a^2}$ **(6)** $Z(a^n \sin n\theta) = \dfrac{az\sin\theta}{z^2 - 2az\cos\theta + a^2}$

7.3.4 SHIFTING PROPERTIES

If Z[{(f(n)}] = F(z), then

(1) $Z[< f(n \pm k) >] = z^{\pm k} F(z)$ *and if* $n \geq 0$, *then*

(2) $Z[< f(n + k) >] = z^n F(z) - \sum_{r=0}^{k-1} f(r) z^{k-r}$

(3) $Z[< f(n-k)>] = z^{-n}F(z) - \sum\limits_{m=1}^{k} f(-m) z^{-k+m}$

PROOF. (1) Consider

$$Z[< f(n \pm k)>] = \sum_{n=-\infty}^{\infty} f(n \pm k) z^{-n}$$

$$= \sum_{n=-\infty}^{\infty} f(n \pm k) z^{-(n \pm k)} \cdot z^{\pm n} = z^{\pm k} \sum_{n=-\infty}^{\infty} f(r) z^{-r}$$

(where $n \pm k = r$)

$$= z^{\pm k} F(z)$$

Hence, $Z[< f(n \pm k)>] = z^{\pm k} F(z)$

(2) Consider

$$Z[< f(n \pm k)>] = \sum_{n=0}^{\infty} f(n+k) z^{-n}$$

$$= \sum_{n=0}^{\infty} f(n+k) z^{-(n+k)} \cdot z^{n} = z^{k} \sum_{r=k}^{\infty} f(r) z^{-r}$$

(where $n + k = r$)

$$= z^{k} \sum_{r=0}^{\infty} f(r) z^{-r} - z^{k} \sum_{r=0}^{k-1} f(r) z^{-r}$$

$$= z^{k} F(z) - \sum_{r=0}^{k-1} f(r) z^{-r} = z^{k} F(z) - \sum_{r=0}^{k-1} f(r) z^{k-r}$$

Hence, $Z[< f(n+k)>] = z^{k} F(z) - \sum\limits_{r=0}^{k-1} f(r) z^{k-r}$

(3) Consider

$$Z[< f(n-k)>] = \sum_{n=0}^{\infty} f(n-k) z^{-n} = \sum_{n=0}^{\infty} f(n-k) z^{-(n+k)} \cdot z^{-k}$$

$$= z^{-k} \sum_{r=0}^{\infty} f(r) z^{-r} + z^{-k} \sum_{r=-k}^{-1} f(1) z^{-r} \qquad \text{(when } n + k = r)$$

$$= z^{-k} F(z) + \sum_{r=-k}^{-1} f(r) z^{-k-r} = z^{-k} F(z) + \sum_{m=1}^{k} f(-m) z^{-k+m}$$

Hence, $Z[< f(n-k)>] = z^{-k} F(z) + \sum\limits_{m=1}^{k} f(-m) z^{-k+m}$

7.3.5 SOME IMPORTANT DEDUCTIONS

(1) $Z[< f(n+1)>] = z F(z) - z f(0) = z [F(z) - f(0)]$

(2) $Z[< f(n+2)>] = z^{2} F(z) - \sum\limits_{r=0}^{1} f(r) z^{2-r}$

$\Rightarrow Z[< f(n+2)>] = z^{2} F(z) - z^{2} f(0) - z f(1) = z^{2} [F(z) - f(0) - z^{-1} f(1)]$

(3) $Z\,[<f(n+3)>] = z^3 F(z) - \sum\limits_{r=0}^{2} f(r)z^{3-r}$

$\Rightarrow Z\,[<f(n+3)>] = z^3 F(z) - z^3 f(0) - z^2 f(2) - z\,f(2)$

$\Rightarrow Z\,[<f(n+3)>] = z^3\,[F(z) - f(0) - z^{-1} f(1) - z^{-2} f(2)]$ and so on.

Shifting to the Right

If $Z(f_n) = F(z)$, then $Z\,(f\,(n-k)) = z^{-k} F(2)$ $(k > 0)$

Proof. By definition, we have

$$Z[f_{n-k}] = \sum\limits_{n=0}^{\infty} f_{n-k} z^{-n} = f_0 z^{-k} + f_1 z^{-(k+1)} + \ldots$$

$$= z^{-k} \sum\limits_{n=0}^{\infty} f_n z^{-n} = z^{-k} F(z)$$

Shifting to the left

If $Z(f_n) = F(z)$, then $Z(f(n+k) = z^k\,[F(z) - f_0 - f_1 z^{-1} - f_2 z^{-2} - \ldots - f_{k-1} z^{-(k-1)}]$

(JNTU–2002)

Proof. We have

$$Z[f_{n+k}] = \sum\limits_{n=0}^{\infty} f_{n+k} z^{-n} = z^k \left[\sum\limits_{n=0}^{\infty} f_n z^{-n} - \sum\limits_{n=0}^{k-1} u_n z^{-n} \right]$$

Hence, $Z(f(n+k) = z^k\,[F(z) - f_0 - f_1 z^{-1} - f_2 z^{-2} - \ldots - f_{k-1} z^{-(k-1)}]$

7.4 Z-TRANSFORM OF nf(n)

If $z\,[<f(x)>] = F(z)$, *then* $Z[<nf(n)>] = -z\dfrac{d\,F(z)}{dz}$

Proof. We have

$$Z[<nf(n)>] = \sum\limits_{n=-\infty}^{\infty} n\,f(n)\,z^{-n} = -z \sum\limits_{n=-\infty}^{\infty} [-nf(n)]z^{-n-1}$$

$$= -z \sum\limits_{n=-\infty}^{\infty} f(n)(-nz^{-k-1}) = -z \sum\limits_{n=-\infty}^{\infty} f(n)\frac{d(z^{-n})}{dz}$$

$$= -z\frac{d}{dz} \sum\limits_{n=-\infty}^{\infty} f(n)z^{-n} = -z\frac{d}{dz} F(z)$$

i.e., $Z[<n\,f(n)>] = -z\dfrac{d}{dz} F(z)$

REMARK

- The above result can be generalized as follows

$$Z[<n^k f(n) > 1] = \left(-z\frac{d}{dz} \right)^k F(z)$$

7.5 Z-TRANSFORM OF f(n)/n

If $z\,[<f(n)>] = F(z)$, *then* $z\left[\dfrac{f(n)}{n} \right] = \int\limits_{z}^{z} \dfrac{1}{z} F(z)\,dz$

Proof. We have

$$z\left[\frac{f(n)}{n}\right] = \sum_{n=-\infty}^{\infty} \frac{f(n)}{n} z^{-n} = \sum_{n=-\infty}^{\infty} f(n)\left(\frac{1}{n}z^{-n}\right) = -\sum_{n=-\infty}^{\infty} f(n)\int_z z^{-n-1} dz$$

$$= -\int_z \sum_{n=-\infty}^{\infty} f(n) z^{-n-1} dz = -\int_z \frac{1}{z} \sum_{n=-\infty}^{\infty} f(n) z^{-n} dz = -\int_z z^{-1} F(z) dz$$

Hence, $Z\left[\left\{\frac{f(n)}{n}\right\}\right] = -\int_z z^{-1} F(z) dz$

7.6 INITIAL VALUE THEOREM

If $Z[<f(n)>] = F(z)$, $n > 0$, then $f(0) = \lim\limits_{z \to \infty} F(z)$

Proof. Here we have

$$Z[< f(n) >] = \sum_{n=0}^{\infty} f(n) z^{-n} = F(z)$$

or $f(0) + f(1)z^{-1} + f(2)z^{-2} + \dots = F(z)$

Taking limits of both the sides, as $z \to \infty$, we get

$$f(0) = \lim_{z \to \infty} F(z)$$

7.7 FINAL VALUE THEOREM

If $Z[<f(n)>] = F(z)$, $n > 0$, then $\lim\limits_{z \to \infty} F(n) = \lim\limits_{z \to 1} (z-1) F(z)$

Proof. We have $Z[< f(n+1) > -f(n)] = \sum_{n=0}^{\infty} [f(n+1) - f(n)] z^{-n}$

or $\quad Z[<f(n+1)>] - Z[<f(n)>] = \sum_{n=0}^{\infty} [f(n+1) - f(n)] z^{-n}$

$$zF(z) - zF(0) - F(z) = \lim_{k \to \infty} \sum_{n=0}^{k} [f(n+1) - f(n)] z^{-n}$$

Taking the limits of both the sides as $z \to 1$, we get

$$\lim_{z \to 1} (z-1) F(z) = f(0) + \lim_{z \to 1} \lim_{k \to \infty} \sum_{n=0}^{k} [f(n+1) - f(n)] z^{-n}$$

$$= f(0) + \lim_{k \to \infty} \sum_{n=0}^{k} \lim_{z \to 1} [f(n+1) - f(n)] z^{-n}$$

$$\text{(On changing the order of limits)}$$

$$= f(0) + \lim_{k \to \infty} \sum_{n=0}^{k} [f(n+1) - f(n)]$$

$$= \lim_{k \to \infty} [f(0) - f(0) + f(1) - f(1) + f(2) - f(2) + \dots + f(n+1) - f(n)]$$

$$= \lim_{k \to \infty} (k+1) = \lim_{k \to \infty} f(k) = \lim_{n \to \infty} f(n)$$

Hence, $\quad\quad \lim\limits_{n \to \infty} f(n) = \lim\limits_{z \to 1} (z-1) F(z)$

7.8 PARTIAL SUM THEOREM

If $Z[\{f(n)\}] = F(z)$, *then* $Z\left[\left\langle \sum\limits_{k=-\infty}^{n} f(k) \right\rangle\right] = \dfrac{F(z)}{1-z^{-1}}$

Proof. Let $\{u(n)\}$ be a sequence such that $u(n) = \sum\limits_{k=-\infty}^{n} f(k)$ ...(1)

We want to find the value of $Z[\{u(n)\}]$.

From (1), we can write $u(n) - u(n-1) = \sum\limits_{k=-\infty}^{n} f(x) - \sum\limits_{k=-\infty}^{n-1} f(x)$

or $u(n) - u(n-1) = f(n)$

Taking Z-transforms of both the sides, we get

$$Z[\{u(n)\}] - Z[\{u(n-1)\}] = Z[\{f(n)\}]$$

or $U(z) - z^{-1} U(z) = F(z)$, where $U(z) = Z[\{u(n)\}]$

$\Rightarrow$ $(1 - z^{-1}) U(z) = F(z)$

$\Rightarrow$ $U(z) = \dfrac{F(z)}{1-z^{-1}}$,   *i.e.*, $Z[\{u(n)\}] = \dfrac{F(z)}{1-z^{-1}}$

Hence, $Z\left[\left\langle \sum\limits_{k=-\infty}^{n} f(k) \right\rangle\right] = \dfrac{F(z)}{1-z^{-1}}$

7.9 CONVOLUTION THEOREM OF Z-TRANSFORMS

Definition : *Let* $< f(n) >$ *and* $< g(n) >$ *be two sequence and let the convolution of* $< f(n) >$ *and* $< g(n) >$ *be* $< h(n) >$ *where* $< h(n) > = < f(n) >^* < g(n) >$, *then* $< h(n) >$ *is defined as*

$$< h(n) > = < f(n) >^* < g(n) >$$

$$= \sum\limits_{n=-\infty}^{\infty} f(n) g(n-k) = \sum\limits_{n=-\infty}^{\infty} g(n) f(n-k) = < g(n) >^* < f(n) >$$

Theorem-1. *If* $Z[<f(n)>] = F(z)$ *and* $Z[< g(n) > = G(z)$, *then* $Z[< h(n) >] = Z[<f(n)>^* < g(n) >] = F(z) G(z)$ *where the region of convergence of* $Z[<h(n)>]$ *is the common region of convergence of* $F(z)$ *and* $G(z)$.

<u>PROOF.</u>   We have, by defintion

$$Z[< h(n) >] = \sum\limits_{n=-\infty}^{\infty} \left[\sum\limits_{n=-\infty}^{\infty} f(k) g(n-k) \right] z^{-n} = \sum\limits_{n=-\infty}^{\infty} \sum\limits_{n=-\infty}^{\infty} f(k) g(n-k) z^{-n}$$

$$= \sum\limits_{k=-\infty}^{\infty} f(k) z^{-k} \sum\limits_{r=-\infty}^{\infty} g(r) z^{-r} \qquad (n-k = r)$$

$$= \sum\limits_{k=-\infty}^{\infty} \hat{f}(k) z^{-k} G(z) = F(z) . G(z)$$

7.10 INVERSE Z-TRANSFORMS

If $F(z)$ is the Z-transform of the sequence $< f(k) >$, then $<f(k)>$ is called the inverse Z-transform of $F(z)$. The operator for inverse Z-transform is denoted by Z^{-1}. Thus, if $F < f(k) > = F(z)$, then

$$Z^{-1}[F(z)] = < f(k) >$$

Some Useful Z-Transforms

Table 1

S.No.	Sequence $f(n)$, $n \geq 0$	Z-transform : F(z)	S.No.	Sequence $f(n)$, $n \geq 0$	Z-transform : F(z)
1.	n	$z/(z-1)^2$	13.	$\sinh n\theta$	$\dfrac{z \sinh\theta}{z^2 - 2z\cosh\theta + 1}$
2.	n^2	$(z^2+z)/(z-1)^3$	14.	$\cosh n\theta$	$\dfrac{z(z - \cosh\theta)}{z^2 - 2z\cosh\theta + 1^2}$
3.	n^p	$-zd/dz\,[Z(n^{p-1})]$, positive integer.	15.	$a^n \sinh n\theta$	$\dfrac{az\sinh\theta}{z^2 - 2az\cosh\theta + a^2}$
4.	$\delta(n) = \begin{cases} 1, & n = 0 \\ 0, & n \neq 0 \end{cases}$	1	16.	$a^n \cosh n\theta$	$\dfrac{z(z - 1\cosh\theta)}{z^2 - 2az\cosh\theta + a^2}$
5.	$f(n) = \begin{cases} 1, & n < 0 \\ 0, & n \geq 0 \end{cases}$	$z/(z-1)$	17.	$a^n f(n)$	$(F(z/a))$
6.	a^n	$z/(z-a)$	18.	$\begin{matrix} f_{n+1} \\ f_{n+2} \\ f_{n+3} \end{matrix}$	$\begin{matrix} z[F(z) - f_0] \\ z^2[F(z) - f_0 - f_1 z^{-1}] \\ z^3[F(z) - f_0 - f_1 z^{-1} - f_2 z^{-2}] \end{matrix}$
7.	na^n	$az/(z-a)^2$	19.	f_{n-k}	$z^{-k} F(z)$
8.	$n^2 a^n$	$(az^2 + a^2 z)/(z-a)^3$	20.	$n f_n$	$-zd/dz\,[F(z)]$
9.	$\sin n\theta$	$\dfrac{z\sin\theta}{z^2 - 2z\cos\theta + 1}$	21.	$\dfrac{1}{n} f_n$	$-\int_0^z z^{-1}[F(z)]dz$
10.	$\cos n\theta$	$\dfrac{z(z - \cos\theta)}{z^2 - 2z\cos\theta + 1}$	22.	f_0	$\displaystyle\lim_{z \to \infty} F(z)$
11.	$a^n \sin n\theta$	$\dfrac{az\sin\theta}{z^2 - 2az\cos\theta + a^2}$	23.	$\displaystyle\lim_{n \to \infty} f_n$	$\displaystyle\lim_{z \to 1}[(z - 1)F(z)]$
12.	$a^n \cos n\theta$	$\dfrac{z(z - a\cos\theta)}{z^2 - 2az\cos\theta + a^2}$			

Table 2

S.No.	Sequence $f(n)$, $n \geq 0$	Z-transform : F(z)	S.No.	Sequence $f(n)$, $n \geq 0$	Z-transform : F(z)
1.	$au_n + bv_n + \ldots$	$aU(z) + bV(z) + \ldots$	8.	c	$c/1 - z$
2.	u_{n+k}	$k = 1,$ $z^{-k}U(z) - \sum\limits_{r=0} u_r z^{r-k}$	9.	n^3	$z(z^2 + uz + 1)/(1 - z)^4$
3.	$y_n = \begin{cases} 0 & (n < k) \\ u_{n-k} & (n \geq k) \end{cases}$	$z^k U(z)$	10.	c^n	$1/(1 - cz)$
4.	$c^{an} u^n$	$U(c^a z)$	11.	$n^2 c^n$	$cz(1 + cz)/(1 - cz)^3$

5.	$\sum_{r=0}^{n} uv_{n-r}$	$U(z)V(z)$	12.	$n^3 c^n$	$cz(c^2 z^2 + 4cz + 1)/(1-cz)^4$
6.	$\Delta u_n = u_{n+1} - u_n$	$z^{-1}[(1-z)U(z) - u_0]$	13.	$c^n n^k$	$z\,du(cz)dz$ with $u_n = n^{k-1}$
7.	$y_n = \begin{cases} 1, & (n=k) \\ 0 & (n \neq k) \end{cases}$	z^k	14.	$\left(\dfrac{n+k}{k}\right)c^n$	$1/(1-cz)^{k+1}$
15.	$\left(\dfrac{b}{n}\right)c^n a^{b-n}$	$(a+cz)^b$	21.	$c^n/n!$ $(n=0,2,4,...)$	$\cosh cz$
16.	c^n/n $(n=1,2,3,...)$	$-\log(1-cz)$	22.	$(\log c)^n/n!$	c^z
17.	c^n/n $(n=1,3,5,...)$	$\dfrac{1}{2}\log\left(\dfrac{1+cz}{1-cz}\right) = \tanh^{-1} cz$	23.	$b^{-an}\sin(nc)$	$\dfrac{z \sin c}{b^a + b^{-a} z^2 - 2z \cos c}$
18.	c^n/n $(n=2,4,6,...)$	$-\dfrac{1}{2}\log(1-c^2 z^2)$	24.	$b^{-an}\cos(nc)$	$\dfrac{b^a - z \cos c}{b^{-a} z^2 + b^a - 2z \cos c}$
19.	$c^n \cdot n!$	e^{cz}	25.	$[y_{ij}(n)]$ (a matrix)	$[y_{ij}(z)]$
20.	$c^n/n!$ $(n=1,3,5,...)$	$\sinh cz$			

Some Useful Inverse Z-Transforms : Table 3

S.No.	F(z)	Inverse Z-transform f(n)	S.No.	F(z)	Inverse Z-transform f(n)
1.	$\dfrac{z}{z-a}$	$a^n u(n)$	4.	$\dfrac{1}{z-a}$	$a^{n-1} u(n-1)$
2.	$\dfrac{z^2}{(z-a)^2}$	$(n+1)a^n u(n)$	5.	$\dfrac{1}{(z-a)^2}$	$(n-1)a^{n-2} u(n-2)$
3.	$\dfrac{z^3}{(z-a)^3}$	$\dfrac{1}{2!}(n+1)(n+2)a^n u(n)$	6.	$\dfrac{1}{(z-a)^3}$	$\dfrac{1}{2}(n-1)(n-2)a^{n-3} u(n-3)$

Solved Examples

EXAMPLE 1. *Find the Z-transforms of the following sequences :*

(i) $<f(n)> = \{15, 10, 7, 4, 1, -1, 0, 3, 6\}$

(ii) $<f(n)> = \{15, 10, 7, 4, 1\}$

(iii) $<f(n)> = 1/3^n$

(iv) $<f(n)> = \dfrac{1}{2^n}, \ -3 \leq n \leq 2$

(v) $<f(n)> = \{a^n\}, n \geq 0$

SOLUTION. (i) We have $Z[<f(n)>]$

$$= F(z) = \sum_{n=-3}^{5} f(n)z^{-n}$$

$$= 15z^3 + 10z^2 + 7z + 4 + \frac{1}{z} - \frac{1}{z^2} + 0 + \frac{3}{z^4} + \frac{6}{z^5}$$

Therefore, $Z[<f(n)>]$

$$= 15z^3 + 10z^2 + 7z + 4 + \frac{1}{z} - \frac{1}{z^2} + 0 + \frac{3}{z^4} + \frac{6}{z^5}$$

(ii) We have $Z[<f(n)>] = F(z)$

$$= \sum_{n=0}^{4} f(n)z^{-n} = 15 + \frac{10}{z} + \frac{7}{z^2} + \frac{4}{z^3} + \frac{1}{z^4}$$

(iii) $Z[<f(n)>] = F(z)$

$$= \sum_{n=-\infty}^{\infty} f(n)z^{-n} = \sum_{n=-\infty}^{\infty} \frac{1}{3^n} z^{-n}$$

$$= \ldots + 27z^3 + 9z^2 + 3z + 1 + \frac{1}{3z} + \frac{1}{9z^2} + \frac{1}{27z^3} + \ldots$$

(iv) $Z[<f(n)>] = F(z) = \sum_{n=-3}^{2} \frac{1}{2^n} z^{-n}$

$$= 8z^3 + 4z^2 + 2z + 1 + \frac{1}{2z} + \frac{1}{4z^2}$$

(v) $Z[\{f(n)\}] = F(z)$

$$= \sum_{n=0}^{\infty} a^n z^{-n} = 1 + \frac{a}{z} + \frac{a^2}{z^2} + \frac{a^3}{z^3} + \ldots$$

$$= \frac{1}{1 - \dfrac{a}{z}} = \frac{z}{z - a}$$

EXAMPLE 2. *Find the Z-transforms of the following sequences :*

    (i) $< f(n) > = < a^{|n|} >$

    (ii) $< f(n) > = \left\{ \dfrac{a^n}{n!} \right\}, n \geq 0$ (SVTU–2009)

    (iii) $< f(n) > = \{C^n \cos(\alpha_n)\}, n \geq 0$ (UPTU–2004)

    (iv) $< f(n) > = \{\sin(3n+5)\}, n \geq 0$ (VTU–2009; Kottayam–2005)

    (v) $< f(n) > = \{C^n \cosh(\alpha_n)\}, n \geq 0$

SOLUTION.   (i) We have

$$Z[< a^{|n|} >] = \sum_{n=-\infty}^{\infty} a^{|n|} z^{-n} = \sum_{n=-\infty}^{-1} a^{-n} z^{-n} + \sum_{n=0}^{\infty} a^n z^{-n}$$

$$= [\ldots + a^3 z^3 + a^2 z^2 + az] + [1 + az^{-1} + a^2 z^{-2} + a^3 z^{-3} + \ldots]$$

$$= \frac{az}{1 - az} + \frac{1}{1 - az^{-1}} = \frac{az}{1 - az} + \frac{z}{z - a}$$

$$= \frac{az(z - a) + z(1 - az)}{(1 - az)(z - a)} = \frac{z(1 - a^2)}{(1 - az)(z - a)}$$

    (ii) $Z\left[\left\langle \dfrac{a^n}{n!} \right\rangle \right] = \sum_{n=0}^{\infty} \left(\dfrac{a^n}{n!} \right) z^{-n} = \sum_{n=0}^{\infty} \dfrac{(az^{-1})}{n!} \cdot z^{-n}$

$$= \sum_{n=0}^{\infty} \frac{(az^{-1})^n}{n!} = 1 + \frac{az^{-1}}{1!} + \frac{(az^{-1})^2}{2!} + \frac{(az^{-1})^3}{3!}$$

$$= e^{az^{-1}} = e^{a/z}$$

(iii) We have $Z\left[\left\langle C^n \cos(\alpha n)\right\rangle\right]$

$$= \sum_{n=0}^{\infty} [C^n \cos(\alpha n)] z^{-n} = \sum_{n=0}^{\infty} C^n \left[\frac{e^{i\alpha n} + e^{-i\alpha n}}{2}\right] z^{-n}$$

$$= \frac{1}{2}\sum_{n=0}^{\infty} (C^n e^{i\alpha n}) z^{-n} + \frac{1}{2}\sum_{n=0}^{\infty} (C^n e^{-i\alpha n}) z^{-n}$$

$$= \frac{1}{2}\sum_{n=0}^{\infty} (C^n e^{i\alpha} z^{-1})^n + \frac{1}{2}\sum_{n=0}^{\infty} (C^n e^{-i\alpha} z^{-1})^n$$

$$= \frac{1}{2}\left[1 + Ce^{i\alpha}z^{-1} + (Ce^{i\alpha}z^{-1})^2 + (Ce^{i\alpha}z^{-1})^3 + ...\right]$$

$$+ \frac{1}{2}\left[1 + Ce^{-i\alpha}z^{-1} + (Ce^{-i\alpha}z^{-1})^2 + (Ce^{-i\alpha}z^{-1})^3 + ...\right]$$

$$= \frac{1}{2}\left[\frac{1}{1 - Ce^{i\alpha}z^{-1}}\right] + \frac{1}{2}\left[\frac{1}{1 - Ce^{-i\alpha}z^{-1}}\right] \qquad [\because |z| > |C|]$$

$$= \frac{1}{2}\left[\frac{1 - Ce^{-i\alpha}z^{-1} + 1 - Ce^{i\alpha}z^{-1}}{(1 - Ce^{i\alpha}z^{-1})(1 - Ce^{-i\alpha}z^{-1})}\right]$$

$$= \frac{1}{2}\left[\frac{2 - Cz^{-1}(e^{i\alpha} + e^{-i\alpha})}{1 - Ce^{-i\alpha}z^{-1} - Ce^{i\alpha}z^{-1})C^2z^{-2}}\right]$$

$$= \frac{1}{2}\left[\frac{2 - Cz^{-1}(e^{i\alpha} + e^{-i\alpha})}{1 - Ce^{-i\alpha}z^{-1} - Ce^{i\alpha}z^{-1} + C^2z^{-2}}\right]$$

$$= \frac{\left[1 - C\left(\dfrac{e^{i\alpha} + e^{-i\alpha}}{2}\right)z^{-1}\right]}{1 - C(e^{i\alpha} + e^{-i\alpha})z^{-1} + C^2z^{-2}}$$

$$= \frac{1 - (C\cos\alpha)z^{-1}}{1 - (2C\cos\alpha)z^{-1} + C^2z^{-2}} = \frac{z^2 - Cz\cos\alpha}{z^2 - 2Cz\cos\alpha + C^2}$$

(iv) We have

$$Z[< \sin(3n + 5) >] = \sum_{n=0}^{\infty} \sin(3n + 5) z^{-n}$$

$$= \sum_{n=0}^{\infty}\left[\frac{e^{i(3n+5)} - e^{-i(3n+5)}}{2!}\right] z^{-n}$$

$$= \frac{1}{2!}\sum_{n=0}^{\infty} e^{i(3n+5)} \cdot z^{-n} - \frac{1}{2!}\sum_{n=0}^{\infty} e^{-i(3n+5)} z^{-n}$$

$$= \frac{e^{5!}}{2!}\sum_{n=0}^{\infty} (e^{3i} - z^{-1})^n - \frac{e^{-5i}}{2!}\sum_{n=0}^{\infty} (e^{-3i}z^{-1})^n$$

$$= \frac{e^{5i}}{2!}[1 + e^{3i}z^{-1} + (e^{3i}z^{-1})^2 + (e^{3i}z^{-1})^3 + ...]$$

$$- \frac{e^{-5i}}{2!}[1 + e^{-3i}z^{-1} + (e^{-3i}z^{-1})^2 + (e^{-3i}z^{-1})^3 + ...]$$

$$= \frac{e^{5i}}{2!}\left[\frac{1}{1-e^{3i}z^{-1}}\right] - \frac{e^{-5i}}{2i}\left[\frac{1}{1-e^{-3i}z^{-1}}\right] \qquad (|Z| > 1)$$

$$= \frac{1}{2!}\left[\frac{e^{5i}(1-e^{-3i}z^{-1}) - e^{-5i}(1-e^{3i}z^{-1})}{(1-e^{3i}z^{-1})(1-e^{-3i}z^{-1})}\right]$$

$$= \frac{1}{2!}\left[\frac{(e^{5i}-e^{-5i}) - e^{2i}z^{-1} + e^{-2i}z^{-1}}{1-e^{3i}z^{-1} - e^{-3i}z^{-1} + z^{-2}}\right]$$

$$= \frac{\left(\dfrac{e^{5i}-e^{-5i}}{2!}\right) - \left(\dfrac{e^{2i}-e^{-2i}}{2!}\right)z^{-1}}{1-(2\cos 3)z^{-1} + z^{-2}}$$

$$= \frac{\sin 5 - (\sin 2)z^{-1}}{1-(2\cos 3)z^{-1}+z^{-2}} = \frac{z^2\sin 5 - z\sin 2}{z^2 - 2z\cos 3 + 1}$$

(v) We have $Z[< C^n\cosh(\alpha n)>]$

$$= \sum_{n=0}^{\infty}[C^n\cosh(\alpha n)]z^{-n}$$

$$= \sum_{n=0}^{\infty}C^n\left[\frac{e^{\alpha n}+e^{-\alpha n}}{2}\right]z^{-n} = \frac{1}{2}\sum_{n=0}^{\infty}(Ce^{\alpha}z^{-1})^n + \frac{1}{2}\sum_{n=0}^{\infty}(Ce^{-\alpha}z^{-1})^n$$

$$= \frac{1}{2}[1 + Ce^{\alpha}z^{-1} + (Ce^{\alpha}z^{-1})^2 + (Ce^{\alpha}z^{-1})^3 + ...]$$

$$+ \frac{1}{2}[1 + Ce^{-\alpha}z^{-1} + (Ce^{-\alpha}z^{-1})^2 + (Ce^{-\alpha}z^{-1})^3 + ...]$$

$$= \frac{1}{2}\left[\frac{1}{1-Ce^{\alpha}z^{-1}}\right] + \frac{1}{2}\left[\frac{1}{1-Ce^{-\alpha}z^{-1}}\right]$$

$$[\because |Z| > |Ce^{\alpha}| \text{ and } |Z| > |Ce^{-\alpha}|]$$

$$= \frac{1}{2}\left[\frac{1-Ce^{-\alpha}z^{-1} + 1 - Ce^{\alpha}z^{-1}}{1-Ce^{\alpha}z^{-1} - Ce^{-\alpha}z^{-1} + C^2z^{-2}}\right]$$

$$= \frac{1}{2}\left[\frac{2-C(e^{\alpha}+e^{-\alpha})z^{-1}}{1-C(e^{\alpha}+e^{-\alpha})z^{-1} + C^2z^{-2}}\right] = \frac{1-(C\cosh\alpha)z^{-1}}{1-(2C\cosh\alpha)z^{-1} + C^2z^{-2}}$$

$$= \frac{z^2 - Cz\cosh\alpha}{z^2 - 2Cz\cosh\alpha + C^2} = \frac{z(z-C\cosh\alpha)}{z^2 - 2Cz\cosh\alpha + C^2}$$

EXAMPLE 3. *Show that* $Z\left(\dfrac{1}{n!}\right) = e^{1/z}$. *Hence, evaluate* $Z\left[\dfrac{1}{(n+1)!}\right]$ *and* $Z\left[\dfrac{1}{(n+2)!}\right]$.

SOLUTION. We have $Z\left[\dfrac{1}{n!}\right] = \sum_{n=0}^{\infty}\dfrac{1}{n!}z^{-n} = 1 + \dfrac{z^{-1}}{1!} + \dfrac{z^{-2}}{2!} + \dfrac{z^{-3}}{3!} + ... = e^{1/z}$

Now, shifting $\left(\dfrac{1}{n!}\right)$ one unit to the left gives

$$Z\left[\frac{1}{(n+1)!}\right] = z\left[z\left(\frac{1}{n!}\right) - 1\right] = z(e^{1/z} - 1)$$

Again, shifting $\left(\dfrac{1}{n!}\right)$ two units to the left gives

$$Z\left[\frac{1}{(n+2)!}\right] = z^2(e^{1/z} - 1 - z^{-1})$$

EXAMPLE 4. *Find the Z-transforms of discrete unit step function* $F(n) = \begin{cases} 1, n < 0 \\ 0, n \geq 0 \end{cases}$

SOLUTION. We have $Z = [\{F(n)\}] = \displaystyle\sum_{n=0}^{\infty} F(n)z^{-n}$

$$= \left[1 + z^{-1} + z^{-2} + z^{-3} + \ldots\right] = \frac{1}{1 - z^{-1}} = \frac{1}{1 - \dfrac{1}{z}} = \frac{z}{z - 1}$$

EXAMPLE 5. *Find the Z-transforms of* $\{f(n)\}$*, where*

$$F(n) = \begin{cases} 5^n, \ n < 0 \\ 3^n, \ n \geq 0 \end{cases}$$

SOLUTION. We have $Z = [\{F(n)\}] = \displaystyle\sum_{n=0}^{-1} 5^n z^{-n} + \sum_{n=0}^{\infty} 3^n z^{-n}$

$$= [\ldots + 5^{-3}z^3 + 5^{-2}z^2 + 5^{-1}z^1] + \left[1 + \frac{3}{z} + \frac{9}{z^2} + \frac{27}{z^3} + \ldots\right]$$

$$= [5^{-1}z + 5^{-2}z^2 + 5^{-3}z^3 + \ldots] + \left[1 + \frac{3}{z} + \frac{9}{z^2} + \ldots\right]$$

$$= \frac{5^{-1}z}{1 - 5^{-1}z} + \frac{1}{1 - \dfrac{3}{z}} = \frac{z}{5 - z} + \frac{z}{3 - z}$$

where $\left|\dfrac{z}{5}\right| < 1$ and $\left|\dfrac{3}{z}\right| < 1$

$$= \frac{z^2 - 3z + 5z - z^2}{(5-z)(z-3)} = \frac{-2z}{z^2 - 2z + 15}$$

Fig. 1

REMARK

- The above two series are convergent in annulus , where $|z| < 3$ and $|z| < 5$.

EXAMPLE 6. *Find the Z-transform of* $\left\{\cos\left(\dfrac{n\pi}{8} + \alpha\right)\right\}$*.*

SOLUTION. We have $Z\left\{\cos\left(\dfrac{n\pi}{8} + \alpha\right)\right\} = \displaystyle\sum \cos\left(\frac{n\pi}{8} + \alpha\right)z^{-n}$

$$= \sum\left[\cos\frac{n\pi}{8}\cos\alpha - \sin\frac{n\pi}{8}\sin\alpha\right]z^{-n}$$

$$= \sum\cos\frac{n\pi}{8}\cos\alpha \, z^{-n} - \sum\sin\frac{n\pi}{8}\sin\alpha z^{-n}$$

$$= \cos\alpha\sum\cos\frac{n\pi}{8}z^{-n} - \sin\alpha\sum\frac{n\pi}{8}z^{-n}$$

$$= \cos\alpha\frac{z^2 - z\cos\dfrac{\pi}{8}}{z^2 - 2z\cos\dfrac{\pi}{8} + 1} - \sin\alpha\frac{z\sin\dfrac{\pi}{8}}{z^2 - 2z\cos\dfrac{\pi}{8} + 1}$$

$$= \frac{\left(z^2 - z\cos\dfrac{\pi}{8}\right)\cos\alpha - z\sin\dfrac{\pi}{8}\sin\alpha}{z^2 - 2z\cos\dfrac{\pi}{8} + 1}$$

$$= \frac{z^2\cos\alpha - z\cos\left(\dfrac{\pi}{8} - \alpha\right)}{z^2 - 2z\cos\dfrac{\pi}{8} + 1}$$

EXAMPLE 7. *Find the Z-transform of* $\cosh\left(\dfrac{n\pi}{2} + \alpha\right)$. *(VTU–2011, UPTU–2008)*

SOLUTION. We have $F(z) = \displaystyle\sum_{n=0}^{\infty} \cosh\left(\dfrac{n\pi}{8} + \alpha\right) z^{-k}$

$$= \sum_{n=0}^{\infty}\left[\frac{e^{(n\pi/2)+\alpha} - e^{-(n\pi/2)+\alpha)}}{2}\right] z^{-n}$$

$$= \frac{1}{2}e^{\alpha}\sum_{n=0}^{\infty} (e^{\pi/2}z^{-1})^n + \frac{1}{2}e^{-\alpha}\sum_{n=0}^{\infty} (e^{-\pi/2}z^{-1})^n$$

$$= \frac{1}{2}e^{\alpha}\left[1 + \left(e^{\pi/2}z^{-1}\right) + \left(e^{\pi/2}z^{-1}\right)^2 + \dots\right]$$
$$+ \frac{1}{2}e^{-\alpha}\left[1 + \left(e^{-\pi/2}z^{-1}\right) + \left(e^{-\pi/2}z^{-1}\right)^2 + \dots\right]$$

[Sum of two G.P.'s]

$$= \frac{1}{2}e^{\alpha}\left[\frac{1}{e^{-\pi/2}z^{-1}}\right] + \frac{1}{2}e^{-\alpha}\left[\frac{1}{1 - e^{-\pi/2}z^{-1}}\right]$$

$$= \frac{1}{2}\left[\frac{e^{\alpha}(1 - e^{-\pi/2}z^{-1}) + e^{-\alpha}(1 - e^{\pi/2}z^{-1})}{(1 - e^{-\pi/2}z^{-1})(1 - e^{-\pi/2}z^{-1})}\right]$$

$$= \frac{\dfrac{e^{\alpha} + e^{-\alpha}}{2} - \dfrac{e^{\alpha - \frac{\pi}{2}} + e^{-\alpha + \frac{\pi}{2}}}{2}}{1 - e^{\pi/2}z^{-1} - e^{-\pi/2}z^{-1} + z^{-2}}$$

$$= \frac{\cosh\alpha - \cosh(\alpha - \pi/2)\cdot z^{-1}}{1 - \left(2\cosh\dfrac{\pi}{2}\right)z^{-1} + z^{-2}}$$

$$= \frac{z^2\cosh\alpha - z\cosh\left(\dfrac{\pi}{2} - \alpha\right)}{z^2 - 2z\cosh\dfrac{\pi}{2} + 1}$$

EXAMPLE 8. *Find the Z-transform of*

(i) ^nC_k (ii) $^{k+n}C_n$ (iii) $^{k+n}C_n\, a^k$

SOLUTION. (i) We have $Z\left[\{^nC_k\}\right] = \displaystyle\sum_{k=0}^{n} {}^nC_k z^{-k}$

$$= 1 + {}^nC_1 z^{-1} + {}^nC_2 z^{-2} + \dots + {}^nC_n z^{-k} = (1 + z^{-1})^n$$

(ii) We have

$$Z\left[\left\{^{k+n}C_n\right\}\right] = \sum_{k=0}^{\infty} {}^{k+n}C_n z^{-k}$$

$$= \sum_{k=0}^{\infty} {}^{k+n}C_k z^{-k} \qquad [\because {}^nC_r = {}^nC_{n-r}]$$

$$= 1 + {}^{n+1}C_1 z^{-1} + {}^{n+2}C_2 z^{-2} + {}^{n+3}C_3 z^{-3} + \ldots$$

$$= 1 + (n+1)z^{-1} + \frac{(n+2)(n+1)}{2!} z^{-2}$$

(iii) We have

$$Z\left[\left\{^{k+n}C_n a^k z^{-k}\right\}\right] = \sum_{k=0}^{\infty} {}^{k+n}C_k a^k z^{-k} \qquad (\because {}^nC_r = {}^nC_{n-r})$$

$$= \sum_{k=0}^{\infty} {}^{k+n}C_k (az^{-1})^k \quad |z| > a = (1 - az^{-1})^{-(n+1)}$$

EXAMPLE 9. *Find Z-transforms of the following sequences by using the change of scale property*

(i) $<a^n>$, $n \geq 0$ (ii) $<C^n \sin(\alpha n)>$, $n \geq 0$

SOLUTION.

(i) We have

$$Z[<1>] = \sum_{n=0}^{\infty} 1.z^{-n} = 1 + \frac{1}{z} + \frac{1}{z^2} + \frac{1}{z^3} + \ldots \qquad \text{(Infinite G.P.)}$$

$$= \frac{1}{1 - \dfrac{1}{z}} = \frac{z}{z-1}$$

Using change of scale property, *i.e.*, if $Z\ [<\{f(n)>] = F(z)$, then $Z\ [<a^n f(n)>] = F(z/a)$, we have

$$Z\left[<a^n>\right] = z\left[<a^n - 1>\right] = \frac{(z/a)}{(z/a) - 1} = \frac{z}{z - a}$$

(ii) We have

$$Z[<\sin\alpha n>] = \sum_{n=0}^{\infty} (\sin\alpha n)z^{-n}$$

$$= \sum_{n=0}^{\infty} \left[\frac{e^{i\alpha n} - e^{-i\alpha n}}{2i}\right] z^{-n}$$

$$= \frac{1}{2i} \sum_{n=0}^{\infty} (e^{i\alpha} z^{-1})^n - \frac{1}{2!} \sum_{n=0}^{\infty} (e^{-i\alpha} z^{-1})^n$$

$$= \frac{1}{2i}\left[1 + e^{i\alpha} z^{-1} + (e^{i\alpha} z^{-1})^2 + \ldots\right]$$

$$\qquad\qquad - \frac{1}{2i}\left[1 + e^{-i\alpha} z^{-1} + (e^{-i\alpha} z^{-1})^2 + \ldots\right]$$

$$= \frac{1}{2i}\left[\frac{1}{1 - e^{i\alpha} z^{-1}}\right] - \frac{1}{2i}\left[\frac{1}{1 - e^{-i\alpha} z^{-1}}\right]$$

$$= \frac{1}{2i}\left[\frac{1 - e^{-i\alpha} z^{-1} - (1 - e^{i\alpha} z^{-1})}{1 - e^{i\alpha} z^{-1} - e^{-i\alpha} z^{-1} + z^{-2}}\right]$$

$$= \frac{1}{2i}\left[\frac{(e^{i\alpha} - e^{-i\alpha})z^{-1}}{1 - (e^{i\alpha} + e^{-i\alpha})z^{-1} + z^{-2}}\right]$$

$$= \frac{(\sin \alpha) z^{-1}}{1 - (2 \cos \alpha) z^{-1} + z^{-2}} = \frac{z \sin \alpha}{z^2 - 2z \cos \alpha + 1}$$

Now, using the change of scale property, we have

$$Z \left[< c^n \sin \alpha n > \right] = \frac{(z/c) \sin \alpha}{(z/c)^2 - (2z/c) \cos \alpha + 1} = \frac{cz \sin \alpha}{z^2 - 2cz \cos \alpha + c^2}$$

EXAMPLE 10. *Find the Z-transforms of the sequence*

$$< f(n) >= \sum_{n=0}^{\infty} z^n \sum_{n=0}^{\infty} 4^n$$

SOLUTION. We know that

$$Z[< 3^n >] = \frac{1}{1 - 3z^{-1}} \qquad \text{and} \qquad Z[< 4^n >] = \frac{1}{1 - 4z^{-1}}$$

Therefore

$$Z[<f(n)>] = Z[<3^n>] \cdot Z[<4^n>] \qquad \text{(By convolution property)}$$

$$= \left(\frac{1}{1 - 3z^{-1}} \right) \left(\frac{1}{1 - 4z^{-1}} \right) = \frac{z^2}{(z-3)(z-4)} = \frac{z^2}{z^2 - 7z + 12}$$

EXAMPLE 11. *If* $F(z) = Z[< f(n) >] = \dfrac{2z^2 + 5z + 14}{(z-1)^4}$*, then find the value of* $f(0)$.

SOLUTION. By initial value theorem, we have

$$f(0) = \lim_{z \to \infty} F(z)$$

$$= \lim_{z \to \infty} \left[\frac{2z^2 + 5z + 14}{(z-1)^4} \right] = \lim_{z \to 0} \frac{1}{z^2} \left[\frac{2 + 5z^{-1} + 14z^{-2}}{(1 - z^{-1})^4} \right] = 0$$

7.11 EVALUATION OF INVERSE Z-TRANSFORMS

(1) **Power Series Method :** This is the simplest method of finding the inverse Z-transforms. If $F(z)$ is expressed as the ratio of two polynomials which cannot be factorized, we simply divide the numerator by the denominator and take the inverse Z-transforms of each term in the quotient.

(2) **Partial Fractions Method :** In this method, we decompose $F(z)/z$ into partial fractions and multiply the resulting expansion by z and then inverting the same. We use this method in a similar fashion as that of finding the inverse Laplace transforms, using partial fractions.

(3) **Inversion Integral Method :** The inverse Z-transforms of $F(z)$ is given by the following formula

$$f(n) = \frac{1}{2\pi i} \int_c F(z) z^{n-1} dz$$

= sum of residue of $F(z) z^{n-1}$ at the poles of $F(z)$ which are inside the contour *c* drawn.

Solved Examples

EXAMPLE 1.   *Using convolution theorem, find*

$$Z^{-1}\left\{\frac{z^2}{(z-a)(z-b)}\right\}$$

SOLUTION.   Since, we know that

$$Z^{-1}\left\{\frac{z}{z-a}\right\} = a^n \qquad \text{and} \qquad Z^{-1}\left\{\frac{z}{z-a}\right\} = b^n$$

Therefore

$$Z^{-1}\left\{\frac{z^2}{(z-a)(z-b)}\right\} = Z^{-1}\left\{\frac{z}{z-a}\frac{z}{z-b}\right\}$$

$$= a^n + b^n$$

$$= \sum_{m=0}^{n} a^m.b^{n-m} = b^n \sum_{m=0}^{n}\left(\frac{a}{b}\right)^m, \text{ which in a G.P.}$$

$$= b^n.\frac{(a/b)^{n+1}-1}{\dfrac{a}{b}-1} = \frac{a^{n+1}-b^{n+1}}{a-b}$$

EXAMPLE 2.   *Find the inverse Z-transforms of* $\log\left(\dfrac{z}{z+1}\right)$ *by power series method.*

SOLUTION.   Putting $z = \dfrac{1}{t}$, $F(z) = \log\left(\dfrac{\dfrac{1}{t}}{\dfrac{1}{t}+1}\right) = -\log(1+t)$

$$= -t + \frac{t^2}{2} - \frac{t^3}{3} + \dots = -z^{-1} + \frac{1}{2}z^{-2} - \frac{1}{3}z^{-3} + \dots$$

Hence, $f(n) = \begin{cases} 0, & \text{for } n = 0 \\ (-1)^n/n, & \text{otherwise} \end{cases}$

EXAMPLE 3.   *Find the inverse Z-transform of* $\dfrac{z}{(z+1)^2}$.

SOLUTION.   We have $F(z) = \dfrac{z}{z^2+2z+1} = z^{-1} - \dfrac{2+z^{-1}}{z^2+2z+1}$ (By actual division)

$$= z^{-1} - 2z^{-2} + \frac{3z^{-1}+2z^{-2}}{z^2+2z+1}$$

$$= z^{-1} - 2z^{-2} + 3z^{-3} - \frac{4z^{-2}+3z^{-3}}{z^2+2z+1}$$

Continuing the process of division, we get an infinite series, *i.e.*,

$$F(z) = \sum_{n=0}^{\infty}(-1)^{n-1}.nz^{-n}$$

Hence, $f(n) = (-1)^{n-1}.n$

EXAMPLE 4. *Find the inverse Z-transform of the following*

(i) $\dfrac{2z^2 + 3z}{(z+2)(z-4)}$ (VTU–2008; SVTU–2007)

(ii) $\dfrac{z^3 - 20z}{(z-2)^3(z-4)}$ (VTU–2011)

SOLUTION.   (i)   We have

$$F(z) = \frac{2z^2 + 3z}{(z+2)(z-4)}$$

$$\Rightarrow \qquad \frac{F(z)}{z} = \frac{2z+1}{(z+2)(z-4)} = \frac{A}{z+2} + \frac{B}{z-4}$$

Using method of partial fraction, we get $A = 1/6$ and $B = 11/6$

Therefore

$$F(z) = \frac{1}{6} \cdot \frac{z}{z+2} + \frac{11}{6} \cdot \frac{z}{z-4}$$

On inversion, we have

$$f(n) = \frac{1}{6}(-2)^n + \frac{11}{6}(4)^n$$

(ii)   We have

$$F(z) = \frac{z^3 - 20z}{(z-2)^3(z-4)}$$

$$\Rightarrow \qquad \frac{F(z)}{z} = \frac{z^2 - 20}{(z-2)^3(z-4)} = \frac{A + Bz + Cz^2}{(z-2)^3} + \frac{D}{z-4}$$

Then, we get $D = \dfrac{1}{2}$

Multiplying throughout by $(z-2)^3 \ (z-4)$, we get

$$z^2 - 20 = (A + Bz + Cz)^2(z-4) + D(z-2)^3$$

Putting $z = 0, 1, -1$ successively and solving the resulting simultaneous equations, we get

$$A = 6, B = 0, C = 1/2$$

Hence, $F(z) = \dfrac{1}{2} \cdot \dfrac{12z + z^3}{(z-2)^3} - \dfrac{z}{z-4}$

$$= \frac{1}{2} \frac{z(z-2)^2 + 4z^2 + 8z}{(z-2)^3} - \frac{z}{z-4}$$

$$= \frac{1}{2}\left\{ \frac{z}{z-2} + 2\frac{2z^2 + 4z}{(z-2)^3} \right\} - \frac{z}{z-4}$$

On inversion, we get

$$f(n) = \frac{1}{2}(2^n + 2 \cdot n^2 2^n) - 4^n$$

$$= 2^{n-1} + n^2 2^n - 4^n$$

EXAMPLE 5.   *Find the inverse Z-transform of* $\dfrac{2(z^2 - 5z + 6.5)}{[(z-2)(z-3)^2]}$ *for* $2 < |z| < 3.$

SOLUTION.   We have

$$F(z) = \frac{2(z^2 - 5z + 6.5)}{[(z-2)(z-3)^2]} = \frac{A}{z-2} + \frac{B}{z-3} + \frac{C}{(z-3)^2}$$

Using the method of partial fraction, we get

$$A = B = C = 1$$

Therefore

$$F(z) = \frac{1}{z-2} + \frac{1}{z-3} + \frac{1}{(z-3)^2}$$

$$= \frac{1}{2}\left(1 - \frac{2}{z}\right)^{-1} - \frac{1}{3}\left(1 - \frac{z}{3}\right)^{-1} + \frac{1}{9}\left(1 - \frac{z}{3}\right)^{-2}$$

$$\left(\frac{2}{z} < 1, \frac{z}{3} < 1, \text{ i.e., } 2 < |z| < 3\right)$$

$$= \frac{1}{2z}\left[1 + \frac{2}{z} + \frac{4}{z^2} + \frac{8}{z^3} + \dots\right] - \frac{1}{3}\left[1 + \frac{z}{3} + \frac{z^2}{9} + \frac{z^3}{27} + \dots\right]$$

$$+ \frac{1}{9}\left[1 + \frac{2z}{3} + \frac{3z^2}{9} + \frac{4z^3}{27} + \dots\right]$$

$$= \left(\frac{1}{z} + \frac{2}{z^2} + \frac{2^2}{z^3} + \frac{2^3}{z^4} + \dots\right) - \left(\frac{1}{3} + \frac{z}{3^2} + \frac{z^2}{3^3} + \frac{z^3}{3^4} + \dots\right)$$

$$+ \left(\frac{1}{3^2} + \frac{2z}{3^3} + \frac{3z^2}{3^4} + \frac{4z^3}{3^5} + \dots\right)$$

$$= \sum_{n=1}^{\infty} 2^{n-1} z^{-n} - \sum_{n=0}^{\infty} \left(\frac{1}{3}\right)^{n-1} z^n + \sum_{n=0}^{\infty} (n+1)\left(\frac{1}{3}\right)^{n+2} \cdot z^n$$

Finally, on inversion, we get

$$f(u) = 2^{n-1}, \; n \geq 1$$

and $$f(n) = -(n+2)3^{n-2}, \; n \leq 0$$

EXAMPLE 6.   *Using the inversion integral method, find the inverse Z-transfrom of* $\dfrac{3z}{(z-1)(z-2)}$

SOLUTION.   Let $$F(z) = \frac{3z}{(z-1)(z-2)}$$

Its poles are at $z = 1$ and $z = 2$.
Using $F(z)$ in the inversion integral, we have

$$f(n) = \frac{1}{2\pi i}\int_c (z)z^{n-1}dz,$$

where c is a circle large enough to enclose both the poles of $F(z)$.

Now, we have   = sum of residues of $F(z)^{n-1}$ at $z = 1$ and $z = 2$

$$\text{res}[F(z). z^{n-1}]_{z=1} = \lim_{z \to 1}\left\{(z-1).\frac{3z^n}{(z-1)(z-2)}\right\} = -3$$

$$\text{res}[F(z). z^{n-1}]_{z=2} = \lim_{z \to 2}\left\{(z-2).\frac{3z^n}{(z-1)(z-2)}\right\} = 3.2^n$$

Hence, the requird inverse Z-transform

$$f(n) = 3(2^n - 1), \quad n = 0, 1, 2, 3, \ldots$$

EXAMPLE 7. *Find the inverse Z-transform of* $\dfrac{2z}{[(z-1)(z^2+1)]}$

SOLUTION. Let $F(z) = \dfrac{2z}{(z-1)(z+i)(z-i)}$

It has three poles at $z = 1, z = \pm i$.

Now, using $F(z)$ in the inversion integral, we have

$$f(n) = \frac{1}{2\pi i} \int_c f(z) z^{n-1} dz,$$

$$= \text{sum of residues of } U(z). z^{n-1} \text{ at } z=1, z= \pm i.$$

Now we have

$$\text{res}[F(z). z^{n-1}]_{z=1} = \lim_{z \to 1} \left\{ (z-1). \frac{2z^n}{(z-1)(z^2+1)} \right\} = 1$$

$$\text{res}[F(z). z^{n-1}]_{z=i} = \lim_{z \to i} \left\{ (z-i). \frac{2z^n}{(z-1)(z+i)(z-i)} \right\} = \frac{(-i)^n}{1+i}$$

$$\text{res}[F(z). z^{n-1}]_{z=-i} = \lim_{z \to -i} \left\{ (z+i). \frac{2z^n}{(z-1)(z+i)(z-i)} \right\} = \frac{(-i)^n}{i-1}$$

Hence, $\qquad f(n) = 1 - \dfrac{(i)^n}{1+i} - \dfrac{(-i)^n}{1-i}$

EXAMPLE 8. *Find* $Z^{-1}\left\{ \dfrac{3z^2 - 18z + 26}{(z-2)(z-3)(z-4)} \right\}.$ (Anna–2005)

SOLUTION. Clearly, the poles are given by

$(z-2)(z-3)(z-4) = 0 \Rightarrow z = 2, 3, 4$

Now, residue at $z = 2$

$$= \left[\frac{(z-2)z^{n-1}(3z^2 - 18z + 26)}{(z-2)(z-3)(z-4)} \right]_{z=2}$$

$$= \left[\frac{3z^{n+1} - 18z^n + 26z^{n-1}}{(z-3)(z-4)} \right]_{z=2}$$

$$= \frac{3.2^{n+1} - 18.2^n + 26.2^{n-1}}{(-1)(-2)}$$

$$= 3.2^n - 9.2^n + 13.2^{n-1} = 2^{n-1}$$

Residue at $z = 3$

$$= \left[\frac{(z-3)z^{n-1}.(3z^2 - 18z + 26)}{(z-2)(z-3)(z-4)} \right]_{z=3}$$

$$= \left[\frac{3z^{n+1} - 18z^n + 26z^{n-1}}{(z-2)(z-4)} \right]_{z=3}$$

$$= \frac{3.3^{n+1} - 18.3^n + 26.3^{n-1}}{1(-1)}$$

and residue at $z = 4$

$$= \left[(z-4) \left[\frac{z^{n-1}(3z^2 - 18z + 26)}{(z-2)(z-3)(z-4)} \right] \right]_{z=4}$$

$$= \left[\frac{3z^{n+1} - 18z^n + 26z^{n-1}}{(z-2)(z-3)} \right]_{z=4}$$

$$= \left[\frac{3.4^{n+1} - 18.4^n + 26.4^{n-1}}{2.1} \right] = 4^{n-1}$$

$$= z^{-1} F(z) = F(n) = \text{sum of residues}$$

$$= 2^{n-1} + 3^{n-1} + 4^{n-1}, n > 0$$

7.12 SOLUTION OF DIFFERENCE EQUATION WITH CONSTANT COEFFICIENTS BY Z-TRANSFORM

WORKING PROCEDURE

For solving a linear difference equation with constant coefficient by Z-transforms, we use the following steps :

Step 1. *Take the Z-transforms of both sides of the given difference equation using the formulae of shifting property and given conditions.*

Step 2. *Transpose all terms without F(z), to the right hand side.*

Step 3. *Divide the coefficient of F(z), getting F(z) as a function of z.*

Step 4. *Express this function in terms of the Z-transforms of known functions and take the inverse Z-transforms of both sides, which gives f(k), i.e., f_k as a function of k which is the required solution.*

Solved Examples

EXAMPLE 1.   *Solve the difference equation by Z-transform.*

$6f_{n+2} - f_{n+1} + f_n = 0$, *with* $f(0) = 0$, $f(1) = 1$,

SOLUTION.   The given difference equation is

$$6f_{n+2} - f_{n+1} + f_n = 0 \qquad \text{...(1)}$$

Taking Z-transform of both sides of (1), we get

$$Z[6f_{n+2} - f_{n+1} + f_n] = 0$$

$$\Rightarrow \qquad Z(6f_{n+2}) - Z(f_{n+1}) + Z(f_n) = 0$$

$$= 6[z^2 F(z) - z^2 f(0) - z f(1)]$$

$$- [zF(z) - zF(0)] + F(z) = 0 \qquad \text{...(2)}$$

Using the given initial condition in (2), we get

$$6z^2 F(z) - 6z - z F(z) + F(z) = 0$$

$$\Rightarrow \qquad (6z^2 - z + 1) \, F(z) = 6z$$

$$\Rightarrow \qquad F(z) = \frac{6z}{6z^2 - z + 1} = \frac{6z}{(3z+1)(2z-1)}$$

$$= \frac{z^{-1}}{\left(1 + \dfrac{z^{-1}}{3}\right)\left(1 - \dfrac{z^{-1}}{2}\right)} = \frac{6/5}{1 - \dfrac{z^{-1}}{2}} - \frac{6/5}{1 + \dfrac{z^{-1}}{3}}$$

$$\Rightarrow \qquad f(n) = Z^{-1}\left[\frac{6/5}{1 - z^{-1}/2}\right] - Z^{-1}\left[\frac{6/5}{1 + z^{-1}/3}\right]$$

$$= \frac{6}{5}\left(\frac{1}{2}\right)^n - \frac{6}{5}\left(-\frac{1}{3}\right)^n = \frac{6}{5}\left[\left(\frac{1}{2}\right)^n - \left(-\frac{1}{3}\right)^n\right]$$

EXAMPLE 2. *Solve by Z-transforms*

$$f_{n+1} + \frac{1}{4}f_n = \left(\frac{1}{4}\right)^4, \ n \geq 0, \ f(0) = 0$$

SOLUTION. The given difference equation is

$$f_{n+1} + \frac{1}{4}f_n = \left(\frac{1}{4}\right)^4 \qquad \qquad \dots(1)$$

Taking the Z-transforms of both sides of (1), we get

$$Z\left[f_{n+1} + \frac{1}{4}f_n\right] = Z\left[\left(\frac{1}{4}\right)^n\right] \quad \Rightarrow \quad Z[f_{n+1}] + z\left(\frac{1}{4}f_n\right) = Z\left[\left(\frac{1}{4}\right)^n\right]$$

$$\Rightarrow zF(z) - zf(0) + \frac{1}{4}F(z) = \frac{1}{1 - \frac{1}{4}z^{-1}}, \text{ where } |Z| > \frac{1}{4}$$

$$\Rightarrow \qquad zF(z) - 0 + \frac{1}{4}F(z) = \frac{1}{1 - \frac{1}{4}z^{-1}} \qquad \qquad [\because f(0) = 0]$$

$$\Rightarrow \qquad \left(z + \frac{1}{4}\right)F(z) = \frac{1}{1 - \frac{1}{4}z^{-1}}$$

Therefore $\qquad F(z) = \dfrac{1}{z + \dfrac{1}{4}} \times \dfrac{1}{1 - \dfrac{1}{4}z^{-1}} = \dfrac{z^{-1}}{1 + \dfrac{1}{4}z^{-1}} + \dfrac{2}{1 - \dfrac{1}{4}z^{-1}}$

$$\Rightarrow \qquad f(n) = Z^{-1}\left[\frac{-2}{1 + \frac{1}{4}z^{-1}}\right] + Z^{-1}\left[\frac{2}{1 - \frac{1}{4}z^{-1}}\right]$$

$$= Z^{-1}\left[-2\left(1 + \frac{1}{4}z^{-1}\right)^{-1}\right] + Z^{-1}\left[2\left(1 - \frac{1}{4}z^{-1}\right)^{-1}\right]$$

$$= -2\left(-\frac{1}{4}\right)^n + 2\left(\frac{1}{4}\right)^n$$

which is the required solution.

EXAMPLE 3. *Solve the difference equation*

$$f_{n+3} - 3f_{n+2} + 3f_{n+1} - f_n = F(n)$$

where $f(0) = f(1) = f(2) = 0$, *by Z-transforms.*

SOLUTION. The given difference equation is

$$f_{n+3} - 3f_{n+2} + 3f_{n+1} - f_n = F(n) \qquad \qquad \dots(1)$$

Taking Z-tranforms of both sides of (1), we get

$$Z(f_{n+3} - 3f_{n+2} + 3f_{n+1} - f_n) = Z[F(n)]$$

$$\Rightarrow Z[f_{n+3}] - 3Z[f_{n+2}] + 3Z[f_{n+1}] - Z[f_n] = Z[F(n)]$$

$$\Rightarrow \quad [z^3 F(z) - z^3 f(0) - z^2 f(1) - zf(2)] - 3[z^2 F(z) - z^2 f(0) - zf(1)]$$

$$+ 3[z F(z) - z f(0)] - F(z) = Z[F(n)] \qquad \ldots(2)$$

Using the given initial conditions in (2), we get

$$z^2 F(z) - 3z^2 F(z) + 3z - F(z) = \frac{1}{1 - z^{-1}}$$

$$\Rightarrow \qquad (z^2 - 3z^2 + 3z - 1) F(z) = \frac{1}{1 - z^{-1}}$$

$$\Rightarrow \qquad (z - 1)^3 F(z) = \frac{1}{1 - z^{-1}}$$

i.e., $\qquad F(z) = \dfrac{1}{(z-1)^3 (1 - z^{-1})} = \dfrac{1}{z^3 (1 - z^{-1})^3 (1 - z^{-1})}$

$$= z^{-3} (1 - z^{-1})^{-4}$$

$$\Rightarrow \qquad f(n) = \text{Ceofficients of } z^{-n} \text{ in } z^{-3} (1 - z^{-1})^{-4}$$

$$= \text{Ceofficients of } z^{-n-3} \text{ in } (1 - z^{-1})^4$$

$$= \frac{(n-2)(n-1)n}{6}, n \geq 3$$

EXAMPLE 4.   Solve $f_n + \dfrac{1}{4} f_{n-1} = F(n) + \dfrac{1}{3} F(n-1)$

SOLUTION.   The given difference equation is

$$f_n + \frac{1}{4} f_{n-1} = F(n) + \frac{1}{3} F(n-1) \qquad \ldots(1)$$

Taking Z-transforms of both sides of (1), we get

$$Z[\{f_n\}] + \frac{1}{4} Z[\{f_{n-1}\}] = Z[\{F(n)\}] + \frac{1}{3} Z[\{F(n-1)\}]$$

$$\Rightarrow \quad F(z) + \frac{1}{4} z^{-1} F(z) = \left[1 + \frac{1}{3} z^{-1} \right] \qquad \Rightarrow \qquad F(z) = \frac{1 + \dfrac{1}{3} z^{-1}}{1 + \dfrac{1}{4} z^{-1}} = \frac{z + \dfrac{1}{3}}{z + \dfrac{1}{4}}$$

Clearly, there is only one simple pole at $z = -\dfrac{1}{4}$.

Consider the contour $|z| > \dfrac{1}{4}$. Then, Residue at $z = -\dfrac{1}{4}$

$$= \left[\left(z + \frac{1}{4} \right) . z^{n-1} . \frac{z + \dfrac{1}{3}}{z + \dfrac{1}{4}} \right]_{z = -\frac{1}{4}} = \left[z^n + \frac{z^{n-1}}{3} \right]_{z = -\frac{1}{4}}$$

$$= \left(-\frac{1}{4} \right)^n + \frac{1}{3} \left(-\frac{1}{4} \right)^{n-1} = -\frac{1}{4} \left(-\frac{1}{4} \right)^{n-1} + \frac{1}{3} \left(-\frac{1}{4} \right)^{n-1} = \frac{1}{12} \left(-\frac{1}{4} \right)^{n-1}$$

Hence, $\qquad f(n) = \text{Residue} = \dfrac{1}{12} \left(-\dfrac{1}{4} \right)^{n-1}$

EXAMPLE 5. *Using Z-transform, solve*
$$f_{n+2} + 4f_{n+1} + 3f_n = 3^n \text{ with } f(0)=0, f(1) = 1. \qquad \text{(UPTU–2003)}$$

SOLUTION. If $\quad Z(f_n) = F(z)$, then $Z[f_{n+1}]=z[F(z)-f(0)]$

$$Z[f_{n+2}] = z^2[F(z)-f(0) - f(1)z^{-1}]$$

Also, $Z(2^n) = \dfrac{z}{z-2}$

Taking Z-transforms of both sides of the given equation, we get

$$z^2[F(z)-f(0) - f(1)z^{-1}]+4z[F(z) - f(0)]+3F(z) = \frac{z}{z-3} \qquad \text{...(1)}$$

Using the given condition, (1) reduces to

$$F(z) = (z^2 + 4z+3) = z + \frac{z}{z+3}$$

Therefore,

$$\frac{F(z)}{z} = \frac{1}{(z+1)(z+3)} + \frac{1}{(z-3)(z+1)(z+3)} = \frac{3}{8} \cdot \frac{1}{z+1} + \frac{1}{24}\frac{1}{z-3} - \frac{5}{12}\frac{1}{z+3}$$

$$\text{(On resolving into partial fractions)}$$

$$\therefore \quad F(z) = \frac{3}{8}\frac{z}{z+1} + \frac{1}{24}\frac{z}{z-3} - \frac{5}{12}\frac{z}{z+3}$$

On inversion, we get

$$f_n = \frac{3}{8}Z^{-1}\left(\frac{z}{z+1}\right) + \frac{1}{24}Z^{-1}\left(\frac{z}{z-3}\right) - \frac{5}{12}Z^{-1}\left(\frac{z}{z+3}\right)$$

$$= \frac{3}{8}(-1)^n + \frac{1}{24}3^n - \frac{5}{12}(-3)^n$$

EXAMPLE 6. *Solve* $f_{n+2} + 6f_{n+1}+9f_n = 2^n$ *with* $f(0) = f(1)=0$ *using Z-tranforms.*
$$\text{(VTU–2011; Anna–2009; SVTU–2009)}$$

SOLUTION. If $\qquad Z(f_n) = F(z)$, then

$$Z(f_{n+1}) = z[F(z) - f(0)]$$

and $\qquad Z(f_{n+2}) = Z[F(z) - f(0) - f(1)z^{-1}]$

Also, $\qquad Z(2^n) = \dfrac{z}{z-2}$

Taking Z-transform of both sides of the given equation, we get

$$z^2[F(z) -f(0)-f(1)Z^{-1}] + 6z[F(z) - f(0)]+ 9F(z) = \frac{z}{z-2} \qquad \text{...(1)}$$

Using the given condition in (1), we get

$$F(z)(z^2 + 6z + 9) = \frac{z}{z-2}$$

$$\Rightarrow \qquad \frac{F(z)}{z} = \frac{1}{(z-2)(z-3)^2}$$

$$= \frac{1}{25}\left[\frac{1}{z-2} - \frac{1}{z+3} - \frac{5}{[z+3]^2}\right]$$

Therefore

$$F(z) = \frac{1}{25}\left[\frac{z}{z-2} - \frac{z}{z+3} - 5.\frac{z}{(z+3)^2}\right] \qquad \text{...(2)}$$

On taking inverse Z-transform of both sides of (2), we get

$$f(n) = \frac{1}{25}\left[Z^{-1}\left(\frac{z}{z-2}\right) - Z^{-1}\left(\frac{z}{z+3}\right) + \frac{5}{3}Z^{-1}\left(-\frac{3z}{(z+3)^2}\right)\right]$$

$$= \frac{1}{25}\left[2^n - (-3)^n + \frac{5}{3}n(-3)^n\right] \qquad \left[\because Z^{-1}\left\{\frac{az}{(z-a)^2}\right\} = na^n\right]$$

Example 7.   *Using the Z-transform, solve*

$$f_{n+2} - 2f_{n+1} + f_n = 3n+5 \qquad\qquad \text{(SVTU–2007)}$$

Solution.   The given difference equation is

$$f_{n+2} - 2f_{n+1} + f_n = 3n+5 \qquad\qquad \text{...(1)}$$

Taking Z-transform of both sides of (1), we get

$$z^2[F(z) - f(0) - f(1)z^{-1}] - 2z[F(z) - f(0)] + F(z)$$

$$= 3.\frac{z}{(z+1)^2} + 5.\frac{z}{z-1}$$

or

$$F(z) = \frac{5z^2 - 2z}{(z-1)^2} + f(0).\frac{z^2 - 2z}{(z-1)^2} + f(1).\frac{z}{(z-1)^2}$$

Taking inverse Z-transform, we get

$$f(n) = Z^{-1}\left[\frac{5z^2 - 2z}{(z-1)^4}\right] + f(0)Z^{-1}\left[\frac{z^2 - 2z}{(z-1)^2}\right]$$

$$+ f(1)Z^{-1}\left[\frac{z}{(z-1)^2}\right] \qquad\qquad \text{...(2)}$$

Notting that

$$Z(1) = \frac{z}{z-1}, \qquad\qquad Z(n) = \frac{z}{(z-1)^2}$$

$$Z(n^2) = \frac{z^2 + z}{(z-1)^3}, \qquad\qquad Z(n^3) = \frac{z^3 + 4z^2 + z}{(z-1)^4}$$

We write

$$\frac{5z^2 - 2z}{(z-1)^4} + A.\frac{z^3 + 4z^2 + z}{(z-1)^4} + B.\frac{z^2 + z}{(z-1)^3} + C.\frac{z}{(z-1)^2} + D.\frac{z}{z-1}$$

Equating coefficient of like powers of z, we find

$$A = \frac{1}{2}, B = 1, C = -\frac{3}{2}, D = 0.$$

Therefore, $Z^{-1}\left\{\dfrac{5z^2 - 2z}{(z-1)^4}\right\} = \dfrac{1}{2}n^3 + n^2 - \dfrac{3}{2}n$

$$= \frac{1}{2}n(n-1)(n+3)$$

also,

$$Z^{-1}\left\{\frac{z^2 - 2z}{(z-1)^2}\right\} = Z^{-1}\left\{\frac{z}{z-1}\right\} - Z^{-1}\left\{\frac{z}{(z-1)^2}\right\}$$

$$= 1 - n$$

and $Z^{-1}\left\{\dfrac{z}{(z-1)^2}\right\} = n$

Putting these values in (2), we get

$$f(n) = \frac{1}{2}n(n-1)(n+3) + f(0)(1-n) + f(1).n$$

$$= \frac{1}{2}n(n-1)(n+3) + C_0 + C_1 n \text{ where } C_0 = f(0),\ C_1 = f(1) - f(0)$$

EXERCISE 7.1

1. Find the Z-transforms of the following functions for $n \geq 0$:

(i) 2^n (ii) $\sin \alpha n$

(iii) $\sin(3n + 5)$ (iv) $\sinh \dfrac{n\pi}{2}$

(v) $\cosh n\theta$ (VTU–2011)

2. Find the Z-transforms of the following functions for $n \geq 0$:

(i) $\sin\left(\dfrac{n\pi}{2} + \alpha\right)$ (ii) $C^n \sinh \alpha n$

(iii) $\cos\left(\dfrac{n\pi}{2} + \dfrac{\pi}{4}\right)$ (iv) $\dfrac{a^n}{n!}$

3. Find the Z-transforms of the following functions for $n > 0$:

(i) $\sin 5n$ (ii) $e^{\alpha n}$

(iii) $3n \cosh 5n$

4. Find the inverse of Z-transforms of the following functions:

(i) $\dfrac{1}{(z-3)(z-2)}$ for $|z| < 2$

(ii) $\dfrac{1}{(z-5)^3}, |z| > 5$

(iii) $\dfrac{3z^2 + 4z}{z^2 - z + 1}, |z| > 1$

(iv) $\dfrac{ze^{-a}}{(z - e^{-a})^2}, |z| > |e^{-a}|$

(v) $\dfrac{z^3 - 20z}{(z-2)^3(z-4)}$ (VTU–2011)

(vi) $\dfrac{4z^2 - 2z}{z^3 - 5z^2 + 8z - 4}$ (VTU–2011)

(vii) $\dfrac{18z^2}{(2z-1)(4z+1)}$ (SVTU–2009)

ANSWERS

1. (i) $\dfrac{z}{z-2}, |z| > 2$ (ii) $\dfrac{z \sin \alpha}{z^2 - 2z \cos \alpha + 1}$ (iii) $\dfrac{z^2 \sin 5 - z \sin 2}{z^2 - 2z \cos 3 + 1}$ (iv) $\dfrac{z \sinh \dfrac{\pi}{2}}{z^2 - 2z \cosh \dfrac{\pi}{2} + 1}$

(v) $\dfrac{z(z - a \cosh \theta)}{z^2 - 2az \cosh \theta + a^2}$ (vi) $2 + (2)^n + 3(n-1)2^n, n > 1$ (vii) $\dfrac{3}{4}\left(\dfrac{1}{2^{n-1}} + \dfrac{1}{(-4)^n}\right)$

2. (i) $\dfrac{z^2 \sin \alpha + z \cos \alpha}{z^2 + 1}, |z| > 1$ (ii) $\dfrac{Cz \sinh \alpha}{z^2 - 2Cz \cosh \alpha + 1}$ (iii) $\dfrac{z^2 - z}{\sqrt{2}(z^2 + 1)}$ (iv) $e^{a/z}$

3. (i) $\dfrac{z \sin 5}{z^2 - 2z \cos 5 + 1}$ (ii) $\dfrac{1}{1 - z^{-1}e^{\alpha n}}$ (iii) $\dfrac{z(z - 3 \cosh 5)}{z^2 - 2z \cosh 5 + 9}$

4. (i) $\left(-\dfrac{1}{3} - \dfrac{z}{3^2} - \dfrac{z^2}{3^3} - \dfrac{z^3}{3^4} - ....\right) + \left(\dfrac{1}{2} + \dfrac{z}{2^2} + \dfrac{z^2}{2^3} + \dfrac{z^3}{2^4} + ...\right)$ (ii) $\begin{cases} \dfrac{(n-2)(n-1)5^{n-3}}{2}, & n \geq 3 \\ 0, & n < 3 \end{cases}$

(iii) $\left[3\left\{\cos\dfrac{n\pi}{3}\right\} + \dfrac{11}{\sqrt{3}}\sin\dfrac{n\pi}{3}\right]F(n)$ (iv) $[ne^{-an}]$ (v) $2^{n-1} + n^2 2^n - 4^n$

Self Assessment Test

1. Solve the following difference equations by Z-transform:

 (i) $f_{n+1} - 2f_n + f_{n-1} = a_n$, $a \neq 1$

 (ii) $f_n - \dfrac{5}{6}f_{n-1} + \dfrac{1}{6}f_{n-2} = F(n)$

 (iii) $f_n + \dfrac{1}{a}f_{n-2} = \left(\dfrac{1}{3}\right)^n \cos\dfrac{n\pi}{2}$; $(n \geq 0)$

 (iv) $6f_{n+2} + 5f_{n+1} - f_n = 6F(n)$

 (v) $6y_{n+2} - y_{n+1} - y_n = 0$, $y(0) = y(1) = 1$

                        (Kottayam–2005)

2. Show that $Z(\sinh n\theta) = \dfrac{z\sinh\theta}{z^2 - 2z\cosh\theta + 1}$

                           (VTU–2011)

3. Show that

$$Z(e^{-an}\sin n\theta) = \dfrac{ze^a\sin\theta}{z^2e^{2a} - 2ze^a\cos\theta + 1}$$

                           (SVTU–2007)

4. Show that $Z(^{n+p}C_p) = \left(1 - \dfrac{1}{z}\right)^{-(p+1)}$.

 Using the damping rule deduce that

$$Z(^{n+p}C_p a^n) = \left(1 + \dfrac{a}{z}\right)^{-(p+1)}$$ (SVTU–2009)

5. Show that $Z\left(\dfrac{1}{n}\right) = z\log\dfrac{z}{z-1}$ (Madras–2003)

Answers

1. (i) $f_n = \dfrac{1}{a}(n+1)F(n) - \dfrac{a}{(a-1)^2}F(n) + \dfrac{a}{(a-1)^2}a^n F(n) + \dfrac{1}{1-a}nF(n-1)$

 (ii) $f_n = \left[3 - \left(\dfrac{1}{2}\right)^n + \left(\dfrac{1}{3}\right)^n\right]F(n)$ (iii) $f_n = \left(\dfrac{n+2}{2}\right)\left(\dfrac{1}{3}\right)^n \cos\dfrac{n\pi}{2}F(n)$

 (iv) $f_n = \left[\dfrac{6}{7}(n+1) - \dfrac{78}{49} + \dfrac{36}{49}\left(-\dfrac{1}{6}\right)^n\right]F(n)$ (v) $y_n = \dfrac{8}{5}\left(\dfrac{1}{2}\right)^n - \dfrac{3}{5}\left(-\dfrac{1}{3}\right)^n$

Objective Evaluations

Fill in the Blanks

1. $Z(a^n \sin n\theta)$ is equal to _____.

2. $Z(\cos n\theta) =$ _____.

3. $Z(n^4) =$ _____.

4. $Z(na^n) =$ _____.

5. $Z(\sin n\theta) =$ _____.

True or False

Write 'T' for True and 'F' for False statement.

1. $Z(1) = \dfrac{Z}{(Z-1)^2}$. **(T/F)**

2. $Z(n^p) = -Z\dfrac{d}{dZ}Z(n^{p-1})$ where p is a positive integer. **(T/F)**

3. If $Z[<f(n)>] = F(z)$, then $Z[<a^{-n}f(n)>] = -F(az)$. **(T/F)**

4. $Z(n^2a^n) = \dfrac{aZ^2 - a^2Z}{(Z-a)^3}$ **(T/F)**

5. If $Z[<f(n)>] = F(z)$, $n > 0$, then
$$\lim_{Z\to\infty} F(n) = \lim_{Z\to\infty} (Z-1)F(Z)$$ **(T/F)**

Multiple Choice Questions

Choose the most appropriate one.

1. $Z\left[\lim_{n\to\infty} f_n\right]$ is equal to :

   (a) $\lim_{n\to 1}[(Z-1)f(Z)]$ (b) $\lim_{n\to 1}[(Z+1)f(Z)]$

   (c) $\lim_{n\to 2}[(Z-1)f(Z)]$ (d) $\lim_{n\to 2}[(Z+1)f(Z)]$

2. Z-transforms of the sequence $< f(n) >= \sum_{n=0}^{\infty} Z^n \sum_{n=0}^{\infty} 4^n$ is :

   (a) $\dfrac{Z}{Z^2 - 7Z + 12}$ (b) $\dfrac{Z}{Z^2 + 7Z + 12}$

   (c) $\dfrac{Z^2}{Z^2 - 7Z + 12}$ (d) $\dfrac{Z^2}{Z^2 + 7Z + 12}$

3. If $F(Z) = Z[<f(n)>] = \dfrac{2Z^2 + 5Z + 14}{(Z-1)^4}$, then the value of $f(0)$ is :

   (a) 0 (b) 1
   (c) 2 (d) 3

4. The inverse Z-transform of $\dfrac{2Z^2 + 3Z}{(Z+2)(Z-4)}$ is :

   (a) $\dfrac{1}{6}(2)^n + \dfrac{11}{6}(4)^n$

   (b) $\dfrac{1}{6}(-2)^n + \dfrac{11}{6}(4)^n$

   (c) $\dfrac{1}{6}(-2)^n - \dfrac{11}{6}(4)^n$

   (d) none of these

5. The value of $f_{n+3} - 3f_{n+2} + 3f_{n+1} - f_n = F(n)$ is :

   (a) $\dfrac{n(n+1)(n+2)}{6}, n \geq 3$

   (b) $\dfrac{n(n-1)(n+2)}{6}, n \geq 3$

   (c) $\dfrac{n(n+1)(n-2)}{6}, n \geq 3$

   (d) $\dfrac{n(n-1)(n-2)}{6}, n \geq 3$

6. Z-transform of 2^n is :

   (a) $\dfrac{Z}{Z-2}, |Z| > 2$ (b) $\dfrac{Z}{Z-1}, |Z| > 2$

   (c) $\dfrac{Z}{Z+2}, |Z| > 2$ (d) $\dfrac{Z}{Z+1}, |Z| > 2$

7. The value of $Z(\sinh n\theta)$ is :

   (a) $\dfrac{Z \sinh\theta}{Z^2 + 2Z\cosh\theta + 1}$

   (b) $\dfrac{Z \sinh\theta}{Z^2 - 2Z\cosh\theta + 1}$

   (c) $\dfrac{Z \sinh\theta}{Z^2 - 2Z\cosh\theta - 1}$

   (d) none of these

ANSWERS

❖ Fill in the Blanks

1. $\dfrac{aZ\sin\theta}{Z^2 - 2aZ\cos\theta + a^2}$

2. $\dfrac{Z(Z - \cos\theta)}{Z^2 - 2Z\cos\theta + 1}$

3. $\dfrac{Z^4 + 11Z^3 + 11Z^2 + Z}{(Z-1)^5}$

4. $\dfrac{aZ}{(Z-a)^2}$

5. $\dfrac{Z\sin\theta}{Z^2 - 2Z\cos\theta + 1}$

❖ True or False

1. T   **2.** T   **3.** F   **4.** F   **5.** T

❖ Multiple Choice Questions

1. (a)   **2.** (c)   **3.** (a)   **4.** (b)   **5.** (d)   **6.** (a)   **7.** (b)

VVVVVV

Chapter 8

Wave, Heat and Laplace Equations

8.1 INTRODUCTION

Many problems in science and engineering, when formulated mathematically, lead to partial differential equations involving one or more unknown functions together with certain prescribed conditions on the functions which arise from the physical situations.

The process of obtaining all solutions of a partial differential equations under given conditions is known as boundary value problem. If time t is regarded as an independent variable and the conditions are stated as $t = 0$, the problem is called initial value problem.

8.2 CLASSIFICATION OF PARTIAL DIFFERENTIAL EQUATIONS

(UPTU–2005, 07)

Consider the equation $\quad A\dfrac{\partial^2 u}{\partial x^2} + B\dfrac{\partial^2 u}{\partial x \partial y} + C\dfrac{\partial^2 u}{\partial y^2} + F(x, y, u, p, q) = 0$ ... (1)

where A, B, C may be constants or functions of x and y.

This equation (1) will be

    (i)   elliptic if $B^2 - 4AC < 0$ (ii) parabolic if $B^2 - 4AC = 0$

    (iii) hyperbolic if $B^2 - 4AC > 0$

  1. Parabolic equation : $\dfrac{\partial u}{\partial t} = \dfrac{\partial^2 u}{\partial x^2}$

  2. Elliptic equation : $\dfrac{\partial^2 u}{\partial x^2} + \dfrac{\partial^2 u}{\partial y^2} = 0$

  3. Hyperbolic equation : The wave equation $\dfrac{\partial^2 u}{\partial t^2} = C^2 \dfrac{\partial^2 u}{\partial x^2}$

8.3 PRODUCT METHOD : SOLUTION OF BOUNDARY VALUE PROBLEMS BY THE METHOD OF SEPARATION OF VARIABLES

Let $\quad A\dfrac{\partial^2 u}{\partial x^2} + B\dfrac{\partial^2 u}{\partial x \partial y} + C\dfrac{\partial^2 u}{\partial y^2} + D\dfrac{\partial u}{\partial x} + E\dfrac{\partial u}{\partial y} + Fu = G$ ... (1)

be the general linear partial differential equation, where A, B, ..., G are functions of x and y. Here, it must be noted that

    (i) If any partial differential equation of second order cannot be put in above form, then it is said to be non-linear.

(ii) If $G = 0$, then equation (i) is called homogeneous. Thus, for non-homogeneous equation $G \neq 0$.

Let $u(x, y) = X(x)\, Y(y)$

$$\ldots(2)$$

be the solution of (1). Here $X(x)$ and $Y(y)$ are respectively the functions of x and y alone. Using (2), we can find the following values

$$\frac{\partial^2 u}{\partial x^2},\ \frac{\partial^2 u}{\partial y^2},\ \frac{\partial^2 u}{\partial x \partial y},\ \frac{\partial u}{\partial x}\ \text{ and }\ \frac{\partial u}{\partial y}\ \text{ in terms of } X \text{ and } Y.$$

Putting all these values in (1), we get $\dfrac{F(D)\,.\,X}{X} = \dfrac{g(D')\,.\,Y}{Y}$, where, $F(D)$ and $g(D')$ are functions of $D = \dfrac{\partial}{\partial x}$ and $D' = \dfrac{\partial}{\partial y}$ respectively.

Thus, we can find a relation in which LHS is a function of x alone while RHS is a function of y alone.

Then putting

$$\frac{F(D)\,.\,X}{X} = \frac{g(D')\,.\,Y}{Y} = \lambda \ \text{ (say)} \qquad \ldots (3)$$

Thus, the solution of (1) reduces to be solution of a pair of ODE given by (3).

Solved Examples

EXAMPLE 1. *Solve the boundary value problem* $\dfrac{\partial u}{\partial x} = 4\dfrac{\partial u}{\partial y}$ *with* $u(0, y) = 8e^{-3y}$ *by the method of separation of variables.*

(UPTU–2008, JNTU–2006)

SOLUTION. The given equation is

$$\frac{\partial u}{\partial x} = 4\frac{\partial u}{\partial y} \qquad \ldots (1)$$

with boundary conditions $u(0, y) = 8e^{-3y}$.

Let $u(x, y) = X(x)\, Y(y)$ (where X and Y are respectively the functions of x and y alone) be the solution of (1).

Putting the value in (1), we get

$$X'Y = 4XY'$$

$$\Rightarrow \qquad \frac{X'}{4X} = \frac{Y'}{Y} \qquad \ldots (2)$$

Here, dashes denote derivatives with respect to the relevant variable. Equation (2) is true only when each side is equal to same constant say λ.

Therefore (2) gives $X' - 4\lambda X = 0$ and $Y' - \lambda Y = 0$.

On solving, we get $X(x) = A e^{4\lambda x}$ and $Y(y) = B e^{\lambda y}$.

Thus, $u(x, y) = ABe^{4\lambda x + \lambda y} = C\, e^{\lambda(4x + y)}$,

where $C\ (= AB)$ is any arbitrary constant.

Using given boundary condition

$$u(0, y) = 8e^{-3y} = C\, e^{\lambda y},$$

we get $\qquad C = 8,\ \lambda = -3$.

Hence, the solution of given boundary value problem is $u(x, y) = 8e^{-3(4x+y)}$

EXAMPLE 2. *Solve by the method of separation of variables* $\dfrac{\partial^2 u}{\partial x^2} - 2\dfrac{\partial u}{\partial x} + \dfrac{\partial u}{\partial y} = 0.$

(UPTU–2005, Q. Bank–2002, PTU–2009 S, Bhopal–2008)

SOLUTION. The given equation is

$$\frac{\partial^2 u}{\partial x^2} - 2\frac{\partial u}{\partial x} + \frac{\partial u}{\partial y} = 0 \qquad \text{... (1)}$$

Let $u(x, y) = X(x)\, Y(y)$ be the solution of this equation.

Putting the value of u in the given equation, we get

$$X''Y - 2X'Y + XY' = 0 \qquad \text{... (2)}$$

The dashes denote derivatives with respect to the relevant variable.

Now, relation (2) can be written as

$$\frac{X'' - 2X'}{X} = \frac{Y'}{Y} \qquad \text{... (3)}$$

This is true if each side is equal to the same constant, say λ. Then, we have

$$X'' - 2X' - \lambda X = 0 \quad \text{and} \quad Y' - \lambda Y = 0 \qquad \text{... (4)}$$

$$\Rightarrow \quad (D^2 - 2D - \lambda)X = 0 \quad \text{and} \quad \frac{dY}{dy} = \lambda Y \qquad \text{... (5)}$$

The auxiliary equation of (4) is given by

$$m^2 - 2m - \lambda = 0,$$

i.e., $\qquad m = 1 \pm \sqrt{(1+\lambda)}.$

Thus, $\qquad X(x) = A e^{[1+\sqrt{(1+\lambda)}]x} + B e^{[1-\sqrt{(1+\lambda)}]x}$

and solution of (5) is given by

$$Y(y) = C e^{\lambda y}$$

Finally, we get

$$u(x, y) = \left\{ ACe^{[1+\sqrt{(1+\lambda)}]x} + BC e^{[1-\sqrt{(1+\lambda)}]x} \right\} e^{\lambda y}$$

$$= \left\{ A_1 e^{[1+\sqrt{(1+\lambda)}]x} + A_2 e^{[1-\sqrt{(1+\lambda)}]x} \right\} e^{\lambda y}$$

EXAMPLE 3. *Use the method of separation of variables to solve the equation* $\dfrac{\partial^2 u}{\partial x^2} - 2\dfrac{\partial u}{\partial x} + \dfrac{\partial u}{\partial y} = 0$

(UPTU–2009, GBTU (AG)–2012)

SOLUTION. Let $u = X \cdot Y$ $\qquad \text{... (1)}$

where X is a function x only and Y is function of y only.

$$\frac{\partial u}{\partial x} = \frac{\partial}{\partial x}(XY) = Y\frac{dX}{dx}$$

$$\frac{\partial u}{\partial y} = \frac{\partial}{\partial y}(XY) = X\frac{dY}{dy}$$

$$\frac{\partial^2 u}{\partial x^2} = \frac{\partial^2}{\partial x^2}(XY) = Y\frac{d^2 X}{dx^2}$$

Putting these values in (1), we get

$$Y\frac{d^2 X}{dx^2} - 2Y\frac{dX}{dx} + X\frac{dY}{dy} = 0$$

$$\Rightarrow \qquad YX'' - 2YX' + XY' = 0 \qquad \Rightarrow \qquad \frac{X'' - 2X'}{X} + \frac{Y'}{Y} = 0$$

$$\Rightarrow \qquad \frac{X'' - 2X'}{X} = -\frac{Y'}{Y} = -P^2 \text{ (say)}$$

(i) $\qquad \dfrac{X'' - 2X'}{X} = -P^2 \qquad \Rightarrow \qquad X'' - 2X' + P^2 X = 0$

A. E. is $m^2 - 2m + p^2 = 0 \qquad \Rightarrow \qquad m = \dfrac{2 \pm \sqrt{4 - 4P^2}}{2} = 1 \pm \sqrt{1 - P^2}$

$\therefore \qquad$ C. F. $= C_1 e^{(1+\sqrt{1-P^2})x} + C_2 e^{(1-\sqrt{1-P^2})x}$

P. I. $= 0$

Hence $X = $ C.F. $+$ P. I. $= C_1 e^{(1+\sqrt{1-P^2})x} + C_2 e^{(1-\sqrt{1-P^2})x}$ ... (2)

(ii) $\qquad \dfrac{-Y'}{Y} = -P^2 \Rightarrow \dfrac{dY}{dy} = p^2 Y \Rightarrow \dfrac{dy}{Y} = p^2 dy$

On integrating, we get $\log Y = p^2 y + \log C_3$

$$Y = C_3 e^{p^2 y}$$

$\therefore \quad u(x, y) = [C_1 e^{(1+\sqrt{1-p^2})x} + C_2 e^{(1-\sqrt{1-p^2})x}] C_3 e^{p^2 y}$ ...(3)

EXERCISE 8.1

1. Use the method of separation of variables to solve the equation $\dfrac{\partial^2 V}{\partial x^2} = \dfrac{\partial V}{\partial t}$.

   Given that $V = 0$ when $t \to \infty$ as well as $v = 0$ at $x = 0$ and $x = l$. (AMIE–1997)

2. Use the method of separation of variables to solve $\dfrac{\partial u}{\partial x} = 2\dfrac{\partial u}{\partial t} + u$, where $u(x, 0) = 6e^{-3x}$.

   (Kerala–2005; UPTU–2006; Kurukshetra–2006; VTU–2009; GBTU(Sum)–2010; GBTU(AG)–2011; AMIETE 1997)

3. Use the method of separation of variables to solve the equation

   $$\frac{\partial^2 u}{\partial x^2} = \frac{\partial u}{\partial y} + 2u.$$

4. Solve by the method of separation of variables $4\dfrac{\partial u}{\partial t} + \dfrac{\partial u}{\partial x} = 3u$, $u = 3e^{-x} - e^{-5x}$ when $t = 0$.

   (SVTU–2008)

5. Use the method of separation of variables, find the solution of the following differential equation:

   $$3\frac{\partial u}{\partial x} + 2\frac{\partial u}{\partial y} = 0, \quad u(x, 0) = 4e^{-x}.$$

   (AMIETE–2000; VTU–2008S)

6. Solve by the method of separation of variables.

   $$py^3 + qx^2 = 0 \quad \text{(VTU–2011; SVTU–2008)}$$

7. Solve by the method of separation of variables

   $$x^2 \frac{\partial u}{\partial x} + y^2 \frac{\partial u}{\partial y} = 0 \qquad \text{(VTU–2008)}$$

8. Find a solution of the equation $\dfrac{\partial^2 u}{\partial x^2} = \dfrac{\partial u}{\partial y} + 2u$ in the form $u = f(x)g(y)$. Solve the equation subject to the conditions $u = 0$ and $\dfrac{\partial u}{\partial x} = 1 + e^{-3y}$, when $x = 0$ for all values of y.

   (Andhra–2000)

9. Solve by method of separation of variables

   $$4\frac{\partial u}{\partial x} + \frac{\partial u}{\partial y} = 3u; u(0, y) = 4e^{-y} - e^{-5y}$$

   (UKTU–2012)

10. Solve by method of separation of variables

    $$y^3 \frac{\partial u}{\partial x} + x^2 \frac{\partial u}{\partial y} = 0 \qquad \text{(GBTU–2011)}$$

11. Solve by method of separation of variables

    $$\frac{\partial^2 u}{\partial x^2} - \frac{\partial u}{\partial y} = 0 \qquad \text{(UPTU (SUM)–2007)}$$

12. Solve by method of separation of variables

    $$\frac{\partial u}{\partial x} = \frac{\partial u}{\partial y} \qquad \text{(GBTU–2012)}$$

13. Solve by method of separation of variables

$$\frac{\partial z}{\partial x} + \frac{\partial^2 z}{\partial y^2} = 0; z(x,0) = 0,$$

$$z(x,\pi) = 0, z(0,y) = 4\sin 3y \quad \text{(MTU–2012)}$$

14. Solve by method of separation of variables

$$\frac{\partial^2 u}{\partial x \partial t} = e^{-t}\cos x$$

given that $u = 0$ when $t = 0$ and $\frac{\partial u}{\partial t} = 0$

when $x = 0$.

(UPTU (SUM)–2008; GBTU–2012; GBTU (AG)–2012)

15. Solve by method of separation of variables

$$u_{xx} = uy + 2u, u(0,y) = 0, \frac{\partial}{\partial x}u(0,y) = 1 + e^{-3y}$$

(GBTU–2010, UPTU–2009)

Hint to Selected Problems

1. Let us assume that $V = XT$, where X is a function of x only and T that of t only.

$$\frac{\partial V}{\partial t} = X\frac{dT}{dt} \text{ and } \frac{\partial^2 V}{\partial x^2} = T\frac{d^2 X}{dx^2}$$

Substitute in $\frac{\partial^2 V}{\partial x^2} = \frac{\partial V}{\partial t}$, we get $X\frac{dT}{dt} = T\frac{d^2 X}{dx^2}$

$$\Rightarrow \quad \frac{1}{T}\frac{dT}{dt} = \frac{1}{X}\frac{d^2 X}{dx^2} \quad \dots (1)$$

Let each side of (1) be equal to constant (p^2)

$\frac{1}{T}\frac{dT}{dt} = -p^2$	$\frac{1}{X}\frac{d^2 X}{dx^2} = -p^2$
$\frac{dT}{dt} + p^2 T = 0$	$\Rightarrow \quad \frac{d^2 X}{dx^2} + p^2 X = 0$
$DT + p^2 T = 0$	$D^2 X + p^2 X = 0$
$\Rightarrow \quad (D + p^2)T = 0$	$(D^2 + p^2)X = 0$
$\Rightarrow \quad D = -p^2$	$D^2 + p^2 = 0$
$\Rightarrow \quad T = C_1 e^{-p^2 t}$	$D^2 = -p^2 \Rightarrow D = \pm ip$
	$X = C_2 \cos px + C_3 \sin px$

$$V = C_1 e^{-p^2 t}(C_2 \cos px + C_3 \sin px) \quad \dots (2)$$

On putting $x = 0$, $V = 0$ in (2), we get

$$0 = C_1 e^{-p^2 t}C_2 \Rightarrow C_2 = 0, \text{ since } C_1 \neq 0.$$

On putting the value of C_2 in (2), we get

$$V = C_1 e^{-p^2 t}C_3 \sin px \quad \dots (3)$$

Since C_3 cannot be zero

$$\sin pl = 0 = \sin n\pi \Rightarrow pl = n\pi$$

$$\Rightarrow \quad p = \frac{n\pi}{l}, \quad n \in \mathbf{Z}$$

On putting the value of P in (3), we get

$$V = C_1 C_3 e^{-\frac{n^2\pi^2 t}{l^2}} \sin\frac{n\pi x}{l}$$

$$V = b_n e^{-\frac{n^2\pi^2 t}{l^2}} \sin\frac{n\pi x}{l}, \text{ where } b_n = C_1 C_2.$$

The most general solution is

$$V = \sum_{m=1}^{\infty} b_n e^{-\frac{n^2\pi^2 t}{l^2}} \sin\frac{n\pi x}{l}.$$

ANSWERS

1. $v = b_n e^{-n^2\pi^2 t/l^2} \sin\frac{n\pi x}{l}$, where $b_n = C_1 C_2$

2. $u = 6e^{-(3x+2t)}$

3. $u(x,y) = (C_1 \cos px + C_2 \sin px) C_3 e^{-(b^2+2)y}$

4. $u(x,t) = 3e^{-x+t} - e^{-5x+2t}$

5. $u = 4e^{-x+3/2y}$

6. $Z = Ce^{4ax^3} \cdot e^{-3ay^4}$

7. $u = Ce^{k(1/y - 1/x)}$

8. $u = \frac{1}{\sqrt{2}}\sinh\sqrt{2x} + e^{-3y}\sin x$

9. $u = 4e^{x-y} - e^{2x-5y}$

10. $u = C_1 C_2 e^{p^2\left(\frac{y^2}{4} - \frac{x^3}{3}\right)}$

11. $u(x,y) = C_1 e^{-p^2 y}(C_2 \cos px + C_3 \sin px)$

12. $u(x,y) = C_1 C_2 e^{k(x+y)}$

13. $z(x,y) = 4e^{ax}\sin 3y$

14. $u(x,t) = \sin x(1 - e^{-t})$

15. $u(x,t) = \frac{1}{\sqrt{2}}\sinh\sqrt{2x} + e^{-3y}\sin x$

8.4 HYPERBOLIC DIFFERENTIAL EQUATIONS

We know that the most important homogeneous hyperbolic differential equation is the wave equation, which can be written as follows :

$$\frac{\partial^2 u}{\partial t^2} = C^2 \frac{\partial^2 u}{\partial x^2}, \text{ where } C \text{ is the wave speed.}$$

In this section, we shall discuss the derivation and solution of wave equation of one, two and three dimensional.

8.5 SOLUTION OF ONE DIMENSIONAL WAVE EQUATION

Consider one dimensional wave equation

$$\frac{\partial^2 u}{\partial x^2} = \frac{1}{C^2} \frac{\partial^2 u}{\partial t^2} \qquad \ldots (1)$$

Suppose that solution of (1) is of the form

$$u(x, t) = X(x) \, T(t) \qquad \ldots (2)$$

Putting the values from (2) in (1), we get

$$X''T = \frac{1}{C^2} XT''$$

$$\Rightarrow \qquad \frac{X''}{X} = \frac{T''}{C^2 T} = \mu \ (say) \qquad \ldots (3)$$

where μ is the separation constant.

From (3), we can deduce that $\quad X'' - \mu X = 0 \qquad \ldots (4)$

and

$$T'' - C^2 \mu T = 0 \qquad \ldots (5)$$

Solving (4) and (5), we get

(i) When $\mu = 0$, $X = a_1 x + a_2$, $T = a_3 t + a_4$

(ii) When $\mu = -\lambda^2$, $(\lambda \neq 0)$ then

$$X = b_1 e^{\lambda x} + b_2 e^{-\lambda x} \text{ and } T = b_3 e^{C\lambda t} + b_4 e^{-C\lambda t}$$

(iii) When $\mu = \lambda^2$, $\lambda \neq 0$ then

$$X = C_1 \cos \lambda x + C_2 \sin \lambda x \text{ and } T = C_3 \cos Cpt + C_4 \sin Cpt$$

Therefore, the various possible solutions are given by

$$u(x, t) = (a_1 x + a_2)(a_3 t + a_4) \qquad \ldots (6)$$

$$u(x, t) = (b_1 e^{\lambda x} + b_2 e^{-\lambda x})(b_3 e^{C\lambda t} + b_4 e^{-C\lambda t}) \qquad \ldots (7)$$

and

$$u(x, t) = (C_1 \cos \lambda x + C_2 \sin \lambda x)(C_3 \cos Cpt + C_4 \sin Cpt) \qquad \ldots (8)$$

Finally, since we are dealing with problems on vibration, $u(x, t)$, must be a periodic function of t.

Thus, $u(x, t)$ must involve trigonometric terms. Hence, the solution given by (8) is the only suitable solution.

8.6 D'ALEMBERT'S SOLUTION OF WAVE EQUATION

Let

$$\frac{\partial^2 \phi}{\partial x^2} = \frac{1}{C^2} \frac{\partial^2 \phi}{\partial t^2} \qquad \ldots (1)$$

be the given wave equation.

Define two new independent variables u and v such that

$$u = x + Ct \text{ and } v = x - Ct \qquad \ldots (2)$$

Now,

$$\frac{\partial \phi}{\partial x} = \frac{\partial \phi}{\partial u}.\frac{\partial u}{\partial x} + \frac{\partial \phi}{\partial v}.\frac{\partial v}{\partial x} = \frac{\partial \phi}{\partial u} + \frac{\partial \phi}{\partial v}$$

$\Rightarrow$

$$\frac{\partial}{\partial x} \equiv \frac{\partial}{\partial u} + \frac{\partial}{\partial v} \qquad \ldots (3)$$

Then

$$\frac{\partial^2 \phi}{\partial x^2} = \frac{\partial}{\partial x}\left(\frac{\partial \phi}{\partial x}\right) = \left(\frac{\partial}{\partial u} + \frac{\partial}{\partial v}\right)\left(\frac{\partial \phi}{\partial u} + \frac{\partial \phi}{\partial v}\right) \qquad \ldots (4)$$

$\Rightarrow$

$$\frac{\partial^2 \phi}{\partial x^2} = \frac{\partial^2 \phi}{\partial u^2} + 2\frac{\partial^2 \phi}{\partial u.\partial v} + \frac{\partial^2 \phi}{\partial v^2}$$

Also,

$$\frac{\partial \phi}{\partial t} = \frac{\partial \phi}{\partial u}.\frac{\partial u}{\partial t} + \frac{\partial \phi}{\partial v}.\frac{\partial v}{\partial t} = C\frac{\partial \phi}{\partial u} - C\frac{\partial \phi}{\partial v} = C\left(\frac{\partial \phi}{\partial u} - \frac{\partial \phi}{\partial v}\right)$$

$\Rightarrow$

$$\frac{\partial}{\partial t} \equiv C\left(\frac{\partial}{\partial u} - \frac{\partial}{\partial v}\right) \qquad \ldots (5)$$

Thus,

$$\frac{\partial^2 \phi}{\partial t^2} = \frac{\partial}{\partial t}\left(\frac{\partial \phi}{\partial t}\right) = C^2\left(\frac{\partial}{\partial u} - \frac{\partial}{\partial v}\right)\left(\frac{\partial \phi}{\partial u} - \frac{\partial \phi}{\partial v}\right)$$

$\Rightarrow$

$$\frac{1}{C^2}\frac{\partial^2 \phi}{\partial t^2} = \frac{\partial^2 \phi}{\partial u^2} - 2\frac{\partial^2 \phi}{\partial u\,\partial v} + \frac{\partial^2 \phi}{\partial v^2} \qquad \ldots (6)$$

Putting the values from (4) and (6) in (1), we get

$$\frac{\partial^2 \phi}{\partial u^2} + 2\frac{\partial^2 \phi}{\partial u\,\partial v} + \frac{\partial^2 \phi}{\partial v^2} = \frac{\partial^2 \phi}{\partial u^2} - 2\frac{\partial^2 \phi}{\partial u\partial v} + \frac{\partial^2 \phi}{\partial v^2}$$

$\Rightarrow$

$$\frac{\partial^2 \phi}{\partial u.\partial v} = 0$$

On integrating, we get

$$\frac{\partial \phi}{\partial u} = F(u) \text{ , where } F(u) \text{ is any arbitrary function of } u.$$

Again integrating w.r.t. u, we get

$$\phi = \int F(u)\,du + g(v) = f(u) + g(v)$$

$\Rightarrow$

$$\phi = f(x + Ct) + g(x - Ct)$$

which is the D'Alembert's solution of wave equation.

8.7 SOLUTION OF TWO DIMENSIONAL WAVE EQUATION

To find the solution of $\dfrac{\partial^2 u}{\partial t^2} = C^2\left(\dfrac{\partial^2 u}{\partial x^2} + \dfrac{\partial^2 u}{\partial y^2}\right)$ subject to the boundary conditions

$u(0, y, t) = u(a, y, t) = u(x, 0, t) = u(x, b, t) = 0$ and initial condition $u(x, y, 0) = f(x, y)$ and

$\left(\dfrac{\partial u}{\partial t}\right)_{t=0} = g(x, y)$.

The given equation is

$$\frac{\partial^2 u}{\partial x^2} + \frac{\partial^2 u}{\partial y^2} = \frac{1}{C^2}\frac{\partial^2 u}{\partial t^2} \qquad \dots (1)$$

with boundary conditions

    (a) $u(0, y, t) = 0$, (b) $u(a, y, t) = 0$, (c) $u(x, 0, t) = 0$, and (d) $u(x, b, t) = 0$.

Also, $u(x, y, 0) = f(x, y)$ $\dots (2)$

and $\left(\dfrac{\partial u}{\partial t}\right)_{t=0} = g(x, y)$ $\dots (3)$

Now, suppose that solution of (1) is of the form

$$u(x, y, t) = X(x)\, Y(y)\, T(t) \qquad \dots (4)$$

Putting the values from (4) in (1), we get

$$X''YT + XY''T = \frac{1}{C^2} XYT''$$

$\Rightarrow$ $\dfrac{X''}{X} + \dfrac{Y''}{Y} = \dfrac{T''}{C^2 T}$ $\dots (5)$

Now, since x, y and t are independent variables, (5) can only be true if each term on each side is equal to a constant, say μ_1.

Let $\dfrac{X''}{X} = \mu_1 \quad \Rightarrow \quad X'' - \mu_1 X = 0$ $\dots (6)$

Now, using (a) and (b), equation (4) gives

$$X(0)\, Y(y)\, T(t) = 0 \quad \text{and} \quad X(a)\, Y(y)\, T(t) = 0 \qquad \dots (7)$$

By assuming $Y(y) \neq 0$, $T(t) \neq 0$, we get

$$X(0) = 0 \;\text{ and }\; X(a) = 0 \qquad \dots (8)$$

Therefore, we want to solve equation (6) subject to the boundary conditions (8). Now, there are following cases :

Case (I). Let $\mu_1 = 0$. In this case, solution of (6) is given by

$$X(x) = Ax + B \qquad \dots (9)$$

Using (8) in (9), we get $B = 0$ and $Aa + B = 0$

$\Rightarrow$ $A = B = 0$ $\Rightarrow$ $X(x) = 0$

$\Rightarrow$ $u = 0$

which does not satisfy (2). Thus, we reject this case.

Case (II). Let $\mu_1 = \lambda_1^2, \lambda_1 \neq 0$. In this case, solution of (6) is given by

$$X(x) = Ae^{x\lambda_1} + Be^{-x\lambda_1} \qquad \dots (10)$$

Using (8) in (10), we get

$$0 = A + B \;\text{ and }\; 0 = Ae^{a\lambda_1} + Be^{-a\lambda_1} \qquad \dots (11)$$

On solving, we get $A = B = 0 \Rightarrow u = 0$

   This case is also rejected.

Case III. Let $\mu_1 = -\lambda_1^2, \lambda_1 \neq 0$. In this case, solution of (6) is given by

$$X(x) = A\cos\lambda_1 x + B\sin\lambda_1 x \qquad \dots (12)$$

Using (8) in (12), we get $\quad 0 = A$ and $\ 0 = A\cos\lambda_1 a + B\sin\lambda_1 a$

$\Rightarrow \qquad\qquad\qquad\qquad A = 0$ and $\sin\lambda_1 a = 0$.

Now, $\sin\lambda_1 a = 0 \ \Rightarrow \ \lambda_1 a = m\pi$

$\Rightarrow \qquad\qquad\qquad \lambda_1 = \dfrac{m\pi}{a}, \quad m = 1, 2, ... \qquad\qquad\qquad\qquad$ ... (13)

Thus, the non-zero solution $X_m(x)$ are given by

$$X_m(x) = B_n \sin\frac{m\pi x}{a}, \quad m = 1, 2, 3, ... \qquad\qquad\qquad \text{... (14)}$$

Further, let $\qquad\qquad \dfrac{Y''}{Y} = \mu_2 \ \Rightarrow \ Y'' - \mu_2 Y = 0 \qquad\qquad\qquad$ ... (15)

Using (c) and (d), (4) gives

$$Y(0) = 0 \ \text{ and } \ Y(b) = 0 \qquad\qquad\qquad\qquad\qquad \text{... (16)}$$

On solving (15) under boundary condition (16), we get

$$Y_n(y) = D_n \sin\left(\frac{n\pi y}{b}\right), \quad n = 1, 2, 3, ... \qquad\qquad\qquad \text{... (17)}$$

where, $\qquad\qquad \mu_2 = -\lambda_2^2 \ \text{ and } \ \lambda_2 = \dfrac{n\pi}{b}, \ n = 1, 2, 3, ... \qquad$ ... (18)

Therefore, (5) reduces to

$$\frac{T''}{C^2 T} = \mu_1 + \mu_2 = -\lambda_1^2 - \lambda_2^2 = -\pi^2\left(\frac{m^2}{a^2} + \frac{n^2}{b^2}\right)$$

$\Rightarrow \qquad\qquad\qquad T'' + \lambda_{mn}^2 T = 0 \qquad\qquad\qquad\qquad\qquad$ ... (19)

where, $\qquad\qquad \lambda_{mn}^2 = C^2 \pi^2 \left(\dfrac{m^2}{a^2} + \dfrac{n^2}{b^2}\right) \qquad\qquad\qquad\qquad$ ... (20)

The solution of (19) is given by

$$T_{mn}(t) = E_{mn} \cos\lambda_{mn}t + F_{mn} \sin\lambda_{mn}t \qquad\qquad \text{... (21)}$$

Hence, $\qquad u_{mn}(x, y, t) = X_m(x) Y_n(y) T_{mn}(t)$

$$= (A_{mn} \cos\lambda_{mn}t + B_{mn} \sin\lambda_{mn}t)\left(\sin\frac{m\pi x}{a}\right)\sin\frac{n\pi y}{b} \qquad \text{... (22)}$$

are solution of (1) satisfying (a) to (d).

Hence, the more general solution is given by

$$u(x, y, t) = \sum_{m=1}^{\infty} \sum_{n=1}^{\infty} (A_{mn} \cos\lambda_{mn}t + B_{mn} \sin\lambda_{mn}t)\left(\sin\frac{m\pi x}{a}\right)\sin\frac{n\pi y}{b}$$

where $\qquad\qquad A_{mn} = B_m D_n E_{mn} \ \text{ and } \ B_{mn} = B_m D_n F_{mn}$.

Solved Examples

EXAMPLE 1. *Find the general solution of one dimensional wave equation* $\dfrac{\partial^2 u}{\partial x^2} = \dfrac{1}{C^2}\dfrac{\partial^2 u}{\partial t^2}$ *and find the particular solution for which* $u = f(x)$, $\dfrac{\partial u}{\partial t} = g(x)$ *at* $t = 0$. (UPTU (CO)–2008)

SOLUTION. The given equation is $\dfrac{\partial^2 u}{\partial x^2} = \dfrac{1}{C^2}\dfrac{\partial^2 u}{\partial t^2} \qquad\qquad\qquad$ ... (1)

Let string be stretched between fixed points $(0, 0)$ and $(a, 0)$. So we are to find $u(x, t)$ subject to the boundary condition

$$u(0, t) = 0, \quad u(a, t) = 0 ; \quad \forall t \qquad \qquad \text{... (2)}$$

and initial conditions

$$u(x, 0) = f(x)$$
$$\left. \left(\frac{\partial u}{\partial t} \right) \right|_{t=0} = u_t(x, 0) = g(x) \qquad \qquad \text{... (3)}$$

Suppose that solution of (1) is of the form

$$u(x, t) = X(x) \, T(t) \qquad \qquad \text{... (4)}$$

Putting the values from (4) in (3), we get

$$X''T = \frac{1}{c^2} XT''$$

$$\Rightarrow \qquad \frac{X''}{X} = \frac{1}{c^2} \frac{T''}{T} = \mu \text{ (say)}, \qquad \qquad \text{... (5)}$$

where μ is a separation constant.

Using (5), we can deduce that

$$X'' - \mu X = 0 \qquad \qquad \text{... (6)}$$

and $\quad T'' - \mu c^2 T = 0 \qquad \qquad \text{... (7)}$

Using (2) and (4), we get

$$X(0) \, T(t) = 0 \text{ and } X(a) \, T(t) = 0 \qquad \qquad \text{... (8)}$$

$$\Rightarrow \qquad X(0) = 0, X(a) = 0 \; [T(t) \neq 0] \qquad \qquad \text{... (9)}$$

Now, we have to solve (6) under (9). There are following cases :

Case (I). Let $\mu = 0$. In this case, solution of (6) is given by

$$X(x) = Ax + B \qquad \qquad \text{... (10)}$$

Using (9) in (10), we get

$$B = 0 \text{ and } Aa + B = 0$$

$$\Rightarrow \qquad A = B = 0 \Rightarrow \; X(x) = 0$$

$$\Rightarrow \qquad u = 0$$

which does not satisfy (3). Thus, we reject this case.

Case (II). Let $\mu = \lambda^2, \lambda \neq 0$. In this case, solution of (6) is given by

$$X(x) = Ae^{\lambda x} + Be^{-\lambda x} \qquad \qquad \text{... (11)}$$

Using (9) in (11), we get

$$0 = A + B \text{ and } 0 = Ae^{\lambda a} + Be^{-\lambda a} \qquad \qquad \text{... (12)}$$

On solving, we get $A = B = 0$

$$\Rightarrow \; X(x) = 0 \qquad \Rightarrow \quad u = 0$$

Hence, we reject this case also.

Case (III). Let $\mu = -\lambda^2, \lambda \neq 0$. In this case, solution of (6) is given by

$$X(x) = A \cos \lambda x + B \sin \lambda x \qquad \qquad \text{... (13)}$$

Using (9) in (13), we get

$$0 = A \text{ and } 0 = A \cos \lambda a + B \sin \lambda a$$

On solving, we get $A = 0$ and $\sin \lambda a = 0$ $\hspace{2cm}$ $(B \neq 0)$

Now, $\hspace{1cm}$ $\sin \lambda a = 0 \implies \lambda a = n\pi$

$\implies$ $\hspace{2cm}$ $\lambda = \dfrac{n\pi}{a}, \quad n = 1, 2, \dots$ $\hspace{2cm}$ ... (14)

Thus, the non-zero solution of (6) is given by

$$X_n(x) = B_n \sin\left(\dfrac{n\pi x}{a}\right) \hspace{2cm} \text{... (15)}$$

Now, using (14) in (7), we get

$$T'' + \left(\dfrac{n^2 \pi^2 C^2}{a^2}\right) T = 0$$

whose general solution is given by

$$T_n(t) = C_n \cos \dfrac{n\pi Ct}{a} + D_n \sin \dfrac{n\pi Ct}{a} \hspace{2cm} \text{... (16)}$$

So, $\hspace{1cm}$ $u_n(x, t) = X_n(x)\, T_n(x) = \left(E_n \cos \dfrac{n\pi Ct}{a} + F_n \sin \dfrac{n\pi Ct}{a}\right) \sin \dfrac{n\pi x}{a}$

are solutions of (1) satisfying (2).

Thus, more general solution is given by

$$u(x, t) = \sum_{n=1}^{\infty} u_n(x, t)$$

$$= \sum_{n=1}^{\infty} \left(E_n \cos \dfrac{n\pi Ct}{a} + F_n \sin \dfrac{n\pi Ct}{a}\right) \sin \dfrac{n\pi x}{a} \hspace{1cm} \text{... (17)}$$

$$\implies \dfrac{\partial u}{\partial t} = \sum_{n=1}^{\infty} \left(-\dfrac{n\pi C E_n}{a} \sin \dfrac{n\pi Ct}{a} + \dfrac{n\pi C}{a} F_n \cos \dfrac{n\pi Ct}{a}\right) \sin \dfrac{n\pi x}{a} \hspace{0.5cm} \text{... (18)}$$

Putting $t = 0$ in (17) and (18) and using (3), we get

$$f(x) = \sum_{n=1}^{\infty} E_n \sin \dfrac{n\pi x}{a} \hspace{2cm} \text{... (19)}$$

and $\hspace{1cm}$ $g(x) = \sum_{n=1}^{\infty} \dfrac{n\pi C F_n}{a} \sin \dfrac{n\pi x}{a} \hspace{2cm} \text{... (20)}$

which are Fourier sine series expansions for $f(x)$ and $g(x)$ respectively. Thus, we get

$$E_n = \dfrac{2}{\pi} \int_0^a f(x) \sin \dfrac{n\pi x}{a}\, dx \hspace{2cm} \text{... (21)}$$

and $\hspace{1cm}$ $\dfrac{n\pi C F_n}{a} = \dfrac{2}{a} \int_0^a g(x) \sin \dfrac{n\pi x}{a}\, dx$

$$\implies F_n = \dfrac{2}{n\pi C} \int_0^a g(x) \sin \dfrac{n\pi x}{a}\, dx \hspace{2cm} \text{... (22)}$$

EXAMPLE 2. *Solve the wave equation* $\dfrac{\partial^2 u}{\partial t^2} = C^2 \dfrac{\partial^2 u}{\partial x^2}$, *where* $u = p_0 \cos pt$ (p_0 *is a constant*)

when $x = l$ *and* $u = 0$ *when* $x = 0$. $\hspace{3cm}$ [AMIE–1997]

SOLUTION.   We know that solution of the given equation is

$$u(x,t) = (C_1 \cos nx + C_2 \sin nx).(C_3 \cos nCt + C_4 \sin nCt) \qquad \ldots (1)$$

Putting $u = 0$ when $x = 0$, we get

$$C_1(C_3 \cos nCt + C_4 \sin nCt) = 0$$

$$\Rightarrow \qquad C_1 = 0$$

Therefore, we have

$$u(x,t) = C_2 \sin nx \, (C_3 \cos nCt + C_4 \sin nCt)$$

$$= \sin nx \cos nCt.b_1 + \sin nx \sin nCt.b_2 \text{ where } b_1 = C_2C_3 \text{ and } b_2 = C_2C_4.$$

Now, using $u = p_0 \cos pt$, when $x = l$, we get

$$p_0 \cos pt = \sin nl \cos pt.b_1 + \sin nl \sin Ct.b_2$$

$$\Rightarrow \qquad p_0 \cos pt = \sin nl \cos nCt.b_1 \quad \text{and}$$

$$0 = \sin nl \sin nCt.b_2 \qquad \Rightarrow \qquad p = nC.$$

Therefore,

$$p_0 \cos pt = \sin nl \cos pt.b_1$$

$$\Rightarrow \qquad b_1 = \frac{p_0}{\sin nl} = \frac{p_0}{\sin(pl/C)}$$

Hence, $u(x,t) = \sin nx \cos nCt.b_1 = \dfrac{p_0}{\sin(pl/C)} \cos pt \, \sin(px/C)$

EXAMPLE 3.   *A string is stretched to two fixed points distance l apart. Motion is started by displacing*

the string in the form $u = a \sin\left(\dfrac{\pi x}{l}\right)$ *from which it is released at time $t = 0$. Show*

that the displacement at any point at a distance x from one end at time t is given by

$$u(x,t) = a \, \sin\left(\frac{\pi x}{l}\right) \cos\left(\frac{\pi Ct}{l}\right) \quad \text{(AMIETE–2003, UPTU–2004, 09, UPTU (Sum)–2009,}$$

UKTU–2011, 12, VTU–2010, SVTU–2008, Kerala–2005)

SOLUTION.   It is known that the vibrations of the string are governed by the equation

$$\frac{\partial^2 u}{\partial x^2} = C^2 \frac{\partial^2 u}{\partial x^2}$$

As per given, the string is fastened at O and A respectively. Therefore,

$$u(0,t) = 0 \quad \text{and} \quad u(l,t) = 0 \, ; \, \forall t \qquad \ldots (2)$$

and $\quad u(x,0) = a\sin\dfrac{\pi x}{l} \quad$ and $\quad \left(\dfrac{\partial u}{\partial t}\right) = 0 \qquad \ldots (3)$

Fig. 1

The solution of (1) is given by

$$u(x,t) = (C_1 \cos nx + C_2 \sin nx) (C_3 \cos Cnt + C_4 \sin Cnt)$$

Now, $u(0,t) = 0$ implies

$$0 = C_1(C_3 \cos Cnt + C_4 \sin Cnt)$$

$$\Rightarrow \qquad\qquad C_1 = 0$$

$$\Rightarrow \qquad\qquad u(x, t) = \sin nx \, (b_1 \cos Cnt + b_2 \sin Cnt)$$

Thus, $\qquad\qquad \dfrac{\partial u}{\partial t} = \sin nx \left[-Cn \sin Cnt \, b_1 + Cn.b_2 \cos Cnt \right]$

$$\Rightarrow \qquad\qquad \left[\dfrac{\partial u}{\partial t} \right]_{t=0} = \sin nx \, [Cn.b_2] = 0 \, ; \; \forall t$$

$$\Rightarrow \qquad\qquad b_2 = 0$$

Thus, (4) reduces to $u(x, t) = \sin x \cos Cnt \, . \, b_1$ ... (5)

Further, $\qquad u(l, t) = 0$ implies $\sin nl \cos Cnt.b_1 = 0 \, ; \; \forall t$

$\Rightarrow$ Either $\quad \sin nl = 0 \;$ or $\; b_1 = 0$.

Since $b_1 = 0$ makes the solution trivial, therefore, we take $\sin nl = 0$.

Now, $\qquad\qquad \sin nl = 0 \Rightarrow nl = m\pi \Rightarrow n = \dfrac{m\pi}{l}, m = 1, 2, ...$

Then, from (5), we get

$$u(x, t) = \sin \dfrac{m\pi}{l} \, x \, \cos\left(\dfrac{Cm\pi}{l}.t \right). \, b_1$$

Since $\qquad u(x, 0) = a \sin \dfrac{\pi x}{l}$, therefore $a \sin \dfrac{\pi x}{l} = \sin \dfrac{m\pi x}{l}.b_1$

which is true only when $m = 1$ and $b_1 = a$.

Hence, $\qquad u(x, t) = a \sin\left(\dfrac{\pi x}{l} \right) \cos\left(\dfrac{\pi Ct}{l} \right)$

EXAMPLE 4. *A string is stretched and fastened to two points distance l apart. Motion is started displacing the string into the form* $y = k(lx - x^2)$ *from which it is released at time t = 0. Find the displacement of any point on the string at a distance of x from one end at time t.* (UPTU Q. Bank 2002; IAS 1990,93,97)

Or

If the string is released from rest in the position $y = \dfrac{4t}{l^2} \, x(l - x)$, *find the displacement y.*

(AMIE–1991,97,2003)

SOLUTION. It is known that the vibrations of the string are governed by the equation

$$\dfrac{\partial^2 y}{\partial t^2} = C^2 \dfrac{\partial^2 y}{\partial x^2} \qquad\qquad ...(1)$$

where $y(0, t) = 0 \;$ and $\; y(l, t) = 0$ ...(2)

Also, the initial transverse velocity of any point of the string is zero. Thus, the initial condition is given by

$$\left. \left(\dfrac{\partial y}{\partial t} \right) \right|_{t=0} = 0$$

$$y(x, 0) = k(lx - x^2) \qquad\qquad ... (3)$$

We know that solution of (1) is given by

$$y(x, t) = (C_1 \cos px + C_2 \sin px) \, (C_3 \cos Cpt + C_4 \sin Cpt) \qquad ... (4)$$

Now $\qquad y(0, t) = 0$

$\Rightarrow \qquad 0 = C_1(C_3 \cos Cpt + C_4 \sin Cpt) \Rightarrow C_1 = 0$

$\therefore \qquad y(x, t) = \sin px \, (C_1' \cos Cpt + C_2' \sin Cpt) \qquad \qquad \dots (5)$

Therefore,

$$\left(\frac{\partial y}{\partial t}\right)_{t=0} = 0 \Rightarrow \quad 0 = C_2' \sin px.(Cp)$$

$$\Rightarrow \quad C_2' = 0 \qquad \qquad \left[\because C_p \neq 0\right]$$

Then, (5) gives $y(x, t) = C_1' \sin px \, (\cos Cpt)$

Now, $\qquad y(l, t) = 0 \Rightarrow C_1' \sin pl \cos Cpt = 0$

$\Rightarrow \qquad C_1' = 0 \quad$ or $\sin pl = 0$

Since $C_1' = 0$ makes the solution trivial, thus, we have

$$\sin pl = \sin n\pi, \ n = 1, 2, \dots \Rightarrow \quad p = \frac{n\pi}{l}$$

$\therefore \qquad y_n(x, t) = C_1' \sin \frac{n\pi x}{l} \cos \frac{n\pi C}{l}.t = b_n \sin \frac{n\pi x}{l} \cos \frac{n\pi Ct}{l}$

Therefore, the complete solution will be of the form

$$f(x, t) = \sum_{n=1}^{\infty} b_n \sin \frac{n\pi x}{l} \cos \frac{n\pi Ct}{l} \qquad \qquad \dots (6)$$

Further, $\qquad f(x, 0) = \sum_{n=1}^{\infty} b_n \sin \frac{n\pi x}{l}$

$\Rightarrow \qquad lx - x^2 = \sum_{n=1}^{\infty} b_n \sin \frac{n\pi x}{l} \qquad \qquad \dots (7)$

which is the expansion of $f(x)$ in the form of Fourier sine series. Thus we have

$$b_n = \frac{2}{l} \int_0^l f(x) \sin \frac{n\pi x}{l} dx$$

$$= \frac{2}{l} \int_0^l (lx - x^2) \sin \frac{n\pi x}{l} dx$$

$$= \frac{2}{l} \left[(lx - x^2)\left(-\cos \frac{n\pi x}{l}\right) \frac{l}{n\pi} - (l - 2x) \right]$$

$$\left[\left(-\sin \frac{n\pi x}{l}\right) \frac{l^2}{n^2\pi^2} + (-2)\left(\cos \frac{n\pi x}{l}\right) \frac{l^3}{n^3\pi^3} \right]_0^l$$

$$= \frac{2}{l} \left[(-1)^{n+1} \frac{2l^3}{n^3\pi^3} + \frac{2l^3}{n^3\pi^3} \right] = \frac{8l^2}{n^3\pi^3}, \text{ if } n \text{ is odd}$$

$$= 0, \text{ if } n \text{ is even.}$$

Putting the value of b_n in (7), we get

$$y(x, t) = \sum \frac{8l^2}{n^3\pi^3} \sin \frac{n\pi x}{l} \cos \frac{n\pi Ct}{l}, \text{ when } n \text{ is odd.}$$

EXAMPLE 5. *A tightly stretched string with fixed end points $x = 0$ and $x = 1$ is initially in a position given by $u = u_0 \sin^3\left(\dfrac{\pi x}{l}\right)$. If it is released from rest from this position, find the displacement $u(x, t)$.*

(AMIE–1997; GBTU (CO)–2011; Rajasthan–2006; VTU–2003; JNTU–2002)

SOLUTION. It is known that the equation of vibrating string is given by

$$\frac{\partial^2 u}{\partial t^2} = C^2 \frac{\partial^2 u}{\partial x^2} \qquad \qquad \dots (1)$$

with boundary conditions $u(0, t) = 0$, $u(l, t) = a$

$$\left(\frac{\partial u}{\partial t}\right)_{t=0} = 0; \quad u(x, 0) = u_0 \sin^3\left(\frac{\pi x}{l}\right)$$

We know that solution of (1) is given by

$$u(x, t) = (C_1 \cos nx + C_2 \sin nx)(C_3 \cos nCt + C_4 \sin nCt) \qquad \dots (2)$$

Now, $\qquad u(0, t) = 0$

$\Rightarrow \qquad\qquad 0 = C_1(C_3 \cos nCt + C_4 \sin nCt)$

$\Rightarrow \qquad\qquad C_1 = 0$

Then, (2) reduces to

$$u(x, t) = C_2 \sin nx \, (C_3 \cos nCt + C_4 \sin nCt) \qquad \dots (3)$$

$\Rightarrow \qquad u(x, t) = \sin nx \, (b_1 \cos nCt + b_2 \sin nCt)$

where, $\qquad b_1 = C_2 C_3, \quad b_2 = C_2 C_4$

Further, $\qquad u(l, t) = 0 \Rightarrow \sin nl \, (b_1 \cos nCt + b_2 \sin nCt) = 0$

$\Rightarrow \qquad \sin nl = 0 \qquad \Rightarrow \qquad \sin nl = \sin n\pi \Rightarrow n = \dfrac{m\pi}{l}$

Therefore, $u(x,t) = \sin\dfrac{m\pi x}{l}\left(b_1 \cos\dfrac{m\pi Ct}{l} + b_2 \sin\dfrac{m\pi Ct}{l}\right) \qquad \dots (4)$

$\Rightarrow \qquad \dfrac{\partial u}{\partial t} = \sin\left(\dfrac{m\pi x}{l}\right)\left(-\dfrac{m\pi C}{l}b_1 \sin\dfrac{m\pi C}{l}.t + b_2 \dfrac{m x C}{l}\cos\dfrac{m\pi C}{l}.t\right)$

Further, $\qquad \dfrac{\partial u}{\partial t} = 0$

$\Rightarrow \qquad 0 = \sin\dfrac{m\pi x}{l}.b_2 \dfrac{m\pi C}{l} \Rightarrow b_2 = 0$

Then, (4) gives $u(x, t) = \sin\dfrac{m\pi C}{l}\cos\dfrac{m\pi Ct}{l}.b_1$

$\Rightarrow \qquad u_m(x, t) = b_m \sin\dfrac{m\pi x}{l}\cos\dfrac{m\pi Ct}{l}$

Now, $\qquad u(x, t) = \displaystyle\sum_{m=1}^{\infty} u_m(x, t)$

$$= \sum_{m=1}^{\infty} b_m \sin\frac{m\pi x}{l}\cos\frac{m\pi Ct}{l} \qquad \dots (5)$$

Also, $\qquad u(x, 0) = u_0 \sin^3\left(\dfrac{\pi x}{l}\right) = \dfrac{u_0}{4}\left(3\sin\dfrac{\pi x}{l} - \sin\dfrac{3\pi x}{l}\right)$

Putting $t = 0$ in (5), we get

$$\sum_{m=1}^{\infty} b_m \sin\frac{m\pi x}{l} = \frac{u_0}{4}\left(3\sin\frac{\pi x}{l} - \sin\frac{3\pi x}{l}\right)$$

where $\qquad b_1 = \dfrac{3u_0}{4}$ and $b_3 = \dfrac{-u_0}{4}$ remaining b_i's are all zero.

Hence, $\qquad u(x,t) = \dfrac{u_0}{4}\left(3\sin\dfrac{\pi x}{l}.\cos\dfrac{C\pi t}{l} - \sin\dfrac{3\pi x}{l}\cos\dfrac{3\pi C t}{l}\right).$

EXAMPLE 6. *Find the solution of the wave equation $\dfrac{\partial^2 y}{\partial t^2} = C^2\dfrac{\partial^2 y}{\partial x^2}$ such that $y = p_0 \cos pt$,*

when $x = l$ and $y = 0$ when $x = 0$ (b_0 is a constant). (IAS–1992)

SOLUTION. The given equation is

$$\frac{\partial^2 y}{\partial t^2} = C^2\frac{\partial^2 y}{\partial x^2} \qquad \qquad ...\ (1)$$

The solution of (1) is given by

$$y(x,t) = (C_1 \cos C\sqrt{k}\,t + C_2 \sin C\sqrt{k}\,t) - (C_3 \cos\sqrt{k}\,x + C_4 \sin\sqrt{k}\,x) \quad ...\ (2)$$

Now, $y(0, t) = 0$ implies

$$0 = (C_1 \cos C\sqrt{k}\,t + C_2 \sin C\sqrt{k}\,t)\,C_3 \Rightarrow C_3 = 0$$

Therefore, (2) becomes

$$y(x,t) = (C_1 \cos C\sqrt{k}t + C_2 \sin C\sqrt{k}t)(C_4 \sin\sqrt{k}x)$$

$$= C_1' \cos C\sqrt{k}\,t \sin\sqrt{k}\,x + C_2' \sin C\sqrt{k}\,t\sin\sqrt{k}\,x \qquad ...\ (3)$$

where, $\qquad C_1' = C_1 C_4,\ C_2' = C_2 C_4$

Further, $\quad y(x, t) = p_0 \cos pt$, when $x = l$.

Also, $\qquad p_0 \cos pt = C_1' \cos C\sqrt{k}\,t \sin\sqrt{k}\,l + C_2' \sin C\sqrt{k}\,t \sin\sqrt{k}\,l$

Equating the coefficients of sin and cos on both the sides, we get

$$p_0 = C_1' \sin\sqrt{k}\,l \quad \Rightarrow \quad C_1' = \frac{p_0}{\sin\sqrt{k}\,l}$$

$$0 = C_2' \sin\sqrt{k}\,l \Rightarrow C_2' = 0$$

Hence, $\quad y(x, t) = \dfrac{p_0}{\sin\sqrt{k}\,l} \cos C\sqrt{k}\,t.\sin\sqrt{k}\,x$

$$= \frac{p_0}{\sin\sqrt{k}\,t} . \cos pt \sin\frac{p}{C}x = \frac{p_0}{\sin(pl\,/\,C)} \cos pt \sin\left(\frac{p}{C}x\right)$$

EXAMPLE 7. *A string is stretched and fastened to two points at a distance l apart. Motion is started by displacing the string in the form $u = a\sin\dfrac{\pi x}{l}$ from which it is released at a time*

$t = 0$. Show that the displacement of any point at a distance x from one end at time t is given by

$$u(x, t) = a\sin\left(\frac{\pi x}{l}\right)\cos\left(\frac{\pi C t}{l}\right).$$

SOLUTION.   It is known that the vibration of the string are governed by the equation

$$\frac{\partial^2 u}{\partial t^2} = C^2 \frac{\partial^2 u}{\partial x^2} \qquad \ldots (1)$$

Since, end points of the string are fixed for all t, therefore, we have

$$u(0, t) = 0 \quad \text{and} \quad u(l, t) = 0$$

Also, the initial transverse velocity of any point of the string is zero, therefore

$$\left(\frac{\partial u}{\partial t}\right)_{t=0} = 0 \qquad \ldots (2)$$

$$u(x, 0) = a \sin \frac{\pi x}{l}$$

Since the vibration of the string is periodic, therefore, solution of (1) is of the following periodic form

$$u(x, t) = (C_1 \cos px + C_2 \sin px)(C_3 \cos Cpt + C_4 \sin Cpt) \qquad \ldots (3)$$

Therefore,

$$u(x, t) = C_2 \sin px \, (C_3 \cos Cpt + C_4 \sin Cpt)$$

$$= \sin px \, (C_3' \cos Cpt + C_4' \sin Cpt) \qquad \ldots (4)$$

$$\Rightarrow \qquad \frac{\partial u}{\partial t} = \sin px \left[C_3'(-Cp \sin Cpt) + C_4'(Cp \cos Cpt) \right] \qquad \ldots (5)$$

$$\Rightarrow \qquad \left(\frac{\partial u}{\partial t}\right)_{t=0} = \sin px \, (C_4' C_p) = 0$$

$$\Rightarrow \qquad C_4' = 0 \Rightarrow u(x, t) = \sin px \, C_3' \cos Cpt \qquad (C_3' \neq 0)$$

Also, $u(l, t) = 0 \Rightarrow \sin pl = 0$

Further, $\sin pl = 0$

$$\Rightarrow \qquad pl = n\pi \Rightarrow p = \frac{n\pi}{l}$$

$$\Rightarrow \qquad u(x, t) = \sin\left(\frac{n\pi x}{l}\right) C_3' \cos\left(\frac{Cn\pi t}{l}\right) \qquad \ldots (6)$$

Also, $u(x, 0) = a \sin \dfrac{\pi x}{l}$

Thus, (6) gives $a \sin \dfrac{\pi x}{l} = \sin \dfrac{n\pi x}{l} \cdot C_3'$

Let us take $C_3' = a$ and $n = 1$.

Hence, the required solution is given by

$$u(x, t) = a \sin \frac{\pi x}{l} \cos\left(\frac{C\pi t}{l}\right).$$

EXAMPLE 8.   *A string is stretched between the fixed points (0, 0) and (1, 0) and released at rest from the position $u = A \sin \pi x$. Obtain the formulae for its subsequent displacement $u(x, t)$.*

SOLUTION.   It is known that the vibration of the string are governed by one-dimensional wave equation $\dfrac{\partial^2 u}{\partial t^2} = C^2 \dfrac{\partial^2 u}{\partial x^2}$ subject to the boundary conditions $u(0, t) = 0$, $u(1, t) = 0$

and initial conditions are

$$u(x, 0) = A \sin \pi x, \quad \left(\frac{\partial u}{\partial t}\right)_{t=0} = 0$$

Therefore, $u(x, t) = \sum_{n=1}^{\infty} C_n \cos nC\pi t \sin \pi x$

where, $\qquad C_n = 2\int_0^1 A \sin \pi x \sin n\pi x \, dx$

$\Rightarrow \qquad C_1 = 2\int_0^1 A \sin \pi x \sin \pi x \, dx$

$$= 2A\int_0^1 \sin^2 \pi x \, dx = A\int_0^1 2\sin^2 \pi x \, dx$$

$$= A\int_0^1 (1 - \cos 2\pi x) \, dx = A$$

All other coefficients $C_2 = C_3 = C_4 = \dots = 0$

Hence, $\qquad u(x, t) = C_1 \cos C\pi t \sin \pi x = A \cos C\pi t \sin \pi x$.

EXAMPLE 9. *A string is stretched between the fixed points (0,0) and (1, 0) and released at rest from the initial deflection given by*

$$f(x) = \frac{2l}{l} \cdot x, \quad \text{when} \ \ 0 < x < l/2 \, ; f(x) = \frac{2l}{l} \cdot (l - x), \quad \text{when} \ \ l/2 < x < l$$

Find the deflection of the string at any time t.   (UPTU Q. Bank–2002)

SOLUTION. We know that the deflection $u(x, t)$ of the string is given by

$$u(x, t) = \sum_{n=1}^{\infty} C_n \cos \frac{Cn\pi}{l} t \sin \frac{n\pi x}{l}$$

where, $\qquad C_n = \frac{2}{l}\int_0^l f(x) \sin \frac{n\pi x}{l} \, dx$

$$= \frac{2}{l}\left[\int_0^{l/2} f(x) \sin \frac{n\pi x}{l} \, dx + \int_{l/2}^l f(x) \sin \frac{n\pi x}{l} dx\right]$$

$$= \frac{2}{l}\left[\int_0^{l/2} \frac{2k}{l} \sin \frac{n\pi x}{l} \, dx + \int_{l/2}^l \frac{2k}{l}(l - x) \sin \frac{n\pi x}{l} dx\right]$$

$$= \frac{4k}{l^2}\left[\int_0^{l/2} x \sin \frac{n\pi x}{l} dx + \frac{4k}{l^2}\int_{l/2}^l (l - x) \sin \frac{n\pi x}{l} dx\right]$$

$$= \frac{4k}{l^2}\left[\left[-x\frac{l}{n\pi}\cos \frac{n\pi x}{l}\right]_0^{l/2} + \frac{l}{\pi x}\int_0^{l/2} 1 . \cos \frac{n\pi x}{l} dx\right.$$

$$\left. + \frac{4k}{l^2}\left[-(1 - x)\frac{l}{n\pi}\cos \frac{n\pi x}{l}\right]_{l/2}^l - \frac{l}{n\pi}\int_{l/2}^l \cos \frac{n\pi x}{l} dx\right]$$

$$= \frac{4k}{l^2}\left[-\frac{l^2}{2n\pi}\cos \frac{n\pi}{2} + \frac{l}{n\pi}\left(\frac{l}{n\pi}\sin \frac{n\pi x}{l}\right)_0^l\right.$$

$$\left. + \frac{4k}{l}\left[-0 + \frac{l}{2}\frac{l}{n\pi}\cos \frac{n\pi}{2} - \frac{l}{n\pi}\left(\frac{l}{n\pi}\sin \frac{n\pi x}{l}\right)_{l/2}^l\right]\right.$$

$$= \frac{4k}{l^2}\left[-\frac{l^2}{2n\pi}\cos\frac{n\pi}{2} + \frac{l^2}{n^2 x^2}\sin\frac{n\pi}{2}\right] + \frac{4k}{l^2}$$

$$\left[\frac{l^2}{2n\pi}\cos\frac{n\pi}{2} - \frac{l^2}{n^2\pi^2}\left(\sin n\pi - \sin\frac{n\pi}{2}\right)\right]$$

$$= \frac{4k}{l^2}\left[-\frac{l^2}{2n\pi}\cos\frac{n\pi}{2} + \frac{l^2}{n^2 x^2}\sin\frac{n\pi}{2}\right.$$

$$\left. + \frac{l^2}{2n\pi}\cos\frac{n\pi}{2} - \frac{l^2}{n^2\pi^2}\sin n\pi + \frac{l^2}{n^2\pi^2}\sin\frac{n\pi}{2}\right]$$

$$= \frac{8k}{n^2\pi^2}\sin\frac{n\pi}{2}$$

Thus, $\quad u(x,t) = \sum_{n=1}^{\infty}\frac{8k}{n^2\pi^2}\sin\frac{n\pi}{2}\cos\frac{Cn\pi}{l}.t.\sin\frac{n\pi x}{l}$

$$= \frac{8k}{\pi^2}\sin\frac{\pi}{2}\cos\frac{C\pi}{l}.t.\sin\frac{\pi x}{l}$$

$$+ \frac{8k}{2^2\pi^2}\sin\pi\cos\frac{2C\pi}{l}.t.\sin\frac{2\pi x}{l}$$

$$+ \frac{8k}{3^2\pi^2}\sin\frac{3\pi}{2}\cos\frac{3C\pi}{l}.t.\sin\frac{3\pi x}{l} + ...$$

$$= \frac{8k}{\pi^2}\left[\sin\frac{\pi x}{l}\cos\frac{\pi C}{l} - \frac{1}{3^2}\sin\frac{3\pi x}{l}\cos\frac{3\pi C}{l}.t + ...\right]$$

EXAMPLE 10. *The point of trisection of a string are pulled aside through a distance h on opposite sides of the position of equilibrium, and the string is released from rest. Find an expression for the displacement of the string at any subsequent time and show that the mid point of the string always remains at rest.* [Kerala–2005, UPTU Q. Bank–2002]

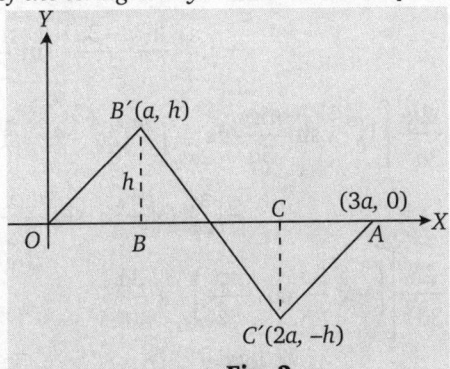

Fig. 2

SOLUTION. Suppose *OBCA* be the position of the equilibrium of the string of length $l = 3a$. Further, let B and C be the points of trisection of the string be pulled through a distance h on opposite sides and then released. It is known that this is the case in which vibrations are governed by one-dimensional wave equation

$$\frac{\partial^2 u}{\partial t^2} = C^3\frac{\partial^2 u}{\partial x^2} \qquad\qquad ... (1)$$

Now, equation of the line joining $O(0, 0)$ to $B'(a, h)$ is given by

$$y - 0 = \frac{h - 0}{a - 0} \quad \Rightarrow \quad y = \frac{h}{a} x$$

Also, the equation of joining $B'(a, h)$ to $C'(2a, -h)$ is given by

$$y - h = \frac{-h - h}{2a - a}(x - a) \Rightarrow y = \frac{h(3a - 2x)}{a}$$

Further, equation of the line joining $C'(2a - h)$ to $A(3a- 0)$ is

$$y - (-h) = \frac{0 - (-h)}{3a - 2a}(x - 2a)$$

$$\Rightarrow \qquad y = \frac{h(x - 3a)}{a}$$

Then, its initial deflection is given by

$$f(x) = \begin{cases} hx / a & ; \ 0 \le x \le a \\ \dfrac{h(3a - 2x)}{a} & ; \ a \le x \le 2a \\ \dfrac{h(x - 3a)}{a} & ; \ 2a \le x \le 3a \end{cases}$$

and initial velocity is $g(x) = 0$.

So, deflection $u(x, t)$ is given by

$$u(x, t) = \sum_{n=1}^{\infty} C_n \cos\frac{Cn\pi}{3a}.t \, \sin\frac{n\pi x}{3a}$$

where, $C_n = \dfrac{2}{3a} \int_0^{3a} f(x) \sin\dfrac{n\pi x}{3a} \, dx$

$$= \frac{2}{3a}\left[\int_0^a \frac{hx}{a} \sin\frac{n\pi x}{3a} dx + \int_a^{2a} \frac{h(3a - 2x)}{a} \sin\frac{n\pi x}{3a} dx \right.$$

$$\left. + \int_{2a}^{3a} \frac{h(x - 3a)}{a} \sin\frac{n\pi x}{3a} . dx \right]$$

or $\qquad C_n = \dfrac{2h}{3a^2}\left[\int_0^a x \sin\dfrac{n\pi x}{3a} dx + \int_a^{2a}(3a - 2x) \sin\dfrac{n\pi x}{3a} dx \right.$

$$\left. + \int_{2a}^{3a}(x - 3a)\sin\frac{n\pi x}{3a} dx \right]$$

$$= \frac{2h}{3a^2}\left[\left(-x.\frac{3a}{n\pi}\cos\frac{n\pi x}{3a}\right)_0^a + \frac{3a}{n\pi}\right.$$

$$\int_0^a \cos\frac{n\pi x}{3a} dx + \left[-(3a - 2x)\frac{3a}{n\pi}\cos\frac{n\pi x}{3a}\right]_a^{2a}$$

$$-\frac{6a}{n\pi}\int_a^{2a}\cos\frac{n\pi x}{3a} dx + \left[-(x - 3a)\frac{3a}{n\pi}\cos\frac{n\pi x}{3a}\right]_{2a}^{3a}$$

$$\left. +\frac{3a}{n\pi}\int_a^{2a}\cos\frac{n\pi x}{3a} dx \right.$$

$$= \frac{2h}{3a^2} \left[\frac{27a^2}{n^2\pi^2} \sin\frac{n\pi}{3} - \frac{27a^2}{n^2\pi^2}\sin\frac{2n\pi}{3} \right] \qquad \text{[Using } \sin n\pi = 0\text{]}$$

$$= \frac{18h}{n^2\pi^2} \left[\sin\frac{n\pi}{3} - \sin\left(n\pi - \frac{n\pi}{3} \right) \right]$$

$$= \frac{18h}{n^2\pi^2} \left[1 + (-1)^n \right] \sin\frac{n\pi}{3}$$

Thus, $C_n \begin{cases} \dfrac{18h}{n^2\pi^2} \cdot 2\sin\dfrac{n\pi}{3} \;; & \text{if } n \text{ is even} \\ 0 & \text{; if } n \text{ is odd} \end{cases}$

Further, let $n = 2m$. Then

$$u(x,t) = \frac{36h}{\pi^2} \sum_{m=1}^{\infty} \frac{1}{(2m)^2} \sin\frac{2m\pi}{3} \cos\frac{2m\pi Ct}{3} \sin\frac{2m\pi x}{3a}$$

$$= \frac{9h}{\pi^2} \sum_{m=1}^{\infty} \frac{1}{m^2} \sin\frac{2m\pi}{3} \sin\frac{2m\pi}{2a} x \cos\frac{2m\pi Ct}{3a}$$

Finally, to get the displacement of the mid point of the string, we put $x = \dfrac{3a}{2}$ in the above expression, then we get

$$u(x,t) = \frac{9h}{\pi^2} \sum_{m=1}^{\infty} \frac{1}{m^2} \sin\frac{2m\pi}{3}$$

$$\sin m\pi \cos\frac{2m\pi Ct}{3} = 0$$

Hence, we conclude that the mid point of the string always remains at rest.

EXERCISE 8.2

1. Find the deflection $u(x,t)$ of the vibrating string whose length is π^2 and $C^2 = 1$ corresponding to zero initial velocity and initial deflection $f(x) = k(\sin x - \sin 2x)$.

(VTU–2011)

2. A string of length l has its ends $x = 0$ and $x = l$ fixed. It is released from rest in the position $y = \dfrac{4\lambda x\,[l-x]}{l^2}$. Find an expression for the displacement of the string at any subsequent time.

3. Solve completely the equation $\dfrac{\partial^2 y}{\partial t^2} = C^2 \dfrac{\partial^2 y}{\partial x^2}$, representing the vibration of a string of length l, fixed at both ends, given that $y(0,t) = 0$, $y(l,t) = 0$; $y(x,0) = f(x)$ and $\dfrac{\partial}{\partial t} y(x,0) = 0$, $0 < x < l$. (UPTU–2005)

4. If a string of length l is initially at rest in equilibrium position and each of its points is given the velocity

$$\left(\frac{\partial y}{\partial t} \right)_{t=0} = b \sin^3\frac{\pi x}{l} , \text{ find the displacement}$$

$y(x,t)$. (UPTU–2001,03,06)

5. The vibration of an elastic string is governed by the P.D.E. $\dfrac{\partial^2 u}{\partial t^2} = \dfrac{\partial^2 u}{\partial x^2}$. The length of the string is π and the ends are fixed. The initial velocity is zero and the initial deflection is $u(x,0) = 2(\sin x + \sin 3x)$ Find the deflection $u(x,t)$ of the vibrating string for $t > 0$.

6. Solve the wave equation

$$\frac{\partial^2 u}{\partial t^2} = a^2\frac{\partial^2 u}{\partial x^2}$$

under the conditions $u = 0$, when $x =$ and $x = \pi$, $\dfrac{\partial u}{\partial t} = 0$, where $t = 0$ an

$u(x,0) = x, \quad 0 < x < \pi$.

7. A string is stretched and fastened to two points apart from a distance l. Motion is started by displacing the string into the form $y = k(lx - x^2)$ from which it is released at time $t = 0$. Find the displacement of any point on the string at a distance x from one end at time t. (UPTU–2002; IAS–1993)

8. Find the deflection $u(x, y, t)$ of the square membrane with $a = b = 1$ and $c = 1$, if the initial velocity is zero and the initial deflection is $f(x, y) = A \sin \pi x \sin 2\pi y$.

(Kurukshetra –2004)

9. Solve the following boundary value problem

$$\frac{\partial^2 u}{\partial t^2} = C^2 \frac{\partial^2 u}{\partial x^2}, \ 0 \le x \le 1, \ t \ge 0$$

subject to the boundary conditions
$u(0, t) = 0, \ t > 0, \ u_x(1, t) = 0, \ t > 0$
and initial conditions

$$u(x, 0) = \begin{cases} x, & 0 < x < 1/4 \\ \frac{1}{2} - x, & 1/4 < x < 1/2 \\ 0, & 1/2 < x < 1 \end{cases}$$

and $u_t(x, 0) = 0, 0 < x < 1$.

10. Show how the wave equation $C^2 \dfrac{\partial^2 y}{\partial x^2} = \dfrac{\partial^2 y}{\partial t^2}$ can be solved by the method of separation of variables. If the initial displacement and velocity of a string stretched between $x = 0$ and $x = l$ are given by $y = f(x)$ and $\partial y / \partial t = g(x)$. Determine the constants in the series solution. (UPTU Q. Bank–2002)

11. A tightly stretched violin string of length l and fixed at both ends is plucked at $x = l / 3$ and assumed initially the shape of a triangle of height a. Find the displacement y at any distance x and any time t after the string is released from rest. (UPTU Q. Bank–2002)

12. Transform the equation $\dfrac{\partial^2 y}{\partial t^2} = C^2 \dfrac{\partial^2 y}{\partial x^2}$ to its normal form using the transformation $u = x + Ct, \ v = x - Ct$ and hence solve it. Show that the solution may be put in the form $y = \dfrac{1}{2}\left[f(x + Ct) + f(x - Ct) \right]$

Assume initial conditions $y = f(x)$ and $(\partial y / \partial t) = 0$ at $t = 0$. (UPTU–2003)

13. A tightly stretched string with fixed end points $x = 0$ and $x = l$ is initially at rest in its equilibrium position. If it is set vibrating by giving to each of its points an initial velocity $\lambda x (1 - x)$, find the displacement of the string at any distance x from one end at any time t. (MTU(SUM)–2011; Bhopal–2008; Madras–2006; JNTU–2005; Anna–2009; UPTU–2002; PTU–2005)

14. Find the deflection of the vibrating string which is fixed at the ends $x = 0$ and $x = 2$ and the motion is started by displacing the string into the form $\sin^3\left(\dfrac{\pi x}{2}\right)$ and releasing it with zero initial velocity at $t = 0$. (MTU–2012)

15. Find the deflection of the vibrating string of unit length whose end points are fixed if the initial velocity is zero and the initial deflection is given by

$$u(x, 0) = \begin{cases} 1 & 0 \le x \le 1/2 \\ -1 & \frac{1}{2} < x \le 1 \end{cases}$$

(GBTU–2012)

Hint to Selected Problems

1. By taking $a = \pi$ and $c^2 = 1$, the required deflection $u(x, t)$ is given by

$$u(x, t) = \sum_{n=1}^{\infty} \cos nt \sin nx \qquad ... (1)$$

where, $E_n = \dfrac{2}{\pi} \int_0^\pi f(x) \sin nx \, dx$

$= \dfrac{2k}{\pi} \int_0^\pi (\sin x - \sin 2x) \sin nx \, dx$

$= \dfrac{2k}{\pi} \left[\int_0^\pi \sin x \sin nx \, dx \right.$

$\left. - \int_0^\pi (\sin x - \sin 2x) \sin nx \, dx \right]$

... (2)

We know that

$\int_0^\pi \sin px \sin qx \, dx = 0$ if $p \ne q$

$= \pi / 2$ if $p = q$.

Using (2) and (3), we get

$E_1 = k, \ E_2 = -k, \ ...., E_n = 0 \ \forall n \ge 3$

Putting all these values in (1), we get

$u(x, t) = k \cos t \sin x - \cos 2t \sin 2x$

2. It is known that the displacement function $y(x, t)$ is the solution of wave equation given by

$$\frac{\partial^2 y}{\partial x^2} = \frac{1}{C^2} \frac{\partial^2 y}{\partial t^2} \qquad ... (1)$$

with boundary conditions

$$y(0,t) = y(l,t) = 0 \quad \text{for all } t \geq 0 \quad \text{... (2)}$$

and initial conditions

$$\left(\frac{\partial y}{\partial t}\right)_{t=0} = 0 \quad \text{for } 0 \leq x \leq l$$

an $\quad y(x,0) = f(x) = \{4\lambda x(l-x)\}/l^2 \quad$ ... (3)

Proceeding same as above, the solution of (1) satisfying the above boundary and initial conditions is

$$y(x,t) = \sum_{n=1}^{\infty} E_n \sin\frac{n\pi x}{l} \cos\frac{n\pi Ct}{l} \quad \text{... (4)}$$

where, $\quad E_n = \frac{2}{l}\int_0^l f(x) \sin\frac{n\pi x}{l} dx \quad$ ... (5)

Putting the values of $f(x)$ given by (3) in (5), we get

$$E_n = \left(\frac{2}{l}\right) \times \left(\frac{4\lambda}{l^2}\right) \int_0^l (lx - x^2) \sin\frac{n\pi x}{l} dx$$

$$= \frac{8\lambda}{l^3}\left[(lx - x^2)\left(-\frac{l}{n\pi}\cos\frac{n\pi x}{l}\right) - (l - 2x)\right.$$

$$\left.\left(-\frac{l^2}{n^2\pi^2}\sin\frac{n\pi x}{l}\right) + (-2)\left(\frac{l^3}{n^3\pi^3}\cos\frac{n\pi x}{l}\right)\right]_0^l$$

$$= \frac{8\lambda}{l^3}\left[-\frac{2l^3(-1)^n}{n^3\pi^3} + 2\frac{l^3}{n^3\pi^3}\right]$$

$$= \frac{16\lambda}{n^3\pi^3}\left[1 - (-1)^n\right] \quad \left[\because \cos n\pi = (-1)^n\right]$$

$$= \begin{cases} 0 & ; \text{ if } n = 2m, \ m = 1,2,3,... \\ \dfrac{32\lambda}{(2m-1)^3\pi^3} & ; \text{ if } n = 2m-1, m = 1,2,... \end{cases}$$

Finally putting the value of E_n in (4), we get

$$y(x,t) = \frac{32\lambda}{\pi^3}\sum_{m=1}^{\infty}\frac{1}{(2m-1)^3}\sin\frac{(2m-1)\pi x}{l}$$

$$\cos(2m-1)\frac{\pi Ct}{l}$$

3. Here, $\qquad \dfrac{\partial^2 y}{\partial t^2} = C^2\dfrac{\partial^2 y}{\partial x^2} \qquad$ ... (1)

Let $\qquad\qquad y = X(x)\,T(t) \qquad$ ... (2)

$$\frac{\partial^2 y}{\partial t^2} = X\frac{d^2 T}{dt^2}$$

$$\frac{\partial^2 y}{\partial x^2} = T\frac{d^2 X}{dx^2}$$

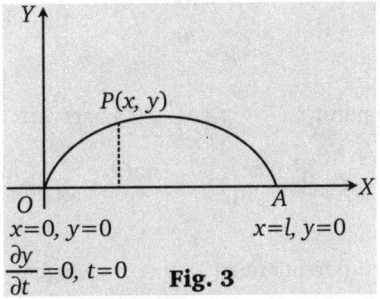

$x=0, y=0 \qquad\qquad x=l, y=0$

$\dfrac{\partial y}{\partial t} = 0, t=0 \qquad$ **Fig. 3**

Equation (1) becomes

$$X\frac{d^2 T}{dt^2} = C^2 T\frac{d^2 X}{dx^2}$$

Separating the variables

$$\frac{1}{X}\frac{d^2 X}{dx^2} = \frac{1}{C^2 T}\frac{d^2 T}{dt^2} = -p^2 \text{ (let)}$$

$$\frac{1}{X}\frac{d^2 X}{dx^2} = -p^2$$

If $\quad \dfrac{1}{C^2 T}\dfrac{\partial^2 T}{\partial t^2} = -p^2$

$$\Rightarrow \qquad \frac{d^2 X}{dx^2} = -Xp^2$$

$$\Rightarrow \quad (D^2 + b^2 C^2)T = 0$$

A.E. is $d^2 + p^2 C^2 = 0$

$$\Rightarrow \qquad \frac{d^2 X}{dx^2} + Xp^2 = 0$$

$$\Rightarrow D^2 = -p^2 C^2 \Rightarrow D = \pm pC_i$$

$$\Rightarrow \qquad (D^2 + p^2)X = 0$$

$$T = (C_2 \cos pCt + C_4 \sin pCt)$$

$$\Rightarrow \qquad X = (C_1 \cos px + C_2 \sin px)$$

Putting the values of X and T in equation (2), we get

$$y = (C_1 \cos px + C_2 \sin px)$$
$$(C_3 \cos pCt + C_4 \sin pCt) \qquad \text{... (3)}$$

Now applying the boundary condition

$$x = 0, \quad y = 0$$

Putting these values in (3), we get

$$0 = C_1(C_3 \cos pCt + C_4 \sin pCt) \Rightarrow C_1 = 0$$

Equation (3) becomes

$$y = C_2 \sin px (C_3 \cos pCt + C_4 \sin pCt) \quad \text{... (4)}$$

Putting $x = l$ and $y = 0$ in (4), we get

$$0 = C_2 \sin pl (C_3 \cos pCt + C_4 \sin pCt)$$

$$\Rightarrow \qquad \sin pl = 0 = \sin n\pi$$

$\Rightarrow \qquad pl = n\pi$

$\Rightarrow \qquad p = \dfrac{n\pi}{l}$

On putting $p = \dfrac{n\pi}{l}$, (4) becomes

$$y = C_2 \sin\frac{n\pi x}{l}\left(C_3 \cos\frac{n\pi Ct}{l} + C_4 \sin\frac{n\pi Ct}{l}\right)$$

... (5)

On differentiating (5) w.r.t. t, we get

$$\frac{\partial y}{\partial t} = C_2 \sin\frac{n\pi x}{l}\left(-\frac{n\pi C}{l}C_3 \sin\frac{n\pi Ct}{l}\right.$$
$$\left. +\frac{n\pi C}{l}C_4\frac{n\pi C}{l}t\right) \qquad \text{... (6)}$$

On putting $\dfrac{\partial y}{\partial t} = 0$ and $t = 0$ in (6), we get

$$0 = C_2 \sin\frac{n\pi x}{l}\cdot\frac{n\pi C}{l}C_4$$

$\Rightarrow \qquad C_4 = 0$

On putting $C_4 = 0$, (5) becomes

$$y = C_2 C_3 \sin\frac{n\pi x}{l}\cos\frac{n\pi Ct}{l} \quad \left[\text{let } C_2 C_3 = b_n\right]$$

... (7)

Now applying $y = f(x)$ and $t = 0$, (7) becomes

$$f(x) = b_n \sin\frac{n\pi x}{l}$$

$C_2 C_3$ can be calculated using Fourier sine series as

$$b_n = \frac{2}{l}\int_0^l f(x)\sin\frac{n\pi x}{l}\,dx$$

The required solution for the given equation is

$$y = b_n \sin\frac{n\pi x}{l}\cos\frac{n\pi Ct}{l}$$

4. The equation for the vibrations of the string is

$$\frac{\partial^2 y}{\partial t^2} = C^2\frac{\partial^2 y}{\partial x^2} \qquad \text{... (1)}$$

The solution of equation (1) is

$$y(x, t) = (C_1 \cos Cpt + C_2 \sin Cpt)$$
$$(C_3 \cos px + C_4 \sin px) \qquad \text{... (2)}$$

Boundary conditions are

$$y(0, t) = 0 \qquad \text{... (3)}$$
$$y(l, t) = 0 \qquad \text{... (4)}$$
$$y(x, 0) = 0 \qquad \text{... (5)}$$
$$\left(\frac{\partial y}{\partial t}\right) = b\sin^3\frac{\pi x}{l} \text{ at } t = 0 \quad \text{... (6)}$$

Putting $x = 0$ and $y = 0$ in (2), we get

$$0 = (C_1 \cos Cpt + C_2 \sin Cpt)C_3 \Rightarrow C_3 = 0$$

Putting the value of C_3 in (2), we get

$$y = (C_1 \cos Cpt + C_2 \sin Cpt)C_4 \sin px \quad \text{... (7)}$$

Putting $x = l$ and $y = 0$ in (7), we get

$$0 = (C_1 \cos Cpt + C_2 \sin Cpt)\,C_4 \sin pl$$

$\Rightarrow \quad \sin pl = 0 = \sin n\pi \; (n \in I) \Rightarrow p = \dfrac{n\pi}{l}$

Putting the value of p in (7), we get

$$y = \left(C_1 \cos\frac{n\pi Ct}{l} + C_2 \sin\frac{n\pi Ct}{l}\right)$$
$$C_4 \sin\frac{n\pi x}{l} \qquad \text{... (8)}$$

Putting $t = 0$ and $y = 0$ in (8), we get

$$0 = C_1 C_4 \sin\frac{n\pi x}{l} \Rightarrow C_1 = 0$$

Putting the value of C_1 in (8), we get

$$y = C_2 C_4 \sin\frac{n\pi Ct}{l}\sin\frac{n\pi x}{l}$$

$$y = b_n \sin\frac{n\pi Ct}{l}\sin\frac{n\pi x}{l}$$

(where $C_2 C_4 = b_n$)

The general solution is

$$y = \sum_1^\infty b_n \sin\frac{n\pi Ct}{l}\sin\frac{n\pi x}{l} \qquad \text{... (9)}$$

On differentiating (9) w.r.t. x^t, we get

$$\frac{\partial y}{\partial t} = \sum_1^\infty b_n \frac{n\pi C}{l}\cos\frac{n\pi Ct}{l}\sin\frac{n\pi x}{l} \quad \text{... (10)}$$

On putting the values of $\dfrac{\partial y}{\partial t} = b\sin^3\dfrac{\pi x}{l}$ and $t = 0$ in (10), we get

$$b\sin^3\frac{\pi x}{l} = \sum_1^\infty b_n \frac{n\pi C}{l}\sin\frac{n\pi x}{l}$$

$$\Rightarrow \frac{b}{4}\left[3\sin\frac{\pi x}{l}\frac{3\pi x}{l}\right] = b_1\frac{\pi C}{l}\sin\frac{\pi x}{l}$$

$$+\frac{2b_2\pi C}{l}\sin^2\frac{\pi x}{l} + 3b_3\frac{\pi C}{l}\sin\frac{3\pi C}{l} + ...$$

Equating the coefficients of $\sin\dfrac{\pi x}{l}$, $\sin\dfrac{2\pi x}{l}$, we get

$$\frac{3b}{4} = b_1\frac{\pi C}{l} \Rightarrow b_1 = \frac{3bl}{4\pi C}$$

$$b_2 = 0$$

and $\dfrac{3b_3\pi C}{l} = \dfrac{b}{4} \Rightarrow b_3 = -\dfrac{bl}{12\pi C}.$

Also, $b_4 = 0 = b_5 = ...$ etc.

Putting the values of b's in equation (9), we get

$$y(x,t) = \frac{3bl}{4\pi C} \sin\frac{\pi Ct}{l} \sin\frac{\pi x}{l}$$

$$- \frac{bl}{12\pi C} \sin\frac{3\pi Ct}{l} \sin\frac{3\pi x}{l}$$

$$= \frac{bl}{12\pi C}\left[9\sin\frac{\pi x}{l}\sin\frac{\pi Ct}{l} - \sin\frac{3\pi x}{l}\sin\frac{3\pi Ct}{l}\right].$$

5. The solution of given equation is

$$u(x,t) = (C_1\cos nx + C_2\sin nx)$$
$$(C_3\cos nt + C_4\sin nt) \qquad \text{... (1)}$$

where, $u(0,t) = 0,\ u(\pi,t) = 0;\ \forall t$

$$u(0,t) = 0 \Rightarrow (C_3\cos nt + C_4\sin nt)C_1 = 0$$

$$\left(\frac{\partial u}{\partial t}\right)_{t=0} = 0;\quad u(x,0) = 2(\sin x + \sin 3x)$$
$$\text{... (2)}$$

which gives $C_1 = 0$.

Thus $u(x,t) = (C_3\cos nt + C_4\sin nt)C_2\sin nx$
$$= (b_1\cos nt + b_2\sin nt)\sin nx$$

Again,
$$u(\pi,t) = 0 \Rightarrow (b_1\cos nt + b_2\sin nt)\sin n\pi = 0$$
$$\Rightarrow \quad \sin n\pi = 0$$
$$\Rightarrow n \text{ must be an integer.}$$

Thus, $u(x,t) = (b_1\cos nt + b_2\sin nt)\sin nx$

Then, $\left(\dfrac{\partial u}{\partial t}\right)_{t=0} = b_2 n\sin x \Rightarrow b_2 = 0$

$$\Rightarrow \quad u(x,t) = b_1\cos nt\sin nx$$

Also, $\quad u(x,0) = 2(\sin x + \sin 3x)$

$$\Rightarrow 2(\sin x + \sin 3x) = \sin nx\, b_1$$
$$\Rightarrow \quad 4\sin 2x\cos x = \sin x . b_1$$

which is true when $b_1 = 4\cos x$ and $n = 2$.

Hence, we have $u(x,t) = 4\cos x\cos 2t\sin 2x$.

6. The solution is of the form

$$u(x,t) = (C_1\cos px + C_2\sin px)$$
$$(C_3\cos apt + C_4\sin apt) \qquad \text{... (1)}$$

On putting $u = 0$ and $x = 0$ in (1), we get

$$0 = C_1(C_3\cos apt + C_4\sin apt) \Rightarrow C_1 = 0.$$

On putting $C_1 = 0$ in (1), we get

$$u(x,t) = C_2\sin px(C_3\cos apt + C_4\sin apt)$$
$$\text{... (2)}$$

On putting $x = \pi$ and $u = 0$ in (2), we get

$$0 = C_2\sin p\pi(C_3\cos apt + C_4\sin apt)$$

$$\Rightarrow \quad \sin px = 0 = \sin n\pi$$

$$\Rightarrow \qquad p = n$$

On putting $p = n$ in equation (2), we get

$$u(x,t) = C_2\sin nx(C_3\cos ant + C_4\sin ant)$$

and $\quad u(x,t) = \sin nx(b_1\cos ant + b_2\sin ant)$

$$[C_2C_3 = b_1 \text{ and } C_2C_4 = b_2] \qquad \text{... (3)}$$

On differentiating w.r.t. t, we get

$$\frac{\partial u}{\partial t} = \sin nx\left[-ab_1 n\sin ant + ab_2 n\cos ant\right]$$
$$\text{... (4)}$$

On putting $\dfrac{\partial u}{\partial t} = 0$ and $t = 0$ in (4), we have

$$0 = \sin n\pi(ab_2 n) \Rightarrow b_2 = 0$$

On putting $b_2 = 0$ in (3), we get

$$u(x,t) = \sin nx\,(b_1\cos ant) \qquad \text{... (5)}$$

General solution is

$$u(x,t) = \sum_{n=1}^{\infty} b_n\sin nx\cos ant \qquad \text{... (6)}$$

On putting $u = x$ and $t = 0$ in (6), we get

$$x = \sum_{n=1}^{\infty} b_n\sin nx\text{, where } b_n = \frac{2}{\pi}\int_0^\pi x\sin nx\, dx.$$

$$= \frac{2}{\pi}\left[x\left(-\frac{\cos nx}{n}\right)(-1)\left(-\frac{\sin nx}{n^2}\right)\right]_0^\pi$$

$$= \frac{2}{\pi}\left[-\frac{\pi}{n}\cos nx\right] = -\frac{2}{n}(-1)^n$$

On putting the value of b_n in (6), we get

$$u(x,t) = -2\sum_{n=1}^{\infty}\frac{(-1)^n}{n}\sin nx\cos nat$$

7. The vibrations of the string are governed by the equation

$$\frac{\partial^2 u}{\partial t^2} = C^2\frac{\partial^2 u}{\partial x^2} \qquad \text{... (1)}$$

The boundary conditions are

$$u(0,t) = 0 \quad \text{and} \quad u(l,t) = 0 \qquad \text{... (2)}$$

Also, the initial transverse velocity of any point of the string is zero, therefore we get

$$\left(\frac{\partial u}{\partial t}\right)_{t=0} = 0\ ;$$

Also, $u(x,0) = k\,(lx - x^2) \qquad \text{... (3)}$

Proceeding same as usual, the solution of (1) under (2) and (3) is given by

$$u(x,t) = (C_1\cos px + C_2\sin px)$$
$$(C_3\cos Cpt + C_4\sin Cpt)$$

Now, $u(0,t) = 0 \Rightarrow (C_1\cos 0 + C_2\sin 0)$

$(C_3 \cos Cpt + C_4 \sin Cpt) = 0$

$\Rightarrow C_1 (C_3 \cos Cpt + C_4 \sin Cpt) = 0 \Rightarrow C_1 = 0$

Therefore,

$u(x, t) = C_2 \sin px (C_3 \cos Cpt + C_4 \sin Cpt)$

$= \sin px \, (C_3' \cos Cpt + C_4' \sin Cpt)$

$\Rightarrow \sin px = 0 = \sin n\pi$, n being integer.

$\Rightarrow \qquad p = \dfrac{n\pi}{l}$.

Therefore,

$u(x, t) = \sin \dfrac{n\pi x}{l}$

$\left[C_3' \cos \dfrac{Cn\pi}{l}.t + C_4' \sin \dfrac{Cn\pi}{l}.t \right]$

$\Rightarrow \dfrac{\partial u}{\partial t} = \sin \dfrac{n\pi x}{l}$

$\left[C_3' \dfrac{Cn\pi}{l} \sin \dfrac{Cn\pi}{l}t + C_4' \dfrac{Cn\pi}{l}.\cos \dfrac{Cn\pi}{l}.t \right]$

$\Rightarrow \left(\dfrac{\partial u}{\partial t} \right)_{t=0} = C_4' \dfrac{Cn\pi}{l} \sin \dfrac{n\pi x}{l} = 0 \Rightarrow C_4' = 0$

$\Rightarrow \quad u(x, t) = C_3' \sin \dfrac{n\pi x}{l} \cos \dfrac{Cn\pi t}{l}$

Hence, the general solution of (1) is given by

$u(x, t) = \overset{\infty}{\underset{n=1}{\Sigma}} u_n(x, t)$

$= \overset{\infty}{\underset{n=1}{\Sigma}} b_n \sin \dfrac{n\pi x}{l} \cos \dfrac{Cn\pi t}{l}$

Also, $u(x, 0) = \overset{\infty}{\underset{n=1}{\Sigma}} b_n \sin \dfrac{n\pi x}{l} = k(lx - x^2)$,

where $C_3' = b_n$

So, $\quad b_n = \dfrac{2}{l} \int_0^l k(lx - x^2) \sin \dfrac{n\pi x}{l} dx$

8. The deflection of the square membrane is given by the two dimensional wave equation

$$\dfrac{\partial^2 u}{\partial t^2} = C^2 \left(\dfrac{\partial^2 u}{\partial x^2} + \dfrac{\partial^2 u}{\partial y^2} \right)$$

The boundary conditions are

$u(x, 0, t) = 0 = u(x, 1, t);$

$u(0, y, t) = 0 = u(1, y, t)$

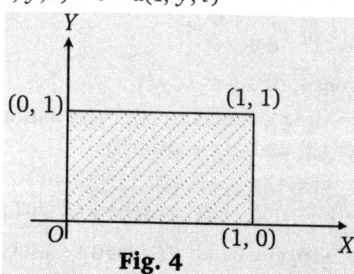

Fig. 4

Also, the initial conditions are

$u(x, y, 0) = f(x, y) = A \sin \pi x \sin 2\pi y$

$\left(\dfrac{\partial u}{\partial t} \right)_{t=0} = 0$

Thus, the deflection is given by

$u(x, y, t) = \overset{\infty}{\underset{m=1}{\Sigma}} \overset{\infty}{\underset{n=1}{\Sigma}} A_{mn} \cos k_{mn} t \sin m\pi x \sin n\pi y$

where,

$A_{mn} = 4 \int_0^1 \int_0^1 f(x, y) \sin m\pi x \sin n\pi y \, dx \, dy$

$\qquad \qquad \qquad \qquad \qquad \dots (1)$

$= 4 \int_0^1 \int_0^1 \sin \pi x \sin 2\pi y \sin m\pi x \sin n\pi y \, dx dy$

On integrating, we find that

$A_{m1} = A_{m3} = A_{m4} = \dots \dots = 0$

But

$A_{m2} = 4A \int_0^1 \int_0^1 \sin \pi x \sin m\pi x \sin^2 2\pi y \, dx \, dy$

$= 2A \int_0^1 \int_0^1 \sin \pi x \sin m\pi x \, (1 - \cos 4\pi y) \, dx \, dy$

$= 2A \int_0^1 \sin \pi x \sin m\pi x \left(y - \dfrac{1}{4\pi} \sin 4\pi y \right)_0^1 dx$

$= 2A \int_0^1 \sin \pi x \sin m\pi x \, dx$

Again, on integrating, we get

$A_{22} = A_{32} = \dots \dots = 0$.

Further, we find that

$A_{12} = 2A \int_0^1 \sin \pi x \sin \pi x \, dx$

$= A \int_0^1 2 \sin^2 \pi x \, dx$

$= A \int_0^1 (1 - \cos 2\pi x) dx$

$= A \left[x - \dfrac{1}{2\pi} \sin 2\pi x \right]_0^1 = A$

Hence, from (1) we get

$u(x, y, t) = A_{12} \cos k_{12} t \sin \pi x \sin 2\pi y$

$= A \cos \sqrt{5} \, \pi t \sin \pi x \sin 2\pi y$

$\left[\because k_{12}^2 = \pi^2 (m^2 + n^2) = \pi^2 (1^2 + 2^2) = \sqrt{5} \, \pi \right]$

ANSWERS

1. $u(x, t) = k \cos t \, \sin x - \cos 2t \, \sin 2x$

2. $y(x, t) = \dfrac{32\lambda}{\pi^3} \displaystyle\sum_{m=1}^{\infty} \dfrac{1}{(2m-1)^3} \sin \dfrac{(2m-1)\pi x}{l} \cos(2m-1) \dfrac{\pi Ct}{l}$

3. $y = b_n \sin \dfrac{n\pi x}{l} \cos \dfrac{n\pi Ct}{l}$, where $b_n = \dfrac{2}{l} \int_0^1 f(x) \sin \dfrac{n\pi x}{l} \, dx$.

4. $y(x, t) = \dfrac{bl}{12\pi c} \left[9 \sin \dfrac{\pi x}{l} \sin \dfrac{\pi ct}{l} - \sin \dfrac{3\pi x}{l} \sin \dfrac{3\pi ct}{l} \right]$

5. $u(x, t) = 4 \cos x \cot 2t . \sin 2x$

6. $u(x, t) = -2 \displaystyle\sum_{n=1}^{\infty} \dfrac{(-1)^n}{n} \sin nx . \cos nat$

7. $u(x, t) = \displaystyle\sum_{n=1}^{\infty} u_n(x, t) = \displaystyle\sum_{n=1}^{\infty} b_n \dfrac{n\pi x}{l} \cos \dfrac{cn\pi t}{l}$; $u(x, 0) = \displaystyle\sum_{n=1}^{\infty} b_n \sin \dfrac{n\pi x}{l} = k(lx - x^2)$

where, $b_n = \dfrac{2}{l} \int_0^1 k(lx - x^2) \sin \dfrac{n\pi x}{l} \, dx$.

8. $u(x, y, t) = A_{12} \cos k_{12} t \, \sin \pi x \, \sin 2\pi y = A \cos \sqrt{5} \, \pi t \, \sin \pi x \, \sin 2\pi y$

9. $u(x, t) = \dfrac{8}{\pi^2} \displaystyle\sum_{n=0}^{\infty} \dfrac{\sin \frac{1}{4}(2n-1)\pi}{(2n-1)^2} \cos \dfrac{(2n-1)\pi Ct}{2} \sin \dfrac{(2n-1)\pi x}{2}$

10. $y(x, t) = \displaystyle\sum_{1} \left(a_n \cos \dfrac{n\pi Ct}{l} + b_n \sin \dfrac{n\pi Ct}{l} \right) \sin \dfrac{n\pi x}{l}$

$a_n = \dfrac{2}{l} \int_0^l f(x) \sin \dfrac{n\pi x}{l} \, dx$; $b_n = \dfrac{2}{n\pi C} \int_0^l g(x) \sin \dfrac{n\pi x}{l} \, dx$

11. $y(x, t) = \dfrac{9a}{\pi^2} \displaystyle\sum_{1}^{\infty} \dfrac{1}{n^2} \sin \dfrac{n\pi}{3} \cos \dfrac{n\pi Ct}{l} \sin \dfrac{n\pi x}{l}$

13. $y(x, t) = \dfrac{8\lambda l^3}{c\pi^4} \displaystyle\sum_{m=1}^{\infty} \dfrac{1}{(2m-1)^4} \sin \dfrac{(2m-1)\pi ct}{l} \sin \dfrac{(2m-1)\pi x}{l}$

8.8 LAPLACE EQUATION

The heat equation is given by $\dfrac{\partial U}{\partial t} = C^2 \nabla^2 U$

If temperature are in steady state (*i.e.,* does not depend upon time *t*), then the heat equation reduces to

$$\nabla^2 u = 0$$

i.e., $\qquad \dfrac{\partial^2 U}{\partial x^2} + \dfrac{\partial^2 U}{\partial y^2} + \dfrac{\partial^2 U}{\partial z^2} = 0$. This is known as Laplace equation.

8.9 LAPLACE EQUATION IN TERMS OF POLAR COORDINATES

Let us suppose that the boundary of the region ∂R is a circle. Then, we use the polar coordinates as follows :

Let $\qquad\qquad\qquad x = r \cos\theta, \ y = r \sin\theta$

Then, $\qquad\qquad\qquad r^2 = x^2 + y^2 \ $ and $\ \theta = \tan^{-1} y / x$

Therefore, we can find $\dfrac{\partial r}{\partial x} = \cos\theta,\ \dfrac{\partial r}{\partial y} = \sin\theta,\ \dfrac{\partial\theta}{\partial x} = \dfrac{-\sin\theta}{r}$ and $\dfrac{\partial\theta}{\partial y} = \dfrac{\cos\theta}{r}$ ... (1)

$$u = u(r,\theta) \Rightarrow \begin{aligned} \dfrac{\partial u}{\partial x} &= \dfrac{\partial u}{\partial r}\dfrac{\partial r}{\partial x} + \dfrac{\partial u}{\partial\theta}\dfrac{\partial\theta}{\partial x} = \left(\dfrac{\partial u}{\partial r}\cos\theta - \dfrac{\partial u}{\partial\theta}\dfrac{\sin\theta}{r}\right) \\ \dfrac{\partial u}{\partial y} &= \dfrac{\partial u}{\partial r}\dfrac{\partial r}{\partial y} + \dfrac{\partial u}{\partial\theta}\dfrac{\partial\theta}{\partial y} = \left(\dfrac{\partial u}{\partial r}\sin\theta + \dfrac{\partial u}{\partial\theta}\dfrac{\cos\theta}{r}\right) \end{aligned} \Bigg] \quad ... (2)$$

Also,

and

$$\begin{aligned} \dfrac{\partial^2 u}{\partial x^2} &= \dfrac{\partial}{\partial x}\left(\dfrac{\partial u}{\partial x}\right) = \dfrac{\partial}{\partial r}\left\{\dfrac{\partial u}{\partial r}\cos\theta - \dfrac{\partial u}{\partial\theta}\dfrac{\sin\theta}{r}\right\}\cos\theta + \dfrac{\partial}{\partial\theta}\left\{\dfrac{\partial u}{\partial r}\cos\theta - \dfrac{\partial u}{\partial\theta}\dfrac{\sin\theta}{r}\right\}\left(\dfrac{-\sin\theta}{r}\right) \\ &= \left(\dfrac{\partial^2 u}{\partial r^2}\cos\theta - \dfrac{\partial^2 u}{\partial\theta\partial r}\dfrac{\sin\theta}{r} + \dfrac{\partial u}{\partial\theta}\dfrac{\sin\theta}{r^2}\right)\cos\theta \\ &\quad + \left(\dfrac{\partial^2 u}{\partial r\partial\theta}\cos\theta - \dfrac{\partial u}{\partial r}\sin\theta - \dfrac{\partial^2 u}{\partial\theta^2}\dfrac{\sin\theta}{r} - \dfrac{\partial u}{\partial\theta}\dfrac{\cos\theta}{r}\right)\left(\dfrac{-\sin\theta}{r}\right) \end{aligned}$$

   ... (3)

Similarly,

$$\begin{aligned} \dfrac{\partial^2 u}{\partial y^2} &= \left(\dfrac{\partial^2 u}{\partial r^2}\sin\theta + \dfrac{\partial^2 u}{\partial r\partial\theta}\dfrac{\cos\theta}{r} - \dfrac{\partial u}{\partial\theta}\dfrac{\cos\theta}{r^2}\right)\sin\theta \\ &\quad + \left(\dfrac{\partial^2 u}{\partial r\partial\theta}\sin\theta + \dfrac{\partial u}{\partial r}\cos\theta + \dfrac{\partial^2 u}{\partial\theta^2}\dfrac{\cos\theta}{r} - \dfrac{\partial u}{\partial\theta}\dfrac{\sin\theta}{r}\right)\left(\dfrac{\cos\theta}{r}\right) \end{aligned}$$

   ... (4)

Adding (3) and (4), we get

$$\dfrac{\partial^2 u}{\partial x^2} + \dfrac{\partial^2 u}{\partial y^2} = \dfrac{\partial^2 u}{\partial r^2} + \dfrac{1}{r}\dfrac{\partial u}{\partial r} + \dfrac{1}{r^2}\dfrac{\partial^2 u}{\partial\theta^2} = 0$$

which is the required Laplace equation in polar coordinates.

REMARK

- The Laplace equation in Cartesian coordinates has constant coefficients only, whereas in polar coordinates, it has variable coefficients.

8.10   LAPLACE EQUATION IN CYLINDRICAL COORDINATES (r, θ, z)

Let (x, y, z) be the cartesian coordinates of the point P whose cylindrical coordinates are given by (r, θ, z) such that

$$x = r\cos\theta, \qquad y = r\sin\theta, \qquad z = z$$

which implies $r^2 = x^2 + y^2$ and $\theta = \tan^{-1} y/x$

Therefore,

$$\dfrac{\partial r}{\partial x} = \dfrac{x}{r} = \cos\theta,\ \dfrac{\partial r}{\partial y} = \dfrac{y}{r} = \sin\theta,\ \dfrac{\partial\theta}{\partial x} = \dfrac{1}{1 + y^2/x^2}\left(-\dfrac{y}{x^2}\right) = \dfrac{-\sin\theta}{r}\ \text{and}\ \dfrac{\partial\theta}{\partial y} = \dfrac{\cos\theta}{r}$$

Now, we have

$$\dfrac{\partial u}{\partial x} = \dfrac{\partial u}{\partial r}\dfrac{\partial r}{\partial x} + \dfrac{\partial u}{\partial\theta}\cdot\dfrac{\partial\theta}{\partial x} + \dfrac{\partial u}{\partial z}\cdot\dfrac{\partial z}{\partial x} = \dfrac{\partial u}{\partial r}\cdot\cos\theta + \dfrac{\partial u}{\partial\theta}\cdot\left(-\dfrac{\sin\theta}{r}\right)$$

$\Rightarrow$

$$\dfrac{\partial}{\partial x} = \cos\theta\,\dfrac{\partial}{\partial r} - \dfrac{\sin\theta}{r}\dfrac{\partial}{\partial\theta}$$

Therefore,

$$\dfrac{\partial^2 u}{\partial x^2} = \dfrac{\partial}{\partial x}\left(\dfrac{\partial u}{\partial x}\right) = \left(\cos\theta\,\dfrac{\partial}{\partial r} - \dfrac{\sin\theta}{r}\dfrac{\partial}{\partial\theta}\right)\left(\cos\theta\,\dfrac{\partial u}{\partial r} - \sin\theta\,\dfrac{\partial u}{\partial\theta}\right)$$

$$= \cos^2\theta \frac{\partial^2 u}{\partial r^2} - 2\frac{\sin\theta\cos\theta}{r}\frac{\partial^2 u}{\partial r \partial\theta} + \frac{2\sin\theta\cos\theta}{r^2}\cdot\frac{\partial u}{\partial\theta}$$

$$+ \frac{\sin^2\theta}{r^2}\frac{\partial u}{\partial r} + \frac{\sin^2\theta}{r^2}\frac{\partial^2 u}{\partial\theta^2} \qquad \dots (1)$$

In a similar manner, we can find

$$\frac{\partial^2 u}{\partial y^2} = \sin^2\theta \frac{\partial^2 u}{\partial r^2} + 2\frac{\sin\theta\cos\theta}{r}\frac{\partial^2 u}{\partial r \partial\theta} - \frac{2\sin\theta\cos\theta}{r^2}\cdot\frac{\partial u}{\partial\theta}$$

$$+ \frac{\cos^2\theta}{r}\frac{\partial u}{\partial r} + \frac{\cos^2\theta}{r^2}\frac{\partial^2 u}{\partial\theta^2} \qquad \dots (2)$$

and
$$\frac{\partial^2 u}{\partial z^2} = \frac{\partial^2 u}{\partial z^2} \qquad \dots (3)$$

On adding (1), (2) and (3), we get

$$\frac{\partial^2 u}{\partial x^2} + \frac{\partial^2 u}{\partial y^2} + \frac{\partial^2 u}{\partial z^2} = \frac{\partial^2 u}{\partial r^2} + \frac{1}{r^2}\frac{\partial^2 u}{\partial\theta^2} + \frac{1}{r}\frac{\partial u}{\partial r} + \frac{\partial^2 u}{\partial z^2} = 0$$

which is the required Laplace equation in cylindrical form.

REMARK

- The above equation can also be written as

$$\frac{1}{r}\frac{\partial}{\partial r}\left(r\frac{\partial u}{\partial r}\right) + \frac{1}{r^2}\frac{\partial^2 u}{\partial\theta^2} + \frac{\partial^2 u}{\partial z^2} = 0$$

8.11 LAPLACE EQUATION IN SPHERICAL COORDINATES

Let (x, y, z) be the cartesian coordinates and (r, θ, ϕ) be the spherical coordinates at P such that
$$x = r\sin\theta\cos\phi, \qquad y = r\sin\theta\sin\phi, \qquad z = r\cos\theta$$

then, clearly we have

$$r^2 = x^2 + y^2 + z^2, \qquad \phi = \tan^{-1}(y/x) \text{ and } \theta = \tan^{-1}\left(\frac{\sqrt{x^2+y^2}}{z}\right)$$

Now,
$$\frac{\partial r}{\partial x} = \frac{x}{r} = \sin\theta\cos\phi, \quad \frac{\partial r}{\partial y} = \frac{y}{r} = \sin\theta\sin\phi, \quad \frac{\partial r}{\partial z} = \frac{z}{r} = \cos\theta,$$

$$\frac{\partial\theta}{\partial x} = \frac{\cos\theta\cos\phi}{r}, \quad \frac{\partial\theta}{\partial y} = \frac{\cos\theta\sin\phi}{r}, \quad \frac{\partial\theta}{\partial z} = \frac{-\sin\theta}{r}, \quad \frac{\partial\phi}{\partial x} = -\frac{\sin\phi}{r\sin\theta}$$

Therefore
$$\frac{\partial u}{\partial x} = \frac{\partial u}{\partial r}\frac{\partial r}{\partial x} + \frac{\partial u}{\partial\theta}\cdot\frac{\partial\theta}{\partial x} + \frac{\partial u}{\partial\phi}\cdot\frac{\partial\phi}{\partial x}$$

$$= \frac{\partial u}{\partial r}\cdot\sin\theta\cos\phi + \frac{\partial u}{\partial\theta}\cdot\frac{\cos\theta\cos\phi}{r} + \frac{\partial u}{\partial\phi}\left(-\frac{\sin\phi}{r\sin\theta}\right)$$

$$\Rightarrow \qquad \frac{\partial}{\partial x} \equiv \sin\theta\sin\phi\frac{\partial}{\partial r} + \frac{\cos\theta\cos\phi}{r}\frac{\partial}{\partial\theta} - \frac{\sin\phi}{r\sin\theta}\frac{\partial}{\partial\phi}$$

Now,
$$\frac{\partial^2 u}{\partial x^2} = \frac{\partial}{\partial x}\left(\frac{\partial u}{\partial x}\right) = \left(\sin\theta\cos\phi\frac{\partial}{\partial r} + \frac{\cos\theta\cos\phi}{r}\frac{\partial}{\partial\theta} - \frac{\sin\phi}{r\sin\theta}\frac{\partial}{\partial\phi}\right)$$

$$\left(\sin\theta\cos\phi\frac{\partial u}{\partial r} + \frac{\cos\theta\cos\phi}{r}\frac{\partial u}{\partial\theta} - \frac{\sin\phi}{r\sin\theta}\frac{\partial u}{\partial\phi}\right)$$

$$= \sin^2\theta\cos^2\phi\,\frac{\partial^2 u}{\partial r^2} + 2\frac{\sin\theta\cos\theta\cos^2\phi}{r}\,\frac{\partial^2 u}{\partial r\,\partial\theta}$$

$$- \frac{2\sin\theta\cos\theta\cos^2\phi}{r^2}\cdot\frac{\partial u}{\partial\theta} - \frac{2\sin\phi\cos\phi}{r}\cdot\frac{\partial^2 u}{\partial r\partial\phi}$$

$$+ \frac{\sin\phi\cos\phi}{r^2}\cdot\frac{\partial u}{\partial\phi} + \frac{\cos^2\theta\cos^2\phi}{r}\cdot\frac{\partial u}{\partial r} + \frac{\cos^2\theta\cos^2\phi}{r^2}\cdot\frac{\partial^2 u}{\partial\theta^2}$$

$$- 2\frac{\cos\theta\sin\phi\cos\phi}{r^2\sin\theta}\cdot\frac{\partial^2 u}{\partial\theta\partial\phi} + \frac{\cos^2\theta\sin\phi\cos\phi}{r^2\sin^2\theta}\cdot\frac{\partial u}{\partial\phi}$$

$$+ \frac{\sin^2\phi}{r}\cdot\frac{\partial u}{\partial r} + \frac{\cos\theta\sin^2\phi}{r^2\sin\theta}\cdot\frac{\partial u}{\partial\theta} + \frac{\sin^2\phi}{r^2\sin^2\theta}\cdot\frac{\partial^2 u}{\partial\phi^2} + \frac{\sin\phi\cos\phi}{r^2\sin^2\theta}\frac{\partial u}{\partial\phi} \quad \ldots(1)$$

Also,
$$\frac{\partial u}{\partial y} = \frac{\partial u}{\partial r}\frac{\partial r}{\partial y} + \frac{\partial u}{\partial\theta}\cdot\frac{\partial\theta}{\partial y} + \frac{\partial u}{\partial\phi}\cdot\frac{\partial\phi}{\partial y}$$

$$= \frac{\partial u}{\partial r}\cdot\sin\theta\sin\phi + \frac{\partial u}{\partial\theta}\cdot\frac{\cos\theta\sin\phi}{r} + \frac{\partial u}{\partial\phi}\left(\frac{\cos\phi}{r\sin\theta}\right)$$

$$\Rightarrow \qquad \frac{\partial}{\partial y} \equiv \sin\theta\sin\phi\frac{\partial}{\partial r} + \frac{\cos\theta\sin\phi}{r}\frac{\partial}{\partial\theta} + \frac{\cos\phi}{r\sin\theta}\frac{\partial}{\partial\phi}$$

which gives
$$\frac{\partial^2 u}{\partial y^2} = \frac{\partial}{\partial y}\left(\frac{\partial u}{\partial y}\right) = \left(\sin\theta\sin\phi\frac{\partial}{\partial r} + \frac{\cos\theta\sin\phi}{r}\frac{\partial}{\partial\theta} + \frac{\cos\phi}{r\sin\theta}\frac{\partial}{\partial\phi}\right)$$

$$\left(\sin\theta\sin\phi\frac{\partial u}{\partial r} + \frac{\cos\theta\sin\phi}{r}\frac{\partial u}{\partial\theta} + \frac{\cos\phi}{r\sin\theta}\frac{\partial u}{\partial\phi}\right)$$

$$= \sin^2\theta\sin^2\phi\,\frac{\partial^2 u}{\partial r^2} + 2\frac{\sin\theta\cos\theta\sin^2\phi}{r}\,\frac{\partial^2 u}{\partial r\,\partial\theta}$$

$$- \frac{2\sin\theta\cos\theta\sin^2\phi}{r^2}\cdot\frac{\partial u}{\partial\theta} + \frac{2\sin\phi\cos\phi}{r}\cdot\frac{\partial^2 u}{\partial r\partial\phi}$$

$$- \frac{\sin\phi\cos\phi}{r^2}\cdot\frac{\partial u}{\partial\phi} + \frac{\cos^2\theta\sin^2\phi}{r}\cdot\frac{\partial u}{\partial r} + \frac{\cos^2\theta\sin^2\phi}{r^2}\cdot\frac{\partial^2 u}{\partial\theta^2}$$

$$+ 2\frac{\cos\theta\sin\phi\cos\phi}{r^2\sin\theta}\cdot\frac{\partial^2 u}{\partial\theta\partial\phi} - \frac{\cos^2\theta\sin\phi\cos\phi}{r^2\sin^2\theta}\cdot\frac{\partial u}{\partial\phi}$$

$$+ \frac{\cos^2\phi}{r}\cdot\frac{\partial u}{\partial r} + \frac{\cos\theta\cos^2\phi}{r^2\sin\theta}\cdot\frac{\partial u}{\partial\theta} + \frac{\cos^2\phi}{r^2\sin^2\theta}\cdot\frac{\partial^2 u}{\partial\phi^2} - \frac{\sin\phi\cos\phi}{r^2\sin^2\theta}\frac{\partial u}{\partial\phi}$$

$$\ldots(2)$$

Now,
$$\frac{\partial u}{\partial z} = \frac{\partial u}{\partial r}\frac{\partial r}{\partial z} + \frac{\partial u}{\partial\theta}\cdot\frac{\partial\theta}{\partial z} + \frac{\partial u}{\partial\phi}\cdot\frac{\partial\phi}{\partial z} = \frac{\partial u}{\partial r}\cdot\cos\theta + \frac{\partial u}{\partial\theta}\left(-\frac{\sin\theta}{r}\right)$$

$$\Rightarrow \qquad \frac{\partial}{\partial z} \equiv \cos\theta\frac{\partial}{\partial r} - \frac{\sin\theta}{r}\frac{\partial}{\partial\theta}$$

Therefore,
$$\frac{\partial^2 u}{\partial z^2} = \frac{\partial}{\partial z}\left(\frac{\partial u}{\partial z}\right) = \left(\cos\theta\frac{\partial}{\partial r} - \frac{\sin\theta}{r}\frac{\partial}{\partial\theta}\right)\left(\cos\theta\frac{\partial u}{\partial r} + \left(\frac{-\sin\theta}{r}\right)\frac{\partial u}{\partial\theta}\right)$$

$$= \cos^2\theta \frac{\partial^2 u}{\partial r^2} - 2\frac{\sin\theta\cos\theta}{r}\frac{\partial^2 u}{\partial r\,\partial\theta}$$

$$+ \frac{2\sin\theta\cos\theta}{r^2}\cdot\frac{\partial u}{\partial\theta} + \frac{\sin^2\theta}{r^2}\cdot\frac{\partial u}{\partial r} + \frac{\sin^2\theta}{r^2}\cdot\frac{\partial^2 u}{\partial\theta^2} \qquad \ldots (3)$$

On adding (1), (2) and (3), we get

$$\frac{\partial^2 u}{\partial x^2} + \frac{\partial^2 u}{\partial y^2} + \frac{\partial^2 u}{\partial z^2} = \frac{\partial^2 u}{\partial r^2} + \frac{2}{r}\frac{\partial u}{\partial r} + \frac{1}{r^2}\frac{\partial^2 u}{\partial\theta^2} + \frac{\cot\theta}{r^2}\frac{\partial u}{\partial\theta} + \frac{1}{r^2\sin^2\theta}\frac{\partial^2 u}{\partial\phi^2} = 0$$

which is the required Laplace equation in spherical coordinates.

REMARK

- The above equation can also be written as

$$\frac{1}{r^2}\frac{\partial}{\partial r}\left(r^2\frac{\partial u}{\partial r}\right) + \frac{1}{r^2\sin\theta}\frac{\partial}{\partial\theta}\left(\sin\theta\frac{\partial u}{\partial\theta}\right) + \frac{1}{r^2\sin^2\theta}\frac{\partial^2 u}{\partial\phi^2} = 0$$

8.12 SOLUTION OF TWO DIMENSIONAL LAPLACE EQUATION : SEPARATION OF VARIABLES

Consider a two-dimensional Laplace equation in cartesian coordinates

$$\nabla^2 u = \frac{\partial^2 u}{\partial x^2} + \frac{\partial^2 u}{\partial y^2} = 0 \qquad \ldots (1)$$

Let
$$u(x, y) = X(x)\,Y(y) \qquad \ldots (2)$$
be the solution of (1).

Using (2) in (1), we get

$$X''Y + Y''X = 0$$

$$\Rightarrow \qquad \frac{X''}{X} = -\frac{Y''}{Y} = \lambda \text{ (say), where } \lambda \text{ is a separation constant.}$$

Now, we have the following cases :

Case (i) - Let $\lambda = \mu^2, \mu$ is real. Then $\dfrac{d^2 X}{dx^2} - \mu^2 X = 0$ and $\dfrac{d^2 Y}{dy^2} + \mu^2 Y = 0$

The solution of above equations are given by

$$X = C_1 e^{\mu x} + C_2 e^{-\mu x} \text{ and } Y = C_3\cos\mu y + C_4\sin\mu y$$

Thus, in this case, the required solution is given by

$$u(x, y) = X(x)\,Y(y) = (C_1 e^{\mu x} + C_2 e^{-\mu x})(C_3\cos\mu y + C_4\sin\mu y)$$

Case (ii) - If $\lambda = 0$

Then the equation reduces to $\dfrac{d^2 X}{dx^2} = 0$ and $\dfrac{d^2 Y}{dy^2} = 0$

Integrating twice, we get $\qquad X = d_1 x + d_2$ and $Y = d_3 y + d_4$

Thus, in this case, the required solution is given by

$$u(x, y) = (d_1 x + d_2)(d_3 y + d_4)$$

Case (iii) - Let $\lambda = -\mu^2$

Proceeding in the same way as in case (i), we get

$$X = e_1\cos\mu x + e_2\sin\mu x \text{ and } Y = e_3 e^{\mu y} + e_4 e^{-\mu y}$$

Hence, in this case, the required solution is given by

$$u(x, y) = (e_1 \cos \mu x + e_2 \sin \mu x)(e_3 e^{\mu y} + e_4 e^{-\mu y}).$$

REMARK

- In all the above cases, C_i, d_i and e_i $(i = 1, ..., 3)$ are integration constants, which can be calculated by using the given boundary conditions.

8.12.1 Some Particular Problems

(1) Interior Dirichlet's Problem for a Rectangle

The interior Dirichlet's problem for a rectangle is defined as follows :

PDE : $\nabla^2 u = 0, \quad 0 \le x \le a, \ 0 \le y \le b$

BCs : $u(x, b) = u(a, y) = 0, u(0, y) = 0, u(x, 0) = f(x)$

(2) The Neumann Problem for a Rectangle

The Neumann problem for a rectangle is defined as follows :

PDE : $\nabla^2 u = 0, 0 \le x \le a, 0 \le y \le b$

BCs : $u_x(0, y) = u_x(a, y) = 0, \ u_y(x, 0) = 0, \ u_y(x, b) = f(x)$

(3) Interior Dirichlet's Problem for a Circle

The interior Dirichlet's problem for a circle is defined as follows :

PDE : $\nabla^2 u = 0, 0 \le r \le a, 0 \le \theta \le 2\pi$

BC : $u(a, \theta) = f(\theta), 0 \le \theta \le 2\pi$

(4) Exterior Dirichlet's Problem for a Circle

The exterior Dirichlet's problem for a circle is defined as follows :

PDE : $\quad \nabla^2 u = 0$

BC : $u(a, \theta) = f(\theta)$

Here, u must be bounded as $r \to \infty$.

8.13 SOLUTION OF LAPLACE EQUATION OF THREE DIMENSIONAL

Consider the three dimensional Laplace equation

$$\frac{\partial^2 u}{\partial x^2} + \frac{\partial^2 u}{\partial y^2} + \frac{\partial^2 u}{\partial z^2} = 0 \qquad \qquad \text{... (1)}$$

Let $\qquad\qquad u(x, y, z) = X(x) \ Y(y) \ Z(z)$ ... (2)

be the solution of (1), where X, Y and Z are functions of x, y and z, respectively.

Putting the value of u [From (2)] in (1), we get

$$\frac{X''}{X} + \frac{Y''}{Y} = -\frac{Z''}{Z} \qquad\qquad \text{... (3)}$$

Further, since x, y and z are independent variables, equation (3) can be true if each term on each side is equal to a constant.

Now, we have the following three cases :

Case (i) . If each term in (3) is zero.

In this case, we have $X'' = 0, Y'' = 0, Z'' = 0$

On integrating each twice, we get

$$X = Ax + B, \quad Y = Cy + D, \quad Z = Ez + F$$

Hence, we get the solution of the form

$$u(x, y, z) = (Ax + B)(Cy + D)(Ez + F) \qquad \text{... (4)}$$

Case (ii). Suppose

$$\frac{X''}{X} = \lambda_1^2, \frac{Y''}{Y} = \lambda_2^2 \text{ such that } \lambda_1^2 + \lambda_2^2 = \lambda^2$$

Then, from equation (3), we can find

$$X'' - \lambda_1^2 X = 0, Y'' - \lambda_2^2 Y = 0, Z'' + \lambda^2 Z = 0$$

On solving, we get

$$X = Ae^{x\lambda_1} + Be^{-x\lambda_1} \; ; Y = Ce^{y\lambda_2} + De^{-y\lambda_2} \text{ and } Z = E\cos\lambda z + F\sin\lambda z$$

Hence, we get the solution of (1) is of the form

$$u(x, y, z) = (Ae^{x\lambda_1} + Be^{-x\lambda_1})(Ce^{y\lambda_2} + De^{-y\lambda_2})(E\cos\lambda z + F\sin\lambda z)$$

Case (iii). In this case, suppose that

$$\frac{X''}{X} = -\lambda_1^2, \frac{Y''}{Y} = -\lambda_2^2, \text{ and } -(\lambda_1^2 + \lambda_2^2) = \lambda^2$$

Then from (3), we can find

$$X'' + \lambda_1^2 X = 0, Y'' + \lambda_2^2 Y = 0, Z'' - \lambda^2 Z = 0$$

On solving, we get

$$X = A\cos\lambda_1 x + B\sin\lambda_1 x$$
$$Y = C\cos\lambda_2 y + D\sin\lambda_2 y$$

and

$$Z = E\,e^{\lambda z} + F\,e^{-\lambda z}$$

Hence, the general solution of (1) is given by

$$u(x, y, z) = \sum_{\lambda_1}\sum_{\lambda_2} (A\cos\lambda_1 x + B\sin\lambda_1 x)(C\cos\lambda_2 y + D\sin\lambda_2 y)(E\,e^{\lambda z} + F\,e^{-\lambda z})$$

Solved Examples

EXAMPLE 1. *If u be a harmonic function in the interior of a rectangle* $0 \le x \le a$, $0 \le y \le b$ *in the XY - plane satisfying Laplace equation*

$$\frac{\partial^2 u}{\partial x^2} + \frac{\partial^2 u}{\partial y^2} = 0 \qquad \text{... (1)}$$

with boundary conditions

$$u(0, y) = 0, u(a, y) = 0 \qquad \text{... (2)}$$
$$u(x, b) = 0, u(x, 0) = f(x) \qquad \text{... (3)}$$

Obtain the solution of above problem. [UPTU–2008]

SOLUTION. By the method of separation of variables, we can find a function $u(x, y)$ such that

$$u(x, y) = X(x)\,Y(y) \qquad \text{... (4)}$$

Putting this value of u in equation (1), we get

$$X''Y + XY'' = 0$$

$$\Rightarrow \qquad \frac{X''}{X} = -\frac{Y''}{Y} \qquad \text{... (5)}$$

For independent x, y, each side of equation (5) must be equal to the same constant say k.

Then (5) reduces to

$$X'' - kX = 0 \qquad \qquad \dots (6)$$

and $\qquad \qquad Y'' + kY = 0 \qquad \qquad \dots (7)$

Now, using the given boundary conditions (2) in (4), we get

$X(0)\, Y(y) = 0 \quad$ and $\quad X(a)\, Y(y) = 0$

$\qquad X(0) = 0 \quad$ and $\qquad \qquad X(a) = 0 \qquad \qquad \dots (8)$

$\qquad \qquad \qquad$ [$Y(y) \neq 0$ have been taken, because otherwise $\quad u = 0$]

Now, we have the following cases :

Case (i). Let $k = 0$. Then solution of (6) is given by

$$X(x) = Ax + B$$

Using the boundary conditions given by (8), we get $B = 0$ and $Aa + B = 0$

i.e., $\Rightarrow \qquad \qquad \qquad A = B = 0 \quad \Rightarrow \quad X(x) = 0$

$\qquad \Rightarrow \qquad \qquad \qquad u = 0$

which does not satisfy the given boundary condition $u(x, 0) = f(x)$.

Hence, we reject this case (i.e., $k = 0$).

Case (ii). Let $k = \lambda^2, \lambda \neq 0$. In this case, solution of equation (6) is given by

$$X(x) = Ae^{\lambda x} + Be^{-\lambda x}$$

Using the boundary conditions given by (6), we get

$$A + B = 0 \ \text{ and } \ Ae^{a\lambda} + Be^{-\lambda a} = 0$$

On solving, we get

$$A = B = 0 \quad \Rightarrow \qquad X(x) = 0 \quad \Rightarrow \qquad u = 0$$

Hence, again, we reject this case.

Case (iii). Let $k = -\lambda^2, \lambda \neq 0$. In this case, the solution of (6) is given by

$$X(x) = A\cos\lambda x + B\sin\lambda x$$

Using the boundary conditions given by (8), we get

$\qquad A = 0 \ \text{ and } \ A\cos\lambda a + B\sin\lambda a = 0$

On solving, we get $A = 0$ and $\sin\lambda a = 0$

Here, we have taken $B \neq 0$ because otherwise $X(x) = 0$.

Further, $\sin\lambda a = 0$

$\Rightarrow \qquad \lambda a = n\pi$, i.e., $\lambda = \dfrac{n\pi}{a}; \quad n = 1, 2, 3, \dots\dots$

Therefore, in this case, non-zero solution $X_n(x)$ of (6) is given by

$$X_n(x) = B_n \sin\left(\frac{n\pi x}{a}\right)$$

Further, using $\mu = -\lambda^2 = -\dfrac{n^2\pi^2}{a^2}$, equation (7) becomes

$$Y'' - \left(\frac{n^2\pi^2}{a^2}\right)Y = 0$$

whose solution is given by

$$Y_n(y) = C_n e^{n\pi y/a} + D_n e^{-n\pi y/a} \qquad \qquad \dots (9)$$

Using the given boundary conditions, we get

$$X(x)Y(b) = 0 \quad \Rightarrow Y(b) = 0$$

$$\Rightarrow \quad Y_n(b) = 0 \qquad \qquad \dots (10)$$

Putting $y = b$ in (9) and using (10), we get

$$0 = C_n e^{n\pi b/a} + D_n e^{-n\pi b/a}$$

$$C_n = -\left. \left\{ D_n e^{-n\pi b/a} \right\} \middle/ e^{n\pi b/a} \right. \qquad \dots (11)$$

Putting this value in (9), we get

$$Y_n(y) = \frac{D_n \left(e^{-n\pi y/a} e^{n\pi b/a} - e^{n\pi y/a} e^{-n\pi b/a} \right)}{e^{n\pi b/a}} = \frac{D_n \left(e^{n\pi(b-y)/a} - e^{-n\pi(b-y)/a} \right)}{e^{n\pi b/a}}$$

which can also be written as

$$Y_n(y) = 2D_n e^{-n\pi b/a} \sinh\{n\pi(b-y)/a\} \qquad \left[\because \sinh\theta = \frac{e^\theta - e^{-\theta}}{2} \right]$$

Therefore, $\quad U_n(x, y) = X_n(x) Y_n(y)$

$$= F_n \sin(n\pi x/a) \sinh\{n\pi(b-y)/a\}$$

Also, the more general solution is given by

$$U(x, y) = \sum_{n=1}^{\infty} F_n \sin\left(\frac{n\pi x}{a}\right) \sinh\left[\frac{n\pi(b-y)}{a}\right]$$

Putting $y = 0$ and using given boundary conditions $u(x, 0) = f(x)$, we get

$$f(x) = \sum_{n=1}^{\infty} \left\{ F_n \sin\left(\frac{n\pi b}{a}\right) \right\} \sin\left[\frac{n\pi x}{a}\right]$$

which is the half range Fourier sine series of $f(x)$ in $(0, a)$. Hence, we get

$$\Rightarrow \quad F_n \sin\frac{n\pi b}{a} = \frac{2}{a} \int_0^a f(x) \frac{n\pi x}{a} dx$$

$$\Rightarrow \quad F_n = \frac{2}{a \sinh\left(\frac{n\pi b}{a}\right)} \int_0^a f(x) \sin\frac{n\pi x}{a} dx$$

EXAMPLE 2. *Find the steady temperature distribution in a thin plate bounded by the lines $x = 0$, $x = a$, $y = 0$, $y = \infty$. Assuming that heat can not escape from either surface; the sides $x = 0$, $x = a$ being kept at temperature zero. The lower edge $y = 0$ is kept at $f(x)$ and the edge $y = \infty$ at temperature zero.*

SOLUTION. For steady state, we know that $\dfrac{\partial u}{\partial t} = 0$

Therefore the heat equation

$$\frac{\partial^2 u}{\partial x^2} + \frac{\partial^2 u}{\partial y^2} = \frac{1}{k}\frac{\partial u}{\partial t}$$

reduces to

$$\frac{\partial^2 u}{\partial x^2} + \frac{\partial^2 u}{\partial y^2} = 0 \qquad \dots (1)$$

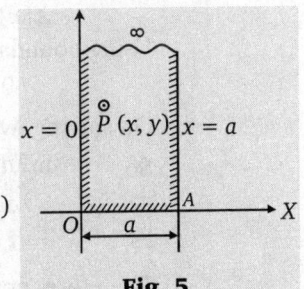

Fig. 5

The given boundary conditions can be written as

$$u(0, y) = 0, \quad u(a, y) = 0 \qquad \dots (2)$$

$$u(x, y) \to 0 \text{ as } y \to \infty \text{ and } u(x, 0) = f(x) \qquad \dots (3)$$

Using the method of separation of variables, let us suppose solution of (1) is of the form

$$u(x, y) = X(x) \ Y(y) \qquad \qquad \dots (4)$$

Putting this value in (1), we get

$$X''Y + XY'' = 0$$

$$\Rightarrow \qquad \frac{X''}{X} = -\frac{Y''}{Y} \qquad \qquad \dots (5)$$

Now, since x and t are independent, therefore each side of (5) must be equal to the same constant say k. Then (5) reduces to

$$X'' - kX = 0 \qquad \qquad \dots (6)$$

and $\quad Y'' + kY = 0 \qquad \qquad \dots (7)$

Using (2) in (4), we get

$$X(0) \ Y(y) = 0 \quad \text{and} \quad X(a) \ Y(y) = 0$$

$$\Rightarrow \qquad X(0) = 0 \quad \text{and} \quad X(a) = 0 \qquad \qquad \text{(By taking } Y(y) \neq 0 \text{)}$$

$$\dots (8)$$

Now, we have to solve equation (6) using the boundary conditions (8) under the following three cases :

Case (i). Let $k = 0$. In this case, solution of (6) is given by

$$u = Ax + B \qquad \qquad \dots (9)$$

Using boundary conditions given by (8) in (9), we get

$$B = 0 \text{ and } Aa + B = 0$$

On solving, we get

$$A = B = 0 \ \Rightarrow \ X(x) = 0 \quad u = 0$$

So, we reject this case.

Case (ii). Let $k = -\lambda^2, \lambda \neq 0$. In this case, solution of (6) is given by

$$X(x) = Ae^{x\lambda} + Be^{-x\lambda}$$

Using boundary conditions (8), we get

$$0 = A + B \text{ and } 0 = Ae^{a\lambda} + Be^{-a\lambda}$$

On solving, we get $A = B = 0$

$$\Rightarrow \qquad X(x) \equiv 0 \ \Rightarrow \ u = 0$$

Hence, we reject $k = \lambda^2$.

Case (iii). Let $k = -\lambda^2, \lambda \neq 0$. In this case, solution of (6) is given by

$$X(x) = A \cos \lambda x + B \sin \lambda x$$

Using boundary conditions (8), we get

$$0 = A \text{ and } 0 = A \cos \lambda a + B \sin \lambda a$$

On solving, we get $A = 0$ and $\sin \lambda a = 0 \qquad \qquad$ (By taking $B \neq 0$)

So, $\quad \sin \lambda a = 0$

$$\Rightarrow \qquad \lambda a = n\pi$$

$$\Rightarrow \qquad \lambda = n\pi / a; \quad n = 1, 2, 3, \dots\dots$$

Therefore, non-zero solutions $X_n(x)$ of (6) are given by

$$X_n(x) = B_n \sin\left(\frac{n\pi x}{a}\right) \qquad \qquad \dots (10)$$

Using $k = -\lambda^2 = -\dfrac{n^2\pi^2}{a^2}$, equation (7) becomes

$$Y'' - \left(\frac{n^2\pi^2}{a^2}\right)Y = 0 \qquad \text{... (11)}$$

On solving (11), we get

$$Y_n(y) = C_n e^{n\pi y/a} + D_n e^{-n\pi y/a} \qquad \text{... (12)}$$

with $\quad Y_n(y) \to 0 \quad as \quad y \to \infty \qquad$ [Using (3)]

Therefore, we must take $C_n = 0$. Then, from (12), we have

$$Y_n(y) = D_n e^{-n\pi y/a}$$

$$\Rightarrow \quad u_n(x, y) = X_n Y_n = E_n \sin\frac{n\pi x}{a} e^{-n\pi y/a} \qquad (E_n = B_n\, D_n)$$

are solutions of (1) satisfying (2) and (3).
The more general solution is given by

$$u(x, y) = \sum_{n=1}^{\infty} E_n \sin\frac{n\pi x}{a} e^{-n\pi y/a}$$

Now, putting $y = 0$ and using $u(x, 0) = f(x)$, we get

$$f(x) = \sum_{n=1}^{\infty} E_n \sin\frac{n\pi x}{a}$$

which is a Fourier sine series and therefore

$$E_n = \frac{2}{a}\int_0^a f(x) \sin\frac{n\pi x}{a} dx .$$

EXAMPLE 3. *A square plate is bounded by the lines $x = 0, y = 0, x = 10$ and $y = 10$. Its faces are insulated. The temperature along the upper horizontal edge is given by $u(x, 10) = x(10 - x)$ while the other three faces are kept at $0°C$. Find the steady state temperature in the plate.*

SOLUTION. The steady state temperature $u(x, y)$ is governed by the equation

$$\frac{\partial^2 u}{\partial x^2} + \frac{\partial^2 u}{\partial y^2} = 0 \qquad \text{... (1)}$$

subject to the boundary conditions

$$u(0, y) = u(10, y) = 0, \ 0 \le y \le a \qquad \text{... (2)}$$
$$u(x, 0) = 0, \quad 0 \le x \le a$$
$$u(x, 10) = 10x - x^2, \ 0 \le x \le a \qquad \text{... (3)}$$

Now proceed same as example (2), and using
$\qquad u(x, 10) = 10x - x^2$ in place of $u(x, b) = 100$
and $\qquad a = b = 10$,

we get $u(x, y) = \displaystyle\sum_{n=1}^{\infty} E_n \sin\frac{n\pi x}{10} \sinh\frac{n\pi y}{10} \qquad \text{... (4)}$

Putting $y = 10$ in (4) and using (3), we get

$$u(x, 10) = 10x - x^2 = \sum_{n=1}^{\infty} (E_n \sinh n\pi) \sin\frac{n\pi x}{10}$$

which is a half range Fourier series of $(10x - x^2)$ in $(0, 10)$
Therefore, we have

$$E_n \sinh n\pi = \frac{2}{10} \int_0^{10} (10x - x^2) \sin \frac{n\pi x}{10} dx$$

$$= \frac{1}{5} \left[(10x - x^2)\left(-\frac{10}{n\pi}\right) \cos \frac{n\pi x}{10} - (10 - 2x)\left(-\frac{100}{n^2\pi^2}\right) \sin \frac{n\pi x}{10} \right.$$

$$\left. + (-2)\left(\frac{1000}{n^3\pi^3}\right) \cos \frac{n\pi x}{10} \right]_0^{10}$$

$$= \frac{1}{5}\left[-\frac{2000(-1)^n}{n^3\pi^3} + \frac{2000}{n^3\pi^3} \right] = \frac{400}{n^3\pi^3}[1 - (-1)^n]$$

Thus, $$E_n = \frac{400\,\mathrm{cosec}\,h\,n\pi}{n^3\pi^3}[1 - (-1)^n]$$

$$E_n = \begin{cases} 0, \\ \quad\quad \text{if} \quad n = 2m \ \text{and} \ m = 1, 2, 3, \dots \\ \dfrac{800\,\mathrm{cosec}\,h\,(2m-1)\pi}{(2m-1)^3\,\pi^3}, \\ \quad\quad \text{if} \quad n = 2m - 1, \ m = 1, 2, 3, \dots \end{cases}$$

Hence, the required temperature is given by

$$u(x, y) = \frac{800}{\pi^3} \sum \frac{1}{(2m-1)^3} \sin \frac{(2m-1)\pi x}{10} \sinh \frac{(2m-1)\pi y}{10} \mathrm{cosec}\,(2m-1)\pi .$$

EXAMPLE 4.   *Find the steady state temperature distribution in a rectangular plate of sides a and b insulated at the lateral surface and satisfying the boundary conditions*
$$u(0, y) = u(a, y) = 0, \quad 0 \le y \le b$$
and $u(x, 0) = 0$ *and* $u(x, b) = f(x), \ 0 \le y \le a$

SOLUTION.   We know that the heat flow in a body for a two dimensional case is governed by the equation

$$\frac{\partial^2 u}{\partial x^2} + \frac{\partial^2 u}{\partial y^2} = \frac{1}{c^2}\frac{\partial u}{\partial t}$$

For steady flow, we have $\dfrac{\partial u}{\partial t} = 0$.

Therefore, we have the following Laplace's equation

$$\frac{\partial^2 u}{\partial x^2} + \frac{\partial^2 u}{\partial y^2} = 0 \qquad\qquad \dots (1)$$

subject to the boundary conditions
$$u(0, y) = u(a, y) = 0, \quad \text{for } 0 \le y \le b \qquad\qquad \dots (2)$$

and $$\left. \begin{array}{l} u(x, 0) = 0, \quad\quad \text{for } 0 \le x \le a \\ u(x, b) = f(x), \ \text{for } 0 \le x \le a \end{array} \right\} \qquad\qquad \dots (3)$$

Now proceeding same as in example (2), and taking $u(x, b) = f(x)$ for $u(x, b) = 100$, we may get

$$u(x, y) = \sum_{n=1}^{\infty} E_n \sin\frac{n\pi x}{a} \sinh\frac{n\pi y}{a} \qquad\qquad \dots (4)$$

Putting $y = b$ in (4) and using (3), we get

$$f(x) = \sum_{n=1}^{\infty} \left(E_n \sinh \frac{n\pi b}{a} \right) \sin \frac{n\pi x}{a}$$

which is a half range Fourier series of $f(x)$ in $(0, a)$.

Hence, $\quad E_n = \dfrac{2}{a \sin\left(\dfrac{n\pi b}{a}\right)} \int_0^a f(x) \, \sin\frac{n\pi x}{a} dx$

EXAMPLE 5. *Find the steady state temperature in a rectangular plate bounded by the lines $x = 0$, $x = a$, $y = 0$ and $y = b$ if the edge $y = 0$ is insulated, the edge $x = 0$ and $x = a$ are kept at $0°C$ and the edge $y = b$ is kept at temperature $f(x)$.*

SOLUTION. We know that the temperature $u(x,y)$ in steady state in two-dimensional plate is governed by

$$\frac{\partial^2 u}{\partial x^2} + \frac{\partial^2 u}{\partial y^2} = 0 \qquad \text{... (1)}$$

As per given, the boundary conditions are

$$u(0, y) = u(a, y) = 0, \text{ for } 0 \le y \le b \qquad \text{....(2)}$$

Also $\quad \left(\dfrac{\partial u}{\partial y} \right)_{y=0} = 0$

(Because the edge $y = 0$ is insulated for $0 \le x \le a$) ... (3a)

and $\quad u(x, b) = f(x)$, for $0 \le x \le a$ ... (3b)

Now proceeding same as in example (4), we get

$$u(x, y) = \sum_{n=1}^{\infty} u_n(x, y) = E_n \sin\left(\frac{n\pi x}{a}\right) \cosh\left(\frac{n\pi y}{a}\right) \qquad \text{... (4)}$$

Putting $y = b$ in (4) and using (3b), we get

$$f(x) = \sum_{n=1}^{\infty} \left(E_n \cosh \frac{n\pi b}{a} \right) \sin \frac{n\pi x}{a}$$

which is a half range Fourier sine series of $f(x)$ in $(0, a)$.

Hence, we get

$$E_n \cos\frac{n\pi b}{a} = \frac{2}{a} \int_0^a f(x) \, \sin\frac{n\pi x}{a} dx$$

$\Rightarrow \quad E_n = \dfrac{2}{a \cosh\left(\dfrac{n\pi b}{a}\right)} \int_0^a f(x) \, \sin\frac{n\pi x}{a} dx.$

EXAMPLE 6. *A rectangular metal plate is bounded by the lines $x = 0$, $x = a$, $y = 0$ and $y = b$. The three sides $x = 0$, $x = a$ and $y = b$ are insulated and the side $y = 0$ is kept at temperature $u_0 \cos\left(\dfrac{\pi x}{a}\right)$. Find the steady state temperature at any point of the plate.*

SOLUTION. The governing equation is given by

$$\frac{\partial^2 u}{\partial x^2} + \frac{\partial^2 u}{\partial y^2} = 0 \qquad \text{... (1)}$$

Since, $x = 0$, $x = a$ and $y = b$ are insulated.

Therefore, we have

$$\left(\frac{\partial u}{\partial x}\right)_{x=0} = 0, \quad \left(\frac{\partial u}{\partial x}\right)_{x=a} = 0 \qquad \qquad \dots (2)$$

$$\left(\frac{\partial u}{\partial y}\right)_{y=0} = 0 \qquad \text{and} \quad u(x, a) \le u_0 \cos\left(\frac{\pi x}{a}\right) \qquad \dots (3)$$

Using the method of separation of variables, suppose (1) has a solution of the form

$$u(x, y) = X(x) \, Y(y) \qquad \qquad \dots (4)$$

Using (4) in (1), we get $\quad X''Y + XY'' = 0$

$$\Rightarrow \qquad \qquad \frac{X''}{X} = -\frac{Y''}{Y} = k \text{ (say)} \qquad \qquad \dots (5)$$

$$\Rightarrow \qquad \qquad X'' - kX = 0 \qquad \qquad \dots (6)$$

and $\qquad \qquad Y'' + kY = 0 \qquad \qquad \dots (7)$

Also, from (4)

$$\frac{\partial u}{\partial x} = X'(x)\, Y(y) \qquad \qquad \dots (8)$$

Using (2) in (8), we get

$$X'(0) = 0 \text{ and } X'(a) = 0 \qquad \qquad \dots (9)$$

Now, we have the following three cases :

Case (i). Let $k = 0$. In this case, solution of equation (6) is given by

$$X(x) = Ax + B \qquad \qquad \dots (10)$$

$$\Rightarrow \qquad \qquad X'(x) = A$$

Using (9), we get $\qquad A = 0.$

Therefore $\qquad \qquad X(x) = B$

Case (ii). Let $k = \lambda^2, \lambda \ne 0$. In this case, the solution of (6) is given by

$$X(x) = Ae^{x\lambda} + Be^{-x\lambda}$$

$$\Rightarrow \qquad \qquad X'(x) = \lambda(Ae^{\lambda x} - Be^{-\lambda x})$$

Using (9) in the above equation, we get

$$\lambda(A - B) = 0 \text{ and } \lambda\,(Ae^{a\lambda} - Be^{-a\lambda}) = 0$$

On solving, we get $A = B = 0$

$$\Rightarrow \qquad \qquad X(x) = 0$$

$$\Rightarrow \qquad \qquad u = 0$$

Case (iii) Let $k = -\lambda^2, \lambda \ne 0$. In this case, solution of (6) is given by

$$X(x) = A\cos\lambda x + B\sin\lambda x \qquad \qquad \dots (10)$$

$$\Rightarrow \qquad \qquad X'(x) = -A\lambda\sin\lambda x + B\lambda\cos\lambda x \qquad \dots (11)$$

Using (9) in (10), we get

$$0 = B\lambda$$

and $\qquad \qquad 0 = -A\lambda\sin\lambda a + B\lambda\cos\lambda a \qquad \qquad \dots (12)$

Now, letting $\lambda \neq 0$ and $A \neq 0$, equation (12) gives

$B = 0$ and $\sin \lambda a = 0$

$B = 0$ and $\quad \lambda = \dfrac{n\pi}{a}, \; n = 1, 2, 3, \ldots$ ... (13)

Therefore, non-zero solutions of (10) is given by

$$X(x) = A \cos\left(\dfrac{n\pi x}{a}\right); \quad n = 1, 2, 3, \ldots \qquad \text{... (14)}$$

∴ All non-zero solution of (6) are given by

$$X_n(x) = A_n \cos\left(\dfrac{n\pi x}{a}\right), \; n = 0, 1, 2, 3, \ldots \qquad \text{... (15)}$$

Now putting $k = -\lambda^2 = -\dfrac{n^2\pi^2}{a^2}$ in (7), we get

$$Y'' - \left(\dfrac{n^2\pi^2}{a^2}\right)Y = 0$$

whose general solution is given by

$$Y_n(y) = C_n e^{n\pi x/a} + D_n e^{-n\pi y/a} \qquad \text{... (16)}$$

Also, from (4), we have

$$\dfrac{\partial u}{\partial y} = X(x)\, Y'(y) \qquad \text{... (17)}$$

Putting $y = b$ in (17) and using (3), we get

$\qquad 0 = X(x)\, Y'(b)$

$\Rightarrow \qquad Y'(b) = 0$ [By letting $X(x) \neq 0$]

$\Rightarrow \qquad Y_n'(b) = 0$ ... (18)

From (16), we have

$$Y_n'(y) = C_n\left(\dfrac{n\pi}{a}\right) e^{n\pi y/a} - D_n\left(\dfrac{n\pi}{a}\right) e^{-n\pi y/a} \qquad \text{... (19)}$$

Putting $y = b$ in (19) and using (18), we get

$$0 = \left(\dfrac{n\pi}{a}\right)(C_n e^{n\pi b/a} - D_n e^{-n\pi b/a})$$

$\Rightarrow \qquad D_n = \left(C_n e^{n\pi b/a}\right) \Big/ e^{-n\pi b/a}$

Putting this value in (16), we get

$$Y_n(y) = C_n\left[\dfrac{(e^{n\pi y/a} e^{-n\pi b/a} + e^{-n\pi y/a} e^{n\pi b/a})}{e^{-n\pi b/a}}\right]$$

$\Rightarrow \qquad Y_n(y) = C_n e^{n\pi b/a}\, \{e^{n\pi(b-y)/a} + e^{-n\pi(b-y)/a})$

$$= 2C_n\, e^{n\pi b/a}\, \cosh\left\{\dfrac{n\pi(b-y)}{a}\right\}$$

Thus, $\quad u_n(x, y) = X_n(x)\, Y_n(y)$

$$= E_n \cos\frac{n\pi x}{a} \cosh\left\{\frac{n\pi(b-y)}{a}\right\}$$

Further, consider the more general solution given by

$$u(x, y) = \sum_{n=0}^{\infty} u_n(x, y) = \sum_{n=0}^{\infty} E_n \cos\left(\frac{n\pi x}{a}\right)\cosh\left\{\frac{n\pi(b-y)}{a}\right\} \qquad \dots (20)$$

Putting $y = b$ and using (3), we get

$$u_0 \cos\left(\frac{\pi x}{a}\right) = \sum_{n=0}^{\infty} E_n \cos\left(\frac{n\pi x}{a}\right)\cosh\left\{\frac{n\pi(b-a)}{a}\right\}$$

$$= E_0 + E_1 \cos\left(\frac{\pi x}{a}\right) \cosh\left\{\frac{\pi(b-a)}{a}\right\}$$

$$+ E_2 \cos\left(\frac{2\pi x}{a}\right) \cosh\left\{\frac{2\pi(b-a)}{a}\right\}$$

$$+ E_3 \cos\left(\frac{3\pi x}{a}\right) \cosh\left\{\frac{3\pi(b-a)}{a}\right\} + \dots$$

On equating the coefficients of like terms, we get $E_0 = 0$.

$$E_1 \cosh\left\{\frac{\pi(b-a)}{a}\right\} = u_0$$

and $E_n = 0$ for $n \geq 1$.

Finally, putting this value in (20), we get

$$u(x, y) = u_0 \operatorname{sech}\left\{\frac{(b-a)\pi}{a}\right\} \cos\left(\frac{\pi x}{a}\right)\cosh\left\{\frac{(b-y)\pi}{a}\right\}.$$

EXAMPLE 7. *An infinitely long uniform plate is bounded by two parallel edges and an end at right angles to them. The breadth is π. The end is maintained at 100°C at all points and the other edges are at 0°C. Find the steady state temperature function $u(x, y)$.*

(UPTU–2008; PTU–2005; JNTU–2005, 08)

SOLUTION. The governing equation for the steady state temperature is given by

$$\frac{\partial^2 u}{\partial x^2} + \frac{\partial^2 u}{\partial y^2} = 0 \qquad \dots (1)$$

with boundary conditions

$$u(0, y) = u(\pi, y) = 0 \text{ for all } y \geq 0 \qquad \dots (2)$$

$$\left.\begin{array}{l} u(x, y) \to 0 \text{ as } y \to \infty, \text{ for } 0 \leq x \leq \pi \\ \text{and } u(x, 0) = f(x) = 100, \quad \text{for } 0 \leq x \leq \pi \end{array}\right\} \qquad \dots (3)$$

Now proceeding same as in example (2) by taking $a = \pi$, we get

$$u(x, y) = \sum_{n=1}^{\infty} E_n \sin nx \, e^{-ny} \qquad \dots (4)$$

where

$$E_n = \frac{2}{\pi}\int_0^{\pi} f(x) \, \sin nx \, dx \qquad \dots (5)$$

$$= \frac{2}{\pi}\int_0^{\pi} 100 \sin nx \, dx = \frac{200}{\pi}[-\cos nx]_0^{\pi} = \frac{200}{\pi}[1 - (-1)^n]$$

$$= \begin{cases} 0 & \text{if } n \text{ is even} \\ \dfrac{400}{(2m-1)\pi}, & \text{if } n = 2m-1, \ m = 1, 2, 3,... \end{cases}$$

Finally, putting all these values in (4), we get

$$u(x, y) = \frac{400}{\pi} \sum_{n=1}^{\infty} \frac{\sin(2m-1)x}{2m-1} e^{-(2m-1)y} .$$

EXAMPLE 8. *A thin rectangular homogeneous thermally conducting plate lies in the xy-plane defined by $0 \leq x \leq a,\ 0 \leq y \leq b$. The edge $y = 0$ is held at its temperature $Tx(x-a)$, where T is a constant, while the remaining edges are held at $0^\circ C$. The other faces are insulated and no internal sources and sinks are present. Find the steady state temperature inside the plate.*

SOLUTION. As per given, we have no heat sources and sinks are present in the plate. Thus, the steady state temperature function is the solution of

$$\frac{\partial^2 u}{\partial x^2} + \frac{\partial^2 u}{\partial y^2} = 0 \qquad\qquad ... (1)$$

with boundary conditions
$$u(0, y) = 0, \quad u(a, y) = 0, \ u(x, b) = 0, \qquad\qquad u(x, 0) = Tx(x-a) \quad ....(2)$$

Now proceeding same as example (1), we get

$$u(x, y) = \sum_{n=1}^{\infty} A_m \sin\left(\frac{n\pi x}{a}\right) \sinh\left(\frac{n\pi(y-b)}{a}\right)$$

where, $\quad A_n \sinh\left(\dfrac{-n\pi b}{a}\right) = \dfrac{2}{a}\int_0^a f(x) \sin\left(\dfrac{n\pi x}{a}\right) dx$

Using the boundary conditions
$$U(x, 0) = Tx(x-a) = f(x)$$

we get

$$A_n \sinh\left(\frac{-n\pi b}{a}\right) = \frac{2}{a}\int_0^a Tx(x-a) \sin\left(\frac{n\pi}{a}x\right) dx$$

$$= \frac{2T}{a}\int_0^a x(x-a) \sin\left(\frac{n\pi x}{a}\right) dx$$

$$= \frac{-a}{n\pi} \cdot \frac{2T}{a}\left[\int_0^a x(x-a) . d\left\{\cos\left(\frac{n\pi x}{a}\right)\right\}\right]$$

$$= -\frac{2T}{n\pi}\left[(x-a)\cos\left(\frac{n\pi x}{a}\right)\right]_0^a - \frac{a}{n\pi}\int_0^a (2x-a)\, d\left\{\sin\left(\frac{n\pi x}{a}\right)\right\}$$

$$= \frac{2aT}{n^2\pi^2}\left[(2x-a)\sin\left(\frac{n\pi x}{a}\right)\right]_0^a - \int_0^a 2\sin\left(\frac{n\pi x}{a}\right) dx$$

$$= \frac{2aT}{n^2\pi^2}\left\{a\sin n\pi + \frac{2a}{n\pi}\left[\cos\left(\frac{n\pi x}{a}\right)\right]_0^a\right\}$$

$$= \frac{2aT}{n^2\pi^2} \cdot \frac{2a}{n\pi}\cos(n\pi - 1) = \frac{4a^2T}{n^3\pi^3}[(-1)^n - 1]$$

Hence, the required temperature function is given by

$$u(x,y) = \sum_{n=1}^{\infty} \frac{4Ta^2}{n^3\pi^3} \operatorname{cosech}\left(-\frac{n\pi b}{a}\right)[(-1)^n - 1]\sin\left(\frac{n\pi x}{a}\right)\sinh\left[\frac{n\pi(y-b)}{a}\right].$$

EXAMPLE 9.   *Solve* $\dfrac{\partial^2 u}{\partial x^2} + \dfrac{\partial^2 u}{\partial y^2} = 0$, $0 \le x \le a$, $0 \le y \le b$ *subject to the boundary conditions*

$$u(0,y) = 0, \; u(x,0) = 0, \; u(x,b) = 0, \frac{\partial u}{\partial x}(a,y) = T\sin^3\frac{\pi y}{a}.$$

SOLUTION.   Proceeding same as example (3), we get

$$u(x,y) = (C_1 e^{\lambda x} + C_2 e^{-\lambda x})(C_3\cos\lambda y + C_4\sin\lambda y)$$

Using $u(x,0) = 0$, we get

$$u(x,0) = 0 = C_4\sin\lambda y\,(C_1 e^{\lambda x} + C_2 e^{-\lambda x})$$

Again, using $u(x,b) = 0$, we get

$$0 = C_4\sin\lambda b\,(C_1 e^{\lambda x} + C_2 e^{-\lambda x})$$

Now, $C_4 \ne 0 \;\Rightarrow\; \sin\lambda b = 0$

$\Rightarrow$ $\lambda b = n\pi \;\Rightarrow\; \lambda = \dfrac{n\pi}{b}; \quad n = 1, 2, 3, \dots\dots$

$\Rightarrow$ $u(x,y) = C_4\sin\left(\dfrac{n\pi y}{b}\right)(C_1 e^{\lambda x} + C_2 e^{-\lambda x})$

which can also be written as   (By renaming the constants)

$$u(x,y) = \sin\left(\frac{n\pi y}{b}\right)\left[A\exp\left(\frac{n\pi x}{b}\right) - B\exp\left(\frac{-n\pi x}{b}\right)\right].$$

$n = 1, 2, 3, \dots$

Now, using $(0,y) = 0$, we get

$$0 = \sin\left(\frac{n\pi y}{b}\right)(A + B)$$

$\Rightarrow$ $A + B = 0$,   *i.e.,* $A = -B$

Therefore,

$$u(x,y) = A\sin\left(\frac{n\pi y}{b}\right)\left[\exp\left(\frac{n\pi x}{b}\right) - \exp\left(\frac{-n\pi x}{b}\right)\right]$$

$$= 2A\sin\left(\frac{n\pi y}{b}\right)\sinh\left(\frac{n\pi x}{b}\right), \quad n = 1, 2, 3,\dots \qquad \dots (1)$$

$\Rightarrow$ $\dfrac{\partial u}{\partial x} = 2A\dfrac{n\pi}{b}\sin\left(\dfrac{n\pi y}{b}\right)\cosh\left(\dfrac{n\pi x}{b}\right)$

Putting $x = a$ and using last boundary condition, we have

$$T\sin^3\frac{\pi y}{a} = 2A\frac{n\pi}{b}\sin\left(\frac{n\pi y}{b}\right)\cosh\left(\frac{n\pi a}{b}\right)$$

$\Rightarrow$ $2A = \dfrac{T\sin^3\dfrac{\pi y}{a}}{\dfrac{n\pi}{b}\sin\left(\dfrac{n\pi y}{b}\right)\cosh\left(\dfrac{n\pi a}{b}\right)}$

Putting this value in (1), the required steady state temperature function is given by

$$u(x, y) = \frac{bT}{n\pi} \sin^3 \frac{\pi y}{a} \operatorname{sech} \frac{n\pi}{b} a \sinh\left(\frac{n\pi x}{b}\right)$$

Hence, the general solution is given by

$$u(x, y) = \sum_{n=1}^{\infty} \frac{bT}{n\pi} \sin^3 \frac{\pi y}{a} \operatorname{sech}\left(\frac{n\pi a}{b}\right) \sinh\left(\frac{n\pi x}{b}\right).$$

EXAMPLE 10. *Find the potential function u(x, y, z) in a rectangular box defined by $0 \le x \le a$; $0 \le y \le b$; $0 \le z \le c$ if the potential is zero on all sides and the bottom, while u = f(x, y) on the top of the box.*

SOLUTION. Since, the potential distribution in the rectangular box satisfies the Laplace equation given by

$$\frac{\partial^2 u}{\partial x^2} + \frac{\partial^2 u}{\partial y^2} + \frac{\partial^2 u}{\partial z^2} = 0 \qquad \text{... (1)}$$

As per given, the boundary conditions are

$$\left. \begin{array}{l} u(0, y, z) = u(a, y, z) = 0 \\ u(x, 0, z) = u(x, b, z) = 0 \\ u(x, y, 0) = 0 \\ \text{and} \quad u(x, y, c) = f(x, y) \end{array} \right\} \qquad \text{... (2)}$$

Fig. 6

Assume that

$$u(x, y, z) = X(x)\ Y(y)\ Z(z) \qquad \text{....(3)}$$

be the form of the solution of (1).

Putting the value from (3) into (1), we get

$$X''(x)\,Y(y)\,Z(z) + X(x)Y''(y)\,Z(z) + X(x)Y(y)Z''(z) = 0$$

$$\Rightarrow \qquad \frac{Y''(y)}{Y(y)} + \frac{Z''(z)}{Z(z)} = -\frac{X''(x)}{X(x)} = \lambda_1^2 \text{ (say)}$$

Now, we get $\qquad X''(x) + \lambda_1^2 X(x) = 0 \qquad \text{... (4)}$

After the second separation, we have

$$\frac{Z''(z)}{Z(z)} = \lambda_3^2, \quad -\frac{Y''(y)}{Y(y)} = \lambda_2^2$$

$$\Rightarrow \qquad Y''(y) + \lambda_2^2 Y(y) = 0 \qquad \text{... (5)}$$

and $\qquad Z''(z) - \lambda_3^2 Z(z) = 0 \qquad \text{... (6)}$

where $\qquad \lambda_3^2 = \lambda_1^2 + \lambda_2^2$

The general solution of (4), (5) and (6) are given by

$$X(x) = C_1 \cos \lambda_1 x + C_2 \sin \lambda_1 x$$

$$Y(y) = C_3 \cos \lambda_2 y + C_4 \sin \lambda_2 y$$

$$Z(z) = C_5 \cosh \lambda_3 z + C_6 \sinh \lambda_3 z$$

Using the boundary condition given by (2), we get

$$X(0) = X(a) = 0\ ;\ Y(0) = Y(b) = 0\ ;\ Z(0) = 0$$

Now, $\qquad X(0) = 0 \ \Rightarrow\ C_1 = 0$

$$X(a) = 0 \ \Rightarrow\ \lambda_1 a = m\pi$$

$$\Rightarrow \quad \lambda_1 = \frac{m\pi}{a}; \quad m = 1, 2, 3, \ldots$$

In a similar manner, we can say that

$$Y(0) = 0 \text{ gives } C_3 = 0$$
$$Y(b) = 0 \text{ gives } \lambda_2 b = n\pi$$

Therefore,

$$\lambda_2 = \frac{n\pi}{b}; \quad n = 1, 2, \ldots$$

and $\quad Z(0) = 0 \Rightarrow C_5 = 0$

Also, $\quad \lambda_3^2 = \lambda_1^2 + \lambda_2^2 = \pi^2 \left(\frac{m^2}{a^2} + \frac{n^2}{b^2} \right) = \lambda_{mn}^2 \quad$ (say)

Then, $\quad \lambda_3 = \pi \sqrt{\dfrac{m^2}{a^2} + \dfrac{n^2}{b^2}} = \lambda_{mn}$

Therefore, the required general solution is given by

$$u(x, y, z) = X(x) \, Y(y) \, Z(z)$$

$$= \sum_{m=1}^{\infty} \sum_{n=1}^{\infty} C_{mn} \sin \frac{m\pi x}{a} \sin \frac{n\pi y}{b} \sinh \lambda_{mn} \cdot z$$

Finally, using the boundary condition $f(x, y) = u(x, y, c)$, we get

$$f(x, y) = \sum \sum C_{mn} \sinh \lambda_{mn} C \sin \frac{m\pi x}{a} \sin \frac{n\pi y}{b}$$

which is a double Fourier sine series.

Hence, $\quad C_{mn} \sinh \lambda_{mn} C = \dfrac{4}{ab} \int_0^a \int_0^b f(x, y) \sin \dfrac{m\pi x}{a} \sin \dfrac{n\pi y}{b} \, dx dy$

EXERCISE 8.3

1. Find the steady state temperature distribution in a thin rectangular plate bounded by the lines $x = 0, x = a, y = 0, y = b$. The edges $x = 0, x = a, y = 0$ are kept at temperature zero while the edge $y = b$ is kept at $100°C$.

2. Evaluate the steady temperature in a rectangular plate of length a and width b, the sides of which are kept at temperature zero, the lower end is kept at temperature $f(x)$ and the upper edge is kept insulated.

3. Show that the velocity potential for an irrational flow of an incompressible fluid satisfies the Laplace equation.

4. Find the steady state temperature distribution in a rectangular plate of sides a and b insulated at the lateral surface and satisfying the boundary conditions

$$u(0, y) = u(a, y) = 0 \text{ for } 0 \le y \le b$$
$$u(x, b) = 0$$

and $u(x, 0) = x(a - x)$ for $0 \le y \le a$

5. A rectangular plate with insulated surface 8 cm wide and so long compared to its width that it can be considered infinite in the length without introducing an appreciable error. If the temperature along the short edge $y = 0$ is given by $u(x, 0) = 100 \sin\left(\dfrac{\pi x}{8} \right)$, in $0 < x < 8$, while the two long edges $x = 0$ and $x = 8$ as well as the other short edges are kept at $0°C$. Find the steady state temperature function $u(x, y)$.

6. By separating the variables, show that $\dfrac{\partial^2 V}{\partial x^2} + \dfrac{\partial^2 V}{\partial y^2} = 0$ has solutions of the form $A \exp(\pm nx \pm ixy)$, where A and n are constants. Deduce that the function of the form

$$V(x, y) = \sum_r A_r \sin\left(\frac{r x y}{a} \right) e^{-(r\pi x)/a},$$
$$x \ge 0, 0 \le y \le \infty$$

where A_r are constants or plane harmonic functions satisfying the conditions $V(x, 0) = V(x, a) = 0$ and $V(x, y) \to 0$ as $x \to \infty$.

7. The diameter of a semi-circular plate of radius a is kept at 0°C and the temperature at the semi-circular boundary is T°C. Show that the steady state temperature in the plate is given by

$$u(r, \theta) = \frac{4T}{\pi} \sum_{n=1}^{\infty} \frac{1}{2n-1} \left(\frac{r}{a}\right)^{2n-1} \sin(2n-1)\theta$$

(GBTU (CO) 2011)

8. Solve: $\dfrac{\partial^2 V}{\partial r^2} + \dfrac{1}{r}\dfrac{\partial V}{\partial r} + \dfrac{1}{r^2}\dfrac{\partial^2 V}{\partial \theta^2} = 0$ with boundary

conditions.

(i) V is finite when $r \to 0$

(ii) $V = \sum C_n \cos n\theta$ on $r = a$ (GBTU–2010)

9. Solve the Laplace equation $\dfrac{\partial^2 u}{\partial x^2} + \dfrac{\partial^2 u}{\partial y^2} = 0$

subject to the conditions

$$u(0, y) = u(l, y) = u(x, 0) = 0$$

and $u(x, a) = \dfrac{\sin n\pi x}{l}$

(VTU–2011; JNTU–2006; Kerala M. Tech.–2005; UPTU–2004)

Hint to Selected Problems

1. In this problem, we consider the steady state temperature, *i.e.*, it does not depend upon time. Thus the flow is governed by the Laplace equation given by

$$\frac{\partial^2 V}{\partial x^2} + \frac{\partial^2 V}{\partial y^2} = 0 \qquad \ldots (1)$$

As per given, the boundary conditions are

$$u(0, y) = 0, \ u(a, y) = 0 \qquad \ldots (2)$$

$$u(x, 0) = 0, \ u(x, b) = 100 \qquad \ldots (3)$$

Using the method of separation of variables, suppose (1) has a solution of the form

$$u(x, y) = X(x)\, Y(y) \qquad \ldots (4)$$

Putting this value in equation (1), we get

$$X''Y + XY'' = 0$$

$$\frac{X''}{X} = -\frac{Y''}{Y} \qquad \ldots (5)$$

Now, for independent x and t, each side of (5) must be equal to the constant, say k. Then we have

$$X'' - kX = 0 \qquad \ldots (6)$$

and $$Y'' + kY = 0 \qquad \ldots (7)$$

Also, from (2) and (4), we have

$$X(0)Y(a) = 0 \text{ and } X(a)\, Y(y) = 0$$

$$X(0) = 0 \text{ and } X(a) = 0 \quad \ldots (8)$$

Here, we must take $Y(y) \neq 0$ because otherwise $u = 0$, which does not satisfy the given conditions.

Now, we shall discuss the following three cases :

Case (I). Let $k = 0$. Then solution of equation (6) is given by

$$X(x) = Ax + B$$

Using boundary conditions (8), we get

$$B = 0 \text{ and } Aa + B = 0$$

$$A = B = 0 \implies X(x) = 0 \implies u = 0$$

Which does not satisfy given condition (3). Thus, we reject this case, *i.e.,.*, $k = 0$.

Case (II). Let $k = \lambda^2, \lambda \neq 0$. In this case, solution of (6) is given by

$$X(x) = A e^{x\lambda} + B e^{-x\lambda}$$

Using the boundary conditions (8), we get

$$A + B = 0 \text{ and } 0 = e^{a\lambda} + e^{-a\lambda}.$$

On solving, we get

$$A = B = 0 \implies X(x) = 0 \implies u = 0$$

Thus, we reject this case also.

Case (III). Let $k = -\lambda^2, \lambda \neq 0$. In this case, the solution of equation (6) is given by

$$X(x) = A \cos \lambda x + B \sin \lambda x$$

Using the boundary conditions (8), we get

$$A = 0 \text{ and } A \cos \lambda a + B \sin \lambda a = 0$$

On solving, we get

$$A = 0 \text{ and } \sin \lambda a = 0$$

[By taking $B \neq 0$]

Further, $\sin \lambda a = 0$

$$\implies \qquad \lambda = \frac{n\pi}{a}; \ n = 1, 2, 3, \ldots$$

Thus, the non-zero solution $X_n(x)$ of (6) is given by

$$X_n(x) = B_n \sin\left(\frac{n\pi x}{a}\right)$$

Further, using $k = -\lambda^2 = -\dfrac{n^2 \pi^2}{a^2}$, equation (7) reduces to

$$Y'' - \left(\frac{n^2\pi^2}{a^2}\right)Y = 0$$

The general solution of this equation is given by

$$Y_n(y) = C_n e^{n\pi y/a} + D_n e^{-n\pi y/a} \quad \dots (9)$$

Using (3) and (4)

$$0 = X(x)\, Y(0)$$

$$\Rightarrow \qquad Y(0) = 0 \quad \text{Now,} \qquad Y(0) = 0$$

$$\Rightarrow \qquad Y_n(0) = 0 \quad \dots (10)$$

Putting $y = 0$ in (9) and using (10), we get

$$0 = C_n + D_n$$

$$D_n = -C_n$$

Then equation (7) becomes

$$Y_n(y) = C_n(e^{n\pi y/a} - e^{-n\pi y/a}) = 2\sinh\left(\frac{n\pi y}{a}\right)$$

Therefore,

$$u(x,y) = X_n(x)\, Y_n(y)$$

$$= E_n \sin(n\pi x/a)\, \sinh(n\pi y/a)$$

where, $E_n = 2B_n C_n$ are arbitrary constants.
The more general solution is given by

$$u(x,y) = \sum_{n=1}^{\infty} E_n \sin\left(\frac{n\pi x}{a}\right) \sinh\left(\frac{n\pi y}{a}\right)$$

$$\dots (11)$$

Putting $y = b$ and using given condition $u(x, b)$ = 100, we get

$$100 = \sum_{n=1}^{\infty} E_n \sin\left(\frac{n\pi x}{a}\right) \sinh\left(\frac{n\pi b}{a}\right)$$

which is a Fourier sine series.

Thus, $E_n \sinh\dfrac{n\pi b}{a} = \dfrac{2}{a}\displaystyle\int_0^a 100.\sin\left(\frac{n\pi x}{a}\right) dx$

$$= \frac{200}{a}\left[-\frac{\cos(n\pi x/a)}{(n\pi/a)}\right]_0^a$$

So, $E_n = \dfrac{200}{n\pi}\left[1-(-1)^n\right]\operatorname{cosech}\left(\dfrac{n\pi b}{a}\right)$

$$= \begin{cases} 400\,\operatorname{cosech}\{(2m-1)\pi b/a & \text{if } n = 2m \\ (2m-1)\pi & \text{if } n = 2m-1 \end{cases}$$

Putting this value in (11), we get
$u(x,y)$

$$= \sum_{m=1}^{\infty} E_{2m-1} \sin\frac{(2m-1)\pi x}{a} \sinh\frac{(2m-1)\pi y}{a}$$

$$\Rightarrow \quad u(x,y) = \frac{400}{\pi}\sum_{m=1}^{\infty}\frac{1}{2m-1}\sin\frac{(2m-1)\pi x}{a}$$

$$\sinh\frac{(2m-1)\pi y}{a}\operatorname{cosech}\frac{(2m+1)\pi b}{a}$$

2. We know that temperature $u(x, y)$ in steady state in two-dimensional plate is governed by the Laplace equation.

$$\frac{\partial^2 u}{\partial x^2} + \frac{\partial^2 u}{\partial y^2} = 0 \qquad \dots (1)$$

According to the given conditions, we have

$$u(0, y) = 0, \ u(a, y) = 0 \qquad \dots (2)$$

$$\left(\frac{\partial u}{\partial y}\right)_{y=b} = 0 \ \text{ and } \ u(x,0) = f(x) \ \dots (3)$$

Using the method of separation of variables, suppose

$$u(x, y) = X(x)\, Y(y) \qquad \dots (4)$$

is solution of (1).
Putting this value of u from (4) in (1), we get

$$X'' Y + XY'' = 0$$

$$\Rightarrow \qquad \frac{X''}{X} = -\frac{Y''}{Y} \qquad \dots (5)$$

Now, since x and t are independent, therefore each side of (5) must be equal to the same constant, say k. Then (5) reduces to

$$X'' - KX = 0 \qquad \dots (6)$$

and $\qquad Y'' + KY = 0 \qquad \dots (7)$

with $\qquad X(0) = 0 \ \text{ and } \ X(a) = 0. \ \dots (8)$

Now, there are following three cases :

Case (I). Let $k = 0$. In this case, solution of (6) is given by

$$X(x) = Ax + B. \qquad \dots (9)$$

Using given conditions (8), we get

$$A = B = 0.$$

Hence, we reject this case.

Case (II). **Let $k = \lambda^2, \lambda \neq 0$**. In this case, solution of (6) is given by

$$X(x) = Ae^{x\lambda} + Be^{-x\lambda} \qquad \dots (10)$$

Again, using boundary conditions (8) in (10), we get

$$A + B = 0 \ \text{ and } \ Ae^{a\lambda} + Be^{-a\lambda} = 0$$

$$\Rightarrow \qquad A = B = 0 \ \Rightarrow \ X(x) = 0 \ \Rightarrow \ u = 0.$$

Hence, we reject this case also.

Case (III). Let $k = -\lambda^2, \lambda \neq 0$. Then solution of (6) is given by

$$X(x) = A\cos\lambda x + B\sin\lambda x$$

Using boundary conditions (8), we get

$$A = 0 \text{ and } A\cos\lambda a + B\sin\lambda a = 0$$

$$A = 0 \text{ and } \sin\lambda a = 0$$

[By taking $B \neq 0$]

Now, $\sin\lambda a = 0$

$$\Rightarrow \quad \lambda a = n\pi \Rightarrow \lambda = \frac{n\pi}{a}; \quad n = 1, 2, 3, \ldots$$

Thus, non-zero solution $X_n(x)$ of (6) is given by

$$X_n(x) = B_n \sin\left(\frac{n\pi x}{a}\right)$$

Now, $\quad k = -\lambda^2 = -\dfrac{n^2\pi^2}{a^2}$,

Then, equation (7) reduces to

$$Y'' - \left(\frac{n^2\pi^2}{a^2}\right)Y = 0$$

whose general solution is given by

$$Y_n(y) = C_n e^{n\pi y/a} + D_n e^{-n\pi y/a} \qquad \ldots (11)$$

From (4), we can find

$$\frac{\partial u}{\partial y} = X(x)\,Y'(y) \Rightarrow \left(\frac{\partial u}{\partial y}\right)_{y=b} = X(x)\,Y'(b)$$

Using (3), we get

$$X(x).Y'(b) = 0$$

$$\Rightarrow \quad Y'(b) = 0 \qquad \ldots (12)$$

Differentiating (11) w.r.t. y, we get

$$Y_n'(y) = \left(\frac{n\pi}{a}\right)C_n\,e^{n\pi y/a} - \left(\frac{n\pi}{a}\right)D_n e^{-n\pi y/a} \qquad \ldots (13)$$

Putting $y = b$ and using (12), we get

$$0 = \left(\frac{n\pi}{a}\right)\left(C_n e^{n\pi b/a} - D_n e^{-n\pi b/a}\right)$$

$$D_n = C_n e^{n\pi b/a} / e^{-n\pi b/a}$$

Putting this value in (11), we get

$$Y_n(y) = \frac{C_n(e^{n\pi y/a} - e^{-n\pi y/a} + e^{-n\pi y/a}e^{n\pi b/a}}{e^{-n\pi b/a}}$$

$$Y_n(y) = C_n e^{n\pi b/a}\left\{e^{n\pi(b-y)/a} + e^{-n\pi(b-y)/a}\right\}$$

$$= 2C_n\,e^{n\pi b/a}\cosh\left\{\frac{n\pi(b-y)}{a}\right\}$$

Thus, $u_n(x,y) = X_n(x).Y_n(y)$

$$= E_n.\sin\frac{n\pi x}{a}\cosh\left\{\frac{n\pi(b-y)}{a}\right\} \qquad \ldots (14)$$

Now, the more general solutions are given by

$$u(x,y) = \sum_{n=1}^{\infty} u_n(x,y)$$

$$= \sum_{n=1}^{\infty} E_n \sin\left(\frac{n\pi x}{a}\right)\cosh\left(\frac{n\pi(b-y)}{a}\right) \ldots (15)$$

Now, putting $y = 0$ in (15) and using (3), we get

$$f(x) = \sum_{n=1}^{\infty}\left\{E_n\cosh\left(\frac{n\pi b}{a}\right)\right\}\sin\left(\frac{n\pi x}{a}\right)$$

which is the half range Fourier sine series of $f(x)$ in $(0, a)$.
Therefore, we get

$$E_n = \frac{2}{a\cosh\left(\dfrac{n\pi b}{a}\right)}\int_0^a f(x).\sin\frac{n\pi x}{a}\,dx$$

3. Let S be a closed surface enclosed a volume V.

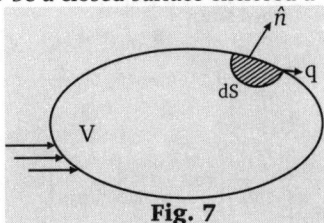

Fig. 7

Let ρ be the density of the fluid. If $\hat{n}$ be a unit normal vector drawn from outside to the surface element ds and $\overline{q}$ be the velocity of the fluid at that point. Then

Inward normal velocity = $-\overline{q}\,\hat{n}$

$\therefore$ Mass of the fluid entering per unit time through the element dS is $(-\overline{q}.\hat{n})\,dS$.

Thus the mass of the fluid entering in the surface S in the unit time is

$$= -\int\int_S \rho\,(\overline{q}.\hat{n})\,dS.$$

Further, the mass of the fluid within S is

$$= \int\int\int_S \rho dV$$

Rate of mass of fluid increasing in S is given by :

$$\frac{\partial}{\partial t}\iiint_V \rho\,dV = \iiint_V \frac{\partial\rho}{\partial t}\,dV.$$

Using the law of conservation of mass, we have

$$\iiint_V \frac{\partial\rho}{\partial t}\,dV = -\iint_S \rho(\overline{q}\,\hat{n})\,dS$$

$$= \iiint_V \nabla(\rho\overline{q})\,dV$$

[By Gauss divergence theorem]

$$\Rightarrow \iiint_V \left[\frac{\partial\rho}{\partial t} + \nabla(\rho\overline{q})\right]dV = 0$$

$\Rightarrow \dfrac{\partial \rho}{\partial t} + \nabla(\rho \overline{q}) = 0$, which is known as equation

of continuity.

For an incompressible fluid :

$$\rho = \text{constant}$$
$$\nabla . \overline{q} = 0$$

Also, if flow is irrotational

$\overline{q} = -\nabla \phi$, where ϕ is a potential function.

$\nabla . q = \nabla . \nabla \phi = \nabla \phi = 0$, which is a Laplace equation.

4. Proceeding same as example (1) and using $f(x) = x(a - x)$, we get the required temperature.

$$u(x, y) = \sum_{n=1}^{\infty} E_n \sin\dfrac{n\pi x}{a} \sinh\dfrac{n\pi(b - y)}{a}$$

where $E_n = \dfrac{2}{a \sinh\left(\dfrac{n\pi b}{a}\right)} \displaystyle\int_0^a f(x) \sin\dfrac{n\pi x}{a} dx$

$= \dfrac{2}{a \sinh\left(\dfrac{n\pi b}{a}\right)} \displaystyle\int_0^a (ax - x^2) \sin\dfrac{n\pi x}{a} dx$

$= \dfrac{2}{a \sinh\left(\dfrac{n\pi b}{a}\right)} \left[(ax - x^2)\left(-\dfrac{a}{n\pi}\right)\cos\dfrac{n\pi x}{a} \right.$

$\left. -(a - 2x)\left(-\dfrac{a^2}{n^2\pi^2}\right)\sin\dfrac{n\pi x}{a} \right.$

$\left. +(-2)\left(\dfrac{a^3}{n^3\pi^3}\right)\cos\dfrac{n\pi x}{a} \right]_0^a$

$= \dfrac{2}{a \sinh\left(\dfrac{n\pi b}{a}\right)} \left[\dfrac{-2a^3(-1)^n}{n^3\pi^3} + \dfrac{2a^3}{n^3\pi^3} \right]$

$= \dfrac{4a^2}{n^3\pi^3}[1 - (-1)^n]\text{cosec}\dfrac{n\pi b}{a}$

$= \begin{cases} 0, \\ \qquad \text{if } n = 2m \text{ (even) and } m = 1, 2, 3, \ldots \\ \left\{\dfrac{8a^2}{\pi^3(2m - 1)^3}\right\} \text{cosech}\left\{\dfrac{(2m - 1)\pi b}{a}\right\}, \\ \qquad \text{if } n = 2m - 1 \text{ and } m = 1, 2, 3, \ldots \end{cases}$

Putting this value in (1), we get

$$u(x, y) = \dfrac{8a^2}{\pi^3} \sum_{n=1}^{\infty} \dfrac{1}{(2m - 1)^3} \sin\dfrac{(2m - 1)\pi x}{a}$$

$$\sinh\dfrac{(2m - 1)(b - y)\pi}{a}\text{cosech}\dfrac{(2m - 1)\pi b}{a}$$

5. Steady state temperature is governed by the following Laplace's equation

$$\dfrac{\partial^2 u}{\partial x^2} + \dfrac{\partial^2 u}{\partial y^2} = 0 \qquad \text{.. (1)}$$

As per given, the boundary conditions are

$$u(0, y) = u(8, y) = 0 \text{ for } 0 < y < \infty \qquad \text{... (3)}$$

$$\left. \begin{array}{l} u(x, y) \to 0 \text{ as } y \to \infty \text{ for } 0 < x < 8 \\[2mm] \text{and} \quad u(x, 0) = 100 \sin\left(\dfrac{\pi x}{8}\right) \text{ for } 0 < x < 9 \end{array} \right\}$$

By taking $a = 8$, we get

$$u(x, y) = \sum_{n=1}^{\infty} E_n \sin\dfrac{n\pi x}{8} e^{-n\pi y/8} \qquad \text{... (4)}$$

Putting $y = 0$ in (4) and using (3), we get

$$100 \sin\left(\dfrac{\pi x}{8}\right) = \sum_{n=1}^{\infty} E_n \sin\left(\dfrac{n\pi x}{8}\right)$$

$$= E_1 \sin\left(\dfrac{\pi x}{8}\right) + E_2 \sin\left(\dfrac{2\pi x}{8}\right) + \ldots$$

Now comparing the coefficients of like terms of both the sides, we get

$$E_1 = 100 \text{ and } E_n = 0 \text{ for } n \geq 2.$$

Hence, the required temperature function is given by

$$u(x, y) = 100 \sin\left(\dfrac{\pi x}{8}\right) e^{-\pi y/8}$$

6. The given equation is

$$\dfrac{\partial^2 V}{\partial x^2} + \dfrac{\partial^2 V}{\partial y^2} = 0 \qquad \text{... (1)}$$

Suppose the solution of (1) is of the form

$$V(x, y) = X(x) Y(y) \qquad \text{... (2)}$$

Putting the value from (2) in (1), we get

$$X''Y + XY'' = 0$$

$$\Rightarrow \qquad \dfrac{X''}{X} = -\dfrac{Y''}{Y} = n^2 \text{ (say)} \qquad \text{... (3)}$$

From (3), we can find

$$X'' - n^2 X = 0 \qquad \text{... (4)}$$

and $\qquad Y'' + n^2 Y = 0 \qquad \text{... (5)}$

whose solutions are given by

$$X = Ae^{nx} + Be^{-nx}$$

and $\qquad Y = C \cos ny + D \sin ny$

Putting these values in (2), we get

$$V(x, y) = (Ae^{nx} + Be^{-nx})(C \cos ny + D \sin ny) \qquad \text{... (6)}$$

$$= A.\exp(\pm nx \pm iny) \qquad \text{... (7)}$$

Here, the given boundary conditions are

$$V(x, y) \to 0 \text{ as } x \to \infty \qquad \ldots (8)$$

and $\quad V(x, 0) = V(x, a) = 0 \qquad \ldots (9)$

Thus, we can write

$$V(x, y) = (E \cos ny + F \sin ny) e^{-nx} \ldots (10)$$

Putting $y = 0$ in (10) and using (9), we get

$$0 = Ee^{-nx} \Rightarrow E = 0$$

Further, putting $y = a$ in (10) and using (9), we get

$$0 = F \sin na e^{-nx} \text{ for all.}$$

$$\Rightarrow \sin na = 0 \quad \Rightarrow \quad n = \frac{r\pi}{a}; \quad r = 1, 2, 3, \ldots$$

Therefore, non-zero solution of (1) are given by (10) in the form

$$V(x, y) = F \sin\left(\frac{r\pi y}{a}\right) e^{-\left(\frac{r\pi x}{a}\right)}, \quad r = 1, 2, 3, \ldots$$

Hence, the more general solution of (1) is of the form

$$V(x, y) = \sum_{r=1}^{\infty} A_r \sin\left(\frac{r\pi y}{a}\right) e^{-\left(\frac{r\pi x}{a}\right)}$$

ANSWERS

1. $u(x, y) = \dfrac{400}{\pi} \sum_{m=1}^{\infty} \dfrac{1}{2m-1} \sin\dfrac{(2m-1)\pi x}{a} \sinh\dfrac{(2m-1)\pi y}{a} \operatorname{cosech}\dfrac{(2m+1)\pi b}{a}$

2. $u(x, y) = \sum_{n=1}^{\infty} u_n(x, y) = \sum_{n=1}^{\infty} E_n \sin\left(\dfrac{n\pi x}{a}\right) \cosh\left\{\dfrac{n\pi(b-y)}{a}\right\}$

$f(x) = \sum_{n=1}^{\infty} \left\{ E_n \cosh\left(\dfrac{n\pi b}{a}\right)\right\} \sin\dfrac{n\pi x}{a}$, where, $E_n = \dfrac{2}{a \cosh\left(\dfrac{n\pi b}{a}\right)} \int_0^a f(x) \sin\dfrac{n\pi x}{a}\, dx$

4. $u(x, y) = \dfrac{8a^2}{\pi} \sum_{m=1}^{\infty} \dfrac{1}{(2m-1)^3} \sin\dfrac{(2m-1)nx}{a} \sinh\dfrac{(2m-1)(b-y)\pi}{a} \operatorname{cosech}\dfrac{(2m-1)\pi b}{a}$

5. $u(x, y) = 100 \sin\left(\dfrac{\pi x}{8}\right) e^{-\pi y/8}$

6. $V(x, y) = \sum_{r=1}^{\infty} A_r \sin\left(\dfrac{rxy}{a}\right) \cdot e^{-\left(\frac{r\pi x}{a}\right)}$

7. $u(r, \theta) = \dfrac{4T}{\pi} \sum_{n=1}^{\infty} \dfrac{1}{2n-1}\left(\dfrac{r}{a}\right)^{2n-1} \sin(2n-1)\theta$

8. $V = \sum C_n \left(\dfrac{r}{a}\right)^n \cos n\theta$

9. $u(x, y) = \dfrac{\sinh(n\pi y / l)}{\sinh(n\pi a / l)} \sin\dfrac{n\pi x}{l}$

8.14 PARABOLIC DIFFERENTIAL EQUATIONS

Here, we shall consider a few problems dealing with the simplest of all parabolic equations, namely the one-dimensional heat equation. Under suitable physical assumptions and choice of units, this equation governs the distribution of temperature on a homogeneous thin rod occupying part of all x-axis, the variable t denoting the time.

8.15 ONE DIMENSIONAL HEAT EQUATION

(UPTU–2007, UPTU(CO)–2009, A.M.I.E.–1995,97)

Let us consider the flow of heat by conduction in a bar OA. Consider an element PQQ'P' of the bar. The temperature u(x, t) of the bar at any point P is function of x and time t.

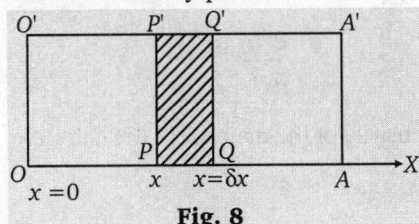

Fig. 8

(i) The position of the bar coincides with the *x*-axis.

(ii) The bar is homogeneous.

(iii) It is sufficiently thin so that the heat is uniformly distributed over its cross-section at a given time *t*.

(iv) The surface of the bar is insulated to prevent any loss of heat through the boundary.

(v) $u(x, t)$ is the temperature at the point x at time t.

(vi) The amount of heat crossing any section of the bar is given by $kA\left(\dfrac{\partial u}{\partial x}\right)\delta t$

where, A = Area of the cross-section of the bar

$$\frac{\partial u}{\partial x} = \text{Temperature gradient at the section}$$

$$\delta t = \text{Time of the flow of heat}$$

$$k = \text{Thermal conductivity of the material of the bar.}$$

The quantity of heat flowing into the element across the section PP' in time δt.

$$= -kA\left(\frac{\partial u}{\partial x}\right)_x \delta t$$

(The negative sign has been taken because heat flows in the direction of decreasing temperature.)

Also, the quantity of heat flowing out of the element across the section QQ' in time δt.

$$= -kA\left(\frac{\partial u}{\partial x}\right)_{x+\delta x} \delta t$$

Therefore the quantity of heat retained by the element is

$$= -kA\left(\frac{\partial u}{\partial x}\right)_x \delta t + kA\left(\frac{\partial u}{\partial x}\right)_{x+\delta x} \delta t = kA\,\delta t \left\{\left(\frac{\partial u}{\partial x}\right)_{x+\delta x} - \left(\frac{\partial u}{\partial x}\right)_x\right\} \quad \dots (1)$$

Now, suppose that this heat raises the temperature of the element by a small quantity δu. Therefore, the same quantity of heat is given by

$$= (\rho A\delta x)\sigma\,\delta u \qquad \dots (2)$$

where σ is specific heat of the bar.

Since (1) and (2) are equal, therefore, we have

$$kA\delta t\left\{u(x+\delta x, t) - u(x, t)\right\} = (\rho A\delta x)\,\sigma\,\delta u$$

$$\Rightarrow \qquad k\,\frac{u(x+\delta x, t) - u(x, t)}{\delta x} = \rho\sigma\frac{\delta u}{\delta t} \qquad \dots (3)$$

As $\delta x \to 0$ and $\delta t \to 0$, we get

$$k\frac{\partial^2 u}{\partial x^2} = \rho\sigma\frac{\partial u}{\partial t}$$

$$\Rightarrow \qquad \frac{\partial u}{\partial t} = k_1\frac{\partial^2 u}{\partial x^2} \qquad \dots (4)$$

where, $k_1 = \dfrac{k}{\rho\sigma}$ is called the diffusivity of the material of the bar. Here, equation (4) is known as one-dimensional heat equation.

Remark

- Heat equation is also known as Diffusion equation.

8.16 SOLUTION OF ONE DIMENSIONAL HEAT EQUATION

(GBTU (CO)–2011, UPTU–2007)

Consider the equation
$$\frac{\partial u}{\partial t} = k\frac{\partial^2 u}{\partial x^2} \qquad ... (1)$$

For the method of separation of variables, let us assume that solution of (1) is of the form
$$u(x, t) = X(x)\, T(t) \qquad ... (2)$$

Putting the values from (2) in (1), we get
$$\frac{X''}{X} = \frac{T'}{kT} = \mu \ \text{(say)}, \text{ a separation constant.} \qquad ... (3)$$

where the dashes denote derivatives with respect to the relevant variable.

From (3), we can find
$$X'' - \mu X = 0 \qquad ... (4)$$

and
$$T' = \mu kT \qquad ... (5)$$

Now, we have the following three cases :

Case (I). Let $\mu = 0$ Then solutions of (4) and (5) are given by
$$X = a_1 x + a_2 \ \text{ and } \ T = a_3 \qquad ... (6)$$

Case (II). Let $\mu = \lambda^2, \lambda \neq 0$. Then (4) and (5) reduce to
$$X'' - \lambda^2 X = 0 \ \text{ and } \ T' = \lambda^2 kT$$

On solving these equations, we get
$$\left. \begin{array}{l} X = b_1 e^{\lambda x} + b_2 e^{-\lambda x} \\ T = b_3 e^{\lambda^2 kt} \end{array} \right] \qquad ... (7)$$

Case (III). Let $\mu = -\lambda^2, \lambda \neq 0$. Then (4) and (5) reduce to
$$X'' + \lambda^2 X = 0 \ \text{ and } \ T' = -\lambda^2 kT$$

On solving, we get
$$\left. \begin{array}{l} X = C_1 \cos \lambda x + C_2 \sin \lambda x \\ T = C_3 e^{-\lambda^2 kt} \end{array} \right] \qquad ... (8)$$

Hence, the various possible solutions are
$$u(x, t) = A_1 x + A_2$$
$$u(x, t) = (B_1 e^{\lambda x} + B_2 e^{-\lambda x}) e^{\lambda^2 kt}$$

and
$$u(x, t) = (C_1 \cos \lambda x + C_2 \sin \lambda x) e^{-\lambda^2 kt}$$

8.17 SOLUTION OF TWO DIMENSIONAL HEAT EQUATION

The two dimensional heat equation is given by
$$\frac{1}{k}\frac{\partial u}{\partial t} = \left(\frac{\partial^2 u}{\partial x^2} + \frac{\partial^2 u}{\partial y^2} \right) \qquad ... (1)$$

Let us suppose (1) has solution of the form

$$u(x, y, t) = X(x) \, Y(y) \, T(t) \qquad \ldots (2)$$

Putting the values from (2) in (1), we get

$$X''YT + XY''T = \frac{1}{k} XYT'$$

$$\Rightarrow \qquad \frac{X''}{X} + \frac{Y''}{Y} = \frac{1}{k} \frac{T'}{T} \qquad \ldots (3)$$

Now, since x, y and t are independent variables, thus (3) is true if each term on each side is equal to a constant such that

$$\frac{X''}{X} = -n^2, \quad \frac{Y''}{Y} = -m^2 \quad \text{and} \quad \frac{T'}{kT} = -p^2 \qquad \ldots (4)$$

with $\qquad n^2 + m^2 = p^2$

The constants may be chosen such that the solution u has the property that $u \to 0$ *as* $t \to \infty$.
Solving (4), we get

$$X_n(x) = A_n \cos nx + B_n \sin nx ; \quad Y_m(y) = C_m \cos my + D_m \sin my$$

and

$$T_p(t) = E_p \, e^{-p^2 kt} = F_{nm} \, e^{-(n^2 + m^2) \, kt}$$

Hence, a suitable solution of (1) is given by

$$u_{nm}(x, y, t) = F_{nm}(A_n \cos nx + B_n \sin nx)(C_m \cos my + D_m \sin my) \, e^{-(n^2 + m^2) kt}$$

8.18 SOLUTION OF THREE DIMENSIONAL HEAT EQUATION

Three dimensional heat equation is given by

$$\frac{\partial^2 u}{\partial x^2} + \frac{\partial^2 u}{\partial y^2} + \frac{\partial^2 u}{\partial z^2} = \frac{1}{k} \frac{\partial u}{\partial t} \qquad \ldots (1)$$

Let solution of (1) be of the form

$$u(x, y, z, t) = X(x) \, Y(y) Z(z) \, T(t) \qquad \ldots (2)$$

where X, Y, Z, T are respectively the function of x, y, z and t alone.
Putting the values from (2) in (1), we get

$$X''YZT + XY''ZT + XYZ''T = \frac{1}{k} XYZT'$$

$$\Rightarrow \qquad \frac{X''}{X} + \frac{Y''}{Y} + \frac{Z''}{Z} = \frac{1}{k} \frac{T'}{T} \qquad \ldots (3)$$

Now, since x, y, z and t are independent variables, equation (3) is true only when each term on each side is a constant such that

$$\frac{X''}{X} = -n^2, \quad \frac{Y''}{Y} = -m^2, \quad \frac{Z''}{Z} = -l^2 \quad \text{and} \quad \frac{T'}{kT} = -p^2 \qquad \ldots (4)$$

with $\qquad n^2 + m^2 + l^2 = p^2$

Further, we have to choose the constants such that solution $u(x, y, z, t)$ has the property that $u \to 0$ *as* $t \to \infty$.
On solving (4), we get

$$X_n(x) = A_n \cos nx + B_n \sin nx$$

$$Y_m(y) = C_m \cos my + D_m \sin my$$

$$Z_l(z) = E_l \cos lz + F_l \sin lz$$

and

$$T_p(t) = G_p \, e^{-p^2 kt} = H_{mnl} \, e^{-(n^2+m^2+l^2)\, kt}$$

which gives

$$u_{nml}(x, y, z, t) = H_{nml}(A_n \cos nx + B_n \sin nx)\,(C_m \cos my + D_m \sin my)$$
$$.(E_l \cos lz + F_l \sin lz)\, e^{-(n^2+m^2+l^2)kt}$$

which are the required suitable solution of (1).

The general solution of (1) can be obtained by putting $u(x,y,z,t) = \sum\limits_{n=1}^{\infty} \sum\limits_{m=1}^{\infty} \sum\limits_{l=1}^{\infty} u_{nml}(x,y,z,t) \cdot$

8.19 TRANSMISSION LINE EQUATIONS

(i) Telegraph Equation : $\dfrac{\partial^2 V}{\partial x^2} = RC\dfrac{\partial V}{\partial t}$ and $\dfrac{\partial^2 i}{\partial x^2} = RC\dfrac{\partial i}{\partial t}$ (UPTU (CO)–2009)

(ii) Radio Equation : $\dfrac{\partial^2 V}{\partial x^2} = LC\dfrac{\partial^2 V}{\partial t^2}$ and $\dfrac{\partial^2 i}{\partial x^2} = LC\dfrac{\partial^2 i}{\partial t^2}$ (UPTU (CO)–2009)

where V = potential, i = current, C = capacitance and L = inductance.

Solved Examples

EXAMPLE 1. *Determine the solution of one dimensional heat equation*

$$\frac{\partial u}{\partial t} = C^2 \frac{\partial^2 u}{\partial x^2}$$

subject to the boundary conditions $u(0, t)=0$, $u(l,t) = 0$ $(t > 0)$ *and the initial condition* $u(x,0) = x$, l *being the length of the bar.* [UPTU–2006]

SOLUTION. We have $\dfrac{\partial u}{\partial t} = C^2 \dfrac{\partial^2 u}{\partial x^2}$ ... (1)

Boundary conditions are
$$u(0, t) = 0$$
$$u(l,t) = 0 \ (t > 0)$$
$$u(x,0) = x$$

On solving (1), we get

$$u = C_1 \, e^{-p^2 C^2 t}(C_2 \cos px + C_3 \sin px) \qquad \text{... (2)}$$

Putting $x = 0$ and $u = 0$ in (2), we get

$$0 = C_1 \, e^{-p^2 C^2 t}(C_2) \ \Rightarrow \ C_2 = 0$$

Putting $C_2 = 0$ in (2), we get

$$u = C_1 e^{-p^2 C^2 t} C_3 \sin px \qquad \text{... (3)}$$

Again putting $x = l$, $u = 0$ in (3), we get

$$0 = C_1 e^{-p^2 C^2 t} C_3 \sin pl \ \Rightarrow \ \sin pl = 0 = \sin n\pi$$

$$\Rightarrow \qquad pl = n\pi \ \Rightarrow \ p = \frac{n\pi}{l}, \ n \text{ is any integer.}$$

Hence, (3) becomes

$$u = C_1 C_3 e^{-\frac{n^2 C^2 \pi^2}{l^2} t} \cdot \sin \frac{n\pi x}{l}$$

$$= b_n e^{-\frac{n^2 C^2 \pi^2}{l^2} t} \sin \frac{n\pi}{l} x \qquad \qquad \ldots (4)$$

On putting $t = 0$ and $u = x$ in (4), we get

$$x = b_n \sin \frac{n\pi}{l} x$$

General solution is $x = \sum_{n=1}^{\infty} b_n \sin \frac{n\pi}{l} x$.

Now, $\quad b_n = \frac{2}{l} x \sin \frac{n\pi x}{l} dx = \frac{2}{l} \left[x \cdot \frac{l}{n\pi} \left(-\cos \frac{n\pi x}{l} \right) - (1) \left(-\frac{l^2}{n^2 \pi^2} \sin \frac{n\pi x}{l} \right) \right]_0^l$

$$= \frac{2}{l} \left[\left(l \cdot \frac{l}{n\pi} (-\cos n\pi) + \frac{l^2}{n^2 \pi^2} \sin n\pi \right) - 0 \right]$$

$$= \frac{2}{l} \left[-\frac{l^2}{n\pi} (-1)^n \right] = (-1)^{n+1} \cdot \frac{2l}{n\pi} .$$

Putting the value of b_n in (4), we get

$$u = \frac{2l}{\pi} \sum_{n=1}^{\infty} \frac{(-1)^{n+1}}{n} \sin \frac{n\pi x}{l} e^{-\frac{n^2 C^2 \pi^2}{l^2} t} .$$

EXAMPLE 2. *An insulated rod of length l has its ends A and B maintained at $0°C$ and $100°C$ respectively until steady state conditions prevail. If B is suddenly reduced to $0°C$ and maintained at $0°C$, find the temperature at a distance x from A at time t.*

[UPTU–2004, 05; GBTU (AG)–2011; UKTU–2011]

SOLUTION. The initial temperature of the rod can be written as

$$u(x, t) = 0 + \frac{100}{l} x = \frac{100}{l} x$$

While in steady state, the temperature distribution can be written as

$$u(x, t) = 0 + \frac{0}{l} x = 0$$

To find u in the intermediate period, calculating time from the instant when the end temperature were changed

$$u = u_1(x) + u_2(x)$$

where $u_2(x)$ is temperature after a sufficient long time and $u_1(x, t)$ is the transient temperature distribution tending to zero as $t \to \infty$. Hence, $u_2(x) = 0$. Also, $u_1(x, t)$ satisfies one-dimensional heat flow

$$C^2 \frac{\partial^2 u}{\partial x^2} = \frac{\partial u}{\partial t}$$

Thus, $\quad u = (C_1 \cos px + C_2 \sin px) e^{-C^2 p^2 t} \qquad \qquad \ldots (1)$

On putting $x = 0, u = 0$ in (1), we get

$$0 = C_1 e^{-p^2 C^2 t} \Rightarrow C_1 = 0$$

On putting $C_1 = 0$ in (1), we get

$$u = C_2 \sin px e^{-C^2 p^2 t} \qquad \qquad \text{... (2)}$$

On putting $x = l, u = 0$ in (2), we get

$$0 = C_2 \sin pl \, e^{-p^2 C^2 t}$$

$\Rightarrow \qquad \qquad \sin pl = 0 = \sin n\pi$

$\Rightarrow \qquad \qquad pl = n\pi \qquad \qquad \Rightarrow \qquad p = \dfrac{n\pi}{l}$

On putting the value of P in (2), we get

$$u = C_2 \sin \frac{n\pi x}{l} e^{-\frac{n^2 \pi^2 C^2}{l^2} t} \qquad \qquad \text{... (3)}$$

On putting $t = 0$, $u = \dfrac{100}{l} x$ in (3), we get

$$\frac{100x}{l} = C_2 \sin \frac{n\pi x}{l}$$

$$C_2 = \frac{2}{l} \int_0^l \frac{100}{l} . x . \sin \frac{n\pi x}{l} dx = \frac{200}{l^2} \int_0^l x \frac{\sin n\pi x}{l} dx$$

$$C_2 = \frac{200}{l^2} \left[-\frac{xl}{n\pi} \cos \frac{n\pi x}{l} - (-1) \frac{l^2}{n^2 \pi^2} \sin \frac{n\pi x}{l} \right]_0^l$$

$$C_2 = \frac{200}{l^2} \left[-\frac{l^2}{n\pi} \cos n\pi \right] \Rightarrow C_2 = -\frac{200}{n\pi} (-1)^n$$

On putting the value of in (3), we get

$$u = -\frac{200}{n\pi} (-1)^n \sin \frac{n\pi x}{l} e^{-\frac{n^2 \pi^2 C^2}{l^2} t}$$

$$u = (-1)^n \frac{200}{n\pi} . \sin \frac{n\pi x}{l} e^{-\frac{n^2 \pi^2 C^2}{l^2} t} .$$

EXAMPLE 3. *Find by method of separation of variables, the solution u(x, t) of separation of variables,*

$$\frac{\partial u}{\partial t} = 3 \frac{\partial^2 u}{\partial x^2}, \ t > 0, \ 0 < x < 2 \ , \ u(0, t) = 0,$$

$$u(2, t) = 0, \ t > 0, \ u(x, 0) = x, \ 0 < x < 2.$$

SOLUTION. The given equation is

$$\frac{\partial u}{\partial t} = 3 \frac{\partial^2 u}{\partial x^2} \qquad \qquad \text{... (1)}$$

Suppose (1) has the solution of the form

$$u = X(x) T(t) \qquad \qquad \text{... (2)}$$

where X and T respectively the function of x and t alone. Putting the value from (2) in (1), we get

$$XT' = 3X''T$$

$$\Rightarrow \quad \frac{X''}{X} = \frac{T'}{3T} = p^2 \text{ (say)} \quad \quad \dots (3)$$

where p is a separation constant.

From (3), we can deduce that

$$\frac{X''}{X} = \frac{T'}{3T} = p^2$$

$$\Rightarrow \quad X'' - p^2 X = 0 \quad \quad \dots (4)$$

$$T' - 3p^2 T = 0 \quad \quad \dots (5)$$

On solving (4) and (5), we get

$$T = C_3 e^{3p^2 t} \text{ and } X = C_1 e^{px} + C_2 e^{-px}$$

which gives $u(x, t) = e^{3p^2 t} (C_1' e^{px} + C_2' e^{-px})$

where $C_1' = C_1 C_3$ and $C_2' = C_2 C_3$

Now, we shall use the given boundary conditions

$$u(0, t) = e^{3p^2 t} (C_1' + C_2') = 0$$

$$\Rightarrow \quad C_2' = -C_1'$$

$$\therefore \quad u(x, t) = C_1' e^{3p^2 t} (e^{-px} - e^{-px}) \quad \quad \dots (6)$$

Further, $u(x, t) = e^{3p^2 t} (C_1' e^{2p} + C_2' e^{-2p}) = 0$

$$\Rightarrow \quad C_1' e^{2p} + C_2' e^{-2p} = 0 \quad \quad \Rightarrow \quad C_1' e^{2p} - C_1' e^{-2p} = 0$$

$$\Rightarrow \quad C_1' = 0, \quad C_2' = 0$$

$$\Rightarrow \quad u(x, t) = 0, \text{ which is absurd.}$$

Further, let $p = 0$, then we have

$$\frac{X''}{X} = \frac{T'}{3T} = 0 \quad \quad \Rightarrow \quad X'' = 0 \text{ and } T' = 0$$

$$\Rightarrow \quad X = C_1 x + C_2, \quad T = C_3$$

$$\Rightarrow \quad u(x, t) = C_3 (C_1 x + C_2) = C_1' x + C_2$$

$$\Rightarrow \quad u(x, t) \text{ is independent of } t, \text{ which is not admissible.}$$

Now, let p be negative, *i.e.*, $p = -k^2$.

Then, we have

$$\frac{X''}{X} = \frac{T'}{3T} = -k^2$$

$$\Rightarrow \quad X'' + k^2 X = 0 \text{ and } T' + 3k^2 T = 0$$

On solving these equations, we have

$$X = C_1 \cos kx + C_2 \sin kx \text{ and } T = C_3 e^{-3k^2 t}$$

$$\Rightarrow \quad u(x, t) = e^{-3k^2 t} (C_1' \cos kx + C_2' \sin kx)$$

Now, $u(x, t) = 0$

$$\Rightarrow \quad 0 = e^{-3k^2 t} C_1' \Rightarrow C_1' = 0$$

$$\Rightarrow \qquad u(x, t) = C_1' \sin kx \, e^{-3k^2 t} \qquad \Rightarrow \qquad u(2, t) = 0 = C_2' \sin 2k e^{-3k^2 t}$$

$$\Rightarrow \qquad \sin 2k = 0$$

$$\Rightarrow \qquad 2k = n\pi; \quad n = 1, 2, \ldots\ldots \quad \Rightarrow \qquad k = \frac{n\pi}{2}; \; n = 1, 2, \ldots\ldots$$

Therefore, $\qquad u(x, t) = C_2' \sin\dfrac{n\pi x}{2} e^{-3n^2\pi^2 t/4}$

$$= b_n \sin\frac{n\pi x}{2} e^{-3n^2\pi^2 t/4} = u_n(x, t) \quad \text{(say)}$$

Hence, the general solution of (1) is given by

$$u(x, t) = \sum_{n=1}^{\infty} b_n \sin\frac{n\pi x}{2} e^{-3n^2\pi^2 t/4}.$$

EXAMPLE 4. *A rod of length l with insulated sides is initially at a uniform temperature μ_0. Its ends are suddenly cooled to 0°C and are kept at that temperature. Show that the temperature function u(x, t) is given by* $u(x, t) = \sum\limits_{n=1}^{\infty} b_n \sin\dfrac{n\pi x}{l} e^{-c^2\pi^2 n^2 t/l^2}$,

where b_n is given by $u_0 = \sum\limits_{n=1}^{\infty} b_n \sin\dfrac{n\pi x}{l}$ \hfill (GBTU–2010, 11)

SOLUTION. The heat equation is given by

$$\frac{\partial u}{\partial t} = C^2 \frac{\partial^2 u}{\partial x^2} \qquad \ldots (1)$$

Suppose that solution of (1) is of the form
$$u(x, t) = X(x) \, T(t) \qquad \ldots (2)$$
where X and T are respectively the functions of x and t alone. Putting the value from (2) in (1), we get

$$X\frac{dT}{dt} = C^2 T \frac{d^2 X}{dx^2}$$

$$\Rightarrow \qquad \frac{1}{C^2 T}\frac{dT}{dt} = \frac{1}{X}\frac{d^2 X}{dx^2} = -p^2 \quad \text{(say)} \qquad \ldots (3)$$

where p is a separation constant.
From (3), we can find

$$\frac{1}{C^2 T}\frac{dT}{dt} = -p^2$$

$$\Rightarrow \qquad \frac{dT}{dt} = -p^2 C^2 T \qquad \ldots (4)$$

and $\qquad \dfrac{1}{X}\dfrac{d^2 X}{dx^2} = -p^2 \qquad \ldots (5)$

Solving (4) and (5), we get

$$T = C_1 e^{-p^2 C^2 t} \text{ and } X = (C_2 \cos px + C_3 \sin px)$$

which gives

$$u(x, t) = e^{-p^2 C^2 t} (C_1' \cos px + C_2' \sin px)$$

Now, $\quad u(0, t) = 0$

$\Rightarrow \qquad\qquad 0 = C_1' e^{-p^2 c^2 t} \Rightarrow C_1' = 0$

$\Rightarrow \qquad\qquad u(x, t) = e^{-p^2 c^2 t} \cdot C_2' \sin px$

Now, $\qquad u(l, t) = 0 \Rightarrow 0 = e^{-p^2 c^2 t}(C_1' \cos pl + C_2' \sin pl)$

$\Rightarrow \qquad\qquad \sin pl = 0 \quad \Rightarrow \quad p = \dfrac{n\pi}{l}; \ n = 1, 2, \ldots\ldots$

Therefore,

$$u(x, t) = e^{-p^2 c^2 t} \sin\frac{n\pi x}{l} C_2' = e^{-n^2 \pi^2 c^2 t / l^2} \sin\frac{n\pi x}{l} C_2'$$

$$= b_n e^{-n^2 \pi^2 c^2 t / l^2} \sin\frac{n\pi x}{l}$$

Hence, the general solution is given by

$$u(x, t) = \sum_{n=1}^{\infty} u_n(x, t) = \sum_{n=1}^{\infty} b_n e^{-n^2 \pi^2 c^2 t / l^2} \sin\frac{n\pi x}{l}$$

Finally, using $u = u_0$ when $t = 0$, we get

$$u_0 = \sum_{n=1}^{\infty} b_n \sin\frac{n\pi x}{l}.$$

EXAMPLE 5. *Find the temperature in a bar of length 2 whose ends are kept at zero and lateral surface insulated if the initial temperature is* $\sin\dfrac{\pi x}{2} + 3\sin\dfrac{5\pi x}{2}$.

[MTU–2011, UPTU (CO)–2007, 09]

SOLUTION. Let $u(x, t)$ be the temperature in the bar. The boundary conditions are
$$u(0, t) = 0 = u(2, t) \text{ for any } t \qquad \ldots (1)$$
The initial condition is
$$u(x, 0) = \sin\frac{\pi x}{2} + 3\sin\frac{5\pi x}{2} \qquad \ldots (2)$$
One dimensional heat flow equation is
$$\frac{\partial u}{\partial t} = C^2 \frac{\partial^2 u}{\partial x^2} \qquad \ldots (3)$$
Its solution is
$$u(x, t) = (C_1 \cos px + C_2 \sin px) C_3 e^{-c^2 p^2 t} \qquad \ldots (4)$$
$$u(0, t) = 0 = C_1 C_3 e^{-C^2 p^2 t} \qquad\qquad \text{(On using (1))}$$
$$\Rightarrow \qquad C_1 = 0$$
$\therefore$ From (4) $\quad u(x, t) = C_2 C_3 \sin px \, e^{-C^2 p^2 t} \qquad \ldots (5)$
$$u_1(2, t) = 0 = C_2 C_3 \sin 2p \, e^{-C^2 p^2 t} \qquad\qquad \text{(On using (1))}$$
$$\Rightarrow \qquad \sin 2p = 0 = \sin n\pi \qquad \therefore \qquad p = \frac{n\pi}{2}, n \in I$$
Hence from (5)
$$u(x, t) = b_n \sin\frac{n\pi x}{2} e^{\frac{-n^2 \pi^2 C^2 t}{4}} \qquad\qquad (\because C_2 C_3 = b_n)$$

The most general solution is

$$u(x,t) = \sum_{n=1}^{\infty} b_n \sin\frac{n\pi x}{2} e^{\frac{-n^2\pi^2 c^2 t}{4}} \qquad \dots(6)$$

$$u(x,0) = \sin\left(\frac{\pi x}{2}\right) + 3\sin\left(\frac{5\pi x}{2}\right) = \sum_{n=1}^{\infty} b_n \sin\frac{n\pi x}{2}$$

$$= b_1 \sin\left(\frac{\pi x}{2}\right) + b_2 \sin\left(\frac{2\pi x}{2}\right) + \dots + b_5 \sin\left(\frac{5\pi x}{2}\right) + \dots$$

Comparing, we get $b_1 = 1$ and $b_5 = 3$
Hence from (6),

$$u(x,t) = \sin\left(\frac{\pi x}{2}\right) e^{-\pi^2 c^2 t/4} + 3\sin\left(\frac{5\pi x}{2}\right) e^{-25\pi^2 c^2 t/4}$$

EXAMPLE 6. *Solve* $\dfrac{\partial^2 u}{\partial x^2} + \dfrac{\partial^2 u}{\partial y^2} = 0$*, which satisfies the conditions* $u(0, y) = u(l, y) = u(x, 0) = 0$

and $u(x,a) = \sin\dfrac{n\pi x}{l}$. (UPTU–2004; (CO)–2009; GBTU (AG)–2012)

SOLUTION. Consider the heat flow in a metal plate of uniform thickness in the directions parallel to length and breadth of the plate. There is no heat flow along the normal to the plane of the rectangle.
Let $u(x,y)$ be the temperature at any point (x,y) of the plate at time t is given by

$$\frac{\partial u}{\partial t} = C^2\left(\frac{\partial^2 u}{\partial x^2} + \frac{\partial^2 u}{\partial y^2}\right)$$

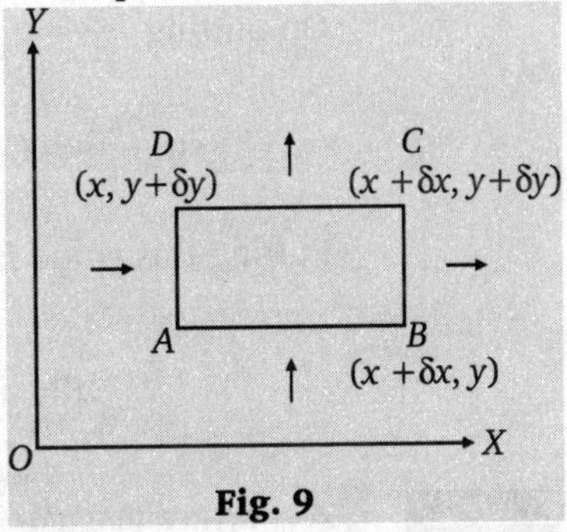

In the steady state, u does not change with t.

$$\frac{\partial u}{\partial t} = 0$$

Let $\qquad u = X(x) \cdot Y(y) \qquad \dots (1)$

Putting the values of $\dfrac{\partial^2 u}{\partial x^2}$ and $\dfrac{\partial^2 u}{\partial y^2}$ in (1),

we have $\quad X''Y + XY'' = 0 \qquad \dots (2)$

Fig. 9

$$\Rightarrow \qquad \frac{X''}{X} = -\frac{-Y''}{Y} = -p^2 \qquad \text{(say)}$$

$$D^2X = -p^2X \quad \Rightarrow \qquad D^2Y = p^2Y$$

$$\Rightarrow \qquad D^2X + p^2X = 0 \quad \Rightarrow \quad D^2Y - p^2Y = 0$$

$$\Rightarrow \qquad (D^2 + p^2)X = 0 \quad \Rightarrow \quad (D^2 - p^2)Y = 0$$

A.E. is $D^2 + p^2 = 0$ A.E. is $\quad D^2 - p^2 = 0$

$$\Rightarrow \qquad D^2 = -p^2 \quad \Rightarrow \qquad D^2 = p^2$$

$$\Rightarrow \qquad D = \pm ip \quad \Rightarrow \qquad D = \pm p$$

$$X = C_1\cos px + C_2 \sin px \qquad Y = C_3 e^{py} + C_4 e^{-py}$$

Putting the values of X and Y in (1), we have

$$u = (C_1 \cos px + C_2 \sin px)(C_3 e^{py} + C_4 e^{-py}) \qquad \text{... (3)}$$

Putting $x = 0$, $u = 0$ in (3), we have

$$0 = C_1(C_3 e^{py} + C_4 e^{-py}) \Rightarrow C_1 = 0$$

(3) is reduced to $u = C_2 \sin px(C_3 e^{py} + C_4 e^{-py})$ ... (4)

On putting $x = l$, $u = 0$ in (4), we have

$$0 = C_2 \sin pl(C_3 e^{py} + C_4 e^{-py}) \Rightarrow C_2 \neq 0$$

$$\therefore \qquad \sin pl = 0 = \sin n\pi \Rightarrow Pl = n\pi$$

or $\qquad P = \dfrac{n\pi}{l}$

Now, (4) becomes

$$u = C_2 \sin\frac{n\pi x}{l}\left[C_3 e^{\frac{n\pi y}{l}} + C_4 e^{-\frac{n\pi y}{l}}\right] \qquad \text{... (5)}$$

On putting $u = 0$ and $y = 0$ in (5), we have $0 = C_2 \sin\dfrac{n\pi x}{l}(C_3 + C_4)$

$$C_3 + C_4 = 0 \qquad \text{or} \qquad C_3 = -C_4$$

(5) becomes $u = C_2 C_3 \sin\dfrac{n\pi x}{l}\left(e^{\frac{n\pi y}{l}} - e^{-\frac{n\pi y}{l}}\right)$

On putting $y = a$ and $u = \sin\dfrac{n\pi x}{l}$ in (6), we have

$$\sin\frac{n\pi x}{l} = C_2 C_3 \sin\frac{n\pi x}{l}\left(e^{\frac{n\pi a}{l}} - e^{-\frac{n\pi a}{l}}\right), \text{ i.e., } C_2 C_3 = \frac{1}{e^{\frac{n\pi a}{l}} - e^{-\frac{n\pi a}{l}}}$$

Putting this value in (6), we have

$$u = \sin\frac{n\pi x}{l}\cdot\frac{e^{\frac{n\pi y}{l}} - e^{-\frac{n\pi y}{l}}}{e^{\frac{n\pi a}{l}} - e^{-\frac{n\pi a}{l}}} \qquad \text{or} \qquad u = \sin\frac{n\pi x}{l}\cdot\frac{\sinh\frac{n\pi y}{l}}{\sin\frac{n\pi a}{l}}.$$

EXAMPLE 7. *A thin rectangular plate whose surface is impervious to heat flow, has at t = 0 an arbitrary distribution of temperature f(x, y). If four edges x = 0, x = a, y = 0, y = b are kept at zero temperature, find the temperature at a point of the plate as t increases.*   [UPTU–2002]

SOLUTION.   We know that the two dimensional heat equation is

$$\frac{\partial u}{\partial t} = C^2\left(\frac{\partial^2 u}{\partial x^2} + \frac{\partial^2 u}{\partial y^2}\right) \qquad \text{... (1)}$$

As per given, the initial temperature of the plate is $f(x, y)$ and the temperature of the four edges of the plate are kept at $0°$.

Therefore, the required boundary conditions are

(i) $u(0, y, t) = 0$,   (ii) $u(a, y, t) = 0$,

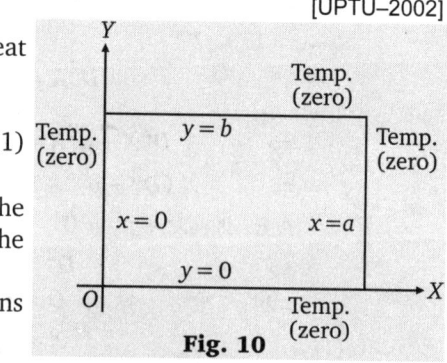

Fig. 10

(iii) $u(x, 0, t) = 0$,   (iv) $u(x, b, t) = 0$.

Also, the initial condition is given by

$$u(x, y, 0) = f(x, y) \qquad \text{... (2)}$$

Suppose that solution of (1) is of the form

$$u(x, y, t) = X(x)\, Y(y)\, T(t) \qquad \text{... (3)}$$

where X, Y and T are respectively the functions of x, y and t alone. Putting the values of (3) in (1), we get

$$\frac{1}{C^2 T} = \frac{1}{X}\frac{d^2 X}{dx^2} + \frac{1}{Y}\frac{d^2 Y}{dy^2} \qquad \text{... (4)}$$

If (3) satisfies (1), we have the following three possibilities :

(a) $\dfrac{1}{X}\dfrac{d^2 X}{dx^2} = 0, \quad \dfrac{1}{Y}\dfrac{d^2 Y}{dy^2} = 0, \quad \dfrac{1}{C^2 T}\dfrac{dT}{dt} = 0$

(b) $\dfrac{1}{X}\dfrac{d^2 X}{dx^2} = p_1^2, \dfrac{1}{Y}\dfrac{d^2 Y}{dy^2} = p_2^2, \dfrac{1}{C^2 T}\dfrac{dT}{dt} = p^2$

(c) $\dfrac{1}{X}\dfrac{d^2 X}{dx^2} = -p_1^2, \dfrac{1}{Y}\dfrac{d^2 Y}{dy^2} = -p_2^2, \dfrac{1}{C^2 T}\dfrac{dT}{dt} = -p^2$

where $p^2 = p_1^2 + p_2^2$.

It can be easily verified that differential equation (c) only gives the solution. In this case, the general solution is given by

$$X = A_1 \cos p_1 x + B_1 \sin p_1 x \,;$$

$$Y = A_2 \cos p_2 y + B_2 \sin p_2 y \,; \quad T = A_3\, e^{-C^2 p^2 t}$$

Therefore

$$u(x, y, t) = (A_1 \cos p_1 x + B_1 \sin p_1 x)\, (A_2' \cos p_2 y + B_2' \sin p_2 y)\, e^{-C^2 p^2 t} \quad \text{... (5)}$$

Using boundary condition (i), we get

$$u(0, y, t) = A_1 (A_2' \cos p_2 y + B_2' \sin p_2 y)\, e^{-C^2 p^2 t} = 0$$

$\Rightarrow \qquad A_1 = 0$

Now, using boundary condition (ii), we get

$$u(a, y, t) = B_1 \sin p_1 a (A_2' \cos p_2 y + B_2' \sin p_2 y)\, e^{-C^2 p^2 t} = 0$$

$\Rightarrow \qquad \sin p_1 a = 0 \Rightarrow p_1 a = m\pi$

$$p_1 = \frac{m\pi}{a}\,; \; m = 1, 2, 3, \dots$$

Similarly, by using boundary condition (iii) and (iv), we get

$$A_2' = 0 \text{ and } P_2 = \frac{n\pi}{b}, \; n = 1, 2, 3, \dots$$

Thus, we have

$$u_{mn}(x, y, t) = A_{mn}\, e^{-C^2 p_{mn}^2 t} \cdot \sin\frac{m\pi}{a} \sin\frac{n\pi}{b} y$$

where, $p^2 = p_{mn}^2 = \pi^2 \left(\dfrac{m^2}{a^2} + \dfrac{n^2}{b^2} \right)$

which gives

$$u(x, y, t) = \sum_{m=1}^{\infty} \sum_{n=1}^{\infty} A_{mn} e^{-C^2 p_{mn}^2 t} . \sin\frac{m\pi}{a} x \sin\frac{n\pi}{b} y$$

Finally, to find the solution which satisfies the initial conditions also, we proceed as follows :

$$u(x, y, 0) = \sum_{m=1}^{\infty} \sum_{n=1}^{\infty} A_{mn} e^{-C^2 p_{mn}^2 t} . \sin\frac{m\pi}{a} x \sin\frac{n\pi}{b} y = f(x, y)$$

The LHS is the double Fourier sine series of $f(x, y)$. Therefore,

$$A_{mn} = \frac{2}{a} . \frac{2}{b} \int_{x=0}^{a} \int_{y=0}^{b} f(x, y) \sin\frac{m\pi}{a} x \sin\frac{n\pi}{b} y \, dx \, dy$$

EXAMPLE 8. *A uniform rod, 20 cm in length is insulated over its sides. Its ends are kept at $0°C$. Its initial temperature is* $\sin\left(\dfrac{\pi x}{20}\right)$ *at a distance x from an end, find temperature $u(x, t)$ at time t. Given that* $\dfrac{\partial u}{\partial t} = a^2\left(\dfrac{\partial^2 u}{\partial x^2}\right)$.

SOLUTION. The given equation is

$$\frac{\partial u}{\partial t} = a^2\left(\frac{\partial^2 u}{\partial x^2}\right) \qquad \text{... (1)}$$

As per given, the boundary conditions are

$$u(0, t) = u(20, t) = 0, \ \forall t \qquad \text{... (2)}$$

$$u(x, 0) = \sin\left(\frac{\pi x}{20}\right) \qquad \text{... (3)}$$

Suppose that the solution of (1) is of the form

$$u(x, t) = X(x) T(t) \qquad \text{... (4)}$$

Putting the values from (4) in (1), we get

$$\frac{X''}{X} = \frac{T'}{a^2 T} = \mu \ (say) \qquad \text{... (5)}$$

where μ is a separation constant.
From (5), we can deduce that

$$X'' - \mu X = 0 \qquad \text{... (6)}$$

and $\qquad\qquad T' = \mu a^2 T \qquad \text{... (7)}$

The new boundary conditions are

$$X(0) = 0 \text{ and } X(20) = 0 \qquad \text{... (8)}$$

Now, we want to solve (6) using boundary conditions (8).
We shall discuss the following three cases :
Case (I). Let $\mu = 0$. In this case, solution of (6) is given by

$$X(x) = Ae^{\lambda x} + Be^{-\lambda x}$$

Using (8), we have $\quad B = 0$
and $\qquad\qquad 20\,A + B = 0 \Rightarrow A = 0$

$\Rightarrow \qquad\qquad X = 0$

$\Rightarrow \quad u = 0$, which does not satisfy (3). Hence, we reject this case.

Case (II). Let $\mu = \lambda^2, \lambda \neq 0$. In this case, solution of (6) is given by
$$X(x) = Ae^{\lambda x} + Be^{-\lambda x}$$
Using boundary condition (8), we have
$$A + B = 0 \quad \text{and} \quad Ae^{20\lambda} + De^{-20\lambda} = 0$$
On solving we get $A = B = 0 \Rightarrow X = 0 \Rightarrow u = 0$.
Therefore, we reject this case also.

Case (III). Let $\mu = -\lambda^2, \lambda \neq 0$. In this case, solution of (6) is given by
Using boundary condition (8), we get
$$A = 0 \quad \text{and} \quad A\cos 20\lambda + B\sin 20\lambda = 0$$

$\Rightarrow \qquad \sin 20\lambda = 0$

$\Rightarrow \qquad 20\lambda = n\pi, \quad n = 1, 2, \ldots\ldots$

$\Rightarrow \qquad \lambda = \dfrac{n\pi}{20}, \quad n = 1, 2, \ldots\ldots$ \hfill ... (9)

Thus, the non-zero solution of equation (6) is given by
$$X_n(x) = B_n \sin\left(\frac{n\pi x}{20}\right)$$
Using (9) in (7), we get
$$\frac{dT}{T} = -\frac{n^2\pi^2 a^2}{400}\,dt$$
whose general solution is given by
$$T_n(t) = D_n\, e^{-(n^2\pi^2 a^2/400)t}$$

Therefore, $\quad u_n(x, t) = X_n(x)\,T_n(t) = E_n \sin\left(\dfrac{n\pi x}{20}\right) e^{-\frac{(n^2\pi^2 a^2)}{400}t}$
are solutions of (1) satisfying (2).
The more general solution of (1) is given by
$$u(x,t) = \sum_{n=1}^{\infty} u_n(x,t) = \sum_{n=1}^{\infty} E_n \sin\left(\frac{n\pi x}{20}\right) e^{-\frac{(n^2\pi^2 a^2)}{400}t}$$
Putting $t = 0$ and using boundary condition (3), we get
$$\sin\frac{\pi x}{20} = \sum_{n=1}^{\infty} E_n \sin\left(\frac{n\pi x}{20}\right)$$
which is a Fourier sine series. Thus the constants E_n are given by
$$E_n = \frac{1}{20}\int_0^{20} \sin\frac{\pi x}{20}\sin\left(\frac{n\pi x}{20}\right)dx, n = 1,2,3,\ldots \hfill ... (10)$$
Here, we discuss the following two cases :

Case I. If $n \neq 1$, then (10) gives
$$E_n = \frac{1}{20}\int_0^{20}\left[\cos\frac{(n-1)\pi x}{20} - \cos\frac{(n+1)\pi x}{20}\right]dx$$
$$= \frac{1}{20}\left[\frac{20}{(n-1)\pi}\sin\frac{(n-1)\pi x}{20} - \frac{20}{(n+1)\pi}\sin\frac{(n+1)\pi x}{20}\right]_0^{20} = 0$$

$\Rightarrow \qquad E_n = 0 \text{ for } n = 2, 3, 4, \ldots$

Case (II). If $n = 1$, then (10) gives

$$E_1 = \frac{1}{20}\int_0^{20}\left(2\sin^2\frac{\pi x}{20}\right)dx = \frac{1}{20}\int_0^{20}\left(1-\cos\frac{2\pi x}{20}\right)dx$$

$$= \frac{1}{20}\left[x - \frac{10}{x}\sin\frac{\pi x}{10}\right]_0^{20} = 1$$

Finally, the general solution is given by

$$u(x, t) = E_1 \sin\frac{\pi x}{20}e^{-\frac{(\pi^2 a^2)}{400}t} = \sin\frac{\pi x}{20}e^{-\frac{(\pi^2 a^2)}{400}t}$$

EXAMPLE 9.   *Solve the one-dimensional heat equation $\dfrac{\partial^2 u}{\partial x^2} = \dfrac{1}{k}\dfrac{\partial u}{\partial t}$ in the region $0 \le x \le \pi$, $t > 0$ when*

   (i) u remains finite as $t \to \infty$,

   (ii) $u = 0$ if $x = 0$ or π, $\forall t$

   (iii) At $t = 0$, $u = x$ for $0 \le x \le \pi/2$ and $u = \pi - x$ for $\dfrac{\pi}{2} < x \le \pi$.

SOLUTION.   The given equation is

$$\frac{\partial u}{\partial t} = a^2\left(\frac{\partial^2 u}{\partial x^2}\right) \qquad \qquad \text{... (1)}$$

As per given, we have

$$u(0, t) = u(20, t) = 0, \ \forall t \ \text{finite quantity as } t \to \infty \qquad \text{... (2)}$$

and boundary conditions are

$$u(0, t) = u(\pi, t) = 0; \ \forall \ t \qquad \qquad \text{... (3)}$$

Also, the initial condition is given by

$$u(x, 0) = \begin{cases} x \ ; \ \text{when } 0 \le x \le \pi/2 \\ \pi - x \ ; \ \text{when } \dfrac{\pi}{2} \le x \le \pi \end{cases} \qquad \text{... (4)}$$

Suppose that solution of (1) is of the form

$$u(x, t) = X(x)\,T(t) \qquad \qquad \text{... (5)}$$

where X and T are respectively the functions of x and t alone. Putting the values from (5) in (1), we get

$$\frac{X''}{X} = \frac{T'}{kT} = \lambda \ \text{(say)} \qquad \qquad \text{... (6)}$$

where λ is a separation constant.

In view of (2), we choose, $\lambda = -n^2$, $n \ne 0$. Then (6) reduces to

$$\frac{X''}{X} = -n^2 \ \Rightarrow \ (D^2 + n^2) X = 0$$

and

$$\frac{T'}{kT} = -n^2 \ \Rightarrow \ \frac{dT}{T} = -n^2 k\,dT$$

On solving, we get

$$X_n(x) = A_n \cos nx + B_n \sin nx \ \text{and} \ T_n(t) = C_n\,e^{-n^2 kt}$$

Therefore

$$u(x, t) = (D_n \cos nx + E_n \sin nx)\, e^{-n^2 kt} \qquad \qquad \dots (7)$$

Putting $x = 0$ and $x = \pi$ in (7) and using (3), we get $0 = D_n$ and

$$0 = D_n \cos n\pi + E_n \sin n\pi$$

$$\Rightarrow \qquad D_n = 0 \text{ and } \sin n\pi = 0$$

Now, $\sin n\pi = 0 \Rightarrow n$ must be an integer.
Therefore, from (7), we have

$$u_n(x, t) = E_n \sin nx \cdot e^{-n^2 kt}, \quad n = 1, 2, 3, \dots \qquad \qquad \dots (8)$$

which are solutions of (1) satisfying the boundary condition (3).
Now, consider the series

$$u(x, t) = \sum_{n=1}^{\infty} u_n(x, t) = \sum_{n=1}^{\infty} E_n \sin nx\, e^{-n^2 kt} \qquad \qquad \dots (9)$$

$$u(x, 0) = \sum_{n=1}^{\infty} E_n \sin nx, \quad n = 1, 2, 3, \dots \qquad \qquad \dots (10)$$

which is a half range Fourier sine series in $(0, \pi)$, therefore E_n is given by

$$E_n = \frac{2}{\pi} \int_0^{\pi} u(x, 0) \sin nx\, dx$$

$$= \frac{2}{\pi} \left[\int_0^{\pi/2} u(x, 0) \sin nx\, dx + \int_{\pi/2}^{\pi} u(x, 0) \sin nx\, dx \right]$$

$$= \int_0^{\pi/2} \frac{2\pi}{x} \sin nx\, dx + \int_{\pi/2}^{\pi} \left(\frac{2}{\pi}\right)(\pi - x)\sin nx\, dx$$

$$= \left[\left(\frac{2x}{\pi}\right)\left(-\frac{\cos nx}{n}\right) - \left(\frac{2}{\pi}\right)\left(-\frac{\sin nx}{x^2}\right) \right]_0^{\pi/2}$$

$$+ \left[\left(\frac{2}{\pi}(\pi - x)\right)\left(-\frac{\cos nx}{n}\right) - \left(-\frac{2}{\pi}\right)\left(-\frac{\sin nx}{n^2}\right) \right]_{\pi/2}^{\pi}$$

$$= -\left(\frac{1}{n}\right)\cos\left(\frac{n\pi}{2}\right) + \left(\frac{2}{\pi n^2}\right)\sin\left(\frac{n\pi}{2}\right) + \frac{1}{n}\cos\left(\frac{n\pi}{2}\right) + \frac{2}{\pi n^2}\sin\left(\frac{n\pi}{2}\right)$$

Thus $E_n = \dfrac{4}{\pi n^2} \sin \dfrac{n\pi}{2}$

$$= \begin{cases} 0 & ; \text{if } n = 2m \text{ and } m = 1, 2, 3, \dots \\[2mm] \dfrac{4(-1)^{m+1}}{\pi(2m-1)^2} & ; \text{if } n = 2m - 1 \text{ and } m = 1, 2, 3, \dots \end{cases}$$

Hence, the required solution is given by

$$u(x, t) = \frac{4}{\pi} \sum_{m=1}^{\infty} \frac{(-1)^{m+1}}{(2m-1)^2} \sin(2m-1)\, x\, e^{(2m-1)^2 kt} \quad m = 1, 2, \dots$$

EXAMPLE 10. *Solve the boundary value problem* $\dfrac{\partial^2 u}{\partial x^2} = \dfrac{l}{k}\dfrac{\partial u}{\partial t}$ *satisfying the condition* $u(0, t) =$ $u(l, t) = 0$ *and* $u(x, 0) = lx - x^2.$

SOLUTION. Proceeding same as in example (4) by taking $a=l$ and $f(x) = u(x, 0) = lx - x^2$, we get

$$u(x, t) = \sum_{n=1}^{\infty} E_n \sin\left(\frac{n\pi x}{l}\right) e^{-(n^2\pi^2 kt)/l^2} \qquad \dots (1)$$

where

$$E_n = \frac{2}{l} \int_0^l f(x) \sin\frac{n\pi x}{l} dx$$

$$= \frac{2}{l} \int_0^l (lx - x^2) \sin\frac{n\pi x}{l} dx$$

$$= \frac{2}{l}\left[(lx - x^2)\left\{ \frac{-\cos(n\pi x / l)}{n\pi / l} \right\} - (l - 2x) \right.$$

$$\left. \left\{ \frac{-\sin(n\pi x / l)}{(n\pi)^2 / l^2} \right\} + (-2)\left\{ \frac{\cos(n\pi x) / l}{(n\pi)^3 / l^3} \right\} \right]_0^l$$

$$= \frac{2}{l}\left[-\left(\frac{2l^3}{n^3\pi^3} \right)\cos n\pi + \left(\frac{2l^3}{n^3\pi^3} \right) \right] = \frac{4l^2}{n^3\pi^3}[1 - (-1)^n]$$

Thus,

$$E_n = \begin{cases} (8l^2) / (2m-1)^3 \pi^3 ; \text{ if } n = 2m - 1, m = 1, 2, 3, \dots \\ 0 ; \text{ if } n = 2m(\text{even}), \text{ where } m = 1, 2, 3, \dots \end{cases}$$

Hence, $u(x, t) = \dfrac{8l^2}{\pi^3} \sum_{m=1}^{\infty} \dfrac{1}{(2m-1)^3} \sin\dfrac{(2m-1)\pi x}{l} e^{-\left[(2m-1)^2\pi^2 kt\right]/l^2}$

EXAMPLE 11. *Solve the boundary value problem* $\dfrac{\partial^2 u}{\partial x^2} = \dfrac{l}{k}\dfrac{\partial u}{\partial t}$ *satisfying the conditions* $u(0, t) =$ $u(l, t) = 0$ *and* $u(x, 0) = x$ *when* $0 \leq x \leq l/2$ $u(x, 0) = (l-x)$ *when* $l/2 \leq x \leq l$.

SOLUTION. Proceeding same as in Example (4) by taking $a = l$ and we get

$$u(x, t) = \sum_{n=1}^{\infty} E_n \sin\left(\frac{n\pi x}{l}\right) e^{-(n^2\pi^2 kt)/l^2} \qquad \dots (1)$$

It is also given that

$$u(x, 0) = \begin{cases} x & ; \text{ when } 0 \leq x \leq l/2 \\ l - x & ; \text{ when } l/2 \leq x \leq l \end{cases} \qquad \dots (2)$$

Putting $t = 0$ in (1), we get

$$u(x, 0) = \sum_{n=1}^{\infty} E_n \sin\frac{n\pi x}{l}$$

which is a half range Fourier sine series in $(0, l)$. Thus, E_n is given by

$$E_n = \frac{2}{l}\int_0^l u(x, 0) \sin\frac{n\pi x}{l} dx$$

$$= \frac{2}{l}\left[\int_0^{l/2} u(x, 0)\sin\frac{n\pi x}{l} dx + \int_{l/2}^l u(x, 0)\sin\frac{n\pi x}{l} dx \right]$$

$$= \int_0^{l/2} \frac{2x}{l}\sin\frac{n\pi x}{l} dx + \int_{l/2}^l \frac{2}{l}(l - x)\sin\frac{n\pi x}{l} dx$$

$$= \left[\left(\frac{2x}{l} \right) \left(-\frac{\cos(n\pi x)/l}{(n\pi)/l} \right) - \left(\frac{2}{l} \right) \left(-\frac{\sin(n\pi x)/l}{(n\pi)^2/l^2} \right) \right]_0^{l/2}$$

$$+ \left[\left(\frac{2(l-x)}{l} \right) \left(-\frac{\cos(n\pi x)/l}{(n\pi)/l} \right) - \left(-\frac{2}{l} \right) \left(-\frac{\sin(n\pi x)/l}{(n\pi)^2/l^2} \right) \right]_{l/2}^{l}$$

$$= -\left(\frac{l}{n\pi} \right) \cos\left(\frac{n\pi}{2} \right) + \left(\frac{2l}{n^2\pi^2} \right) \sin\left(\frac{n\pi}{2} \right) + \left(\frac{l}{n\pi} \right) \cos\left(\frac{n\pi}{2} \right) + \left(\frac{2l}{n^2\pi^2} \right) \sin\frac{n\pi}{2}$$

Thus, $E_n = \frac{4l}{n^2\pi^2} \sin\frac{n\pi}{2}$

$$= \begin{cases} 0 \ ; \text{if } n = 2m \text{ and } m = 1,2,3,\ldots \\ \dfrac{4l}{(2m-1)^2 \pi^2} \ ; \text{ if } n = 2m-1, \ m = 1,2,\ldots \end{cases}$$

Hence, by (1), we have

$$u(x,t) = \frac{4l}{\pi^2} \sum_{m=1}^{\infty} \frac{1}{(2m-1)^2} \sin\frac{(2m-1)\pi x}{l} e^{-\left[(2m-1)^2\pi^2 kt\right]/l^2}$$

EXAMPLE 12. *Find the solution of one dimensional heat equation satisfying the following boundary conditions*

(i) *T is bounded as* $t \to \infty$ (ii) $\dfrac{\partial T}{\partial x}\Big|_{x=0} = 0 \ ; \ \forall t$

(iii) $\dfrac{\partial T}{\partial x}\Big|_{x=a} = 0 \ ; \ \forall t$ (iv) $T(x,0) = x(a-x); \ 0 < x < a$

SOLUTION. The general acceptable solution of the given partial differential equation is :

$$T(x,t) = \exp(-\alpha\lambda^2 t)(A\cos\lambda x + B\sin\lambda x)$$

$$\Rightarrow \quad \frac{\partial T}{\partial x} = \exp(-\alpha\lambda^2 t)(-A\lambda\sin\lambda x + B\lambda\cos\lambda x) \qquad \ldots (1)$$

Using boundary condition (ii) in (1), we get $B = 0$.

By using boundary condition (iii), we get

$$\sin\lambda a = 0 \quad \Rightarrow \quad \lambda a = n\pi \ ; \ n = 0, 1, 2,\ldots$$

Therefore, the more general solution is given by

$$T(x,t) = \sum A_n \exp(-\alpha\lambda^2 t)\cos\lambda x$$

$$= \sum_{n=0}^{\infty} A_n \exp\left[-\alpha\left(\frac{n\pi}{a} \right)^2 t \right] \cos\left(\frac{n\pi}{a} \right) x$$

Using boundary condition (iv), we have
$$T(x,0) = x(a-x)$$

$$= A_0 + \sum_{n=1}^{\infty} A_n \exp\left[-\alpha\left(\frac{n\pi}{a} \right)^2 t \right] \cos\left(\frac{n\pi}{a} \right) x$$

where, $A_0 = \dfrac{2}{\pi}\int_0^a (ax-x^2)dx = \dfrac{a^2}{6}$

$$A_n = \frac{2}{\pi}\int_0^a (ax-x^2)\cos\left(\frac{n\pi x}{a}\right)dx$$

$$= \frac{2a^2}{n^2\pi^2}(1+\cos n\pi) = \frac{2a^2}{n^2\pi^2}[1+(-1)^n]$$

$$\Rightarrow \qquad A_n = \begin{cases} -\dfrac{4a^2}{n^2\pi^2}\;;\; \text{if } n \text{ is even} \\ 0 \qquad\; ;\; \text{if } n \text{ is odd} \end{cases}$$

Hence, the required solution is given by

$$T(x,t) = \frac{a^2}{6} - \frac{4a^2}{\pi^2}$$

$$\sum_{n=2,4,\dots}^{\infty} \frac{1}{n^2}\cos\left(\frac{n\pi}{a}\right)x.\exp\left[-\alpha\left(\frac{n\pi}{a}\right)^2 t\right]$$

EXAMPLE 13. *Find the current i and voltage V in a transmission line of length l, t seconds after the ends are suddenly grounded given that* $i(x,0)=i_0$, $V(x,0)=V_0\sin\left(\dfrac{\pi x}{l}\right)$ *and that R and G are negligible.* [UPTU–2008]

SOLUTION. We have $\dfrac{\partial^2 V}{\partial x^2} = LC\dfrac{\partial^2 V}{\partial t^2}$

Let $V = XT$, where X and T are the functions of x and t respectively.

$$\frac{\partial^2 V}{\partial x^2} = T\frac{\partial^2 X}{\partial x^2} \text{ and } \frac{\partial^2 V}{\partial t^2} = X\frac{d^2 T}{dt^2}$$

or $\qquad \dfrac{\dfrac{d^2 X}{dx^2}}{X} = LC\dfrac{\dfrac{d^2 T}{dt^2}}{T} = -p^2$ (say)

Since the initial conditions suggest the values of V and i are periodic functions.

$$X = C_1\cos px + C_2\sin px$$

and $\qquad T = C_3\cos\dfrac{pt}{\sqrt{LC}} + C_4\sin\dfrac{pt}{\sqrt{LC}}$

so $\qquad V = (C_1\cos px + C_2\sin px)$

$$\left(C_3\cos\frac{pt}{\sqrt{LC}} + C_4\sin\frac{pt}{\sqrt{LC}}\right) \qquad\qquad \dots (1)$$

where, $t = 0$, $V = V_0\sin\dfrac{\pi x}{l}$

$$V_0 \sin\frac{\pi x}{l} = (C_1 \cos px + C_2 \sin px) C_3 \qquad \ldots (2)$$

On equating the coefficients, we get

$$C_1 C_3 = 0 \Rightarrow C_1 = 0 \text{ and } C_2 C_3 = V_0, \ p = \pi/l$$

becomes (1)

$$V = \sin\frac{\pi x}{l}\left[V_0 \cos\frac{pt}{\sqrt{LC}} + C_2 C_4 \sin\frac{pt}{\sqrt{LC}}\right] \qquad \ldots (3)$$

Now, when $t = 0, \ i = i_0$ (constant).

Hence,

$$\frac{\partial i}{\partial x} = 0$$

$$\frac{\partial i}{\partial x} = -C\frac{\partial V}{\partial t}$$

$\therefore$

$$\frac{\partial V}{\partial t} = 0 \quad \text{when} \quad t = 0$$

Now,

$$\frac{\partial V}{\partial t} = \sin\frac{\pi x}{l}\left(\frac{p}{\sqrt{LC}}\right)\left[-V_0 \sin\frac{pt}{\sqrt{LC}} + C_2 C_4 \cos\frac{pt}{\sqrt{LC}}\right] \qquad \ldots (4)$$

On putting $\dfrac{\partial V}{\partial t} = 0$ and $t = 0$ in (4), we get

$$C_2 C_4 = 0 \Rightarrow C_4 = 0$$

Now (3) becomes, $V = V_0 \sin\dfrac{\pi x}{l}\cos\dfrac{\pi t}{\sqrt{LC}}$

$$\frac{\partial V}{\partial x} = \frac{\pi}{l}V_0 \cos\frac{\pi x}{l}\cos\frac{\pi t}{\sqrt{LC}} = -L\frac{\partial i}{\partial t} \qquad \ldots (5)$$

$$\frac{\partial V}{\partial t} = -\frac{V_0 \pi}{l\sqrt{LC}}\sin\frac{\pi x}{l}\sin\frac{\pi t}{l\sqrt{LC}} = -\frac{1}{C}\frac{\partial t}{\partial x} \qquad \ldots (6)$$

Integrating (5) and (6), we get

$$i = -V_0\sqrt{\frac{C}{L}}\cos\frac{\pi x}{l}\sin\frac{\pi t}{l\sqrt{LC}} + f(x)$$

and

$$i = -V_0\sqrt{\frac{C}{L}}\cos\frac{\pi x}{l}\sin\frac{\pi t}{l\sqrt{LC}} + F(t)$$

$\therefore \ f(x)$ and $F(t)$ must be constant only, since $i = i_0$ when $t = 0$

$\therefore$ $\qquad$ Constant $= \ i_0 = f(x)$

Hence, $\qquad i = i_0 - V_0\sqrt{\dfrac{C}{L}}\cos\dfrac{\pi x}{l}\sin\dfrac{\pi t}{l\sqrt{LC}}.$

EXERCISE 8.4

1. A rod of length l with insulated sides, is initially at a uniform temperature u_0. Its ends are suddenly cooled at $0°C$ and are kept at that temperature. Find the temperature distribution $u(x, t)$.

2. Find temperature distribution $y(x, t)$ in a uniform bar of unit length, whose one end is kept at $10°C$ and the other end is insulated. It is given that $y(x, 0) = 1 - x$, $0 < x < 1$.
[UPTU Q. Bank–2002]

3. The faces $x = 0$ and $x = a$ of an infinite slab are maintained at zero temperature. Given that the temperature $u(x, t) = f(x)$ at $t = 0$. Find the temperature at time t.

4. Find the solution of one dimensional heat equation $\dfrac{\partial\theta}{\partial t} = a^2 \dfrac{\partial^2\theta}{\partial x^2}$ under the boundary conditions $\theta(0, t) = 0, \theta(l, t) = 0$. When $t > 0$ and the initial condition $\theta(x, 0) = x$ when $0 < x < l$, l being the length of bar.

5. The four edges of a thin square plate of area π^2 are kept at temperature zero and the faces are perfectly insulated. The initial temperature is assumed to be $u(x, y, 0) = xy(n - x)(\pi - y)$. Find the temperature $u(x, y, t)$ in the plate.

6. Obtain the temperature $u(x, t)$ in a slab where ends $x = 0$ and $x = l$ are kept at temperature zero and whose initial temperature is given by
$$f(x) = \begin{cases} A \;; & \text{when } 0 < x < l/2 \\ 0 \;; & \text{when } l/2 < x < l \end{cases}$$
[GBTU–2010; GBTU (CO)–2010]

7. Solve $\dfrac{\partial^2 u}{\partial x^2} = h^2\left(\dfrac{\partial u}{\partial t}\right)$ when $u(0, t) = u(l, t) = 0$ and $u(x, 0) = \sin\dfrac{\pi x}{l}$.

8. Solve the one dimensional heat equation
$$\dfrac{\partial^2 u}{\partial x^2} = \dfrac{1}{k}\dfrac{\partial u}{\partial t} \quad \text{in the range } 0 \le x \le 2\pi, t \ge 0,$$
subject to the boundary condition $u(x, 0) = \sin^3 x$, for $0 \le x \le 2\pi$ and $u(0, t) = u(2\pi, t) = 0$ for $t \ge 0$.

9. Solve $\dfrac{\partial^2 z}{\partial x^2} = \dfrac{1}{k}\dfrac{\partial z}{\partial t}$ with the condition $z = 0$ when $x = 0$ and $x = 1$ for all values of t.

10. Find a solution of $\dfrac{\partial V}{\partial t} = k\dfrac{\partial^2 V}{\partial x^2}$ such that $V \neq \infty$ if $t \to \infty$, $V = 100$ if $x = 0$ or π for all values of t; $V = 0$ if $t = 0$ for all values of x between 0 and π.

11. The heat flow equation in a homogeneous rod is $\dfrac{\partial^2 T}{\partial x^2} = \dfrac{1}{\alpha^2}\dfrac{\partial T}{\partial t}$, where T is the temperature and α^2 thermal diffusivity. The rod is of length L with insulated sides. Solve it with boundary conditions $T(x, 0) = f(x)$, $0 < x < L, T_x(0, t) = T_x(L, t) = 0$. What is the steady temperature of the rod.

12. A conducting bar of uniform cross-section lies among the x-axis with ends at $x = 0$ and $x = L$. It is kept initially at temperature 0^0 and its lateral surface is insulated. There are no heat sources in the bar. The end $x=0$ is kept at 0^0 and heat is suddenly applied at the end $x = L$ so that there is a constant flux q_0 at $x = L$. Find the temperature distribution in the bar for $t > 0$.

13. Show that the solution of $\dfrac{\partial u}{\partial t} = k\dfrac{\partial^2 u}{\partial x^2}$ subject to the conditions

   (i) u is not infinite for $t \to \infty$.

   (ii) $\dfrac{\partial u}{\partial x} = 0$ for $x = 0$ and $x = l$.

   (iii) $u = lx - x^2$ for $t = 0$ between $x = 0$ and $x = l$ is
   $$u = \dfrac{1}{6}l^2 - \dfrac{l^2}{\pi^2}\sum_{n=1}^{\infty}\dfrac{1}{n^2}\cos\dfrac{n\pi x}{l}e^{-(4\pi^2 n^2 kt)/l}$$

14. A rectangular plate bounded by the lines $x = 0, y = 0, x = a, y = b$ has an initial distribution given by $V = A\sin\left(\dfrac{\pi x}{a}\right)\sin\left(\dfrac{\pi y}{b}\right)$.
The edges are kept at zero temperature and the plane faces are impervious to heat. Find the temperature at any point. [UPTU–2002]

15. A square plate with sides of unit length has its faces insulated and its sides kept at $0°C$. If the initial temperature is specified, determine the subsequent temperature at any point of the plate.

16. The temperature distribution in a bar of length. π which is perfectly insulated at ends $x = 0$ and $x = \pi$ is governed by partial differential equation.

$$\frac{\partial u}{\partial t} = \frac{\partial^2 u}{\partial x^2}$$

Assuming the initial temperature distribution as $u(x, 0) = f(x) = \cos 2x$, find the temperature distribution at any instant of time.

[MTU–2011]

17. Solve $\dfrac{\partial u}{\partial t} = a\dfrac{\partial^2 u}{\partial x^2}$; a constant, subject to the boundary conditions $u(0, t) = 0$, $u(\pi, t) = 0$ and the initial condition $u(x, 0) = \sin 2x$.

[MTU–2012]

18. Find the temperature distribution in a rod of length 2m whose end points are fixed at temperature zero and the initial temperature distribution is $f(x) = 100x$.   [GBTU–2012]

19. Find the temperature $u(x, t)$ in a homogeneous bar of heat conducting material of length l cm with its ends kept at zero temperature and initial temperature given by $\dfrac{x(L - x)d}{L^2}$.

[UPTU (SUM)–2009]

Hint to Selected Problems

3. The temperature $u(x, t)$ in the given solid is governed by the one dimensional heat equation

$$C\frac{\partial^2 u}{\partial x^2} = \frac{\partial u}{\partial t} \qquad \ldots (1)$$

As per given, the boundary conditions are

$$u(0, t) = 0, \ u(a_1 - 1) = 0, \ \forall t \quad \ldots (2)$$

$$u(x, 0) = f(x) \qquad \ldots (3)$$

Now, suppose that (1) has the solution of the form

$$u(x, t) = X(x)\, T(t) \qquad \ldots (4)$$

where X and T are respectively the functions of x and t alone. Using the values of (4) in (1), we get

$$\frac{X''}{X} = \frac{T'}{CT} = \mu \ \text{(say)} \qquad \ldots (5)$$

where μ is a separation constant.

From equation (5), we can deduce that

$$X'' - \mu X = 0 \qquad \ldots (6)$$

and

$$T' = \mu CT \qquad \ldots (7)$$

Using (2) in (4), we get

$$X(0) = 0 \ \text{and} \ X(a) = 0 \qquad \ldots (8)$$

Now, we want to solve (6) subject to the boundary condition (8). Here, we have the following cases :

Case (I). Let $\mu = 0$. Then solution of (6) is given by

$$X(x) = Ax + B \qquad \ldots (9)$$

Using (8), we get

$$A = B = 0$$

$$X(x) = 0$$

$u = 0$, which does not satisfy (3).

Case (II). Let $\mu = \lambda^2, \lambda \neq 0$: In this case, solution of (6) is given by

$$X(x) = Ae^{\lambda x} + Be^{-\lambda x}.$$

Using (8), we get $A + B = 0$ and

$$0 = Ae^{0\lambda} + De^{-a\lambda}$$

$$\Rightarrow \qquad A = B = 0$$

$$\Rightarrow \qquad X = 0$$

$$\Rightarrow \qquad u = 0$$

Thus, we reject this case also.

Case (III). Let $\mu = -\lambda^2, \lambda \neq 0$. In this case, solution of (6) is given by

$$X(x) = A\cos \lambda x + B\sin \lambda x$$

Using (8), we get

$$A = 0$$

and $A\cos \lambda a + B\sin \lambda a = 0$

Let $B \neq 0$, then $\sin \lambda a = 0$

$$\lambda a = n\pi, \ n = 1, 2, \ldots$$

$$\lambda = \frac{n\pi}{a}, \ n = 1, 2, \ldots$$

Therefore, non-zero solution of (6) is given by

$$X_n(x) = B_n \sin\left(\frac{n\pi x}{a}\right) \qquad \ldots (10)$$

Putting $\lambda = \dfrac{n\pi}{a}$ in (7), we get

$$\frac{dT}{T} = -\frac{n^2\pi^2 C}{a^2}dt$$

$$\Rightarrow \qquad \frac{dT}{T} = -C_n^2 dt$$

whose solution is given by

$$T_n(t) = D_n e^{-C_n^2 t} \qquad \ldots (11)$$

Thus, we have

$$u_n(x,t) = X_n(x)\, T_n(t) = E_n \sin\left(\frac{n\pi x}{a}\right) e^{-C_n^2 t}$$

... (12)

The more general solution of (1) is given by

$$u(x,t) = \sum_{n=1}^{\infty} u_n(x,t) = \sum_{n=1}^{2} E_n \sin\left(\frac{n\pi x}{a}\right) e^{-C_n^2 t}$$

... (13)

Putting $t = 0$ in (13) and using (3), we get

$$f(x) = \sum_{n=1}^{\infty} E_n \sin\left(\frac{n\pi x}{a}\right)$$

which is a Fourier sine series, thus the constants E_n are given by

$$E_n = \frac{2}{a}\int_0^a f(x)\sin\left(\frac{n\pi x}{a}\right) dx, \quad n = 1,2,3,\ldots$$

4. Proceeding same as example 3, we get

$$u(x,y,t) = \sum_{m=1}^{\infty}\sum_{n=1}^{\infty} F_{mn}\sin mx \sin ny \; e^{-\lambda_{mn}^2 t}$$

... (1)

where, $\quad F_{mn} = \dfrac{4}{\pi^2}\displaystyle\int_{x=0}^{\pi}\int_{y=0}^{\pi} xy\,(\pi - x)$

$$(\pi - y)\sin mx \sin ny\, dx\, dy$$

$$= \frac{4}{\pi^2}\left[\int_0^{\pi}(\pi x - x^2)\sin mx\, dx\right]$$

$$\left[\int_0^{\pi}(\pi y - y^2)\sin ny\, dy\right] \;\ldots\;(2)$$

Using chain rule of integration by parts, we get

$$\int_0^{\pi}(\pi x - x^2)\sin mx\, dx = \left[(\pi x - x^2)\left(-\frac{\cos mx}{m}\right)\right.$$

$$\left. -(\pi - 2x)\left(-\frac{\sin mx}{m^2}\right) + (-2)\left(\frac{\cos mx}{m^2}\right)\right]_0^{\pi}$$

$$= \frac{2}{m^3}\left[1 - (-1)^m\right]$$

In a similar manner, we get

$$\int_0^{\pi}(\pi y - y^2)\sin ny\, dy = \frac{2}{n^3}\left[1 - (-1)^n\right]$$

Thus,

$$F_{mn} = \frac{16}{\pi^2 m^3 n^3}\left[1-(-1)^m\right]\left[1-(-1)^n\right]$$

$$= \begin{cases} 0, & \text{when } m = 2p \text{ or } n = 2q \\[2mm] \dfrac{64}{\pi^2(2p-1)^3(2q-1)^3}, \\[2mm] & \text{when } m = 2p-1,\; n = 2q-1 \end{cases}$$

Hence, the required solution is given by

$$u(x,y,t) = \sum_{p=1}^{\infty}\sum_{q=1}^{\infty} F_{pq}\sin\left[(2p-1)x\right]$$

$$\sin\left[(2q-1)y\right]e^{-\lambda_{pq}^2 t}$$

where, $\quad \lambda_{pq}^2 = C^2\left[(2p-1)^2 + (2q-1)^2\right]$

and $\quad F_{pq} = \dfrac{64}{\pi^2}(2p-1)^3(2q-1)^3.$

8. Proceeding same as example (3) by taking $a = 2\pi$ and $f(x) = 5m^3 x$, we get

$$u(x,t) = \sum_{n=1}^{\infty} E_n \sin\left(\frac{nx}{2}\right) e^{-(n^2 kt/4)}, \quad n = 1,2,3,\ldots$$

... (1)

Putting $t = 0$ in (1) and using $u(x,0) = \sin^3 x$, we get

$$\sum_{n=1}^{\infty} E_n \sin\left(\frac{nx}{2}\right) = \sin^3 x = \frac{3}{4}\sin x - \frac{1}{4}\sin 3x\,.$$

Because $\quad \sin 3x = 3\sin x - 4\sin^3 x$

$$\Rightarrow \qquad \sin^3 x = \frac{1}{4}(3\sin x - \sin 3x)$$

Putting these values in (1), we get the required solution

$$u(x,t) = \frac{1}{4}(3\sin xe^{-kt} - \sin 3x\, e^{-9kt})$$

12. The given initial boundary value problem is given as follows :

$$\frac{\partial T}{\partial t} = \alpha\frac{\partial^2 T}{\partial x^2}$$

Subject to the boundary conditions

$$T(0,t) = 0,\; t > 0,\quad \frac{\partial T}{\partial x}(L,t) = q_0,\; t > 0$$

with initial conditions

$$T(x,0) = 0,\quad 0 \le x \le L.$$

Let $\quad T(x,t) = T_s(x) + T_1(x,t)$

Where, T_s is a steady part and T is the transient part of the solution.

Thus, we have

$$\frac{\partial^2 T_s}{\partial x^2} = 0 \;\Rightarrow\; T_s = Ax + B.$$

Using $x = 0$, $T_s = 0$, we get $B = 0$, therefore,

$$T_s = Ax.$$

Further using the boundary condition $\dfrac{\partial T_s}{\partial x} = q_0$, we get $A = q_0$.

Thus the steady solution is given by
$$T_s = q_0 x$$

For the transient part, the boundary and initial condition can be redefined as follows :

(i) $T_1(0,t) = T(0,t) - T_s(0) - 0 - 0 = 0$

(ii) $\dfrac{\partial T_1(x,t)}{\partial x} = \dfrac{\partial T(x,t)}{\partial x} - \dfrac{\partial T_s(L,t)}{\partial x}$

$$= q_0 - q_0 = 0$$

(iii) $T_1(x,0) = T(x,0) - T_s(x)$

$$= -q_0 x, \, 0 < x < L.$$

Therefore, for the transient part, we have to solve the given PDE subject to these conditions. We know that the solution of given PDE is

$$T_1(x,t) = e^{-a\lambda^2 t} \, (A \cos \lambda x + B \sin \lambda x)$$

Using (i), we get $A = 0$

$$T_1(x,t) = B e^{-a\lambda^2 t} \sin \lambda x$$

Again using boundary conditions (ii), we get

$$\left.\frac{\partial T}{\partial x}\right|_{x=L} = B\lambda \, e^{-a\lambda^2 t} \cos \lambda L = 0 \, .$$

$$\Rightarrow \qquad \lambda L = (2n-1)\frac{\pi}{2}, \, n = 1, 2, ...$$

Thus, the general solution is given by

$$T_1(x,t) = \sum_{n=1}^{\infty} B_n \exp\left[-\alpha \left\{ \frac{(2n-1)}{L} \right\}^2 \pi^2 t \right]$$

$$\sin\left(\frac{2n-1}{2L} \pi x \right)$$

Applying initial condition (iii), we get

$$T_1(x,0) = -q_0 x = \sum_{n=1}^{\infty} B_n \sin\left(\frac{2n-1}{2L} \pi x \right).$$

Now, multiplying both sides by $\sin\left(\dfrac{2n-1}{2L} \pi x \right)$,

integrating between 0 to L and using

$$\int_0^L B_n \sin\left(\frac{2n-1}{2L} \right) \pi x \sin\left(\frac{2m-1}{2L} \pi x \right) dx$$

$$= \begin{cases} 0, & n \neq m \\ \dfrac{B_m L}{2}, & n = m \end{cases}$$

We get

$$-q_0 \frac{4L^2}{(2m-1)^2 \pi^2}\left[\sin\left(\frac{2m-1}{2}\pi \right) \right] = B_m \cdot \frac{L}{2} \, .$$

$$\Rightarrow \quad -q_0 \frac{4L^2}{(2m-1)^2 \pi^2}(-1)^{m-1} = B_m \cdot \frac{L}{2}$$

$$\Rightarrow \quad B_m = \frac{(-1)^m 8 L q_0}{(2m-1)^2 \pi^2}$$

Hence, the required temperature distribution is given by

$$T(x,t) = q_0 x + \frac{8 L q_0}{\pi^2} \sum_{m=1}^{\infty}\left[\frac{(-1)^m}{(2m-1)^2} \cdot \right.$$

$$\exp\left[-\alpha \left\{ (2m-1)/2 \right\}^2 \pi^2 t \right] \cdot \sin\left(\frac{2m-1}{2L} \pi x \right)$$

ANSWERS

1. $u(x,t) = x + 2 + \sum\limits_{n=1}^{\infty} E_n \sin n\pi x \, e^{-n^2 \pi^2 c^2 t}$;

2. $y(x,t) = 10 + \sum\limits_{n=1}^{\infty} E_n \sin \dfrac{(2n-1)}{2}\pi x \, e^{-C_n^2 t}$

3. $u(x,t) = \sum\limits_{n=1}^{\infty} u_n(x,t) = \sum\limits_{n=1}^{\infty} E_n \sin\left(\dfrac{n\pi x}{a} \right) e^{-C_n^2 t}$; $f(x) = \sum\limits_{n=1}^{\infty} E_n \sin \dfrac{n\pi x}{a}$,

where $E_n = \dfrac{2}{a} \int_0^a f(x) \sin\left(\dfrac{n\pi x}{a} \right) dx, \, n = 1, 2, 3, ...$

4. $E_n = \dfrac{2}{l}\int_0^l x \sin \dfrac{n\pi x}{l} dx = \begin{cases} \dfrac{2l}{n\pi} & ; \, n \text{ is odd} \\ -\dfrac{2l}{n\pi} & ; \, n \text{ is even} \end{cases}$

5. $u(x,y,t) = \sum\limits_{p=1}^{\infty} \sum\limits_{q=1}^{\infty} F_{pq} \sin[(2p-1)x] \sin[(2q-1)y] e^{-\lambda p^2 qt}$

where $\lambda_{pq}^2 = C^2 \left[(2p-1)^2 + (2q-1)^2 \right]$ and $F_{pq} = \dfrac{64}{\pi^2}(2p-1)^3 (2q-1)^3$

6. $u(x,l) = \dfrac{4A}{\pi} \displaystyle\sum_{n=1}^{\infty} \dfrac{1}{n} \sin^2 \dfrac{n\pi}{4} \sin \dfrac{n\pi x}{l} e^{-(n^2\pi^2 kt)/l^2}$

7. $u(x,t) = \sin\left(\dfrac{\pi x}{l}\right) e^{-\pi^2 t/h^2 l^2}$

8. $u(x,t) = \dfrac{1}{4}\left[3\sin x e^{-kt} - \sin 3x e^{-akt}\right]$

12. $T(x,t) = q_0 x + \dfrac{8Lq_0}{\pi^2} \displaystyle\sum_{m=1}^{\infty}\left[\dfrac{(-1)^m}{(2m-1)^2}\exp\left[-\alpha\{(2m-1)/2\}^2 \pi^2 t\right]\sin\left(\dfrac{2m-1}{2L}\pi x\right)\right]$

14. $u(x,y,t) = A\sin\left(\dfrac{\pi x}{a}\right)\sin\left(\dfrac{\pi y}{a}\right)e^{-\pi^2 kt\left(\frac{1}{a^2}+\frac{1}{b^2}\right)}$

16. $u(x,t) = e^{-4t}\cos 2x$

17. $u(x,t) = \sin 2x e^{-4at}$

18. $u(x,t) = \dfrac{-400}{\pi}\displaystyle\sum_{n=1}^{\infty}\dfrac{\cos n\pi}{n}\dfrac{\sin n\pi x}{2}e^{-\left(\frac{c^2 n^2 \pi^2 t}{4}\right)}$

19. $u(x,t) = \dfrac{8d}{\pi^3}\displaystyle\sum_{n=1}^{\infty}\dfrac{1}{(2n-1)^3}\dfrac{\sin(2n-1)\pi x}{L}e^{\frac{-(2n-1)^2\pi^2 C^{2t}}{L^2}}$

Self Assessment Test

1. Find the temperature in a thin metal rod of length L with both ends insulated (so that there is no passage of heat through the ends) and with initial temperature $\sin\dfrac{\pi x}{L}$ in the rod. [UPTU (SUM)–2009]

2. A homogeneous rod of conducting material of length 'l' has its ends kept at zero temperature. The temperature at the centre is T and falls uniformly to zero at the two ends. Find the temperature distribution. [UKTU–2012]

3. An infinitely long plane uniform plate is bounded by two parallel edges and an end at right angles to them. The breadth is π. This end is maintained at temperature u_0 at all points and the other edges are at zero temperature. Determine the temperature at any point of the plate in the steady state. [GBTU–2012]

4. Solve $\dfrac{\partial^2 u}{\partial x^2}+\dfrac{\partial^2 u}{\partial y^2}=0, 0<x<\pi, 0<y<\pi$

   which satisfies the conditions:

   $$u(0,y)=u(\pi,y)=u(x,\pi)=0$$

   and $u(x,0)=\sin^2 x$ UKTU–2011]

5. Solve the Laplace equation $\dfrac{\partial^2 u}{\partial x^2}+\dfrac{\partial^2 u}{\partial y^2}=0$

   in a rectangle in the xy-plane with $u(x,0)=0$, $u(x,b)=0$, $u(0,y)=0$ and $u(a,y)=f(y)$ plarallel to y-axis. [UPTU (SUM)–2008]

6. Find the steady state temperature distribution in a rectangular thin plate with its two surfaces insulated and with the conditions $u(0,y)=0$, $u(x,0)=0$, $u(a,y)=g(y)$, $u(x,b)=f(x)$. [UPTU (SUM)–2007]

7. Solve the following Laplace equation $\dfrac{\partial^2 u}{\partial x^2}+\dfrac{\partial^2 u}{\partial y^2}=0$ in a rectangle with $u(0,y)=0$, $u(a,y)=0$, $u(x,b)=0$ and $u(x,0)=f(x)$ along x-axis. [UPTU–2008]

8. Solve the boundary value problem.

   $$\dfrac{\partial^2 u}{\partial x^2}+\dfrac{\partial^2 u}{\partial y^2}=0, 0\le x\le a, 0\le y\le b$$

   with the boundary conditions

   $$u_x(0,y)=u_x(a,y)=u_y(x,0)=0$$

   and $u_y(x,b)=f(x)$.   [MTU (SUM)–2011]

9. Neglecting R and G, find the emf $v(x,t)$ in a line of length l, t seconds after the ends were suddenly grounded, given that $i(x,0)=i_0$ and $v(x,0)=e_1\sin\dfrac{\pi x}{l}+e_5\sin\dfrac{5\pi x}{l}$ [SVTU–2008, MTU–2012]

10. Solve $\dfrac{\partial^2 V}{\partial x^2}=LC\dfrac{\partial^2 V}{\partial t^2}$ assuming that the initial voltage is $V_0\sin\dfrac{\pi x}{l}$; $V_t(x_0)=0$ and $V=0$ at the ends $x=0$ and $x=l$ for all t. [UPTU (SUM)–2007]

11. A steady voltage distribution of 20 volts at the sending end and 12 volts at the receiving end is maintained in a telephone wire of length l. A time $t=0$, the receiving end is grounded. Find the voltage and current t sec later. Neglect leakance and inductance. [MTU–2011, UPTU (SUM)–2008]

12. A homogeneous rod of conducting malerial of length 100 cm has its ends kept at zero temperature and the temperature initially is

    $$u(x,0)=x \qquad 0\le x\le 50$$
    $$=100-x \quad 50\le x\le 100$$

    Find the temperature $u(x,t)$ at any time. [Bhopal–2007, SVTU–2007, Kurukshetra–2006]

ANSWERS

1. $u(x,t)=\dfrac{2}{\pi}-\dfrac{4}{\pi}\displaystyle\sum_{m=1}^{\infty}\dfrac{1}{(4m^2-1)}\cos\left(\dfrac{2m\pi x}{L}\right)e^{\dfrac{-4m^2\pi^2 C^2 t}{L^2}}$

2. $u(x,t)=\dfrac{8T}{\pi^2}\displaystyle\sum_{m=1}^{\infty}\dfrac{(-1)^{m+1}}{(2m-1)^2}\sin(2m-1)\pi x e^{-[(2m-1)^2\pi^2 c^2 t]}$

3. $u(x,t) = \dfrac{4u_0}{\pi} \displaystyle\sum_{n=1}^{\infty} \dfrac{1}{(2n-1)} \sin(2n-1)x\, e^{-(2n-1)y}$

4. $u(x,t) = \dfrac{-8}{\pi} \displaystyle\sum_{m=1,2,3,\dots}^{\infty} \dfrac{\sin(2m-1)x \sinh(2m-1)(\pi-y)}{(2m-1)\{(2m-1)^2-4\}\sinh(2m-1)\pi}$

5. $u(x,t) = \displaystyle\sum_{n=1}^{\infty} b_n \sin\dfrac{n\pi y}{b} \sinh\dfrac{n\pi x}{b}$ where, $b_n = \dfrac{2}{b\sinh\left(\dfrac{n\pi a}{b}\right)} \displaystyle\int_0^b f(y) \sin\dfrac{n\pi y}{b}\, dy$

6. $u(x,y) = \displaystyle\sum_{n=1}^{\infty}\left[b_n \sin\left(\dfrac{n\pi x}{a}\right)\sinh\left(\dfrac{n\pi y}{a}\right) + B_n \sin\left(\dfrac{n\pi y}{b}\right)\sinh\left(\dfrac{n\pi x}{b}\right)\right]$

where $b_n = \dfrac{2}{a\sinh\left(\dfrac{n\pi b}{a}\right)}\displaystyle\int_0^a f(x)\sin\dfrac{n\pi x}{a}\, dx$ and $B_n = \dfrac{2}{b\sinh\left(\dfrac{n\pi a}{b}\right)}\displaystyle\int_0^a g(y)\sin\dfrac{n\pi y}{b}\, dy$

7. $u(x,y) = \displaystyle\sum_{n=1}^{\infty} B_n \sin\left(\dfrac{n\pi x}{a}\right)\sinh\dfrac{n\pi}{a}(b-y)$, where, $B_n = \dfrac{2}{a\sinh\left(\dfrac{n\pi b}{a}\right)}\displaystyle\int_0^a f(x)\sin\dfrac{n\pi x}{a}\, dx$

8. $u(x,y) = \displaystyle\sum_{n=1}^{\infty} b_n \cos\dfrac{n\pi x}{a}\left(e^{\frac{n\pi y}{a}} - e^{\frac{-n\pi y}{a}}\right)$ where $b_n = \dfrac{1}{n\pi\cosh\dfrac{n\pi b}{a}}\displaystyle\int_0^a f(x)\cos\dfrac{n\pi x}{a}\, dx$

9. $V = e_1 \sin\dfrac{\pi x}{l}\cos\dfrac{\pi t}{l\sqrt{2C}} + e_5 \sin\dfrac{5\pi x}{l}\cos\dfrac{5\pi t}{l\sqrt{LC}}$ **10.** $V(x,t) = V_0 \sin\dfrac{\pi x}{l}\cos\dfrac{\pi t}{l\sqrt{LC}}$

11. $V(x,t) = \dfrac{20(l-x)}{l} + \dfrac{24}{\pi}\displaystyle\sum_{n=1}^{\infty} \dfrac{(-1)^{n+1}}{n}\sin\dfrac{n\pi x}{l}\, e^{-n^2\pi^2 t/RCl^2}$

$i(x,t) = \dfrac{20}{lR} + \dfrac{24}{lR}\displaystyle\sum_{n=1}^{\infty} (-1)^n \cos\dfrac{n\pi x}{l}\, e^{-n^2\pi^2 t/RCl^2}$

12. $u(x,t) = \dfrac{400}{\pi^2}\displaystyle\sum_{n=1}^{\infty} \dfrac{(-1)^n}{(2n+1)^2} - e^{-[(2n+1)c\pi/100]^2 t}\sin\dfrac{(2n+1)\pi x}{100}$

 Objective Evaluations

❖ Multiple Choice Questions

Choose the most appropriate one.

1. The general second order linear differential equation in two independent variables x, y is given by

$$Au_{xx} + Bu_{xy} + Cu_{yy} + Du_x + Eu_y + Fu = G,$$

where A, B, C, D, E and F are given functions of x and y or constant. Then equation is hyperbolic if :

(a) $B^2 - 4AC > 0$ (b) $B^2 - 4AC = 0$

(c) $B^2 - 4AC < 0$ (d) none of these

2. The general second order linear differential equation in two independent variables x, y is given by

$$Au_{xx} + Bu_{xy} + Cu_{yy} + Du_x + Eu_y + Fu = G,$$

where A, B, C, D, E and F are given functions of x and y or constant. Then equation is parabolic if :

(a) $B^2 - 4AC > 0$ (b) $B^2 - 4AC = 0$

(c) $B^2 - 4AC < 0$ (d) none of these

3. The general second order linear differential equation in two independent variables x, y is given by

$$Au_{xx} + Bu_{xy} + Cu_{yy} + Du_x + Eu_y + Fu = G,$$

where A, B, C, D, E and F are given functions of x and y or constant. Then equation is elliptic if :

(a) $B^2 - 4AC > 0$ (b) $B^2 - 4AC = 0$

(c) $B^2 - 4AC < 0$ (d) none of these

4. The general second order linear differential equation in two independent variables x, y is given by

$$Au_{xx} + Bu_{xy} + Cu_{yy} + Du_x + Eu_y + Fu = G,$$

where A, B, C, D, E and F are given functions of x and y or constant. Then characteristic equation is given by :

(a) $\dfrac{dy}{dx} = (B \pm \sqrt{B^2 - 4AC})$

(b) $\sqrt{B^2 - 4AC}$

(c) $\dfrac{dy}{dx} = \dfrac{1}{2A}(B \pm \sqrt{B^2 - 4AC})$

(d) none of these

5. In Dirichlet conditions :

(a) u is prescribed by each point of a boundary ∂D of a domain D.

(b) where value of normal derivative $\dfrac{\partial u}{\partial n}$, in the boundary ∂D are specified.

(c) $\left(\dfrac{\partial u}{\partial n} + au\right)$ is specific on ∂D

(d) none of these

6. In Neumann conditions :

(a) u is prescribed by each point of a boundary ∂D of a domain D.

(b) where value of normal derivative $\dfrac{\partial u}{\partial n}$, in the boundary ∂D are specified.

(c) $\left(\dfrac{\partial u}{\partial n} + au\right)$ is specific on ∂D.

(d) none of these

7. Wave equation is represented by :

(a) $u_{tt} - c^2 \nabla^2 u = 0$, where c is a constant.

(b) $u_t - k\nabla^2 u = 0$, where k is a constant.

(c) $\nabla^2 u = 0$

(d) $\nabla^2 u = f(x, y, z)$

8. Heat equation is represented by :

(a) $u_{tt} - c^2 \nabla^2 u = 0$, where c is a constant.

(b) $u_t - k\nabla^2 u = 0$, where k is a constant.

(c) $\nabla^2 u = 0$

(d) $\nabla^2 u = f(x, y, z)$

9. Laplace equation is represented by :

(a) $u_{tt} - c^2 \nabla^2 u = 0$, where c is a constant.

(b) $u_t - k\nabla^2 u = 0$, where k is a constant.

(c) $\nabla^2 u = 0$

(d) $\nabla^2 u = f(x, y, z)$

10. Poisson equation is represented by :

   (a) $u_{tt} - c^2 \nabla^2 u = 0$, where c is a constant.

(b) $u_t - k\nabla^2 u = 0$, where k is a constant.

(c) $\nabla^2 u = 0$

(d) $\nabla^2 u = f(x, y, z)$

ANSWERS

1. (a)   **2.** (b)   **3.** (b)   **4.** (c)   **5.** (a)   **6.** (b)   **7.** (a)   **8.** (b)   **9.** (a)

10. (a)

VVVVVV

Chapter 9

Fourier Series

9.1 INTRODUCTION

In this section, we shall study a special type of functional series extensively studied by Joseph Fourier. Joseph Fourier represented expansions in trigonometrical series in connection with boundary value problem in conduction of heat. Although such expansions had been studied earlier, these series bear the name 'Fourier series' because of the major contributions of Fourier in this field.

9.2 PERIODIC FUNCTIONS

(UPTU–2002)

A function $f(x)$ which satisfies the relation $f(x + T) = f(x)$ for all real x and some fixed T is called a periodic function. The smallest positive number T, for which this relation holds, is called the period of $f(x)$.

If T is the period of $f(x)$. Then

$$f(x) = f(x + T) = f(x + 2T) = ... = f(x + nT) = ...$$

Also,

$$f(x) = f(x - T) = f(x - 2T) = ... = f(x - nT) = ...$$

$\therefore$

$$f(x) = f(x \pm nT), \text{ where } n \text{ is a positive integer.}$$

For example: Consider the function $f(x) = \sin x$. We have

$$\sin x = \sin (x + 2\pi) = \sin (x + 4\pi) = ...$$

Here, $f(x) = \sin x$ is a periodic function with period 2π. This function is also called sinusoidal periodic function.

We have studied about the Macluarian's theorem which is used to expand a function provided the function's derivative are continuous. Now, the need arise to expand functions which have discontinuities in their derivatives. By Fourier series, we can expand both types of functions under certain conditions as an infinite series of sine and cosine of x and it's integral multiple of a function $f(x)$ is defined in the interval $c < x < c + 2\pi$.

Then, Fourier series of $f(x)$ is given by

$$f(x) = \frac{a_0}{2} + \sum_{n=1}^{\infty} a_n \cos nx + \sum_{n=1}^{\infty} b_n \sin nx \qquad ...(1)$$

where a_0, a_n and b_n are called Fourier coefficient of $f(x)$ and their values are given as :

$$a_0 = \frac{1}{\pi} \int_c^{c+2\pi} f(x)\,dx \qquad ...(2)$$

$$a_n = \frac{1}{\pi} \int_c^{c+2\pi} f(x) \cos nx\,dx \qquad ...(3)$$

$$b_n = \frac{1}{\pi} \int_c^{c+2\pi} f(x) \sin nx\,dx \qquad ...(4)$$

The series (1) with coefficients a_0, a_n and b_n given by (2), (3) and (4) respectively is called the Fourier series of $f(x)$ and the coefficients a_0, a_n and b_n are called the Fourier coefficients corresponding to $f(x)$.

(i) When $c = 0$, the interval becomes $0 < x < 2\pi$ and formula for a_0, a_n, b_n is obtained by putting $c = 0$.

(ii) When $c = -\pi$, then interval becomes $-\pi < x < \pi$. In this interval, the formula for a_0, a_n and b_n becomes as under :

(a) When $f(x)$ is an odd function, then

$$a_0 = \frac{1}{\pi} \int_{-\pi}^{\pi} f(x)dx = 0 \cdot \quad a_n = \frac{1}{\pi} \int_{-\pi}^{\pi} f(x)\cos nx\, dx = 0$$

[By property of definite integral]

$$b_n = \frac{1}{\pi} \int_{-\pi}^{\pi} f(x)\sin nx\, dx = \frac{2}{\pi} \int_{0}^{\pi} f(x)\sin x\, dx$$

Hence, if function $f(x)$ is odd, its Fourier expansion contains only sine series,

i.e., $\qquad f(x) = \sum_{n=1}^{\infty} b_n \sin nx, \text{where } b_n = \frac{2}{\pi} \int_{0}^{\pi} f(x)\sin nx\, dx.$

(b) When $f(x)$ is even function, then formula for a_0, a_n and b_n are given by

$$a_0 = \frac{1}{\pi} \int_{-\pi}^{\pi} f(x)dx = \frac{2}{\pi} \int_{0}^{\pi} f(x)dx,$$

$$a_n = \frac{1}{\pi} \int_{-\pi}^{\pi} f(x)\cos nx\, dx = \frac{2}{\pi} \int_{0}^{\pi} f(x)\cos nx\, dx$$

and $\qquad b_n = \frac{1}{\pi} \int_{-\pi}^{\pi} f(x)\sin nx\, dx = 0$ $\qquad$ [$\because f(x) \sin nx$ is odd.]

Hence, if a periodic function $f(x)$ is even, its Fourier expansion contains only cosine terms, i.e., $f(x) = \frac{a_0}{2} + \sum_{n=1}^{\infty} \int_{0}^{\pi} f(x)dx$, where

$$a_0 = \frac{2}{\pi} \int_{0}^{\pi} f(x)dx \text{ and } a_n = \frac{2}{\pi} \int_{0}^{\pi} f(x).\cos nx\, dx$$

9.3 SOME IMPORTANT RESULTS

The following results are useful in the Fourier series :

(i) $\sin n\pi = 0, \cos n\pi = (-1)^n, \cos\left(n + \frac{1}{2}\right)\pi = 0,$ where $n \in Z.$

(ii) $\int uv = uv_1 - u'v_2 + u''v_3 - u'''v_4 + ...,$ where $u' = \dfrac{du}{dx}, u'' = \dfrac{d^2u}{dx^2}, ...$

$$v_1 = \int v\, dx, v_2 = \int v_1 dx, ...$$

(iii) $\int\limits_{0}^{2\pi} \sin nx \, dx = 0$ (iv) $\int\limits_{0}^{2\pi} \cos nx \, dx = 0$ (v) $\int\limits_{0}^{2\pi} \sin^2 nx \, dx = \pi$

(vi) $\int\limits_{0}^{2\pi} \cos^2 nx \, dx = \pi$ (vii) $\int\limits_{0}^{2\pi} \sin nx . \sin mx \, dx = 0$ (viii) $\int\limits_{0}^{2\pi} \cos nx . \cos mx \, dx = 0$

(ix) $\int\limits_{0}^{2\pi} \sin nx . \cos nx \, dx = 0$ (x) $\int\limits_{0}^{2\pi} \sin nx . \cos nx \, dx = 0$

(xi) $\int e^{ax} \sin bx \, dx = \dfrac{e^{ax}}{a^2 + b^2}(a \sin bx - b \cos bx) + c$

(xii) $\int e^{ax} \cos bx \, dx = \dfrac{e^{ax}}{a^2 + b^2}(a \cos bx - b \sin bx) + c$

9.4 DETERMINATION OF FOURIER COEFFICIENTS: EULER'S FORMULAE

The fourier series is given by

$$f(x) = \frac{a_0}{2} + a_1 \cos x + a_2 \cos 2x + \dots + a_n \cos nx + b_1 \sin x + \dots + b_2 \sin 2x + \dots + b_n \sin nx + \dots$$
$$\dots(i)$$

or $f(x) = \dfrac{a_0}{2} + \sum\limits_{n=1}^{\infty} a_n \cos nx + \sum\limits_{n=1}^{\infty} b_n \sin nx.$

To find a_0 : Integrating both sides of equation (1) from $x = c+0, x = c+2\pi$

$$\int\limits_{c}^{c+2\pi} f(x)dx = \frac{a_0}{2} \int\limits_{c}^{c+2\pi} dx + \int\limits_{c}^{c+2\pi} \left(\sum\limits_{n=1}^{\infty} a_n \cos nx \right) dx + \int\limits_{c}^{c+2\pi} \left(\sum\limits_{n=1}^{\infty} b_n \sin nx \right) dx$$

$$= \frac{a_0}{2}(c + 2\pi - c) + 0 + 0 = a_0 \pi$$

$$\Rightarrow \qquad a_0 = \frac{1}{\pi} \int\limits_{c}^{c+2\pi} f(x) dx .$$

To find a_n : Multiplying each side of equation (1) by cos nx and integrate w.r.t. x., between the limit c to $c+2\pi$.

$$\int\limits_{c}^{c+2\pi} f(x) \cos nx \, dx = \frac{a_0}{2} \int\limits_{c}^{c+2\pi} \cos nx \, dx + \int\limits_{c}^{c+2\pi} \left(\sum\limits_{n=1}^{\infty} a_n \cos nx \right) \cos nx \, dx$$

$$+ \int\limits_{c}^{c+2\pi} \left(\sum\limits_{n=1}^{\infty} b_n \sin nx \right) \cos nx \, dx$$

$$= 0 + a_n \pi + 0 = a_n \pi$$

$$\Rightarrow \qquad a_n = \frac{1}{\pi} \int\limits_{c}^{c+2\pi} f(x) \cos nx \, dx .$$

To find b_n : Multiplying each side of equation (1) by sin nx and integrate w.r.t. x between the limit c to $c + 2\pi$.

$$\int_{c}^{c+2\pi} f(x)\sin nx \, dx = \frac{a_0}{2} \int_{c}^{c+2\pi} \sin nx \, dx + \int_{c}^{c+2\pi} \left(\sum_{n=1}^{\infty} a_n \cos nx \right) \sin nx \, dx +$$

$$\int_{c}^{c+2\pi} \left(\sum_{n=1}^{\infty} b_n \sin nx \right) \sin nx \, dx$$

$$= 0 + 0 + b_n \pi = b_n \pi$$

$$\Rightarrow \qquad b_n = \frac{1}{\pi} \int_{c}^{c+2\pi} f(x)\sin nx \, dx$$

These values of a_0, a_n and b_n are called Euler's formulae.

9.5 DIRICHLET'S CONDITIONS

Any function $f(x)$ can be expressed as a Fourier series $\dfrac{a_0}{2} + \displaystyle\sum_{n=1}^{\infty} a_n \cos nx + \sum_{n=1}^{\infty} b_n \sin nx$, where a_0, a_n and b_n are constants.

   (i) $f(x)$ is finite and single valued in the interval $c < x < c + 2\pi$.

   (ii) $f(x)$ is periodic with period 2π.

   (iii) $f(x)$ and $f'(x)$ are piecewise continuous in the interval $c < x < c + 2\pi$.

The Fourier series with its coefficients converge to

   (a) $f(x)$ if x is a point of continuity.

   (b) $\dfrac{f(x+0)+f(x-0)}{2}$, if x is a point of discontinuity.

The conditions (i), (ii) and (iii) imposed on $f(x)$ are sufficient but not necessary, *i.e.*, if the conditions are satisfied, the convergence is guranted. However, if they are not satisfied the series may or may not converge.

Solved Examples

EXAMPLE 1.   *Expand the function* $f(x) = x \sin x$ *as a Fourier series in interval* $-\pi \le x \le \pi$. *Deduce that* $\dfrac{1}{1.3} - \dfrac{1}{3.5} + \dfrac{1}{5.7} - \dfrac{1}{7.9} + ... = \dfrac{\pi - 2}{4}$

(UPTU–2001, 2005, 2008; Q.Bank–2001; SVTU–2009; Bhopal–2009; Rohtak–2006; Agra-1990)

SOLUTION.   Since $x \sin x$ is an even function of x, so $b_n = 0$, then Fourier series is given by

$$f(x) = x \sin x = \frac{a_0}{2} + \sum_{n=1}^{\infty} a_n \cos nx,$$

where

$$a_0 = \frac{2}{\pi} \int_{0}^{\pi} x \sin x \, dx = \frac{2}{\pi} \left[-x \cos x + \sin x \right]_{0}^{\pi}$$

$$= \frac{2}{\pi} (-\pi \cos \pi) = 2$$

$$a_n = \frac{2}{\pi} \int_{0}^{\pi} x \sin x \cos nx \, dx = \frac{1}{\pi} \int_{0}^{\pi} x.2 \cos nx \sin x \, dx$$

$$= \frac{1}{\pi} \int_{0}^{\pi} x \{ \sin(n+1)x - \sin(n-1)x \} \, dx$$

$$= \frac{1}{\pi}\left[x\left\{\frac{-\cos(n+1)x}{n+1} + \frac{\cos(n-1)x}{n-1}\right\}\right] - 1\left\{\frac{-\sin(n+1)x}{(n+1)^2} + \frac{\sin(n-1)x}{(n+1)^2}\right\}\right]_0^\pi$$

$$= \frac{1}{\pi}\left[x\left\{\frac{-\cos(n+1)\pi}{n+1} + \frac{\cos(n-1)\pi}{n-1}\right\}\right] = \frac{\cos(n-1)\pi}{n-1} - \frac{\cos(n+1)\pi}{n+1} ; n \neq 1$$

$$= \begin{cases} \dfrac{1}{n-1} - \dfrac{1}{n+1} = \dfrac{2}{n^2-1} & \text{if } n \text{ is odd } n \neq 1 \\[3mm] \dfrac{-1}{n-1} + \dfrac{1}{n+1} = \dfrac{-2}{n^2-1} & \text{if } n \text{ is even} \end{cases}$$

When $n = 1$, then

$$a_1 = \frac{2}{\pi}\int_0^\pi x\sin x\cos x\,dx = \frac{1}{\pi}\int_0^\pi x\sin 2x\,dx$$

$$= \frac{1}{\pi}\left[x\left(\frac{-\cos 2x}{2}\right) - \left(\frac{-\sin 2x}{4}\right)\right]_0^\pi = \frac{1}{\pi}\left[\frac{-\pi\cos 2\pi}{2}\right] = -\frac{1}{2}$$

$$\therefore \quad x\sin x = 1 - \frac{1}{2}\cos x - 2\left[\frac{\cos 2x}{2^2-1} - \frac{\cos 3x}{3^2-1} + \frac{\cos 4x}{4^2-1} - \frac{\cos 5x}{5^2-1} + ...\right]$$

Putting $x = \dfrac{\pi}{2}$, we get

$$\frac{\pi}{2} = 1 - 2\left(\frac{-1}{2^2-1} + \frac{1}{4^2-1} - \frac{1}{6^2-1} + ...\right)$$

$$\Rightarrow \quad \frac{\pi}{2} - 1 = 2\left(\frac{1}{3} - \frac{1}{15} + \frac{1}{35} - ...\right) \Rightarrow \frac{\pi-2}{4} = \left(\frac{1}{1.3} - \frac{1}{3.5} + \frac{1}{5.7} - ....\right).$$

EXAMPLE 2. *Find the Fourier series to represent e^{ax} in interval $-\pi < x < \pi$.* (SVTU-2005)

SOLUTION. Let $\quad f(x) = e^{ax} = \dfrac{a_0}{2} + \sum_{n=1}^{\infty} a_n\cos nx + \sum_{n=1}^{\infty} b_n\sin nx$

$$a_0 = \frac{1}{\pi}\int_{-\pi}^{\pi} f(x)\,dx = \frac{1}{\pi}\int_{-\pi}^{\pi} e^{ax}\,dx = \frac{1}{\pi}\left[\frac{e^{ax}}{a}\right]_{-\pi}^{\pi}$$

$$= \frac{1}{a\pi}(e^{a\pi} - e^{-a\pi}) = \frac{2\sinh a\pi}{\pi a}$$

$$a_n = \frac{1}{\pi}\int_{-\pi}^{\pi} f(x)\cos nx\,dx = \frac{1}{\pi}\int_{-\pi}^{\pi} e^{ax}\cos nx\,dx$$

$$= \left[\frac{e^{ax}}{\pi(a^2+n^2)}a\cos nx + a\sin nx\right]_{-\pi}^{\pi}$$

$$= \frac{a\cos n\pi(e^{a\pi} - e^{-ia\pi})}{\pi(a^2+n^2)} = \frac{2a(-1)^n\sinh a\pi}{\pi(a^2+n^2)}$$

Similarly, we can get

$$b_n = \frac{2n(-1)^n \sinh a\pi}{\pi(a^2 + n^2)}$$

$$\therefore \quad e^{ax} = \frac{\sinh a\pi}{a\pi} + \sum_{n=1}^{\infty} \frac{2a(-1)^n \sinh a\pi}{\pi(a^2 + n^2)} \cos nx + \sum_{n=1}^{\infty} \frac{2n(-1)^n \sinh a\pi}{\pi(a^2 + n^2)} \sin n\pi$$

$$= \frac{2\sinh a\pi}{\pi} \left[\frac{1}{2a} - a\left(\frac{\cos x}{a^2 + 1^2} - \frac{\cos 2x}{a^2 + 2^2} + \frac{\cos 3x}{a^2 + 3^2} - \cdots \right) \right.$$

$$\left. - \left(\frac{\sin x}{a^2 + 1^2} - \frac{2\sin 2x}{a^2 + 2^2} + \frac{3\sin 3x}{a^2 + 3^2} - \cdots \right) \right]$$

EXAMPLE 3. *Obtain the Fourier series for the function $f(x) = x^2$, $-\pi < x < \pi$. Sketch the graph f function $f(x)$. Hence, show that* (PTU–2009; Bhopal–2008; BPTU–2006; Kanpur–2014)

(i) $\dfrac{1}{1^2} + \dfrac{1}{2^2} + \dfrac{1}{3^2} + \dfrac{1}{4^2} + \cdots = \displaystyle\sum_{n=1}^{\infty} \dfrac{1}{n^2} = \dfrac{\pi^2}{6}$

(UPTU(Q. Bank)–2001, Anna–2009, PTU–2009, Osmania–2003, Mumbai–2009, SVTU–2008)

(ii) $\dfrac{1}{1^2} - \dfrac{1}{2^2} + \dfrac{1}{3^2} - \dfrac{1}{4^2} + \cdots = \dfrac{\pi^2}{12}$ (UPTU–2004, 08; SVTU–2008)

(iii) $\dfrac{1}{1^2} + \dfrac{1}{3^2} + \dfrac{1}{5^2} + \cdots = \displaystyle\sum_{n=1}^{\infty} \dfrac{1}{(2n-1)^2} = \dfrac{\pi^2}{8}$

SOLUTION. Since, $f(x) = x^2$ is an even function, therefore $b_n = 0$

Now $f(x) = x^2 = \dfrac{a_0}{2} + \displaystyle\sum_{n=1}^{\infty} a_n \cos nx$. Then

$$a_0 = \frac{2}{\pi} \int_0^{\pi} f(x)\,dx = \frac{2}{\pi} \int_0^{\pi} x^2\,dx = \frac{2}{\pi} \left[\frac{x^3}{3} \right]_0^{\pi} = \frac{2}{3}\pi^2$$

$$a_n = \frac{2}{\pi} \int_0^{\pi} f(x) \cos nx\,dx = \frac{2}{\pi} \int_0^{\pi} x^2 \cos nx\,dx$$

$$= \frac{2}{\pi} \left[x^2 \left(\frac{\sin nx}{n} \right) - 2x \left(\frac{-\cos nx}{n^2} \right) + 2 \left(\frac{-\sin nx}{n^2} \right) \right]_0^{\pi}$$

$$= \frac{2}{\pi} \left[2x \frac{\cos n\pi}{n^2} \right] = 4 \frac{(-1)^n}{n^2}$$

$$\therefore \quad x^2 = \frac{\pi^2}{3} - 4 \left(\frac{\cos x}{1^2} - \frac{\cos 2x}{2^2} + \frac{\cos 3x}{3^2} - \frac{\cos 4x}{4^2} + \cdots \right)$$

$$\Rightarrow \quad x^2 = \frac{\pi^2}{3} + 4 \sum_{n=1}^{\infty} \frac{(-1)^n}{n^2} \cos nx \qquad \cdots (1)$$

Put $x = \pi$ in (1), we get

$$\pi^2 = \frac{\pi^2}{3} - 4 \left(-\frac{1}{1^2} - \frac{1}{2^2} - \frac{1}{3^2} - \frac{1}{4^2} - \cdots \right)$$

$$\Rightarrow \quad \frac{2\pi^2}{3} = -4\left(-\frac{1}{1^2} - \frac{1}{2^2} - \frac{1}{3^2} - \frac{1}{4^2} - \ldots\right)$$

$$\therefore \quad \frac{1}{1^2} + \frac{1}{2^2} + \frac{1}{3^2} + \frac{1}{4^2} \ldots = \frac{\pi^2}{6} \quad \ldots (2)$$

Put $x = 0$ in (1), we get

$$0 = \frac{\pi^2}{3} - 4\left(-\frac{1}{1^2} - \frac{1}{2^2} - \frac{1}{3^2} - \frac{1}{4^2} - \ldots\right)$$

$$\therefore \quad \frac{1}{1^2} - \frac{1}{2^2} - \frac{1}{3^2} - \frac{1}{4^2} - \ldots = \frac{\pi^2}{12} \quad \ldots (3)$$

Adding (2) and (3), we get

$$\frac{\pi^2}{4} = 2\left(\frac{1}{1^2} + \frac{1}{3^2} + \frac{1}{5^2} + \ldots\right)$$

$$\therefore \quad \frac{1}{1^2} + \frac{1}{3^2} + \frac{1}{5^2} + \ldots = \frac{\pi^2}{8}.$$

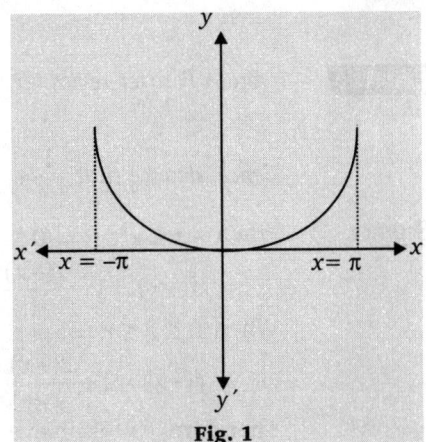

Fig. 1

EXAMPLE 4. *Obtain the Fourier series for $f(x) = e^{-x}$ in the interval $0 < x < 2\pi$.*

(UPTU(Q.Bank)–2001, SVTU–2007)

SOLUTION. Let $f(x) = e^{-x}$. The Fourier series of $f(x)$ can be written as

$$f(x) = e^{-x} = \frac{a_0}{2} + \sum_{n=1}^{\infty} a_n \cos nx + \sum_{n=1}^{\infty} b_n \sin nx$$

Then, $\quad a_0 = \frac{1}{2}\int_0^{2\pi} f(x)dx = \frac{1}{\pi}\int_0^{2\pi} e^{-x}dx = \frac{1}{\pi}\cdot\left[-e^{-x}\right]_0^{2\pi} = \frac{1-e^{-2\pi}}{\pi}$

$$a_n = \frac{1}{\pi}\int_0^{2\pi} f(x)\cos nx\, dx = \frac{1}{\pi}\int_0^{2\pi} e^{-x}\cos nx\, dx$$

$$= \frac{1}{\pi(1+n^2)}[e^{-x}(-\cos nx + n\sin nx)]_0^{2\pi} = \frac{1-e^{-2\pi}}{\pi(1+n^2)}$$

$$b_n = \frac{1}{\pi}\int_0^{2\pi} f(x)\sin nx\, dx = \frac{1}{\pi}\int_0^{2\pi} e^{-x}\sin nx\, dx$$

$$= \frac{1}{\pi(1+n^2)}[-\sin nx - n\cos nx]_0^{2\pi} = \frac{1-e^{-2\pi}}{\pi}\cdot\frac{n}{1+n^2}$$

$$\therefore \quad e^{-x} = \frac{1-e^{-2\pi}}{\pi}\left[\frac{1}{2} + \left(\frac{1}{2}\cos x + \frac{1}{5}\cos 2x + \frac{1}{10}\cos 3x + \ldots\right)\right]$$

$$+ \left(\frac{1}{2}\sin x + \frac{2}{5}\sin 2x + \frac{3}{10}\sin 3x + \ldots\right)$$

$$= \frac{1-e^{-2\pi}}{2\pi} + \frac{1-e^{-2\pi}}{\pi}\sum_{n=1}^{\infty}\frac{\cos nx}{1+n^2} + \frac{1-e^{-2\pi}}{\pi}\sum_{n=1}^{\infty}\frac{n\sin nx}{1+n^2}$$

EXAMPLE 5.   *Obtain Fourier series for the function f(x), given by* $f(x) = \begin{cases} 1 + \dfrac{2x}{\pi}; & -\pi \le x \le 0 \\ 1 - \dfrac{2x}{\pi}; & 0 \le x \le \pi \end{cases}$

Hence, deduce that $\dfrac{1}{1^2} + \dfrac{1}{3^2} + \dfrac{1}{5^2} + ... = \dfrac{\pi^2}{8}$.   (VTU–2010, Mumbai–2007)

SOLUTION.   When $-\pi \le x \le 0 \Rightarrow 0 \le -x \le \pi$

$\therefore \qquad f(-x) = 1 - \dfrac{2(-x)}{\pi} = 1 + \dfrac{2x}{\pi} = f(x)$

When $0 \le x \le \pi \Rightarrow -\pi \le -x \le 0$

$\therefore \qquad f(-x) = 1 + \dfrac{2(-x)}{\pi} = 1 - \dfrac{2x}{\pi} = 1 - \dfrac{2x}{\pi} = f(x)$

Therefore, $f(x)$ is an even function of x in the interval $[-\pi, \pi]$. Hence $b_n = 0$.

Now, Fourier series of $f(x)$ is given by

$$f(x) = \frac{a_0}{2} + \sum_{n=1}^{\infty} a_n \cos nx$$

Then,

$$a_0 = \frac{2}{\pi} \int_0^\pi f(x)\,dx = \frac{2}{\pi} \int_0^\pi \left(1 - \frac{2x}{\pi}\right) dx = \frac{2}{\pi}\left[x - \frac{x^2}{\pi}\right]_0^\pi = 0$$

$$a_n = \frac{2}{\pi} \int_0^\pi f(x) \cos nx\,dx = \frac{2}{\pi}\int_0^\pi \left(1 - \frac{2x}{\pi}\right)\cos nx\,dx$$

$$= \frac{2}{\pi}\left[\left(1 - \frac{2x}{\pi}\right)\frac{\sin nx}{n} - \left(-\frac{2}{\pi}\right)\left(-\frac{\cos nx}{n^2}\right)\right]_0^\pi$$

$$= \frac{2}{\pi}\left[-\frac{2\cos n\pi}{\pi n^2} + \frac{2}{\pi n^2}\right] = \frac{4}{\pi^2 n^2}[1 - (-1)^n]$$

$\Rightarrow \qquad f(x) = \dfrac{4}{\pi^2} \sum_{n=1}^{\infty} [1 - (-1)^n] \dfrac{\cos nx}{n^2}$

$$= \frac{4}{\pi^2}\left(\frac{2\cos x}{1^2} + \frac{2\cos 3x}{3^2} + \frac{2\cos 5x}{5^2} + ...\right)$$

$$= \frac{8}{\pi^2}\left(\frac{\cos x}{1^2} + \frac{\cos 3x}{3^2} + \frac{\cos 5x}{5^2} + ...\right).$$

Putting $x = 0$, we get $\dfrac{1}{1^2} + \dfrac{1}{3^2} + \dfrac{1}{5^2} + ... = \dfrac{\pi^2}{8}$.

                                               [Since $f(0) = 1$)]

EXAMPLE 6.   *Find a Fourier series to represent* $x - x^2$ *from* $x = -\pi$ *to* $x = \pi$.

Deduce that $\dfrac{1}{1^2} - \dfrac{1}{2^2} + \dfrac{1}{3^2} - \dfrac{1}{4^2} + ... = \dfrac{\pi^2}{12}$.   (VTU–2011, Madras–2006)

SOLUTION.   The Fourier series for $f(x)$ in $(-\pi, \pi)$ is

$$f(x) = a_0 + \sum_{n=1}^{\infty} a_n \cos nx + \sum_{n=1}^{\infty} b_n \sin nx$$

Here,

$$a_0 = \frac{1}{2\pi} \int_{-\pi}^{\pi} (x - x^2) dx = \frac{1}{2\pi} \left[\frac{x^2}{2} - \frac{x^3}{3} \right]_{-\pi}^{\pi} = -\frac{\pi^2}{3}$$

$$a_n = \frac{1}{\pi} \int_{-\pi}^{\pi} (x - x^2) \cos nx \, dx$$

$$= \frac{1}{\pi} \left[(x - x^2) \frac{\sin nx}{n} - (1 - 2x) \left(-\frac{\cos nx}{n^2} \right) + (-2) \left(-\frac{\sin nx}{n^3} \right) \right]_{-\pi}^{\pi} = \frac{-4(-1)^n}{n^2}$$

and $b_n = \frac{1}{\pi} \int_{-\pi}^{\pi} (x - x^2) \sin nx \, dx$

$$= \frac{1}{\pi} \left[(x - x^2) \left(-\frac{\cos nx}{n} \right) - (1 - 2x) \cdot \left(-\frac{\sin nx}{n^2} \right) + (-2) \left(\frac{\cos nx}{n^3} \right) \right]_{-\pi}^{\pi}$$

$$= \frac{(-2)(-1)^n}{n}$$

∴ The required Fourier series is

$$x - x^2 = -\frac{\pi^2}{3} + 4 \left[\frac{\cos x}{1^2} - \frac{\cos 2x}{2^2} + \frac{\cos 3x}{3^2} - \frac{\cos 4x}{4^2} + ... \right]$$

$$+ 2 \left[\frac{\sin x}{1} - \frac{\sin 2x}{2} + \frac{\sin 3x}{3} - \frac{\sin 4x}{4} + ... \right] \qquad ...(1)$$

Deduction. Putting $x = 0$ in (1), we get

$$0 = -\frac{\pi^2}{3} + 4 \left(\frac{1}{1^2} - \frac{1}{2^2} + \frac{1}{3^2} - \frac{1}{4^2} + ... \right)$$

or $\qquad \frac{1}{1^2} - \frac{1}{2^2} + \frac{1}{3^2} - \frac{1}{4^2} + ... = \frac{\pi^2}{12}.$

EXAMPLE 7. *Find the Fourier series of the function defined as*

$$f(x) = \begin{cases} x + \pi & ; & 0 \le x \le \pi \\ -x - \pi & ; & -\pi \le x \le 0 \end{cases}$$

and $f(x + 2\pi) = f(x).$ (UPTU–2006, Q. Bank–2001)

SOLUTION. Let $\qquad f(x) = \frac{a_0}{2} + \sum_{n=1}^{\infty} a_n \cos nx + \sum_{n=1}^{\infty} b_n \sin nx$

Then, $\qquad a_0 = \frac{1}{\pi} \int_{-\pi}^{\pi} f(x) dx = \frac{1}{\pi} \int_{-\pi}^{0} f(x) dx + \frac{1}{\pi} \int_{0}^{\pi} f(x) dx$

$$= \frac{1}{\pi} \int_{-\pi}^{0} (-x - \pi) dx + \frac{1}{\pi} \int_{0}^{\pi} (x + \pi) dx = \frac{1}{\pi} \left[\left(-\frac{x^2}{2} - \pi x \right)_{-\pi}^{0} + \left(\frac{x^2}{2} + \pi x \right)_{0}^{\pi} \right]$$

$$= \frac{1}{\pi} \left\{ \left(\frac{\pi^2}{2} - \pi^2 \right) + \left(\frac{\pi^2}{2} + \pi^2 \right) \right\} = \pi$$

$$a_n = \frac{1}{\pi} \int_{-\pi}^{\pi} f(x) \cos nx \, dx = \frac{1}{\pi} \int_{-\pi}^{0} f(x) . \cos nx \, dx + \frac{1}{\pi} \int_{0}^{\pi} f(x) . \cos nx \, dx$$

$$= \frac{1}{\pi} \int_{-\pi}^{0} (-x - \pi) \cos nx \, dx + \frac{1}{\pi} \int_{0}^{\pi} (x + \pi) \cos nx \, dx$$

$$= \frac{1}{\pi} \left[(-x - \pi) \frac{\sin nx}{n} - (-1) \left\{ -\frac{\cos nx}{n^2} \right\} \right]_{-\pi}^{0}$$

$$+ \frac{1}{\pi} \left[(x + \pi) \frac{\sin nx}{n} - (-1) \left\{ -\frac{\cos nx}{n^2} \right\} \right]_{0}^{\pi}$$

$$= \frac{1}{\pi} \left[-\frac{1}{n^2} + \frac{(-1)^n}{n^2} \right] + \frac{1}{\pi} \left[\frac{(-1)^n}{n^2} - \frac{1}{n^2} \right]$$

$$= \frac{2}{n^2 \pi} [(-1)^n - 1] = \begin{cases} -\dfrac{4}{n^2 \pi} & ; \quad \text{if } n \text{ is odd} \\ 0 & ; \quad \text{if } n \text{ is even} \end{cases}$$

Also $$b_n = \frac{1}{\pi} \int_{-\pi}^{\pi} f(x) \sin nx \, dx$$

$$= \frac{1}{\pi} \left\{ \int_{-\pi}^{0} f(x) . \sin nx \, dx + \int_{0}^{\pi} f(x) . \sin nx \, dx \right\}$$

$$= \frac{1}{\pi} \left\{ \int_{-\pi}^{0} (-x - \pi) \sin nx \, dx + \int_{0}^{\pi} (x + \pi) \sin nx \, dx \right\}$$

$$= \frac{1}{\pi} \left[(-x - \pi) \left(-\frac{\cos nx}{n} \right) - (-1) \left\{ -\frac{\sin nx}{n^2} \right\} \right]_{-\pi}^{0}$$

$$+ \frac{1}{\pi} \left[(x + \pi) \left(-\frac{\cos nx}{n} \right) - (-1) \left\{ -\frac{\sin nx}{n^2} \right\} \right]_{0}^{\pi}$$

$$= \frac{1}{\pi} \left[\frac{\pi}{n} \right] + \frac{1}{\pi} \left[\frac{-2\pi}{n} (-1)^n + \frac{\pi}{n} \right] = \frac{1}{n} [1 - 2(-1)^n + 1] = \frac{2}{n} [1 - (-1)^n]$$

$$= \begin{cases} \dfrac{4}{n} & , \quad \text{if } n \text{ is odd} \\ 0 & , \quad \text{if } n \text{ is even} \end{cases}$$

The required Fourier series is given by

$$f(x) = \frac{a_0}{2} + a_1 \cos x + a_2 \cos 2x + \ldots + b_1 \sin x + b_2 \sin 2x + \ldots$$

$$= \frac{\pi}{2} - \frac{4}{\pi} \left(\frac{\cos x}{1^2} + \frac{\cos 3x}{3^2} + \ldots \right) + 4 \left(\frac{\sin x}{1} + \frac{\sin 3x}{3} + \ldots \right)$$

EXAMPLE 8. *Find the Fourier series for the function $f(x) = x + x^2$, $-\pi < x < \pi$. Hence, show that*

(Kurukshetra–2005, UPTU–2003)

(i) $$\frac{\pi^2}{6} = 1 + \frac{1}{2^2} + \frac{1}{3^2} + \frac{1}{4^2} + \ldots$$

(UPTU–2003)

(ii) $\dfrac{\pi^2}{12} = \dfrac{1}{1^2} - \dfrac{1}{2^2} + \dfrac{1}{3^2} - \dfrac{1}{4^2} + \dots$ 　　　　　　　　(Agra-1992, 93; 2014)

SOLUTION.　　Let the Fourier series be

$$x + x^2 = \dfrac{a_0}{2} + \sum_{n=1}^{\infty} a_n \cos nx + \sum_{n=1}^{\infty} b_n \sin nx \qquad \dots (1)$$

Here,　　$a_0 = \dfrac{1}{\pi} \int_{-\pi}^{\pi} (x + x^2)dx = \dfrac{1}{\pi}\left[\int_{-\pi}^{\pi} x\, dx + \int_{-\pi}^{\pi} x^2\, dx\right] = \dfrac{2}{\pi}\int_0^{\pi} x^2 dx = \dfrac{2}{3}\pi^2$

$$a_n = \dfrac{1}{\pi}\int_{-\pi}^{\pi}(x + x^2)\cos nx\, dx$$

$$= \dfrac{1}{\pi}\left[\int_{-\pi}^{\pi} x\cos nx\, dx + \int_{-\pi}^{\pi} x^2 \cos nx\, dx\right] = \dfrac{2}{\pi}\int_0^{\pi} x^2 \cos nx\, dx$$

$$= \dfrac{2}{\pi}\left[\left(x^2 \dfrac{\sin nx}{n}\right)_0^{\pi} - \int_0^{\pi} 2x \cdot \dfrac{\sin nx}{n}dx\right] = -\dfrac{4}{n\pi}\int_0^{\pi} x\sin nx\, dx$$

$$= -\dfrac{4}{n\pi}\left[\left\{x\left(-\dfrac{\cos nx}{n}\right)\right\}_0^{\pi} - \int_0^{\pi} 1\cdot\left(-\dfrac{\cos nx}{n}\right)dx\right]$$

$$= -\dfrac{4}{n\pi}\left(-\dfrac{\pi}{n}\cos nx\right) = \dfrac{4}{n^2}\cos n\pi = \dfrac{4}{n^2}(-1)^n$$

and　　$b_n = \dfrac{1}{\pi}\int_{-\pi}^{\pi}(x + x^2)\sin nx\, dx$

$$= \dfrac{2}{\pi}\int_0^{\pi} x\sin nx\, dx + \dfrac{2}{\pi}\int_0^{\pi} x^2 \sin nx\, dx \qquad \left[\because \int_0^{\pi} x^2 \sin nx\, dx = 0\right]$$

$$= \dfrac{2}{\pi}\left(-\dfrac{\pi}{n}\cos n\pi\right) = -\dfrac{2}{n}(-1)^n.$$

From (1),　$x + x^2 = \dfrac{\pi^2}{3} + 4\sum_{n=1}^{\infty}\dfrac{(-1)^n}{n^2}\cos nx - 2\sum_{n=1}^{\infty}\dfrac{(-1)^n}{n}\sin nx$

$$f(x) = \dfrac{\pi^2}{3} + 4\left[-\dfrac{1}{1^2}\cos x + \dfrac{1}{2^2}\cos 2x - \dfrac{1}{3^2}\cos 3x + \dots\right]$$

$$-2\left[-\dfrac{1}{1}\sin x + \dfrac{1}{2}\sin 2x - \dfrac{1}{3}\sin 3x + \dots\right]. \qquad \dots(2)$$

We observe that the series on the R.H.S. given by equation (2), always represents $x + x^2$ for all values of x except the end points $-\pi$ or π.

At the point of discontinuity

$$f(-\pi) = \dfrac{1}{2}(\text{L.H.L.} + \text{R.H.L.})$$

$$= \dfrac{1}{2}[f(-\pi - 0) + f(-\pi + 0)] = \dfrac{1}{2}[f(\pi - 0) + f(-\pi + 0)]$$

$$= \dfrac{1}{2}[\pi + \pi^2 + (-\pi) + (-\pi)^2] = \pi^2 .$$

Putting $x = -\pi$ in equation (2), we get

$$\pi^2 = \frac{\pi^2}{3} + 4\left[\frac{1}{1^2} + \frac{1}{2^2} + \frac{1}{3^2} + \frac{1}{4^2} + \ldots\right]$$

Therefore, $\quad \dfrac{\pi^2}{6} = 1 + \dfrac{1}{2^2} + \dfrac{1}{3^2} + \dfrac{1}{4^2} + \ldots$ ... (3)

Again, putting $x = 0$ in equation (2), we get

$$0 = \frac{\pi^2}{3} + 4\left[-\frac{1}{1^2} + \frac{1}{2^2} - \frac{1}{3^2} + \frac{1}{4^2} - \ldots\right]$$

$$\Rightarrow \qquad \frac{\pi^2}{12} = \frac{1}{1^2} - \frac{1}{2^2} + \frac{1}{3^2} - \frac{1}{4^2} \ldots$$

EXAMPLE 9.   *Express* $f(x) = |x|$, $-\pi < x < \pi$, *as Fourier series. Hence, show that*

$$\frac{1}{1^2} + \frac{1}{3^2} + \frac{1}{5^2} + \ldots = \frac{\pi^2}{8}.$$

(UPTU(Q.Bank)–2001; SVTU–2005, 09; Kerala–2005; PTU–2005)

SOLUTION.   Here, $f(-x) = |-x| = |x| = f(x)$

$\therefore \quad f(x)$ is an even function and hence $b_n = 0$.

Let $\qquad f(x) = |x| = \dfrac{a_0}{2} + \displaystyle\sum_{n=1}^{\infty} a_n \cos nx$

Then, $\qquad a_0 = \dfrac{2}{\pi}\int_0^{\pi} f(x)dx = \dfrac{2}{\pi}\int_0^{\pi}|x|\,dx = \dfrac{2}{\pi}\int_0^{\pi} x\,dx = \dfrac{2}{\pi}\left[\dfrac{x^2}{2}\right]_0^{\pi} = \pi$

and $\qquad a_n = \dfrac{2}{\pi}\int_0^{\pi} f(x)\cos nx\,dx$

$$= \frac{2}{\pi}\int_0^{\pi}|x|.\cos nx\,dx = \frac{2}{\pi}\int_0^{\pi} x\cos nx\,dx$$

$$= \frac{2}{\pi}\left[x\left(\frac{\sin nx}{n}\right) - 1\left(-\frac{\cos nx}{n^2}\right)\right]_0^{\pi} = \frac{2}{\pi}\left[\frac{\cos nx}{n^2} - \frac{1}{n^2}\right]$$

$$= \frac{2}{\pi n^2}[(-1)^n - 1] = \begin{cases} 0 & , \text{ if } n \text{ is even} \\ -\dfrac{4}{\pi n^2} & , \text{ if } n \text{ is odd} \end{cases}$$

Hence, $\qquad |x| = \dfrac{\pi}{2} - \dfrac{4}{\pi}\left(\cos x + \dfrac{\cos 3x}{3^2} + \dfrac{\cos 5x}{5^2} + \ldots\right)$ ... (1)

Deduction. Putting $x = 0$, in equation (1), we get $\dfrac{1}{1^2} + \dfrac{1}{3^2} + \dfrac{1}{5^2} + \ldots = \dfrac{\pi^2}{8}$.

EXAMPLE 10.   *Find the Fourier series expansion of* $f(x)$, *if* $f(x) = \begin{cases} -\pi & , & -\pi < x < 0 \\ x & , & 0 < x < \pi \end{cases}$

Deduce that $\dfrac{1}{1^2} + \dfrac{1}{3^2} + \dfrac{1}{5^2} + \ldots = \dfrac{\pi^2}{8}$.

SOLUTION.   Let the Fourier series be

$$f(x) = \frac{a_0}{2} + \sum_{n=1}^{\infty} a_n \cos nx + \sum_{n=1}^{\infty} b_n \sin nx$$

Then, $a_0 = \frac{1}{2\pi} \int_{-\pi}^{\pi} f(x)dx = \frac{1}{2\pi}\left[\int_{-\pi}^{0}(-\pi)dx + \int_0^{\pi} x\, dx \right]$

$$= \frac{1}{\pi}\left[-\pi(x)_{-\pi}^{0} + \left(\frac{x^2}{2}\right)_0^{\pi} \right] = \frac{1}{\pi}\left[-\left(\pi^2 + \frac{\pi^2}{2}\right)\right] = -\frac{\pi}{2}$$

$$a_n = \frac{1}{\pi}\int_{-\pi}^{\pi} f(x)\cos nx\, dx = \frac{1}{\pi}\left[\int_{-\pi}^{0}(-\pi)\cos nx\, dx + \int_0^{\pi} x\cos nx\, dx\right]$$

$$= \frac{1}{\pi}\left[-\pi\left(\frac{\sin nx}{n}\right)_{-\pi}^{0} + \left(\frac{x\sin nx}{n} + \frac{\cos nx}{n^2}\right)_0^{x}\right]$$

$$= \frac{1}{\pi}\left[0 + \frac{1}{n^2}\cos n\pi - \frac{1}{n^2}\right] = \frac{1}{\pi n^2}(\cos n\pi - 1)$$

and

$$b_n = \frac{1}{\pi}\int_{-\pi}^{\pi} f(x)\sin nx\, dx = \frac{1}{\pi}\left[\int_{-\pi}^{0}(-\pi)\sin nx\, dx + \int_0^{\pi} x\sin nx\, dx\right]$$

$$= \frac{1}{\pi}\left[\left(\frac{\pi\cos nx}{n}\right)_{-\pi}^{0} + \left(-\frac{\cos nx}{n} + \frac{\sin nx}{n^2}\right)_0^{\pi}\right]$$

$$= \frac{1}{\pi}\left[\frac{\pi}{n}(0 - \cos n\pi) - \frac{\pi}{n}\cos n\pi\right] = \frac{1}{n}(1 - 2\cos n\pi)$$

The required Fourier series is

$$f(x) = -\frac{\pi}{4} - \frac{2}{\pi}\left(\cos x + \frac{\cos 3x}{3^2} + \frac{\cos 5x}{5^2} + ...\right)$$

$$+ \left(3\sin x - \frac{\sin 2x}{2} + \frac{3\sin 3x}{3} - \frac{\sin 4x}{4} + ...\right) \qquad ... (1)$$

Deduction. Putting $x = 0$ in (1) , we get

$$f(0) = \frac{\pi}{4} - \frac{2}{\pi}\left(1 + \frac{1}{3^2} + \frac{1}{5^2} + ...\right) \qquad ... (2)$$

But $f(x)$ is continuous at $x = 0$, and we have $f(0-0) = -\pi$ and $f(0+0) = 0$

$\therefore \qquad f(0) = \frac{1}{2}[f(0-0) + f(0+0)] = -(\pi/2) \qquad ...(3)$

Hence, from (2) and (3), we have

$$-\frac{\pi}{2} = -\frac{\pi}{4} - \frac{2}{\pi}\left[\frac{1}{1^2} + \frac{1}{3^2} + \frac{1}{5^2} + ...\right] \text{ or } \frac{1}{1^2} + \frac{1}{3^2} + \frac{1}{5^2} + ... = \frac{\pi^2}{8}.$$

Additional Solved Examples

EXAMPLE 1.   *Express $f(x) = \frac{1}{2}(\pi - x)$ in a Fourier series in the interval $0 < x < 2\pi$.*

SOLUTION.   Let $f(x) = \frac{1}{2}(\pi - x) = \frac{a_2}{2} + \sum_{n=1}^{\infty} a_n \cos nx + \sum_{n=1}^{\infty} b_n \sin nx$

$$a_0 = \frac{1}{\pi}\int_0^{2\pi} f(x)\,dx = \frac{1}{\pi}\int_0^{2\pi}\frac{1}{2}(\pi-x)\,dx = \frac{1}{2\pi}\left[\pi x - \frac{x^2}{2}\right]_0^{2\pi} = 0$$

$$a_n = \frac{1}{\pi}\int_0^{2\pi} f(x)\cos nx\,dx = \frac{1}{\pi}\int_0^{2\pi}\frac{1}{2}(\pi-x)\cos nx\,dx$$

$$= \frac{1}{2\pi}\left[(\pi-x)\frac{\sin nx}{n} - (-1)\frac{\cos nx}{n^2}\right]_0^{2\pi}$$

$$b_n = \frac{1}{\pi}\int_0^{2\pi} f(x)\sin nx\,dx = \frac{1}{\pi}\int_0^{2x}\frac{1}{2}(\pi-x)\sin nx\,dx$$

$$= \frac{1}{2\pi}\left[(\pi-x)\left(\frac{-\cos nx}{n}\right) - (-1)\left(\frac{-\sin nx}{n^2}\right)\right]_0^{2\pi} = \frac{1}{2\pi}\left[\frac{2\pi}{n}\right] = \frac{1}{n}$$

$$f(x) = \sum_{n=1}^{\infty}\frac{\sin nx}{n}$$

EXAMPLE 2. Find the Fourier series to represent the function $f(x) = |\sin x|$, $-\pi < x < \pi$.

SOLUTION. $f(-x) = |(-\sin x)| = |\sin x| = f(x)$ (UPTU-2001)

$f(x)$ is an even function and hence $b_n = 0$

Let $f(x) = |\sin x| = \dfrac{a_0}{2} + \displaystyle\sum_{n=1}^{\infty} a_n \cos n x$

Let $a_0 = \dfrac{2}{\pi}\int_0^{\pi} f(x)\,dx = \dfrac{2}{\pi}\int_0^{\pi}|\sin x|\,dx$

$$= \frac{2}{\pi}[-\cos x]_0^{\pi} = \frac{2}{\pi}[-\cos x + \cos 0] = \frac{4}{\pi}$$

$$a_n = \frac{2}{\pi}\int_0^{\pi} f(x)\cos nx\,dx$$

$$= \frac{2}{\pi}\int_0^{\pi}|\sin x|\cos nx\,dx = \frac{2}{\pi}\int_0^{\pi}\sin x\cos nx\,dx = \frac{1}{\pi}\int_0^{\pi} 2\cos nx\sin x\,dx$$

$$= \frac{1}{\pi}\int_0^{\pi}[\sin(n+1)x - \sin(n-1)x]\,dx = \frac{1}{\pi}\left[\frac{-\cos(n+1)x}{(n+1)} + \frac{\cos(n-1)x}{n-1}\right]_0^{\pi}$$

$$= \frac{1}{\pi}\left[\frac{-\cos(n+1)\pi}{n+1} + \frac{\cos(n-1)\pi}{n-1} + \frac{1}{n+1} - \frac{1}{n-1}\right]$$

$$= \frac{1}{\pi}\left[\frac{-(-1)^{n+1}}{n+1} + \frac{(-1)^{n-1}}{n-1} + \frac{1}{n+1} - \frac{1}{n-1}\right]$$

$$= \frac{1}{\pi}\left[\begin{array}{ll} 0, & \text{when } n \text{ is odd } i.e.,\ n = 3,5,7 \\[2mm] -\dfrac{4}{n^2-1}, & n \text{ is even} \end{array}\right]$$

$$|\sin x| = \frac{2}{\pi} - \frac{4}{\pi}\left(\frac{\cos 2x}{3} + \frac{\cos 4x}{15} + \frac{\cos 2nx}{4n^2-1}\right)$$

EXAMPLE 3. *Obtain the Fourier series to represent* $f(x) = \dfrac{1}{4}(\pi - x)^2, \ 0 < x < 2\pi$. *Hence obtained the following:*

(i) $\dfrac{1}{1^2} + \dfrac{1}{2^2} + \dfrac{1}{3^2} + \dfrac{1}{4^2} + \ldots = \dfrac{\pi^2}{6}$

(ii) $\dfrac{1}{1^2} - \dfrac{1}{2^2} + \dfrac{1}{3^2} - \dfrac{1}{4^2} + \ldots + = \dfrac{\pi^2}{12}$

(iii) $\dfrac{1}{1^2} + \dfrac{1}{3^2} + \dfrac{1}{5^2} + \ldots + = \dfrac{\pi^2}{8}$

SOLUTION.

(i) Consider

$$f(x) = \frac{1}{4}(\pi - x)^2$$

$$a_0 = \frac{1}{\pi}\int_0^{2\pi}\frac{1}{4}(\pi - x)^2\,dx = \frac{1}{4\pi}\int_0^{2\pi}(\pi - x)^2\,dx$$

$$= \frac{1}{4\pi}\left[\frac{(\pi - x)^3}{-3}\right]_0^{2\pi} = \frac{1}{-12\pi}[-\pi^3 - \pi^3] = \frac{\pi^2}{6}$$

$$a_n = \frac{1}{\pi}\int_0^{2\pi}f(x)\cos nx\,dx = \frac{1}{\pi}\int_0^{2\pi}\frac{1}{4}(\pi - x)^2\cos nx\,dx$$

$$= \frac{1}{4\pi}\int_0^{2\pi}(\pi - x)^2\cos nx\,dx$$

$$= \frac{1}{4\pi}\left[(\pi - x)^2\frac{\sin nx}{n} - \left\{-2(\pi - x)\left(-\frac{\cos nx}{n^2}\right) + 2\left(-\frac{\sin nx}{n^3}\right)\right\}\right]_0^{2\pi}$$

$$= \frac{1}{4\pi}\left[\frac{2\pi}{n^2} + \frac{2\pi}{n^2}\right] = \frac{1}{n^2}$$

$$b_n = \frac{1}{\pi}\int_0^{2\pi}\frac{1}{4}(\pi - x)^2\sin nx\,dx$$

$$= \frac{1}{4x}\left[(\pi - x)^2\left(-\frac{\cos nx}{n}\right) - \left\{2(\pi - x)(-1)\left(-\frac{\sin nx}{n^2}\right) + 2\left(\frac{\cos nx}{n^3}\right)\right\}\right]_0^{2\pi}$$

$$b_n = 0$$

$$f(x) = \frac{\pi^2}{12} + \sum_{n=1}^{\infty}\frac{\cos nx}{n^2} = \frac{\pi^2}{12} + \frac{\cos x}{1^2} + \frac{\cos 2x}{2^2} + \frac{\cos 3x}{3^2} + \ldots \qquad (1)$$

(i) Putting $x = 0$ in eq^n (1)

$$\frac{\pi^2}{4} = \frac{\pi^2}{12} + \left(\frac{1}{1^2} + \frac{1}{2^2} + \frac{1}{3^2} + \frac{1}{4^2} + \frac{1}{5^2} + \ldots\right)$$

$$\frac{1}{1^2} + \frac{1}{2^2} + \frac{1}{3^2} + \frac{1}{4^2} = \frac{\pi^2}{6}$$

(ii) Putting $x = \pi$ in eq^n. (1)

$$\frac{1}{4}(\pi - \pi)^2 = \frac{\pi^2}{12} + \left[\left(-\frac{1}{1^2} + \frac{1}{2^2} - \frac{1}{3^2} + \frac{1}{4^2} + \ldots\right)\right]$$

$$0 = \frac{\pi^2}{12} + \left[(-1) + \frac{1}{2^2} + \left(-\frac{1}{3^2}\right) + \frac{1}{4^2} + \ldots\right]$$

$$\frac{\pi^2}{12} = \frac{1}{1^2} - \frac{1}{2^2} + \frac{1}{3^2} - \frac{1}{4^2} + \ldots$$

(iii) Adding (i) and (ii)

$$\frac{\pi^2}{6} + \frac{\pi^2}{12} = 2\left(\frac{1}{1^2} + \frac{1}{3^2} + \frac{1}{5^2} + \ldots\right)$$

$$\frac{\pi^2}{4} = 2\left(\frac{1}{1^2} + \frac{1}{3^2} + \frac{1}{5^2} + \ldots\right)$$

$$\frac{\pi^2}{8} = \frac{1}{1^2} + \frac{1}{3^2} + \frac{1}{5^2} + \ldots$$

EXAMPLE 4.   *Expand in a Fourier series the function $f(x) = x$ in the interval $0 < x < 2\pi$.*

(UPTU-2001)

SOLUTION.   Here, we have $f(x) = x$

$$a_0 = \frac{1}{\pi}\int_0^{2\pi} f(x)\,dx = \frac{1}{\pi}\int_0^{\pi} x\,dx = \frac{1}{\pi}\left[\frac{x^2}{2}\right] = \frac{4\pi^2}{2\pi} = 2\pi$$

$$a_n = \frac{1}{\pi}\int_0^{2\pi} x\cos nx\,dx = \frac{1}{\pi}\left[\frac{x\sin nx}{n} - \left(-\frac{\cos nx}{n^2}\right)\right]_0^{2\pi} = 0$$

$$b_n = \frac{1}{\pi}\int_0^{2\pi} x\sin nx\,dx = \frac{1}{\pi}\left[x\left(-\frac{\cos nx}{n}\right) - \left(-\frac{\sin nx}{n^2}\right)\right]_0^{2\pi} = \frac{1}{\pi}\left[-\frac{2\pi}{n}\right] = -\frac{2}{n}$$

$$f(x) = \pi - 2\sum_{n=1}^{\infty} \frac{\sin nx}{n}$$

EXAMPLE 5.   *Show that for $-\pi < x < \pi$*

$$\sin ax = \frac{2\sin a x}{\pi}\left(\frac{\sin x}{1^2 - a^2} - \frac{2\sin 2x}{2^2 - a^2} + \frac{3\sin 3x}{3^2 - a^2}\right)$$

SOLUTION.   We have $f(x) = \sin ax$

$$a_0 = \frac{1}{\pi}\int_{-\pi}^{\pi} f(x)\,dx = \frac{2}{\pi}\int_0^{\pi} \sin ax\,dx = \frac{2}{\pi}\left[-\frac{\cos ax}{a}\right]_0^{2\pi} = 0$$

$$a_n = \frac{2}{\pi}\int_0^{\pi} \sin ax\cos nx\,dx$$

$$= \frac{1}{\pi}\int_0^{\pi} [\sin(n+a)x - \sin(n-a)x]\,dx$$

$$= \frac{1}{\pi}\left[-\frac{\cos(n+a)x}{(n+a)} + \frac{\cos(n-a)x}{(n-a)}\right]_0^{\pi} = 0$$

$$b_n = \frac{2}{\pi}\int_0^{\pi} \sin ax\sin nx\,dx$$

$$= \frac{1}{\pi}\left[\int_0^{\pi}[[\cos(n-a)x - \cos(n+a)x]\,dx\right]$$

$$= \frac{1}{\pi} \left[\frac{\sin(n-a)x}{(n-a)} - \frac{\sin(n+a)x}{(n+a)} \right]_0^\pi$$

$$= \frac{1}{\pi} \left[\frac{\sin(n-a)\pi}{(n-a)} - \sin\frac{(n+a)\pi}{(n+a)} - 0 + 0 \right]$$

$$= \frac{1}{\pi} \left[\frac{(-1)^n(-\sin a\pi)}{(n-a)} - \frac{(-1)^n \sin a\pi}{(n+a)} \right]$$

$$= \frac{(-1)^{n+1} \sin a\pi}{\pi} \left[\frac{1}{n-a} + \frac{1}{n+a} \right] = \frac{(-1)^{n+1} 2n \sin a\pi}{\pi(n^2 - a^2)}$$

$$f(x) = \frac{a_0}{2} + \sum_{n=1}^\infty a_n \cos nx + \sum_{n=1}^\infty b_n \sin nx$$

$$\sin ax = \frac{2\sin a\pi}{\pi} \sum \frac{(-1)^{n+1}}{n^2 - a^2} \sin nx$$

$$\sin ax = \frac{2\sin ax}{\pi} \left[\frac{\sin x}{1^2 - a^2} - \frac{2\sin 2x}{2^2 - a^2} + \frac{3\sin 3x}{3^2 - a^2} + ... \right]$$

EXAMPLE 6. *Obtain a Fourier expansion for $\sqrt{1-\cos x}$ in the interval $-\pi < x < \pi$.*

SOLUTION. We have $f(x) = \sqrt{1 - \cos x} = \sqrt{2} \sin\frac{x}{2}$

Since $\sqrt{2} \sin\frac{x}{2}$ is an even function of x, so $b_n = 0$.

$$a_0 = \frac{\sqrt{2}}{\pi} \int_{-\pi}^\pi \sin\frac{x}{2} dx = \frac{2\sqrt{2}}{\pi} \int_0^\pi \sin\frac{x}{2} dx$$

$$= \frac{2\sqrt{2}}{\pi} \left[2\left(-\cos\frac{x}{2}\right) \right]_0^\pi = \frac{4\sqrt{2}}{\pi}$$

$$a_n = \frac{2\sqrt{2}}{\pi} \int_0^\pi \sin\frac{x}{2} \cos nx\, dx = \frac{\sqrt{2}}{\pi} \int_0^\pi 2\cos nx \sin\frac{x}{2} dx$$

$$= \frac{\sqrt{2}}{\pi} \int_0^\pi \left[\sin\left(n + \frac{1}{2}\right)x - \sin\left(n - \frac{1}{2}\right)x \right]$$

$$= \frac{\sqrt{2}}{\pi} \left[\frac{-\cos\left(n + \frac{1}{2}\right)x}{\left(n + \frac{1}{2}\right)} + \frac{\cos\left(n - \frac{1}{2}\right)x}{\left(n - \frac{1}{2}\right)} \right]_0^\pi$$

$$= \frac{\sqrt{2}}{\pi} \left\{ \frac{-\cos\left(n + \frac{1}{2}\right)\pi}{\left(n + \frac{1}{2}\right)} + \frac{\cos\left(n - \frac{1}{2}\right)\pi}{\left(n - \frac{1}{2}\right)} + \frac{1}{\left(n + \frac{1}{2}\right)} - \frac{1}{\left(n - \frac{1}{2}\right)} \right\}$$

$$\left\{ \text{because } \cos\left(n + \frac{1}{2}\right)\pi = 0 \right\}$$

$$= \frac{\sqrt{2}}{\pi} \left\{ \frac{1}{\left(n+\frac{1}{2}\right)} - \frac{1}{\left(n-\frac{1}{2}\right)} \right\} = \frac{\sqrt{2}}{\pi} \left\{ \frac{\left[\left(n-\frac{1}{2}\right) - \left(n+\frac{1}{2}\right)\right]}{\left(n^2 - \frac{1}{4}\right)} \right\} = -\frac{4\sqrt{2}}{\pi(4n^2 - 1)}$$

$$f(x) = \frac{a_0}{2} + \sum_{n=1}^{\infty} a_n \cos nx$$

$$\sqrt{1 - \cos x} = \frac{2\sqrt{2}}{\pi} - \frac{4\sqrt{2}}{\pi} \sum_{n=1}^{\infty} \frac{\cos nx}{4n^2 - 1}$$

EXAMPLE 7. *Obtain a Fourier series to represent e^{-ax} from $x = -\pi$ to $x = \pi$. Hence derive series for*
$$\frac{\pi}{\sin hx}.$$

SOLUTION. Let $\quad f(x) = e^{-ax} = \frac{a_0}{2} + \sum_{n=1}^{\infty} a_n \cos nx + \sum_{n=1}^{\infty} b_n \sin nx$

$$a_0 = \frac{1}{\pi} \int_{-\pi}^{\pi} f(x) dx = \frac{1}{\pi} \int_{-\pi}^{\pi} e^{-ax} dx$$

$$= \frac{1}{\pi} \left[\frac{e^{-ax}}{-a} \right]_{-\pi}^{\pi} = \frac{1}{a\pi} (e^{a\pi} - e^{-a\pi}) = \frac{2 \sinh a\pi}{a\pi}$$

$$a_n = \frac{1}{\pi} \int_{-\pi}^{\pi} e^{-ax} \cos nx \, dx$$

$$= \frac{1}{\pi} \frac{e^{-ax}}{(a^2 + n^2)} [-a \cos nx + n \sin nx]_{-\pi}^{\pi}$$

$$= \frac{1}{\pi(a^2 + n^2)} [e^{-a\pi}(-a \cos nx) + e^{a\pi}(a \cos nx)]$$

$$= \frac{a \cos n\pi}{\pi(a^2 + n^2)} (e^{a\pi} - e^{-a\pi}) = \frac{(-1)^n 2a \sinh a\pi}{\pi(a^2 + n^2)} = \frac{2a(-1)^n \sinh a\pi}{\pi(a^2 + n^2)}$$

Now, $\quad b_n = \frac{1}{\pi} \int_{-\pi}^{\pi} e^{-ax} \sin nx \, dx = \frac{1}{\pi} \frac{e^{-ax}}{\pi(a^2 + n^2)} [(-a \sin nx - n \cos nx)]_{-\pi}^{\pi}$

$$= \frac{(-1)^n 2a \sinh a\pi}{\pi(a^2 + n^2)}$$

$\therefore \quad e^{-ax} = \frac{\sinh a\pi}{a\pi} + \sum_{n=1}^{\infty} \frac{2a(-1)^n \sinh a\pi}{\pi(a^2 + n^2)} \cdot \cos nx + \sum_{n=1}^{\infty} \frac{2n(-1)^n \sinh a\pi}{\pi(a^2 + b^2)} \cdot \sin nx$

$$= \frac{2 \sinh a\pi}{\pi} \left[\frac{1}{2a} - a \left(\frac{\cos x}{a^2 + 1^2} - \frac{\cos 2x}{a^2 + 2^2} + \frac{\cos 3x}{a^2 + 3^2} \right) \right.$$

$$\left. - \left(\frac{\sin x}{a^2 + 1^2} - \frac{2 \sin 2x}{a^2 + b^2} + \frac{3 \sin 3x}{a^2 + 3^2} + \ldots \right) \right]$$

EXAMPLE 8. *Find a series of sines and cosines to multiples of x which will represent* $\dfrac{\pi}{\sinh \pi} e^x$ *in the interval* $-\pi < x < \pi$.

SOLUTION. Consider, $f(x) = \dfrac{\pi}{2\sinh \pi} e^x$, $-\pi < x < \pi$

Now, Fourier series is given by

$$f(x) = \frac{a_0}{2} + \sum_{n=1}^{\infty} (a_n \cos nx + b_n \sin nx) \qquad ....(1)$$

where $\quad a_0 = \dfrac{1}{\pi} \int_{-\pi}^{\pi} f(x)\,dx = \dfrac{1}{\pi} \int_{-\pi}^{\pi} \dfrac{\pi}{2\sin n\pi} e^x\,dx \ = \dfrac{1}{\pi} \cdot \dfrac{1}{2\sinh \pi} \int_{-\pi}^{\pi} e^x\,dx$

$\dfrac{1}{2\sinh \pi} [e^x]_{-\pi}^{\pi} = \dfrac{1}{2\sinh \pi} (e^\pi - e^{-\pi})$

$\qquad\qquad = \dfrac{2\sinh \pi}{2\sinh \pi} = 1 \qquad\qquad \left[\because \sinh \pi = \dfrac{e^\pi - e^{-\pi}}{2} \right]$

$a_n = \dfrac{1}{\pi} \int_{-\pi}^{\pi} f(x) \cos nx\,dx = \dfrac{1}{\pi} \int_{-\pi}^{\pi} \dfrac{\pi}{2\sinh \pi} e^x \cos nx\,dx$

$\qquad = \dfrac{1}{2\sinh \pi} \int_{-\pi}^{\pi} e^x \cos nx\,dx$

$\qquad = \dfrac{1}{2\sinh \pi} \left[\dfrac{e^x}{1+n^2} (\cos nx + n \sin nx) \right]_{-\pi}^{\pi}$

$\qquad = \dfrac{1}{2\sinh \pi} \cdot \dfrac{1}{1+n^2} [e^\pi (-1)^n - e^{-\pi} (-1)^n]$

$\qquad\qquad\qquad\qquad\qquad\qquad [\because \cos n\pi = (-1)^n, \ \sin n\pi = 0]$

$\qquad = \dfrac{(-1)^n}{1+n^2} \cdot \dfrac{1}{2\sinh \pi} \cdot 2\sinh h\pi = \dfrac{(-1)^n}{1+n^2}$

Now, $\quad b_n = \dfrac{1}{\pi} \int_{-\pi}^{\pi} f(x) \sin nx\,dx = \dfrac{1}{\pi} \int_{-\pi}^{\pi} \dfrac{\pi}{2\sinh \pi} e^x \sin nx\,dx$

$\qquad = \dfrac{1}{\pi} \cdot \dfrac{\pi}{2\sinh \pi} \int_{-\pi}^{\pi} e^x \sin nx\,dx$

$\qquad = \dfrac{1}{2\sinh \pi} \left[\dfrac{e^x}{1+n^2} (\sin nx - n \cos nx) \right]_{-\pi}^{\pi}$

$\qquad = \dfrac{1}{2\sinh \pi} \cdot \dfrac{1}{1+n^2} [e^\pi (-n)(-1)^n - e^{-\pi} (-n).(-1)^n]$

$\qquad = -(-1)^n \dfrac{n}{1+n^2} \cdot \dfrac{2\sinh \pi}{2\sinh \pi} = -(-1)^n \dfrac{n}{1+n^2}$

Putting all these values in eq^n (1), we get

$$f(x) = \frac{1}{2} + \sum_{n=1}^{\infty} \frac{(-1)^n}{n^2+1} (\cos nx - n \sin nx)$$

$\qquad = \dfrac{1}{2} + \dfrac{(-1)^1}{1^2+1} (\cos x - 1.\sin x) + \dfrac{(-1)^n}{2^2+1} (\cos 2x - 2\sin 2x) + ...$

$\therefore \quad \dfrac{\pi}{2\sinh \pi} e^x = \dfrac{1}{2} - \dfrac{1}{2} (\cos x - \sin x) + \dfrac{1}{5} (\cos 2x - 2\sin 2x) + ...$

EXAMPLE 9. *Prove that* $x^2 = \dfrac{\pi^3}{3} + 4\sum_{n=1}^{\infty}(-1)^n \dfrac{\cos nx}{n^2}$. (UPTU-2004)

SOLUTION. Since x^2 is even function. So $b_n = 0$.

Now, $f(x) = x^2$

$$a_0 = \frac{1}{\pi}\int_{-\pi}^{\pi}x^2\,dx = \frac{2}{\pi}\left[\frac{x^3}{3}\right]_{-\pi}^{\pi} = \frac{2\pi^3}{3}$$

$$a_n = \frac{2}{\pi}\int_0^{\pi}x^2\cos nx\,dx$$

$$= \frac{2}{\pi}\left[\frac{x^2\sin nx}{n} - 2x\left(-\frac{\cos nx}{n^2}\right) + 2\left(-\frac{\sin nx}{n^3}\right)\right]_0^{\pi} = \frac{4\cos n\pi}{n^2} = \frac{4(-1)^n}{n^2}$$

$$f(x) = \frac{2\pi^3}{2.3} + 4\sum_{n=1}^{\infty}\frac{(-1)^n\cos nx}{n^2}$$

EXAMPLE 10. *Prove that in the interval* $x\cos x = -\dfrac{2}{\pi} - \dfrac{1}{2}\sin x + 2\sum_{n=2}^{\infty}\dfrac{n(-1)^n\sin nx}{n^2-1}$ (UPTU-2001)

SOLUTION. We have, $a_0 = \dfrac{2}{\pi}\int_0^{\pi}x\cos x\,dx = \dfrac{2}{\pi}[x\sin x - 0)\{-\cos x\}]_0^{\pi}$

$$\frac{2}{\pi}[\cos\pi - \cos 0] = -\frac{4}{\pi}$$

$$a_n = \frac{2}{\pi}\int_0^{\pi}x\cos x\cos nx\,dx = \frac{1}{\pi}\int_0^{\pi}x.2\cos nx\cos x\,dx$$

$$= \frac{1}{\pi}\int_0^{\pi}[x\{\cos(n+1)x + \cos(n-1)x\}\,dx]$$

$$= \frac{1}{\pi}\left\{x\{\sin(n+1)x + \sin(n-1)x\} - (1) - \left\{\frac{-\cos(n+1)x}{(n+1)^2} - \frac{\cos(n-1)x}{(n-1)^2}\right\}\right]_0^{\pi}$$

$$= \frac{1}{\pi}\left\{\pi\{\sin(n+1)\pi + \sin(n-1)\pi\} + \frac{\cos(n+1)\pi}{(n+1)^2} + \frac{\cos(n-1)\pi}{(n-1)^2} - 0\right\} = 0$$

$$b_n = \frac{2}{\pi}\int_0^{\pi}x\cos x\sin nx\,dx = \frac{1}{\pi}\int_0^{\pi}x.2\sin nx\cos x\,dx$$

$$= \frac{1}{\pi}\int_0^{\pi}x\{\sin(n+1) + \sin(n-1)x\}\,dx$$

$$= \frac{1}{\pi}\left[x\left\{-\frac{\cos(n+1)x}{(n+1)} - \frac{\cos(n-1)x}{(n-1)}\right\} - 1\left\{\frac{\sin(n+1)x}{(n+1)^2} - \frac{\sin(n-1)x}{(n-1)^2}\right\}\right]_0^{\pi}$$

$$= \frac{1}{\pi}\left[\pi\left\{-\frac{\cos(n+1)\pi}{n+1} - \frac{\cos(n-1)\pi}{n-1}\right\}\right]$$

$$= \left[\frac{1}{\pi}.\pi\left\{\frac{-(-1)^{n+1}}{n+1} - \frac{(-1)^{n-1}}{n-1}\right\}\right]$$

$$= \begin{cases} \dfrac{2}{n^2-1}; & \text{when } n \text{ is even} \\[3mm] -\dfrac{2n}{n^2-1}; & \text{when } n \text{ is odd} \end{cases}$$

when $n = 1$, $\quad a_1 = \dfrac{2}{\pi}\int_0^\pi x\cos x \sin x\, dx = \dfrac{1}{\pi}\int_0^\pi x\cos x \sin x\, dx$

$$= \dfrac{1}{\pi}\left[x\left(-\dfrac{\cos 2x}{2}\right) - 1\left(-\dfrac{\sin 2x}{4}\right) \right]_0^\pi = \dfrac{1}{\pi}\left[\dfrac{-\pi\cos 2\pi}{2} \right] = -\dfrac{1}{2}$$

Hence, $\quad f(x) = -\dfrac{2}{\pi} - \dfrac{1}{2}\sin x + 2\sum_{n=2}^\infty (-1)^n n \sin nx$

9.6 FOURIER SERIES FOR DISCONTINUOUS FUNCTIONS

At the point of discontinuity, the value of function for Fourier series is obtained by the average of left hand limit and right hand limit of function at that point of discontinuity.

Solved Examples

EXAMPLE 1. *Obtain Fourier series for the function*

$$f(x) = \begin{cases} x & ; & -\pi < x < 0 \\ -x & ; & 0 < x < \pi \end{cases}$$

and hence show that $\dfrac{1}{1^2} + \dfrac{1}{3^2} + \dfrac{1}{5^2} + \ldots = \dfrac{\pi^2}{8}$. $\hspace{2cm}$ (UPTU–2002)

SOLUTION. We know that

$$f(x) = \dfrac{a_0}{2} + \sum_{n=1}^\infty a_n \cos nx + \sum_{n=1}^\infty b_n \sin nx \hspace{2cm} \ldots(1)$$

$$a_0 = \dfrac{1}{\pi}\int_{-\pi}^\pi f(x)\, dx = \dfrac{1}{\pi}\left[\int_{-\pi}^0 x\, dx + \int_0^\pi -x\, dx \right]$$

$$= \dfrac{1}{\pi}\left[\left(\dfrac{x^2}{2}\right)_{-\pi}^0 - \left(\dfrac{x^2}{2}\right)_0^\pi \right] = \dfrac{1}{\pi}\left[0 - \dfrac{\pi^2}{2} - \dfrac{\pi^2}{2} \right] = -\pi$$

$$a_n = \dfrac{1}{\pi}\int_{-\pi}^\pi f(x)\cos nx\, dx = \dfrac{1}{\pi}\left[\int_{-\pi}^0 x\cos nx\, dx + \int_0^\pi -x\cos nx\, dx \right]$$

$$= \dfrac{1}{\pi}\left[\left(\dfrac{x\sin nx}{n}\right)_{-\pi}^0 - \int_{-\pi}^0 \dfrac{\sin nx}{n}\, dx + \left(-x\dfrac{\sin nx}{n}\right)_0^\pi - \int_0^\pi (-1)\dfrac{\sin nx}{n}\, dx \right]$$

$$= \dfrac{1}{\pi}\left[\dfrac{1}{n^2}(\cos nx)_{-\pi}^0 - \dfrac{1}{n^2}(\cos nx)_0^\pi \right] = \dfrac{1}{\pi}\left[\left\{\dfrac{1-(-1)^n}{n^2}\right\} - \left\{\dfrac{(-1)^n-1}{n^2}\right\} \right]$$

$$= \frac{1}{\pi}\left[\frac{2\{1-(-1)^n\}}{n^2}\right] = \frac{2}{\pi n^2}[1-(-1)^n] = \begin{cases} 0 & ; \text{ if } n \text{ is even} \\ \frac{4}{\pi n^2} & ; \text{ if } n \text{ is odd} \end{cases}$$

and $b_n = \frac{1}{\pi}\int_{-\pi}^{\pi} f(x)\sin nx\, dx$

$$= \frac{1}{\pi}\left[\int_{-\pi}^{0} x\sin nx\, dx + \int_0^{\pi} -x\sin nx\, dx\right]$$

$$= \frac{1}{\pi}\left[\left(x\frac{-\cos nx}{n}\right)_{-\pi}^0 - \int_{-\pi}^0 \frac{-\cos nx}{n}\, dx + \left(x\frac{\cos nx}{n}\right)_0^{\pi} - \int_0^{\pi}(-1)\frac{-\cos nx}{n}\, dx\right]$$

$$= \frac{1}{\pi}\left[\frac{-\pi}{n}(-1)^n + \frac{1}{n}(-1)^n\right] = 0$$

From (1)

$$f(x) = -\frac{\pi}{2} + \frac{4}{\pi}\left(\frac{\cos x}{1^2} + \frac{\cos 3x}{3^2} + \frac{\cos 5x}{5^2} + \dots\right)$$

At the point of discontinuity

$$f(0) = \frac{1}{2}[f(0^-) + f(0^+)] = \frac{1}{2}[0-0] = 0$$

Putting, $x = 0$ in (2), we get $0 = -\frac{\pi}{2} + \frac{4}{\pi}\left(\frac{1}{1^2} + \frac{1}{3^2} + \frac{1}{5^2} + \dots\right)$

Hence, $\frac{1}{1^2} + \frac{1}{3^2} + \frac{1}{5^2} + \dots = \frac{\pi^2}{8}$.

EXAMPLE 2.   *Obtain the Fourier series to represent $f(x)$ given as follows :*

$$f(x) = \begin{cases} x & ; \text{ for } 0 \leq x \leq \pi \\ 2\pi - x & ; \text{ for } \pi \leq x \leq 2\pi \end{cases}$$ (SVTU–2008, BPTU–2005S)

SOLUTION.   Let $f(x) = \frac{a_0}{2} + \sum_{n=1}^{\infty} a_n \cos nx + \sum_{n=1}^{\infty} b_n \sin nx, \; 0 \leq x \leq 2\pi$ ... (1)

where $a_0 = \frac{1}{\pi}\int_0^{2\pi} f(x)\, dx$

$$= \frac{1}{\pi}\left[\int_0^{\pi} x\, dx + \int_{\pi}^{2\pi}(2\pi - x)\, dx\right] = \frac{1}{\pi}\left[\left(\frac{x^2}{2}\right)_0^{\pi} + \left(2\pi x - \frac{x^2}{2}\right)_{\pi}^{2\pi}\right]$$

$$= \frac{1}{\pi}\left[\frac{\pi^2}{2} + 2\pi(2\pi - x) - \frac{1}{2}(4\pi^2 - \pi^2)\right] = \frac{1}{\pi}(\pi^2) = \pi$$

$$a_n = \frac{1}{\pi}\int_0^{2\pi} f(x)\cos nx\, dx = \frac{1}{\pi}\left[\int_0^{\pi} x\cos nx\, dx + \int_0^{2\pi}(2\pi - x)\cos nx\, dx\right]$$

$$= \frac{1}{\pi}\left[\left\{\frac{x\sin nx}{n} + \frac{\cos nx}{n^2}\right\}_0^{\pi} + \left\{(2\pi - x)\frac{\sin nx}{n} - \frac{\cos nx}{n^2}\right\}_{\pi}^{2\pi}\right]$$

$$= \frac{1}{\pi}\left[\left(\frac{\cos n\pi - 1}{n^2}\right) - \left(\frac{1 - \cos n\pi}{n^2}\right)\right]$$

$$= \frac{2}{n^2\pi}[(-1)^n - 1] = \begin{cases} 0 & , \text{ if } n \text{ is even} \\ -\dfrac{4}{n\pi^2} & , \text{ if } n \text{ is odd} \end{cases}$$

Again $\quad b_n = \frac{1}{\pi}\int_0^{2\pi} f(x).\sin nx\, dx$

$$= \frac{1}{\pi}\left[\int_0^{\pi} x \sin nx\, dx + \int_0^{2\pi}(2\pi - x)\sin nx\, dx\right]$$

$$= \frac{1}{\pi}\left[\left\{-\frac{x\cos nx}{n} + \frac{\sin nx}{n^2}\right\}_0^{\pi} + \left\{-(2\pi - x)\frac{\cos nx}{n} - \frac{\sin nx}{n^2}\right\}_\pi^{2\pi}\right]$$

$$= \left(\frac{-\pi\cos n\pi}{n} + \frac{\pi\cos n\pi}{n}\right) = 0$$

Therefore, $f(x) = \dfrac{\pi}{2} - \dfrac{4}{\pi}\left[\cos x + \dfrac{\cos 3x}{3^2} + \dfrac{\cos 5x}{5^2} + ...\right], 0 \le x \le 2\pi$

which is the required Fourier series for $f(x)$.

EXAMPLE 3. If $f(x) = \begin{cases} 0 & , -\pi \le x \le 0 \\ \sin x & , \quad 0 \le x \le \pi \end{cases}$

Prove that $f(x) = \dfrac{1}{\pi} + \dfrac{1}{2}\sin x - \dfrac{2}{\pi}\sum_{n=1}^{\infty}\dfrac{\cos 2nx}{4n^2 - 1}$. (Agra-1991, Meerut-1990, Kanpur-1988)

Hence, show that

(i) $\dfrac{1}{1.3} + \dfrac{1}{3.5} + \dfrac{1}{5.7} + ... = \dfrac{1}{2}$

(ii) $\dfrac{1}{1.3} - \dfrac{1}{3.5} + \dfrac{1}{5.7} - ... = \dfrac{\pi - 2}{4}$ (Bhopal–2008, Mumbai–2005S, Rohtak–2005)

SOLUTION. Let $\quad f(x) = \dfrac{a_0}{2} + \sum_{n=1}^{\infty} a_n \cos nx + \sum_{n=1}^{\infty} b_n \sin nx$

Then, $\quad a_0 = \dfrac{1}{\pi}\int_{-\pi}^{\pi} f(x)\,dx = \dfrac{1}{\pi}\left[\int_{-\pi}^{0} 0.dx + \int_0^{\pi}\sin x\,dx\right] = \dfrac{2}{\pi}$

$$a_n = \frac{1}{\pi}\int_{-\pi}^{\pi} f(x).\cos nx\, dx = \frac{1}{\pi}\left[\int_{-\pi}^{0} 0.dx + \int_0^{\pi}\sin x \cos nx\, dx\right]$$

$$= \frac{1}{2\pi}\int_0^{2\pi} 2\cos nx.\sin x\, dx = \frac{1}{2\pi}\int_0^{\pi}[\sin(n+1)x - \sin(n-1)x]dx$$

$$= \frac{1}{2\pi}\left[-\frac{\cos(n+1)x}{n+1} + \frac{\cos(n-1)x}{n-1}\right]_0^{\pi}, n \ne 1$$

$$= \frac{1}{2\pi}\left[-\frac{\cos(n+1)\pi}{n+1} + \frac{\cos(n-1)\pi}{n-1} + \frac{1}{n+1} - \frac{1}{n-1}\right]$$

$$= \frac{1}{2\pi}\left[-\frac{(-1)^{n+1}}{n+1} + \frac{(-1)^{n-1}}{n-1} + \frac{1}{n+1} - \frac{1}{n-1}\right]$$

$$= \begin{cases} \dfrac{1}{2\pi}\left(-\dfrac{1}{n+1}+\dfrac{1}{n-1}+\dfrac{1}{n+1}-\dfrac{1}{n-1}\right) & , \quad \text{when } n \text{ is odd} \\[3mm] \dfrac{1}{2\pi}\left(\dfrac{1}{n+1}-\dfrac{1}{n-1}+\dfrac{1}{n+1}-\dfrac{1}{n-1}\right) & , \quad \text{when } n \text{ is even} \end{cases}$$

$$= \begin{cases} 0 & , \quad \text{when } n \text{ is odd}, i.e., n = 3,5,7,... \\[3mm] -\dfrac{2}{\pi(n^2-1)} & , \quad \text{when } n \text{ is even} \end{cases}$$

When $n=1$, we have

$$a_1 = \frac{1}{\pi}\int_0^\pi \sin x \cos x\, dx = \frac{1}{2\pi}\int_0^\pi \sin 2x\, dx = \frac{1}{2\pi}\left[-\frac{\cos 2x}{2}\right]_0^\pi = 0$$

and $\quad b_n = \dfrac{1}{\pi}\int_{-\pi}^\pi f(x)\sin nx\, dx = \dfrac{1}{\pi}\left[\int_{-\pi}^0 0.dx + \int_0^\pi \sin x \sin nx\, dx\right]$

$$= \frac{1}{2\pi}\int_0^\pi 2\sin nx \sin x\, dx = \frac{1}{2\pi}\int_0^\pi [\cos(n-1)x - \cos(n+1)x]\, dx$$

$$= \frac{1}{2\pi}\left[\frac{\sin(n-1)x}{(n-1)} - \frac{\sin(n+1)x}{(n+1)}\right]_0^\pi = 0, n \neq 1$$

When $n = 1$, we have $b_1 = \dfrac{1}{\pi}\int_0^\pi \sin x \sin x\, dx$

$$= \frac{1}{2\pi}\int_0^\pi (1-\cos 2x)dx = \frac{1}{2\pi}\left[x - \frac{\sin 2x}{2}\right]_0^\pi = \frac{1}{2}$$

$$\therefore \; f(x) = \frac{1}{\pi} - \frac{2}{\pi}\left[\frac{\cos 2x}{2^2-1} + \frac{\cos 4x}{4^2-1} + \frac{\cos 6x}{6^2-1} + ...\right] + \frac{1}{2}\sin x$$

$$= \frac{1}{\pi} + \frac{1}{2}\sin x - \frac{2}{\pi}\sum_{n=1}^\infty \frac{\cos 2nx}{(2n)^2-1}$$

Putting $x = 0$ in equation (1), we have

$$0 = \frac{1}{\pi} - \frac{2}{\pi}\sum_{n=1}^\infty \frac{1}{4n^2-1}$$

$$\frac{1}{2} = \sum_{n=1}^\infty \frac{1}{4n^2-1} = \sum_{n=1}^\infty \frac{1}{(2n-1)(2n+1)} = \frac{1}{1.3} + \frac{1}{3.5} + \frac{1}{5.7} + ...$$

Putting $x = \pi/2$ in equation (1), we have,

$$1 = \frac{1}{\pi} + \frac{1}{2} - \frac{2}{\pi}\sum_{n=1}^\infty \frac{\cos n\pi}{4n^2-1}$$

$$\Rightarrow \frac{1}{2} - \frac{1}{\pi} = -\frac{2}{\pi}\sum_{n=1}^\infty \frac{(-1)^n}{4n^2-1}$$

$$\Rightarrow \frac{\pi-2}{4} = -\sum_{n=1}^\infty \frac{(-1)^n}{(2n-1)(2n+1)} = -\left(-\frac{1}{1.3} + \frac{1}{3.5} - \frac{1}{5.7} + ...\right)$$

$$\Rightarrow \frac{1}{1.3} - \frac{1}{3.5} + \frac{1}{5.7} - ... = \frac{\pi-2}{4}$$

Additional Solved Examples

EXAMPLE 1. Find the Fourier series for the following function:

$$f(x) = \begin{cases} x^2, & 0 \le x \le \pi \\ -x^2, & -\pi \le x \le 0 \end{cases}$$

(Mumbai-2009; Hissar-2007)

SOLUTION.

$$a_0 = \frac{1}{\pi} \int_{-\pi}^{\pi} f(x)\,dx$$

$$= \frac{1}{\pi} \left[\int_{-\pi}^{0} (-x^2)\,dx + \int_{0}^{\pi} x^2\,dx \right] = \frac{1}{\pi} \left\{ \left[-\frac{x^3}{3} \right]_{-\pi}^{0} + \left[\frac{x^3}{3} \right]_{0}^{\pi} \right\} = 0$$

$$a_n = \frac{1}{\pi} \int_{-\pi}^{0} f(x)\cos nx\,dx + \frac{1}{\pi} \int_{0}^{\pi} f(x)\cos nx\,dx$$

$$= \frac{1}{\pi} \int_{-\pi}^{0} (-x^2)\cos nx\,dx + \frac{1}{\pi} \int_{0}^{\pi} x^2 \cos nx\,dx$$

$$= \frac{1}{\pi} \left[(-x^2) \left(\frac{\sin nx}{n} \right) - (-2x) \left(-\frac{\cos nx}{n^2} \right) + (-2) \left(\frac{-\sin nx}{n^3} \right) \right]_{-\pi}^{0}$$

$$+ \frac{1}{\pi} \left[x^2 \left(\frac{\sin nx}{n} \right) - 2x \left(\frac{-\cos nx}{n^2} \right) + 2 \left(-\frac{\sin nx}{n^3} \right) \right]_{0}^{\pi}$$

$$b_n = \frac{1}{\pi} \int_{-\pi}^{\pi} f(x)\sin nx\,dx$$

$$= \frac{2}{\pi} \int_{-\pi}^{0} -x^2 \sin nx\,dx + \frac{2}{\pi} \int_{0}^{\pi} x^2 \sin nx\,dx \quad \therefore \int_{0}^{\pi} x^2 \sin nx\,dx = 0$$

$$= \frac{2}{\pi} \int_{-\pi}^{0} -x^2 \sin nx\,dx$$

$$= \frac{2}{\pi} \left[(-x)^2 \left(-\frac{\cos nx}{n} \right) - (-2x) \left(-\frac{\sin nx}{n^2} \right) + (-2) \left(\frac{\cos nx}{n^3} \right) \right]_{-\pi}^{0}$$

$$= \frac{2}{\pi} \left[-\frac{2}{n^3} - \frac{\pi^2 \cos n\pi}{n} + \frac{2\cos n\pi}{n^3} \right] = \frac{2}{\pi} \left[\frac{-2}{n^3} - \frac{\pi^2(-1)^n}{n} + \frac{2(-1)^n}{n^3} \right]$$

$$= \begin{cases} -\dfrac{2\pi}{n}, & \text{when } n \text{ is even} \\ \dfrac{2}{n} \left(\pi - \dfrac{4}{\pi n^2} \right), & \text{when } n \text{ is odd} \end{cases}$$

$$f(x) = 2\left[\pi - \frac{4}{\pi} \right] \sin x - \frac{2\pi}{2} \sin 2x + \frac{2}{3} \left[\pi - \frac{4}{9\pi} .. \right] \sin 3x + ...$$

EXAMPLE 2. Find the Fourier series to represent the function

$$f(x) = \begin{cases} -k, & \text{when } -\pi < x < 0 \\ k, & \text{when } 0 < x < \pi \end{cases}$$

SOLUTION .

$$a_0 = \frac{1}{\pi} \int_{-\pi}^{0} (-k) \, dx + \frac{1}{\pi} \int_{0}^{\pi} k \, dx$$

$$= \frac{1}{\pi} [-kx]_{-\pi}^{0} + \frac{1}{\pi} [kx]_{0}^{\pi} = 0$$

$$a_n = \frac{1}{\pi} \int_{-\pi}^{0} -k \cos nx \, dx + \frac{1}{\pi} \int_{0}^{\pi} k \cos nx \, dx$$

$$= \frac{1}{\pi} \left[\frac{k \cos nx}{n} \right]_{-\pi}^{0} + \frac{1}{\pi} \left[\frac{-k \cos nx}{n} \right]_{0}^{\pi} = 0$$

$$b_n = \frac{1}{\pi} \int_{-\pi}^{0} -k \sin nx \, dx + \frac{1}{\pi} \int_{0}^{\pi} k \sin nx \, dx$$

$$= \frac{1}{\pi} \left[\frac{k \cos nx}{n} \right]_{-\pi}^{0} + \frac{1}{\pi} \left[\frac{-k \cos nx}{n} \right]_{0}^{\pi}$$

$$= \frac{1}{\pi} \left[\frac{2k}{n} - \frac{2 \cos n\pi}{n} \right] = \frac{2}{\pi} \left[\frac{k}{n} - \frac{(-1)^n}{n} \right]$$

$$= \begin{cases} \dfrac{2}{\pi} \left(\dfrac{k}{n} - \dfrac{(-1)^n}{n} \right) = 0, & \text{when } n \text{ is even} \\[3mm] \dfrac{4k}{\pi n}, & \text{when } n \text{ is odd } n = 1, 3, 5, \dots \end{cases}$$

Hence, $\quad f(x) = \dfrac{4k}{\pi} \left(\dfrac{\sin x}{1} + \dfrac{\sin 3x}{3} + \dfrac{\sin 5x}{5} + \dots \right)$

EXAMPLE 3.   *Find the Fourier series for the function:*

$$f(x) = \begin{cases} -1 & , \quad -\pi < x < -\pi/2 \\ 0 & , \quad -\pi/2 < x < \pi/2 \\ 1 & , \quad \pi/2 < x < \pi \end{cases}$$

(UPTU-2004, 05)

SOLUTION .   We have $\quad a_0 = \dfrac{1}{\pi} \int_{-\pi}^{-\pi/2} (-1) \, dx + \dfrac{1}{\pi} \int_{-\pi/2}^{\pi/2} 0 \, dx + \dfrac{1}{\pi} \int_{\pi/2}^{\pi} 1 \, dx = 0$

$$a_n = \frac{1}{\pi} \int_{-\pi}^{-\pi/2} (-1) \cos nx \, dx + \frac{1}{\pi} \int_{-\pi/2}^{\pi/2} 0 \cos nx \, dx + \frac{1}{\pi} \int_{\pi/2}^{\pi} 1 \cos nx \, dx = 0$$

$$b_n = \frac{1}{\pi} \int_{-\pi}^{-\pi/2} (-1) \sin nx \, dx + \frac{1}{\pi} \int_{-\pi/2}^{\pi/2} 0 \sin nx \, dx + \frac{1}{\pi} \int_{\pi/2}^{\pi} 1 \sin nx \, dx$$

$$= \frac{2}{n\pi} \left[\cos \frac{n\pi}{2} - \cos n\pi \right]$$

$$b_1 = \frac{1}{\pi} \int_{-\pi}^{-\pi/2} (-1) \sin x \, dx + \frac{1}{\pi} \int_{-\pi/2}^{\pi/2} 0 . \sin x \, dx + \frac{1}{\pi} \int_{\pi/2}^{\pi} \sin x \, dx$$

$$= \frac{1}{\pi} [\cos x]_{-\pi}^{-\pi/2} + 0 + \frac{1}{\pi} [-\cos x]_{\pi/2}^{\pi} = \frac{2}{\pi}$$

$$b_2 = -\frac{2}{\pi}, \quad b_3 = \frac{2}{3\pi}$$

$$f(x) = b_1 \sin x + b_2 \sin 2x + b_3 \sin 3x + \ldots$$

$$= \frac{2}{\pi} \sin x - \frac{2}{\pi} \sin 2x + \frac{2}{3\pi} \sin 3x + \ldots$$

$$= \frac{2}{\pi} \left[\sin x - \sin 2x + \frac{\sin 3x}{3} + \ldots \right]$$

EXAMPLE 4. *Find the Fourier series expansion for*

$$f(x) = \begin{cases} -\pi \; , & -\pi < x < 0 \\ x \; , & 0 < x < \pi \end{cases}$$

(Bhopal-2008)

Deduce that $\dfrac{1}{1^2} + \dfrac{1}{3^2} + \dfrac{1}{5^2} + \ldots = \dfrac{\pi^2}{8}$.

SOLUTION . Let the Fourier series be

$$f(x) = \frac{a_0}{2} + \sum_{n=1}^{\infty} a_n \cos nx + \sum_{n=1}^{\infty} b_n \sin nx$$

where $a_0 = \dfrac{1}{\pi} \int_{-\pi}^{\pi} f(x)\,dx = \dfrac{1}{\pi} \int_{-\pi}^{0} (-\pi)\,dx + \dfrac{1}{\pi} \int_{0}^{\pi} x\,dx$

$$= \frac{1}{\pi} [-\pi x]_{-\pi}^{0} + \frac{1}{\pi} \left[\frac{x^2}{2} \right]_{0}^{\pi} = \frac{1}{\pi} \left[\left(-\pi^2 + \frac{\pi^2}{2} \right) \right] = -\frac{\pi}{2}$$

$$a_n = \frac{1}{\pi} \int_{-\pi}^{0} (-\pi) \cos nx\,dx + \frac{1}{\pi} \int_{0}^{\pi} x \cos nx\,dx$$

$$= \frac{1}{\pi} \left[-\frac{\pi \sin nx}{n} \right]_{-\pi}^{0} + \frac{1}{\pi} \left[\frac{x \sin nx}{n} - \left(-\frac{\cos nx}{n^2} \right) \right]_{0}^{\pi}$$

$$= 0 + \frac{1}{\pi} \left[\frac{\cos n\pi}{n^2} - \frac{1}{n^2} \right] = \frac{1}{\pi n^2} [\cos n\pi - 1]$$

and $b_n = \dfrac{1}{\pi} \int_{-\pi}^{0} (-\pi) \sin nx\,dx + \dfrac{1}{\pi} \int_{0}^{\pi} x \sin nx\,dx$

$$= \frac{1}{\pi} \left[\frac{\pi \cos nx}{n} \right]_{-\pi}^{0} + \frac{1}{\pi} \left[x \frac{\cos nx}{n} - \frac{\sin nx}{n^2} \right]_{0}^{\pi}$$

$$= \frac{1}{\pi} \left[\frac{\pi}{n} (1 - \cos n\pi) - \frac{\pi}{n} \cos n\pi \right] = \frac{1}{n} (1 - 2\cos n\pi) = \frac{1}{n} [1 - 2(-1)^n]$$

$$= \begin{cases} \frac{1}{2} [1 - 2(-1)^n] = 3 \text{ when } n \text{ is even} \\ -1/2 \qquad \text{when } n \text{ is odd} \end{cases}$$

The required series is

$$f(x) = -\frac{\pi}{4} - \frac{2}{\pi} \left(\cos x + \frac{\cos 3x}{3^2} + \frac{\cos 5x}{5^2} + \ldots \right)$$

$$+ \left(3\sin x - \frac{\sin 2x}{2} + \frac{3\sin 3x}{3} - \frac{\sin 4x}{4} + \ldots \right) \qquad \ldots(1)$$

Putting $x = 0$ in (1) $\Rightarrow f(0) = -\dfrac{\pi}{4} - \dfrac{2}{\pi} \left(1 + \dfrac{1}{3^2} + \dfrac{1}{5^2} + \ldots \right)$ $\qquad \ldots(2)$

$f(x)$ is continuous at $x = 0$

$$f(0-0) = -\pi \qquad f(0+0) = 0$$

$$f(0) = \frac{1}{2}[f(0-0) + f(0+0)] = -\frac{\pi}{2}$$

...(3)

Hence from (2) and (3)

$$-\frac{\pi}{2} = -\frac{\pi}{4} - \frac{2}{\pi}\left(1 + \frac{1}{3^2} + \frac{1}{5^2} + ...\right)$$

$$1 + \frac{1}{3^2} + \frac{1}{5^2} + ... = \frac{\pi^2}{8}$$

EXAMPLE 5.   *Find the Fourier expansion of the function defined in one period by the relations:*

$$f(x) = \begin{cases} 1 & , \quad 0 < x < \pi \\ 2 & , \quad \pi < x < 2\pi \end{cases}$$

(Kottayam–2005)

SOLUTION.   Consider $a_0 = \frac{1}{\pi}\int_0^\pi 1\,dx + \frac{1}{\pi}\int_\pi^{2\pi} 2\,dx = \frac{1}{\pi}[x]_0^\pi + \frac{2}{\pi}[x]_\pi^{2\pi} = 3$

$$a_n = \frac{1}{\pi}\int_0^\pi 1 \cdot \cos nx\,dx + \frac{1}{\pi}\int_\pi^{2\pi} 2\cos nx\,dx$$

$$= \frac{1}{\pi}\left[\frac{\sin nx}{n}\right]_0^\pi + \frac{2}{\pi}\left[\frac{\sin nx}{n}\right]_\pi^{2\pi} = 0$$

$$b_n = \frac{1}{\pi}\int_0^\pi 1 \cdot \sin nx\,dx + \frac{1}{\pi}\int_\pi^{2\pi} 2\sin nx\,dx$$

$$= \frac{1}{\pi}\left[-\frac{\cos nx}{n}\right]_0^\pi + \frac{2}{\pi}\left[-\frac{\cos nx}{n}\right]_\pi^{2\pi}$$

$$= \frac{1}{\pi}\left[-\frac{\cos n\pi}{n} + \frac{1}{n}\right] + \frac{2}{\pi}\left[-\frac{\cos 2n\pi}{n} + \frac{\cos n\pi}{n}\right]$$

$$= \frac{1}{\pi}\left[\frac{(-1)^n}{n} + \frac{1}{n}\right] + \frac{2}{\pi}\left[\frac{(-1)^{2n}}{n} + \frac{(-1)^n}{n}\right]$$

$$= \begin{cases} 0 & \text{when } n \text{ is even} \\ -\dfrac{2}{\pi} & \text{when } n \text{ is odd} \end{cases}$$

$$f(x) = \frac{3}{2} - \frac{2}{\pi}\left(\sin x + \frac{\sin 3x}{3} + \frac{\sin 5x}{5} + ...\right)$$

EXAMPLE 6.   *An alternating current after passing through a rectifier has the form* $i = \begin{cases} I_0 \sin x & \text{for} \quad 0 \le x < \pi \\ 0 & \text{for} \quad \pi \le x \le 2\pi \end{cases}$ *where I_0 is the maximum current and the period is 2π.*

Express i as a Fourier series.   (VTU–2007, Calicut–2005, UPTU(Q.Bank)–2001)

SOLUTION.   We have $a_0 = \frac{1}{\pi}\int_0^{2\pi} f(x)\,dx$

$$a_0 = \frac{1}{\pi}\int_0^\pi I_0 \sin x\,dx + \frac{1}{\pi}\int_\pi^{2\pi} 0\,dx = \frac{I_0}{\pi}[-\cos x]_0^\pi = \frac{I_0}{\pi}[-\cos \pi + \cos 0] = \frac{2I_0}{\pi}$$

$$a_n = \frac{1}{\pi} \int_0^\pi I_0 \sin x \cos nx \, dx + \frac{1}{\pi} \int_\pi^{2\pi} 0 \cos nx \, dx$$

$$\frac{2I_0}{2\pi} \int_0^\pi \sin x \cos nx \, dx = \frac{I_0}{2\pi} \int_0^\pi 2 \cos nx \sin x \, dx$$

$$= \frac{I_0}{2\pi} \int_0^\pi [\sin(n+1)x - \sin(n-1)x]$$

$$= \frac{I_0}{2\pi} \left[\frac{-\cos(n+1)x}{n+1} + \frac{\cos(n-1)x}{n-1} \right]_0^\pi$$

$$\frac{I_0}{2\pi} \left[-\frac{\cos(n+1)\pi}{n+1} + \frac{\cos(n-1)\pi}{n-1} + \frac{1}{n+1} - \frac{1}{n-1} \right]$$

$$\frac{I_0}{2\pi} \left[\frac{-(-1)^{n+1}}{n+1} + \frac{(-1)^{n-1}}{n-1} + \frac{1}{n+1} - \frac{1}{n-1} \right]$$

$$= \begin{cases} \dfrac{I_0}{2\pi} \left(-\dfrac{4}{n^2-1} \right), & \text{when } n \text{ is even} \\ 0, & \text{when } n \text{ is odd} \end{cases}$$

$$b_n = \frac{1}{\pi} \int_0^\pi I_0 \sin x \sin nx \, dx + \frac{1}{\pi} \int_\pi^{2\pi} 0 \sin nx \, dx$$

$$= \frac{I_0}{2\pi} \int_0^\pi 2 \sin nx \sin x \, dx + 0$$

$$= \frac{I_0}{2\pi} \int_0^\pi [\cos(n-1)x - \cos(n+1)x] \, dx$$

$$= \frac{I_0}{2\pi} \left[\frac{\sin(n-1)x}{n-1} - \frac{\sin(n+1)x}{n+1} \right]_0^\pi = 0$$

$$f(x) = \frac{I_0}{\pi} - \frac{2I_0}{\pi} \left(\frac{\cos 2x}{2^2-1} + \frac{\cos 4x}{4^2-1} + \frac{\cos 6x}{6^2-1} + \dots \right)$$

9.7 CHANGE OF INTERVAL

In many problems, the interval of Fourier expansion is $2l$ and not 2π. In order to apply this theory, this interval must be transformed into an interval of length 2π.

Consider a periodic function $f(x)$ defined in the interval $c < x < c + 2l$. To change the interval into one of length 2π, we put

$$\frac{x}{l} = \frac{z}{\pi} \text{ or } z = \frac{\pi x}{l} \text{ so that at } x = c, z = \frac{\pi c}{l} = d(\text{say})$$

When $\qquad x = c + 2l, z = \dfrac{\pi(c+2l)}{l} = \dfrac{\pi c}{l} + 2\pi = d + 2\pi$

Thus, the function $f(x)$ of period $2l$ in $(c, c+2l)$ is transformed to the function $f\left(\dfrac{lz}{\pi}\right) = F(z)$

say, or period in $(d, d+2\pi)$ and then function $F(z)$ can be expressed as a Fourier series such that

$$F(z) = \frac{a_0}{2} + \sum_{n=1}^{\infty} a_n \cos nz + \sum_{n=1}^{\infty} b_n \sin nz \qquad \ldots(1)$$

where
$$a_0 = \frac{1}{\pi}\int_d^{d+2\pi} F(z)dz; a_n = \frac{1}{\pi}\int_d^{d+2\pi} F(z)\cos nz\, dz$$

and
$$b_n = \frac{1}{\pi}\int_d^{d+2\pi} F(z)\sin nz\, dz$$

Now, making the inverse substitution $z = \frac{\pi x}{l}, dz = \frac{\pi}{l}dx$, when $z = d$, $x = c$ and when $z = d + 2\pi, x = c + 2l$. The expression (1) becomes

$$F(z) = F\left(\frac{\pi x}{l}\right) = F(x) = \frac{a_0}{2} + \sum_{n=1}^{\infty} a_n \cos\frac{n\pi x}{l} + \sum_{n=1}^{\infty} b_n \sin\frac{n\pi x}{l} \qquad \ldots(2)$$

The coefficient a_0, a_n, b_n in (2) becomes

$$a_0 = \frac{1}{l}\int_c^{c+2l} f(x)dx, \ a_n = \frac{1}{l}\int_c^{c+2l} f(x)\cos\frac{n\pi x}{l}dx, \ b_n = \frac{1}{l}\int_c^{c+2\pi} f(x)\sin\frac{n\pi x}{l}dx$$

REMARKS

* If $c = 0$, the interval become $0 < x < 2l$ and the a_0, a_n, b_n are given by

$$a_0 = \frac{1}{l}\int_0^{2l} f(x)dx, a_n = \frac{1}{l}\int_0^{2l} f(x)\cos\frac{n\pi x}{l}dx, \ b_n = \frac{1}{l}\int_0^{2l} f(x)\sin\frac{n\pi x}{l}dx \ .$$

* If $c = -l$, the interval become $-l < x < l$ and a_0, a_n, b_n are given by

$$a_0 = \frac{1}{l}\int_{-l}^{l} f(x)dx, \ a_n = \frac{1}{l}\int_{-l}^{l} f(x)\cos\frac{n\pi x}{l}dx, \ b_n = \frac{1}{l}\int_{-l}^{l} f(x)\sin\frac{n\pi x}{l}dx.$$

Solved Examples

EXAMPLE 1.   *Find the Fourier series to represent* $f(x) = x^2 - 2$ *when* $-2 \le x \le 2$.   (SVTU–2006)

SOLUTION.   Here, $b_n = 0$ because $f(x)$ is an even function

Let $\quad f(x) = x^2 - 2 = \frac{a_0}{2} + \sum_{n=1}^{\infty} a_n \cos\frac{n\pi x}{2}$ $[\because 2l = 4 \Rightarrow l = 2]$

Then, $\quad a_0 = \frac{2}{2}\int_0^2 (x^2 - 2)dx = \left[\frac{x^3}{3} - 2x\right]_0^2 = \frac{8}{3} - 4 = -\frac{4}{3}$

and $\quad a_n = \frac{2}{2}\int_0^2 (x^2 - x)\cos\frac{n\pi x}{2}dx$

$$= \left[(x^2 - 2)\frac{\sin n\pi x/2}{(n\pi/2)} - 2x\right.$$

$$\left.\left(-\frac{\cos\dfrac{n\pi x}{2}}{(n^2\pi^2/4)} + 2\left(\frac{\sin\dfrac{n\pi x}{2}}{(n^3\pi^3/8)}\right)\right)\right]_0^2$$

$$= \frac{16 \cos n\pi}{n^2 \pi^2} = \frac{16(-1)^n}{n^2 \pi^2}.$$

$$\therefore \quad f(x) = (x^2 - 2) = -\frac{2}{3} + \frac{16}{\pi^2} \sum \frac{(-1)^n}{n^2} \cos \frac{n\pi x}{2}$$

$$= -\frac{2}{3} - \frac{16}{\pi^2} \left(\cos \frac{\pi x}{2} - \frac{1}{4} \cos \pi x + \frac{1}{9} \cos \frac{3\pi x}{2} - \cdots \right)$$

EXAMPLE 2. *Obtain the Fourier series for the function*

$$f(x) = \begin{cases} \pi x & ; \quad 0 \le x \le 1 \\ \pi(2 - x) & ; \quad 1 \le x \le 2 \end{cases}$$

(UPTU–2001, VTU–2011, Bhopal–2008, Mumbai–2007)

SOLUTION. Here, $2l = 2 \Rightarrow l = 1$.

Let $f(x) = \frac{a_0}{2} + \sum_{n=1}^{\infty} a_n \cos n\pi x + \sum_{n=1}^{\infty} b_n \sin n\pi x$

where $a_0 = \int_0^2 f(x)\,dx = \int_0^1 \pi x\,dx + \int_1^2 \pi(2 - x)\,dx = \pi \left[\frac{x^2}{2} \right]_0^1 + \pi \left[2x - \frac{x^2}{2} \right]_1^2$

$$= \pi \left(\frac{1}{2} \right) + \pi \left[(4 - 2) - \left(2 - \frac{1}{2} \right) \right] = \pi$$

$$a_n = \int_0^2 f(x) \cos n\pi x\, dx$$

$$= \int_0^1 \pi x \cos n\pi x\, dx + \int_1^2 \pi(2 - x) \cos n\pi x\, dx$$

$$= \left[\pi x \frac{\sin n\pi x}{n\pi} - \pi \left(-\frac{\cos n\pi x}{n^2 \pi^2} \right) \right]_0^1 + \left[\pi(2 - x) \frac{\sin n\pi x}{n\pi} - (-\pi) \left(-\frac{\cos n\pi x}{n^2 \pi^2} \right) \right]_1^2$$

$$= \left(\frac{\cos n\pi}{n^2 \pi} - \frac{1}{n^2 \pi} \right) + \left[-\frac{\cos 2n\pi}{n^2 \pi} + \frac{\cos n\pi}{n^2 \pi} \right] = \frac{2}{n^2 \pi} (\cos n\pi - 1)$$

$$= \frac{2}{n^2 \pi} [(-1)^n - 1] = \begin{cases} 0 & ; \quad \text{if } n \text{ is even} \\ -\dfrac{4}{n^2 \pi} & ; \quad \text{if } n \text{ is odd} \end{cases}$$

and $b_n = \int_0^2 f(x) \sin n\pi x\, dx$

$$= \int_0^1 \pi x \sin n\pi x\, dx + \int_1^2 \pi(2 - x) \sin n\pi x\, dx$$

$$= \left[\pi x \left(\frac{-\cos n\pi x}{n\pi} \right) - \pi \left(-\frac{\sin n\pi x}{n^2 \pi^2} \right) \right]_0^1 + \left[\pi(2 - x) \frac{\cos n\pi x}{n\pi} - (-\pi) \left(-\frac{\sin n\pi x}{n^2 \pi^2} \right) \right]_1^2$$

$$= \left[-\frac{\cos n\pi}{n} \right] + \left[\frac{\cos n\pi}{n} \right] = 0$$

Hence, $f(x) = \frac{\pi}{2} - \frac{4}{\pi} \left(\frac{\cos \pi x}{1^2} + \frac{\cos 3\pi x}{3^2} + \frac{\cos 5\pi x}{5^2} + \cdots \right)$

EXAMPLE 3. *Expand $f(x) = e^{-x}$ as a Fourier series in the interval $(-l, l)$.* (Kerala–2005, VTU–2004)

SOLUTION. Let $f(x) = e^{-x} = \dfrac{a_0}{2} + \sum\limits_{n=1}^{\infty} a_n \dfrac{\cos n\pi x}{l} + \sum\limits_{n=1}^{\infty} b_n \dfrac{\sin n\pi x}{l}$

Then, $a_0 = \dfrac{1}{l}\int_{-l}^{l} e^{-x}dx = \dfrac{1}{l}\left[-e^{-x}\right]_{-l}^{l} = -\dfrac{1}{l}(e^l - e^{-l}) = \dfrac{2\sinh l}{l}$

$$a_n = \frac{1}{l}\int_{-l}^{l} e^{-x}\cos\frac{n\pi x}{l}dx = \frac{1}{l}\left[\frac{e^{-x}}{1+\left(\dfrac{n\pi}{l}\right)^2}\left(-\cos\frac{n\pi x}{l} + \frac{n\pi}{l}\sin\frac{n\pi x}{l}\right)\right]_{-l}^{l}$$

$$= \frac{1}{l^2 + (n\pi)^2}\left[-e^{-l}\cos n\pi + e^l\cos n\pi\right]$$

$$= -\frac{2l\cos n\pi}{l^2 + (n\pi)^2}\left(\frac{e^l - e^{-l}}{2}\right) = \frac{2l(-1)^n \sinh l}{l^2 + (n\pi)^2}$$

$$b_n = \frac{1}{l}\int_{-l}^{l} e^{-x}\sin\frac{n\pi x}{l}dx = \frac{1}{l}\left[\frac{e^{-x}}{1+\left(\dfrac{n\pi}{l}\right)^2}\left(-\sin\frac{n\pi x}{l} - \frac{n\pi}{l}\cos\frac{n\pi x}{l}\right)\right]_{-l}^{l}$$

$$= -\frac{1}{l^2 + (n\pi)^2}\left[\frac{n\pi}{l}(e^{-l} - e^l)\cos n\pi\right]$$

$$= \frac{2n\pi\cos n\pi}{l^2 + (n\pi)^2}\left(\frac{e^l - e^{-l}}{2}\right) = \frac{2n\pi(-1)^n \sinh l}{l^2 + (n\pi)^2}$$

Hence, $e^{-x} = \sinh l\left[\dfrac{1}{l} - 2l\left(\dfrac{1}{l^2 + \pi^2}\cos\dfrac{\pi x}{l}\right.\right.$

$$\left. - \frac{1}{l^2 + 2^2\pi^2}\cos\frac{2\pi x}{l} + \frac{1}{l^2 + 3^2\pi^2}\cos\frac{3\pi x}{l} - \cdots\right)$$

$$\left.-2\pi\left(\frac{1}{l^2 + \pi^2}\sin\frac{\pi x}{l} - \frac{2}{l^2 + 2^2\pi^2}\sin\frac{2\pi x}{l} + \frac{3}{l^2 + 3^2\pi^2}\sin\frac{3\pi x}{l} - \cdots\right)\right].$$

EXAMPLE 4. *Prove that $\dfrac{l}{2} - x = \dfrac{l}{\pi}\sum\limits_{n=1}^{\infty}\dfrac{1}{n}\sin\dfrac{2n\pi x}{l}, 0 < x < l.$*

SOLUTION. Let $f(x) = \dfrac{l}{2} - x, 0 < x < l$

The Fourier series for $f(x)$ in the interval $(0, l)$ is

$$f(x) = \frac{a_0}{2} + \sum_{n=1}^{\infty}\left[a_n\cos\frac{n\pi x}{l/2} + b_n\sin\frac{n\pi x}{l/2}\right]$$

Here,

$$a_0 = \frac{1}{(l/2)}\int_0^l f(x)dx = \frac{2}{l}\int_0^l\left(\frac{1}{2} - x\right)dx = \frac{2}{l}\left[\frac{lx}{2} - \frac{x^2}{2}\right]_0^l = 0$$

$$a_n = \frac{1}{(l/2)}\int_0^l f(x)\cos\frac{n\pi x}{(l/2)}dx = \frac{2}{l}\int_0^l\left(\frac{l}{2} - x\right)\cos\frac{2n\pi x}{l}dx$$

$$= \frac{2}{l}\left[\left(\frac{l}{2} - x\right)\frac{1}{2n\pi}\sin\frac{2n\pi x}{l} + (-1)\frac{l^2}{4n^2\pi^2}\cos\frac{2n\pi x}{l}\right]_0^l$$

$$= \frac{2}{l}\left[-\frac{l^2}{4n^2\pi^2}\cos 2n\pi + \frac{l^2}{4n^2\pi^2}\right] = \frac{2}{l}\cdot\frac{l^2}{4n\pi^2}(-\cos 2n\pi + 1)$$

$$= \frac{1}{2n^2\pi^2}(-1+1) = 0$$

and $\qquad b_n = \frac{1}{(l/2)}\int_0^l f(x)\cdot\frac{\sin n\pi x}{(l/2)}dx = \frac{2}{l}\int_0^l\left(\frac{l}{2} - x\right)\sin\frac{2n\pi x}{l}dx$

$$= \frac{2}{l}\left[\left(\frac{l}{2} - x\right)\cdot\left(-\frac{1}{2n\pi}\cos\frac{2n\pi x}{l}\right) - (-1)\left(-\frac{l^2}{4n^2\pi^2}\sin\frac{2n\pi x}{l}\right)\right]_0^l$$

$$= \frac{2}{l}\left[\frac{l}{2}\cdot\frac{1}{2n\pi}\cos 2n\pi + \frac{l}{2}\cdot\frac{1}{2n\pi}(l)\right] = \frac{2}{l}\left[\frac{l^2}{2n\pi}\right] = \frac{l}{n\pi}$$

The required Fourier series is

$$f(x) = \sum_{n=1}^{\infty}\frac{1}{n\pi}\frac{\sin 2n\pi x}{l} \quad \text{or} \quad \frac{1}{2} - x = \frac{1}{\pi}\sum_{n=1}^{\infty}\frac{1}{n}\sin\frac{2n\pi x}{l}.$$

EXAMPLE 5. *Find the Fourier expansion for the function* $f(x) = x - x^2; -1 < x < 1$.

(UPTU(Q.Bank)–2001)

SOLUTION. Let $\quad f(x) = \frac{a_0}{2} + \sum_{n=1}^{\infty} a_n\cos n\pi x + \sum_{n=1}^{\infty} b_n\sin n\pi x$

Then, $\quad a_0 = \int_{-1}^{1}(x - x^2)dx = \int_{-1}^{1}x\,dx - \int_{-1}^{1}x^2\,dx$

$$= 0 - 2\int_0^1 x^2\,dx = -2\left[\frac{x^3}{3}\right]_0^1 = -\frac{2}{3}$$

$$a_n = \int_{-1}^{1}(x - x^2)\cos n\pi x\,dx$$

$$= \int_{-1}^{1}x\cos n\pi x\,dx - \int_{-1}^{1}x^2\cos n\pi x\,dx = 0 - 2\int_0^1 x^2\cos n\pi x\,dx$$

$$= -2\left[x^2\frac{\sin n\pi x}{n\pi} - 2x\left(-\frac{\cos n\pi x}{n^2\pi^2}\right) + 2\left(-\frac{\sin n\pi x}{n^3\pi^3}\right)\right]_0^1$$

$$= -2\left[\frac{2\cos n\pi}{n^2\pi^2}\right] = -\frac{4(-1)^n}{n^2\pi^2}$$

and $\quad b_n = \int_{-1}^{1}(x - x^2)\sin n\pi x\,dx = \int_{-1}^{1}x\sin n\pi x\,dx$

$$= -1\int_{-1}^{1}x^2 n\pi x\,dx = 2\int_0^1 x\sin n\pi x\,dx - 0$$

$$= 2\left[x\left(-\frac{\cos n\pi x}{n\pi}\right) - 1\left(-\frac{\sin n\pi x}{n^2\pi^2}\right)\right]_0^1$$

$$= 2\left[-\frac{\cos n\pi}{n\pi}\right] = -2\frac{(-1)^n}{n\pi}$$

$$\therefore x - x^2 = -\frac{1}{3} + \frac{4}{\pi^2} + \left(\frac{\cos \pi x}{1^2} - \frac{\cos 2\pi x}{2^2} + \frac{\cos 3\pi x}{3^2} - \cdots\right)$$

$$+ \frac{2}{\pi}\left(\frac{\sin \pi x}{1} - \frac{\sin 2\pi x}{2} + \frac{\sin 3\pi x}{3} - \cdots\right).$$

Additional Solved Examples

EXAMPLE 1. Develop $f(x)$ in a Fourier series in the interval $(0, 2)$ if

$$f(x) = \begin{cases} x, & 0 < x < 1 \\ 0, & 1 < x < 2 \end{cases}$$

SOLUTION. Here the interval is $(0, 2)$ and so $l = 1$

Let $$f(x) = \frac{a_0}{2} + \sum_{n=1}^{\infty}\left[a_n \cos\frac{n\pi x}{l} + bn\frac{\sin n\pi x}{l}\right]$$

$$a_0 = \frac{1}{1}\int_0^2 f(x)\,dx = \int_0^1 x\,dx + \int_1^2 0\,dx = \left[\frac{x^2}{2}\right]_0^1 = \frac{1}{2}$$

$$a_n = \int_0^2 f(x)\cos n\pi x\,dx = \int_0^1 x\cos n\pi x\,dx + \int_1^2 0\cos n\pi x\,dx$$

$$= \left[x\left(\frac{\sin n\pi x}{n\pi}\right) - 1\left(-\frac{\cos n\pi x}{n^2\pi^2}\right)\right]_0^1 = \left[\frac{(-1)^n}{n^2\pi^2} - \frac{1}{n^2\pi^2}\right]$$

$$= \begin{cases} 0 & \text{when } n \text{ is even} \\ -\dfrac{2}{n^2\pi^2} & \text{when } n \text{ is odd} \end{cases}$$

$$b_n = \int_0^2 f(x)\sin n\pi x\,dx = \int_0^1 x\sin n\pi x\,dx + \int_1^2 0\sin n\pi x\,dx$$

$$= \left[x\left(\frac{-\cos n\pi x}{n\pi}\right) - 1\left(\frac{-\sin n\pi x}{n^2\pi^2}\right)\right]_0^1$$

$$= \left[-\frac{\cos n\pi}{n\pi} + \frac{\sin n\pi}{n^2\pi^2} + 0 - 0\right]$$

$$= \frac{(-1)^n}{n\pi} = \begin{cases} \dfrac{1}{n\pi} & \text{when } n \text{ is even} \end{cases}$$

$$f(x) = \frac{1}{4} - \frac{2}{\pi^2}\left(\cos \pi x + \frac{\cos 3\pi x}{3^2} + \frac{\cos 5\pi x}{5^2} + \cdots\right)$$

$$+ \frac{1}{\pi}\left(\sin \pi x - \frac{\sin 2\pi x}{2} + \frac{\sin 3\pi x}{3} - \cdots\right)$$

EXAMPLE 2. Given $f(x) = \begin{cases} 0, & 0 < x < c \\ 1, & c < x < 2c \end{cases}$ expand $f(x)$ in a Fourier series of period $2c$.

SOLUTION. Consider $$a_0 = \frac{1}{c}\int_0^{2c} f(x)\,dx$$

$$= \frac{1}{c}\int_0^c 0\,dx + \frac{1}{c}\int_c^{2c} 1\,dx$$

$$= \frac{1}{c}[x]_c^{2c} = 1$$

$$a_n = \frac{1}{c}\int_0^c 0\frac{\cos n\pi x}{c}dx + \frac{1}{c}\int_c^{2c}\frac{1\cos n\pi x}{c}dx$$

$$= \frac{1}{c}\left[\frac{c}{n\pi}\frac{\sin n\pi x}{c}\right]_c^{2c} = \frac{1}{n\pi}[\sin 2n\pi - \sin n\pi] = 0$$

$$b_n = \frac{1}{c}\int_0^{2c} f(x)\frac{\sin n\pi x}{c}dx$$

$$= \frac{1}{c}\int_0^c 0.\frac{\sin n\pi x}{c}dx + \frac{1}{c}\int_c^{2c}\frac{\sin n\pi x}{c}dx$$

$$= \frac{1}{c}\left[\frac{c}{n\pi}\left(-\frac{\cos n\pi x}{c}\right)\right]_c^{2c} = -\frac{1}{n\pi}[\cos 2n\pi - \cos n\pi]$$

$$= -\frac{1}{n\pi}[1-(-1)^n] = \begin{cases}-\dfrac{2}{n\pi} & \text{when } n \text{ is odd} \\ 0 & \text{when } n \text{ is even}\end{cases}$$

$$f(x) = \frac{1}{2} - \frac{2}{\pi}\left(\sin\frac{n\pi x}{c} + \frac{1}{3}\frac{\sin 3\pi x}{c} + ...\right)$$

EXAMPLE 3. *Expand f(x) in a Fourier series in the interval (–2, 2) when* $f(x) = \begin{cases}0 & , & -2 < x < 0 \\ 1 & , & 0 < x < 2\end{cases}$.

(Kanpur-2013)

SOLUTION. Consider $a_0 = \frac{1}{2}\int_{-2}^2 f(x)\,dx$

$$\frac{1}{2}\int_{-2}^0 0\,dx + \frac{1}{2}\int_0^2 1.dx = \frac{1}{2}[x]_0^2 = 1$$

$$a_n = \frac{1}{2}\int_{-2}^0 0\frac{\cos n\pi x}{2}dx + \int_0^2\frac{\cos n\pi x}{2}dx = \left[\frac{2}{n\pi}\frac{\sin n\pi x}{2}\right]_0^2 = 0$$

$$b_n = \frac{1}{2}\int_{-2}^0\frac{0\sin n\pi x}{2}dx + \frac{1}{2}\int_0^2\frac{\sin n\pi x}{2}dx$$

$$= \frac{1}{2}\left[\frac{2}{n\pi}\left(-\cos\frac{n\pi x}{2}\right)\right]_0^2 = \frac{1}{2}\left[-\frac{2}{n\pi}.\cos\frac{2n\pi}{2}+\frac{2}{n\pi}\right] = \frac{1}{2}\left[-\frac{2}{n\pi}(-1)^n+\frac{2}{n\pi}\right]$$

$$= \begin{cases}0, & \text{when } n \text{ is even} \\ \dfrac{2}{n\pi}, & \text{when } n \text{ is odd}, n = 1,3,5,7,...\end{cases}$$

$$f(x) = \frac{1}{2} + \frac{2}{\pi}\left(\frac{\sin \pi x}{2} + \frac{1}{3}\frac{\sin 3\pi x}{2} + \frac{1}{5}\frac{\sin 5\pi x}{2} + ...\right)$$

EXAMPLE 4. *Find a Fourier series for the function given by* $f(t) = \begin{cases}t & , & 0 < t < 1 \\ 1-t & , & 1 < t < 2\end{cases}$ *where* $l = 1$.

SOLUTION. Consider

$$a_0 = \int_0^1 t\,dt + \int_1^2 (1-t)dt$$

$$= \left[\frac{t^2}{2}\right]_0^1 + \left[t - \frac{t^2}{2}\right]_1^2 = 0$$

$$a_n = \int_0^1 t\cos n\pi t\, dt + \int_1^2 (1-t)\cos n\pi t\, dt$$

$$= \left[t\left(\frac{\sin n\pi t}{n\pi}\right) - \left(-\frac{\cos n\pi t}{n^2\pi^2}\right)\right]_0^1 + \left[(1-t)\left(\frac{-\sin n\pi t}{n\pi}\right) - (-1)\left(-\frac{\cos n\pi t}{n^2\pi^2}\right)\right]_1^2$$

$$= \left[\frac{\sin n\pi}{n\pi} + \frac{\cos n\pi}{n^2\pi^2} - \frac{1}{n^2\pi^2}\right] + \left[(-1)\left(-\frac{\sin 2n\pi}{n\pi}\right) - \frac{\cos 2n\pi}{n^2\pi^2} + 0 + \frac{\cos n\pi}{n^2\pi^2}\right]$$

$$= \left[\frac{2(-1)^n}{n^2\pi^2} - \frac{(-1)^{2n}}{n^2\pi^2} - \frac{1}{n^2\pi^2}\right] = \begin{cases} 0 & ,\text{when } n \text{ is even} \\ -\dfrac{4}{n^2\pi^2}, & \text{when } n \text{ is odd} \end{cases}$$

$$b_n = \int_0^1 \sin n\pi t\, dt + \int_1^2(1-t)\sin n\pi t\, dt$$

$$= \left[t\left(-\frac{\cos n\pi t}{n\pi}\right) - 1\left(-\frac{\sin n\pi t}{n^2\pi^2}\right)\right]_0^1 + \left[(1-t)\left(-\frac{\cos n\pi t}{n\pi}\right) - (-1)\left(-\frac{\sin n\pi t}{n^2\pi^2}\right)\right]_1^2$$

$$= \left[-\frac{\cos n\pi}{n\pi} + \frac{\sin n\pi}{n^2\pi^2} + 0 - 0\right] + \left[(-1)\left(-\frac{\cos 2n\pi}{n^2\pi^2}\right) - \frac{\sin 2n\pi}{n^2\pi^2} - 0 + \frac{\sin n\pi}{n^2\pi^2}\right]$$

$$= \left\{-\frac{(-1)^n}{n\pi} + \frac{(-1)^{2n}}{n\pi}\right\} = \begin{cases} 0, & \text{when } n \text{ is even} \\ \dfrac{2}{n\pi}, & \text{when } n \text{ is odd} \end{cases}$$

$$f(x) = -\frac{4}{\pi^2}\left(\cos \pi t + \cos\frac{3\pi t}{3^2} + \cos\frac{5\pi t}{5^2} + \dots\right) + \frac{2}{\pi}\left(\sin \pi t + \frac{\sin 3\pi t}{3} + \dots\right)$$

EXAMPLE 5. *Find a Fourier series corresponding to the function $f(x)$ defined in $(-2, 2)$ as follows;*

$$f(x) = \begin{cases} 2 & , \text{ if } -2 \le x \le 0 \\ x & , \text{ if } \quad 0 < x < 2 \end{cases}$$

SOLUTION.   Consider $a_0 = \dfrac{1}{2}\int_{-2}^2 f(x)\,dx = \dfrac{1}{2}\int_{-2}^0 2\,dx + \dfrac{1}{2}\int_0^2 x\,dx$

$$= \frac{1}{2}[2x]_{-2}^0 + \frac{1}{2}\left[\frac{x^2}{2}\right]_0^2 = 3$$

$$a_n = \frac{1}{2}\int_{-2}^0 \frac{2\cos n\pi x}{2}\,dx + \frac{1}{2}\int_0^2 \frac{x\cos n\pi x}{2}\,dx$$

$$= \frac{1}{2}\left[\frac{4}{n\pi}\left(\sin\frac{n\pi x}{2}\right)_{-2}^0 + \left(\frac{x^2}{n\pi}\sin\frac{n\pi x}{2} + \frac{4}{n^2\pi^2}\cos\frac{n\pi x}{2}\right)\right]_0^2$$

$$= \frac{1}{2}\left[\frac{4}{n^2\pi^2}\cos n\pi - \frac{4}{n^2\pi^2}\right]$$

$$= \frac{2}{n^2\pi^2}[(-1)^n - 1] = \begin{cases} -\dfrac{4}{n^2\pi^2}, & \text{when } n \text{ is odd} \\ 0, & \text{when } n \text{ is even} \end{cases}$$

$$b_n = \frac{1}{2}\left[\int_{-2}^{0} 2\sin\frac{n\pi x}{2}\,dx + \int_{0}^{2} x\sin\frac{n\pi x}{2}\,dx\right]$$

$$= \frac{1}{2}\left[2\left(-\frac{2}{n\pi}\cos\frac{n\pi x}{2}\right)\right]_{-2}^{0} + \frac{1}{2}\left[x\left(-\frac{2}{n\pi}\cos\frac{n\pi x}{2}\right) + \frac{4}{n^2\pi^2}\sin\frac{n\pi x}{2}\right]_{0}^{2}$$

$$= \frac{1}{2}\left[-\frac{4}{n\pi} + \frac{4}{n\pi}\cos n\pi\right] + \frac{1}{2}\left[-\frac{4}{n\pi}\cos n\pi + \frac{4}{n^2\pi^2}\sin n\pi\right]$$

$$= \frac{1}{2}\left[-\frac{4}{n\pi}\right] = -\frac{2}{n\pi}$$

Hence, $f(x) = \dfrac{3}{2} - \dfrac{4}{\pi^2}\left\{\dfrac{1}{1^2}\cos\dfrac{\pi x}{2} + \dfrac{1}{3^2}\cos\dfrac{3\pi x}{2}....\right\}$

$$-\frac{2}{\pi}\left\{\sin\frac{\pi x}{2} + \frac{1}{2}\sin\frac{2\pi x}{2} + \frac{1}{3}\sin\frac{3\pi x}{3} +...\right\}$$

EXAMPLE 6. *Find a Fourier series for the function*

$$f(x) = \begin{cases} 0, & \text{when } -2 < x < -1 \\ k, & \text{when } -1 < x < 1 \\ 0, & \text{when } 1 < x < 2 \end{cases}$$

SOLUTION. Here $2l = 4 \Rightarrow l = 2$

$$a_0 = \frac{1}{l}\int_{-l}^{l} f(x)\,dx \qquad \Rightarrow \qquad a_0 = \frac{1}{2}\int_{-2}^{2} f(x)\,dx$$

$$\frac{1}{2}\int_{-2}^{-1} 0\,dx + \frac{1}{2}\int_{-1}^{1} k\,dx + \frac{1}{2}\int_{1}^{2} 0\,dx = \frac{1}{2}[kx]_{-1}^{1} = k$$

$$a_n = \frac{1}{2}\int_{-2}^{-1} 0\cos\frac{n\pi x}{2}\,dx + \frac{1}{2}\int_{-1}^{1} k\cos\frac{n\pi x}{2}\,dx + \frac{1}{2}\int_{1}^{2} 0\cos\frac{n\pi x}{2}\,dx$$

$$= \frac{1}{2}\int_{-1}^{1} k\cos\frac{n\pi x}{2}\,dx$$

$$\frac{k}{2}\left[\frac{2}{n\pi}\sin\frac{n\pi x}{2}\right]_{-1}^{1} = 0$$

$$b_n = \frac{1}{2}\int_{-2}^{-1} 0\sin\frac{n\pi x}{2}\,dx + \frac{1}{2}\int_{-1}^{1} k\sin\frac{n\pi x}{2}\,dx + \frac{1}{2}\int_{1}^{2} 0\sin\frac{n\pi x}{2}\,dx$$

$$= \frac{k}{2}\left[\frac{2}{n\pi}\left(-\cos\frac{n\pi x}{2}\right)\right]_{-1}^{1} = \frac{k}{2n\pi}[-\cos n\pi - \cos n\pi]$$

$$= \frac{k}{2n\pi}\{-2\cos n\pi\} = \frac{k}{2n\pi}\{-2(-1)^n\}$$

$$= \begin{cases} \dfrac{2k}{2n\pi}, & \text{when } n \text{ is odd} \\[2mm] -\dfrac{k}{n\pi}, & \text{when } n \text{ is even} \end{cases}$$

$$f(x) = \frac{k}{2} + \frac{2k}{2\pi}\left(\sin\frac{\pi x}{2} + \frac{1}{3}\sin\frac{3\pi x}{2} + \frac{1}{5}\sin\frac{5\pi x}{2} + \ldots\right)$$

9.8 HALF RANGE SERIES

When we require to expand a function $f(x)$ in the range $(0, \pi)$ in a Fourier series of period 2π or more generally in the range $(0, l)$ in a Fourier series of period $2l$, a function $f(x)$ defined over the interval $0 < x < l$ is capable of two distinct half range series.

The half range cosine series is $f(x) = \dfrac{a_0}{2l} + \displaystyle\sum_{n=1}^{\infty} a_n \cos\frac{n\pi x}{l}$

where, $\qquad a_0 = \dfrac{2}{l}\int_0^l f(x)\,.dx, \text{ where } a_n = \dfrac{2}{l}\int_0^l f(x)\cos\frac{n\pi x}{l}dx$

The half range sine series is

$$f(x) = \sum_{n=1}^{\infty} b_n \sin\frac{n\pi x}{l}, \text{ where } b_n = \frac{2}{l}\int_0^l f(x)\sin\frac{n\pi x}{l}dx$$

Solved Examples

EXAMPLE 1. If $f(x) = \begin{cases} x & ; & 0 < x < \pi/2 \\ \pi - x & ; & \pi/2 < x < \pi \end{cases}$ *Show that*

(i) $f(x) = \dfrac{4}{\pi}\left[\sin x - \dfrac{\sin 3x}{3^2} + \dfrac{\sin 5x}{5^2} - \ldots\right]$ (Mumbai–2008, SVTU–2008, VTU–2004)

(ii) $f(x) = \dfrac{\pi}{4} - \dfrac{2}{\pi}\left[\dfrac{\cos 2x}{1^2} + \dfrac{\cos 6x}{3^2} + \dfrac{\cos 10x}{5^2} + \ldots\right]$ (VTU–2011)

SOLUTION. (i) Half range sine series, we have $l = \pi$ so

$$f(x) = \sum_{n=1}^{\infty} b_n \sin\frac{n\pi x}{\pi} = \sum_{n=1}^{\infty} b_n \sin nx$$

$$b_n = \frac{2}{\pi}\int_0^\pi f(x)\sin nx\,dx$$

$$= \frac{2}{\pi}\left[\int_0^{\pi/2} x\sin nx\,dx + \int_{\pi/2}^\pi (\pi - x)\sin nx\,dx\right]$$

$$= \frac{2}{\pi}\left[x\left(-\frac{\cos nx}{n}\right) - 1\left(-\frac{\sin nx}{n^2}\right)\right]_0^{\pi/2}$$

$$+ \frac{2}{\pi}\left[(\pi - x)\left(-\frac{\cos nx}{nx}\right) - (-1)\left(-\frac{\sin nx}{n^2}\right)\right]_0^\pi$$

$$= \frac{2}{\pi}\left[-\frac{\pi}{2n}\cos\frac{n\pi}{2} + \frac{1}{n^2}\sin\frac{n\pi}{2}\right] + \frac{2}{\pi}\left[\frac{\pi}{2n}\cos\frac{n\pi}{2} + \frac{1}{n^2}\sin\frac{n\pi}{2}\right]$$

$$= \frac{2}{\pi}\left[\frac{2}{n^2}\sin\frac{n\pi}{2}\right] = \frac{4}{\pi n^2}\sin\frac{n\pi}{2}$$

Hence, $f(x) = \frac{4}{\pi}\left[\sin x - \frac{\sin 3x}{3^2} + \frac{\sin 5x}{5^2} - ...\right].$

(ii) Half range cosine series

Let $f(x) = \frac{a_0}{2} + \sum_{n=1}^{\infty} a_n \cos nx$

Then, $a_0 = \frac{2}{\pi}\int_0^\pi f(x)dx = \frac{2}{\pi}\int_0^{\pi/2} x\,dx + \int_{\pi/2}^\pi (\pi - x)dx$

$$= \frac{2}{\pi}\left[\frac{x^2}{2}\right]_0^{\pi/2} + \left[\pi x - \frac{x^2}{2}\right]_{\pi/2}^\pi = \frac{2}{\pi}\left[\frac{\pi^2}{8} + \left(\pi^2 - \frac{\pi^2}{2}\right) - \left(\frac{\pi^2}{2} - \frac{\pi^2}{8}\right)\right]$$

$$= \frac{2}{\pi}\left[\frac{\pi^2}{4}\right] = \frac{\pi}{2}$$

and

$$a_n = \frac{2}{\pi}\int_0^\pi f(x)\cos x\,dx$$

$$= \frac{2}{\pi}\left[\int_0^{\pi/2} x\cos nx\,dx + \int_{\pi/2}^\pi (\pi - x)\cos nx\,dx\right]$$

$$= \frac{2}{\pi}\left[\frac{x\sin nx}{n} - 1\left(-\frac{\cos nx}{n^2}\right)\right]_0^{\pi/2} + \frac{2}{\pi}\left[(\pi - x)\frac{\sin nx}{n} - (-1)\left(\frac{\cos nx}{n^2}\right)\right]_{\pi/2}^\pi$$

$$= \frac{2}{\pi}\left[\frac{\pi}{2n}\sin\frac{n\pi}{2} + \frac{1}{n^2}\cos\frac{n\pi}{2} - \frac{1}{n^2}\right] + \frac{2}{\pi}\left[\frac{\cos n\pi}{n^2} - \frac{\pi}{2n}\sin\frac{n\pi}{2} + \frac{1}{n^2}\cos\frac{n\pi}{2}\right]$$

$$= \frac{2}{\pi}\left[\frac{2}{n^2}\cos\frac{n\pi}{2} - \frac{\cos n\pi}{n^2} - \frac{1}{n^2}\right] = \frac{2}{\pi n^2}\left[2\cos\frac{n\pi}{2} - \cos n\pi - 1\right]$$

Put $n = 0, 1, 2, 3, ...$ in equation (1), we get

$$a_1 = 0, a_2 = \frac{2}{\pi . 2^2}(2\cos\pi - \cos 2\pi - 1) = \frac{-2}{1^2 . \pi}$$

$$a_3 = 0, a_4 = 0, a_5 = 0, a_6 = \frac{2}{6^2\pi}(2\cos 3\pi - \cos 6\pi - 1) = \frac{-2}{3^2\pi}$$

$$a_7 = a_8 = a_9 = 0, a_{10} = \frac{2}{10^2.\pi}(2\cos 5\pi - \cos 10\pi - 1) = \frac{-2}{5^2\pi}$$

Hence,

$$f(x) = \frac{\pi}{4} - \frac{2}{\pi}\left[\frac{\cos 2x}{1^2} + \frac{\cos 6x}{3^2} + \frac{\cos 10x}{5^2} + ...\right].$$

EXAMPLE 2. *Develop the* $\sin\frac{\pi x}{l}$ *in half range cosine series in range* $0 < x < l.$ (UPTU–2001)

SOLUTION.   Let $\sin\dfrac{\pi x}{l} = \dfrac{a_0}{2} + \sum\limits_{n=1}^{\infty} a_n \cos\dfrac{n\pi x}{l}$

where, $a_0 = \dfrac{2}{l}\int_0^l \sin\dfrac{\pi x}{l}dx = \dfrac{2}{l}\left[-\dfrac{\cos(\pi x/l)}{\pi/l}\right]_0^l = \dfrac{2}{\pi}[\cos\pi - 1] = \dfrac{4}{\pi}$

and $a_n = \dfrac{2}{l}\int_0^l \sin\dfrac{\pi x}{l}\cos\dfrac{n\pi x}{l}dx = \dfrac{1}{l}\int_0^l\left[\sin(n+1)\dfrac{\pi x}{l} - \sin(n-1)\dfrac{\pi x}{l}\right]dx$

$$= \dfrac{1}{l}\left[-\dfrac{\cos(n+1)\dfrac{\pi x}{l}}{(n+1)\pi/l} + \dfrac{\cos(n+1)\dfrac{\pi x}{l}}{(n-1)\pi/l}\right]_0^l$$

$$= \dfrac{1}{\pi}\left[-\dfrac{(-1)^{n+1}}{n+1} + \dfrac{(-1)^{n-1}}{n+1} + \dfrac{1}{n+1} - \dfrac{1}{n-1}\right]$$

(i)   When n is odd

$$a_n = \dfrac{1}{\pi}\left[-\dfrac{1}{n+1} + \dfrac{1}{n-1} + \dfrac{1}{n-1} - \dfrac{1}{n-1}\right] = 0$$

(ii) When n is even

$$a_n = \dfrac{1}{\pi}\left[\dfrac{1}{n+1} - \dfrac{1}{n-1} + \dfrac{1}{n-1} - \dfrac{1}{n-1}\right] = \dfrac{2}{\pi}\left[\dfrac{1}{n+1} - \dfrac{1}{n-1}\right]$$

$$= \dfrac{-4}{\pi(n+1)(n-1)}, n\neq 1$$

∴ $\sin\dfrac{\pi x}{l} = \dfrac{2}{\pi} - \dfrac{4}{\pi}\left[\dfrac{\cos\dfrac{2\pi x}{l}}{1.3} + \dfrac{\cos\dfrac{4\pi x}{l}}{3.5} + \dfrac{\cos\dfrac{6\pi x}{l}}{5.7} + ...\right].$

EXAMPLE 3.   *Obtain the half range sine series for function $f(x) = x^2$ in the interval $0 < x < 3$.*

(UPTU–2001(Sp), 2002)

SOLUTION .   The Fourier half range sine series in the interval $(0, c)$ is given by

$$f(x) = \sum\limits_{n=1}^{\infty} b_n \sin nx \qquad\qquad ...(1)$$

where, $b_n = \dfrac{2}{c}\int_0^c f(x)\sin\dfrac{n\pi x}{c}dx$

Here, $c = 3$ and $f(x) = x^2$

∴ $b_n = \dfrac{2}{3}\int_0^3 x^2 \sin\dfrac{n\pi x}{3}dx$

$$= \dfrac{2}{3}\left[x^2\left(\dfrac{-3}{n\pi}\right)\left(\cos\dfrac{n\pi x}{3}\right) + 2x\left(\dfrac{3}{n\pi}\right)\left(\dfrac{3}{n\pi}\right)\sin\dfrac{n\pi x}{3}\right.$$

$$\left.-2\left(\dfrac{3}{n\pi}\right)\left(\dfrac{3}{n\pi}\right)\left(\dfrac{3}{n\pi}\right)\cos\dfrac{n\pi x}{3}\right]_0^3$$

$$= \dfrac{2}{3}\left[\left\{-\dfrac{27}{n\pi}(-1)^n - \dfrac{54}{n^3\pi^3}(-1)^n\right\} + \dfrac{54}{n^3\pi^3}\right]$$

$$= \frac{2}{3} \left[\frac{54}{n^3 \pi^3} \{1 - (-1)^n\} - \frac{27}{n\pi}(-1)^n \right]$$

$$= \begin{cases} \dfrac{2}{3}\left(\dfrac{108}{n^3\pi^3} + \dfrac{27}{n\pi} \right), & \text{if } n \text{ is odd} \\[3mm] -\dfrac{18}{n\pi}, & \text{if } n \text{ is even} \end{cases}$$

Hence, the required half range sine series is given by

$$f(x) = b_1 \sin x + b_2 \sin 2x + b_3 \sin 3x + \ldots$$

$$= \frac{2}{3}\left[\frac{108}{\pi^3}\left(\frac{\sin x}{1^3} + \frac{\sin 3x}{3^3} + \frac{\sin 5x}{5^3} + \ldots \right) + \frac{27}{\pi}\left(\frac{\sin x}{1} + \frac{\sin 3x}{3} + \frac{\sin 5x}{5} + \ldots \right) \right.$$

$$\left. - \frac{18}{\pi}\left(\frac{\sin 2x}{2} + \frac{\sin 4x}{4} + \ldots \right) \right].$$

EXAMPLE 4. (i) *Express* $f(x) = x$ *as a half range sine series in* $0 < x < 2$,   (UPTU–2004)

(ii) *Express* $f(x) = x$ *as a half-range cosine series in* $0 < x < 2$.

(SVTU–2009, Bhopal–2007, Mumbai–2006)

SOLUTION . (i) The Fourier sine series for $F(x)$ in $(0, 2)$ is

$$f(x) = \sum_{n=1}^{\infty} b_n \sin \frac{n\pi x}{2}$$

where $$b_n = \frac{2}{2} \int_0^2 f(x) \sin \frac{n\pi x}{2} dx = \int_0^2 x \sin \frac{n\pi x}{2} dx$$

$$= \left[-\frac{2x}{n\pi} \cos \frac{n\pi x}{2} + \frac{4}{n^2 \pi^2} \sin \frac{n\pi x}{2} \right]_0^2 = \frac{-4(-1)^n}{n\pi}$$

$$\Rightarrow \qquad b_1 = 4/\pi_1, b_2 = -4/2\pi, b_3 = 4/3\pi, \quad b_4 = -4/4\pi, \text{etc.}$$

Required half range Fourier sine series is

$$f(x) = \frac{4}{\pi}\left[\sin \frac{\pi x}{2} - \frac{1}{2}\sin \frac{2\pi x}{2} + \frac{1}{3}\sin \frac{3\pi x}{2} - \frac{1}{4}\sin \frac{4\pi x}{2} + \ldots \right]$$

(ii) The Fourier cosine series for $f(x)$ in $(0, 2)$ is

$$f(x) = \frac{a_0}{2} + \sum_{n=1}^{\infty} a_n \cos \frac{n\pi x}{2}$$

where $$a_0 = \frac{2}{2} \int_0^2 f(x) dx = \int_0^2 x \, dx = 2$$

and $$a_n = \frac{2}{2} \int_0^2 f(x) \cos \frac{n\pi x}{2} dx = \int_0^2 x \cos \frac{n\pi x}{2} dx$$

$$= \left[\frac{2x}{n\pi} \sin \frac{n\pi x}{2} + \frac{4}{n^2 \pi^2} \cos \frac{n\pi x}{2} \right]_0^2 = \frac{4}{n^2 \pi^2}[(-1)^n - 1]$$

$$\Rightarrow \qquad a_1 = -8/\pi^2, a_2 = 0, a_3 = -8/3^2\pi^2, a_4 = 0, \quad a_5 = -8/5^2\pi^2$$

Required half range Fourier series is given by

$$f(x) = 1 - \frac{8}{\pi^2}\left[\frac{\cos \pi x/2}{1^2} + \frac{\cos 3\pi x/2}{3^2} + \frac{\cos 5\pi x/2}{5^2} + \ldots \right].$$

EXAMPLE 5.   *Find a series of cosines of multiples of x which will represent x sin x in the interval* (0, π) *and show that*

$$\frac{1}{1.3} - \frac{1}{3.5} + \frac{1}{5.7} - \dots = \frac{\pi - 2}{4}.$$ (UPTU–2002, VTU–2003, Anna–2001)

SOLUTION.   Let

$$x \sin x = \frac{a_0}{2} + \sum_{n=1}^{\infty} a_n \cos nx$$

Then

$$a_0 = \frac{2}{\pi} \int_0^\pi x \sin x \, dx = \frac{2}{\pi} [x(-\cos x) - 1.(-\sin x)]_0^\pi$$

$$= \frac{2}{\pi} [-\pi \cos x] = 2$$

and

$$a_n = \frac{2}{\pi} \int_0^\pi x \sin x \cos nx \, dx = \frac{1}{\pi} \int_0^\pi x (2 \cos nx \sin x) dx$$

$$= \frac{1}{\pi} \int_0^\pi x [\sin(n+1)x - \sin(n-1)x] dx$$

$$= \frac{1}{\pi} \left[x \left\{ -\frac{\cos(n+1)x}{n+1} + \frac{\cos(n-1)x}{n-1} \right\} \right.$$

$$\left. -1 \left\{ -\frac{\sin(n+1)x}{(n+1)^2} - \frac{\sin(n-1)\pi}{(n-1)^2} \right\} \right]_0^\pi$$

$$= \frac{1}{\pi} \left[-\frac{\pi \cos(n+1)\pi}{n+1} + \frac{\pi \cos(n-1)\pi}{(n-1)} \right], \text{ when } n \neq 1$$

$$= \frac{-(-1)^{n+1}}{n+1} + \frac{(-1)^{n-1}}{n-1} = (-1)^n \left[\frac{1}{n-1} - \frac{1}{n+1} \right] = \frac{2(-1)^{n-1}}{(n-1)(n+1)}$$

When n=1, we have

$$a_1 = \frac{2}{\pi} \int_0^\pi x \sin x \cos x \, dx = \frac{1}{\pi} \int_0^\pi x \sin 2x \, dx$$

$$= \frac{1}{\pi} \left[x \left(-\frac{\cos 2x}{2} \right) - 1 \left(-\frac{\sin 2x}{2^2} \right) \right]_0^\pi = \frac{1}{\pi} \left[-\frac{\pi \cos 2x}{2} \right] = -\frac{1}{2}$$

∴ $$x \sin x = 1 - \frac{1}{2} \cos x - 2 \left(\frac{\cos 2x}{1.3} - \frac{\cos 3x}{2.4} + \frac{\cos 4x}{3.5} - \dots \right)$$

Putting $x = \frac{\pi}{2}$, we get

$$\frac{\pi}{2} = 1 - 2 \left(-\frac{1}{1.3} + \frac{1}{3.5} - \frac{1}{5.7} - \dots \right)$$

∴ $$1 + \frac{2}{1.3} - \frac{2}{3.5} + \frac{2}{5.7} - \dots = \frac{\pi}{2}$$

⇒ $$\frac{2}{1.3} - \frac{2}{3.5} + \frac{2}{5.7} - \dots = \frac{\pi}{2} - 1$$

Hence, $$\frac{1}{1.3} - \frac{1}{3.5} + \frac{1}{5.7} - \dots = \frac{\pi - 2}{4}.$$

EXAMPLE 6. *Obtain the half range sine series for e^x in $0 < x < 1$.*

SOLUTION. Let

$$e^x = \sum_{n=1}^{\infty} b_n \sin n\pi x \qquad\qquad [\because l = 1]$$

Then,

$$b_n = 2\int_0^l e^x \sin n\pi x \, dx = 2\left[\frac{e^x}{1+(n\pi)^2} (\sin n\pi x - n\pi \cos n\pi x) \right]_0^l$$

$$= 2\left[\frac{e}{1+(n\pi)^2}(-n\pi \cos n\pi x) - \frac{1}{1+(n\pi)^2}(-n\pi) \right]$$

$$= \frac{2}{1+n^2\pi^2}[-en\pi(-1)^n + n\pi] = \frac{2n\pi}{1+n^2\pi^2}[1 - e(-1)^n]$$

Hence,

$$e^x = 2\pi \sum_{n=1}^{\infty} \frac{n[1 - e(-1)^n]}{1+n^2\pi^2}$$

$$= 2\pi\left[\frac{1+e}{1+\pi^2}\sin \pi x + \frac{2(1-e)}{1+4\pi^2}\sin 2\pi x + \frac{3(1+e)}{1+9\pi^2}\sin 3\pi x + ... \right].$$

EXAMPLE 7. *Expand* $f(x) = \begin{cases} \dfrac{1}{4} - x & ,if \quad 0 < x < \dfrac{1}{2} \\ x - \dfrac{3}{4} & ,if \quad \dfrac{1}{2} < x < 1 \end{cases}$ *as the Fourier series of sine terms.*

(UPTU–2001, VTU–2011, Andhra–2000)

SOLUTION. The Fourier sine series for $f(x)$ in $(0, 1)$ is

$$f(x) = \sum_{n=1}^{\infty} b_n \sin n\pi x$$

where, $b_n = \dfrac{2}{l}\int_0^1 f(x)\sin n\pi x \, dx$

$$= 2\left[\int_0^{1/2}\left(\frac{1}{4} - x\right)\sin n\pi x \, dx + \int_{1/2}^1\left(x - \frac{3}{4}\right)\sin n\pi x \, dx \right]$$

$$= 2\left| -\left(\frac{1}{4} - x\right)\frac{\cos n\pi x}{n\pi} - \frac{\sin n\pi x}{n\pi} \right|_0^{1/2} + 2\left| -\left(x - \frac{3}{3}\right)\frac{\cos n\pi x}{n\pi} + \frac{\sin n\pi x}{n^2\pi^2} \right|_{1/2}^1$$

$$= 2\left[\frac{1}{4n\pi}\cos \frac{n\pi}{2} + \frac{1}{4n\pi} - \frac{\sin n\pi / 2}{n^2\pi^2} \right]$$

$$+ 2\left[-\frac{1}{4n\pi}\cos n\pi - \frac{1}{4n\pi}\cos \frac{n\pi}{2} - \frac{\sin n\pi / 2}{n^2\pi^2} \right]$$

$$= \frac{1}{2n\pi}[1 - (-1)^n] - \frac{4\sin n\pi / 2}{n^2\pi^2}$$

$\Rightarrow$ $b_1 = \dfrac{1}{\pi} - \dfrac{4}{\pi^2}, b_2 = 0, b_3 = \dfrac{1}{3\pi} + \dfrac{4}{3^2\pi^2},$

$$b_4 = 0, b_5 = \frac{1}{5} - \frac{4}{5^2\pi^2}, b_6 = 0 \quad \text{etc.}$$

Hence, the required Fourier series is

$$f(x) = \left(\frac{1}{\pi} - \frac{4}{\pi^2}\right)\sin \pi x + \left(\frac{1}{3\pi} + \frac{4}{3^2\pi^2}\right)\sin 3\pi x + \left(\frac{1}{5\pi} - \frac{4}{5^2\pi^2}\right)\sin 5\pi x + \ldots$$

EXAMPLE 8.   *Find the half range cosine series of the function*

$$f(t) = \begin{cases} 2t & ; \quad 0 < t < 1 \\ 2(2-t) & ; \quad 1 < t < 2 \end{cases}$$

SOLUTION.   The Fourier half range cosine series in interval $(0, C)$ is

$$f(t) = \frac{a_0}{2} + a_1 \cos\frac{\pi t}{c} + a_2 \cos\frac{2\pi t}{c} + a_3 \cos\frac{3\pi t}{c} + \ldots$$

Here, $c = 2$ we have   ...(1)

$$a_0 = \frac{2}{c}\int_0^c f(t)dt = \frac{2}{2}\int_0^1 2t\,dt + 2\int_1^2 2(2-t)dt$$

$$= [t^2]_0^1 + \left[2\left(2t - \frac{t^2}{2}\right)\right]_1^2 = 1 + [(4t-t)]_1^2$$

$$= 1 + (8 - 4 - 4 + 1) = 2$$

and $a_n = \dfrac{2}{c}\int_0^c f(t)\cos\dfrac{n\pi t}{c}dt = \dfrac{2}{2}\int_0^1 2t\cos\dfrac{n\pi t}{2}dt + \dfrac{2}{2}\int_1^2 2(2-t)\cos\dfrac{n\pi t}{2}dt$

$$= \left[2t\left(\frac{2}{n\pi}\sin\frac{n\pi t}{2}\right) - (2)\left(-\frac{4}{n^2\pi^2}\cos\frac{n\pi t}{2}\right)\right]_0^1$$

$$+ \left[(4-2t)\left(\frac{2}{n\pi}\sin\frac{n\pi t}{2}\right) - (2)\left(-\frac{4}{n^2\pi^2}\cos\frac{n\pi t}{2}\right)\right]_1^2$$

$$= \left[\frac{4}{n\pi}\sin\frac{n\pi}{2} + \frac{8}{n^2\pi^2}\cos\frac{n\pi}{2} - \frac{8}{n^2 + \pi^2}\right]$$

$$+ \left[0 + \frac{8}{n^2\pi^2}\cos\frac{n\pi}{2} - \frac{4}{n\pi}\sin\frac{n\pi}{2} + \frac{8}{n^2\pi^2}\cos\frac{n\pi}{2}\right]$$

$$= \frac{8}{n^2\pi^2}\cos\frac{n\pi}{2} - \frac{8}{n^2\pi^2} - \frac{4}{n\pi}\sin\frac{n\pi}{2} = \frac{8}{n^2\pi^2}\left[\cos\frac{n\pi}{2} - 1 - \frac{n\pi}{2}\sin\frac{n\pi}{2}\right]$$

If $n = 1$, $a_1 = \dfrac{8}{\pi^2}\left[0 - 1 - \dfrac{\pi}{2}\right] = \dfrac{-8}{\pi^2} - \dfrac{4}{\pi}$

If $n = 2$, $a_2 = \dfrac{8}{4\pi^2}[-1-1] = \dfrac{-16}{4\pi^2} = -\dfrac{4}{\pi^2}$

If $n = 3$, $a_3 = \dfrac{8}{9\pi^2}\left[0 - 1 + \dfrac{3\pi}{2}\right] = \dfrac{-8}{9\pi^2} + \dfrac{4}{3\pi}$

Putting these values of $a_0, a_1, a_2, a_3, \ldots$ in equation (1), we get

$$f(1) = 1 - \left(\frac{8}{\pi^2} + \frac{4}{\pi}\right)\cos\frac{\pi t}{2} - \frac{4}{\pi^2}\cos\frac{2\pi t}{2} + \left(-\frac{8}{9\pi^2} + \frac{4}{3\pi}\right)\cos\frac{4\pi t}{2} + \ldots$$

Additional Solved Examples

EXAMPLE 1. *Find the Fourier half range series expansion of the function*

$$f(x) = (-x/l) + 1, 0 \le x \le l.$$

SOLUTION. Here $c = l$, function is even, $b_n = 0$

$$a_0 = \frac{2}{l}\int_0^l f(x)\,dx = \frac{2}{l}\int_0^l \left(-\frac{x}{l} + 1\right)dx$$

$$= \frac{2}{l}\left[-\frac{x^2}{2l} + x\right]_0^l = \frac{2}{l}\left[-\frac{l^2}{2l} + l\right] = 1$$

$$a_n = \frac{2}{l}\int_0^l f(x)\cos\frac{n\pi x}{l}\,dx = \frac{2}{l}\int_0^l \left(-\frac{x}{l} + 1\right)\cos\frac{n\pi x}{l}\,dx$$

$$= \frac{2}{l}\left[\left(-\frac{x}{l} + 1\right)\left(\frac{l}{n\pi}\frac{\sin n\pi x}{l}\right) - \left(-\frac{1}{l}\right)\left(-\frac{l^2}{n^2\pi^2}\cos\frac{n\pi x}{l}\right)\right]_0^l$$

$$= \frac{2}{l}\left[0 - \frac{l^2}{n^2\pi^2}\cos n\pi + \frac{1}{n^2\pi^2}\right]$$

$$= \frac{2}{n^2\pi^2}\left[1 - (-1)^n\right] = \begin{cases} \dfrac{4}{n^2\pi^2}, & \text{when } n \text{ is odd} \\ 0, & \text{when } n \text{ is even} \end{cases}$$

Hence, $f(x) = \dfrac{1}{2} + \dfrac{4}{\pi^2}\left[\dfrac{1}{1^2}\cos\dfrac{\pi x}{l} + \dfrac{1}{3^2}\cos\dfrac{3\pi x}{l} + ...\right]$

EXAMPLE 2. *Find a series of sines of multiples of x which will represent f(x) in the interval (0, π), where*

$$f(x) = \begin{cases} \dfrac{1}{3}\pi, & 0 < x < \dfrac{1}{3}\pi \\ 0, & \dfrac{1}{3}\pi < x < \dfrac{2}{3}\pi \\ -\dfrac{1}{3}\pi, & \dfrac{2}{3}\pi < x < \pi \end{cases}$$

SOLUTION. Fourier sine series for $f(x)$ in $(0, \pi)$ is

$$f(x) = \sum_{n=1}^{\infty} bn\sin\frac{n\pi x}{\pi}$$

$$b_n = \frac{2}{\pi}\int_0^\pi f(x)\sin\frac{n\pi x}{\pi}$$

$$= \frac{2}{\pi}\left[\int_0^{\pi/3} f(x)\,dx + \int_{\pi/3}^{2\pi/3} f(x)\,dx + \int_{2\pi/3}^\pi f(x)\,dx\right]$$

$$= \frac{2}{\pi}\left[\int_0^{\pi/3}\frac{1}{3}\pi\sin nx\,dx + \int_{\pi/3}^{2\pi/3}0\sin nx\,dx + \int_{2\pi/3}^\pi\left(-\frac{\pi}{3}\right)\sin nx\,dx\right]$$

$$= \frac{2}{3}\left[-\cos\frac{nx}{n}\right]_0^{\pi/3} + 0 - \frac{2}{3}\left[-\cos\frac{nx}{n}\right]_{2\pi/3}^{\pi}$$

$$= -\frac{8}{3}\sin\frac{n\pi}{6}\sin\frac{n\pi}{3}.\cos\frac{n\pi}{2}$$

$$f(x) = -\frac{8}{3}\sum_{n=1}^{\infty}\frac{1}{n}\sin\frac{n\pi}{6}\sin\frac{n\pi}{3}\cos\frac{n\pi}{2}\sin n\pi$$

$$= \frac{1}{2}\left[\frac{1}{2}\sin 2x + \frac{1}{4}\sin 4x + \frac{1}{8}\sin 8x + \frac{1}{10}\sin 10x + ...\right]$$

EXAMPLE 3. *Find the half range cosine series for the function* $f(x) = (x-1)^2$ *in the interval* $0 < x < 1$. (VTU–2010, JNTU–2006)

Hence show that

(i) $\dfrac{1}{1^2} + \dfrac{1}{2^2} + \dfrac{1}{3^2} + \dfrac{1}{4^2} + ... = \dfrac{\pi^2}{6}$,

(ii) $\dfrac{1}{1^2} - \dfrac{1}{2^2} + \dfrac{1}{3^2} - \dfrac{1}{4^2} + ... = \dfrac{\pi^2}{12}$,

(iii) $\dfrac{1}{1^2} + \dfrac{1}{3^2} + \dfrac{1}{5^2} + \dfrac{1}{7^2} + ... = \dfrac{\pi^2}{8}$.

SOLUTION. The Fourier half range cosine series in interval (0, 1) is

$$f(x) = \frac{a_0}{2} + \sum_{n=1}^{\infty} a_n \cos\frac{n\pi x}{1}$$

$$a_0 = \frac{2}{1}\int_0^1(x-1)^2 dx = 2\left[\frac{(x-1)^3}{3}\right]_0^1 = \frac{2}{3}$$

$$a_n = 2\int_0^1(x-1)^2 \cos n\pi x\, dx$$

$$= 2\left[(x-1)^2\left(\frac{\sin n\pi x}{n\pi}\right) - 2(x-1).\left(-\cos\frac{n\pi x}{n^2\pi^2}\right) + 2(1)\left(-\frac{\sin n\pi x}{n^3\pi^3}\right)\right]_0^1$$

$$= 2\left[\frac{2}{n^2\pi^2}\right] = \frac{4}{n^2\pi^2}$$

$$f(x) = \frac{1}{3} + \frac{4}{\pi^2}\left\{\cos\pi x + \frac{\cos 2\pi x}{2^2} + \frac{\cos 3\pi x}{3^2} + ...\right\} \qquad ...(1)$$

$$(x-1)^2 = \frac{1}{3} + \frac{4}{\pi^2}\left\{\cos\pi x + \cos\frac{2\pi x}{2^2} + \frac{\cos 3\pi x}{3^2} + ...\right\}$$

Putting $x = 0$ in (1)

$$1 = \frac{1}{3} + \frac{4}{\pi^2}\left\{1 + \frac{1}{2^2} + \frac{1}{3^2} + ...\right\}$$

$$1 - \frac{1}{3} = \frac{4}{\pi^2}\left(1 + \frac{1}{2^2} + \frac{1}{3^2} + ...\right)$$

$$\frac{\pi^2}{6} = 1 + \frac{1}{2^2} + \frac{1}{3^2} + ... \qquad ...(2)$$

Putting $x = 1$ in eq^n. (1)

$$0 = \frac{1}{3} + \frac{4}{\pi^2}\left(-1 + \frac{1}{2^2} - \frac{1}{3^2} + \frac{1}{4^2} - \ldots\right)$$

$$-\frac{1}{3} = \frac{4}{\pi^2}\left(-1 + \frac{1}{2^2} - \frac{1}{3^2} + \frac{1}{4^2} - \ldots\right)$$

$$\frac{\pi^2}{12} = 1 - \frac{1}{2^2} + \frac{1}{3^2} - \frac{1}{4^2} + \ldots \qquad \ldots(3)$$

Adding (2) and (3), we get

$$\frac{\pi^2}{8} = \frac{1}{1^2} + \frac{1}{3^2} + \frac{1}{5^2} + \frac{1}{7^2} + \ldots$$

EXAMPLE 4. *If* $f(x) = mx, \qquad 0 \le x \le \pi/2$
$\qquad = m(\pi - x), \quad \pi/2 \le x \le \pi$

Then show that $f(x) = \dfrac{4m}{\pi}\left[\dfrac{\sin x}{1^2} - \dfrac{\sin 3x}{3^2} + \dfrac{\sin 5x}{5^2} - \ldots\right].$

SOLUTION. Half range sine series, we have $l = \pi$, so

$$f(x) = \sum_{n=1}^{\infty} b_n \sin\frac{n\pi x}{\pi} = \sum_{n=1}^{\infty} b_n \sin nx$$

$$b_n = \frac{2}{\pi}\int_0^{\pi} f(x)\sin nx\, dx$$

$$= \frac{2}{\pi}\int_0^{\pi/2} mx\sin nx\, dx + \frac{2}{\pi}\int_{\pi/2}^{\pi} m(\pi - x)\sin nx\, dx$$

$$= \frac{2m}{\pi}\left[x\left(-\cos\frac{nx}{n}\right) - 1\left(-\sin\frac{nx}{n^2}\right)\right]_0^{\pi/2} + \frac{2m}{\pi}\left[(\pi - x)\left(-\frac{\cos nx}{n}\right) - (-1)\left(-\frac{\sin nx}{n^2}\right)\right]_{\pi/2}^{\pi}$$

$$= \frac{2m}{\pi}\left[-\frac{\pi}{2}\cos\frac{n\pi}{2} + \frac{1}{n^2}\sin\frac{n\pi}{2}\right] + \frac{2m}{\pi}\left[\frac{\pi}{2n}\cos\frac{n\pi}{2} + \frac{\sin}{n^2}\frac{n\pi}{2}\right]$$

$$= \frac{2m}{\pi}\left[\frac{2}{n^2}\sin\frac{n\pi}{2}\right] = \frac{4}{\pi n^2}\sin\frac{n\pi}{2}$$

$$f(x) = \frac{4}{\pi}\left[\sin x - \frac{\sin 3x}{3^2} + \frac{\sin 5x}{5^2} + \ldots\right]$$

EXAMPLE 5. *Expand* $\pi x - x^2$ *as a half range sine series in the interval* $(0, \pi)$ *upto first three terms. where* $l = \pi$.

SOLUTION. Half range sine series

$$f(x) = \sum_{n=1}^{\infty} b_n \frac{\sin n\pi x}{l}$$

$$b_n = \frac{2}{\pi}\int_0^\pi (\pi x - x^2)\frac{\sin n\pi x}{\pi}\,dx = \frac{2}{\pi}\int_0^\pi (\pi x - x^2)\sin nx\,dx$$

$$= \frac{2}{\pi}\left[(\pi x - x^2)\left(-\cos\frac{nx}{n}\right) - (\pi - 2x)\left(-\sin\frac{nx}{n^2}\right) + (-2)\left(\frac{\cos nx}{n^3}\right)\right]_0^\pi$$

$$\frac{2}{\pi}\left[-\frac{2\cos n\pi}{n^3} + \frac{2}{n^3}\right] = \frac{2}{\pi}\left[\frac{-2(-1)^n}{n^3} + \frac{2}{n^3}\right]$$

$$= \frac{1}{\pi n^3}[1 - (-1)^n] = \begin{cases} 0, & \text{when } n \text{ is even} \\ \dfrac{8}{\pi n^3}, & \text{when } n \text{ is odd} \end{cases}$$

$$\pi x - x^2 = \frac{8}{\pi}\left(\sin x + \frac{\sin 3x}{3^3} + \frac{\sin 5x}{5^3} + \dots\right)$$

EXAMPLE 6.   Obtain a half range cosine series for

$$f(x) = \begin{cases} kx & , \quad \text{for} \quad 0 \le x \le l/2 \\ k(l - x) & , \quad \text{for} \quad l/2 \le x \le l \end{cases}$$

(Bhopal–2008, VTU–2008)

SOLUTION.   Consider $a_0 = \dfrac{2}{l}\int_0^l f(x)\,dx$

$$\frac{2}{l}\left[\int_0^{l/2} kx\,dx + \int_{l/2}^l k(l - x)\,dx\right] = \frac{kl}{2}$$

$$a_n = \frac{2}{l}\int_0^l f(x)\cos\frac{n\pi x}{l}\,dx$$

$$\frac{2}{l}\int_0^{l/2} kx\cos\frac{n\pi x}{l}\,dx + \int_{l/2}^l k(l - x)\cos\frac{n\pi x}{l}\,dx$$

$$= \frac{2kl}{n^2\pi^2}\left[2\cos\frac{n\pi}{2} - 1 - \cos n\pi\right]$$

when n is odd $\cos\dfrac{n\pi}{2} = 0$ and $\cos n\pi = -1$

$$a_n = 0 \quad \text{when } n = 1, 3, 5, \dots$$

n is even $a_2 = \dfrac{2kl}{2^2\pi^2}[2\cos\pi - 1 - \cos 2\pi] = \dfrac{8kl}{2^2\pi^2}$

$$a_4 = \frac{2kl}{4^2\pi^2}[2\cos 2\pi - 1 - \cos 4\pi] = 0$$

$$a_6 = \frac{2kl}{6^2\pi^2}[2\cos 3\pi - 1 - \cos 6\pi] = -\frac{8kl}{6^2\pi^2}$$

$$f(x) = \frac{kl}{4} - \frac{8kl}{\pi^2}\left[\frac{1}{2^2\pi^2}\cos\frac{2\pi x}{l} + \frac{1}{6^2}\cos\frac{6\pi x}{l} + \dots\right] \qquad \dots(1)$$

Putting $x = 1$, $f(x) = 0$
then from (1), we have

$$0 = \frac{kl}{4} - \frac{8kl}{\pi^2}\left(\frac{1}{2^2} + \frac{1}{6^2} + ...\right)$$

$$\frac{1}{2^2} + \frac{1}{6^2} + ... = \frac{\pi^2}{32}$$

$\Rightarrow \qquad \frac{1}{2^2}\left(\frac{1}{1^2} + \frac{1}{3^2} + ...\right) = \frac{\pi^2}{32}$

$\Rightarrow$ Hence, $\qquad \frac{1}{1^2} + \frac{1}{3^2} = \frac{\pi^2}{8}$

9.9 PARSEVEL'S IDENTITY FOR FOURIER SERIES

Consider the Fourier series $\frac{a_0}{2} + \sum\limits_{n=1}^{\infty}(a_n \cos nx + b_n \sin nx)$ .If $f(x)$ converges uniformly to $f(x)$ at

every point of the interval $(0, 2\pi)$, then

$$\frac{1}{\pi}\int_0^{2\pi}\{f(x)\}^2 dx = \frac{a_0^2}{2} + \sum\limits_{n=1}^{\infty}(a_n^2 + b_n^2).$$

Proof. Let the series $\frac{a_0}{2} + \sum\limits_{n=1}^{\infty}(a_n \cos nx + b_n \sin nx)$ represents the Fourier series of $f(x)$. Also,

let this series converges uniformly to $f(x)$ at every point of the interval $(0, 2\pi)$ so that

$$f(x) = \frac{a_0}{2} + \sum\limits_{n=1}^{\infty}(a_n \cos nx + b_n \sin nx) \qquad \qquad ...(1)$$

and that term by term integration is possible.

To prove that $\frac{1}{\pi}\int_0^{2\pi}\{f(x)\}^2 dx = \frac{a_0^2}{2} + \sum(a_n^2 + b_n^2)$

We have $\qquad a_n = \frac{1}{\pi}\int_0^{2\pi} f(x).\cos nx\, dx \quad (n = 0, 1, 2, 3, ...)$

$$b_n = \frac{1}{\pi}\int_0^{2\pi} f(x).\sin nx\, dx \quad (n = 0, 1, 2, 3 ...)$$

Multiplying (1) by $f(x)$ and then integrating from $x=0$ to $x=2\pi$, we get

$$\int_0^{2\pi}\{f(x)\}^2 dx = \frac{a_0}{2} + \int_0^{2\pi} f(x)\, dx + \sum\limits_{n=1}^{\infty}\left(a_n\int_0^{2\pi} f(x).\cos nx dx + b_n\int_0^{2\pi} f(x)\sin nx dx\right)$$

$$= \frac{a_0}{2}.\pi a_0 + \sum\limits_{n=1}^{\infty}(\pi a_n^2 + \pi b_n^2)$$

Dividing by π, we get $\frac{1}{\pi}\int_0^{2\pi}\{f(x)\}^2 dx = \frac{a_0^2}{2} + \sum\limits_{n=1}^{\infty}(a_n^2 + b_n^2).$

Solved Examples

EXAMPLE 1. *Obtain the Fourier series expansion of* $f(x) = x^2$ *in* $-\pi < x < \pi$ *and prove that*

$$\sum_{n=1}^{\infty} \frac{1}{n^4} = \frac{\pi^4}{90} \text{ by using Parsevel's theorem.}$$

SOLUTION. Since $f(x) = x^2$ is even function so $b_n = 0$

Let the Fourier series expansion of $f(x)$ is given by

$$f(x) = x^2 = \frac{a_0}{2} + \sum_{n=1}^{\infty} a_n \cos nx \qquad \ldots(1)$$

where $\qquad a_0 = \frac{2}{\pi} \int_0^{\pi} f(x)dx = \frac{2}{\pi} \int_0^{\pi} x^2 dx = \frac{2\pi^2}{3}$

$$a_n = \frac{2}{\pi} \int_0^{\pi} f(x) \cos nx \, dx = \frac{2}{\pi} \int_0^{\pi} x^2 \cos nx \, dx$$

$\Rightarrow \qquad a_n = \frac{2}{\pi} \left[x^2 \frac{\sin nx}{n} + 2x \cdot \frac{\cos nx}{n^2} - 2 \frac{\sin nx}{n^2} \right]_0^{\pi} = \frac{\pi(-1)^2}{n^2}$

$\therefore$ (1) becomes,

$$x^2 = \frac{\pi^2}{3} + 4 \sum_{n=1}^{\infty} \frac{(-1)^n \cos nx}{n^2} \qquad \ldots(2)$$

which is the required Fourier expansion.

Now, by Parsevel's theorem, we have

$$\int_{-\pi}^{\pi} \{f(x)\}^2 dx = \pi \left[\frac{a_0^2}{2} + \sum_{n=1}^{\infty} (a_n^2 + b_n^2) \right]$$

Hence, $\int_{-\pi}^{\pi} x^4 dx = \pi \left[\frac{4\pi^4}{2.9} + \sum_{n=1}^{\infty} \frac{16}{n^4} \right] \Rightarrow \left(\frac{x^5}{5} \right)_{-\pi}^{\pi} = \frac{2\pi^5}{9} + \pi \sum_{n=1}^{\infty} \frac{16}{n^4}$

or $\quad \frac{2\pi^5}{5} - \frac{2\pi^5}{9} = \pi \sum_{n=1}^{\infty} \frac{16}{n^4} \Rightarrow \frac{\pi^4}{90} = \sum_{n=1}^{\infty} \frac{1}{n^4}.$

EXAMPLE 2. *By using the sine series for* $f(x) = 1$ *in* $0 < x < \pi$, *show that*

$$\frac{\pi^2}{8} = 1 + \frac{1}{3^2} + \frac{1}{5^2} + \frac{1}{7^2} + \ldots .$$

SOLUTION. The Fourier sine series for $f(x) = 1$ in $(0, \pi)$ is $f(x) = \Sigma b_n \sin nx$, where,

$$b_n = \frac{2}{\pi} \int_0^{\pi} f(x) \sin nx \, dx = \frac{2}{\pi} \int_0^{\pi} (1) . \sin nx dx$$

$$= \frac{2}{\pi} \left(-\frac{\cos nx}{n} \right)_0^{\pi}$$

$$= -\frac{2}{n\pi} (\cos n\pi - 1) = -\frac{2}{n\pi} [(-1)^n - 1]$$

$$= \begin{cases} \dfrac{4}{n\pi} & , \quad \text{if } n \text{ is odd} \\ 0 & , \quad \text{if } n \text{ is even} \end{cases}$$

The Fourier sine series is

$$1 = \frac{4}{\pi}\sin x + \frac{4}{3\pi}\sin 3x + \frac{4}{5\pi}\sin 5x + \frac{4}{7\pi}\sin 7x + \dots$$

By Parsevel's formula, we get

$$\int_0^\pi [f(x)]^2 dx = \frac{c}{2}[b_1^2 + b_2^2 + b_3^2 + b_4^2 + b_5^2 + \dots]$$

$$\Rightarrow \qquad \int_0^\pi (1)^2 dx = \frac{\pi}{2}\left[\left(\frac{4}{\pi}\right)^2 + \left(\frac{4}{3\pi}\right)^2 + \left(\frac{4}{5\pi}\right)^2 + \left(\frac{4}{7\pi}\right)^2 + \dots\right]$$

$$\Rightarrow \qquad [x]_0^\pi = \left(\frac{\pi}{2}\right)\left(\frac{16}{\pi^2}\right)\left[1 + \frac{1}{3^2} + \frac{1}{5^2} + \frac{1}{7^2} + \dots\right]$$

$$\Rightarrow \qquad \pi = \frac{\pi}{2}\left(\frac{16}{\pi^2}\right)\left[1 + \frac{1}{3^2} + \frac{1}{5^2} + \frac{1}{7^2} + \dots\right]$$

Hence, $\qquad \dfrac{\pi^2}{8} = 1 + \dfrac{1}{3^2} + \dfrac{1}{5^2} + \dfrac{1}{7^2} + \dots$

EXAMPLE 3. If $f(x) = \begin{cases} \pi x & , \quad 0 < x < 1 \\ \pi(2-x) & , \quad 1 < x < 2 \end{cases}$

Using half range cosine series, show that $\dfrac{1}{1^4} + \dfrac{1}{3^4} + \dfrac{1}{5^4} + \dots = \dfrac{\pi^4}{96}$.

SOLUTION The half range cosine series for $f(x)$ in $(0, c)$ is

$$f(x) = \frac{a_0}{2} + \sum_{n=1}^{\infty} a_n \cos\frac{n\pi x}{c}$$

Here, $a_0 = \dfrac{2}{c}\int_0^c f(x)dx = \dfrac{2}{2}\left[\int_0^1 \pi x\, dx + \int_1^2 \pi(2-x)dx\right]$

$$= \pi\left[\frac{x^2}{2}\right]_0^1 + \pi\left[2x - \frac{x^2}{2}\right]_0^1 = \frac{\pi}{2} + \pi\left[(4-2) - \left(2 - \frac{1}{2}\right)\right] = \pi$$

$$a_n = \frac{2}{c}\int_0^c f(x).\cos\frac{n\pi x}{c}dx = \frac{2}{2}\left[\int_0^1 \pi x \cos\frac{n\pi x}{2}dx + \int_1^2 \pi(2-x)\cos\frac{n\pi x}{2}dx\right]$$

$$= \pi\left[\frac{x\sin\dfrac{n\pi x}{2}}{\dfrac{n\pi}{2}} - \left(-\frac{\cos\dfrac{n\pi x}{2}}{\dfrac{n^2\pi^2}{4}}\right)\right]_0^1 + \pi\left[(2-x)\frac{\sin\dfrac{n\pi x}{2}}{\dfrac{n\pi}{2}} - (-1)\left(-\frac{\cos\dfrac{n\pi x}{2}}{\dfrac{n^2\pi^2}{4}}\right)\right]_1^2$$

$$= \pi \left[\frac{2}{n\pi} \sin \frac{n\pi}{2} + \frac{4}{n^2\pi^2} \cos \frac{n\pi}{2} - \frac{4}{n^2\pi^2} \right]$$

$$+ \pi \left[0 - \frac{4}{n^2\pi^2} \cos n\pi - \frac{2}{n\pi} \sin \frac{n\pi}{2} + \frac{4}{n^2\pi^2} \cos \frac{n\pi}{2} \right]$$

$$= \pi \left[\frac{8}{n^2\pi^2} \cos \frac{n\pi}{2} - \frac{4}{n^2\pi^2} - \frac{4}{n^2\pi^2} \cos n\pi x \right]$$

$$= \frac{4}{n^2\pi} \left[2\cos \frac{n\pi}{2} - 1 - \cos n\pi \right]$$

Putting $n = 1, 2, 3, \ldots$, we get

$$a_1 = 0, a_2 = \frac{-4}{\pi}, a_3 = 0, a_4 = 0, a_5 = 0, a_6 = -\frac{4}{9\pi}$$

By Parsevel's formula, we get

$$\int_0^c \{f(x)\}^2 dx = \frac{c}{2} \left[\frac{a_0^2}{2} + a_1^2 + a_2^2 + a_3^2 + \ldots \right]$$

$$\int_0^1 (\pi x)^2 dx + \int_1^2 \pi^2 (2-x)^2 dx = \frac{2}{2} \left[\frac{\pi^2}{2} + \frac{16}{\pi^2} + \frac{16}{81\pi^2} + \ldots \right]$$

$$\pi^2 \left[\frac{x^3}{3} \right]_0^1 - \pi^2 \left[\frac{(2-x)^3}{3} \right]_1^2 = \frac{\pi^2}{2} + \frac{16}{\pi^2} + \frac{16}{81\pi^2} + \ldots$$

$$\Rightarrow \qquad \frac{\pi^2}{3} - \pi^2 \left(0 - \frac{1}{3} \right) = \frac{\pi^2}{3^2} + \frac{16}{\pi^2} \left[1 + \frac{1}{81} + \ldots \right]$$

$$\Rightarrow \qquad \frac{2\pi^2}{3} - \frac{\pi^2}{2} = \frac{16}{\pi^2} \left[1 + \frac{1}{3^4} + \frac{1}{5^4} + \ldots \right]$$

$$\Rightarrow \qquad \frac{\pi^2}{6} = \frac{16}{\pi^2} \left[1 + \frac{1}{3^4} + \frac{1}{5^4} + \ldots \right]$$

$$\Rightarrow \qquad \frac{\pi^4}{96} = 1 + \frac{1}{3^4} + \frac{1}{5^4} + \ldots$$

9.10 COMPLEX FORM OF FOURIER SERIES

In complex notations, the Fourier series is written as

$$f(x) = \sum_{n=-\infty}^{+\infty} C_n e^{inx} \qquad\qquad (n = 0, 1, 2, \ldots)$$

$$f(x) = C_0 + \sum_{n=1}^{\infty} C_n e^{inx} + \sum_{n=1}^{\infty} C_{-n} e^{-inx} \qquad\qquad \ldots(1)$$

With $\qquad C_n = \frac{1}{2\pi} \int_C^{C+2\pi} f(x) e^{-inx} dx$ and $C_{-n} = \frac{1}{2\pi} \int_C^{C+2\pi} f(x) e^{inx} dx \quad (n = 0, 1, 2, \ldots)$

$f(x)$ being defined in the interval $[C, C + 2\pi]$.

Here,
$$C_0 = \frac{1}{2\pi} \int_C^{C+2\pi} f(x)\, dx = \frac{a_0}{2}.$$

$$C_n = \frac{1}{2\pi} \int_C^{C+2\pi} f(x)\, e^{-inx}\, dx = \frac{1}{2\pi} \int_C^{C+2\pi} f(x)(\cos nx - i\sin nx)dx$$

$$= \frac{1}{2\pi} \int_C^{C+2\pi} f(x) \cos nx\, dx - i\frac{1}{2\pi} \int_C^{C+2\pi} f(x) \sin nx\, dx = \frac{a_n - ib_n}{2}$$

Similarly,
$$C_{-n} = \frac{1}{2\pi} \int_C^{C+2\pi} f(x)\, e^{inx}\, dx = \frac{a_n + ib_n}{2}.$$

Now, (1) becomes

$$f(x) = \frac{a_0}{2} + \sum_{n=1}^{\infty} \left\{ \left(\frac{a_n - ib_n}{2} \right) \cos nx + \left(\frac{a_n + ib_n}{2} \right) \sin nx \right\} \qquad \ldots(2)$$

with
$$a_n = \frac{1}{\pi} \int_C^{C+2\pi} f(x).\cos nx\, dx \quad \text{and} \quad b_n = \frac{1}{\pi} \int_C^{C+2\pi} f(x) \sin nx\, dx$$

Solved Example

EXAMPLE. *Obtain the complex form of the Fourier series of $f(x) = e^{-x}$ in $-1 \le x \le 1$.*

(Mumbai–2005S, Madras–2000S)

SOLUTION. The complex form of Fourier series for the given function $f(x)$ is

$$f(x) = \sum_{n=-\infty}^{\infty} C_n\, e^{inx}.$$

Here,
$$C_n = \int_{-1}^{1} e^{-x}\, e^{-inx\pi} dx = \frac{1}{2} \int_{-1}^{1} e^{-(1+in\pi)x} dx$$

$$= \frac{1}{2} \left[\frac{e^{-(1+in\pi)x}}{-(1+in\pi)} \right]_{-1}^{1} = \frac{e^{1+inx} - e^{-(1+inx)}}{2(1+inx)}$$

$$= \frac{e(\cos n\pi + i\sin n\pi) - e^{-1}(\cos n\pi - i\sin n\pi)}{2(1+in\pi)}$$

$$= \frac{e - e^{-1}}{2}(-1)^n \frac{1 - in\pi}{1+n^2\pi^2} = \frac{(-1)^n(1 - in\pi)\sinh 1}{1+n^2\pi^2}$$

Hence, the required complex form of the Fourier series is

$$e^{-x} = \sum_{n=-\infty}^{\infty} \frac{(-1)^n(1 - in\pi)}{1+n^2\pi^2} \sinh 1.e^{in\pi x}.$$

Miscellaneous Solved Examples

EXAMPLE 1. *Show that the Fourier series for the function $f(x) = x^2$ in $-\pi < x < \pi$ is*

$$\frac{\pi^2}{3} - 4\left[\frac{\cos x}{1^2} - \frac{\cos 2x}{2^2} + \frac{\cos 3x}{3^2} + \ldots \right]$$

(Purvanchal-1991)

SOLUTION. Here, $f(x) = x^2$ and $f(-x) = (-x)^2 = x^2$, hence $f(x)$ is even then $b_n = 0$
We know that

$$f(x) = \frac{a_0}{2} + \sum_{n=1}^{\infty} a_n \cos nx + \sum_{n=1}^{\infty} b_n \sin nx \qquad \ldots(1)$$

Since

$$a_0 = \frac{1}{2\pi} \int_{-\pi}^{\pi} f(x)\,dx = \frac{1}{2\pi} \int_{-\pi}^{\pi} x^2 dx \qquad [\because f(x) = x^2]$$

$$= \frac{1}{2\pi}\left[\frac{1}{3}x^3\right]_{-\pi}^{\pi} = \frac{1}{6\pi}[\pi^3 - (-\pi)^3] = \frac{1}{6\pi}(2\pi^3) = \frac{\pi^2}{3} \qquad ...(2)$$

Now, $\qquad a_n = \frac{1}{\pi}\int_{-\pi}^{\pi} f(x)\cos nx\,dx = \frac{1}{\pi}\int_{-\pi}^{\pi} x^2 \cos nx\,dx$

On integrating by parts, we get

$$= \frac{1}{\pi}\left[\left\{x^2\left(\frac{\sin nx}{n}\right)\right\}_{-\pi}^{\pi} - \int_{-\pi}^{\pi} 2x\left(\frac{\sin nx}{n}\right)dx\right].$$

$$= \frac{1}{\pi}\left[0 - \frac{2}{n}\int_{-\pi}^{\pi} x \sin nx\,dx\right] = -\frac{2}{n\pi}\int_{-\pi}^{\pi} x \sin nx\,dx$$

Again, integrating by parts, we get

$$= \frac{-2}{n\pi}\left[\left\{x\left(-\frac{\cos nx}{n}\right)\right\}_{-\pi}^{\pi} - \int_{-\pi}^{\pi} 1\left(-\frac{\cos nx}{n}\right)dx\right]$$

$$= \frac{2}{\pi n^2}\left[\{x\cos nx\}_{-\pi}^{\pi} - \left\{\frac{\sin nx}{n}\right\}_{-\pi}^{\pi}\right]$$

$$= \frac{2}{\pi n^2}[2\pi\cos n\pi - 0] = \frac{4}{n^2}(-1)^n \qquad ...(3)$$

Now, substituting the value of eq^n (2) and (3) in eq^n (1), we get

$$f(x) = \frac{\pi^2}{3} + \sum_{n=1}^{\infty}\left[\frac{4}{n^2}(-1)^n \cos nx\right]$$

$$= \frac{\pi^2}{3} + 4\left[\frac{(-1)^2}{1^2}\cos x + \frac{(-1)^2}{2^2}\cos 2x + \frac{(-1)^3}{3^2}\cos 3x + ...\right]$$

$$= \frac{\pi^2}{3} - 4\left[\frac{\cos x}{1^2} - \frac{\cos 2x}{2^2} + \frac{\cos 3x}{3^2} - ....\right].$$

EXAMPLE 2.   *Show that* $\dfrac{\pi}{4}\cos x = \dfrac{2}{1.3}\sin(2x) + \dfrac{4}{3.5}\sin(4x) + \dfrac{6}{5.7}\sin(6x) + ...$

where $0 \le x \le \pi.$ (SVTU–2004)

SOLUTION.   Consider, $\qquad f(x) = \frac{\pi}{4}\cos x = \sum_{n=1}^{\infty} b_n(\sin nx) \qquad ...(1)$

Now, $\qquad b_n = \frac{2}{\pi}\int_0^{\pi} f(x)\sin(nx)\,dx$

$\Rightarrow \qquad b_n = \frac{2}{\pi}\int_0^{\pi} \frac{\pi}{4}(\cos x)\sin(nx)\,dx$

$\Rightarrow \qquad 4b_n = \int_0^{\pi}[\sin(n+1)x + \sin(n-1)x]\,dx$

$$= \left[-\frac{\cos(n+1)x}{n+1} - \frac{\cos(n-1)x}{n-1} \right]_0^x$$

$$= -\frac{1}{(n+1)}\{\cos(n+1)\pi - 1\} - \frac{1}{n-1}\{\cos(n-1)\pi - 1\}$$

$$= -\frac{1}{(n+1)}\{-\cos n\pi - 1\} - \frac{1}{(n-1)}\{-\cos n\pi - 1\}$$

$$= \left\{ \frac{1}{n+1} + \frac{1}{n-1} \right\} + \left(\frac{1}{n+1} + \frac{1}{n-1} \right)\cos(n\pi)$$

$$= \frac{2n}{n^2-1}(1 + \cos n\pi)$$

$\Rightarrow \qquad b_n = \left(\frac{n}{n^2-1} \right)\frac{1}{2}[1 + (-1)^n].$

$\Rightarrow \qquad b_n = 0 \qquad$ for $n = 1, 3, 5, 7, \dots\dots$

and $\qquad b_n = \dfrac{n}{n^2-1} \qquad$ for $n = 2, 4, 6, \dots.$

Putting this values in eq^n. (1), we get

$$\frac{\pi}{4}\cos x = \sum_{n=1}^{\infty} b_n \sin(nx)$$

$$\frac{\pi}{4}\cos x = \sum_{n=1}^{\infty} \left(\frac{n}{n^2-1} \right)\sin(nx) \quad \text{for } n = 2, 4, 6, \dots.$$

$$= \frac{2}{1.3}\sin(2x) + \frac{4}{3.5}\sin(4x) + \frac{6}{5.7}\sin(6x) + \dots$$

EXAMPLE 3. *Find Fourier series for $f(x) = x$ in the interval $(-\pi, \pi)$.* (Purvanchal-1990)

SOLUTION. Here, $\qquad f(x) = a_0 + \sum_{n=1}^{\infty}(a_n \cos nx + b_n \sin nx)$ ...(1)

Since, $\qquad a_0 = \dfrac{1}{\pi}\int_{-\pi}^{\pi} f(x)\,dx = \dfrac{1}{\pi}\int_{-\pi}^{\pi} x\,dx \qquad [\because f(x) = x]$

$a_0 = 0$, because intergrand is odd function of x.

Now, $\qquad a_n = \dfrac{1}{\pi}\int_{-\pi}^{\pi} f(x)\cos nx\,dx = \dfrac{1}{\pi}\int_{-\pi}^{\pi} x \cos nx\,dx \qquad [\because f(x) = x]$

$a_n = 0 \qquad$ ...(3), because integrand is odd function of x.

and $\qquad b_n = \dfrac{1}{\pi}\int_{-\pi}^{\pi} f(x)\sin nx\,dx = \dfrac{1}{\pi}\int_{-\pi}^{\pi} x \sin nx\,dx = \dfrac{2}{\pi}\int_{0}^{\pi} x \sin nx\,dx$

On integrating by parts, we get

$$= \frac{2}{\pi}\left[\left\{ x\left(-\frac{\cos nx}{n} \right) \right\}_0^{\pi} - \int_0^{\pi} 1.\left(-\frac{\cos nx}{n} \right)dx \right]$$

$$= \frac{2}{\pi}\left[-\frac{\pi}{n}\cos n\pi + \frac{1}{n}\int_0^{\pi} \cos nx\,dx \right]$$

$$= -\frac{2}{n}\cos n\pi + \frac{2}{\pi n}\left(\frac{\sin nx}{n}\right)_0^\pi = -\frac{2}{n}(-1)^n = (-1)^{n+1}\frac{2}{n}.$$

Putting the value of eq^n (2), (3) and (4) in eq^n (1), we get

$$f(x) = \sum_{n=1}^{\infty}\left[(-1)^{n+1}\frac{2}{n}\sin nx\right]$$

$$x = (-1)^2.\frac{2}{1}\sin x + (-1)^3.\frac{2}{2}\sin 2x + (-1)^4.\frac{2}{3}\sin 3x + ...$$

$$x = 2\left[\sin x - \frac{1}{2}\sin 2x + \frac{1}{3}\sin 3x + ...\infty\right].$$

EXAMPLE 4.   *Obtain the Fourier series for e^x in the interval $[-\pi, \pi]$. What is the sum of the series for $x = \pm \pi$?*

(SVTU-2004)

SOLUTION.   Here, we have

$$f(x) = e^x = \frac{a_0}{2} + \sum_{n=1}^{\infty}\left[a_n\cos(nx) + b_n\sin(nx)\right] \qquad ...(1)$$

Since, $$a_0 = \frac{1}{\pi}\int_{-\pi}^{\pi}e^x dx = \frac{1}{\pi}(e^\pi - e^{-\pi}) = \frac{2}{\pi}\sinh(\pi) \qquad ...(2)$$

$$a_n = \frac{1}{\pi}\int_{-\pi}^{\pi}f(x)\cos(nx)dx = \frac{1}{\pi}\int_{-\pi}^{\pi}e^x\cos(nx)dx$$

$$\Rightarrow \qquad \pi a_n = \frac{1}{(n^2+1)}\left[e^x\{\cos(nx) + n\sin(nx)\}\right]_{x=-\pi}^{\pi}$$

$$\Rightarrow \quad \pi(n^2+1)a_n = e^\pi\{\cos(n\pi) + n\sin(n\pi)\} - e^{-\pi}\{\cos(-n\pi) + n\sin(-n\pi)\}$$

$$= e^{-x}\cos(n\pi) - e^{-\pi}(\cos n\pi)$$

$$= 2\sinh(\pi).\cos n\pi$$

$$\Rightarrow \qquad a_n = \frac{2}{\pi(n^2+1)}\sinh(\pi).\cos(n\pi) \qquad ...(3)$$

$$b_n = \frac{1}{\pi}\int_{-\pi}^{\pi}f(x)\sin(nx)dx = \frac{1}{\pi}\int_{-\pi}^{\pi}e^x\sin(nx)dx$$

$$\Rightarrow \quad \pi(n^2+1)b_n = \left[e^x\{\sin(nx) - n\cos(nx)\}\right]_{x=-\pi}^{\pi}$$

$$= e^\pi\{\sin(n\pi) - n\cos(n\pi)\} - e^{-\pi}\{\sin(-n\pi) - n\cos(-n\pi)\}$$

$$= n\cos(n\pi)(e^{-\pi} - e^\pi) = -2n\sinh(\pi).\cos(n\pi)$$

$$\Rightarrow \qquad b_n = -\frac{2n}{\pi(n^2+1)}.\sinh(\pi).\cos(n\pi).$$

Using eq^n (2), (3) and (4) in eq^n (1), we get

$$e^x = \frac{2\sinh(\pi)}{\pi}\left[\frac{1}{2} + \sum_{n=1}^{\infty}\frac{\cos(n\pi)}{(n^2+1)}\{\cos(nx) - n\sin(nx)\}\right].$$

EXAMPLE 5.   *Obtain Fourier sine series of the following function*

$$f(x) = e^{ax}, \; 0 < x < \pi.$$

(Kanpur-1990)

SOLUTION. Consider $f(x) = e^{ax} = \sum\limits_{n=1}^{\infty} b_n \sin nx$

where $b_n = \dfrac{2}{\pi}\int_0^{\pi} f(x)\sin nx\, dx = \dfrac{2}{\pi}\int_0^{\pi} e^{ax}\sin nx\, dx$

$= \dfrac{2}{\pi}\left[\dfrac{e^{ax}}{a^2+n^2}(a\sin nx - n\cos nx)\right]_0^{\pi}$

$= \dfrac{2}{\pi}\left[\dfrac{e^{a\pi}}{a^2+n^2}(a\sinh n\pi - n\cos n\pi) + \dfrac{n}{a^2+n^2}\right]$

$= \dfrac{2}{\pi}\cdot\dfrac{n}{a^2+n^2}[-(-1)^n e^{a\pi}+1]$ $[\because \cos n\pi = (-1)^n \text{ and } \sin n\pi = 0]$

$\therefore$ $b_n = \dfrac{2n}{(a^2+n^2)\pi}[1-(-1)^n e^{a\pi}]$

Putting $n = 1, 2, 3, \ldots$, we have

$b_1 = \dfrac{2(1+e^{a\pi})}{(a^2+1^2)\pi}$; $b_2 = \dfrac{2.2(1-e^{a\pi})}{(a^2+2^2)\pi}$

.... etc.

Now, the required Fourier sine series is given by

$e^{ax} = \dfrac{2}{\pi}\left[\dfrac{1+e^{ax}}{a^2+1^2}\sin x + \dfrac{2(1-e^{a\pi})}{a^2+2^2}\sin 2x + \ldots\right].$

EXAMPLE 6. *Find Fourier cosine series for* $f(x) = e^x$, $0 < x < \pi$. (SVTU–2006)

SOLUTION. Consider $f(x) = e^x + \dfrac{a_0}{2} + \sum\limits_{n=1}^{\infty} a_n \cos(nx)$ ...(1)

Since $a_0 = \dfrac{2}{\pi}\int_0^{\pi} f(x)\,dx = \dfrac{2}{\pi}\int_0^{\pi} e^x\,dx = \dfrac{2}{\pi}(e^{\pi}-1)$

$a_n = \dfrac{2}{\pi}\int_0^{\pi} f(x)\,dx = \dfrac{2}{\pi(n^2+1)}\left[e^x\{\cos(nx)+n\sin(nx)\}\right]_{x=0}^{\pi}$

$= \dfrac{2}{\pi(n^2+1)}[e^{\pi}\{\cos(n\pi)+n\sin(n\pi)\}-\{\cos 0 + n\sin 0\}]$

$= \dfrac{2}{\pi(n^2+1)}[e^{\pi}\cos(n\pi)-1]$

Putting this value in eq^n (1), we get

$e^x = \dfrac{1}{\pi}(e^{\pi}-1) + \sum\limits_{n=1}^{\infty}\dfrac{2[e^{\pi}\cos(n\pi)-1]\cos(nx)}{\pi(n^2+1)}$

EXAMPLE 7. *Find the Fourier's series for* $f(x)$ *in the interval* $(-\pi, \pi)$, *where*

$f(x) = \begin{cases} \pi+x, & -\pi < x < 0 \\ \pi-x, & 0 < x < \pi \end{cases}$ (SVTU-2004, Agra-1968, Purvanchal-1989, 91)

SOLUTION. We know that $f(x) = \dfrac{a_0}{2} + \sum\limits_{n=1}^{\infty}(a_n\cos nx + b_n\sin nx)$

Then
$$a_0 = \frac{1}{\pi}\int_{-\pi}^{\pi} f(x)\,dx = \frac{1}{\pi}\left[\int_{-\pi}^{0} f(x)\,dx + \int_{0}^{\pi} f(x)\,dx\right]$$

$$= \frac{1}{\pi}\left[\int_{-\pi}^{0}(\pi+x)\,dx + \int_{0}^{\pi}(\pi-x)\,dx\right]$$

$$= \frac{1}{\pi}\left[\left(\pi x + \frac{x^2}{2}\right)_{x=-\pi}^{0} + \left(\pi x - \frac{x^2}{2}\right)_{x=0}^{\pi}\right]$$

$$= \frac{1}{\pi}\left[\pi^2 - \frac{\pi^2}{2} + \pi^2 - \frac{\pi^2}{2}\right] = \pi$$

Now
$$a_n = \frac{1}{\pi}\left[\int_{-\pi}^{0} f(x)\cos nx\,dx + \int_{0}^{\pi} f(x)\cos nx\,dx\right]$$

$$= \frac{1}{\pi}\left[\int_{-\pi}^{0}(\pi+x)\cos nx\,dx + \int_{0}^{\pi}(\pi-x)\cos nx\,dx\right]$$

$$= \frac{1}{\pi}\left[\left(\frac{\pi\sin nx}{n} + \frac{x\sin nx}{n} + \frac{\cos nx}{n^2}\right)_{x=-\pi}^{0}\right]$$

$$+ \frac{1}{\pi}\left[\frac{\pi\sin nx}{n} - \frac{x\sin nx}{n} - \frac{\cos nx}{n^2}\right]_{x=0}^{\pi}$$

$$= \frac{1}{\pi}\left[\frac{1-\cos n\pi}{n^2} - \frac{\cos n\pi}{n^2} - 1\right]$$

$$= \frac{2}{\pi}\left[\frac{1-\cos n\pi}{n^2}\right] = \frac{2}{\pi}\left[\frac{1-(-1)^n}{n^2}\right].$$

$a_n = 0$ if n is even.

$a_n = \dfrac{4}{n^2\pi}$ if n is odd.

and
$$b_n = \frac{1}{\pi}\left[\int_{-\pi}^{0} f(x)\sin nx\,dx + \int_{0}^{\pi} f(x)\sin nx\,dx\right]$$

$$= \frac{1}{\pi}\left[\int_{-\pi}^{0}(\pi+x)\sin nx\,dx + \int_{\pi}^{0}(\pi-x)\sin nx\,dx\right]$$

$$= \frac{1}{\pi}\left[\frac{-\pi\cos nx}{n} - \frac{x\cos nx}{n} + \frac{\sin nx}{n^2}\right]_{-\pi}^{0}$$

$$+ \frac{1}{\pi}\left[\frac{-\pi\cos nx}{n} + \frac{x\cos nx}{n} - \frac{\sin nx}{n^2}\right]_{0}^{\pi}$$

$$= \frac{1}{\pi}\left[\frac{-\pi+\pi\cos n\pi - \pi\cos n\pi}{n} + \left(\frac{-\pi\cos n\pi + \pi\cos n\pi + \pi}{n}\right)\right]$$

Putting the value of a_0, a_n and b_n in eq^n. (1), we get

$$f(x) = \frac{\pi}{2} + \frac{4}{\pi}\left(\cos x + \frac{1}{3^2}\cos 3x + \frac{1}{5^2}\cos 5x + ...\right)$$

EXAMPLE 8. *Find Fourier series for* $f(x) = x$ *in* $(0, \pi)$. (SVTU-2006)

SOLUTION. Consider
$$f(x) = x = \frac{a_0}{2} + \sum_{n=1}^{\infty} \{a_n \cos nx + b_n \sin nx\}$$

Now,
$$a_0 = \frac{1}{\pi} \int_0^{\pi} f(x) \, dx = \frac{1}{\pi} \int_0^{\pi} x \, dx = \frac{\pi}{2}$$

$$a_n = \frac{1}{\pi} \int_0^{\pi} f(x) \cos(nx) \, dx = \frac{1}{\pi} \int_0^{\pi} x \cos(nx) \, dx$$

$$\pi a_n = \left\{\frac{x}{n} \sin(ax)\right\}_0^{\pi} - \frac{1}{n} \int_0^{\pi} \sin(nx) \, dx$$

$$\Rightarrow \qquad \pi a_n = (0 - 0) + \frac{1}{n^2}\{\cos(nx)\}_0^{\pi} = \frac{\cos(n\pi) - 1}{n^2}$$

$$\Rightarrow \qquad a_n = \frac{1}{\pi n^2}[(-1)^n - 1].$$

$a_n = 0$ for even numbers.

and
$$b_n = \frac{1}{\pi} \int_0^{\pi} f(x) \sin(nx) \, dx = \frac{1}{\pi} \int_0^{\pi} x \sin(nx) \, dx$$

$$\Rightarrow \qquad \pi b_n = \left\{-\frac{x}{n} \cos(nx)\right\}_0^{\pi} + \frac{1}{n} \int_0^{\pi} \cos(nx) \, dx$$

$$-\frac{\pi}{n} \cos(n\pi) + \frac{1}{n^2}\{\sin(nx)\}_0^{\pi} = -\frac{\pi}{n} \cos(n\pi) - 0$$

$$\Rightarrow \qquad b_n = -\frac{1}{n} \cos(n\pi)$$

Putting the value of a_0, a_n and b_n in eq^n. (1), we get

$$x = \frac{\pi}{4} + \frac{1}{\pi} \sum_{n=1}^{\infty} \left\{\frac{(-1)^n - 1}{n^2} \cos(nx) + \sum_{n=1}^{\infty} \frac{(-1)^{n+1}}{n} \sin(nx)\right\}$$

$$\Rightarrow \qquad x = \frac{\pi}{4} - \frac{2}{\pi}\left[\frac{1}{1^2}\cos x + \frac{1}{3^2}\cos(3x) + \frac{1}{5^2}\cos(5x) + ...\right]$$

$$+ \left[\sin x - \frac{1}{2}\sin(2x) + \frac{1}{3}\sin(3x) - \frac{1}{4}\sin(4x) + ...\right]$$

EXAMPLE 9. *Find Fourier series for* $f(x) = x$ *in* $(-\pi, \pi)$.

SOLUTION. Consider
$$f(x) = x = \frac{a_0}{2} + \sum_{n=1}^{\infty} a_n \cos(nx) + \sum_{n=1}^{\infty} b_n \sin(nx)$$

Now,
$$a_0 = \frac{1}{\pi} \int_{-\pi}^{\pi} f(x) \, dx = \frac{1}{\pi} \int_{-\pi}^{\pi} x \, dx = 0$$

$$a_n = \frac{1}{\pi} \int_{-\pi}^{\pi} x \cos(nx) \, dx = 0$$

and
$$b_n = \frac{1}{\pi} \int_{-\pi}^{\pi} x \sin(nx) \, dx = \frac{2}{\pi} \int_0^{\pi} x \sin(nx) \, dx$$

$$= \frac{2}{\pi} \cdot \pi \frac{(-1)^n}{n} \Rightarrow b_n = \frac{2}{n}(-1)^{n+1}.$$

Putting the value of a_0, a_n and b_n in eq^n. (1), we get

$$x = 0 + 0 + \sum_{n=1}^{\infty} \frac{2}{n}(-1)^{n+1} \sin(nx)$$

$$x = 2\left[\sin x - \frac{1}{2}\sin(2x) + \frac{1}{3}\sin 3(x) - \frac{1}{4}\sin(4x)...\right]$$

EXAMPLE 10. *Show that the series*

$$2\left[\sin x + \frac{1}{2}\sin 2x + \frac{1}{3}\sin 3x +...\right]$$

represents $(\pi-x)$ in the interval $0 < x < 2\pi$. What is the sum of the series when $x = 0$ and $x = 2\pi$?

(SVTU-2004)

SOLUTION. Consider

$$f(x) = \frac{a_0}{2} + \sum_{n=1}^{\infty} (a_n \cos nx + b_n \sin nx) \qquad ...(1)$$

where

$$f(x) = \pi - x.$$

Since

$$a_0 = \frac{1}{\pi}\int_0^{2\pi} f(x)\,dx = \frac{1}{\pi}\int_0^{2\pi}(\pi - x)\,dx = \frac{1}{\pi}\left[\pi x - \frac{x^2}{2}\right]_0^{2\pi}$$

$$= \frac{1}{\pi}(2\pi^2 - 2\pi^2) = 0$$

Now,

$$a_n = \frac{1}{\pi}\int_0^{2\pi} f(x)\cos nx\,dx = \frac{1}{\pi}\int_0^{2\pi}(\pi - x)\cos nx\,dx$$

$$= \frac{1}{\pi}\left[\frac{(\pi - x)}{n}\sin nx + \int\frac{\sin nx}{n}\,dx\right]_0^{2\pi}$$

$$= \frac{1}{\pi}\left[\frac{(\pi - x)}{n}\sin nx - \frac{1}{n^2}(\cos nx)\right]_0^{2\pi} = 0$$

and

$$b_n = \frac{1}{\pi}\int_0^{2\pi} f(x)\sin nx\,dx = \frac{1}{\pi}\int_0^{2\pi}(\pi - x)\sin nx\,dx$$

$$= \frac{1}{\pi}\left[-\frac{(\pi - x)}{n}\cos nx - \int\frac{\cos nx}{n}\,dx\right]_0^{2\pi}$$

$$= \frac{1}{\pi}\left[-\frac{(\pi - x)}{n}\cos nx - \frac{\sin nx}{n^2}\right]_0^{2\pi}$$

Putting the value of a_0, a_n and b_n in eq^n. (1), we get

$$f(x) = 2\sum_{n=1}^{\infty}\frac{1}{n}\sin nx$$

$$\pi - x = 2\left(\sin x + \frac{1}{2}\sin 2x + \frac{1}{3}\sin 3x +...\right).$$

EXAMPLE 11. *Obtain a series of sines and cosines of multiples of x which will represent f(x) in the interval $-\pi < x < \pi$, when*

$$f(x) = \begin{cases} 0, & -\pi < x < 0 \\ \dfrac{\pi x}{4}, & 0 < x < \pi \end{cases}$$

(Purvanchal-1990; Agra-2000)

and hence deduce that

$$\frac{\pi^2}{8} = 1 + \frac{1}{3^2} + \frac{1}{5^2} +...$$

SOLUTION. Consider,
We know that,

$$f(x) = \begin{cases} 0, & -\pi < x < 0 \\ \dfrac{\pi x}{4}, & 0 < x < \pi \end{cases}$$

$$f(x) = \frac{a_0}{2} + \sum_{n=1}^{\infty} (a_n \cos nx + b_n \sin nx) \qquad \qquad ...(1)$$

Now,

$$a_0 = \frac{1}{\pi}\left[\int_{-\pi}^{0} f(x)\,dx + \int_{0}^{\pi} f(x)\,dx\right]$$

$$= \frac{1}{\pi}\left[\int_{-\pi}^{0} 0\cdot dx + \int_{0}^{\pi}\frac{1}{4}\pi x\,dx\right] = \frac{1}{\pi}\left[0 + \frac{\pi}{4}\left(\frac{x^2}{2}\right)_{0}^{\pi}\right] = \frac{\pi^2}{8}.$$

$$a_n = \frac{1}{\pi}\left[\int_{-\pi}^{0} f(x)\cos nx\,dx + \int_{0}^{\pi} f(x)\cos nx\,dx\right]$$

$$= \frac{1}{\pi}\left[\int_{-\pi}^{0} 0\cdot\cos nx\,dx + \int_{0}^{\pi}\frac{1}{4}\pi x\cos nx\,dx\right]$$

$$= \frac{1}{\pi}\left[0 + \frac{\pi}{4}\left(\frac{x}{n}\sin nx + \frac{\cos nx}{n^2}\right)_{x=0}^{\pi}\right]$$

$$= \frac{1}{4}\left[\frac{\cos n\pi - 1}{n^2}\right] = \frac{1}{4}\left[\frac{(-1)^n - 1}{n^2}\right]$$

$$a_n = \begin{cases} 0, & \text{when } n \text{ is even} \\ -\dfrac{1}{2n^2}, & \text{when } n \text{ is odd.} \end{cases}$$

and

$$b_n = \frac{1}{\pi}\left[\int_{-\pi}^{0} f(x)\sin nx\,dx + \int_{0}^{\pi} f(x)\sin nx\,dx\right]$$

$$= \frac{1}{\pi}\left[\int_{-\pi}^{0} 0\cdot\sin nx\,dx + \int_{0}^{\pi}\frac{\pi}{4}\cdot x\sin nx\,dx\right]$$

$$= \frac{1}{\pi}\left[0 + \frac{\pi}{4}\left(-\frac{x\cos nx}{n} + \frac{\sin nx}{n^2}\right)_{x=0}^{\pi}\right]$$

$$= \frac{-\pi\cos n\pi}{4n} = \frac{\pi}{4n}(-1)^{n+1}.$$

Putting the values of a_0, a_n and b_n in eqn. (1), we get

$$f(x) = \frac{\pi^2}{16} - \frac{1}{2}\left(\cos x + \frac{1}{3^2}\cos 3x + \frac{1}{5^2}\cos 5x + ...\right)$$

$$+ \frac{\pi}{4}\sum_{n=1}^{\infty}\frac{(-1)^{n+1}\sin nx}{n}$$

Now, putting $x = 0$, we get

$$0 = \frac{\pi^2}{16} - \frac{1}{2}\left(1 + \frac{1}{3^2} + \frac{1}{5^2} + ...\right) + \frac{\pi}{4}\sum_{n=1}^{\infty}\frac{(-1)^{n+1}(0)}{n}$$

$$\Rightarrow \qquad \frac{2\pi^2}{16} = 1 + \frac{1}{3^2} + \frac{1}{5^2} + \dots$$

$$\Rightarrow \qquad \frac{\pi^2}{8} = 1 + \frac{1}{3^2} + \frac{1}{5^2} + \dots$$

EXAMPLE 12. *Find a Fourier series for the function defined by the equation,*

$$f(x) = \begin{cases} -1, & -\pi < x < 0 \\ 0, & x = 0 \\ 1, & 0 < x < \pi \end{cases} \qquad \text{(Kanpur-1983, 2000, Agra-1992)}$$

Hence prove that, $\dfrac{\pi}{4} = 1 - \dfrac{1}{3} + \dfrac{1}{5} - \dfrac{1}{7} + \dfrac{1}{9} - \dots$

SOLUTION.   Here, we have

$$f(x) = \begin{cases} -1, & -\pi < x < 0 \\ 0, & x = 0 \\ 1, & 0 < x < \pi \end{cases}$$

We know that

$$f(x) = \frac{a_0}{2} + \sum_{n=1}^{\infty} (a_n \cos nx + b_n \sin nx) \qquad \dots(1)$$

Now,

$$a_0 = \frac{1}{\pi} \int_{-\pi}^{\pi} f(x)\,dx = \frac{1}{\pi} \int_{-\pi}^{0} f(x)\,dx + \int_{0}^{\pi} f(x)\,dx$$

$$= \frac{1}{\pi}\left[\int_{-\pi}^{0}(-dx) + \int_{0}^{\pi} dx \right] + \frac{1}{\pi}\left[(-x)_{-\pi}^{0} + (x)_{\pi}^{0} \right]$$

$$= \frac{1}{\pi}\left[-(0+\pi) + (\pi - 0) \right] = 0$$

$$a_n = \frac{1}{\pi} \int_{-\pi}^{\pi} f(x)\cos nx\,dx = \frac{1}{\pi}\left[\int_{-\pi}^{0}(-\cos nx)\,dx + \int_{0}^{\pi}(\cos nx)\,dx \right]$$

$$= \frac{1}{n\pi}\left[-(\sin nx)_{-\pi}^{0} + (\sin nx)_{0}^{\pi} \right] = 0$$

and

$$b_n = \frac{1}{\pi} \int_{-\pi}^{\pi} f(x)\sin nx\,dx = \frac{1}{\pi}\left[\int_{-\pi}^{0}(-\sin nx)\,dx + \int_{0}^{\pi}\sin nx\,dx \right]$$

$$= \frac{1}{n\pi}\left[(\cos nx)_{-\pi}^{0} - (\cos nx)_{0}^{\pi} \right]$$

$$= \frac{1}{n\pi}\left[(1 - \cos n\pi) + (1 - \cos n\pi) \right] = \frac{2}{n\pi}\left[1 - (-1)^n \right]$$

$$b_n = \begin{cases} 0, & \text{if } n \text{ is even} \\ \dfrac{4}{n\pi}, & \text{if } n \text{ is odd.} \end{cases}$$

Putting the value of a_0, a_n and b_n in eq^n (1), we get

$$f(x) = \frac{4}{\pi}\left[\sin x + \frac{1}{3}\sin 3x + \frac{1}{5}\sin 5x + \frac{1}{7}\sin 7x + \dots \right]$$

Now, putting $x = \pi/2$, we get

$$1 = \frac{4}{\pi}\left[1 - \frac{1}{3} + \frac{1}{5} - \frac{1}{7} + \frac{1}{9} - \ldots\right]$$

$$\Rightarrow \qquad \frac{\pi}{4} = 1 - \frac{1}{3} + \frac{1}{5} - \frac{1}{7} + \frac{1}{9} - \ldots$$

EXAMPLE 13. *Find a function which shall be equal to cos x for values of x between 0 and π equal to $-\cos x$ for values between $-\pi$ and 0.* (Agra-1999)

SOLUTION. Consider,

$$f(x) = \begin{cases} -\cos x, & \pi \le x \le 0 \\ \cos x, & 0 \le x \le \pi \end{cases}$$

We know that

$$f(x) = \frac{a_0}{2} + \sum_{n=1}^{\infty} (a_n \cos nx + b_n \sin nx) \qquad \ldots(1)$$

Then,

$$a_0 = \frac{1}{\pi}\int_{-\pi}^{\pi} f(x)\,dx = \frac{1}{\pi}\left[\int_{-\pi}^{0} f(x)\,dx + \int_{0}^{\pi} f(x)\,dx\right]$$

$$= \frac{1}{\pi}\left[\int_{-\pi}^{0}(-\cos x)\,dx + \int_{0}^{\pi}\cos x\,dx\right] = \frac{1}{\pi}\left[(-\sin x)_{-\pi}^{0} + (-\sin x)_{\pi}^{0}\right] = 0$$

Now,

$$a_n = \frac{1}{\pi}\int_{-\pi}^{\pi} f(x)\cos nx\,dx$$

$$= \frac{1}{\pi}\left[\int_{-\pi}^{0}(-\cos x)\cos nx\,dx + \int_{0}^{\pi}\cos x\cos nx\,dx\right]$$

$$= \frac{1}{\pi}\left[\int_{\pi}^{0}\cos y\cos ny\,dy + \int_{0}^{\pi}\cos x\cos nx\,dx\right] \qquad [\because\ -x = y]$$

$$= \frac{1}{\pi}\left[-\int_{0}^{\pi}\cos x\cos nx\,dx + \int_{0}^{\pi}\cos x\cos nx\,dx\right]$$

$$= \frac{1}{\pi}\left[-\int_{0}^{\pi}\cos x\cos nx\,dx + \int_{0}^{\pi}\cos x\cos nx\,dx\right]$$

$$= 0 \qquad \left[\int_{a}^{b} f(t)\,dt = \int_{a}^{b} f(x)\,dx\right]$$

and

$$b_n = \frac{1}{\pi}\int_{-\pi}^{\pi} f(x)\sin nx\,dx$$

$$= \frac{1}{\pi}\left[\int_{-\pi}^{0}-\cos x\sin nx\,dx + \int_{0}^{\pi}\cos x\sin nx\,dx\right]$$

Let $x = -y$ in first integral

$$= \frac{1}{\pi}\left[\int_{0}^{\pi}\cos y\sin ny\,dy + \int_{0}^{\pi}\cos x\sin nx\,dx\right]$$

$$= \frac{1}{\pi}\left[\int_{0}^{\pi}\cos x\sin nx\,dx + \int_{0}^{\pi}\cos x\sin nx\,dx\right]$$

$$= \frac{2}{\pi}\int_{0}^{\pi}\cos x\sin nx\,dx = -\frac{1}{\pi}\left[\frac{\cos(n+1)x}{n+1} + \frac{\cos(n-1)x}{n-1}\right]_{x=0}^{\pi}$$

$$= -\frac{1}{\pi}\left[\frac{\cos(n+1)\pi - 1}{n+1} + \frac{\cos(n-1)\pi - 1}{n-1}\right]$$

$$= -\frac{1}{\pi}\left[\frac{-\cos n\pi - 1}{n+1} + \frac{-\cos n\pi - 1}{n-1}\right] \qquad [\because \cos(\pi \pm \theta) = -\cos\theta]$$

$$= \frac{(1+\cos n\pi)}{\pi}\left(\frac{1}{n+1} + \frac{1}{n-1}\right) = \frac{2n}{n^2-1}[1+(-1)^n]$$

Here b_1 is infinite, thus again calculate it.

$$b_1 = \lim_{n\to 1}\frac{2n(1+\cos n\pi)}{n^2-1}$$

$$= \lim_{n\to 1}\frac{2[(1+\cos n\pi) - n\pi\sin n\pi]}{2n}$$

$$= \frac{2[(1+\cos\pi) - \sin\pi]}{2} = 0$$

$$b_n = \frac{2n}{n^2-1}[1+(-1)^n],\ n = 2,3,4,5,....$$

$$b_n = \begin{cases} 0, & \text{if } n \text{ is odd} \\ \dfrac{4n}{n^2-1}, & \text{if } n \text{ is even} \end{cases}$$

Also, $b_1 = 0$, $a_0 = 0$, $a_n = 0$ for $n = 1, 2, 3, 4, ...$
Putting the value of a_0, a_n and b_n in eq^n (1), we get

$$f(x) = 4\left[\frac{2}{3}\sin 2x + \frac{4}{15}\sin 4x + \frac{6}{35}\sin 6x + ...\right]$$

MISCELLANEOUS EXERCISE

1. Find the Fourier series to represent the periodic function

$$f(x) = \begin{cases} x, & -\dfrac{\pi}{2} < x < \dfrac{\pi}{2} \\ \pi - x, & \dfrac{\pi}{2} < x < \dfrac{3\pi}{2} \end{cases}$$

(UPTU-2001)

2. If $f(x) = \cos\omega x$, $-\pi < x < \pi$, where ω is a function as a Fourier series, prove that

$$\cot\theta = \frac{1}{\theta} + \frac{2\theta}{\theta^2 - \pi^2} + \frac{2\theta}{\theta^2 - 4\pi^2} + ...$$

(UPTU–2001)

3. Find the Fourier series expansion of $f(x) = 2x - x^2$ in (0, 3) and hence show that

$$\frac{1}{1^2} - \frac{1}{2^2} + \frac{1}{3^2} - \frac{1}{4^2} + ... - \infty = \frac{\pi}{12}$$

(Mumbai-2005)

4. If $f(x) = \begin{cases} \dfrac{hx}{a}, & 0 < x < a \\ \dfrac{h(l-x)}{l-a}, & a < x < l \end{cases}$

Prove that for all values of x between 0 and l

$$f(x) = \frac{2hl^2}{a(l-a)\pi^2}\left[\begin{array}{l}\sin\dfrac{\pi a}{l}\sin\dfrac{\pi x}{l} + \\ \dfrac{1}{2^2}\sin\dfrac{2\pi a}{l}\sin\dfrac{2\pi x}{l} + ...\end{array}\right]$$

(UPTU–2001)

5. Prove that in $0 < x < c$,

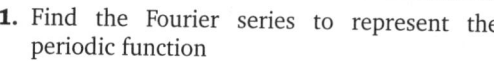

$$x = \frac{c}{2} - \frac{4c}{\pi^2}\left(\begin{array}{l}\cos\dfrac{\pi x}{c} + \dfrac{1}{3^2}\cos\dfrac{3\pi x}{c} \\ + \dfrac{1}{5^2}\dfrac{\cos 5\pi x}{c} + ...\end{array}\right)$$

and deduce that

(i) $\dfrac{1}{1^4} + \dfrac{1}{3^4} + \dfrac{1}{5^4} + ... = \dfrac{\pi^4}{96}$

(ii) $\dfrac{1}{1^4} + \dfrac{1}{2^4} + \dfrac{1}{3^4} + \dfrac{1}{4^4} + ... = \dfrac{\pi^4}{90}$

6. Find the complex form of the Fourier series of $f(x) = \cos ax$, $-\pi < x < \pi$.

(Anna-2009, Mumbai–2009)

7. Find the complex form of the Fourier series of $f(x) = e^{ax}$, $-l < x < l$

(Madras-2003)

8. Find the complex form of the Fourier series of
$$f(x) = \begin{cases} 0, & \text{if } 0 < x < l \\ a, & \text{if } 0 < x < 2l \end{cases}$$

9. Find the complex form of the Fourier series of $f(x) = \cosh 3x + \sinh 3x$ in $(-3, 3)$.

(Mumbai–2008)

10. Prove that in $0 < x < C$,

$$x = \frac{C}{2} - \frac{4C}{\pi^2}\left(\cos\frac{\pi x}{C} + \frac{1}{3^2}\cos\frac{3\pi x}{C} + \frac{1}{5^2}\cos\frac{5\pi x}{C} + ...\right)$$

and deduce that

(i) $\dfrac{1}{1^4} + \dfrac{1}{3^4} + \dfrac{1}{5^4} + ... = \dfrac{\pi^4}{96}$

(ii) $\dfrac{1}{1^4} + \dfrac{1}{2^4} + \dfrac{1}{3^4} + \dfrac{1}{4^4} + ... = \dfrac{\pi^4}{90}$

11. Find the complex form of the Fourier series of $f(x) = \cos ax,\ -\pi < x < \pi$.

(Anna–2009, Mumbai–2009)

12. Find the complex form of the Fourier series of $f(x) = e^{ax},\ -l < x < l$. (Madras–2003)

13. Find the complex form of the Fourier series of
$$f(x) = \begin{cases} 0, & \text{if } 0 < x < l \\ a, & \text{if } 0 < x < 2l \end{cases}.$$

14. Find the complex form of the Fourier series of $f(x) = \cosh 3x + \sinh 3x$ in $(-3, 3)$.

(Mumbai–2008)

$\mathcal{ANSWERS}$

1. $f(x) = \dfrac{4}{\pi}\left[\dfrac{\sin x}{1^2} - \dfrac{\sin 3x}{3^2} + \dfrac{\sin 5x}{5^2} - ...\right]$

3. $f(x) = -\displaystyle\sum_{n=1}^{\infty}\dfrac{9}{n^2\pi^2}\cos\dfrac{2h\pi x}{3} + \sum_{n=1}^{\infty}\dfrac{3}{n\pi}\sin\dfrac{2n\pi x}{3}$

6. $\dfrac{a}{\pi}\sin a\pi \displaystyle\sum_{n=-\infty}^{\infty}\dfrac{(-1)^n e^{inx}}{a^2 - n^2}$

7. $\dfrac{2}{\pi} - \dfrac{2}{\pi}\left[\dfrac{e^{2it} + e^{-2it}}{1.3} + \dfrac{e^{4it} + e^{-4it}}{3.5} + \dfrac{3^{6it} + e^{-6it}}{5.7} + ...\right]$

8. $\dfrac{a}{2} - \dfrac{a}{\pi}\left[(e^u - e^{-u}) + \dfrac{1}{3}(e^{3u} - e^{-3u}) + \dfrac{1}{5}(e^{5u} - e^{-5u}) + ...\right]$, where $u = \dfrac{i\pi x}{l}$

9. $\sinh 9 \displaystyle\sum_{-\infty}^{\infty}\dfrac{(-1)^n(9 + n\pi i)}{81 + (n\pi)^2}e^{n\pi ix/3}$.

11. $\dfrac{a}{\pi}\sin a\pi \displaystyle\sum_{n=-\infty}^{\infty}\dfrac{(-1)^n e^{inx}}{a^2 - n^2}$

12. $\dfrac{2}{\pi} - \dfrac{2}{\pi}\left[\dfrac{e^{2it} + e^{-2it}}{1.3} + \dfrac{e^{4it} + e^{-4it}}{3.5} + \dfrac{e^{6it} + e^{-6it}}{5.7} + ...\right]$

13. $\dfrac{a}{2} - \dfrac{a}{\pi}\left[(e^u - e^{-u}) + \dfrac{1}{3}(e^{3u} - e^{-3u}) + \dfrac{1}{5}(e^{5u} - e^{-5u}) + ...\right]$, where $u = \dfrac{i\pi x}{l}$

14. $\sinh 9 \displaystyle\sum_{-\infty}^{\infty}\dfrac{(-1)^n(9 + n\pi i)}{81 + (n\pi)^2}e^{n\pi ix/3}$

Self Assessment Test

1. Find the Fourier series for $f(x) = 1-x^2$, when $-1 \leq x \leq 1$.

2. Find the Fourier series of

   (i) $f(x) = \begin{cases} 0, & -5 < x < 0 \\ 3, & 0 < x < 5 \end{cases}$

   (ii) $f(x) = \begin{cases} 0, & -l < x < 0 \\ E \sin \omega x, & 0 < x < l \end{cases}$

3. Find a series of sines of multiples of x, when function is defined as

   $$f(x) = \begin{cases} \dfrac{\pi x}{4}, & 0 \leq x \leq \pi/2 \\ \dfrac{\pi(\pi - x)}{4}, & \dfrac{\pi}{2} < x \leq \pi \end{cases}$$

4. Find the Fourier series to represent $f(x)$, where

   $$f(x) = \begin{cases} -a, & -l < x < 0 \\ a, & 0 < x < l \end{cases}$$

5. $f(x) = \begin{cases} -x+1, & -\pi \leq x \leq 0 \\ x+1, & 0 \leq x \leq \pi \end{cases}$ and deduce the

   value of $\dfrac{1}{1^2} + \dfrac{1}{3^2} + \dfrac{1}{5^2} + ...$

6. $f(x) = |\cos x|$ in $(-\pi, \pi)$.

7. $f(x) = \sin ax$ in $(-\pi, \pi)$, where a is not an integer.

8. Find the Fourier series of the function :

   $$f(x) = \begin{cases} -2, & -4 < x < -2 \\ x, & -2 < x < 2 \\ 2, & 2 < x < 4 \end{cases}$$

9. Find the Fourier series expansion for the function $f(x) = x - x^3$ in the interval $-1 < x < 1$.

10. Find Fourier series if $f(x) = \begin{cases} -x, & -\pi < x < 0 \\ 0, & 0 < x \leq \pi \end{cases}$.

11. Find Fourier series if $f(x) = \begin{cases} 2x, & 0 \leq x \leq \pi \\ x, & -\pi < x \leq 0 \end{cases}$

12. Find Fourier series if $f(x) = \begin{cases} 1, & -\pi < x < 0 \\ -2, & 0 < x \leq \pi \end{cases}$

Answers

1. $f(x) = \dfrac{2}{3} + \dfrac{4}{\pi^2}\left(\cos \pi x - \dfrac{\cos 2\pi x}{2^2} + \dfrac{\cos 3\pi x}{3^2} + ...\right)$

2. (i) $f(x) = \dfrac{E}{\pi} + \dfrac{E}{2}\sin \omega x - \dfrac{2E}{\pi}\left(\dfrac{1}{1.3}\cos 2\omega x + \dfrac{1}{1.3}\cos 4\omega x + ...\right)$

   (ii) $f(x) = \dfrac{3}{2} + \displaystyle\sum_{n=1}^{\infty} \dfrac{3[1-(-1)^n]}{n\pi}\sin \dfrac{n\pi x}{5}$

3. $f(x) = \dfrac{\pi}{4}\left(\dfrac{4}{\pi}\right)\left[\sin x - \dfrac{1}{3^2}\sin(3x) + \dfrac{1}{5^2}\sin(5x) - \dfrac{1}{7^2}\sin(7x) + ...\right]$

4. $f(x) = \dfrac{4l}{\pi}\left[\sin \dfrac{\pi x}{l} + \dfrac{1}{3}\sin \dfrac{3\pi x}{l} + \dfrac{1}{5}\sin \dfrac{5\pi x}{l} + ...\right]$

5. Even function, $f(x) = \dfrac{\pi}{2} + 1 - \dfrac{4}{\pi}\left(\cos x + \dfrac{1}{3^2}\cos 3x + \dfrac{1}{5^2}\cos 5x + ...\right)$

6. Even function, $f(x) = \dfrac{2}{\pi} + \dfrac{4}{\pi}\displaystyle\sum_{n=1}^{\infty} \dfrac{\cos \dfrac{n\pi}{2}}{1-n^2}\cos nx.$

7. Odd function, $f(x) = \dfrac{2\sin a\pi}{\pi}\displaystyle\sum_{n=1}^{\infty} \dfrac{(-1)^n . n}{a^2 - n^2}\sin nx.$

8. $f(x) = \dfrac{4}{\pi} + \dfrac{8}{\pi^2} \sin \dfrac{\pi x}{4} - \dfrac{2}{\pi} \sin \pi x + \left(\dfrac{4}{3\pi} - \dfrac{8}{3^2 \pi} \right) \sin \dfrac{3\pi x}{4} - \dfrac{2}{2\pi} \sin \pi x + \ldots$

9. $f(x) = \dfrac{12}{\pi^3} \left(\dfrac{\sin \pi x}{1^3} - \dfrac{\sin 2\pi x}{2^3} + \dfrac{\sin 3\pi x}{3^3} + \ldots \right)$

10. $f(x) = \dfrac{\pi}{4} - \dfrac{2}{\pi} \displaystyle\sum_{n=0}^{\infty} \dfrac{\cos(2n+1)x}{(2n+1)^2} - \displaystyle\sum_{n=1}^{\infty} (-1)^{n-1} \dfrac{\sin nx}{n}$

11. $f(x) = \dfrac{\pi}{4} - \dfrac{2}{\pi} \left(\dfrac{\cos x}{1^2} + \dfrac{\cos 3x}{3^2} + \dfrac{\cos 5x}{5^2} + \ldots \right) + 3 \left(\dfrac{\sin x}{1} + \dfrac{\sin 2x}{2} + \dfrac{\sin 3x}{3} + \ldots \right)$

12. $f(x) = -\dfrac{1}{2} - \dfrac{6}{\pi} \displaystyle\sum_{n=0}^{\infty} \dfrac{\sin(2n+1)x}{(2n+1)}$

Objective Evaluations

❖ Fill in the Blanks

1. The double sequence $\left\langle \dfrac{1}{mn} \right\rangle$ converges to _____ .

2. The double sequence $\left\langle (-1)^{m+n} \right\rangle$ oscillates _____ .

3. The double sequence $\left\langle (-1)^{m+n}(m+n) \right\rangle$ oscillates _____ .

4. A necessary condition for convergence of a double series $\sum\limits_{m,n} a_{mn}$ that $\lim\limits_{m,n\to\infty} a_{mn} =$ _____ .

5. Every absolutely convergent double series is _____ .

6. If $f(x)$ is periodic function with period 2π, then $\int_{-\pi}^{\pi} f(x)\,dx = \int_{\lambda}^{\lambda+\underline{\quad}} f(x)\,dx$, where λ is any arbitrary number.

7. If $f(x) = \dfrac{1}{2}a_0 + \sum\limits_{n=1}^{\infty}(a_n\cos nx + b_n \sin nx)$ be the Fourier series of f, then the coefficient a_n, b_n are called the _____ of f.

8. If $f(x)$ is a periodic function with period T, then $f(x+nT) =$ _____ for all $n \in I$, the set of all integers.

❖ True / False

Write 'T' for True and 'F' for False statement.

1. $\int_{-\pi}^{\pi} \sin nx\, dx = 0 = \int_{-\pi}^{\pi}\cos nx\, dx \, \forall n$. **(T/F)**

2. The inequality $\sum\limits_{n=1}^{\infty}(a_n^2 + b_n^2) \le \dfrac{1}{\pi}\int_{-\pi}^{\pi} f^2 dx$ is called Bessel's inequality. **(T/F)**

3. If f is integrable on $[-\pi, \pi]$ and has period 2π then $\int_{-\pi}^{\pi} f(t)dt = \int_{-\pi}^{\pi} f(t+a)dt$ for any $a \in R$. **(T/F)**

4. $\int_{0}^{\infty} \dfrac{\sin t}{T} dt = \pi$ **(T/F)**

5. Fourier series has period 2π. **(T/F)**

6. If the trigonometric series $\dfrac{1}{2}a_0 + \sum\limits_{n=1}^{\infty}(a_n\cos nx + b_n \sin nx)$ is uniformly convergent on $[-\pi, \pi]$ and if $f(x)$ is its sum; then it is the Fourier series of $f(x)$. **(T/F)**

7. A constant function is always periodic. **(T/F)**

8. The functions $f(x) = \sin x$ and $g(x) = \cos x$ are periodic functions, the period of each function is π. **(T/F)**

❖ Multiple Choice Questions

Choose the most appropriate one

1. A function $f(x)$ is called periodic if it is defined for real $x \in R$ and:

(a) if there is any positive number p such that $f(x+p) = f(x)+f(p)$

(b) if there is any positive number p, such that $f(x+p) > f(x)$

(c) if there is any positive number p such that $f(x+p) = f(r)$,

(d) none of the above.

2. In Fourier series

$$f(x) = a_0 + \sum_{n=1}^{\infty}(a_n \cos nx + b_n \sin nx),$$

Euler formula is :

(a) $a_0 = \dfrac{1}{\pi}\int_{-\pi}^{\pi} f(x)dx$

(b) $a_0 = \dfrac{1}{2\pi}\int_{-\pi}^{\pi} f(x)dx$

(c) $a_0 = \dfrac{1}{2\pi}\int_{-\pi}^{\pi} f(x)\sin x\, dx$

(d) none of the above

3. In Fouries series

$$f(x) = a_0 + \sum_{n=1}^{\infty}(a_n \cos nx + b_n \sin nx),$$

Euler formula is:

(a) $a_n = \frac{1}{\pi} \int_{-\pi}^{\pi} f(x) dx$

(b) $a_n = \frac{1}{\pi} \int_{-\pi}^{\pi} f(x) \sin nx \, dx$

(c) $a_n = \frac{1}{\pi} \int_{-\pi}^{\pi} f(x) \cos nx \, dx$

(d) none of the above

4. If a function $f(x)$ of period $p = 2l$ has Fourier series, then in series

$$f(x) = a_0 + \sum_{n=1}^{\infty} (a_n \cos nx + b_n \sin nx),$$ the

Fourier coefficients a_n is given by ;

(a) $\int_{-l}^{l} f(x) . \frac{\sin n\pi x}{l} dx$

(b) $\frac{1}{l} \int_{-l}^{l} f(x) . \frac{\cos n\pi x}{l} dx$

(c) $\int_{-l}^{l} f(x) . \frac{\cos n\pi x}{l} dx$

(d) none of the above

5. The Fourier series of an even function of period 2π is a:

(a) Fourier cosine series

(b) Fourier sine series

(c) Fourier complex series

(d) none of the above

6. The Fourier series of an odd function of period $2l$ is given by:

(a) $f(x) = a_0 + \sum_{n=1}^{\infty} a_n \frac{\cos n\pi}{l}$, where

$a_0 = \frac{1}{l} \int_0^l f(x) dx$, $a_n = \frac{2}{l} \int_0^l f(x) . \frac{\cos n\pi x}{l} dx$

(b) $f(x) = \sum_{n=1}^{\infty} b_n \sin \frac{n\pi x}{l}$,

where $b_n = \frac{2}{l} \int_0^l f(x) \sin \frac{n\pi x}{l} dx$

(c) $f(x) = \sum_{n=-\infty}^{\infty} C_n e^{inx}$, where

$C_n = \frac{1}{2\pi} \int_{-\pi}^{\pi} f(x) e^{-inx} dx, n = 1, 2, ...$

(d) none of the above

7. The Fourier series of an odd function of period $2l$ is given by:

(a) Fourier cosine series

(b) Fourier sine series

(c) Fourier complex series

(d) none of the above

8. The Fourier series for an odd function $f(x)$ in the interval $-\pi < x < \pi$ is given by :

(a) $f(x) = \sum_{n=1}^{\infty} b_n \sin nx$

(b) $f(x) = \sum_{n=1}^{\infty} (a_n \cos nx + b_n \sin nx)$

(c) $f(x) = \frac{1}{2} a_0 + \sum_{n=1}^{\infty} a_n \cos nx$

(d) none of these

9. If the Fourier series of the function $f(x) = x^2$ is

$$x^2 = \frac{1}{2} a_0 + \sum_{n=1}^{\infty} (a_n \cos nx + b_n \sin nx), \text{ then the}$$

Fourier coefficient $b_n (n = 1, 2, 3,...)$ is :

(a) $4/n^2$ (b) 0

(c) $-4/n^2$ (d) none of thee

10. If a function f is bounded and integrable on the interval $[-\pi, \pi]$ and if a_n, b_n its Fourier coefficients, then the series $\sum_{n=1}^{\infty} (a_n^2 + b_n^2)$ is :

(a) convergent

(b) divergent

(c) conditionally convergent

(d) none of these

11. If the Fourier series of function $f(x) = x^2$ is

$$x^2 = \frac{1}{2} a_0 + \sum_{n=1}^{\infty} (a_n \cos nx + b_n \sin nx), \quad \text{then}$$

the Fourier coefficient a_0 is equal to :

(a) $\frac{\pi^2}{3}$ (b) $\frac{2\pi^2}{3}$

(c) $\frac{4\pi^2}{3}$ (d) none of these

12. The function $f(x) = x \sin x$ is an :

(a) even function

(b) odd function

(c) neither odd nor even function

(d) none of these

13. The finite Fourier transform of $f(x) = x$ is :

(Bundelkhand-2005, 06)

(a) $\frac{\pi}{p^2}(-1)^{p+1}$ (b) $\frac{\pi}{p^2}(-1)^p$

(c) $\frac{\pi}{p}(-1)^{p+1}$ (d) $\frac{\pi}{p}(-1)^p$

14. Sine transform of e^{-x} is : (Bundelkhand–2006)

(a) $\dfrac{1}{1+p^2}$ (b) $\dfrac{p}{1+p^2}$

(c) $\dfrac{1}{1-p^2}$ (d) $\dfrac{p}{1-p^2}$

15. The infinite Fourier series transform of e^{-x} is :

(a) $\dfrac{1}{p}\sqrt{\dfrac{2}{\pi}}$ (b) $\dfrac{1}{p^2}\sqrt{\dfrac{2}{\pi}}$

(c) $\dfrac{1}{1+p^2}\sqrt{\dfrac{2}{\pi}}$ (d) $\dfrac{p}{1+p^2}\sqrt{\dfrac{2}{\pi}}$

16. If $f_s(p) = \dfrac{p}{1+p^2}$, then $f(x)$ is : (Agra-2004)

(a) e^{-x} (b) e^x

(c) e^{-x^2} (d) e^{x^2}

17. The Fourier cosine transform of $F(x) = e^{-x}$ is :

(Agra-2004)

(a) $\dfrac{1}{1-p^2}$ (b) $\dfrac{1}{1+p^2}$

(c) $\dfrac{p}{1-p^2}$ (d) $\dfrac{p}{1+p^2}$

18. If Fourier sine transform is e^{-ap}, then $f(x)$ is equal to :

(a) $\dfrac{2x}{\pi(a^2+x^2)}$ (b) $\dfrac{2\pi ax}{a^2+x^2}$

(c) $\dfrac{2\pi x}{a^2+x^2}$ (d) $\dfrac{2a}{\pi(a^2+x^2)}$

19. What is $f(x)$ if its infinite Fourier sine transform is e^{-ap}/p ?

(a) $\sqrt{\dfrac{2}{\pi}}\tan^{-1}\dfrac{2a}{x}$ (b) $\sqrt{\dfrac{2}{\pi}}\tan^{-1}\dfrac{2x}{a}$

(c) $\sqrt{\dfrac{2}{\pi}}\tan^{-1}\dfrac{a}{x}$ (d) $\sqrt{\dfrac{2}{\pi}}\tan^{-1}\dfrac{x}{a}$

20. If $\tilde{f}(p)$ is the complex Fourier transform of $f(x)$, then the complex Fourier transform of $f(ax)$ is :

(a) $\tilde{f}\left(\dfrac{p}{a}\right)$ (b) $\dfrac{1}{a}\tilde{f}(p)$

(c) $\dfrac{1}{a}\tilde{f}\left(\dfrac{p}{a}\right)$ (d) none of these

21. The Fourier sine transform of $4/\pi$ is :

(a) $\dfrac{(-1)^{p+1}}{p}$ (b) $\dfrac{(-1)^{p-1}}{p}$

(c) $\dfrac{(-1)^p}{p^2}$ (d) $\dfrac{(-1)^{p+1}}{p^2}$

22. If $f(x)$ is continuous and $f'(x)$ is sectionally continuous, then $F_s\{f'(x)\}$ is :

(a) $-F_c\{f(x)\}$ (b) $-pF_c\{f(x)\}$

(c) $-p^2 F_c\{f(x)\}$ (d) $F_c\{f(x)\}$

ANSWERS

❖ Fill in the Blanks

1. 0   **2.** finitely   **3.** infinitely   **4.** 0   **5.** convergent

6. 2π **7.** Fourier coefficient   **8.** $f(x)$

❖ True/False

1. T   **2.** T   **3.** T   **4.** F   **5.** T   **6.** T   **7.** T   **8.** F

❖ Multiple Choice Questions

1. (c)   **2.** (c)   **3.** (a)   **4.** (b)   **5.** (a)   **6.** (b)   **7.** (b)   **8.** (a)   **9.** (b)

10. (a)   **11.** (b)   **12.** (a)   **13.** (c)   **14.** (b)   **15.** (d)   **16.** (a)   **17.** (b)   **18.** (a)

19. (d)   **20.** (a)   **21.** (a)   **22.** (b)

vvvvvv

Bessel's Function : $J_n(t) = \dfrac{t^n}{2^n \, \Gamma(n+1)} \left[1 - \dfrac{t^2}{2(2n+2)} + \dfrac{t^4}{2.4(2n+2)(2n+4)} \cdots \right].$

Change of Scale Property : If $L\{F(t)\} = f(p)$, then $L\{F(at)\} = \dfrac{1}{a} f\left(\dfrac{p}{a}\right)$.

Convolution : If $L^{-1}\{f(p)\} = F(t)$ and $L^{-1}\{g(p)\} = G(t)$, where $F(t)$ and $G(t)$ are two functions of class A, then $L^{-1}\{f(p).g(p)\} = \int_0^t F(u)\,G(t-u)\,du = F * G$.

Convolution Theorem of Z-Transforms : Let $<f(n)>$ and $<g(n)>$ be two sequences and let the convolution of $<f(n)>$ and $<g(n)>$ be $<h(n)>$ where $<h(n)> = <f(n)>*$ $<g(n)>$, then $<h(n)>$ is defined as

$$<h(n)> = <f(n)>* <g(n)>$$

$$= \sum_{n=-\infty}^{\infty} f(n)g(n-k) = \sum_{n=-\infty}^{\infty} g(n) f(n-k) = <g(n)> * <f(n)>$$

Error Function : The error function denoted by $erf\,(t)$, is defined by $erf(t) = \dfrac{2}{\sqrt{\pi}} \int_0^t e^{-u^2} du$.

Final Value Theorem : If $Z[<f(n)>] = F(z)$, $n > 0$, then $\lim\limits_{z \to \infty} F(n) = \lim\limits_{z \to 1} (z-1) F(z)$

Finite Fourier Cosine Transform : Let $f(x)$ be a function which is sectionally continuous over some finite interval $]0, l[$ of the variable x. Then the finite Fourier cosine transform of $f(x)$ on this interval is defined as

$$\tilde{f}_c(p) = \int_0^l f(x) \cos\dfrac{p\pi x}{l} dx \text{ , where } p \text{ is any integer.}$$

Finite Fourier Sine Transforms : Let $f(x)$ be a function, which is sectionally continuous over some finite interval $]0, 1[$ for the variable x. Then finite Fourier sine transforms of $f(x)$ on this interval is given by $\tilde{f}(s) = \int_0^1 f(x) \sin\dfrac{p\pi x}{l} dx$, where p is any integer.

If the end points of the interval become $x = 0$ and $x = \pi$, then $\tilde{f}_s(p) = \int_0^\pi f(x) \sin px\, dx$.

First Translation or Shifting Theorem : If $f(p)$ is the Laplace transform of $F(t)$, then $f(p - a)$ is the Laplace transform of $e^{at}F(t)$, i.e., if $L\{F(t)\} = f(p)$, when $p > a$, then $L\{e^{at} F(t)\} = f(p-a), p > a$.

Fourier Cosine Transform : The infinite Fourier cosine transform of $f(x)$, $0 < x < \infty$ is denoted by $F_c[f(x)]$ or $\tilde{f}_c(p)$ and is defined by

$$\tilde{f}_c[f(x)] = \tilde{f}_c(p) = \sqrt{\dfrac{2}{\pi}} \int_0^\infty f(x) \cos px\, dx$$

Fourier Sine Transform : The infinite Fourier sine transform of the function $f(x)$, $0 < x < \infty$ is defined by $F_s[f(x)]$ or $\tilde{f}_s(p)$ and defined by

$$\tilde{f}_s(p) = F_s[f(x)] = \sqrt{\dfrac{2}{\pi}} \int_0^\infty f(x) \sin px\, dx$$

Fourier Transform : If a function $f(x)$ defined on the interval $]-\infty, \infty[$, and piecewise continuous in each finite partial interval and absolutely integrable in $]-\infty, \infty[$, then

$$F(f(x)) = \tilde{f}(p) = \int_{-\infty}^{\infty} e^{ipx} f(x) dx$$

is defined as Fourier transform of $f(x)$.

Heaviside's Unit Function : The unit step function or heaviside's unit function denoted by $H(t-a)$ is defined by

$$H(t-a) = \begin{cases} 0, & t < a \\ 1, & t \geq a \end{cases}.$$

Initial Value Theorem : If $Z[<f(n)>] = F(z)$, $n > 0$, then $f(0) = \lim\limits_{z \to \infty} F(z)$

Inverse Fourier Transform : The inverse formula for Fourier transform is given by

$$f^{-1}[\tilde{f}(p)] = f(x) = \frac{1}{2\pi} \int_{-\infty}^{\infty} \tilde{f}(p) e^{-ipx} dp .$$

Inverse Infinite Fourier Cosine Transform : The inversion formula for Fourier cosine transform is given by

$$f(x) = \sqrt{\frac{2}{\pi}} \int_0^{\infty} \tilde{f}_c(p) \cos pxdx .$$

Inverse Z-transforms : If $F(z)$ is the Z-transform of the sequence $< f(k) >$, then $<f(k)>$ is called the inverse Z-transform of $F(z)$. The operator for inverse Z-transform is denoted by Z^{-1}. Thus, if $F < f(k) > = F(z)$, then

$$Z^{-1} [F(z)] = < f(k) >$$

Inversion Formula for Cosine Transform : If $\tilde{f}_c(p)$ is the finite Fourier cosine transform of $f(x)$ over the interval $]0, l[$ then the inversion formula for cosine transform is given by

$$f(x) = \frac{1}{l} \tilde{f}_c(0) + \frac{2}{l} \sum_{p=1}^{\infty} \tilde{f}_c(p) \cos \frac{p\pi x}{l} ,$$

where $\tilde{f}_c(0) = \int_0^{\pi} f(x) dx$.

Inversion Formula for Sine transform : If $\tilde{f}_s(p)$ is the finite Fourier sine transform of $f(x)$ over the interval $]0, l[$, then the inversion formula for sine transform is given by

$$f(x) = \frac{2}{l} \sum_{p=1}^{\infty} \tilde{f}_s(p) \sin \frac{p\pi x}{l}$$

Laplace Transform : If $F(t)$ be a function of t defined for all values of t, then Laplace transform of $f(t)$, denoted by $L\{F(t)\}$ or $f(p)$ is defined by

$$L \{F(t)\} = f(p) = \int_0^{\infty} e^{-pt} F(t) \, dt$$

Linearity property of Fourier Transforms : Let $\tilde{f}(p)$ and $\tilde{g}(p)$ are Fourier transform of $f(x)$ and $g(x)$ respectively. Then $F\{af(x) + bg(x)\} = a\tilde{f}(p) + b\tilde{g}(p)$ where a and b constants.

Modulation Theorem : If $\tilde{f}(p)$ is the complex Fourier transform of $f(x)$ then, the Fourier transform of $f(x) \cos ax$ is $\frac{1}{2}[\tilde{f}(p-a) + \tilde{f}(p+a)]$.

Multiple Fourier Transforms : Let $f(x, y)$ be a function of two variables x and y. Temporarily, as a function of x, its Fourier transform is given by

$$\tilde{f}(p, y) = \frac{1}{\sqrt{2\pi}} \int_{-\infty}^{\infty} f(x, y) e^{ipx} dx$$

Null Function : A function $N(t)$ of t such that $\int_0^t N(t)\, dt = 0,\ \forall\ t > 0$ is called null function.

Parsevel's Identity for Fourier series : Consider the Fourier series

$\dfrac{a_0}{2} + \sum\limits_{n=1}^{\infty} (a_n \cos nx + b_n \sin nx)$. If $f(x)$ converges uniformly to $f(x)$ at every point of the interval

$(0, 2\pi)$, then

$$\frac{1}{\pi}\int_0^{2\pi} \{f(x)\}^2 dx = \frac{a_0^2}{2} + \sum_{n=1}^{\infty} (a_n^2 + b_n^2).$$

Partial Sum Theorem : If $Z[\{f(n)\}] = F(z)$, then $Z\left[\left\langle \sum\limits_{k=-\infty}^{n} f(k) \right\rangle\right] = \dfrac{F(z)}{1 - z^{-1}}$

Periodic Function : A function $f(x)$ which satisfies the relation $f(x + T) = f(x)$ for all real x and some fixed T is called a periodic function.

Second Translation or Heaviside's Shifting Theorem : If $L\{F(t)\} = f(p)$ and

$G(t) = \begin{cases} F(t-a), & t > a \\ 0, & t < a \end{cases}$ then, $L\{G(t)\} = e^{-ap} f(p)$.

Shifting Property : If $\tilde{f}(p)$ is the complex Fourier transform of $f(x)$, then complex Fourier transform of $f(x - a)$ is $e^{ipa} f(p)$.

The Gamma Function : If $n > 0$, the gamma function is defined by $\Gamma(n) = \int_0^{\infty} u^{n-1} e^{-u} du$.

The Heaviside Expansion Formula : If $F(p)$ and $G(p)$ are polynomials in P, the degree of $F(p)$ being less than that of $G(p)$ and if

$$G(p) = (p - \alpha_1)(p - \alpha_2)...(p - \alpha_n)$$

where, $\alpha_1, \alpha_2, ..., \alpha_n$ are distinct constants, real or complex, then

$$L^{-1}\left\{\frac{F(p)}{G(p)}\right\} = \sum_{r=1}^{n} \frac{F(\alpha_r)}{G'(\alpha_r)} e^{\alpha_r . t}$$

The Inverse Infinite Fourier Sine Transform : The inverse formula for infinite Fourier sine transform is given by

$$f(x) = F_s^{-1}[\tilde{f}_s(p)] = \sqrt{\frac{2}{\pi}} \int_0^{\infty} \tilde{f}_s(p) \sin px\, dx \ .$$

The Inverse Laplace Transform : If the Laplace transform of a function $F(t)$ is $f(p)$, *i.e.*, if $L\{F(t)\} = f(p)$. Then, $F(t)$ is known as inverse Laplace transform of $F(p)$.

The sine and cosine integrals : The sine and cosine integrals, which are denoted by $S_i(t)$ and $C_i(t)$ respectively are defined by $S_i(t) = \int_0^t \dfrac{\sin u}{u}\, du$ and $C_i(t) = \int_t^{\infty} \dfrac{\cos u}{u}\, du$.

Unit Step Function : The unit step function, denoted by $H(t - a)$ is defined by

$$H(t - a) = \begin{cases} 0, & t < a \\ 1, & t > a \end{cases}$$

Z-Transforms : If the function f_n is defined for discrete values ($n = 0, 1, 2, ...$) and $f_n = 0$, for $n < 0$ then its Z-transform is defined by

$$Z(f^n) = F(z) = \sum_{n=-\infty}^{\infty} f(n) z^{-n} = \sum_{n=-\infty}^{\infty} \frac{f(n)}{z^n}$$

whenever the infinite series converges. Here z is a complex number. Z is an operator and $F(z)$ is the Z-transforms of $\{f(n)\}$.

vvvvvv

 **Comprehensive Table for Areas of Application of Integral Transforms**

S. No.	Area of Application	Integral Transform Used
1.	Summation of Series	Laplace and Inverse Laplace
2.	Evaluation of improper integrals	Laplace and Mellin
3.	Linear ordinary differential equation with constant and variable coefficients	Laplace, Mellin and Euler
4.	Evaluation of coefficient of asymptotic expansion	Laplace and Mellin
5.	System of linear ordinary differential equation	Laplace
6.	Linear equation of mathematical physics	Laplace, Fourier and Hankel
7.	Linear integral equation	Laplace, Fourier and Mellin
8.	Linear integral equation	Laplace, Fourier and Mellin
9.	Linear difference equation	Laplace
10.	Linear differential equation	Laplace
11.	Linear integro-differential equation	Laplace and Fourier

Bibliography

1.	**Antimirov, M.Ya**	*Applied Integral Transforms,* American Mathematician Society, Providence, Rhode Island, (1993)
2.	**Bateman, H. and Erdelyi, A.**	*Tables of Integral Transforms,* Vols. 1 and 2, McGraw-Hill, New York (1954)
3.	**Bellman, R. and Roth, R.**	*The Laplace Transform,* World Scientific Publishing Co., Singapore (1984)
4.	**Boetsch, G.**	*Introduction to the Theory and Application of Laplace Transform,* Springer-Verlag, Berlin (1974)
5.	**Davis, B.**	*Integral Transforms and Their Applications,* Springer-Verlag, New York (1978)
6.	**Krantz, S.G.**	*Handbook of Complex Variables,* Birkhauser, Boston (1999)
7.	**Milles, J.W.**	*Integral Transforms in Applied Mathematics,* Cambridge University Press, Cambridge (1971)
8.	**Pinkus, A. and Zafrany, S.**	*Fourier Series and Integral Transforms,* Cambridge University Press, Cambridge (1997)
9.	**Polyanin, A.D.**	*Handbook of Linear Partial Differential Equations for Engineers and Scientists,* Chapman & Hall|CRC Press, Boca Roton (2002)
10.	**Sneddon, I.**	*Fourier Transforms,* Dover Publications, New York (1995)
11.	**Sneddon, I.**	*The Use of Integral Transforms,* McGraw-Hill, New York (1972)
12.	**Titchmarch, E.C.**	*Introduction to the Theory of Fourier Integrals,* 3rd Edition, Chelsea Publishing, New York (1986)
13.	**Zwillinger, D.**	*Handbook of Differential Equations,* 3rd Edition, Academic Press, Boston (1977)

Index